PROCESSING OF BIOMASS WASTE

PROCESSING OF BIOMASS WASTE

Technological Upgradation and Advancement

Edited by

ANGANA SARKAR
Department of Biotechnology and Medical Engineering, National Institute of Technology Rourkela, Rourkela, Odisha, India

ULLA LASSI
Research Unit of Sustainable Chemistry, Faculty of Technology, University of Oulu, Oulu, Finland; Unit of Applied Chemistry, University of Jyvaskyla, Kokkola University Consortium Chydenius, Kokkola, Finland

ELSEVIER

Elsevier
Radarweg 29, PO Box 211, 1000 AE Amsterdam, Netherlands
125 London Wall, London EC2Y 5AS, United Kingdom
50 Hampshire Street, 5th Floor, Cambridge, MA 02139, United States

Copyright © 2024 Elsevier Inc. All rights are reserved, including those for text and data mining, AI training, and similar technologies.

Publisher's note: Elsevier takes a neutral position with respect to territorial disputes or jurisdictional claims in its published content, including in maps and institutional affiliations.

No part of this publication may be reproduced or transmitted in any form or by any means, electronic or mechanical, including photocopying, recording, or any information storage and retrieval system, without permission in writing from the publisher. Details on how to seek permission, further information about the Publisher's permissions policies and our arrangements with organizations such as the Copyright Clearance Center and the Copyright Licensing Agency, can be found at our website: www.elsevier.com/permissions.

This book and the individual contributions contained in it are protected under copyright by the Publisher (other than as may be noted herein).

MATLAB® is a trademark of The MathWorks, Inc. and is used with permission. The MathWorks does not warrant the accuracy of the text or exercises in this book. This book's use or discussion of MATLAB® software or related products does not constitute endorsement or sponsorship by The MathWorks of a particular pedagogical approach or particular use of the MATLAB® software.

Notices
Knowledge and best practice in this field are constantly changing. As new research and experience broaden our understanding, changes in research methods, professional practices, or medical treatment may become necessary.

Practitioners and researchers must always rely on their own experience and knowledge in evaluating and using any information, methods, compounds, or experiments described herein. In using such information or methods they should be mindful of their own safety and the safety of others, including parties for whom they have a professional responsibility.

To the fullest extent of the law, neither the Publisher nor the authors, contributors, or editors, assume any liability for any injury and/or damage to persons or property as a matter of products liability, negligence or otherwise, or from any use or operation of any methods, products, instructions, or ideas contained in the material herein.

ISBN: 978-0-323-95179-1

For Information on all Elsevier publications
visit our website at https://www.elsevier.com/books-and-journals

Publisher: Candice Janco
Senior Acquisition Editor: Anita Koch
Editorial Project Manager: Kyle Gravel
Production Project Manager: Rashmi Manoharan
Cover Designer: Christian Bilbow

Typeset by MPS Limited, Chennai, India

Contents

List of contributors   xi

1. Introduction to waste biomass processing and valorization   1
ULLA LASSI AND ANGANA SARKAR

1.1 What is waste biomass?   1
1.2 Waste biomass valorization   1
1.3 Challenges of waste biomass valorization   5
1.4 Opportunities of waste biomass valorization   5
1.5 Conclusions   6
References   7

2. Assessment of wastes for future bioprospecting   9
JEETESH KUSHWAHA, JYOTI RANI, MADHUMITA PRIYADARSINI, KAILASH PATI PANDEY AND ABHISHEK S. DHOBLE

2.1 Introduction   9
2.2 Classification of biomass waste   9
2.3 Techniques for the extraction of value-added products from biowaste   11
2.4 High-value bio-products from biomass waste   15
2.5 Metabolic engineering approaches   17
2.6 Current challenges and future prospects   18
References   18

3. Recent advances in pretreatment of waste biomass   21
ASISH BISAI AND MADHUMITA PATEL

3.1 Introduction   21
3.2 Types of biomasses   21
3.3 Different pretreatment processes   23
3.4 Typical biomass component   24
3.5 Goals of pretreatment process   24
3.6 Biomass torrefaction   25
3.7 Ozonolysis pretreatment process   26
3.8 Ionic liquid pretreatment process   29
3.9 Alkaline pretreatment process   30
3.10 Organic solvent pretreatment process   32
3.11 Steam explosion pretreatment process   35
3.12 Conclusion   37
References   38

4. Emerging technologies for waste biomass pretreatment: pros and cons   41
SUBHRAJIT ROY AND SAIKAT CHAKRABORTY

4.1 Introduction   41
4.2 Pretreatment of waste biomass for bioenergy production   42
4.3 Pretreatment technologies for waste biomass: pros   47
4.4 Pretreatment technologies for waste biomass: cons   48
4.5 Advance pretreatment in biorefineries   50
4.6 Technological upgradation and Scale-up   51
4.7 Conclusions   51
References   52

5. Design and operation of advanced waste biomass processing system   55
NITIN KUMAR, JAYDEV KUMAR MAHATO AND SUNIL KUMAR GUPTA

5.1 Introduction   55
5.2 Global trend of waste generation and compositional variation in household organic waste   56
5.3 Sources of solid organic waste (SOW)   56
5.4 Classification of technologies for processing of solid organic waste   58
5.5 SWOT analysis of various technologies   65
5.6 Summary and conclusions   66
References   66

6. Application of cutting-edge molecular biotechnological techniques in waste valorization   71
POULOMI SARKAR AND ANGANA SARKAR

6.1 Introduction   71
6.2 Status of waste generation in India and world   71
6.3 Sources and types of wastes   72
6.4 Waste remedial approaches and associated challenges   74
6.5 Waste bioprocessing as a valuable tool for resource recovery   74
6.6 Recent advances in molecular biotechnology tools for waste valorization   75
6.7 Future perspectives   77
References   78

7. Microwave application in bioenergy production from waste biomass — 83
REJETI VENKATA SRINADH AND NEELANCHERRY REMYA

7.1 Introduction — 83
7.2 Need for bioenergy — 83
7.3 Biomass materials — 84
7.4 Conventional bioenergy production methods — 88
7.5 Microwave application in bioenergy production — 90
7.6 Lab-scale to large-scale opportunities — 100
References — 102

8. Bioprospecting of algal biomass for value generation from municipal waste — 105
RENUPAMA BHOI, SHUBHAM AGRAWAL, JASMITA AND ANGANA SARKAR

8.1 Introduction — 105
8.2 Composition of municipal wastewater — 106
8.3 Biorefinery approaches of microalgal cultivation in municipal wastewater — 109
8.4 Cultivation of microalgae in municipal wastewater — 110
8.5 Challenges in microalgal cultivation in municipal wastewater — 112
8.6 Future perceptive — 112
References — 113

9. Bioenergy from waste biomass — 115
SUBHRAJIT ROY AND SAIKAT CHAKRABORTY

9.1 Introduction — 115
9.2 Waste biomass: sources, classification, and characterization — 116
9.3 Sources and classification of bioenergy obtained from waste biomass — 119
9.4 Catalytic conversion of waste biomass to bioenergy — 122
9.5 Biochemical conversion of waste biomass to bioenergy — 123
9.6 Thermochemical conversion of waste biomass to bioenergy — 124
9.7 Production of transportable bioenergy from waste biomass — 126
9.8 Bioeconomy and biorefinery — 129
9.9 Technological upgradation and scale-up — 130
9.10 Conclusions — 130
References — 131

10. Lignocellulosic wastes: different dimensions to a sustainable 2 G bioethanol production — 135
SUBHODEEP BANERJEE, SUBHARA DEY, ANUSHA AND RINTU BANERJEE

10.1 Introduction — 135
10.2 Diversity and potentiality of lignocellulosic residue — 135
10.3 Biochemistry of lignocellulose — 136
10.4 Pretreatment methods of biomass for 2 G bioethanol — 137
10.5 Combined pretreatment — 138
10.6 Biological pretreatment method treat above combined — 139
10.7 Enzymatic pretreatment — 139
10.8 Different enzymes used for the pretreatment and hydrolysis of lignocellulosic biomass — 140
10.9 Saccharification — 140
10.10 Fermentation — 142
10.11 Pilot plant: case study — 142
10.12 Why bioethanol has not yet been commercialized in accordance with fermentation techniques? Major challenges that lie with commercialization of lignocellulosic bioethanol — 143
10.13 Supply chain management of lignocellulosics bioethanol — 144
10.14 Challenges of 2 G bioethanol production: biorefinery concept — 145
10.15 Future scope — 146
10.16 Conclusion — 147
References — 148

11. Biodiesel from lipid-rich wastes: prospects and challenges in commercialization — 151
SARVESHWARAN SARAVANABHUPATHY, SWAGATA DUTTA AND RINTU BANERJEE

Abbreviations — 151
11.1 Introduction — 151
11.2 Biodiesel market and Indian scenario — 152
11.3 Potential of lipid-rich feedstocks for large-scale biodiesel production — 153
11.4 Different techniques for large-scale production of biodiesel — 156
11.5 Scale-up of biodiesel production and technoeconomic analysis — 158
11.6 Commercialization of biodiesel — 159
11.7 Challenges and major bottlenecks of biodiesel commercialization — 160
11.8 Conclusion — 161
References — 162

12. Biogas from organic wastes — 165
JOSEPH SEKHAR SANTHAPPAN, RAJALINGAM ARUMUGANAINAR, GODWIN GLIVIN AND V.M. JAGANATHAN

12.1 Introduction — 165
12.2 Materials and methods — 170
12.3 Results and discussion — 177

12.4 Conclusion	182	16.5 Different methods of pigment production	236
References	182	16.6 Chemical methods	237
		16.7 Biological methods	237

13. Hydrogen from waste and biowaste materials: production, separation, purification, and use — 185
PREM KUMAR SEELAM, PUTRAKUMAR BALLA AND SIMONA LIGUORI

- 13.1 Introduction: H_2 from biomass-derived biowastes — 185
- 13.2 Hydrogen production processes from biowastes and industrial waste gases — 187
- 13.3 Hydrogen separation and purification — 193
- 13.4 Hydrogen distribution and end use — 196
- 13.5 Conclusions and summary — 197
- References — 197

- 16.8 Production of pigments from agrowastes without using microorganisms — 237
- 16.9 Production of pigment from microorganisms using agrowastes as substrate — 238
- 16.10 Process of pigment production from agrowaste — 240
- 16.11 Commercially available pigments from agrowaste and drawbacks of using agrowastes for pigments production — 240
- 16.12 Challenges and future aspects — 241
- References — 241

14. A biorefinery route to treat waste water through extremophilic enzymes: an innovative approach to generate value from waste — 201
TUHIN SUBHRA BISWAS, THAMIZVANI K., KASTURI BIDKAR, KAVYA SINGH, CHANDUKISHORE T. AND ASHISH A. PRABHU

- 14.1 Bioprospecting of novel and industrially relevant enzymes — 201
- 14.2 Extremophilic enzymes — 204
- 14.3 Enzyme modification using immobilization — 207
- 14.4 Metagenomics: a source of enzyme discovery: introduction — 210
- 14.5 Conclusion — 213
- References — 213

17. Plastic waste as a novel substrate for industrial biotechnology — 245
RAJLAKSHMI, PRIYADHARSHINI JAYASEELAN AND RINTU BANERJEE

- 17.1 Introduction — 245
- 17.2 Plastics and their classification — 246
- 17.3 Problems associated with conventional plastic usage and its disposals — 249
- 17.4 Biodegradable plastics: limitations and its market studies — 249
- 17.5 Strategies for management of plastic waste (reduction, replacement and reuse) — 250
- 17.6 Reduction strategies — 251
- 17.7 Micro-organisms in plastic reduction — 251
- 17.8 Enzymes in plastic reduction — 252
- 17.9 Mealworm in plastic reduction — 252
- 17.10 Replacement of conventional plastics using biotechnology — 252
- 17.11 Starch-based polymer — 253
- 17.12 Polylactic acid-based polymer — 253
- 17.13 Modification of plastics for waste management — 253
- 17.14 Conversion strategies using biotechnology — 254
- 17.15 Renewal process for utilization of multiple use plastics waste — 254
- 17.16 Pre-treatment of plastic waste by thermochemical depolymerization of plastic waste — 255
- 17.17 Pyrolysis — 255
- 17.18 Hydrothermal treatment — 256
- 17.19 Incineration — 257
- 17.20 Solvothermal process — 257
- 17.21 Plastic waste as feedstock for potential product formation — 257
- 17.22 Synergistic approach for plastic waste conversion — 258
- 17.23 Ecology and environment: modified plastic usage and its future — 259

15. Waste as a substrate for the production of organic acids and solvents — 215
KAWINHARSUN DHODDURAJ, DURGA ASHOK BURANDE, NIVEDHITHA ULAGANATHAN AND ASHISH A. PRABHU

- 15.1 Introduction — 215
- 15.2 Organic acids — 215
- 15.3 Organic solvent — 221
- References — 228

16. Pigments and paints from wastes — 233
KUMARI GUDDI, G. VIJAY CHITHRA, R. BHAVANI, SAMBIT NAIK AND ANGANA SARKAR

- 16.1 Introduction — 233
- 16.2 Different sources of agrowastes — 234
- 16.3 Different forms of agrowaste — 234
- 16.4 Types of pigments produced from agrowastes — 235

17.24 Conclusion	259	21. Sewage waste as substrate for value	305
References	260	RAHUL RANJAN, ROHIT RAI, VIKASH KUMAR AND PRODYUT DHAR	
18. Catalytic conversion of biomass-based wastes: upgrading and valorization to value-added intermediates and platform molecules	**263**	21.1 Introduction	305
		21.2 Value-added products from sewage sludge	319
PUTRAKUMAR BALLA, SATYA KAMAL CHIRAURI, SRINIVASARAO GINJUPALLI, RAJENIDRAN RAJESH, PRATHAP CHALLA, SUNGTAK KIM AND SEELAM PREM KUMAR		21.3 Applications of sewage sludge	329
		21.4 Conclusion	331
		Acknowledgment	331
		References	332
18.1 Introduction	263	**22. Converting biomass waste to water treatment chemicals**	**341**
18.2 Biomass conversion technologies	264		
18.3 Utilization of biological-waste to value-added products	266	TATIANA SAMARINA, VARSHA SRIVASTAVA, OUTI LAATIKAINEN AND SARI TUOMIKOSKI	
18.4 Efficient conversion process of biomass derived platform molecules into fine chemicals and fuels	267	22.1 Introduction	341
		22.2 Biomass feedstock for recovery of chemicals	342
18.5 Conclusions	273	22.3 Recovery of biochemicals from biomass	344
References	273	22.4 Technology readiness level and application of bio-based chemicals in water and wastewater treatment practices	350
19. Thermal digestion process—a novel technique for converting solid organic waste into nutrient-rich organic fertilizer	**275**	22.5 Challenges and future perspective	351
		22.6 Conclusions	352
NITIN KUMAR AND SUNIL KUMAR GUPTA		Acknowledgments	352
		References	352
19.1 Introduction	275	**23. Waste-biomass-derived potential catalyst materials for water reclamation**	**361**
19.2 Qualitative and quantitative analysis of SOW	276		
19.3 Existing technologies on the nutrient recovery from solid organic waste	276	VARSHA SRIVASTAVA, ANNE HEPONIEMI, SARI TUOMIKOSKI, RIIKKA KUPILA, DAVIDE BERGNA AND ULLA LASSI	
19.4 Thermal digestion process	279	Highlights	361
19.5 Comparative assessment with various technologies	281	Abbreviations	361
		23.1 Introduction	362
19.6 Summary and conclusions	282	23.2 Biomass-derived catalyst materials	364
References	282	23.3 Carbon-based catalyst preparation and properties	368
20. Waste biomass conversion to energy storage material	**285**	23.4 Application of biomass-derived catalysts for water treatment	370
GLAYDSON SIMÕES DOS REIS, SARI TUOMIKOSKI, DAVIDE BERGNA, SYLVIA LARSSON, MIKAEL THYREL, HELINANDO PEQUENO DE OLIVEIRA, PALANIVEL MOLAIYAN AND ULLA LASSI		23.5 Challenges and future perspectives	377
		23.6 Conclusions	378
		Acknowledgments	379
		References	379
20.1 Introduction	285	**24. Technoeconomic feasibility analysis of waste bioprocessing**	**385**
20.2 Suitable sources of biomass for electrode preparation	286		
20.3 Production and properties of biomass-derived anode materials	288	V.M. JAGANATHAN, JOSEPH SEKHAR SANTHAPPAN, RAJALINGAM ARUMUGANAINAR, M. EDWIN AND GODWIN GLIVIN	
20.4 Applications of biomass-derived carbon materials for energy storage systems	290	24.1 Introduction	385
20.5 Conclusions, challenges, and prospects	300	24.2 Materials and methods	389
References	300	24.3 Results and discussion	396

24.4 Conclusion	402	25.3 Case study	409
References	403	25.4 Discussion and conclusions	415
		Acknowledgments	415
25. Sustainability assessment method for waste biomass processing	**405**	References	415
ANUSHA AIRI, SARI PIIPPO AND EVA PONGRACZ		Index	417
25.1 Introduction	405		
25.2 Sustainability	406		

List of contributors

Shubham Agrawal Department of Biotechnology and Medical Engineering, National Institute of Technology Rourkela, Rourkela, Odisha, India

Anusha Airi University of Oulu, Water, Energy and Environmental Engineering Research Unit, Oulu, Finland

Anusha Agricultural & Food Engineering Department, Indian Institute of Technology Kharagpur, Kharagpur, West Bengal, India

Rajalingam Arumuganainar Engineering Department, College of Engineering and Technology, University of Technology and Applied Sciences, Shinas, Oman

Durga Ashok Burande Bioprocess Development Laboratory, Department of Biotechnology, National Institute of Technology Warangal, Warangal, Telangana, India

Putrakumar Balla Department of Chemical Engineering and Applied Chemistry, Chungnam National University, Daejeon, Republic of Korea

Rintu Banerjee Microbial Biotechnology and Downstream Processing Laboratory, Agricultural & Food Engineering Department, Indian Institute of Technology Kharagpur, Kharagpur, West Bengal, India; P K Sinha Centre for Bioenergy and Renewables, Indian Institute of Technology Kharagpur, Kharagpur, West Bengal, India; Agricultural & Food Engineering Department, Indian Institute of Technology-Kharagpur, Kharagpur, West Bengal, India

Subhodeep Banerjee Advanced Technology Development Centre, Indian Institute of Technology Kharagpur, Kharagpur, West Bengal, India; P K Sinha Centre for Bioenergy and Renewables, Indian Institute of Technology Kharagpur, Kharagpur, West Bengal, India

Davide Bergna Research Unit of Sustainable Chemistry, Faculty of Technology, University of Oulu, Oulu, Finland; Unit of Applied Chemistry, University of Jyvaskyla, Kokkola University Consortium Chydenius, Kokkola, Finland

R. Bhavani Department of Biotechnology and Medical Engineering, National Institute of Technology, Rourkela, Odisha, India

Renupama Bhoi Department of Biotechnology and Medical Engineering, National Institute of Technology Rourkela, Rourkela, Odisha, India

Kasturi Bidkar Bioprocess Development Laboratory, Department of Biotechnology, National Institute of Technology, Warangal, Telangana, India

Asish Bisai Adani Airport Holding Limited, Ahemadabad, Gujarat, India

Tuhin Subhra Biswas Bioprocess Development Laboratory, Department of Biotechnology, National Institute of Technology, Warangal, Telangana, India

Saikat Chakraborty Biological Systems Engineering, Plaksha University, Mohali, Punjab, India

Prathap Challa Energy & Environmental Engineering Department, CSIR-Indian Institute of Chemical Technology, Hyderabad, Telangana, India

Satya Kamal Chirauri Micron Technology Operations LLP, Hyderabad, Telangana, India

Helinando Pequeno de Oliveira Institute of Materials Science, Federal University of São Francisco Valley, Juazeiro, BA, Brazil

Subhara Dey P K Sinha Centre for Bioenergy and Renewables, Indian Institute of Technology Kharagpur, Kharagpur, West Bengal, India

Prodyut Dhar School of Biochemical Engineering, Indian Institute of Technology (BHU), Varanasi, Uttar Pradesh, India

Abhishek S. Dhoble School of Biochemical Engineering, Indian Institute of Technology (BHU), Varanasi, Uttar Pradesh, India

Kawinharsun Dhodduraj Bioprocess Development Laboratory, Department of Biotechnology, National Institute of Technology Warangal, Warangal, Telangana, India

Glaydson Simões Dos Reis Department of Forest Biomaterials and Technology, Swedish University of Agricultural Sciences, Biomass Technology Centre, Umeå, Sweden

Swagata Dutta Agricultural & Food Engineering Department, Indian Institute of Technology, Kharagpur, Kharagpur, West Bengal, India

M. Edwin Department of Mechanical Engineering, University College of Engineering Nagercoil, Anna University Constituent College, Nagercoil, Tamil Nadu, India

Srinivasarao Ginjupalli Department of Applied Science, University of Technology and Applied Science, Muscat, Sultanate of Oman

Godwin Glivin Department of Energy and Environment, National Institute of Technology, Tiruchirappalli, Tamil Nadu, India

Kumari Guddi Department of Biotechnology and Medical Engineering, National Institute of Technology, Rourkela, Odisha, India

Sunil Kumar Gupta Department of Environmental Science & Engineering, Indian Institute of Technology (Indian School of Mines), Dhanbad, Jharkhand, India

Anne Heponiemi Research Unit of Sustainable Chemistry, Faculty of Technology, University of Oulu, Oulu, Finland

V.M. Jaganathan Department of Energy and Environment, National Institute of Technology, Tiruchirappalli, Tamil Nadu, India

Jasmita Department of Biotechnology, College of Commerce, Arts, and Science Patna (Bihar) Patliputra University, Patna, Bihar, India

Priyadharshini Jayaseelan Microbial Biotechnology and Downstream Processing Laboratory, Agricultural & Food Engineering Department, Indian Institute of Technology Kharagpur, Kharagpur, West Bengal, India

Thamizvani K. Bioprocess Development Laboratory, Department of Biotechnology, National Institute of Technology, Warangal, Telangana, India

Sungtak Kim Department of Chemical Engineering and Applied Chemistry, Chungnam National University, Daejeon, Republic of Korea

Nitin Kumar Department of Environmental Science & Engineering, Indian Institute of Technology (Indian School of Mines), Dhanbad, Jharkhand, India

Vikash Kumar School of Biochemical Engineering, Indian Institute of Technology (BHU), Varanasi, Uttar Pradesh, India

Riikka Kupila Research Unit of Sustainable Chemistry, Faculty of Technology, University of Oulu, Oulu, Finland

Jeetesh Kushwaha School of Biochemical Engineering, Indian Institute of Technology (BHU), Varanasi, Uttar Pradesh, India

Outi Laatikainen School of Engineering, Kajaani University of Applied Sciences, Kajaani, Finland

Sylvia Larsson Department of Forest Biomaterials and Technology, Swedish University of Agricultural Sciences, Biomass Technology Centre, Umeå, Sweden

Ulla Lassi Research Unit of Sustainable Chemistry, Faculty of Technology, University of Oulu, Oulu, Finland; Unit of Applied Chemistry, University of Jyvaskyla, Kokkola University Consortium Chydenius, Kokkola, Finland

Simona Liguori Department of Chemical & Biomolecular Engineering, Clarkson University, Potsdam, NY, United States

Jaydev Kumar Mahato Department of Environmental Science & Engineering, Indian Institute of Technology (Indian School of Mines), Dhanbad, Jharkhand, India

Palanivel Molaiyan Research Unit of Sustainable Chemistry, Faculty of Technology, University of Oulu, Oulu, Finland

Sambit Naik Fakir Mohan University, Balasore, Odisha, India

Kailash Pati Pandey School of Biochemical Engineering, Indian Institute of Technology (BHU), Varanasi, Uttar Pradesh, India

Madhumita Patel Department of Environmental Science and Engineering, IIT (ISM) Dhanbad, Dhanbad, Jharkhand, India

Sari Piippo Finnish Environment Institute, Syke, Oulu, Finland

Eva Pongracz University of Oulu, Water, Energy and Environmental Engineering Research Unit, Oulu, Finland

Ashish A. Prabhu Bioprocess Development Laboratory, Department of Biotechnology, National Institute of Technology, Warangal, Telangana, India

Madhumita Priyadarsini School of Biochemical Engineering, Indian Institute of Technology (BHU), Varanasi, Uttar Pradesh, India

Rohit Rai School of Biochemical Engineering, Indian Institute of Technology (BHU), Varanasi, Uttar Pradesh, India

Rajenidran Rajesh Energy & Environmental Engineering Department, CSIR-Indian Institute of Chemical Technology, Hyderabad, Telangana, India

Rajlakshmi Microbial Biotechnology and Downstream Processing Laboratory, Agricultural & Food Engineering Department, Indian Institute of Technology Kharagpur, Kharagpur, West Bengal, India

Jyoti Rani School of Biochemical Engineering, Indian Institute of Technology (BHU), Varanasi, Uttar Pradesh, India

Rahul Ranjan School of Biochemical Engineering, Indian Institute of Technology (BHU), Varanasi, Uttar Pradesh, India

Neelancherry Remya School of Infrastructure, Indian Institute of Technology Bhubaneswar, Bhubaneswar, Odisha, India

Subhrajit Roy Biological Systems Engineering, Plaksha University, Mohali, Punjab, India

Tatiana Samarina School of Engineering, Kajaani University of Applied Sciences, Kajaani, Finland; Research Unit of Sustainable Chemistry, Faculty of Technology, University of Oulu, Oulu, Finland

Joseph Sekhar Santhappan Engineering Department, College of Engineering and Technology, University of Technology and Applied Sciences, Shinas, Oman

Sarveshwaran Saravanabhupathy Agricultural & Food Engineering Department, Indian Institute of Technology-Kharagpur, Kharagpur, West Bengal, India

Angana Sarkar Department of Biotechnology and Medical Engineering, National Institute of Technology Rourkela, Rourkela, Odisha, India

Poulomi Sarkar Division of Plant Physiology, ICAR-Indian Agricultural Research Institute, New Delhi, India

Prem Kumar Seelam Sustainable Chemistry Research Unit, Faculty of Technology, University of Oulu, Oulu, Finland

Kavya Singh Bioprocess Development Laboratory, Department of Biotechnology, National Institute of Technology, Warangal, Telangana, India

Rejeti Venkata Srinadh School of Infrastructure, Indian Institute of Technology Bhubaneswar, Bhubaneswar, Odisha, India

Varsha Srivastava Research Unit of Sustainable Chemistry, Faculty of Technology, University of Oulu, Oulu, Finland

Chandukishore T. Bioprocess Development Laboratory, Department of Biotechnology, National Institute of Technology, Warangal, Telangana, India

Mikael Thyrel Department of Forest Biomaterials and Technology, Swedish University of Agricultural Sciences, Biomass Technology Centre, Umeå, Sweden

Sari Tuomikoski Research Unit of Sustainable Chemistry, Faculty of Technology, University of Oulu, Oulu, Finland

Nivedhitha Ulaganathan Bioprocess Development Laboratory, Department of Biotechnology, National Institute of Technology Warangal, Warangal, Telangana, India

G. Vijay Chithra Department of Biotechnology and Medical Engineering, National Institute of Technology, Rourkela, Odisha, India

CHAPTER 1

Introduction to waste biomass processing and valorization

Ulla Lassi[1,2] and Angana Sarkar[3]

[1]Research Unit of Sustainable Chemistry, Faculty of Technology, University of Oulu, Oulu, Finland [2]Unit of Applied Chemistry, University of Jyvaskyla, Kokkola University Consortium Chydenius, Kokkola, Finland [3]Department of Biotechnology and Medical Engineering, National Institute of Technology Rourkela, Rourkela, Odisha, India

1.1 What is waste biomass?

Forest and pulping industry and agriculture produce large amounts of waste biomass. Global annual generation of biomass waste is estimated to be around 140 Gt. Biomass wastes primarily contain forestry residues, agricultural wastes, animal wastes, industrial wastes, municipal solid wastes (MSWs), and food processing wastes (Tripathi et al., 2019). Main types of waste biomass are wood and agricultural products, solid waste, and landfill municipal wastes. There are several options for waste biomass valorization. The use of waste biomass for electric power or energy (fuels) is widely reported. In addition to traditional energy use, waste biomass can be used to produce polymers and other platform chemicals and construction materials. Recently, also the modification of waste biomass to functional materials or composites is done.

Management of waste biomass is normally done by mechanical, biological, thermal, and/or chemical treatments or combination of these. However, waste biomass is a heterogeneous material, which presents significant management problems and can have even negative environmental impacts. Sustainable biomass processing requires that technoeconomic, societal, and environmental aspects are considered. This chapter introduces the processing of lignocellulosic and agricultural waste biomass.

1.2 Waste biomass valorization

1.2.1 General

Waste biomass valorization refers to the conversion of organic waste materials into usable products such as fuel, chemicals, and fertilizer. This can include materials such as agricultural waste, food waste, and sewage sludge. The process of converting waste biomass into usable products is known as biomass conversion, and it can include several valorization methods. The goal of waste biomass processing is to turn these materials into a valuable resource, while also reducing the amount of waste that ends up in landfills (Sharma et al., 2024).

There are several technologies that can be utilized for waste biomass valorization. Some of the most common include thermal, (bio)chemical, mechanical, and other methods, which are shown in Table 1.1.

Thermal methods for waste biomass management are combustion, gasification, and pyrolysis. Combustion is the burning of biomass to produce heat and electricity. This can be done in a stand-alone furnace or in a combined heat and power (CHP) system. During pyrolysis, waste biomass is heated in the absence of oxygen to produce a liquid bio-oil or biochar. Gasification is the process of converting biomass into a gaseous fuel, such as syngas, by heating it in the presence of a limited amount of oxygen (El-Emam, 2012).

TABLE 1.1 Methods of waste biomass valorization.

Techniques	Treated waste biomass	Product	References
Thermal methods			
a) Combustion b) Pyrolysis c) Gasification	Solid waste, agricultural straws, animal slurries, lignocellulosic feedstock, food waste, aquatic plants, and energy crops, etc.	Mechanical/electrical energy Bio-oil Biochar Syngas	Lombardi, Carnevale, and Corti (2015), Laird et al. (2009), Luterbacher et al. (2009), El-Emam (2012)
Biochemical methods			
a) Anaerobic digestion b) Fermentation c) Composting d) Biochemical conversion	Organic wastewater, lignocellulosic biomass, kitchen waste, household waste, construction waste, street waste, commercial waste, and microalgae, etc.	Methane-rich biogas Bioethanol Biofertilizer Bioplastics	Singh, Szamosi, and Siménfalvi (2020), Beyene, Werkneh, and Ambaye (2018), Duan et al. (2019), Lee et al. (2019)
Mechanical Methods			
a) Supercritical fluid method	Agricultural feedstock	Biodiesel	Lee et al. (2019)
Others			
a) Transesterification b) Photosynthetically fuel cell	Cellulosic biomass, agricultural residue and forest residue	Biodiesel Electrical energy	Lee et al. (2019)

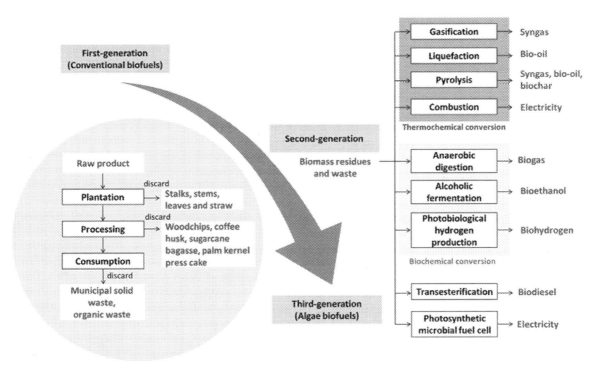

FIGURE 1.1 Diagram of the development of biofuel generation with highlights on the second generation biofuels produced by biomass residues and waste and their conversion pathways to produce a wide variety of bioenergy (Lee et al., 2019).

Bioprocessing methods include the use of microorganisms and/or enzymes to produce value-added platform chemicals, as shown in Fig. 1.1. These include anaerobic digestion (in the absence of oxygen) and biochemical conversion using enzymes or microorganisms to convert biomass into useful products such as chemicals, plastics, or animal feed. The resulting biogas from anaerobic digestion can be used as a fuel source. Composting and fermentation are also considered as bioprocessing methods enabling production of nutrient-rich soil amendments and ethanol fuel, respectively.

These are the most common technologies, but new technologies are continuously being developed, and it's important to keep informed. It is also important to keep in mind that the choice of technology will depend on the type and characteristics of the waste biomass, as well as the desired product.

1.2.2 Biochemical methods

Biochemical methods for waste biomass valorization include anaerobic digestion, fermentation, composting, and biochemical conversion. Anaerobic digestion is a biological process that uses microorganisms to break down organic materials in the absence of oxygen. In this process, organic biomass is converted into methane-rich biogas through various metabolic reactions such as hydrolysis, acidogenesis, acetogenesis, and methanogenesis steps (Singh et al., 2020). The process occurs in an anaerobic reactor, which can be a simple tank or a more complex system.

The process of anaerobic digestion can be divided into four stages, i.e., hydrolysis, acidogenesis, acetogenesis, and methanogenesis. Hydrolysis is the stage in which complex organic compounds are broken down into simpler sugars and amino acids by enzymes secreted by microorganisms. During acidogenesis, the sugars and amino acids are further broken down by acidogenic bacteria into organic acids, such as acetic acid, and alcohols. In the acetogenesis stage, the organic acids are converted into acetic acid and hydrogen by acetogenic bacteria. At the final stage, during the methanogenesis, methanogenic bacteria convert the acetic acid and hydrogen into methane (CH_4) and carbon dioxide (CO_2). This biogas can be captured and used as a source of energy.

Anaerobic digestion can be used to process a wide range of organic materials, including agricultural waste, food waste, sewage sludge, and other types of organic waste biomass. The end products of the process are biogas, which can be used as a source of energy, and a nutrient-rich liquid or solid called digestate, which can be used as a fertilizer. Anaerobic digestion plays an essential role in reducing organic waste, recovering the energy from biomass, and generating energy and biofuels (Fan et al., 2018). Anaerobic digestion process has many advantages, such as reducing greenhouse gas emissions, reducing the volume of waste sent to landfills, and producing a renewable source of energy. However, it can also pose some challenges, such as the need for consistent and high-quality feedstock and the management of the digestate.

Fermentation is an anaerobic process, producing mainly liquid biofuels instead of biogas. Fermentation of biomass residues containing fermentable sugars is produced when cellulose and hemicellulose components of biomass are transformed into fermentable sugars in the presence of yeast or bacteria. Fermentation of waste biomass involves the breakdown of sucrose into fructose and glucose through hydrolysis or using enzymes followed by subsequent conversion into ethanol (Beyene et al., 2018). Typically, fermentation is used to manufacture liquid biofuels and other value-added products from biomass wastes with high sugar concentrations, such as domestic wastes (Matsakas et al., 2014), food and food processing wastes (Sabater et al., 2020; Sadh et al., 2018), and waste from animal products. Fermentation has been extensively used in biorefining, particularly for the transformation of LCB-derived sugars into bulk products such as acetone and butanol.

Composting is an aerobic process in solid waste management. It is chemical-free and produces useful value-added products that are beneficial to improve poor conditions of soil (Wojnowska-Baryła et al., 2020). The production of compost includes different types of feedstocks such as household wastes, commercial wastes, kitchen wastes, street wastes, and construction wastes (Asquer et al., 2017; Awasthi et al., 2020). Microorganisms play an important role in composting that stabilizes the organic debris in temperate and humid conditions. As microbes are involved in composting, the end product formed will be stable and toxic-free (Lee, 2018). Composting process is widely used in recycling of nutrients such as Nitrogen, Phosphorous, and Potassium, which are the most vital elements for the conversion of organic debris into compost (Soobhany, 2019). High density of nitrogen content in organic solid waste found in cattle manure, food wastes, and low density of nitrogen is found in dried leaves, rice bran, residues, newsprints and helps in the production of fertilizer of fine quality (Duan et al., 2019).

Biochemical conversion involves breaking down biomass using microorganisms or their enzymes (see Fig. 1.2). This process utilizes yeast or specific bacteria for transformation of biomass residues into useful energy. Extensively used biochemical technologies for waste transformation include anaerobic digestion, photobiological process, and fermentation. Through biochemical conversion, various types of biofuels are produced Lee et al. (2019).

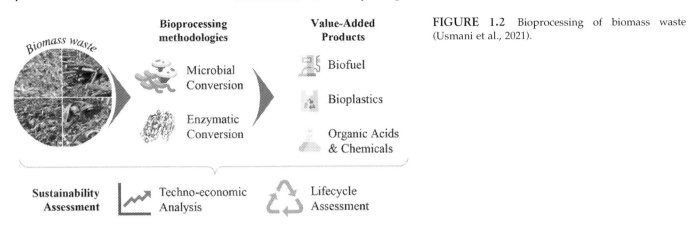

FIGURE 1.2 Bioprocessing of biomass waste (Usmani et al., 2021).

1.2.3 Thermal methods

Waste biomass combustion simply means burning organic material. For millennia, humans have used this basic technology to create heat and, later, to generate power through steam. While wood is the most used biomass feedstock, a wide range of materials can be burned effectively. In a direct combustion system, biomass waste is burned in a combustor or furnace to generate hot gas, which is fed into a boiler to generate steam, which is expanded through a steam turbine or steam engine to produce mechanical or electrical energy (Lombardi et al., 2015).

Waste biomass gasification is a thermochemical process used to convert organic substances into valuable gas (so-called syngas, a mixture of CO and H_2). This gas is used for heat and power production in CHP units. Temperature, equivalent ratio, and pressure impact the syngas composition (El-Emam, 2012). Different types of gasifiers exist, for example, fixed-bed gasifiers and fluidized-bed gasifiers. The type of gasifier (reactor) and its configuration are the most critical factors affecting the reactions and products (Arena, 2012). In all gasification processes, however, the phenomena of pyrolysis followed by partial oxidation of the residual carbon are prevalent (Mohammadi & Anukam, 2022).

Pyrolysis is the thermochemical transformation of waste biomass into various value-added products such as bio-oil, gases, and biochars in the absence of oxygen (Laird et al., 2009). Organic materials are broken down into solid, liquid, and gas mixtures during the pyrolysis process. Pyrolysis is used to valorize several types of biomasses including agricultural residues and wood wastes. Further, syngas can be converted by anaerobic bacteria into biochemicals and biofuels. Pyrolysis can be fast or slow, depending on the heating rate of the waste biomass in inert (or oxygen-poor environment). Fast pyrolysis occurs at high temperature range of 300°C–700°C at a fast-heating rate of 10°C–200°C/s, with a short solid resistance time of 0.5–10 s and with fine particle size (<1 mm) feedstock. Main product of fast pyrolysis is pyrolytic oil, which can be further upgraded. Quite often, pyrolysis oil is still used in energy production by burning. Slow pyrolysis also takes place in an oxygen-poor environment, but the temperature range is 350°C–700°C and heating rate is slow, only 0.1°C–1°C/s. It is one of the most feasible production processes, which is typically carried out at atmospheric pressure, and process heat is often provided by an external source, usually from combustion of generated gas or through partial combustion of biomass feedstock (Laird et al., 2009). Slow pyrolysis primarily targets biochar (up to about 60 wt.%), along with 25–30 wt.% bio-oil and the remaining product as gas. Obtained biochar can further be upgraded, and several applications for biochar exist and are under development. These are more discussed in Section 4.1.

Abovementioned thermal treatment methods are suitable for waste biomass, which has low moisture content. For the wet waste biomass, biochemical processing is often used. It is also possible to treat wet waste biomass in pressurized conditions via hydrothermal carbonization (HTC). HTC, also referred to as "aqueous carbonization at elevated temperature and pressure," is a chemical process for the conversion of organic compounds to structured carbons. It can be used to make a wide variety of nanostructured carbons, simple production of brown coal substitute, synthesis gas, liquid petroleum precursors, and humus from biomass with release of energy (Luterbacher et al., 2009). As mentioned, hydrothermal carbonization of waste biomass is typically done for wet waste biomass material (Laird et al., 2009). Obtained biochar can be upgraded, and this is discussed in Chapter 4.

1.3 Challenges of waste biomass valorization

1.3.1 General

Waste biomass processing can be challenging for several reasons, e.g., variation of feedstock quality, contamination, high operational or capital costs, lack of infrastructure, due to the environmental concerns and regulations, and limited market opportunities. Further, processing of waste biomass should be competitive with other potential treatment methods. Waste biomass can be highly variable in terms of composition, quality, and moisture content, which can make it difficult to process consistently. Waste biomass can also be contaminated with nonorganic materials such as plastic, metal, or glass, which can damage processing equipment or reduce the quality of the product. Contamination is a challenge for transportation as well. Processing equipment and facilities can be expensive to build and maintain, which increases capital and operating costs, and can make it difficult to achieve profitability. This is valid not only for waste biomass processing, but also there is often a lack of infrastructure in place to collect, transport, and process waste biomass, which can make it difficult to scale up operations (Okolie et al., 2022; Tawfik et al., 2022).

Waste biomass processing can generate emissions that can be harmful to air and water quality and could also generate odors. Therefore, there might be environmental concerns and strict regulations that need to be followed when dealing with waste biomass and the end products, which can add complexity and cost to the process. The market for waste biomass products can be limited, which can make it difficult to find buyers for the end products. Waste biomass processing must compete with other waste management options such as landfilling, incineration, or export for energy recovery (Tawfik et al., 2022).

Complete valorization of biomass has been emphasized as an ideal for the future biorefinery, which is in accordance with the advancement of a circular bioeconomy and the principles of green chemistry (Erythropel et al., 2018). Overall, waste biomass processing can be a complex and challenging endeavor, but with the right technology and management, it can be a valuable way to turn waste into a valuable resource.

1.3.2 How to improve treatment and processing of waste biomass?

There are several ways to improve waste biomass processing, including e.g., pretreatment of waste biomass (Liu et al., 2012), standardization of feedstock, improved collaboration between waste generators, processors, and end users. By standardizing the composition feedstock, quality, and moisture content of the waste biomass, it will be easier to process consistently and with fewer problems. Pretreatment of the waste biomass can help to remove contaminants, reduce the moisture content, and make it easier to handle. For this, automation and control systems can help to improve the efficiency and consistency of the processing operations (Guragain & Vadlani, 2021; Wang & Tester, 2023).

Investing in advanced technology, such as high-efficiency anaerobic digesters or gasifiers, can help to improve performance and reduce the costs of waste biomass processing. Collaboration between waste generators, processors, and end users can help to improve the efficiency and effectiveness of waste biomass processing. Government support in terms of grants, subsidies, and favorable regulations can help to reduce the costs and increase the profitability of the process. Market development is also critical for waste biomass processing. Developing new markets for the end products can help to increase demand and make the process more economically viable. Finally, investing in research and development can help to find new and more efficient ways to process the waste biomass and find new applications for the end products. Educating and raising awareness about the benefits of waste biomass processing can help to increase acceptance and support for the process (Wang & Tester, 2023).

Ultimately, improving waste biomass processing requires a multidisciplinary approach that involves a combination of technical, economic, and social factors. By identifying and addressing the challenges, and by implementing best practices, it is possible to improve performance and reduce the costs of the process.

1.4 Opportunities of waste biomass valorization

Waste biomass can be valued for different new products, and there are several options for the treatment of waste biomass, as shown earlier in Chapter 3. In addition to traditional biomass-based fuels and chemicals, some new options exist. These include, for example, upgrading waste biomass to new functional materials and composite materials.

1.4.1 Functional materials from biomass waste

Biomass waste can also contain functional groups, which give unique properties and enable the direct reuse of biomass waste. Especially, the use of biomass waste as adsorbent material for water and gas treatment and as construction material is widely studied (Feizi, 2015; Lee, 2018). Biomass waste is also potential precursor material for functional carbon-based materials, which are produced via hydrothermal treatment or slow pyrolysis followed by physical or chemical activation. The most studied functional materials are biochar, activated carbon (AC), and graphitic carbon, and the properties of them are dependent on the feedstock quality, thermal treatment conditions. Similarly, the properties of functional materials are also dependent on the applications (Tomczyk et al., 2020).

Biochar has shown great application potentials in water and wastewater treatment, soil amendment, and catalytic oxidation of organic pollutants (Liu et al., 2019). However, if processing is incomplete, some pollutants will be formed and released, impacting the environment and health (Fang et al., 2014). AC can be produced from waste biomass through pyrolysis at a temperature above 700°C using chemical or physical activation. AC has a larger specific surface area and smaller microporous structure compared with biochar. AC can be used in electrochemical applications, such as in double-layer capacitors and batteries (dos Reis et al., 2023a; He, 2017; Sankar et al., 2019). Graphitic carbon is commonly obtained from soft carbons (e.g., coke) via heating above 2100°C. The in-plane structure of graphene layers in the graphitic carbon is almost like that in graphite. It is produced by various synthetic methods to increase the degree of graphitization, including direct heating of porous carbons at 2500°C–3000°C, and catalytic graphitization where in-situ graphitic nanostructure was obtained by metal catalysts (Kim et al., 2019). These carbon-based materials not only possess high specific surface area, porous structure, and abundant surface functional groups but also exhibit favorable chemical stability, great performance, and regeneration capacity (dos Reis et al., 2023b; Zhu et al., 2019). Many efforts have been made to study their potential applications as adsorbents, catalysts, electrode materials, and functional composites.

Adsorbents are used in water and gas treatment. Functional groups of adsorbents including carboxyl, hydroxyl, and amide of biomass wastes play a significant role in adsorbing contaminants from liquid phase or natural gas. Utilizing the biomass waste not only lowers the production cost of adsorbents but also improves the recovery efficiency of biomass wastes (Kilpimaa et al., 2014, 2015; Lee, 2018; Mondal, 2009; Monteiro et al., 2016). However, there are some drawbacks of direct use of biomass waste as adsorbent. For example, the removal rate of heavy metals is relatively low due to the slow diffusion or limited surface-active sites. It is also difficult to separate the absorbents from solutions and recycle them for multiple usage. As a result, the conversion of biomass wastes into different forms of carbon-based materials with better adsorption capacities has been investigated (Sattar et al., 2019).

Biochar from thermal treatment is widely used as an adsorbent for removing heavy metals from wastewater. Biochar exhibits a microporous structure and many oxygen-containing functional groups. Biochar also shows higher removal rates than those reported for untreated biomass waste. Biomass-derived AC also can be used as adsorbent for environmental remediation (Bergna et al., 2022). The adsorption capacity of AC usually depends on the precursor, activation methods, physical and chemical pretreatments, type of activating agents, and gasification time (Mehrvarz et al., 2017; Njoku et al., 2014). Especially, after thermal or chemical pretreatments, the surface functional groups and physical and chemical structures of AC can be improved, resulting in better adsorption performance (Georgin et al., 2016).

Overall, the direct use of biomass wastes as adsorbents is a simple and cost-effective way to remove the heavy metals. Through modifying the structure or surface functional groups of biomass wastes, the adsorption capacity could be improved. Additionally, carbon-based adsorbents obtained from biomass wastes usually have higher adsorption capacities due to their larger surface area and higher porosity. However, these processes increased the cost, and the recyclability of absorbents needs to be improved.

1.5 Conclusions

Global annual generation of waste biomass is estimated to be around 140 Gt. Main types of waste biomass are wood and agricultural products, solid waste, and landfill municipal wastes. In addition to traditional energy use, waste biomass can be used to produce polymers, chemicals, and construction materials. Added value of waste biomass is obtained by conversion of it to functional materials or composites. These include, for example, water treatment chemicals and carbon-based materials for energy storage applications.

Management of waste biomass is normally done by mechanical, biological, thermal, and/or chemical treatments or combination of these. The use of waste biomass for electric power or energy (fuels) is widely reported. However, waste biomass is a heterogeneous material, which presents significant management problems and can have even negative environmental impacts. Sustainable biomass processing requires to include technoeconomic, societal, and environmental aspects as foremost priority.

References

Arena, U. (2012). Process and technological aspects of municipal solid waste gasification. A review. *Waste Management, 32*, 625–639. Available from https://doi.org/10.1016/j.wasman.2011.09.025.

Asquer, C., et al. (2017). Biomass ash reutilisation as an additive in the composting process of organic fraction of municipal solid waste. *Waste Management, 69*, 127–135. Available from https://doi.org/10.1016/j.wasman.2017.08.009.

Awasthi, M. K., et al. (2020). Refining biomass residues for sustainable energy and bio-products: An assessment of technology, its importance, and strategic applications in circular bio-economy. *Renewable and Sustainable Energy Reviews, 127*, 109876. Available from https://doi.org/10.1016/j.rser.2020.109876.

Bergna, D., et al. (2022). Activated carbon from hydrolysis lignin: Effect of activation method on carbon properties. *Biomass and Bioenergy, 159*, 106387. Available from https://doi.org/10.1016/j.biombioe.2022.106387, February.

Beyene, H. D., Werkneh, A. A., & Ambaye, T. G. (2018). Current updates on waste to energy (WtE) technologies: A review. *Renewable Energy Focus, 24*, 1–11. Available from https://doi.org/10.1016/j.ref.2017.11.001.

dos Reis, G. S., et al. (2023a). Biomass-derived carbon–silicon composites (C@Si) as anodes for lithium-ion and sodium-ion batteries: A promising strategy towards long-term cycling stability: A mini review. *Electrochemistry Communications, 153*. Available from https://doi.org/10.1016/j.elecom.2023.107536, July.

dos Reis, G. S., et al. (2023b). Preparation of highly porous nitrogen-doped biochar derived from birch tree wastes with superior dye removal performance. *Colloids and Surfaces A: Physicochemical and Engineering Aspects, 669*, 131493. Available from https://doi.org/10.1016/j.colsurfa.2023.131493, February.

Duan, Y., et al. (2019). Evaluation of integrated biochar with bacterial consortium on gaseous emissions mitigation and nutrients sequestration during pig manure composting. *Bioresource Technology, 291*, 121880. Available from https://doi.org/10.1016/j.biortech.2019.121880.

El-Emam. (2012). Energy and exergy analyses of an integrated SOFC and coal gasification system. *International Journal of Hydrogen Energy, 37*, 1689–1697. Available from https://doi.org/10.1016/j.ijhydene.2011.09.139.

Erythropel, H. C., et al. (2018). The Green ChemisTREE: 20 years after taking root with the 12 principles. *Green Chemistry, 20*(9), 1929–1961. Available from https://doi.org/10.1039/C8GC00482J.

Fan, Y. V., et al. (2018). Anaerobic digestion of municipal solid waste: Energy and carbon emission footprint. *Journal of Environmental Management, 223*, 888–897. Available from https://doi.org/10.1016/j.jenvman.2018.07.005.

Fang, J., Leavey, A., & Biswas, P. (2014). Controlled studies on aerosol formation during biomass pyrolysis in a flat flame reactor. *Fuel, 116*, 350–357. Available from https://doi.org/10.1016/j.fuel.2013.08.002.

Feizi. (2015). Removal of heavy metals from aqueous solutions using sunflower, potato, canola and walnut shell residues. *Journal of the Taiwan Institute of Chemical Engineers*, 125–136. Available from https://doi.org/10.1016/j.jtice.2015.03.027, Elsevier Ltd.

Georgin, J., et al. (2016). Preparation of activated carbon from peanut shell by conventional pyrolysis and microwave irradiation-pyrolysis to remove organic dyes from aqueous solutions. *Journal of Environmental Chemical Engineering, 4*(1), 266–275. Available from https://doi.org/10.1016/j.jece.2015.11.018.

Guragain, Y. N., & Vadlani, P. V. (2021). Renewable biomass utilization: A way forward to establish sustainable chemical and processing industries. *Clean Technologies, 3*(1), 243–259. Available from https://doi.org/10.3390/cleantechnol3010014.

He, N. (2017). Engineering biorefinery residues from loblolly pine for supercapacitor applications. *Carbon* (pp. 304–312). Elsevier Ltd. Available from https://doi.org/10.1016/j.carbon.2017.05.056.

Kilpimaa, S., et al. (2014). Removal of phosphate and nitrate over a modified carbon residue from biomass gasification. *Chemical Engineering Research and Design, 92*(10), 1923–1933. Available from https://doi.org/10.1016/j.cherd.2014.03.019.

Kilpimaa, S., et al. (2015). Physical activation of carbon residue from biomass gasification: Novel sorbent for the removal of phosphates and nitrates from aqueous solution. *Journal of Industrial and Engineering Chemistry, 21*, 1354–1364. Available from https://doi.org/10.1016/j.jiec.2014.06.006.

Kim, M. H., et al. (2019). Porous graphitic activated carbon sheets upcycled from starch-based packing peanuts for applications in ultracapacitors. *Journal of Alloys and Compounds, 805*, 1282–1287. Available from https://doi.org/10.1016/j.jallcom.2019.05.359.

Laird, D. A., et al. (2009). Review of the pyrolysis platform for coproducing bio-oil and biochar. *Biofuels, Bioproducts and Biorefining, 3*(5), 547–562. Available from https://doi.org/10.1002/bbb.169.

Lee. (2018). Persimmon leaf bio-waste for adsorptive removal of heavy metals from aqueous solution. *Journal of Environmental Management*, 382–392. Available from https://doi.org/10.1016/j.jenvman.2017.12.080, Elsevier Ltd.

Lee, S., et al. (2019). Waste to bioenergy: A review on the recent conversion technologies. *BMC Energy*, 1–22. Available from https://doi.org/10.1186/s42500-019-0004-7.

Liu, C., et al. (2019). From rice straw to magnetically recoverable nitrogen doped biochar: Efficient activation of peroxymonosulfate for the degradation of metolachlor. *Applied Catalysis B: Environmental, 254*, 312–320. Available from https://doi.org/10.1016/j.apcatb.2019.05.014, April.

Liu, X., et al. (2012). Effect of thermal pretreatment on the physical and chemical properties of municipal biomass waste. *Waste Management, 32*(2), 249–255. Available from https://doi.org/10.1016/j.wasman.2011.09.027.

Lombardi, L., Carnevale, E., & Corti, A. (2015). A review of technologies and performances of thermal treatment systems for energy recovery from waste. *Waste Management, 37*, 26–44. Available from https://doi.org/10.1016/j.wasman.2014.11.010.

Luterbacher, J. S., et al. (2009). Hydrothermal gasification of waste biomass: Process design and life cycle asessment. *Environmental Science and Technology*, 43(5), 1578–1583. Available from https://doi.org/10.1021/es801532f.

Matsakas, L., et al. (2014). Utilization of household food waste for the production of ethanol at high dry material content. *Biotechnology for Biofuels*, 7(1), 1–9. Available from https://doi.org/10.1186/1754-6834-7-4.

Mehrvarz, E., Ghoreyshi, A. A., & Jahanshahi, M. (2017). Surface modification of broom sorghum-based activated carbon via functionalization with triethylenetetramine and urea for CO_2 capture enhancement. *Frontiers of Chemical Science and Engineering*, 11(2), 252–265. Available from https://doi.org/10.1007/s11705-017-1630-6.

Mohammadi, A., & Anukam, A. (2022). In P. Vizureanu (Ed.), *The technical challenges of the gasification technologies currently in use and ways of optimizing them: A review*. Rijeka: IntechOpen. Available from https://doi.org/10.5772/intechopen.102593.

Mondal, M. K. (2009). Removal of Pb(II) ions from aqueous solution using activated tea waste: Adsorption on a fixed-bed column. *Journal of Environmental Management*, 90(11), 3266–3271. Available from https://doi.org/10.1016/j.jenvman.2009.05.025.

Monteiro, R. J. R., et al. (2016). Sustainable approach for recycling seafood wastes for the removal of priority hazardous substances (Hg and Cd) from water. *Journal of Environmental Chemical Engineering*, 4(1), 1199–1208. Available from https://doi.org/10.1016/j.jece.2016.01.021.

Njoku, V. O., et al. (2014). Preparation of activated carbons from rambutan (Nephelium lappaceum) peel by microwave-induced KOH activation for acid yellow 17 dye adsorption. *Chemical Engineering Journal*, 250, 198–204. Available from https://doi.org/10.1016/j.cej.2014.03.115.

Okolie, J. A., et al. (2022). Waste biomass valorization for the production of biofuels and value-added products: A comprehensive review of thermochemical, biological and integrated processes. *Process Safety and Environmental Protection*, 159, 323–344. Available from https://doi.org/10.1016/j.psep.2021.12.049.

Sabater, C., et al. (2020). Valorization of vegetable food waste and by-products through fermentation processes. *Frontiers in Microbiology*, 11, 1–11. Available from https://doi.org/10.3389/fmicb.2020.581997, October.

Sadh, P. K., et al. (2018). Fermentation: A boon for production of bioactive compounds by processing of food industries wastes (by-products). *Molecules (Basel, Switzerland)*, 23(10). Available from https://doi.org/10.3390/molecules23102560.

Sankar, S., et al. (2019). Biomass-derived ultrathin mesoporous graphitic carbon nanoflakes as stable electrode material for high-performance supercapacitors. *Materials and Design*, 169, 107688. Available from https://doi.org/10.1016/j.matdes.2019.107688.

Sattar, M. S., et al. (2019). Comparative efficiency of peanut shell and peanut shell biochar for removal of arsenic from water. *Environmental Science and Pollution Research*, 26(18), 18624–18635. Available from https://doi.org/10.1007/s11356-019-05185-z.

Sharma, S., et al. (2024). Unlocking the potential: A paradigm-shifting approach for valorizing lignocellulosic waste biomass of constructed wetland enabling environmental and societal sustainability. *Industrial Crops and Products*, 207(P1), 117709. Available from https://doi.org/10.1016/j.indcrop.2023.117709.

Singh, B., Szamosi, Z., & Siménfalvi, Z. (2020). Impact of mixing intensity and duration on biogas production in an anaerobic digester: A review. *Critical Reviews in Biotechnology*, 40(4), 508–521. Available from https://doi.org/10.1080/07388551.2020.1731413.

Soobhany, N. (2019). Insight into the recovery of nutrients from organic solid waste through biochemical conversion processes for fertilizer production: A review. *Journal of Cleaner Production*, 241. Available from https://doi.org/10.1016/j.jclepro.2019.118413, 118413.

Tawfik, A., et al. (2022). Sustainable microalgal biomass valorization to bioenergy: Key challenges and future perspectives. *Chemosphere*, 296, 133812. Available from https://doi.org/10.1016/j.chemosphere.2022.133812, February.

Tomczyk, A., Sokołowska, Z., & Boguta, P. (2020). Biochar physicochemical properties: Pyrolysis temperature and feedstock kind effects. *Reviews in environmental science and biotechnology* (pp. 191–215). Netherlands: Springer. Available from https://doi.org/10.1007/s11157-020-09523-3.

Tripathi, N., et al. (2019). Biomass waste utilisation in low-carbon products: Harnessing a major potential resource. *npj Climate and Atmospheric Science*, 2(1). Available from https://doi.org/10.1038/s41612-019-0093-5.

Usmani, Z., et al. (2021). Bioprocessing of waste biomass for sustainable product development and minimizing environmental impact. *Bioresource Technology*, 322, 124548. Available from https://doi.org/10.1016/j.biortech.2020.124548, December 2020.

Wang, K., & Tester, J. W. (2023). Sustainable management of unavoidable biomass wastes. *Green Energy and Resources*, 1(1), 100005. Available from https://doi.org/10.1016/j.gerr.2023.100005.

Wojnowska-Baryła, I., Kulikowska, D., & Bernat, K. (2020). Effect of bio-based products on waste management. *Sustainability (Switzerland)*, 12(5), 1–12. Available from https://doi.org/10.3390/su12052088.

Zhu, L., et al. (2019). Turning biomass waste to a valuable nitrogen and boron dual-doped carbon aerogel for high performance lithium-sulfur batteries. *Applied Surface Science*, 489, 154–164. Available from https://doi.org/10.1016/j.apsusc.2019.05.333, April.

CHAPTER 2

Assessment of wastes for future bioprospecting

Jeetesh Kushwaha, Jyoti Rani, Madhumita Priyadarsini, Kailash Pati Pandey and Abhishek S. Dhoble

School of Biochemical Engineering, Indian Institute of Technology (BHU), Varanasi, Uttar Pradesh, India

2.1 Introduction

Globally, around 17 billion tonnes of solid garbage is produced annually, and 27 billion tonnes is anticipated by the year 2050 (Karak et al., 2012). The cities around the world produce about 1.3 billion tonnes of municipal solid waste, and by 2025, it is predicted that this amount could rise to 2.2 billion tonnes, primarily as a result of population growth, accelerating urbanization, and the improvement of living standards of households in low and middle-income nations (Hoornweg & Bhada-Tata, 2012). The production of waste material is an inevitable by-product of human activity, and poor waste management harms both environmental and human health. Experts describe Municipal Solid Waste (MSW) as garbage thrown from the residential and commercial sectors and materials that are no longer valuable to the user (Varma & Kalamdhad, 2017; Vergara & Tchobanoglous, 2012). The composition of the waste stream is influenced by several variables, including "climate, geographic location, and cultural norms," in addition to economic development (Hoornweg & Bhada-Tata, 2012). Depletion of the ozone layer, environmental issues, human health risks, ecological degradation, including climate change, and the loss of abiotic resources, are all exacerbated by the inappropriate management of MSW via open dumping, open burning, and unhygienic landfilling (Laurent et al., 2014).

Energy demand is increasing worldwide, and crude oil is the only energy source for most countries. Worldwide oil output would drop from around 25 billion barrels per day to 5 billion barrels per day. The global attention to finding a different non-petroleum-oriented energy source has been sparked by this inevitable decline of the world's petroleum supplies in the next years (Zheng et al., 2009). Using waste for biofuel is an excellent alternative energy source to combat the energy crisis head-on. The first-generation biofuels are produced from corn and sugarcane. The second generation of biofuels may also consist of bioethanol made from cash crops like jatropha, maize, and hemp, as well as mixed paper waste that has been segregated from municipal solid trash. Microbes, mostly algae, can be used to make third-generation biofuels. Biodiesel is a fourth-generation biofuel made from vegetable oil (Byadgi & Kalburgi, 2016). Biofuels may be made from various materials, including organic waste, cellulosic biomass, and used chicken feathers. The energy in biomass is produced by organically fixing carbon. Wood, food trash, garden waste, leaves, textiles, rubber, metal (ferrous or non-ferrous metals), leather, and glass comprise municipal solid waste. All these components participate in the process of conversion from waste to energy. Anaerobic digestion, recycling, thermo-chemical waste-to-energy (WTE) technologies, composting, including pyrolysis, incineration, gasification, and ultimate waste disposal practices like landfilling are some of the strategies used for the management of waste (Bundhoo, 2018). According to composition, the percentage of recyclables in MSW is shown in Fig. 2.1.

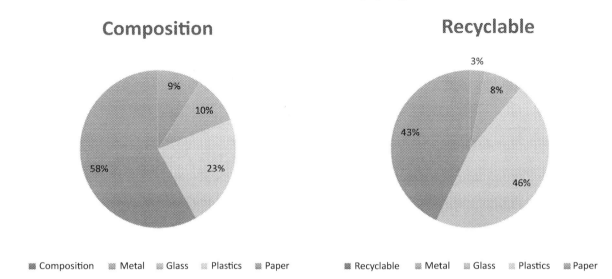

FIGURE 2.1 Pie chart for municipal solid waste (MSW) composition and recyclable waste.

TABLE 2.1 Types of biomass wastes and their composition.

Type of waste	Composition	References
Agricultural waste	Straw (rice, wheat, maize, etc.), Husk, leaves, seeds, kernels, pits, nut shells, vegetables, and fruits (peels, rotten, spoiled, pressings, etc.)	Merlin (2013)
Municipal waste	Solid waste (plastics, papers, glasses, metals), food waste, liquid waste (household used water, sanitary waste), etc.	Barampouti et al. (2019)
Animal Husbandry waste	Pig manure, cow manure, Dairy cattle manure, urinary waste, water waste, slaughterhouse waste etc.	Dhanya et al. (2020)
Algal biomass	Algae and micro-algae	Voloshin et al. (2016)

2.2 Classification of biomass waste

In the scientific community, the concept of biomass is recognized and accepted as organic material generated from live or recently existing lifeforms. Understanding the overall biomass problem is made more accessible by the classification of biomass. Based on various criteria, they can be classified. Based on origin, biomass waste can be segregated into algal biomass, municipal wastes, agricultural wastes, wastes produced from livestock farming, etc. as shown in Table 2.1 (Ochieng et al., 2022).

2.2.1 Agricultural biomass waste

After agricultural harvesting, the remaining residues and tree pruning in the croplands are referred to as agricultural biomass. Crop residues such as straws, kernels, husks, olive pits, walnut pits, nut shells, etc., are among the significant agricultural residues. Agricultural residues can be classified into two broad categories. The first is field residues, which include the leftover parts after crop harvesting, and the second is the process residues, which include the remnants after crop processing. The waste produced by agriculture and agroindustries is rising by 5%–10% each year globally (Wang et al., 2016). According to some data, the United States generates 998 million tonnes of agricultural biomass waste annually (Khan et al., 2015). A very little of this produced agricultural biomass is used as animal fodder. More proportions are burnt in the farmyard. This combustion step can be overcome by switching to green energy production using this leftover agricultural biomass. The utilization of these crop residues might support a large percentage of energy production.

2.2.2 Municipal waste

Municipal wastes provide an enticing biomass source. The most pervasive and significant societal producer of garbage is municipal waste. It is garbage generated by people and households on a day-to-day basis. Each community could be a significant supplier, although contemporary urban cultures provide far more than rural ones. Urban garbage is produced in large quantities. Both liquid and solid wastes are included in municipal waste. Wastes from urban and rural areas that are homogenous or heterogeneous are included in solid waste (Getahun et al., 2012). The majority of substantial trash, including paper, plastic, glass, and metal, can be recycled; however, it must first be segregated before or after collection. After decomposing, the other organic leftovers might be utilized to improve soil nutrients. Municipal liquid waste contains a lot of organic nutrients. These biowastes can be fed to digesters to produce high-value by-products. There will always be a part that cannot be recycled or degraded but can be used to produce energy and requires special technology and management.

2.2.3 Biomass waste from animal husbandry

Biomass waste from animal husbandry comprises the three primary types of livestock waste. They are solid wastes (manure produced from the farm), liquid manure (urinary waste), and water waste. The European agricultural sector produces 1500 Mt of animal manure annually (Koul et al., 2022). Cattle manure and feces comprise 1284 million tonnes of these wastes, whereas pig excrement makes up 295 million tonnes (Holm-Nielsen et al., 2009). Manure amounts rose in a few concentrated places as a result of the movement of livestock and poultry production toward larger and more specialized production units. Untreated manure poses a major hazard to the soil, air, water, animals, and poultry. Pathogen contamination, nutrient-rich liquid manure, and solid waste leaching are potential causes of surface water pollution and soil contamination. Using biogas technologies, up to 57% of methane can be produced from the excrements. So, utilizing these livestock farming wastes for the production of some value-added products like biogas can effectively address some of these environmental issues.

2.2.4 Algal biomass

Algae are becoming more popular as a source of fuel because of their high light-to-biomass conversion efficiency. Microalgae typically complete a developmental cycle in only a few days, making them extremely fast-growing photosynthesizing microorganisms. Some algae have an oil content of up to 50% of their mass. Due to high oil content and rapid biomass synthesis, microalgae are long being considered as possible viable sources for biofuel generation. Microalgae contain different amounts of lipids and fatty acids depending on the growing conditions. Many microalgae produce more oil than the highest-yielding oil crops. According to some estimations, diatom algae potentially might generate 40 tonnes of oil per ha per year. This production is 7–31 times higher than palm oil plants which produce the most vegetable oils and 200 times more than soybean plants. Research scientists, industrialists, and entrepreneurs have recently revived their interest in using microalgae as an alternative biodiesel source. An alternative being researched is biodiesel made from microalgae. As a result, research into the manufacturing of algal oil for biodiesel is picking up momentum.

2.3 Techniques for the extraction of value-added products from biowaste

Organic solid waste management is now a matter of concern, as its improper management may lead to environmental and health problems. Currently, 2 billion tons of solid waste is produced every year, and it is estimated to reach up to 3 billion tons every year.

2.3.1 Biochemical methods

This menace can be converted into a valuable resource through different methods. Anaerobic and aerobic digestion are biochemical methods through which this waste can be converted into wealth.

2.3.1.1 Anaerobic digestion

Among various methods, anaerobic digestion seems to be the most promising approach to converting this waste into wealth. In anaerobic digestion, different metabolic reactions named hydrolysis, acidogenesis,

TABLE 2.2 Comparison of different methods of waste management.

Waste management methods	Nature of method	Value-added products	Drawbacks
Anaerobic digestion	Biological	Biogas and manure	Time-consuming, bad odor
Aerobic digestion	Biological	Manure	Energy consuming
Gasification	Thermochemical	Syngas and char	High maintenance cost
Hydrothermal liquification	Thermochemical	Gas, Biocrude oil, aqueous products	Difficult product separation
Incineration	Thermochemical	Thermal energy	Harmful gases and incineration residue
Pyrolysis	Thermochemical	Biochar, Syn gas, and bio-oil	Economically unviable

acetogenesis, and methanogenesis occur. Greenhouse gases such as methane and carbon dioxide are released from landfills as a result of anaerobic digestion and affect the environment adversely. If the same process is performed under controlled conditions, value-added products like biofuel and manure can be produced by anaerobic digestion. Unarguably methane and hydrogen are cleaner than fossil fuels.

Anaerobic digestion occurs in four steps. The first step is hydrolysis which involves the transformation of insoluble organic materials into soluble organic materials through enzyme-mediated reaction. Anaerobes such as clostridia, bacteroides, and facultative microbes like streptococci are involved in this step. These microorganisms secrete extracellular enzymes to convert larger and more complex molecules into smaller and simpler molecules, making these molecules readily available for absorption. Some of these microorganisms secrete different enzymes, allowing them to break a wide range of substrates, while some are specific and can break only specific substrates.

The second step is Acidogenesis where various facultative and obligatory anaerobic bacteria intake the monomers produced in hydrolysis and degrade them further into short-chain organic acids such as propanoic acids, acetic acids, butyric acids, carbon dioxide, hydrogen, and alcohols.

The third step is Acetogenesis in which Products of acidogenesis are used as, substrates of acetogenesis. Products of acidogenesis that methanogenic bacteria cannot directly convert into methane are transformed into methanogenic substrates. Alcohols and volatile fatty acids (VFA) oxidize and produce acetate and hydrogen.

The last step is Methanogenesis where methanogenic bacteria use the intermediate products and convert them into methane and carbon dioxide under strictly anaerobic conditions.

Different types of biomasses are used as a substrate for the process of anaerobic digestion, such as animal manure, agricultural residues, digestible organic wastes from food and agro-industries, organic fraction of municipal waste, sewage sludge, etc. In addition to substrate, the process of anaerobic digestion depends on various factors. A brief summary of different factors and their effect on the whole process is presented in Table 2.2.

2.3.1.2 Aerobic digestion

In Aerobic digestion, organic matter is broken down in the presence of oxygen. In the presence of oxygen, bacteria degrade the substrate, and carbon dioxide is produced. As substrate decreases, the bacteria die, and it is consumed as a substrate by other bacteria, called endogenous respiration, and the number of participants lowers at this stage. Activated sludge is a process in which organic matter is consumed in the presence of oxygen, and carbon dioxide is produced.

In aerobic digestion, a dense microbial culture is maintained in an aeration tank to oxidize the organic matter present in the wastewater in low hydraulic residence time.

In the activated sludge, wastewater is aerated in the presence of aerobic microbes. After the aeration is settled, biological solids are back into the aerated wastewater, and after that, the removal of biological solids is done through sedimentation.

Aerobic microbial communities are advantageous as they are low maintenance, easy to control and operate, and have relatively low odor. But it has some drawbacks too, such as it produces more sludge, high electricity costs, and no gain in terms of energy.

2.3.2 Thermochemical methods

In general, the following thermochemical methods are used for waste treatment:

2.3.2.1 Gasification

Gasification is incomplete thermal oxidation, resulting in a higher proportion of gaseous products such as hydrogen, carbon monoxide, etc., and a low proportion of solids like char, and ash; other than this, condensable compounds like tar and oil are also produced. Produced gases can be purified and used for different purposes, such as power generation. Thus, gasification produces value-added products from low-value substrates by transforming them into fuels and other products.

Gasification can be divided into the following steps. Drying is performed between 100°C–200°C and reduces the moisture content of the biomass. The second step, pyrolysis is the breakdown of the substrate in absence of oxygen by high temperature. Pyrolysis reduces the volatile matter present in biomass (substrate) and releases harmful gases from the biomass. Now the biomass is reduced to solid charcoal. Condensation of generated hydrocarbon gases can be done for the production of liquid tar. In oxidation, the reaction between oxygen present in the air and carbonized solid biomass occurs, which releases heat, and CO_2 and water are generated. If oxygen is absent, the reduction occurs between 800°C–1000°C temperature.

2.3.2.2 Hydrothermal liquification

Hydrothermal liquefaction (HTL) is a type of thermo-chemical conversion technology that can be used for the conversion of wet fuel stock into energy. In HTL, biomass is hydrolyzed into smaller fragments, after which it's degraded by dehydration, decarboxylation, deoxygenation, and dehydrogenation. Because of the complexity of hydrothermal liquefaction products and feedstock, the exact mechanism of the process is unclear.

A comparative study was done by Beckman and Elliott (1985) yields and properties of obtained oil. Correlation for various parameters like time, pressure, and temperature was developed, and they concluded that fluctuation in key parameters shows a great impact on the product output.

2.3.2.3 Incineration for heat and electricity

In developing countries, incineration is an economical form of waste management. Incineration is combusting the waste in an incinerator, reducing the waste by 80%–90%. Incineration is done at the temperature range of 800°C to 1000°C. The heat produced during the process is used to boil the water in the boiler, and produced steam is used to run the turbine generator for electricity production. The gases produced by incineration contain pollutants. Therefore, proper treatment of the air is required. Fly and bottom ash are obtained as end products of incineration used in the cement industry and for road construction.

2.3.3 Mechanical methods

Some of the non-conventional mechanical methods for the extraction of value-added compounds from biomass waste have been described below:

2.3.3.1 Pulse electric field

Pulse electric field (PEF) is a non-thermal technique that works on the principle of selective disintegration of biological membranes through electroporation and dielectric breakdown to release components of interest. The organic waste to be treated is kept between two electrodes in an electric field, and as the transmembrane potential of the cells exceeds the threshold value (Ec), pores are created, which increase permeability and may rupture the membrane (Kumar et al., 2011). These pores may stay temporarily or permanently. The strength of the electric field for the process varies from 10–80 kV/cm depending upon the period of exposure, frequency of repetition, energy input, and the number of pulses. PEF uses high-intensity voltage that may cause ohmic heating of the sample, increasing the temperature to not more than 40°C, which is undesirable for heat-labile samples. External cooling of the sample is required as a mitigation step.

PEF has been extensively used in the pre-treatment of organic wastes to achieve improved solubilization and the hydrolysis step, which is a rate-limiting step in anaerobic digestion (Capodaglio et al., 2016). PEF has also been investigated in phosphorous recovery from cells by breaking the phospholipid bilayer, and it has shown better efficiency as compared to microwave irradiation (Hu et al., 2018). PEF is also used in sustainable biorefineries for the treatment of organic waste to produce biogas, biodiesel, and bioethanol and extract valuable compounds like polysaccharides, antioxidants, proteins, lipids, and pigments (Eppink et al., 2019).

2.3.3.2 High voltage electric discharge

High voltage electric discharge (HVED) is an efficient technique that may serve as an alternative to conventional methods of extraction (Li et al., 2019). HVED extraction may be carried out in three modes: batch mode, continuous mode, and circulating mode. HVED has been used to treat organic municipal waste and food industry waste. The waste to be treated is kept in an aqueous solution, and energy is injected directly through electrodes submerged into it. A needle electrode with a high voltage of up to 40 kV and a current of up to 10 kA is used to excite electrons of water molecules present in the aqueous phase (Rani et al., 2019). There are two phases of electric discharge in an aqueous solution, namely corona streamer (pre-breakdown) and arc discharge (breakdown). As the breakdown occurs, other events like bubble cavitation, liquid turbulence, overpressure shock waves, and the formation of reactive chemical species also occur. The shock wave propagation and the explosion of cavitation bubbles created through electric discharges result in the fragmentation of biological materials present in the waste (Boussetta et al., 2013). HVED may have applications in the extraction of several bio compounds present in the waste. HVED has been used in extracting polyphenols and proteins from vine shoots, oils from sesame seeds, isothiocyanates from rapeseed press-cake, flavonols, stilbenes, and polyphenols from grape stems, reducing sugars and phenolic compounds from orange and pomegranate peels and flavonoids and polyphenols from peanut shells and spent coffee grounds (Li et al., 2019).

2.3.3.3 Supercritical fluid extraction

Supercritical fluid extraction (SFE) is an emerging method for the extraction of valuable compounds from biomass waste. SFE is an eco-friendly and environmentally benign process for the extraction of value-added products. Any substance can be called a supercritical (SC) fluid at a pressure and temperature above its critical point (known as a triple point) in which different liquid and gas phases do not exist. Carbon dioxide (CO_2) is the most preferred used SC fluid because it is inexpensive, non-flammable, and non-toxic, and it has a critical temperature and pressure of 31°C and 7.4 MPa, respectively (Rani et al., 2019). The low viscosity and minimal surface tension can help CO_2 penetrate the matrix and improve the efficiency of extraction. SC-CO_2 can also be used to extract heat-sensitive products from organic waste because of its adequate critical temperature.

The process is carried out in two steps. First, the cellular matrix is broken, and the SC fluid penetrates the matrix and disturbs the cellular components; second, compounds soluble in SC fluids are extracted. After the completion of the process, the SC fluid is removed (Rani et al., 2019).

Although, SC CO_2 has been used extensively and is a very promising solvent but co-solvents like ethanol, dimethyl ether, propane, etc., have been used to increase the efficiency of extraction of different compounds. Phenolic compounds like osthol from sour orange peel, ferulic acid, vanillic acid, cinnamic acid, and naringenin from pomegranate seeds, quercetin, gallic acid, and mangiferin from mango, resveratrol from passion fruit seeds, and protocatechuic acid, *p*-hydroxybenzoic acid from mangosteen pericarp have been extracted using supercritical fluid extraction (Chai et al., 2021).

2.3.3.4 Ultrasound-assisted extraction

Ultrasound-assisted extraction (UAE) meets the objective of green chemistry extraction. UAE is a very fast method and can be completed in minutes with minimal use of solvents yielding higher-purity products. UAE can be used either as a standalone method or as a step in pre-treatment to enhance extraction efficiency. UAE works on the principle of the cavitation effect that creates high sheer forces in the sample under consideration. The ultrasonic waves produce energy that is transmitted to the aqueous medium, displacing it and generating cycles of low-pressure called rarefaction and high-pressure called compression. The rarefaction cycle creates cavitation bubbles that increase in size on absorbing energy, and when they can no longer absorb energy, the implosive collapse of the bubbles known as cavitation takes place (Rani et al., 2019). This collapse results in the creation of macro-turbulences and micro-mixing that increase the extraction index. The frequency of ultrasound ranges from 20–25 kHz.

The most commonly used sonicators are the bath and probe-type. A probe-type sonicator produces 100 times greater frequency than a bath-type sonicator. UAE has been used to extract antioxidants from pomegranate peels and Zyzyphus lotus fruit, carotenoids (lycopene and β-carotene) from tomato pomace, phenols, antioxidants, and anthocyanins from jaboticaba peels and grape seeds (Chemat et al., 2017).

2.3.3.5 Microwave-assisted extraction

Microwave-assisted extraction (MAE) is widely used in the extraction of bioactive products from organic matter present in fruits, vegetables, and plant waste. The commercially used microwave systems may exceed 100 W when operated at 100% power in the temperature range of 200°C–300°C, which is high enough to rupture cells and extract intracellular compounds. The basic principle behind MAE is based on the ionic conductivity and dipole rotation caused by the microwaves. On application of electromagnetic field, the ions migrate electrophoretically, which is known as ionic conduction that resists the flow of ions producing heat. Then, dipole rotation realigns the dipoles in the electric field (Sparr Eskilsson & Björklund, 2000). The extraction takes place by using microwaves to heat the solvents in which the samples are mixed. The analytes are then partitioned in the solvent from the sample matrix. The MAE process reduces the usage of solvent by almost 10 times, and extraction from multiple samples can be carried out at once.

Bioactive compounds like gallic acid, ferulic acid, flavonoids, anthocyanins, phenolics, and carotenoids have been extracted from peels of apples, bananas, mangoes, blueberries, guava, carrots, and pomegranates using microwave-assisted extraction. These bioactive compounds have also been extracted from vegetable peels and waste from seafood (Alvi et al., 2022).

2.4 High-value bio-products from biomass waste

Biomass waste is generated in huge volumes every year all around the globe (Perea-Moreno et al., 2019). As the world's population grows and urbanization expands, the quantity of biomass waste produced by forestry and agricultural operations, food & beverages, and other industries expands as shown in Fig. 2.2. Sugarcane bagasse, wheat straw, maize straw, rice straw, and rice husk yield 181, 354, 204, 731, and 110 Mt (million tons) per year, consecutively (Sarkar et al., 2012).

2.4.1 Biofuels

Any fuel that is made from biomass, plants, algae, or livestock manure, is referred to as biofuel. Due to its easy acquisition of such feedstock stuff, biofuel is viewed as a renewable energy source compared to fossil fuels like natural gas, coal, and petroleum.

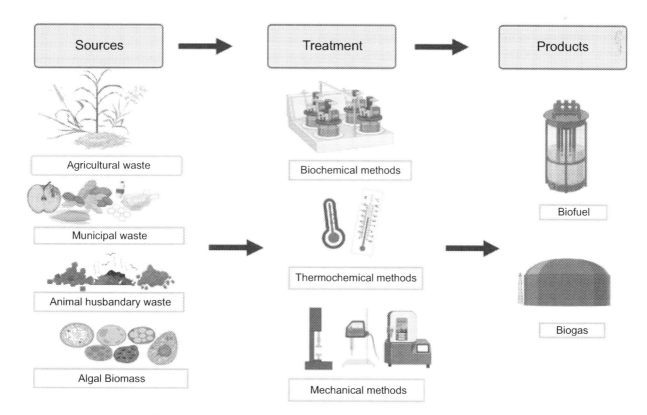

FIGURE 2.2 Schematic representation of the conversion of biowaste into value-added products.

2.4.1.1 Biogas

An ideal answer to the rising energy needs, pollution, fuel costs, and wastewater treatment, biogas generation uses anaerobic co-digestion of food biomass waste with activated sludge to create a more environmentally friendly system (Deena et al., 2022). It has been discovered that co-digestion and pre-treatment of activated sludge and food waste are essential for the effective production of biogas.

The Environmental Protection Agency (EPA) claims that after enteric fermentation and natural petroleum and gas systems, these landfills' municipal organic waste is the largest source of methane emissions (Li et al., 2021). Food waste is composted and naturally digested in a gradual but effective process. Anaerobic digestion may produce biogas, which can then be turned into biofuel and employed to provide heat and electricity (Awasthi et al., 2018). This environmentally friendly method not only produces electricity but also creates a system for managing waste that is both affordable and sustainable, encourages positive health and safety by reducing food waste, avoids water and soil pollution, and aids in environmental protection (Ren et al., 2022).

2.4.1.2 Bioethanol

The concept of recycling waste products from the food industry into very valuable substances like bioethanol has gained popularity in recent years (Hijosa-Valsero et al., 2019). In addition to soluble sugars, protein, polysaccharides, lipids, and lignocellulosic components, including lignin, hemicellulose, and cellulose, food industry wastes are generated in huge numbers across the world (Akbas & Stark, 2016; Hijosa-Valsero et al., 2019). Corn starch from the United States and sugar (sucrose) of sugarcane from Brazil serve as the primary raw materials for bioethanol manufacturing (Chum et al., 2014). In contrast to cereal grains, which account for 60% of this output, sugarcane or its resulting stuff accounts for 40% of the world's total ethanol production (mainly in India, Thailand, and Brazil) (Akbas & Stark, 2016).

Compared to traditional liquid fuels, bioethanol offers benefits. It has been demonstrated that using ethanol instead of gasoline outcomes in an 80% reduction in carbon emissions (Akbas & Stark, 2016). Since the 1970s oil catastrophe, more ethanol has been produced for use in conveyance. Bioethanol production reached over 24 billion gallons in 2014, with the United States and Brazil producing roughly 25% (6.2 billion gallons) and 60% (14.3 billion gallons) of the total (Akbas & Stark, 2016).

2.4.1.3 Biodiesel

Transesterification is one of the procedures that has been developed so far and is well recognized as an economical and environmentally acceptable process of transforming waste cooking oil into biodiesel, a renewable energy source (Tabatabaei et al., 2019). According to stoichiometry, transesterification is the reaction that results in the formation of 3 mol of esters and 1 mol of glycerol from the reaction of 1 mol of triglyceride and 3 mol of light alcohols in the existence of a catalyst (Sinha et al., 2008).

A total of 95% of the world's biodiesel is made with first feedstock, or food crops like rapeseed, wheat, corn, and soy since they are readily available and simple to convert (Prasad et al., 2020). However, it has been suggested to produce biodiesel using second-generation feedstocks, such as animal fats, waste cooking oil, and non-edible vegetable oils, due to worries about the effects of first-generation biodiesel on food security and the environment (Andreo-Martínez et al., 2020). The feedstock's value contributes 70%–95% of the overall cost of making biodiesel, further highlighting the advantageous economic characteristics of waste cooking oil biodiesel due to the cheap price of this feedstock (Amid et al., 2020).

2.4.2 Industrial biocatalyst

For the first time, a novel, environmentally friendly, integrated method for making biodiesel was used, which used enzymatic transesterification facilitated by dried fermented solids with lipase activity and low-value biocatalysts and oil. The dried fermentation solids generated from *Rhizomucor miehei* growth by solid-state fermentation in macaúba cake achieving 92% of ester concentration is used as a biocatalyst in the transesterification processes between macaúba acid oil and alcohol (ethanol or methanol). When implemented in the industry, procedures like the one described here will unquestionably support the expansion of the regional bioeconomy. Similarly, developing a palm biorefinery using three by-products is suggested to obtain a biocatalyst from palm oil cake and fiber in the form of dried fermented solids and utilize it in the production of biodiesel from palm oil deodorizer distillate. The economic assessment of this approach evaluates the cost to produce the biocatalyst to be 25 US dollars per kilogram (Collaço et al., 2021).

2.4.3 Biosurfactants

Biosurfactants have a low level of toxicity, are adaptable, and are acceptable to biological systems. Surfactant is desirable for bioremediation because of their properties, including foaming, dispersion, wetting, coating, de-emulsification, and emulsification (Bustamante et al., 2012). Four hexachlorocyclohexane (HCH) isomers, including alpha (α-HCH), beta (β-HCH), gamma (γ-HCH), and delta (δ-HCH), showed improved solubility, bioavailability, and degradation when biosurfactants trehalolipid *(Rhodococcus erythropolis B7g)*, sophorolipid *(Candida bombicola)*, and rhamnolipid *(Pseudomonas aeruginosa)* used. The biosurfactant dosage of 40–60 g/mL led to the highest solubilization of HCH (Manickam et al., 2012). Biosurfactants can increase hydrocarbon mobility, which can increase oil recovery. Emulsification, mobilization, solubilization, and expanding the hydrocarbon surface area are potential mechanisms for biosurfactant-improved oil recovery. Crude oil is transported by pipelines across a large distance from the site of production to the port of shipment.

2.4.4 Biocontrol agents

Biocontrol Agents employ biological insecticides, which are often made by microbial fermentation. These pesticides are a good substitute for chemical ones since they haven't been linked to any human diseases, Allen and Levy (2013) haven't been proven to pollute ecosystems, and don't generate insect resistance (Smalling et al., 2013). The global crop protection market was worth 49.9 billion dollars in 2016 and had been growing steadily since 2007. Since 2011, values have hovered between 45 to 50 billion dollars.

2.4.5 Microbial pigments

Agricultural items have been industrially processed, which has produced agro-industrial waste. Leaves, Straw, peel, husk, stem, stalk, bagasse, legumes, discarded grains, and other year-round goods are examples of products resulting from agricultural activity. Different agricultural waste sources generate various pigments through microorganisms such as mold, bacteria, fungi, and yeast. It has been reported that microbes from the genera *Monascus, Penicillium, Rhodotorula,* and *Aspergillus* can produce bio pigments (Panesar et al., 2015).

2.4.6 Biopolymers

Different waste biomass obtained from animals, microbes, and plants is used to make a biodegradable polymer, which is a biopolymer. Biopolymers are diverse and essential to humans; they show unique characteristics, have relevance with varied uses, and possess outstanding qualities (Arthington-Skaggs et al., 1999). Biopolymers are a popular idea in industrial applications due to their various characteristics and the process employing limitless resources. In past years, interest in using biodegradable packing materials, health, agriculture, and other purposes has grown. These polymers contain cell attachment property, which is its main benefit and cellulose is the most prevalent kind of biopolymer. It covers 33% of all plant components and is the most prevalent organic substance (Jha & Kumar, 2019).

2.5 Metabolic engineering approaches

The deliberate alteration of cellular metabolism to produce desired molecules is known as metabolic engineering. Recombinant DNA technology allows for the manipulation of different organisms' metabolic pathways. To boost productivity, a number of engineering techniques are used, including metabolic engineering, inverse metabolic engineering, genetic engineering, evolutionary engineering, engineering of different pathways etc. Applications of metabolic engineering are motivated by a goal to deliver a more ecologically friendly method than conventional chemical procedures by reducing costs, increasing productivity, improving efficiency.

The "omics" has shed more light on how microalgae have evolved and adapted to different environmental challenges, reinforcing the involvement of both conserved and unique pathways in lipid synthesis, and the creation of products with added value. The transgenic microalgae produced after metabolic engineering differ in morphology, lipid content, and physiology. Microalgae have been shown to produce more lipids when certain genes related to lipid synthesis are expressed or silenced. Some literature evinced the improvement in lipid production in microalgae post-genetic engineering. Arora et al. (2019) studied the physiological and metabolic flexibility of *Scenedesmus sp.* IITRIND2 under the influence of high salt concentration conditions and got increase in the lipids with negative charge by accumulating sugar and proline upregulation. Malic enzyme's contribution to lipid accumulation in *Phaeodactylum tricornutum* has been examined by Xue et al. (2015). They came to the conclusion that the transgenic *P. tricornutum* had significantly higher levels of malic enzyme overexpression and enzymatic activity, as well as the total lipid content increased by 2 fivefold (Xue et al., 2015).

2.6 Current challenges and future prospects

The use of starch biomass for the manufacturing of biofuel offers a green solution to the world's declining energy resources and rising consumption levels. Starch biomass can also be converted into various value-added products, like acids, solvents, fermentable sugars, beverage softeners, and biofuels. Using bioethanol derived from waste for transportation potentially participates in developing a cleaner and green environment. However, due to a number of technical limitations, it is a difficult process. Developing new, enhanced strains with greater substrate resistance and better production kinetics should be the main emphasis of genetic engineering techniques. The poor productivity of the thermophilic fermentation process is a significant barrier to the industrial use of enzymes and metabolites from these organisms. The physiology of specific microorganisms participating in the process must be researched to increase efficiency. The design of bioprocesses and media, novel bioreactors, and increased production in mesophilic hosts at the molecular level can all be key approaches.

References

Akbas, M. Y., & Stark, B. C. (2016). Recent trends in bioethanol production from food processing byproducts. *Journal of Industrial Microbiology and Biotechnology*, 43(11), 1593–1609. Available from https://doi.org/10.1007/s10295-016-1821-z, http://www.springerlink.com/app/home/journal.asp.

Allen, M. T., & Levy, L. S. (2013). Parkinsons disease and pesticide exposure - A new assessment. *Critical Reviews in Toxicology*, 43(6), 515–534. Available from https://doi.org/10.3109/10408444.2013.798719.

Alvi, T., Asif, Z., & Iqbal Khan, M. K. (2022). Clean label extraction of bioactive compounds from food waste through microwave-assisted extraction technique-A review. *Food Bioscience*, 46, 101580. Available from https://doi.org/10.1016/j.fbio.2022.101580.

Amid, S., Aghbashlo, M., Tabatabaei, M., Hajiahmad, A., Najafi, B., Ghaziaskar, H. S., Rastegari, H., Hosseinzadeh-Bandbafha, H., & Mohammadi, P. (2020). Effects of waste-derived ethylene glycol diacetate as a novel oxygenated additive on performance and emission characteristics of a diesel engine fueled with diesel/biodiesel blends. *Energy Conversion and Management*, 203. Available from https://doi.org/10.1016/j.enconman.2019.112245, https://www.journals.elsevier.com/energy-conversion-and-management.

Andreo-Martínez, P., Ortiz-Martínez, V. M., García-Martínez, N., de los Ríos, A. P., Hernández-Fernández, F. J., & Quesada-Medina, J. (2020). Production of biodiesel under supercritical conditions: State of the art and bibliometric analysis. *Applied Energy*, 264. Available from https://doi.org/10.1016/j.apenergy.2020.114753, https://www.journals.elsevier.com/applied-energy.

Arora, N., Laurens, L. M., Sweeney, N., Pruthi, V., Poluri, K. M., & Pienkos, P. T. (2019). Elucidating the unique physiological responses of halotolerant Scenedesmus sp. cultivated in sea water for biofuel production. *Algal Research*, 37, 260–268.

Arthington-Skaggs, B. A., Jradi, H., Desai, T., & Morrison, C. J. (1999). Quantitation of ergosterol content: Novel method for determination of fluconazole susceptibility of Candida albicans. *Journal of Clinical Microbiology*, 37(10), 3332–3337. Available from https://doi.org/10.1128/jcm.37.10.3332-3337.1999, http://jcm.asm.org.

Awasthi, S. K., Joshi, R., Dhar, H., Verma, S., Awasthi, M. K., Varjani, S., Sarsaiya, S., Zhang, Z., & Kumar, S. (2018). Improving methane yield and quality via co-digestion of cow dung mixed with food waste. *Bioresource Technology*, 251, 259–263. Available from https://doi.org/10.1016/j.biortech.2017.12.063, http://www.elsevier.com/locate/biortech.

Barampouti, E. M., Mai, S., Malamis, D., Moustakas, K., & Loizidou, M. (2019). Liquid biofuels from the organic fraction of municipal solid waste: A review. *Renewable and Sustainable Energy Reviews*, 110, 298–314. Available from https://doi.org/10.1016/j.rser.2019.04.005, https://www.journals.elsevier.com/renewable-and-sustainable-energy-reviews.

Beckman, D., & Elliott, D. C. (1985). Comparisons of the yields and properties of the oil products from direct thermochemical biomass liquefaction processes. *The Canadian Journal of Chemical Engineering*, 63(1), 99–104.

Boussetta, N., Lesaint, O., & Vorobiev, E. (2013). A study of mechanisms involved during the extraction of polyphenols from grape seeds by pulsed electrical discharges. *Innovative Food Science and Emerging Technologies*, 19, 124–132. Available from https://doi.org/10.1016/j.ifset.2013.03.007.

Bundhoo, Z. M. A. (2018). Solid waste management in least developed countries: current status and challenges faced. *Journal of Material Cycles and Waste Management*, 20(3), 1867–1877. Available from https://doi.org/10.1007/s10163-018-0728-3, http://link.springer.de/link/service/journals/10163/index.htm.

Bustamante, M., Durán, N., & Diez, M. C. (2012). Biosurfactants are useful tools for the bioremediation of contaminated soil: A review. *Journal of Soil Science and Plant Nutrition*, 12(4). Available from https://doi.org/10.4067/s0718-95162012005000024, http://www.scielo.cl/pdf/jsspn/v12n4/aop2412.pdf.

Byadgi, Shruti A., & Kalburgi, P. B. (2016). Production of bioethanol from waste newspaper. *Procedia Environmental Sciences*, 35, 555–562. Available from https://doi.org/10.1016/j.proenv.2016.07.040.

References

Capodaglio, A. G., Ranieri, E., & Torretta, V. (2016). Process enhancement for maximization of methane production in codigestion biogas plants. *Management of Environmental Quality: An International Journal*, 27(3), 289–298. Available from https://doi.org/10.1108/MEQ-04-2015-0059, http://www.emeraldinsight.com/info/journals/meq/meq.jsp.

Chai, Y. H., Yusup, S., Kadir, W. N. A., Wong, C. Y., Rosli, S. S., Ruslan, M. S. H., Chin, B. L. F., & Yiin, C. L. (2021). Valorization of tropical biomass waste by supercritical fluid extraction technology. *Sustainability (Switzerland)*, 13(1), 1–24. Available from https://doi.org/10.3390/su13010233, https://www.mdpi.com/2071-1050/13/1/233/pdf.

Chemat, F., Rombaut, N., Sicaire, A. G., Meullemiestre, A., Fabiano-Tixier, A. S., & Abert-Vian, M. (2017). Ultrasound assisted extraction of food and natural products. Mechanisms, techniques, combinations, protocols and applications. A review. *Ultrasonics Sonochemistry*, 34, 540–560. Available from https://doi.org/10.1016/j.ultsonch.2016.06.035, http://www.elsevier.com/inca/publications/store/5/2/5/4/5/1.

Chum, H. L., Zhang, Y., Hill, J., Tiffany, D. G., Morey, R. V., Eng, A. G., & Haq, Z. (2014). Understanding the evolution of environmental and energy performance of the US corn ethanol industry: evaluation of selected metrics. *Biofuels, Bioproducts and Biorefining*, 8(2), 224–240. Available from https://doi.org/10.1002/bbb.1449.

Collaço, A. C. A., Aguieiras, E. C. G., Cavalcanti, E. D. C., & Freire, D. M. G. (2021). Development of an integrated process involving palm industry co-products for monoglyceride/diglyceride emulsifier synthesis: Use of palm cake and fiber for lipase production and palm fatty-acid distillate as raw material. *LWT*, 135. Available from https://doi.org/10.1016/j.lwt.2020.110039, https://www.journals.elsevier.com/lwt.

Deena, S. R., Vickram, A. S., Manikandan, S., Subbaiya, R., Karmegam, N., Ravindran, B., Chang, S. W., & Awasthi, M. K. (2022). Enhanced biogas production from food waste and activated sludge using advanced techniques – A review. *Bioresource Technology*, 355. Available from https://doi.org/10.1016/j.biortech.2022.127234, http://www.elsevier.com/locate/biortech.

Dhanya, B. S., Mishra, A., Chandel, A. K., & Verma, M. L. (2020). Development of sustainable approaches for converting the organic waste to bioenergy. *Science of the Total Environment*, 723. Available from https://doi.org/10.1016/j.scitotenv.2020.138109, http://www.elsevier.com/locate/scitotenv.

Eppink, M. H. M., Olivieri, G., Reith, H., van den Berg, C., Barbosa, M. J., & Wijffels, R. H. (2019). *From current algae products to future biorefinery practices: A review*. Advances in Biochemical Engineering/Biotechnology (pp. 99–123). Netherlands: Springer Science and Business Media Deutschland GmbH. Available from http://www.springer.com/series/10.

Getahun, T., Mengistie, E., Haddis, A., Wasie, F., Alemayehu, E., Dadi, D., Van Gerven, T., & Van Der Bruggen, B. (2012). Municipal solid waste generation in growing urban areas in Africa: Current practices and relation to socioeconomic factors in Jimma, Ethiopia. *Environmental Monitoring and Assessment*, 184(10), 6337–6345. Available from https://doi.org/10.1007/s10661-011-2423-x.

Hijosa-Valsero, M., Garita-Cambronero, J., Paniagua-García, A. I., & Díez-Antolínez, R. (2019). Tomato waste from processing industries as a feedstock for biofuel production. *Bioenergy Research*, 12(4), 1000–1011. Available from https://doi.org/10.1007/s12155-019-10016-7, http://www.springer.com/life+sci/plant+sciences/journal/12155.

Holm-Nielsen, J. B., Al Seadi, T., & Oleskowicz-Popiel, P. (2009). The future of anaerobic digestion and biogas utilization. *Bioresource Technology*, 100(22), 5478–5484. Available from https://doi.org/10.1016/j.biortech.2008.12.046.

HoornwegD., & Bhada-TataP. (2012). *What a waste: A global review of solid waste management*. 29–43. Available from https://doi.org/10.1201/9781315593173-4.

Hu, P., Liu, J., Bao, H., Wu, L., Jiang, L., Zou, L., Wu, Y., Qian, G., & Li, Y. Y. (2018). Enhancing phosphorus release from waste activated sludge by combining high-voltage pulsed discharge pretreatment with anaerobic fermentation. *Journal of Cleaner Production*, 196, 1044–1051. Available from https://doi.org/10.1016/j.jclepro.2018.06.153, https://www.journals.elsevier.com/journal-of-cleaner-production.

Jha, A., & Kumar, A. (2019). Biobased technologies for the efficient extraction of biopolymers from waste biomass. *Bioprocess and Biosystems Engineering*, 42(12), 1893–1901. Available from https://doi.org/10.1007/s00449-019-02199-2, https://rd.springer.com/journal/449.

Karak, T., Bhagat, R. M., & Bhattacharyya, P. (2012). Municipal solid waste generation, composition, and management: The world scenario. *Critical Reviews in Environmental Science and Technology*, 42(15), 1509–1630. Available from https://doi.org/10.1080/10643389.2011.569871.

Khan, S., Waqas, M., Ding, F., Shamshad, I., Arp, H. P. H., & Li, G. (2015). The influence of various biochars on the bioaccessibility and bioaccumulation of PAHs and potentially toxic elements to turnips (Brassica rapa L.). *Journal of Hazardous Materials*, 300, 243–253. Available from https://doi.org/10.1016/j.jhazmat.2015.06.050, http://www.elsevier.com/locate/jhazmat.

Koul, B., Yakoob, M., & Shah, M. P. (2022). Agricultural waste management strategies for environmental sustainability. *Environmental Research*, 206. Available from https://doi.org/10.1016/j.envres.2021.112285, http://www.elsevier.com/inca/publications/store/6/2/2/8/2/1/index.htt.

Kumar, P., Barrett, D. M., Delwiche, M. J., & Stroeve, P. (2011). Pulsed electric field pretreatment of switchgrass and wood chip species for biofuel production. *Industrial and Engineering Chemistry Research*, 50(19), 10996–11001. Available from https://doi.org/10.1021/ie200555u.

Laurent, A., Bakas, I., Clavreul, J., Bernstad, A., Niero, M., Gentil, E., Hauschild, M. Z., & Christensen, T. H. (2014). Review of LCA studies of solid waste management systems - Part I: Lessons learned and perspectives. *Waste Management*, 34(3), 573–588. Available from https://doi.org/10.1016/j.wasman.2013.10.045.

Li, Y., Ni, J., Cheng, H., Zhu, A., Guo, G., Qin, Y., & Li, Y.-Y. (2021). Methanogenic performance and microbial community during thermophilic digestion of food waste and sewage sludge in a high-solid anaerobic membrane bioreactor. *Bioresource Technology*, 342, 125938. Available from https://doi.org/10.1016/j.biortech.2021.125938.

Li, Z., Fan, Y., & Xi, J. (2019). Recent advances in high voltage electric discharge extraction of bioactive ingredients from plant materials. *Food Chemistry*, 277, 246–260. Available from https://doi.org/10.1016/j.foodchem.2018.10.119, http://www.elsevier.com/locate/foodchem.

Manickam, N., Bajaj, A., Saini, H. S., & Shanker, R. (2012). Surfactant mediated enhanced biodegradation of hexachlorocyclohexane (HCH) isomers by Sphingomonas sp. NM05. *Biodegradation*, 23(5), 673–682. Available from https://doi.org/10.1007/s10532-012-9543-z.

Merlin, G. H. (2013). *Anaerobic digestion of agricultural waste: State of the Art and Future Trends Analysis of different techniques used for improvement of biomethanation process: A review* Meena Krishania Variat ion in Anaerobic Digest ion: Need for Process Monit oring.pdf. Nova Science Publishers.

Ochieng, R., Gebremedhin, A., & Sarker, S. (2022). Integration of waste to bioenergy conversion systems: A critical review. *Energies*, 15(7). Available from https://doi.org/10.3390/en15072697, https://www.mdpi.com/1996-1073/15/7/2697/pdf.

Panesar, R., Kaur, S., & Panesar, P. S. (2015). Production of microbial pigments utilizing agro-industrial waste: A review. *Current Opinion in Food Science*, 1(1), 70–76. Available from https://doi.org/10.1016/j.cofs.2014.12.002, http://www.journals.elsevier.com/current-opinion-in-food-science/.

Perea-Moreno, M. A., Samerón-Manzano, E., & Perea-Moreno, A. J. (2019). Biomass as renewable energy: Worldwide research trends. *Sustainability (Switzerland)*, 11(3). Available from https://doi.org/10.3390/su11030863, https://www.mdpi.com/2071-1050/11/3/863/pdf.

Prasad, Shiv, Singh, Anoop, Korres, Nicholas E., Rathore, Dheeraj, Sevda, Surajbhan, & Pant, Deepak (2020). Sustainable utilization of crop residues for energy generation: A life cycle assessment (LCA) perspective. *Bioresource Technology*, 303, 122964. Available from https://doi.org/10.1016/j.biortech.2020.122964.

Rani, J., Indrajeet., Rautela, A., & Kumar, S. (2019). Biovalorization of winery industry waste to produce value-added products. *Biovalorisation of Wastes to Renewable Chemicals and Biofuels* (pp. 63–85). India: Elsevier. Available from https://www.sciencedirect.com/book/9780128179512.

Ren, Yuanyuan, Wang, Chen, He, Ziang, Qin, Yu, & Li, Yu-You (2022). Enhanced biomethanation of lipids by high-solid co-digestion with food waste: Biogas production and lipids degradation demonstrated by long-term continuous operation. *Bioresource Technology*, 348, 126750. Available from https://doi.org/10.1016/j.biortech.2022.126750.

Sarkar, N., Ghosh, S. K., Bannerjee, S., & Aikat, K. (2012). Bioethanol production from agricultural wastes: An overview. *Renewable Energy*, 37(1), 19–27. Available from https://doi.org/10.1016/j.renene.2011.06.045.

Sinha, S., Agarwal, A. K., & Garg, S. (2008). Biodiesel development from rice bran oil: Transesterification process optimization and fuel characterization. *Energy Conversion and Management*, 49(5), 1248–1257. Available from https://doi.org/10.1016/j.enconman.2007.08.010.

Smalling, K. L., Kuivila, K. M., Orlando, J. L., Phillips, B. M., Anderson, B. S., Siegler, K., Hunt, J. W., & Hamilton, M. (2013). Environmental fate of fungicides and other current-use pesticides in a central California estuary. *Marine Pollution Bulletin*, 73(1), 144–153. Available from https://doi.org/10.1016/j.marpolbul.2013.05.028.

Sparr Eskilsson, C., & Björklund, E. (2000). Analytical-scale microwave-assisted extraction. *Journal of Chromatography A*, 902(1), 227–250. Available from https://doi.org/10.1016/S0021-9673(00)00921-3.

Tabatabaei, M., Aghbashlo, M., Dehhaghi, M., Panahi, H. K. S., Mollahosseini, A., Hosseini, M., & Soufiyan, M. M. (2019). Reactor technologies for biodiesel production and processing: A review. *Progress in Energy and Combustion Science*, 74, 239–303. Available from https://doi.org/10.1016/j.pecs.2019.06.001, https://www.journals.elsevier.com/progress-in-energy-and-combustion-science.

Varma, V. S., & Kalamdhad, A. S. (2017). *Solid waste* (pp. 337–368). Informa UK Limited. Available from http://doi.org/10.1201/b22171-12.

Vergara, S. E., & Tchobanoglous, G. (2012). Municipal solid waste and the environment: A global perspective. *Annual Review of Environment and Resources*, 37, 277–309. Available from https://doi.org/10.1146/annurev-environ-050511-122532.

Voloshin, R. A., Rodionova, M. V., Zharmukhamedov, S. K., Nejat Veziroglu, T., & Allakhverdiev, S. I. (2016). Review: Biofuel production from plant and algal biomass. *International Journal of Hydrogen Energy*, 41(39), 17257–17273. Available from https://doi.org/10.1016/j.ijhydene.2016.07.084, http://www.journals.elsevier.com/international-journal-of-hydrogen-energy/.

Wang, Bin, Dong, Faqin, Chen, Mengjun, Zhu, Jingping, Tan, Jiangyue, Fu, Xinmei, Wang, Youzhi, & Chen, Shu (2016). Advances in recycling and utilization of agricultural wastes in China: Based on environmental risk, crucial pathways, influencing factors, policy mechanism. *Procedia Environmental Sciences*, 31, 12–17. Available from https://doi.org/10.1016/j.proenv.2016.02.002.

Xue, J., Niu, Y. F., Huang, T., Yang, W. D., Liu, J. S., & Li, H. Y. (2015). Genetic improvement of the microalga Phaeodactylum tricornutum for boosting neutral lipid accumulation. *Metabolic Engineering*, 27, 1–9. Available from https://doi.org/10.1016/j.ymben.2014.10.002, http://www.elsevier.com/inca/publications/store/6/2/2/9/1/3/index.htt

Zheng, Y., Pan, Z., & Zhang, R. (2009). Overview of biomass pretreatment for cellulosic ethanol production. *International Journal of Agricultural and Biological Engineering*, 2(3), 51–68. Available from https://doi.org/10.3965/j.issn.1934-6344.2009.03.051-068, http://www.ijabe.org/index.php/ijabe/article/view/168/83.

CHAPTER 3

Recent advances in pretreatment of waste biomass

Asish Bisai[1] and Madhumita Patel[2]

[1]Adani Airport Holding Limited, Ahemadabad, Gujarat, India [2]Department of Environmental Science and Engineering, IIT (ISM) Dhanbad, Dhanbad, Jharkhand, India

3.1 Introduction

The demand of energy in the world is increasing year by year. Fossil fuel is the limited source of energy, and it is unsustainable. Therefore this issue can be resolved only by diversification of the energy resources and by introducing various renewable or sustainable energy resources. Global warming and climate changes are also the emerging issues nowadays due to use of these fossil fuels. Therefore the aim is now to promote the clean or sustainable fuels and low carbon fuels. In this regard, biomass has a special role. Biomass is mainly the organic material. The main sources of biomass are wood, crops, food wastes, agricultural waste, municipal waste, etc. Biomass is a renewable resource, and it can be used in the different forms such as pellet, char, ethanol, biodiesel, and biogas. For the production of energy, we should not depend on the one type of biomass sources. Like the generation of energy only from the food base fuel may affect in the food supply in many of the countries. Therefore lignocellulosic biomass is the alternate source of the energy. Lignocellulosic biomass is the most abundant material available on the planet that can be used for the production of energy. Hardwood (oak, eucalyptus), softwood (pine, spruce), grasses, agricultural wastes (wheat straw, rice husks) etc., are the sources of the lignocellulosic biomass. Despite their wide availability, they have various drawbacks, including compositional variance, low density, low energy density, a moisture- absorbing characteristic, etc. To solve these shortcomings of the lignocellulosic biomass, pretreatment is necessary. Usage of biomass has great advantages. It will revitalize rural and agricultural societies. It will lessen the CO_2 emissions by reducing fossil fuels consumption. Besides promoting sustainability, it will increase the safety of the energy and fuel supply (Fig. 3.1). In the pretreatment process, first biomass is collected, processed, stored, left to dry, and decreased in size and followed by hydrolysis process using acid, alkali, or enzymes to convert biomass into hemicellulose and cellulose. This produces sugar, which is fermented to yield a dilute substance through which ethanol is obtained. This produced ethanol is then used as fuel. Lignin is a by-product of hydrolysis process, and it is used to make transportation fuel as well as specialty and commercial chemicals. Fig. 3.1 shows the schematic of biomass pretreatment process to break the bonding between the cellulose, hemicellulose and lignin.

3.2 Types of biomasses

Biomass includes a diverse range of organic sources such as wood, wood residues, straw, grass, agriculture residue, wastes from factories, industries, and kitchen. These sources can be gathered through harvesting, cutting, and clearance from forest management and also can be collected from by-products of food or sawmill waste. Once gathered, they must be transported to a location where they can be pretreated (Fig. 3.2). There are various

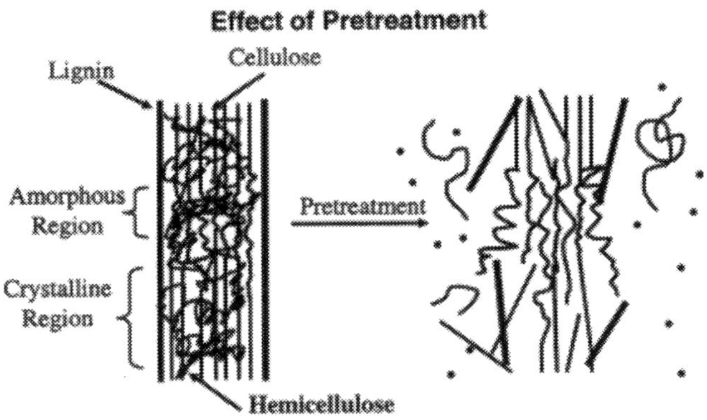

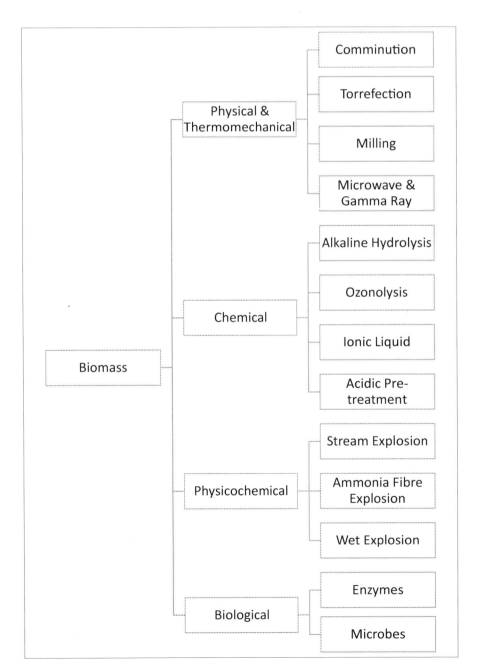

FIGURE 3.1 Schematic illustration of biomass pretreatment process (Mosier et al., 2005). Source: *From Mosier, N. Wyman, C., Dale, B., Elander, R., Lee, Y. Y., Holtzapple, M., Ladisch, M. (2005). Features of promising technologies for pretreatment of lignocellulosic biomass. Bioresource Technology 96(6):673–686, available from: https://doi.org/10.1016/j.biortech.2004.06.025, http://www.elsevier.com/locate/biortech.*

FIGURE 3.2 Pretreatment process of lignocellulosic biomass (Ning et al., 2021).

methods of pretreatment, such as physical, chemical, biological, and thermochemical procedures. Fig. 3.2 shows the general pretreatment method for biomass.

3.3 Different pretreatment processes

3.3.1 Physical pretreatment process

In order to break down the biomass and reduce its size so that cellulose can be easily hydrolyzed, various physical pretreatment techniques are performed. Comminution, milling, extrusion, pyrolysis, and irradiation are some of the main physical pretreatment techniques. The biomass is broken down into smaller pieces during this physical pretreatment, increasing the biomass surface area. By crushing, grinding, cutting, and vibrating, the biomass size is decreased during the comminution pretreatment process. This method aids in the transportation and handling of biomass (Sun & Cheng, 2002). Depending on the qualities of the feedstock, several chemicals are utilized during the extrusion pretreatment process. In addition to thermal pretreatment, other methods include gamma ray, microwave, laser, and electron beam. Due of its rapid heating and minimal energy use, microwave therapy is becoming more popular nowadays. It is simple to use with both acid and alkaline treatments (Eskicioglu et al., 2007).

The refractory component of biomass cannot often be removed by any physical pretreatment procedure. However, the major purpose of physical pretreatment is to lower the size, which assists later when we begin biological or chemical treatment.

There are several physical techniques that can help to reduce the size of the biomass, including chipping, milling, and shredding. The size of the biomass can be reduced to 10–30 mm by chipping, then to 0.2–2 mm by milling and grinding. Compared with chipping process, the milling and grinding processes lower the size of the biomass. The size and nature of the biomass determine the energy needed for physical pretreatment. Compared with agricultural biomass, wood uses more energy. By different experiments, it has been shown that the milling process boosts the generation of bioethanol. But it takes a lot of energy at the business level. However, it has been demonstrated that milling, a physical pretreatment method, may be employed even after biological or chemical pretreatment (Zhu et al., 2010).

3.3.2 Biological pretreatment process

Fungi are mostly employed in biological pretreatment procedure. They exude enzymes that break down the biomass lignin, cellulose, and hemicellulose. It is common to utilize brown-rot, white-rot, and soft-rot fungi that break down cellulose, lignin, and hemicellulose. Utilizing certain fungus, specific biomass can be delignified. *Phanerochaete chrysosporium, Dichomitus squalens*, etc., are employed for materials such as wood and wheat straw (Hatakka, 1994).

However, there are a number of drawbacks to biological pretreatment, such as the lengthy procedure. It might take 10–14 days to complete the treatment. The development of fungus is also crucial in this the pretreatment in order to carry out the test. Fungi can only grow in that particular environment. Microorganisms also eat some of the biomass's carbohydrate-containing components. But by adding some chemicals, we can speed up this biological process. Itoh et al. investigated this biological pretreatment with an organic solvent on beech wood (Itoh et al., 2003). They came to the conclusion that it is possible to cut power use by up to 15% (Itoh et al., 2003).

3.3.3 Chemical pretreatment process

For the treatment of biomass, various types of chemicals are also used such as alkali, acids, organic solvent, ionic and liquids. In an alkali pretreatment, NaOH, KOH, Ca[OH]$_2$ are mainly used, and they help in swelling the feedstocks and also increase surface areas. Delignification of lignocellulosic biomasses mainly occurs by alkaline pretreatment process. This alkali pretreatment process becomes more useful for the low-lignin-containing biomass.

Additionally, the biomass hydrolysis in the acid–base pretreatment is mostly carried out using diluted hydrochloric, phosphoric, and sulfuric acids. When using acids, the corrosive property of the equipment is examined, and acid recovery attempts are made as well.

3.3.4 Physicochemical pretreatment process

The physical and chemical properties of the biomass are impacted by this pretreatment procedure. Lignin and hemicellulose are eliminated during the physicochemical pretreatment. By altering the working conditions, particularly the temperature and pressure, cellulose is also disrupted. This procedure can be carried out with or without the use of chemicals. The most common physicochemical pretreatments are moist oxidation, carbon dioxide explosion, ammonia fiber explosion, and stream explosion.

3.4 Typical biomass component

3.4.1 Cellulose

The most readily accessible polymer in nature is cellulose. It is an organic substance. Its molecular structure is $(C_6H_{10}O_5)_n$. This is made up of a long, linear chain of 1,4-glycosylated linkages. It serves as the foundation of the plant cell wall. It is also found in bacteria, algae, etc. It has been seen that wood only contains up to 40%–50% cellulose, whereas cotton fiber has 90% cellulose content. Cellulose is found in both structured and amorphous manner in plants. Additionally, cellulose offers excellent resilience to acidic and alkaline treatments.

3.4.2 Hemicellulose

Hemicellulose is also one of the most available polymers in the biomass. Normally the portion of hemicellulose in the biomass varies from 20% to 50%. Hemicellulose has lesser molecular weight than cellulose. Hemicellulose also consists of xylose, arabinose, glucose, mannose, galactose, etc.

Hemicellulose composition also varies in the softwood and agricultural waste. Like in agricultural waste, it contains mainly xylan, and in softwood, it contains mainly glucomannan. This xylan is easily removed by either acidic or alkaline pretreatment (Balaban & Uçar, 1999). It has also seen that the degradation of hemicellulose is easier than cellulose. Hemicellulose mainly covers the cellulose of the plant cell wall. Therefore for the degradation of cellulose, hemicellulose must be removed. Also there is chance to form various inhibitory products during hemicellulose degradation.

3.4.3 Lignin

Following cellulose and hemicellulose in terms of availability, lignin is one of the most common polymers in the world. This is mostly present in the cell walls of plants. The major component is phenylpropane. The lignin structure is complex. The plant cell wall is made strong, impermeable, and resistant to microbial action due to the presence of lignin. It functions as a binder and provides structural support to the plant. Also, several research studies have shown that as the lignin component of the biomass is removed, the rate of lignocellulosic biomass digestion rises. There are different pretreatment processes to remove lignin. Therefore in order to remove lignin from a given biomass, we must select a certain pretreatment procedure. By removing the lignin, the biomass begins to swell, and its surface area also expands. Compared with raw biomass, the pretreated biomass is easier to digest in further hydrolysis process.

3.5 Goals of pretreatment process

There are various methods for the pretreatment of lignocellulosic biomass. But to make pretreatment commercially acceptable, the process must have low capital cost as well as low operational cost. Besides that, the pretreatment process should not produce any inhibitory products; otherwise it will affect the further process. There are also different expenses such as the cost of organic solvent, ionic liquid, cost of glucose yield, processing cost, and waste generation. By analyzing all these factors, we have to choose the particular pretreatment process for a particular feedstock. The ideal pretreatment is that in which no inhibitory products form. To analyze the pretreatment process, there is a "severity factor." This factor defines the combined effect of temperature, pressure, duration of the pretreatment, etc. This also helps to compare different pretreatment processes. Pretreatment analysis is done based on the amount of glucose yield, pretreated liquid fermentation, etc.

3.6 Biomass torrefaction

3.6.1 Torrefaction

Torrefaction is a type of thermal pretreatment process. By applying high temperature, ranges between 200°C and 300°C and at atmospheric pressure, the moisture is removed from the biomass, and by this, energy density of the biomass is also increased. Table 3.1 shows the temperature ranges for biomass pretreatment during rorrefaction and its chemcial and physical changes. This process is done in the absence of oxygen. The torrefaction process has the following steps. Firstly, unprocessed wood chips are collected and stored to be used as biomass fuel. Then, the wood chips are dried using a closed-loop belt dryer. Next, the wood chips are heated using microwave technology, creating charcoal-like substances. Then, torrefied wood is made into pellets that produce up to 10% more energy than wood. The purpose of this process is to make a product that resembles coal from a biomass such as wood. Currently, all the countries' energy production depends on coal and natural gas. This coal combustion is the biggest air pollution source and also while burning, large quantities of water have been produced including ash, sludge, and other toxic chemicals. Normal biomass has a lower energy content and has high moisture content, which lowers efficiency. To achieve similar energy input, it requires large amount of mass, which increases the transportation costs. However, torrefied biomass is one alternative to energy production. Energy content is increased since the torrefaction process decreases the weight by 30% while only 10% of energy is lost to the dwarf action. Lastly in the palletization process, making transfers is as easy as coal. The torrefied biomass can be used in resisting coal-fired power plants. There are basically two processes of torrefaction, that is, wet and dry torrefaction. Elevated temperature has an effect on the physical and chemical properties of biomass. The application of elevated heat also changes the color of biomass from brown to black gradually.

After the torrefaction, black color biomass is formed, which is used in the energy production. Besides that, acids, alcohols are formed and gases such as CO_2, methane are also formed. After the torrefaction, water content of the biomass is reduced due to loss of water and volatile components. Due to loss of these compounds, bulk density of the materials is also reduced. But energy density of the biomass is increased significantly. Biomass is also fibrous in nature. By torrefaction, biomass losses this property. This also improves the grindability and also the power consumption is reduced when biomass is torrefied (Bergman & Kiel, 2005).

The variation of the feedstock types, climatic condition, quality of the biomass affect the palletization process. Lignin is the binding component of the lignocellulosic biomass. Therefore palletization depends on the lignin. By torrefaction, these lignin sites came out as hemicellulose is dissolved by torrefaction, and this helps in binding more easily.

3.6.2 Wet torrefaction

Wet torrefaction (WT) is also a type of hydrothermal conversion process. In this process, temperature is applied in between 180°C and 265°C for about 1 hour (van der Stelt et al., 2011; Xiao et al., 2012). Pressure is applied in between 1 and 250 kPa. This is also called as hot water extraction. In this torrefaction step, mainly the

TABLE 3.1 Changes in physical and chemical properties of biomass in different temperature range of torrefaction (Tumuluru et al., 2011).

Torrefaction temperature range (°C)	Features
50–150	- This temperature range is the nonreactive zone as here such changes are not seen in the chemical and physical properties of biomass. - Mainly moisture is released from biomass.
120–150	- In this step dissolution of lignin is started and acts like a binder.
150–200	- This temperature range is the reactive drying zone. - Chemical properties start to change here as C-H bond breaks and from this stage initial structure can't be regained.
200–300	- This stage is known as the destructive drying as here the loss of mass is more. - The entire cell structure of the biomass is disrupted, and its fibrous characteristic is also depleted and behaves like a brittle material. - Lignin, cellulose, hemicellulose all of these decompose and forms char like substances.

hemicellulose and cellulose components are degraded, and further by hydrolysis process, it is transformed to smaller particles. But lignin remains totally unaffected during this pretreatment process. In this process, no chemical or toxic substances are used as input material. The biomass is treated only with water in the inert environment with the help of N_2. This WT process is also cheaper than dry torrefaction. It has been seen that biomass containing more water content is good for this pretreatment. Here water is used in large quantity. The ratio of water and biomass is 6:1. This water is treated at 200°C–265°C. In the output we find solid product, char, carbon dioxide, sugar, etc. (Yan et al., 2010). After this process, char is segregated by filtration and used for palletization.

3.6.3 Dry torrefaction

Dry torrefaction is a thermochemical transformation process. It is performed at the temperature range in between 200°C and 300°C. Also, inert gas is used in the system, and test is performed for 30–60 minutes (van der Stelt et al., 2011).

It has been seen that degradation of hemicellulose only occurs in this process. This makes biomass to repel water. It also breaks down cellulose and from this carboxyl group forms and also water and CO_2 came out. Due to this torrefaction, the color in the lignocellulosic biomass also changes to dark brown. This process reduces the size of the biomass, but the porosity of the biomass is not changed much. Torrefaction process also retains most of the biomass's energy. The reduction of the size occurs in the biomass due to the elevated temperature in each stage of the torrefaction process. Chen and Kuo (2011) experimented on different temperatures, and weight losses are checked in different temperature. The torrefied biomass energy totally depends on the temperature range, reaction time, size and quality of the feedstock used, etc. (van der Stelt et al., 2011). By this process, the microbial resistance, stability of the biomass will enhance. Still more research is required for the commercialization of this process.

3.6.4 Conclusion

Torrefaction is a pretreatment method for biomass that results in breaking down the biomass into several components. Water, volatile compounds, and gases emission take place. The by-products that are derived from the torrefaction have strong fuel characteristics and are also less recalcitrant. When compared with dry torrefaction, WT has a substantial impact on the grass biomass's attributes, such as heating value and energy content. This torrefaction also helps to increase the porosity of biomass and the surface area for further hydrolysis process.

To reduce climate change, the majority of nations want to achieve sustainable development. To achieve this, biomass transformation into ethanol and other products is the foundation of the efficient biorefinery. Additionally, experiments are going on different pretreatment methods to lessen the biomass's recalcitrance in an economical and ecologically safe way. Numerous pretreatment techniques, including physical, chemical, biological, and physicochemical ones, have been employed. Ionic solution, organic solvent pretreatment techniques have recently come to light as viable pretreatment options for the large-scale usage of biomass wastes.

3.7 Ozonolysis pretreatment process

We need to utilize the waste biomass effectively to lessen our existing reliance on fossil fuels and to cut greenhouse gas emissions. For this, lignocellulosic biomass can be used to produce a variety of goods such as biofuel and methane. The first step in the manufacturing process is pretreatment of the biomass, followed by hydrolysis, and finally, fermentation operations.

This pretreatment step modifies biomass structural components to increase enzymes accessibility and to boost the sugar yield. Additionally, the pretreatment procedure lessens the lignocellulosic materials inherent resistance. This ozonolysis pretreatment is a type of chemical pretreatment process. Here ozone is used as an oxidant. Ozone is also easily accessible. The lignin polymer of biomass is degraded during the ozonolysis pretreatment process, while the cellulose and hemicellulose are unaffected. The pH is reduced from 6.5 to 2, when ozone destroys lignin and produces organic acids. Main advantages of the ozonolysis pretreatment are: (1) this treatment can be performed at ambient temperature and pressure and (2) ozonolysis process forms organic acids, which are easily digested by decomposer.

3.7.1 Application of ozonolysis

- Ozone pretreatment is used in pulp paper industry for bleaching.
- Neely performed this ozonolysis test. According to them, the water content in between 25% and 35% and ozone concentration in between 4% and 6% are the optimum conditions for ozonolysis pretreatment (Neely, 1984).
- Vidal and Molinier performed ozonolysis pretreatment process using sawdust as a biomass. They examined that the lignin content decreases and hydrolysis of the biomass increases in this ozonolysis pretreatment process (Vidal & Molinier, 1988).
- Gracia-Cubero et al. performed ozonolysis on wheat, rye, barley, etc. This ozonolysis enhanced the hydrolysis yield up to 90%, and this process also does not produce any inhibitors as well as loss of cellulose is minimum compared with other processes (García-Cubero et al., 2009).
- This ozonolysis pretreatment was tested on wheat straw for 3 hours by Schultz-Jensen et al., and it reduced the lignin level. Additionally, ethanol output is 52%. Before starting the fermentation, the sample has to be cleaned since phenolic and carboxylic chemicals might hinder it. These inhibitors are mostly by-products of lignin (Schultz-Jensen et al., 2011).
- Bule et al. investigated wheat straw pretreatment by ozonolysis and studied that ozone targets lignin either through structural modification or by opening the aromatic ring to reduce recalcitrance for enzymatic hydrolysis (Bule et al., 2013).
- Silverstein et al. used ozone pretreatment method for the conversion of cotton stalks to ethanol by continuously sparging ozone gas over the mixture of cotton stalks and water at 4°C for 30, 60, and 90 minutes. Lignin content is affected much than cellulose and hemicellulose. Ozone did not affect much may be because of an insufficient reaction time, low ozone concentration (Silverstein et al., 2007).

3.7.2 Influence of different process parameters

Ozonolysis is one of the best pretreatment processes. Here ozone is used as solvent. Ozone mainly removes lignin, and thus the structure of biomass is also damaged. Then from the biomass sugar is released by the hydrolysis process. This sugar is used as a source of fuel production, mainly ethanol. During this process, various inhibitory products are also formed such as carboxylic acids. Also, this ozonolysis process depends on various parameters such as design of reactor, water content, pH level, size of the biomass particle, concentration of the ozone used in the process, consumption of the ozone during the process, sugar yields, etc. It has been found that temperature has no effect on ozonolysis pretreatment process. Though this ozonolysis process is simple and easy, but still much research is still required to make this industrial alternative.

3.7.2.1 Effect of reactor design

The reactor is also one of the key factors for ozonolysis. There are different types of reactors used in the ozonolysis process; among them (1) batch reactor and (2) fixed-bed reactors are most common. In batch reactor, ozone concentration remains same throughout the reactor. But in the fixed-bed reactor, the concentration of ozone degrades gradually from the inflow to outflow.

Also, various other types of reactors are used such as rotating cylinder, batch tank reactor, layer fixed bed, bubble column, fixed-bed column are used (Vidal & Molinier, 1988).

Neely also experimented on different types of reactors such as rotating horizontal cylinder and fixed-bed column. After the experiment, they found that rotating cylinder is the most useful among these reactors (Neely, 1984). Vidal and Molinier also experimented on sawdust biomass by using different reactors such as fixed-bed reactor and stirred reactor. They compared these two reactors and found that fixed-bed reactor is most useful than stirred reactor because the stirred reactor consumes less ozone (Vidal & Molinier, 1988).

3.7.2.2 Effect of moisture content

Moisture content is also one of the important parameters in this pretreatment process. The ozonolysis reaction occurs by the transformation of ozone from the gaseous stage to liquid phase. Then from this mobile water phase to bounded phase of water and lastly reaction happens between the ozone and biomasses (Li et al., 2015). Therefore initially there is no ozone present in the water for reaction as all of ozone present in the mobilized water. The time when ozone reaches the immobile water phase, reaction happens.

Neely experimented on different biomass, and they found that different biomass has different moisture content (Neely, 1984). If the water content is low, then ozone can't be able to react with all the biomasses. Also, if the water content is more, then a layer will form, which will delay the reaction process of ozone. It is also observed that this will lead to the more ozone consumption.

The experiment is done on maize by Li et al. (2015). They have seen that water content and the size of the particles are interconnected to each other. According to them, this optimum moisture content totally depends on the ratio of mobile and immobile water.

Due to better ability to collect and hold water, agricultural wastes often have higher water contents. Additionally, the production of inhibitory chemicals in subsequent processes is decreased as moisture levels rise.

3.7.2.3 Effect of particle size

For Ozonolysis pretreatment, particle size is also an important factor. The particle size is related to the surface area, if the particle size reduces, then available surface area increases. The particle size can be reduced by milling process, and this also increases the process cost. However, in this process, an optimum size of particle is also required. It is observed that the reduction of particle size increases the lignin removal.

However, Garcia-Cubero et al. did the experiment with reduced particle size of wheat and poplar sawdust (García-Cubero et al., 2009; Vidal & Molinier, 1988). They concluded that the reduction of particle size does not influence the delignification efficiency. But Souza-Corrêa et al. examined in sugarcane bagasse, and they concluded that reduction of particle size has no effect on lignin removal, but it improves the cellulose conversion (Souza-Corrêa et al., 2014).

3.7.2.4 Effect of pH

In this ozonolysis process, water is used as a moisturizing media. Therefore the pH of water is to be adjusted to control the biomass moisture content. During this ozonolysis process, pH also decreases due to formation of organic acids (Yu et al., 2011). Binder et al. examined ozonolysis pretreatment in wheat straw in alkaline medium. It has been shown that utilization of ozone is more in alkaline medium than in neutral medium. Ozonolysis pretreatment is experimented in rye and wheat biomass with 20% NaOH solution and water. It is observed that 20% NaOH solution decreases the lignin removal and increases the carbohydrate breakdown more than that of the water solution (Oh et al., 2015).

Another experiment is done in pine, and this ozonolysis pretreatment is performed at pH 2, and maximum lignin removal is observed at this pH 2 (Yu et al., 2011). The pH level drops throughout the ozonolysis reaction as it produces carboxylic acids, by-products of lignin degradation.

3.7.3 Conclusion

Although biomass pretreated with ozone is of the best chemical pretreatment process, but further study is needed in this regard. To enhance the efficiency of ozone reactions, it is important to determine the optimum pretreatment condition for different feedstock such as reaction time and ozone consumption. The usage of the ozone should be controlled to use this pretreatment process in the large scale. A cost-effective and efficient process should be made for the production of biofuel. The primary research issue today is to create microbes that can ferment all types of sugars efficiently, even in the presence of inhibitory materials. Also, we have to prepare enzymes that can supply a high sugar yield. Large amounts of ozone are needed for this ozonolysis pretreatment, and this makes ozonolysis commercially unviable (Sun & Cheng, 2002).

However, ozone production is becoming less expensive, and in the future, this pretreatment process could be profitable. Researchers experimented on wheat straw by this ozonolysis process. They concluded that valuable compounds, such as butanol, are also produced during the sugar fermentation step. Butanol is a fuel that may be utilized in vehicles. It has high energy content such as gasoline. It has lower corrosive property. It can be utilized at elevated temperatures.

After the degradation of lignin and hemicellulose, value-added by-products such as biofuel production have been developed from this. Schultz-Jensen et al. examined the release of various degradation products such as phenols, benzene, fatty acids during treatment (Schultz-Jensen et al., 2011). Lignin-derived goods are also tried to recover. Utilizing lignin degradation components adds a significant component to the goal of ozonolysis. It provides cellulose and hemicellulose to microorganisms.

3.8 Ionic liquid pretreatment process

Ionic liquids consist of anion and cations. Ionic liquid breaks down cellulose, starch, and lignin at low-to-moderate temperature under atmospheric pressure. Ionic liquid has advantages over other solvents as it is nonflammable, noncorrosive, thermally stable, no measurable vapor pressure. Also, the properties of ionic liquids can be controlled easily. Ionic liquids have low melting point and also has higher viscosities. Ionic liquids can be recovered after the pretreatment process. Many research studies are carried out to find the suitable ionic liquids for the biomasses. It has been found that 1-butyl-3-methylimidazolium chloride and 1-ethyl-3-methylimidazolium acetate dissolve cellulose at a temperature in between 90°C and 130°C in 24 hours (Oh et al., 2015). Tables 3.2 and 3.3 summerize the different types of ionic liquids used in the ionic liquid pretreatment process and recent literature on the ionic pretreatment, respectively.

3.8.1 Ionic liquid application

These liquids are useful for chemical separation and extraction from aqueous solvents. Ionic liquids are used as electrolytes in batteries. Biofuel is also produced from lignocellulosic biomass by dissolving lignin. Cellulose is densely packed. Lignin forms ester bonds with hemicellulose and hydrogen bonds with cellulose. These bonds make the structure tough to break. Therefore it requires pretreatment otherwise direct hydrolysis process would be very slow. Pretreatment using ionic liquids, mainly cations, has the potential to reduce material recalcitrance

TABLE 3.2 Different types of ionic liquids used in the ionic liquid pretreatment process.

Ionic liquids	Properties	References
Task-specific ionic liquid	• Specific functional groups are added with ionic liquids for specific tasks. • If melting point is greater than 100°C called as task-specific onium salt. • Mainly OH^+, $NH2^+$, and Cl^- are added as the functional groups • TSIL also exhibit negligible vapor pressure.	Davis (2004)
Chiral ionic liquid	• Contains at least one asymmetric/chiral carbon. • Used for biofuel research and as a solvent in medicines.	Tran (2007)
Protic ionic liquid	• This ionic liquid consists of hydrogen ion. • Acidic in nature • Used as fuel in the cell electrolytes and as protein stabilizing agent.	Angell et al. (2007)
Metal salt ionic liquid	• Transition metals, p and f block metals are used. • Examples: $[Cu(Cl)_4]$, $[Co(CO)_4]$, $[NiCl_4]_2$	Lin and Vasam (2005)
Changeable polarity solvent	• Solvent has capacity to change the polarity from high to low and vice versa. • Like for the synthesis of carbamate salts, secondary amines are used as solvents and CO_2 as the activator.	—

TABLE 3.3 Current studies on ionic liquid pretreatment.

Researchers	Feedstock used	Ionic liquid used	Process
Mood et al. (2013)	Barley straw	1-ethyl-3-methyl imidazolium acetate,1-ethyl-3-methyl imidazolium diethyl phosphate, etc.	200 mg of barley straw is mixed with 4.8 g of ionic liquid 110°C for 90 min is heated and continuous stirring is done. Then mixture is allowed to cool and 50 mL of water is added to precipitate out the dissolve cellulose. Precipitates are dried for at least 2 days before the next hydrolysis process.
Cox and Ekerdt (2013)	Yellow pine	1-H-3-methylimidazolium chloride	Temperature is used 110°C–150°C for 5 h. Separation of lignin and hemicellulose occurs from the cell wall of pine.
Fu and Mazza (2011)	Wheat straw	1-ethyl-3-methylimidazolium acetate	By using this ionic liquid, they find that temperature at 158°C, reaction time 3.6 h and initial concentration of 49.5% is the optimum condition for the wheat straw pretreatment.

and enhance the hydrolysis process for a variety of biomass. Also, the high concentration of chlorine mineralizes the cellulose. This process is also depended on various factors such as pretreatment time, temperature, and moisture content. Ionic liquids also react with the aromatic components of biomass, that is, lignin through their cations. BF_4 and PF_6 are good to dissolve lignin and make cellulose available for hydrolysis. Various lignocellulosic biomasses, including Eucalyptus, Pinus radiata, maple wood flour, and sugarcane bagasse, are pretreated with 1-Ethyl-3 methyl imidazolium acetate. Following the pretreatment with ionic solutions, the sugar production is decreased in all of these biomasses. The drop in sugar production varies among various biomasses. These might have occurred because the cellulose in a particular biomass is crystalline in nature, which makes it less soluble in ionic liquids and causes a little drop in sugar. The highest decrease in sugar occurs when the cellulose's crystallinity is extremely low because it has strong solubility in ionic liquid (Uju et al., 2012).

3.8.2 Conclusion

Currently all the countries are demanding for sustainable energy sources and for this lignocellulosic biomass are the great source of it. Therefore the main aim nowadays is to develop an effective and economic pretreatment process. By this pretreatment process, first this lignocellulosic biomass is transferred to sugars and then from sugars to biofuels or other biobased products. The pretreatment of biomass using ionic liquids has been viewed as a potential approach. Ionic liquids are excellent cellulose solvents because of their special chemical and physical characteristics. Though, ionic liquid pretreatment still needs a lot of study to be commercially practical.

3.9 Alkaline pretreatment process

Lignocellulosic biomass is a huge natural source that may be used to make environment friendly goods such as bioethanol and biodiesel. Research has been going on to use these products as an alternative of fossil fuels. There are different types of pretreatment processes such as biological process using enzymes and microbes, also in chemical pretreatment process, various reagents and catalysts are used for different substrates. The methods are stream exploration, acid treatment, alkali treatment, etc. Among these pretreatment processes, alkali pretreatment has emerged out as it has desirable features such as it can be done milder conditions.

Biomass can be divided into different categories depending on where it comes from, such as forest area, agricultural area, municipal waste, and industrial waste. From all these areas, biomass can be collected, and by pretreatment, biofuel can be generated. Agricultural wastes include grains, wheat, barley straws, corn stover, sugarcane bagasse, etc. Biomass is mainly consisted of cellulose, hemicellulose, lignin, and other substances. Cellulose in particular is extremely crystalline, which causes cellulose breakdown by enzymes bit challenging. Additionally, lignin inhibits enzyme activity and microbiological decomposition. Alkali technologies use ammonia fiber explosion, sodium and calcium hydroxide treatment, soaking in liquid ammonia, low liquid ammonia, low-moisture ammonia, etc.

The digestibility of lignocellulosic biomass may be efficiently enhanced by this pretreatment as this pretreatment can effectively eliminate chemical and physical obstacles. Pretreatment at elevated temperature and faster reaction time increases biomass degradability. But this high pressure is produced due to this elevated temperature. Therefore to perform this test also equipment should be well designed. These increase the capital and operating expenses. Various inhibitory products are also formed due to the degradation of biomass by this pretreatment method. Therefore, for the degradation of specific biomass, we must have to choose carefully the specific pretreatment process.

3.9.1 Various alkaline pretreatment technologies

Alkaline pretreatment procedures use a variety of chemicals to increase the biomass's enzymatic digestion. The primary chemicals utilized are calcium hydroxide, sodium hydroxide, sodium carbonate, ammonia, and lime.

In alkaline pretreatment techniques, chemical reagents may also be recovered and used again. Lignin and hemicellulose dissolve during the alkaline pretreatment process.

According to Kim and Holtzapple, the energy needed to delignify agricultural residue such as maize stover by using an alkaline pretreatment was significantly lower than the energy needed to delignify wood using the Kraft method (Kim & Holtzapple, 2006). Therefore, compared with woody species, the alkaline pretreatment is far more successful for delignifying grass.

3.9.1.1 Pretreatment with sodium hydroxide, calcium hydroxides, and sodium carbonate

There are various alkalis used in the pretreatment of lignocellulosic biomass. Among these, sodium hydroxide, calcium hydroxide and ammonia are most common. Calcium hydroxide can remove the lignin most. Therefore calcium hydroxide is generally used in the high-lignin-containing biomasses. Also, calcium hydroxide has several advantages such as it can be easily recycled, it is less expensive, and it has no harmful effect as well (Pinkert et al., 2011).

Kraft process is also a common alkaline pretreatment process for the pretreatment of biomasses. In this process, mainly sodium hydroxide is used. Along with sodium hydroxide, Na_2S is also used, which helps to remove lignin. After the reaction with biomass, NaHS and hydrogen sulfide are formed. This NaOH alkali also helps to remove lignin from biomass. Due to removal of lignin, the biomass starts to swell, so surface area of the biomass also increases. As it removes lignin, so reaction also starts with cellulose and hemicellulose.

Pretreatment with sodium carbonate solution is also an alkali treatment process. (1) To perform this pretreatment, eucalyptus is used as a feedstock. In this process, mainly soaking and percolation of sodium carbonate solution are examined on the biomass. Temperature maintained for percolation and soaking process is 150°C and 60°C, respectively. It is observed that delignification of biomass after this soaking, and percolation process is 8.7% and 21.7%, respectively. From this, we can conclude that the pretreatment process is affected due to temperature and pressure. It is seen than percolation is more effective than soaking pretreatment (Park & Kim, 2012).

3.9.1.2 Ammonia pretreatment

Pretreatment with ammonia is also one of the most used alkaline treatment processes. It increases the digestion of cellulose. Also, the solubilization of lignin increases at higher temperatures. This ammonia pretreatment has several advantages such as ammonia can be reused and recycled. Also, in this process, temperature requirement is also less compared with other process, and it also takes lesser time for reaction. Ammonia fiber explosion method is a type of alkaline pretreatment method. In this process, liquid ammonia is used. In this process, high pressure is applied, but temperature is kept in medium range. When this high pressure is removed, then dissolution of cellulose and lignin occurs. In this ammonia fiber explosion, biomass and ammonia are used in 1:1 ratio. Temperature is kept in between 60°C and 90°C and pressure applied up to 3 MPa for 1 hour in a closed vessel. After half an hour, the pressure is released. This helps to solubilize cellulose and hemicellulose. The structure of the lignin is also changed (Alizadeh et al., 2005). Different types of ammonia pretreatment method has beed summerized in the Table 3.4.

TABLE 3.4 Different types of ammonia pretreatment methods.

Ammonia pretreatment methods	Process parameter condition	Usage of ammonia (gram/gram of biomass)	Usage of water (gram/gram biomass)	Ethanol yield (%)	References
Ammonia fiber explosion method	• Temperature used is 60°C–90°C. • Pressure kept at 3 MPa for 1 hour in a closed vessel.	1:1	—	—	—
Ammonia recycle percolation	• Temperature is kept 170°C for 10 min retention time. • 3.3 g of ammonia is used per gram biomass.	0.5	2.8	71	Kim et al. (2007)
Soaking in aqueous ammonia	• 60°C temperature is kept for 24 hours. • 6 g of ammonia is used per gram biomass.	0.9	5.1	70	Kim and Lee (2005)
Low liquid ammonia	• 30°C temperature is kept for 4 weeks. • 2 g of ammonia is used per gram biomass.	0.5	1.5	70	Li and Kim (2011)
Low moisture anhydrous ammonia	• Optimum temperature is 80°C for 84–96 h.	0.1	1–2.3	89	Yoo et al. (2011)

Ammonia recycle percolation is also a type of ammonia base pretreatment process. In this process, liquid ammonia is used at a temperature range in between 150°C and 180°C for 90 minutes (Kim et al., 2000). In this process, removal of lignin may go up to 80%. During this process, 50% of the hemicellulose is also removed.

As most of the hemicellulose sugar is not degraded in ammonia recycle percolation technique, so a new process is used named as Soaking Aqueous Ammonia (SAA) (Kim & Lee, 2005). This test is performed by Kim et al. on corn at temperature in between 60°C for about 24 hours.

To decrease the ammonia input, Li and Kim et al. performed a test called low liquid ammonia. In this pretreatment process, ammonia input is lower than remaining process (Li & Kim, 2011). Though it uses low ammonia, still it gives same ethanol yield. In all these processes, further washing is needed to recover ammonia.

To resolve this washing process and to reduce the cost of ammonia pretreatment process, another technique is used called as low-moisture anhydrous ammonia. Temperature is kept about 80°C for 3–4 days. The optimum ammonium input that is examined 0.1 g of ammonia per gram of biomass. It is also studied that ethanol yield for this process goes up to 89% (Yoo et al., 2011).

Among all the alkaline pretreatments, ammonia-based treatments are used most extensively for various reasons such as it is easily recoverable, noncorrosive, and nontoxic. It is also inexpensive for production of fertilizer. It is also widely used in the chemical, pharmaceutical, and food industries. It can be easily used and also can be recovered easily because it offers various processing techniques (Kim et al., 2007). Removal or modification of lignin, increase of surface area are the main function of the ammonia pretreatment process.

3.9.1.3 Lime pretreatment

Lime pretreatment of lignocellulosic biomass is the cheapest alkaline pretreatment process. In this pretreatment process, digestion of hemicellulose and lignin occurs at very low temperature and pressure. (Biomass pretreatment: Fundamentals toward application) Chang et al. (1998). Although the lime pretreatment can be used in mild reaction conditions, there are other ways to use it (1) at high temperature for short time period with or without oxygen, such as 100°C–160°C for 6 hours; (2) at moderate temperatures for long time period such as 55°C–65°C for 2–8 weeks; or (3) pretreatment for 1 hour in warm water, without air or oxygen (Sierra et al., 2009).

This lime pretreatment process mainly removes lignin and acetyl groups. The removal of lignin also depends on the type of biomasses. Lignin content is more in the woody biomasses than that of the agricultural plants. Therefore this pretreatment process is helpful for woody biomasses (Pinkert et al., 2011). Besides low cost, there are various advantages of lime pretreatment process such as lime is nontoxic, and it can be reused and recycled. Also, this process is performed at low temperature, so huge energy can be saved at the plant.

3.9.2 Conclusion

There are various advantages of using alkali pretreatment process. Like in this process, heat consumption is less. The treated biomass by alkalis also easily digestible. This process has also less corrosion effect on the equipment. Mainly alkaline treatment helps in the delignification of biomass. But this process has ability to breakdown the biomass in different components such as cellulose, hemicellulose, and lignin. Therefore alkali pretreatment can be used in biorefineries as well.

3.10 Organic solvent pretreatment process

The shortage in fossil fuel and effects of greenhouse gas emission have caused an alternate energy research. Lignocellulosic biomass is a huge source of energy. It has the highest amounts of accessible resources. Many different methods have been created to turn different types of biomasses into ethanol. Pretreatment process is required to decrease cellulose content, to reduce size, to improve hydrolysis and porosity of the feedstock. The properties of biomass are another factor that influences the pretreatment method selection. There are various methods such as physical, chemical, physicochemical, and biological processes. However, the pretreatment of biomass with organic solvents is still under developed and highly recommended due to following reasons.

- Organic solvent pretreatment separates cellulose as a solid. The lignin and hemicellulose present in the biomass dissolve with organic solvent. This delignification increases surface area of cellulose. This helps in further hydrolysis operation.

- Lignin can also be separated by this organic solvent pretreatment of biomass. Good-quality organosolv lignin contains less sulfur, and it has different applications such as sticky substances for coatings, plywood, and grease (Arato et al., 2005).
- Hemicellulose distillation is more effectively performed with the use of organic solvents. Then, hemicellulose can be transformed to ethanol.
- The organic solvent used in this process can be recovered, and it can be reused during pretreatment again.

3.10.1 Application of organic solvent

Initially wood is pretreated with ethanol and HCl to remove lignin. Formic and acetic acids were used for the removal of lignin from biomass. But as time goes on, more and more organic solvents such as acetone, amine, formaldehyde, and alcohol are being utilized to break down biomass into cellulose, hemicellulose, and lignin. Organic solvent is used to treat biomass for a period of time at high pressures and temperatures. By this, lignin and hemicellulose break down to smaller units. Cellulose is also filtered out, and from this ethanol is made. By dilution and precipitation, lignin-containing liquid is separated. Acid catalysts are also applied to fasten lignin removal and lower treatment temperature. In this pretreatment procedure, HCl and H_2SO_4 are utilized as catalysts (Sun & Cheng, 2002). Alcohols with low-boiling points, such as ethanol and methanol, are preferred due to the inexpensive prices. Glycerol and glycol, which have a high boiling point, are also utilized because they work well at low pressure and temperatures.

3.10.2 Pretreatment using alcohol as organic solvent

Alcohols are generally used as solvents for the biomass pretreatments. Among the different types of alcohols, methanol and ethanol are found to be the best alcohols as an organic solvent. These are comparatively cheap and mix with water easily.

3.10.2.1 The use of methanol as an organic solvent

The pretreatment of lignocellulosic biomass can be done using methanol as an organic solvent which is given in Table 3.5. Methanol solvent can be used with or without using catalysts. As a catalyst, liquid acid, magnesium and calcium chloride, hydrogen chloride can be used. If this pretreatment is performed without catalyst, then temperature is increased up to 170°C–200°C. To perform the test without catalyst, it requires high temperature. This pretreatment process mainly helps to remove the cellulose fiber.

There are various advantages to use methanol as an organic solvent like we can recover this methanol easily by distillation, and it is cheaper than other organic solvent. But there are some disadvantages also as methanol is harmful chemical, and it is inflammable in nature. Therefore we have to take care while using this as organic solvent.

3.10.2.2 The use of ethanol as organic solvent

In 1983, Neilson et al. investigated on cotton wood by ethanol pretreatment process, and they observed that the sugar yield is improved up to 2.5 times than untreated biomass (Shafuadeh & Neilson, 1981). Later on, lignol process is developed. This is also a type of organic solvent process, and here also ethanol is used as solvent. Here hemicellulose and lignin are separated first and then from cellulose fraction glucose is produced. And from this glucose, again cellulose is prepared (Arato et al., 2005).

TABLE 3.5 Pretreatment of biomass using methanol as an organic solvent.

Biomass used	Reaction condition	Reaction time	Temperature used	Result	References
Pine wood	60%–80% of methanol with 0.2% of HCl	45 min	170°C	• This process removes 75% of lignin • Hemicellulose is removed	Zhao et al. (2009)
Beach wood	50% aqueous methanol with 0.1% of HCl	45 min	160°C 1°C	• 90% of lignin is removed • Hemicellulose is also removed easily	Zhao et al. (2009)
Poplar and wheat straw	Uses methanol along with a bacteria named *Clostridium thermocellum*	—	175°C–235°C	• In this test mainly the comparison of ethanol yield is done • It has found that wheat straw has more ethanol yield than poplar	Hörmeyer et al. (1988)

Goh et al. experimented in palm fruit by using ethanol pretreatment process (Zhang et al., 2016). H_2SO_4 is used as the catalyst in this test. They performed this test to check the effect of various parameters in this test. They found that the concentration of ethanol is the most important parameter for the glucose yield.

Koo et al. performed ethanol pretreatment process using H_2SO_4 and NaOH as catalyst (Zhang et al., 2016). They performed the test and compared the efficiency of the catalysts. They examined that NaOH as a catalyst can be used in this test at the lower temperature than H_2SO_4. H_2SO_4 prehydrolysis before this organic pretreatment helps to remove more lignin and cellulose during hydrolysis. Also, the usage of ultrasound technique after ethanol pretreatment process increases removal rate of lignin.

Ethanol pretreatment can be done at less solvent concentration. Ethanol is also easily recovered by distillation process. And this will reduce the cost of the pretreatment. But ethanol is a low-boiling point solvent so more pressure is required to use ethanol as the organic solvent. Due to this high pressure requirement, the cost of the equipment may increase. But ethanol is less harmful than methanol.

3.10.3 Pretreatment by organic acids

For the pretreatment of biomass, formic acid and acetic acid are the most commonly used. These two organic solvents are easily soluble in the lignin. The digestion of the biomass increases after this acetic acid pretreatment. But the digestion ability of the acetic acid pretreatment is lesser than ethanol pretreatment.

In this organic acid pretreatment as a catalyst, H_2SO4 and HCl are generally used. These catalysts help in the pretreatment process. This test can be performed at normal temperature. Then the solids that are formed are filtered. Then it is treated with organic solvent. The generated black liquor is then evaporated to remove the organic solvent. Puls et al. used acetic acid solvent as a catalyst (Zhang et al., 2016). In this process temperature is ranged in between 150°C and 200°C, and the reaction time for the treatment used for 2–5 hours. Also, hydrogen peroxide is used as a catalyst to remove lignin content from biomass (Zhang et al., 2016).

Sun et al. treated wheat straw with 90% acetic acid and 4% H_2SO_4(Zhang et al., 2016). They observed that this process helps in the removal of hemicellulose and lignin and getting high cellulose content. Gong et al. experimented acetic acid pretreatment with the help of microwave. They found that 25% of acetic acid and 230 watt of power for 3 minutes retention time are the optimum conditions for this treatment (Zhang et al., 2016). Kootstra et al. experimented on wheat straw by using different acids such as fumaric and sulfuric acids (Zhang et al., 2016). They concluded that for dry wheat straw, we can use fumaric acid as an alternative of sulfuric acid.

3.10.4 Uses of peracetic acid as organic solvent

This is also an oxidizing agent. This acid is made by mixture of hydrogen peroxide and acetic acid. H_2SO_4 is used as a catalyst. This test can be performed at normal temperature. It takes more reaction time. The reaction time can be lessened by using high temperature. Raw biomass can be treated at 80°C temperature for the retention time of 2 hours (Zhao et al., 2009). By this process, cellulose can be easily converted to glucose. Other researchers also concluded that this peracetic acid can be performed easily at low temperature and pressures (Zhao et al., 2009).

3.10.5 Acetone as organic solvent

Acetone pretreatment process is useful mainly for the cellulose-containing biomass. This is performed at 180°C–220°C with sulfuric acid and hydrochloric acid as a catalyst. From cellulose-containing biomass, more glucose is produced (Zhao et al., 2009).

3.10.6 Conclusion

The life cycle assessment method is used to evaluate the viability of goods and processes in light of their effects on the environment and their economic viability. This assessment suggests that biomass biorefineries can reduce greenhouse gas emissions up to 60% when compared with conventional energy systems. Giarola et al. experimented LCA assessment of organic solvent–based biorefineries (Zhang et al., 2016). They concluded that the growth of the sustainable biorefineries depends on the supply chain of biomass, the site of the biorefinery, and quality and amount of the biomass. Inventory cost is also crucial for this biorefinery process.

Kautto et al. performed a test on hardwood (Zhang et al., 2016). They pretreated hardwood with organic solvent and diluted acid for transformation of hardwood to bioethanol. They found that the organic solvent procedure also yields pure lignin fraction and coproducts in addition to bioethanol.

To improve hydrolysis and accomplish bioethanol generation, biomass is pretreated using a variety of techniques. But the choice of the pretreatment method also depends on the biomass characteristics, commercial viability, and environmental effect. Hydrolysis of cellulose is easily done when lignin and hemicellulose are removed. This process also helps to increase the surface area. It has been observed that organic pretreatments have a greater ability to eliminate lignin than other pretreatment methods. The process of decrystallizing cellulose using an organic solvent relies on the organic solvent type. Organic pretreatment also transforms biomass to high-purity components, which helps in easy utilization and recovery of products. Therefore this process helps to form biofuels and also different by-products and biochemicals. Also, hemicellulose is converted to xylitol. From lignin, heat and electricity are generated. Lignin further transforms to phenolics and styrene, which helps to generate concrete plasticizers, brake products, grease, etc. But no commercial-level application of organic solvent treatment of biomass exists. For commercialization of organic solvent pretreatment, various recommendations have to be considered such as: (1) optimization of pretreatment condition such as temperature, reaction time, organic solvent concentration, cost of the process; (2) optimization of lignin base products; and (3) continuous development of pretreatment system for different biomass feedstocks are required.

3.11 Steam explosion pretreatment process

An organized biomass has a strong resistance to biological breakdown. Therefore pretreatment procedure is needed before proceeding to the next breakdown phase. This pretreatment ensures the improvement of biological conversion efficiency and leads to high ethanol formation. Table 3.6 summarizes the different steam explosion pretreatment methods. The good pretreatment increases sugar production, decreasing carbohydrate reduction. Also makes process cheaper. There are several pretreatment techniques, each with its own set of advantages and disadvantages for decomposing biomass and enhancing chemical degradability. Also, this pretreatment should have minimum ecological impact, cost-effective, and no usage of toxic substances. In this process, a screw reactor is employed to transmit the biomass and to generate stress. The equipment's end progressively releases this tension. This continuous stream explosion offers a number of benefits, including improved heat transfer processes and a decrease in the undesirable degradation by-products. By continuously pumping pressure and water vapor into a reactor at temperatures of 220°C and ensuring small retention times, biomass is quickly heated in this method. Water vapor has a role in the biomass first thermal processing. In this technique, high pressures are employed. Water vapor or steam disperses and penetrates in the biomass, contributing in the autohydrolysis reaction. This promotes the dissolution of hemicellulose.

The moisture that has accumulated inside the biomass evaporates when the pressure is abruptly released, disrupting the biomass. The size of the pretreated biomass is likewise reduced by this sudden decompression. This pretreatment results in the production of slurry. Slurry is a mixture of an insoluble solid and a liquid water component. The liquid portion contains hemicellulose sugars as well as solubilized carbohydrates. In the insoluble solid fraction, cellulose, hemicellulose, and lignin are present. This pretreatment step causes hemicellulose to be

TABLE 3.6 Different steam explosion pretreatment methods (Duque et al., 2016).

Various stream explosion method	Catalyst used	Catalyst recovery	Temperature (°C)	Pressure (bar)	Reaction time (min)
Autohydrolysis	–	–	180–230	10–27	2–15
Two-step process	H_2SO_4, SO_2	–	190–210	12	2
Acid catalyst process	H_2SO_4, SO_2, H_3PO_4	–	180–220	10–12	2–10
Ammonia fiber explosion	NH_3	99%	40–140	17–21	2–30
Wet explosion	O_2, H_2O_2, air	–	170–210	15–35	1–30

Adapted from Duque, A., Manzanares, P., Ballesteros, I., Ballesteros, M. (2016). Steam explosion as lignocellulosic biomass pretreatment. Biomass Fractionation Technologies for a Lignocellulosic Feedstock Based Biorefinery *(pp. 349–368). Elsevier Inc., Spain, available from: https://doi.org/10.1016/B978-0-12-802323-5.00015-3, http://www.sciencedirect.com/science/book/9780128023235.*

hydrolyzed and soluble in the liquid fraction, depending on the process parameters. This makes the biomass more porous, which facilitates the digestion of the biomass.

3.11.1 Autohydrolysis process

Because water acts as an acid at very high temperatures, biomass auto-hydrolyzes if no catalytic substance is added to the steam explosion. These acidic conditions cause biomass to produce organic acids. From hemicelluloses acetic acid is formed, which is the main catalyst for the hydrolysis of the biomass. Four main factors such as cellulose recovery from the insoluble solid fraction, cellulose sensitivity to hydrolysis, hemicellulosic sugar restoration, and minimal formation of breakdown chemicals determine the maximum substrate utilization.

3.11.2 Variables effecting steam explosion pretreatment process

Particle size, temperature, water content, residence duration, and biomass composition are all parameters that influence steam explosion pretreatment process.

3.11.2.1 Effect of particle size and water content

Heat transfer and steam utilization are significantly influenced by the biomass's water content and particle size. Heat can move through the reactor more efficiently and quickly in smaller particle sizes. However, large particle size may lead to surface burning, the production of by-products, and partial hydrolysis. Smaller-size particles are difficult to employ in these batch reactors due to their lower density, though. If the biomass has more water, the pores will hold the extra water, which will lower the working temperature and prolong the hydrolysis reaction.

3.11.2.2 Effect of temperature and retention period

Temperature has a significant impact on how a steam explosion process operates. Temperature aids in the recovery of xylose and cellulose but has little impact on the recovery of glucose. As the temperature rises, so does the amount of cellulose that is hydrolyzed by enzymes. Higher temperatures typically result in lower total carbohydrate production. High temperatures can lead to a short retention period. Temperature and retention time are interconnected, therefore a single variable might potentially be used to reflect both of their efficacy. In order to relate temperature and retention time, researchers developed the severity factor. The severity factor evaluates various pretreatment parameters and tries to compare with the restoration rate of solid and liquid fractions of the pretreated biomass. However, researchers found that because the severity factor only takes the boiling phase into account and not the explosion, it may not adequately describe the process of a stream explosion. In order to better understand the stream explosion pretreatment process, they created a new variable called explosion power density, which takes into account the intensity of the explosion phase and supplements this severity component.

3.11.3 Different approaches for steam explosion treatment

3.11.3.1 Two-step pretreatment process

In the first step of the two-stage pretreatment, lower temperature is used to solubilize the hemicellulosic sugars first. Then the segregation of liquid and solid fractions is done from the slurry. From the solid fraction, cellulose hydrolyzes at temperatures over 210°C during the second stage of the pretreatment. Thus, it increases its chemical degradability. In these processes, catalyst as acid is used to enhance hemicellulose solubilization. The two-step pretreatment method has the advantages of increasing ethanol output, employing fewer enzymes, and fully utilizing the biomass. Söderström et al. investigated this two-step pretreatment of softwood treated with SO_2 with the goal of creating the most hexoses. With a temperature of 190°C and a retention period of 2 minutes, the first stage produced the maximum amount of mannose, which was 65% of theoretical. The second steam explosion pretreatment was performed at 220°C for 5 minutes on the solid fraction of the first pretreatment. In this case, the glucose and mannose outputs contributed more than 80% of the theoretical. When compared, two-step pretreatment is found to be more advantageous than one-step pretreatment (Söderström et al., 2002).

3.11.3.2 Acid-catalyzed steam explosion process

In this pretreatment procedure, acid catalysts are required to increase the yield of cellulose hydrolysis and hemicellulose solubility. When softwood is pretreated, acids are an efficient additive that minimizes the harshness of the pretreatment with other organic wastes. The biomass is processed with the chosen catalyst before being added to the steam explosion reactor. The use of acid catalysts has various benefits, including faster breakdown, complete hemicellulose removal, and operation at lower temperatures and shorter reaction periods. H_2SO_4 and SO_2 are more often utilized among the various catalysts. Gaseous sulfur dioxide is often preferred over H_2SO_4 because it is simpler to apply and allows for an equal dispersion of the material across biomass. Some researchers compared the performance of steam explosions catalyzed by SO_2 and H_2SO_4. They found that using H_2SO_4 resulted in more steam being needed to reach the desired temperature. But SO_2 fixed the problem and made it easier to extract more lignin from the pretreated biomass. The usage of H_2SO_4 produced the largest glucose content but the least sugar release. On the other hand, pretreatments with SO_2 did not result in much sugar loss. Between the two, sulfur dioxide typically gives the highest hemicellulosic sugar recovery, resulting in a better fermentability of the sugarcane bagasse. As a result, SO_2 is regarded as the finest option. H_3PO_4 is also utilized as a catalyst in steam explosions. In comparison to H_2SO_4, H_3PO_4 as a catalyst on maize stover gives a low sugar output but produces fewer degradation products. The usage of H_3PO_4 instead of H_2SO_4 has the advantage of being a less corrosive and harmful acid.

3.11.3.3 Ammonia fiber explosion process

Basically, this pretreatment procedure combines alkaline pretreatment with stream explosion. This uses liquid ammonia as a catalyst. A temperature range of 40–140 °C and a pressure range of 250–300 psi are employed. A 30-minute reaction time is maintained. The biomass fibers explode when the imposed pressure is removed. When pressure is released, ammonia begins to evaporate; it is captured and can also be recycled. For materials such as rice straw, maize fiber, rye grass, and switchgrass, this method is highly successful. We can recover all of the consumed ammonia through this procedure, which is quite beneficial.

3.11.3.4 Wet explosion process

Wet explosion is also a type of steam explosion pretreatment process. In this process, thermal hydrolysis and wet oxidation are combined, and this process solubilizes more lignin than normal steam explosion process. Addition of H_2O_2, the oxidizing agent, enhances the decomposition of cellulose, hemicellulose, and lignin. Also, the addition of the oxygen lessens the cost of operation, reducing the furfural production.

3.12 Conclusion

The pretreatment of biomass is a crucial step in the production of biofuels. The biomass disintegrates into its constituent parts through this process. This aids both the reaction process and the production of greater products such as ethanol and biofuel. The most affordable and environmentally friendly pretreatment procedure for the conversion of the biomass has been the subject of research. Additionally, there are several techniques such as stream explosion technique, ozonolysis, and organic solvent treatment. The stream explosion procedure is shown to be the most effective method for removing lignin out of all these methods. Additionally, this procedure produces a lot of glucose. Ionic liquid and ozonolysis processes, however, may extract lignin from biomass up to 88.2% and 84.9%, respectively. These procedures cost significantly more than the stream explosion procedure. However, following the test, this ionic liquid can be retrieved. Additionally, more energy is needed during the ozonolysis process to produce ozone. Since each form of biomass has unique elements, so we are unable to choose a single pretreatment for them all. The kinds and qualities of the biomass determine how effectively each pretreatment method works. We discover that a variety of by-products, including bioethanol, biofuel, and biodiesel, are produced after the pretreatment. It has been discovered that while biohydrogen production is higher in some processes, bioethanol yield is better in some processes. As a result, this area needs additional investigation. In the end, we must concentrate on a pretreatment procedure that lowers operational costs, manufacturing costs, and also improves the output of biofuel, so that ultimately, we may progress toward sustainable growth.

References

Alizadeh, H., Teymouri, F., Gilbert, T. I., & Dale, B. E. (2005). Pretreatment of switchgrass by ammonia fiber explosion (AFEX). *Applied Biochemistry and Biotechnology - Part A Enzyme Engineering and Biotechnology*, 124(1–3), 1133–1141. Available from https://doi.org/10.1385/ABAB:124:1-3:1133, 02732289 United States.

Angell, C. A., Byrne, N., & Belieres, J. P. (2007). Parallel developments in aprotic and protic ionic liquids: Physical chemistry and applications. *Accounts of Chemical Research*, 40(11), 1228–1236. Available from https://doi.org/10.1021/ar7001842.

Arato, C., Pye, E. K., & Gjennestad, G. (2005). The lignol approach to biorefining of woody biomass to produce ethanol and chemicals. *Applied Biochemistry and Biotechnology - Part A Enzyme Engineering and Biotechnology*, 123(1–3), 871–882. Available from https://doi.org/10.1385/abab:123:1-3:0871, 02732289 Humana Press Canada, http://www.humanapress.com/JournalTOC.pasp?issn = 0273-2289.

Balaban, M., & Uçar, G. (1999). The effect of the duration of alkali treatment on the solubility of polyoses. *Turkish Journal of Agriculture and Forestry*, 23(6), 667–671. Available from http://journals.tubitak.gov.tr/agriculture/index.htm.

Bergman, P. C., Kiel, J. H. (2005). Torrefaction for biomass upgrading. *14th European Biomass Conference & Exhibition*, Paris, France.

Bule, M. V., Gao, A. H., Hiscox, B., & Chen, S. (2013). Structural modification of lignin and characterization of pretreated wheat straw by ozonation. *Journal of Agricultural and Food Chemistry*, 61(16), 3916–3925. Available from https://doi.org/10.1021/jf4001988.

Chen, W. H., & Kuo, P. C. (2011). Torrefaction and co-torrefaction characterization of hemicellulose, cellulose and lignin as well as torrefaction of some basic constituents in biomass. *Energy*, 36(2), 803–811. Available from http://www.elsevier.com/inca/publications/store/4/8/3/.

Cox, B. J., & Ekerdt, J. G. (2013). Pretreatment of yellow pine in an acidic ionic liquid: Extraction of hemicellulose and lignin to facilitate enzymatic digestion. *Bioresource Technology*, 134, 59–65. Available from http://www.elsevier.com/locate/biortech.

Davis, J. H. (2004). Task-specific ionic liquids. *Chemistry Letters*, 33(9), 1072–1077. Available from https://doi.org/10.1246/cl.2004.1072.

Duque, A., Manzanares, P., Ballesteros, I., & Ballesteros, M. (2016). Steam explosion as lignocellulosic biomass pretreatment. *Biomass Fractionation Technologies for a Lignocellulosic Feedstock Based Biorefinery* (pp. 349–368). Spain: Elsevier Inc. Available from http://www.sciencedirect.com/science/book/9780128023235.

Eskicioglu, C., Terzian, N., Kennedy, K. J., Droste, R. L., & Hamoda, M. (2007). A thermal microwave effects for enhancing digestibility of waste activated sludge. *Water Research*, 41(11), 2457–2466. Available from http://www.elsevier.com/locate/watres.

Fu, D., & Mazza, G. (2011). Optimization of processing conditions for the pretreatment of wheat straw using aqueous ionic liquid. *Bioresource Technology*, 102(17), 8003–8010. Available from https://doi.org/10.1016/j.biortech.2011.06.023.

García-Cubero, M. T., González-Benito, G., Indacoechea, I., Coca, M., & Bolado, S. (2009). Effect of ozonolysis pretreatment on enzymatic digestibility of wheat and rye straw. *Bioresource Technology*, 100(4), 1608–1613. Available from https://doi.org/10.1016/j.biortech.2008.09.012.

Hatakka, A. (1994). Lignin-modifying enzymes from selected white-rot fungi: production and role from in lignin degradation. *FEMS Microbiology Reviews*, 13(2–3), 125–135. Available from https://doi.org/10.1016/0168-6445(94)90076-0.

Hörmeyer, H. F., Tailliez, P., Millet, J., Girard, H., Bonn, G., Bobleter, O., & Aubert, J.-P. (1988). Ethanol production by *Clostridium thermocellum* grown on hydrothermally and organosolv-pretreated lignocellulosic materials. *Applied Microbiology and Biotechnology*, 29(6), 528–535. Available from https://doi.org/10.1007/bf00260980.

Itoh, H., Wada, M., Honda, Y., Kuwahara, M., & Watanabe, T. (2003). Bioorganosolve pretreatments for simultaneous saccharification and fermentation of beech wood by ethanolysis and white rot fungi. *Journal of Biotechnology*, 103(3), 273–280. Available from http://www.elsevier.com/locate/jbiotec.

Kim, S., & Holtzapple, M. T. (2006). Delignification kinetics of corn stover in lime pretreatment. *Bioresource Technology*, 97(5), 778–785. Available from https://doi.org/10.1016/j.biortech.2005.04.002.

Kim, S. B., Yum, D. M., & Park, S. C. (2000). Step-change variation of acid concentration in a percolation reactor for hydrolysis of hardwood hemicellulose. *Bioresource Technology*, 72(3), 289–294. Available from https://doi.org/10.1016/S0960-8524(99)00081-4.

Kim, T. H., Lee, Y., Sunwoo, C., & Kim, S. J. (2007). Pretreatment of corn stover by low-liquid ammonia recycle percolation process. *Radiation Oncology*, 69(4), 1167–1172.

Kim, T. H., & Lee, Y. Y. (2005). Pretreatment of corn stover by soaking in aqueous ammonia. *Applied Biochemistry and Biotechnology - Part A Enzyme Engineering and Biotechnology*, 124, 1119–1131. Available from https://doi.org/10.1385/ABAB:124:1-3:1119, 02732289 1–3 United States.

Li, C., Wang, L., Chen, Z., Li, Y., Wang, R., Luo, X., Cai, G., Li, Y., Yu, Q., & Lu, J. (2015). Ozonolysis pretreatment of maize stover: The interactive effect of sample particle size and moisture on ozonolysis process. *Bioresource Technology*, 183, 240–247. Available from http://www.elsevier.com/locate/biortech.

Li, X., & Kim, T. H. (2011). Low-liquid pretreatment of corn stover with aqueous ammonia. *Bioresource Technology*, 102(7), 4779–4786. Available from https://doi.org/10.1016/j.biortech.2011.01.008.

Lin, I. J. B., & Vasam, C. S. (2005). Metal-containing ionic liquids and ionic liquid crystals based on imidazolium moiety. *Journal of Organometallic Chemistry*, 690(15), 3498–3512. Available from https://doi.org/10.1016/j.jorganchem.2005.03.007.

Mood, S. H., Golfeshan, A. H., Tabatabaei, M., Abbasalizadeh, S., & Ardjmand, M. (2013). Comparison of different ionic liquids pretreatment for barley straw enzymatic saccharification. *3 Biotech*, 3(5), 399–406. Available from https://doi.org/10.1007/s13205-013-0157-x.

Mosier, N., Wyman, C., Dale, B., Elander, R., Lee, Y. Y., Holtzapple, M., & Ladisch, M. (2005). Features of promising technologies for pretreatment of lignocellulosic biomass. *Bioresource Technology*, 96(6), 673–686. Available from https://doi.org/10.1016/j.biortech.2004.06.025.

Neely, W. C. (1984). Factors affecting the pretreatment of biomass with gaseous ozone. *Biotechnology and Bioengineering*, 26(1), 59–65. Available from https://doi.org/10.1002/bit.260260112.

Ning, P., Yang, G., Hu, L., Sun, J., Shi, L., Zhou, Y., Wang, Z., & Yang, J. (2021). Recent advances in the valorization of plant biomass. *Biotechnology for Biofuels*, 14(1). Available from https://doi.org/10.1186/s13068-021-01949-3.

Oh, Y. H., Eom, I. Y., Joo, J. C., Yu, J. H., Song, B. K., Lee, S. H., Hong, S. H., & Park, S. J. (2015). Recent advances in development of biomass pretreatment technologies used in biorefinery for the production of bio-based fuels, chemicals and polymers. *Korean Journal of Chemical Engineering*, 32(10), 1945–1959. Available from http://www.springerlink.com/content/120599/.

Park, Y. C., & Kim, J. S. (2012). Comparison of various alkaline pretreatment methods of lignocellulosic biomass. *Energy*, 47(1), 31–35. Available from http://www.elsevier.com/inca/publications/store/4/8/3/.

Pinkert, A., Goeke, D. F., Marsh, K. N., & Pang, S. (2011). Extracting wood lignin without dissolving or degrading cellulose: Investigations on the use of food additive-derived ionic liquids. *Green Chemistry*, *13*(11), 3124–3136. Available from https://doi.org/10.1039/c1gc15671c.

Schultz-Jensen, N., Kádár, Z., Thomsen, A. B., Bindslev, H., & Leipold, F. (2011). Plasma-assisted pretreatment of wheat straw for ethanol production. *Applied Biochemistry and Biotechnology*, *165*(3–4), 1010–1023. Available from https://doi.org/10.1007/s12010-011-9316-x.

Shafuadeh, R. G. K., & Neilson, M. J. (1981). Evaluation of organosolv pulp as a suitable substrate for rapid enzymatic hydrolysis. *Biotechnology and Bioengineering*, *25*, 609–612.

Sierra, R., Granda, C. B., & Holtzapple, M. T. (2009). *Lime pretreatment* (vol. 581, pp. 115–124). Springer Science and Business Media LLC. Available from http://doi.org/10.1007/978-1-60761-214-8_9.

Silverstein, R. A., Chen, Y., Sharma-Shivappa, R. R., Boyette, M. D., & Osborne, J. (2007). A comparison of chemical pretreatment methods for improving saccharification of cotton stalks. *Bioresource Technology*, *98*(16), 3000–3011. Available from https://doi.org/10.1016/j.biortech.2006.10.022.

Souza-Corrêa, J. A., Oliveira, C., Nascimento, V. M., Wolf, L. D., Gómez, E. O., Rocha, G. J. M., & Amorim, J. (2014). Atmospheric pressure plasma pretreatment of sugarcane bagasse: The influence of biomass particle size in the ozonation process. *Applied Biochemistry and Biotechnology*, *172*(3), 1663–1672. Available from https://doi.org/10.1007/s12010-013-0609-0.

Söderström, J., Pilcher, L., Galbe, M., & Zacchi, G. (2002). *Two-step steam pretreatment of softwood with SO_2 impregnation for ethanol production* (vol 98–100, pp. 5–11). Springer nature. Available from http://doi.org/10.1007/978-1-4612-0119-9_1.

Sun, Y., & Cheng, J. (2002). Hydrolysis of lignocellulosic materials for ethanol production: A review. *Bioresource Technology*, *83*(1), 1–11. Available from http://www.elsevier.com/locate/biortech.

Tran, C. D. (2007). Ionic liquids for and by analytical spectroscopy. *Analytical Letters*, *40*(13), 2447–2464. Available from https://doi.org/10.1080/00032710701583417.

Tumuluru, J. S., Sokhansanj, S., Hess, J. R., Wright, C. T., & Boardman, R. D. (2011). A review on biomass torrefaction process and product properties for energy applications. *Industrial Biotechnology*, *7*(5), 384–401.

Uju., Shoda, Y., Nakamoto, A., Goto, M., Tokuhara, W., Noritake, Y., Katahira, S., Ishida, N., Nakashima, K., Ogino, C., & Kamiya, N. (2012). Short time ionic liquids pretreatment on lignocellulosic biomass to enhance enzymatic saccharification. *Bioresource Technology*, *103*(1), 446–452. Available from https://doi.org/10.1016/j.biortech.2011.10.003.

van der Stelt, M. J. C., Gerhauser, H., Kiel, J. H. A., & Ptasinski, K. J. (2011). Biomass upgrading by torrefaction for the production of biofuels: A review. *Biomass and Bioenergy*, *35*(9), 3748–3762. Available from https://doi.org/10.1016/j.biombioe.2011.06.023.

Vidal, P. F., & Molinier, J. (1988). Ozonolysis of lignin - Improvement of in vitro digestibility of poplar sawdust. *Biomass*, *16*(1), 1–17. Available from https://doi.org/10.1016/0144-565(88)90012-1.

Xiao, L. P., Shi, Z. J., Xu, F., & Sun, R. C. (2012). Hydrothermal carbonization of lignocellulosic biomass. *Bioresource Technology*, *118*, 619–623. Available from https://doi.org/10.1016/j.biortech.2012.05.060.

Yan, W., Hastings, J. T., Acharjee, T. C., Coronella, C. J., & Vásquez, V. R. (2010). Mass and energy balances of wet torrefaction of lignocellulosic biomass. *Energy and Fuels*, *24*(9), 4738–4742. Available from https://doi.org/10.1021/ef901273n, 15205029 United States.

Yoo, C. G., Nghiem, N. P., Hicks, K. B., & Kim, T. H. (2011). Pretreatment of corn stover using low-moisture anhydrous ammonia (LMAA) process. *Bioresource Technology*, *102*(21), 10028–10034. Available from https://doi.org/10.1016/j.biortech.2011.08.057.

Yu, Z., Jameel, H., Chang, Hm, & Park, S. (2011). The effect of delignification of forest biomass on enzymatic hydrolysis. *Bioresource Technology*, *102*(19), 9083–9089. Available from https://doi.org/10.1016/j.biortech.2011.07.001.

Zhang, K., Pei, Z., & Wang, D. (2016). Organic solvent pretreatment of lignocellulosic biomass for biofuels and biochemicals: A review. *Bioresource Technology*, *199*, 21–33. Available from http://www.elsevier.com/locate/biortech.

Zhao, X., Cheng, K., & Liu, D. (2009). Organosolv pretreatment of lignocellulosic biomass for enzymatic hydrolysis. *Applied Microbiology and Biotechnology*, *82*(5), 815–827. Available from https://doi.org/10.1007/s00253-009-1883-1.

Zhu, J. Y., Pan, X., & Zalesny, R. S. (2010). Pretreatment of woody biomass for biofuel production: Energy efficiency, technologies, and recalcitrance. *Applied Microbiology and Biotechnology*, *87*(3), 847–857. Available from https://doi.org/10.1007/s00253-010-2654-8.

CHAPTER

4

Emerging technologies for waste biomass pretreatment: pros and cons

Subhrajit Roy and Saikat Chakraborty

Biological Systems Engineering, Plaksha University, Mohali, Punjab, India

4.1 Introduction

4.1.1 Waste biomass as a potential substrate for biorefinery

The production of bioenergy from waste biomass has the potential to substantially decrease greenhouse gas emissions. Identifying substrates that don't compete with food sources is therefore crucial. Waste biomass can be used as an abundant and sustainable feedstock for biorefinery operations, including agricultural residues, forestry residues, food waste, and municipal solid waste. It is possible to effectively optimize the biofuel production process to achieve a desired product distribution during scale-up and commercialization, by figuring out the reaction mechanism and the pretreatment method for the conversion of such waste biomass into bioenergy precursors (Roy & Chakraborty, 2023). Waste biomass has a great deal of potential as a feedstock for biorefineries. Biorefineries work to transform biomass into a range of useful goods, including biofuels, biochemicals, bioplastics, and other things.

Large amounts of waste biomass are produced as a by-product of multiple industries and human activities. Because of its accessibility, it is a suitable feedstock for biorefineries because it provides a steady supply. By preventing waste biomass from ending up in landfills and lowering the greenhouse gas emissions connected with its disposal, using it as a feedstock has a positive impact on the environment. By turning waste into useful products, it encourages a circular economy. Given that disposing of waste biomass can be expensive, it is frequently accessible for little or even no cost. The overall cost of acquiring feedstock for biorefineries can be decreased by using waste biomass as a substrate. It can be converted via a variety of techniques, such as biological (Paul & Chakraborty, 2019), thermochemical (Gaikwad & Chakraborty, 2014), and hybrid ones. By turning waste biomass into goods that are in great demand, biorefineries give people the chance to get the most value out of their waste biomass. This promotes resource efficiency and the growth of a circular, sustainable bioeconomy. By developing new businesses, employment opportunities, and value chains in the waste management and bio-based sectors, the construction of biorefineries that use waste biomass can encourage regional economic development. However, it's crucial to keep in mind that using waste biomass in biorefineries may present difficulties such feedstock heterogeneity, compositional unpredictability, and the necessity for effective pretreatment technology. Continuous research, technical development, and partnerships among stakeholders in the biomass supply chain and biorefinery sectors are required to tackle these challenges.

4.1.2 Role of pretreatment in generating bioenergy from waste biomass

Waste biomass is converted into bioenergy by a series of sequential processes such as pretreatment, hydrolysis, and fermentation. Pretreatment is the most crucial first step among these. Prior to hydrolysis and fermentation, key components of the substrate are separated during pretreatment (Dutta & Chakraborty, 2015). The production of bioenergy from waste biomass depends heavily on pretreatment. By breaking the complex structure, improving accessibility, and boosting reactivity, it gets the biomass ready for subsequent conversion procedures. Waste biomass, such as lignocellulosic materials, has a complex and hard structure made of cellulose, hemicellulose, and lignin (Roy & Chakraborty, 2019). By eliminating or degrading lignin and partially depolymerizing hemicellulose, pretreatment techniques try to disturb this structure and make the biomass more susceptible to enzymatic or microbial hydrolysis.

Pretreatment improves the efficiency of enzymatic hydrolysis, the process in which enzymes break down complex carbohydrates (cellulose and hemicellulose) into simpler sugars (glucose, xylose, etc.). By removing barriers such as lignin and opening up the biomass structure, pretreatment increases the accessibility of enzymes to the carbohydrate components, leading to higher sugar yields. Pretreatment liberates the encapsulated sugars present in biomass by breaking down the complex bonds holding them within the cellulosic matrix (Dutta & Chakraborty, 2016). This facilitates the subsequent fermentation process, where microorganisms can convert the released sugars into bioenergy products such as bioethanol, biogas, or other advanced biofuels.

Waste biomass can contain inhibitory compounds, such as phenolics, organic acids, and furans, which can hinder enzymatic activity or microbial fermentation (Paul & Chakraborty, 2019). Certain pretreatment methods can help remove or degrade these inhibitory compounds, enhancing the overall efficiency of bioconversion processes. Effective pretreatment enhances the overall efficiency of bioenergy generation by increasing sugar yields, reducing processing time, and optimizing subsequent conversion processes. This leads to higher bioenergy production rates, improved economics, and a more sustainable utilization of waste biomass. The choice of pretreatment method depends on the biomass feedstock, target bioenergy product, process economics, and environmental considerations. Pretreatment is an essential step in unlocking the energy potential of waste biomass and enabling a more sustainable and efficient bioenergy production.

4.2 Pretreatment of waste biomass for bioenergy production

Waste lignocellulosic biomass has a high energy content, is inexpensive, and has a high concentration of cellulose (50%–60%), making it an attractive feedstock for biochemical conversion to biofuel (Malherbe & Cloete, 2003). Lignocelluloses are polymers made of carbohydrates that are present in plant cell walls. They are made up of cellulose, hemicellulose, and lignin, three distinct polymers, which are resiliently bonded to one another (Armstrong et al., 2016). Lignocelluloses are resistant to microbial attack because of their tertiary structure (Himmel et al., 2007). A crucial stage in recycling waste lignocellulosic biomass into biofuels is pretreatment. In contrast to lignin, which creates covalent bonds with hemicelluloses, cellulose forms hydrogen bonds with the hemicellulose matrix. To remove lignin and hemicellulose from cellulose during the pretreatment of lignocellulose, these intermolecular hydrogen bonds and covalent bonds are broken down. For catalytic and enzymatic hydrolysis, pretreatment increases the accessibility of cellulose to catalysts and enzymes (Mosier et al., 2005). The product yields are reduced by the addition of lignin and hemicelluloses (Vandenbossche et al., 2014). The yields of biofuel precursors can therefore be increased by pretreating lignocellulosic biomass prior to catalytic and enzymatic hydrolyses. Here, we discuss several pretreatment techniques used prior to the hydrolysis of waste lignocellulosic biomasses.

4.2.1 Mechanical pretreatment

Mechanical pretreatment reduces particle size, crystallinity, and Degree of Polymerization (DP) and increases the porosity and surface area of waste lignocellulosic biomass. In mechanical pretreatment, gamma rays, pulsed electric fields, electric beam irradiation, and thermogravimetric techniques are employed. These unconventional heating techniques successfully break down lignocellulosic intra and intermolecular structures, increasing product yields. Compared with conventional pretreatment approaches, microwave and electron beam irradiations reduce the reaction time by 5–10 times (Binod et al., 2010) and increase product yields by 10%–25% (Sarkar et al., 2012). However, the process is capital-intensive due to significant energy usage and equipment expenditures.

4.2.1.1 Drying

Physical pretreatment methods such as densification, pelletization, and drying are utilized to improve the characteristics of waste biomass meant for conversion to thermochemical processes through heat initiation.

However, these methods have limitations, including an inability to remove the lignin content of waste lignocellulosic biomasses, which restricts access to the cellulose content of the material. Additionally, physical pretreatment techniques are energy-intensive and costly to implement on a commercial scale. Studies suggest that high energy consumption associated with lignin removal may be a significant factor affecting the energy efficiency of physical pretreatment techniques and, ultimately, the overall energy efficiency of biorefinery processes (Anukam & Berghel, 2021).

4.2.1.2 Milling

Mechanical pretreatment procedures, such as grinding ball milling, cryo-milling, and compression milling, aid in reducing the reaction time and increasing product yields. The reaction time is reduced by 20%−60%, and product yields improve by 5%−20% depending on biomass, duration, and milling/grinding method employed (Kumar et al., 2008). However, these pretreatment procedures are incredibly energy-intensive and inappropriate for lignin removal (Elgharbawy et al., 2016). The presence of lignin reduces product yields by 30%−35% when employing various pretreatment procedures. As a result, mechanical pretreatment is not recommended for lignin-rich biomasses. Depending on the nature of the waste biomass, other milling processes such as hammer milling, colloid milling, and two-roll milling are also utilized as mechanical pretreatment procedures (Chang & Holtzapple, 2000; Lachke & Rayali, 2009).

4.2.1.3 Sonication

Sonication, also known as ultrasound irradiation, has gained popularity as a green pretreatment method for enhancing the production of biofuel from waste lignocellulosic biomasses. Above 20 kHz sound waves are used in ultrasound, which propagates as compressions and rarefactions through a material (Khanal et al., 2007). Bubbles of gas and vapor start to form as the waves pass through low-pressure zones; these bubbles continue to grow until they reach a critical diameter and implode, a process known as cavitation (Pilli et al., 2011). The idea behind utilizing ultrasonic irradiation as a pretreatment is that circumstances of high temperature and pressure, known as "hot spots," are produced by the implosion of gas bubbles (Hogan et al., 2004). In addition to heating effects, the implosion of bubbles creates powerful shearing forces that may aid in the breakdown of cell walls and membranes, the defibrillation of lignocelluloses, and the oxidation of organic materials. Oxidative radicals created by cavitation may also help in the solubilization of materials. By combining these effects, lignocellulose's structure is altered, the crystalline character of cellulose molecules is broken down, cellulose is depolymerized and solubilized, cell walls and membranes are lysed, and the solubilization of organic materials is enhanced (Bundhoo & Mohee, 2018).

4.2.2 Physicochemical pretreatment

A blend of physical and chemical pretreatment processes is often used in physicochemical pretreatment. This sort of pretreatment helps to disrupt the structural bonds of lignin and hemicelluloses, making the cellulose more accessible to catalysts and enzymes. The most critical physicochemical pretreatment procedures are steam explosion, ammonia fiber explosion (AFEX), liquid hot water (LHW), and CO_2 explosion (Badiei et al., 2014).

Chemical pretreatments are the most often utilized pretreatment procedures for lignocellulosic biomass biochemical conversion. Compared with other pretreatment methods, this method provides a better approach for cellulose depolymerization at low temperature, low pressure, and short reaction time.

4.2.2.1 Acid pretreatment

Acid pretreatment is typically performed by adding 0.2−2.5 wt.% concentrated or dilute acids to dried waste biomass. In the continuous stirring mode, dilute acid hydrolysis is generally performed at high temperatures ($>160°C$) with low biomass loading (Naseeruddin et al., 2013). This procedure employs low concentrations (1 wt.%) of phosphoric acid, nitric acid, and sulfuric acid. Dilute acid dissolves hemicellulose and lignin by dissolving the intermolecular structure of lignocellulose. The high solubility of lignin and hemicellulose in acid benefits this approach. This pretreatment procedure results in a high sugar yield after hydrolysis (Kumar et al., 2009). Low acid concentrations are often employed for pretreatment of cellulose-rich lignocellulosic biomass. The pretreatment of municipal solid trash with dilute H_2SO_4, HNO_3, and HCl and that the sugar yield from pretreated solid waste is acid concentration−dependent (Li et al., 2007). When diluted H_3PO_4 is applied to potato peels, the total sugar production is 82.5% (Lenihan et al., 2010).

Concentrated acids such as H_2SO_4 and HCl are introduced to waste biomass at low temperatures (30°C–60°C) and low pressure in concentrated acid hydrolysis. Acid concentration, temperature, duration, and acid-to-substrate ratio impact the efficiency of the process. At low temperatures, a high acid concentration efficiently destroys the lignocellulose's covalent and hydrogen bonds. It also improves the porosity and accessibility of catalysts and enzymes to biomass (Groenestein et al., 2007). At 50°C, lignocellulose is pretreated with a highly concentrated H_2SO_4 of 70% (v/v) (Liu et al., 2012). Concentrated H_3PO_4 is used to pretreat biomass in water at low temperatures, resulting in a high sugar yield following hydrolysis. However, recovering the acids employed in this process is extremely difficult, necessitating the deployment of corrosion-resistant equipment.

Furthermore, inhibitory compounds are produced during acid pretreatment, reducing process efficiency. After acid pretreatment, the solid substrates must be washed and neutralized many times. Organic solvents such as DMSO and acetone must also be added to remove the particles from the supernatant before catalytic or enzymatic hydrolysis (Shamsudin et al., 2012).

4.2.2.2 Alkali pretreatment

In the alkaline pretreatment technique, waste lignocellulosic biomasses are mixed in alkaline solutions such as NaOH, KOH, and NH_4OH at particular temperatures. Alkali pretreatment of biomass promotes swelling, lowers crystallinity, decreases DP, and increases substrate surface area (Conde-Mejia et al., 2012; Zhao et al., 2009). Hemicelluloses and lignin are extracted from carbohydrates (Zhang et al., 2007). At high temperatures (80°C–100°C) and atmospheric pressure, low alkali concentrations (2–5 wt.%) are commonly utilized. Temperature, pH, and time impact the process. Low alkali concentration pretreatment of biomass favors enzymatic hydrolysis, especially for materials with low lignin content (Tutt et al., 2012). High alkaline concentrations induce cellulose breakdown and a decrease in lignin removal (6–20 wt.%). By increasing the NaOH dosage from 3 to 9 wt.% boosts the delignification of dry bagasse from 52.3% to 75.5% (Zhao et al., 2009). However, alkali salts are difficult to remove after the reaction, prompting repeated washing to remove the salts and other residues. Inhibitory compounds are formed during pretreatment with a high alkali concentration, decreasing product yields (Harmsen et al., 2010).

Alkaline peroxide pretreatment is often preferred over other chemical pretreatment techniques. This procedure is divided into two phases (Yang et al., 2002). In the first phase, the aqueous alkali solution eliminates lignin, acetate, and extractives from lignocellulosic biomass. Hemicellulose is extracted from the biomass using dilute peroxide acid in the second phase. In the alkaline peroxide pretreatment process, oxidants such as hydrogen peroxide, sulfur trioxide, and chlorine dioxide are employed as peroxides (Harmsen et al., 2010). The aromatic ring of lignin reacts with oxidizing chemicals such as hydrogen peroxide and sodium peroxide, resulting in oxidative delignification of the biomass (Banerjee et al., 2011). The oxidant dosage, reaction duration, and temperature impact this process. Sugar yields from lettuce and water hyacinth are increased by alkaline peroxide treatment with NaOH and H_2O_2 (Mishima et al., 2006). At 90°C, alkaline peroxide treatment of waste softwood eliminates 90% lignin in 45 minutes (Gould, 1985).

4.2.2.3 Ionic liquid pretreatment

Ionic liquids (ILs) have received a lot of interest recently because of some of their distinctive and beneficial features that can be used in a variety of industries. Due to their low vapor pressure, they are nonflammable and nonvolatile, which reduces the possibility of harmful emissions (Kolbeck et al., 2010). They exhibit strong polarity, great thermal stability (decomposition temperatures are in the range of 300°C–500°C), good conductivity, and are redox-robust while not being coordinating. The structure and content of ILs, which are made solely of ions, determine their conductivity and viscosity properties (Pinkert et al., 2011). ILs are typically viscous liquids, and the viscosity varies depending on the type of anion. Their capacity to form hydrogen bonds and temperature are additional crucial factors that influence viscosity.

In any electrochemical process, conductivity is crucial. As ILs are made up of ions, there are a lot of charge carriers in each unit of volume. The conductivity of the ILs and viscosity are inversely related (Every et al., 2004). In addition to viscosity, corelated ionic movements, anionic charge delocalization, and ion size and aggregation also affect the conductivity of ILs (Zhang et al., 2006). The oxygen balance, density, and ion energy content of the ILs are determined by their anionic component, which also affects how well they work. In contrast, the cationic component regulates the viscosity, surface tension, and melting point of the ILs to control their stability (Dong et al., 2014). ILs have a strong polarity. Protic ionic liquids (PILs) (Fig. 4.1) often have more polarity than aprotic ionic liquids (APILs). Due to their hydrophobic nature, PILs form hydrogen bond networks and react with metal salts significantly more effectively than APILs (Guan et al., 2017). PILs increase ion–water interactions at high water concentrations, whereas APILs prefer low water concentrations (Ab Rani et al., 2011).

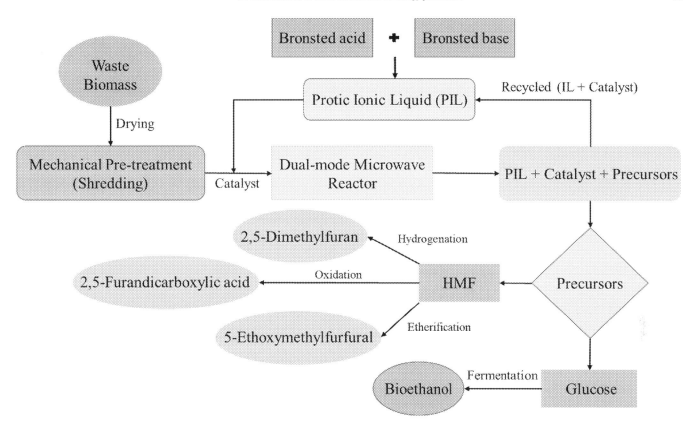

FIGURE 4.1 Process flowsheet of bioenergy precursor production from waste biomass using PIL.

ILs are widely used in a variety of processes, including those involving solvents, electrolytes, liquid crystals, lubricants, lubricant additives, analytics, and electroelastic materials (Greaves & Drummond, 2015). ILs dissolve biomass containing cellulose, hemicellulose, and lignin very efficiently without releasing any volatile organic solvents. As ILs display a wide variety of polarities, another key benefit that may be taken advantage of is their ability to dissolve a range of biomasses.

4.2.2.4 Deep eutectic solvent pretreatment

Deep Eutectic Solvents (DESs) offer a more energy-efficient alternative compared with established processes. They have the ability to dissolve lignin and enhance the availability of cellulose for hydrolysis, all the while operating at lower temperatures and pressures. DESs primarily consist of a fluid mixture composed of two or three ionic compounds that can self-associate to form a eutectic mixture (Dai et al., 2013). Physicochemical properties of DESs resemble those of ionic liquids, including the capability to dissolve lignin (Gorke et al., 2008). However, DESs are more environmentally friendly and cost-effective than ionic liquids. Research has shown that the energy requirement per unit of biomass is approximately half that of steam explosion when using DESs (Kumar et al., 2016). Furthermore, DES pretreatment results in low or negligible concentrations of inhibitors such as hydroxymethyl furfural and acetic acid furfural (Procentese et al., 2018). Glycerol-choline chloride is a promising DES for waste lignocellulosic materials. Typically, lignocellulosic biomass pretreatment studies have focused on a single biomass, pretreatment temperature, and processing time. However, research also indicates that solvent consumption and recovery are crucial parameters for the industrial development of DES pretreatment.

4.2.2.5 Steam explosion

Waste lignocellulosic biomasses are processed with steam at high pressure and temperature in the steam explosion technique (Sánchez & Cardona, 2008). In this process, the reaction temperature is typically kept between 160°C and 200°C. After a few minutes, the reaction is stopped by an immediate pressure release. The waste biomass is blasted into a fine powder by rapidly decompressing to atmospheric pressure. At 210°C, the steam explosion of softwood waste biomass yields a maximum sugar yield of 50% (Mtui, 2009).

The saturated liquid (water) is utilized as feed in the LHW technique at temperatures ranging from 170°C to 180°C, followed by rapid breakdown to atmospheric pressure (Laser et al., 2002). Compared with the steam explosion technique, the quantity of biomass loading, reaction temperature, and reaction pressure required for the LHW method are minimized (Mosier, 2005). In the AFEX process, lignocellulosic materials are treated with concentrated ammonia at high pressure before quickly reducing pressure (Sun & Cheng, 2002). The efficient recovery of ammonia is a critical aspect of making this process economically viable (Teymouri et al., 2004; Wyman et al., 2005).

4.2.2.6 Supercritical CO_2

Supercritical fluids are those that are above their critical temperature and pressure but below the pressure required for condensation (Schacht et al., 2008). The critical point is achieved when the gas and liquid phases coexist and are no longer able to be distinguished. Due to its accessibility, low toxicity, low flammability, low cost, and its ability to not contaminate products, supercritical CO_2 (scCO_2) is a good substitute for typical solvents in green chemistry (Rogalinski et al., 2008). In addition, compared with other solvents, CO_2 has comparatively low energy requirements for supercritical fluid conditions. Due to its high quadrupole moment and low polarizability, CO_2 is a poor solvent for polar molecules but a good solvent for nonpolar materials (Morais et al., 2015). Because it may be used to dissolve monomers but not polymers, it is appropriate for the impregnation and separation of waste lignocellulosic materials. To be suitable for commercial processes within the biorefinery concept, scCO_2 must nevertheless be economically viable.

4.2.3 Biological pretreatment

Biological pretreatment is a low-energy process that relies on incubating biomass with specially selected microorganisms that produce enzymes that modify the biomass, making it more suitable for biological or thermochemical processing. The most common type of biological pretreatment is biopulping using indigenous microorganisms, although previous research has shown that the process can be accelerated and controlled by adding fungal organisms. Fungal microorganisms, particularly white rot fungi, are preferred as inocula due to their effective delignification activities. However, the time required for the process ranges from 28 to 60 days, making biological pretreatments economically unviable for biofuel production (Mielenz, 2020).

4.2.3.1 Bacterial/microbial pretreatment

The only factor used to choose microbial strains for pretreatment is their level of microbial activity. During the pretreatment stage, fungi strains such as *Gloeophyllum trabeum* create enzymes that aid in the breakdown of cellulose and hemicellulose as well as a modified form of lignin that leaves behind brown residues (Gao et al., 2012). According to research, white-rot fungi are particularly good at accelerating enzymatic hydrolysis and completely decomposing lignin. Interestingly, compared with utilizing a single microbe for pretreatment, the employment of a consortium of microorganisms greatly improves delignification, xylan breakdown, and increases the accessibility of cellulose surfaces in biomass (Sindhu et al., 2015). In the pretreatment process, facultative anaerobes such as *Enterobacter sp.* generate a mixture of acids, alcohols, CO_2, and H_2 from carbohydrates, lipids, and proteins. Obligate anaerobes such as *Methanococcus* and *Methanobacterium* take part in the conversion of acetate and related intermediates into methane. In total, 60% of bacterial species and 90% of archaeal species can be effectively used for anaerobic digestion (Mishra et al., 2018).

4.2.3.2 Enzymatic pretreatment

Enzymes are biological catalysts used in the biochemical reactions. They are produced from living cells, microorganisms, and plants (Foreman et al., 2003). Cellulase enzyme is widely used for the enzymatic hydrolysis of cellulose (Wood, 1985). Celluclast, Celtic Ctec2, Celtic Ctec3 are the newly invented enzymes, which are used for the conversion of waste lignocelluloses to sugars (Rodrigues et al., 2015; Weiss et al., 2013). These enzymes are produced from a variety of microbial sources, such as *Trichoderma viride* (Griffin et al., 1974), *Trichoderma ressei* (Hohn & Sahm, 1983), *Trichoderma longibrachiatum* (Pachauri et al., 2017), *Trichoderma koningii* (Wood & Mccrae, 1978), *Aspergillus niger* (Mrudula & Murugammal, 2011), *Penicillium funiculosum* (De Castro et al., 2010), *Clostridium thermocellum* (Olson et al., 2010), *Bacteroides cellulosilyticus* (Robert et al., 2007), *Bacillus cereus* (Kim et al., 1988), and *Bacillus subtilis* (Deka et al., 2013).

The cellulose must be hydrolyzed effectively using numerous enzymes. The three categories listed below represent the traditional taxonomy of enzymes, which is how the cellular enzymes are organized. Cellobiose is broken down into glucose by three different enzymes: exo-1,4-D-glucanases, also known as cellobiohydrolases (CBH), endo-1,4-D-glucanases, which randomly hydrolyze internal glycosidic bonds in the cellulose chain, and 1,4-D-glucosidases, which hydrolyze cellobiose to glucose. By providing each other with newly accessible sites, all of these enzymes collaborate to hydrolyze cellulose (Coughlan & Ljungdahi, 1988).

4.3 Pretreatment technologies for waste biomass: pros

Pretreatment is required to separate the essential components from waste biomass. By choosing a proper pretreatment technique, it is possible to increase the cost-effectiveness and digestibility and decrease the development of undesirable byproducts. Additionally, pretreatment processes can significantly affect the design, effectiveness, and expense of subsequent procedures. Also, knowing the basics of different pretreatment technologies might aid in choosing the ideal pretreatment approach for a specific waste biomass feedstock. The properties of waste biomass, availability, financial resources, and low adverse impact on the environment can all be taken into account when choosing the most effective pretreatment method (Rezania et al., 2017).

4.3.1 Mechanical pretreatment

Mechanical pretreatment procedures can increase the efficacy, sustainability, and commercial feasibility of waste biomass conversion processes. They are simple yet essential steps in turning waste biomass into useful goods (Zadeh et al., 2020). The biomass's composition can be broken down via pretreatment procedures including grinding, milling, and shredding, which will make it easier for subsequent treatment procedures to access. This may improve the effectiveness of subsequent procedures such as thermal conversion or enzymatic hydrolysis. These procedures can manage huge quantities of waste biomass because they don't require chemicals, which results in relatively minimal inhibitor production. A denser, more energy-dense feedstock that is simpler to transport and store can also be produced via mechanical pretreatment procedures such as size reduction or pelletization. This could result in the recovery of energy from waste biomass using thermal conversion techniques such as combustion or gasification (Tan et al., 2021). A more consistent feedstock for following processing steps might result from reducing the biomass's particle size, which can enhance process control and consistency.

4.3.2 Physicochemical pretreatment

Physicochemical pretreatment processes of waste biomass have several advantages over physical pretreatment processes, including accessibility of acid or alkali treatment, steam explosion, and microwave irradiation can break down the waste biomass structure and increase its accessibility to downstream processing steps. This can lead to higher yields of valuable products such as biofuels or other platform chemicals (Nanda et al., 2013).

These processes can also improve the quality of the final product by reducing impurities and enhancing the purity of the target compound. Some physicochemical pretreatment processes such as steam explosion can reduce the energy required for downstream processing steps by increasing the reactivity of the biomass. This can lead to lower energy consumption and reduced operating costs. It can also increase the biodegradability of the biomass, making it more suitable for use in anaerobic digestion or composting. This can lead to increased energy recovery and reduced waste volumes (Chandel et al., 2022).

Physicochemical pretreatment processes can be tailored to the specific characteristics of the biomass feedstock, allowing for greater process flexibility and the optimization of process conditions to achieve desired product yields and quality.

4.3.3 Biological pretreatment

Biological pretreatment processes of waste biomass have several advantages, including low-energy inputs requirements compared with physical and physicochemical pretreatment processes, as they rely on microorganisms to break down the biomass (Fig. 4.2). This can lead to lower operating costs and energy consumption.

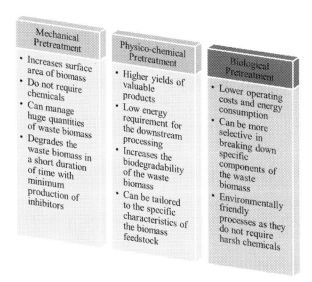

FIGURE 4.2 Pretreatment technologies for waste biomass: Pros.

These processes are environmentally friendly, as they do not require harsh chemicals or produce harmful byproducts. This makes them a more sustainable option for waste biomass conversion (Bhatt & Shilpa, 2015).

Biological pretreatment processes can be more selective in breaking down specific components of the biomass, such as lignin or hemicellulose. This can lead to higher purity products and reduced waste. It can also improve the quality of the final product by removing impurities and enhancing the purity of the target compound (Miskat et al., 2020).

Biological pretreatment processes can also result in the production of value-added products, such as enzymes, biofuels, or other bioproducts, depending on the specific microorganisms and conditions used. These processes can be easily integrated with other processing methods such as anaerobic digestion or composting, leading to increased energy recovery and reduced waste volumes (Achinas et al., 2017).

4.4 Pretreatment technologies for waste biomass: cons

Moisture and lignin in the feedstocks, which may adversely affect the subsequent processes, are the key bottleneck in pretreatment methods to treat waste biomass (Fig. 4.3). Therefore, one of the most important steps in the development of various bioenergy production processes is the pretreatment of waste biomass. Waste biomass for the generation of bioenergy has not yet found a single treatment method. Although combination treatment methods have, to some extent, achieved some satisfactory results, more in-depth studies still need to be conducted to improve moisture (Roy & Chakraborty, 2019) and lignin (Tan et al., 2021) removal in a cost-effective and environmentally friendly way. Here are a few drawbacks of the current pretreatment procedures.

4.4.1 Mechanical pretreatment

While mechanical pretreatment processes of waste biomass offer advantages, they also have some disadvantages to consider. Mechanical pretreatment processes often require significant energy inputs, especially for size reduction methods such as grinding or milling. This can increase operational costs and carbon emissions, making the process less environmentally friendly. These processes also involve the use of machinery and equipment, which are subject to wear and tear due to the abrasive nature of biomass. This can result in frequent maintenance requirements and higher operational costs (Zadeh et al., 2020).

Mechanical pretreatment processes typically do not have high selectivity in breaking down specific components of biomass. While they can increase accessibility and reduce particle size, they do not specifically target certain compounds such as lignin or hemicellulose, which may require further treatment steps (Miskat et al., 2020). Some of these pretreatment methods, such as thermal processes such as drying or size reduction through high

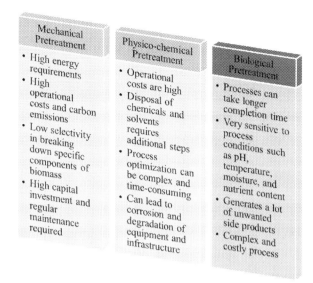

FIGURE 4.3 Pretreatment technologies for waste biomass: Cons.

temperatures, can lead to the loss of volatile compounds (Tan et al., 2021). This can affect the composition and quality of the final product.

Depending on the specific physical pretreatment method used, there may be associated environmental impacts (Chandel et al., 2022). These processes may not be suitable for all types of biomasses. Some biomass feedstocks, such as those with high moisture content or high lignin content, may pose challenges for effective physical pretreatment. It's important to consider these disadvantages alongside the advantages when evaluating the feasibility and suitability of mechanical pretreatment processes for a given waste biomass conversion scenario.

4.4.2 Physicochemical pretreatment

Physicochemical pretreatment processes of waste biomass also have some disadvantages to consider. Some of these processes often require the use of chemicals such as acids, alkalis, or solvents. These chemicals can be costly to procure and handle, and their disposal may require additional steps to ensure environmental safety. The use of chemicals in physicochemical pretreatment processes can have negative environmental impacts. Improper handling or disposal of the chemicals used can lead to soil, water, or air pollution, contributing to environmental degradation (Sharma et al., 2023).

Physicochemical pretreatment processes often require specific reaction conditions such as temperature, pressure, and duration. Optimizing these conditions for different biomass feedstocks can be complex and time-consuming, adding to the process complexity and potential operational challenges. Some of these pretreatment methods, such as acid hydrolysis, can generate inhibitory compounds during the process (Zadeh et al., 2020). These compounds can negatively affect downstream processing steps, such as enzymatic hydrolysis or microbial fermentation, and reduce overall process efficiency.

The use of chemicals in physicochemical pretreatment processes can lead to corrosion and degradation of equipment and infrastructure (Tan et al., 2021). This can increase maintenance requirements, repair costs, and overall operational complexity. These processes involving the use of hazardous chemicals can pose safety risks if proper safety protocols and precautions are not followed. Handling corrosive or toxic substances requires careful training and adherence to safety guidelines. It is important to assess the specific advantages and disadvantages of physicochemical pretreatment processes in the context of the particular waste biomass and the desired end products, considering both the technical and environmental aspects of the process.

4.4.3 Biological pretreatment

Biological pretreatment processes of waste biomass also have some disadvantages. These processes can take longer compared with physical or physicochemical methods. This is because they rely on the activity of microorganisms or enzymes to break down the biomass, and the biological processes often require more time for completion (Bhatt & Shilpa, 2015).

Biological pretreatment processes can be sensitive to process conditions such as pH, temperature, moisture content, and nutrient availability. Any deviations from the optimal conditions can impact the efficiency and effectiveness of the process. Maintaining and controlling these conditions can add complexity to the overall process. These processes involve the use of microorganisms, which can exhibit variability in terms of their activity, growth rates, and metabolic capabilities (Chandel et al., 2022). There is a risk of contamination by unwanted microorganisms that may negatively affect the process or the quality of the final product.

Microorganisms used in biological pretreatment processes may require specific nutrients for their growth and metabolic activities. Providing these nutrients can add to the overall process cost and complexity. Certain types of biomasses may be less amenable to biological pretreatment due to their composition or structural characteristics. Also, these processes can generate secondary waste streams such as spent microbial biomass or fermentation residues (Nanda et al., 2013). Proper management and disposal of these waste streams may be required, adding to the overall waste management challenges.

4.5 Advance pretreatment in biorefineries

The most significant difference between a biorefinery and a petroleum refinery is the source of raw materials. The circular economy, which is based on the idea that waste lignocellulosic materials, which were used to produce bio-based products, can be recovered, and recycled, can be developed by converting waste biomass in the biorefinery into a variety of chemicals and energy carriers. Biorefining is defined as "the sustainable processing of biomass into a spectrum of marketable bio-based products (chemicals, materials) and bioenergy (biofuels, power, heat)" by the International Energy Agency Bioenergy (Galbe & Wallberg, 2019). However, when a large-scale production plant is taken into consideration, the various kinds of raw materials create a significant barrier.

The rate of polymerization and crystallinity of the cellulose in waste biomass cannot be effectively reduced using regular pretreatment procedures. Enzymatic hydrolysis still faces obstacles from the formation of inhibitory chemicals, inadequate digestion of waste biomass, and the production of lignin monomer compounds. The development of nanotechnology has a huge impact on the biorefinery sectors in order to address all of these problems. Pretreatment techniques based on nanotechnology are currently being researched extensively (Yadav et al., 2019). The main component of this method is the ability of nanoparticles to enter the cell membrane of waste biomass. Under extreme conditions, the interaction of nanoparticles with the other essential components can release the desired compounds, such as hemicellulose or lignin (Chandel et al., 2022). This method generates a significant amount of shearing in the reactor, which breaks down the recalcitrance and improves catalyst action. Utilizing magnetic nanoparticles is a beneficial pretreatment approach for waste lignocellulosic biomass that may also provide us with additional advantages. This procedure is economical as the immobilized enzyme is easy to retrieve and to reuse. Due to their important characteristics, including their nanoscale size, high surface-area-to-volume ratio, and formation of numerous active sites that participate in various reactions, high reactivity, thermal stability, chemical stability, high specificity, high catalytic efficiency, high rate of crystallinity, and high adsorption capacity, nanoparticles have captured the attention of researchers (Suhara et al., 2012).

The science of nanotechnology is expanding significantly. Across the world, their major applications are used often. In several industries, including healthcare, food, agriculture, and bioenergy, nanotechnology has great potential. There is a need for a renewable energy source because these nonrenewable resources are extremely detrimental for the environment. Environmental pollution has been reduced thanks to renewable and sustainable energy, which also satisfies global energy demand (Gou et al., 2020). This approach is more affordable, environmentally benign, and sustainable because of the use of waste biomass feedstocks. For the manufacturing of biofuels, various conventional pretreatment techniques were applied, but they had some drawbacks, including environmental contamination, time consumption, and high chemical costs (Chandel et al., 2022).

The use of nanoparticles in the processing of waste lignocellulosic biomass helps to overcome all these limitations and makes the process environmentally safe, cost-effective, and sustainable. With the help of waste lignocellulosic biomass, various nanomaterials, including nanoparticle, nano-biocatalyst, and magnetic-based nanomaterial (Verma et al., 2013), have been used to produce various renewable bioenergy (biogas, bioethanol, biodiesel, bio-oil, and biohydrogen). The breakdown of the biomass is significantly influenced by the nanoscale properties of nanoparticles. They quickly enter the cell wall, interact with the components, and release a lot of sugars as a result. The production of renewable energy is improved by further processing these released elements with an enzyme immobilized on a nanomaterial.

4.6 Technological upgradation and Scale-up

To transition toward a society that relies less on fossil fuels, it is essential to consider several key factors in the biorefinery concept. The future integration of livestock, food production, chemical manufacturing, fuel production, and energy generation heavily relies on the advancement of biorefineries. The synthesis of proteins and platform chemicals such as hydroxymethylfurfural (Roy & Chakraborty, 2020) will increasingly depend on the combination of physical and biotechnological processes. By replacing fossil fuels, waste biomass can contribute to mitigating the excessive levels of carbon dioxide in the atmosphere. Furthermore, for many countries, biorefineries may play a crucial role in ensuring domestic energy sources and the availability of chemicals. Additionally, the potential for developing new industries in rural areas is another important factor, as numerous major biorefineries are expected to be established there.

The bulk of waste lignocellulosic feedstock presents a problem because it is typically far more challenging and recalcitrant than the starchy materials currently employed in, for example, the bioethanol sector (Paul & Chakraborty, 2018). However, over the past 10–15 years, several pilot plant facilities have been operational. Several businesses have examined the possibility of producing bioenergy by converting cellulose into second-generation sugar compounds that may then be converted into ethanol and other chemicals. Many older ventures, though, are currently dormant or shifting their priorities. The information acquired from pilot plants is crucial as it has produced significant quantities of engineering design data that will be required for the full-scale design of upcoming biorefineries.

The deployment of successful future biorefineries faces a number of obstacles. Researchers suggest that the first step toward a big capital investment is scaling up a successful small-scale operation to a large-scale biorefinery. Therefore, there needs to be a very strong economic incentive to support the investment. The fact that the regulations governing biofuels and biochemicals are not yet long-term stable, which means that the necessary financial ground is not yet fully established for significant investments, also causes investors to perceive the low return on invested capital and an unstable future position (Galbe & Wallberg, 2019). Additionally, they draw the conclusion that some biotechnology firms that had been operating an effective early-stage procedure had encountered serious issues as a result of scaling-up difficulties and consequences of construction, testing, and operation delays.

One of the crucial factors to consider is the availability of waste biomass, including a consistent supply of waste lignocellulosic biomass throughout the year, and access to similar types of biomasses that can be used in a biorefinery under specific operating conditions. For the possibility of continuous operation, it is essential to have all the necessary logistics and a fully functional supply chain in place, including transportation options for low-density materials such as straw and bagasse, as well as suitable storage facilities. The major expenses associated with the biorefinery concept are primarily related to the processing and transportation of the feedstock. In addition to the elements mentioned earlier, there are other significant challenges that need to be addressed, such as mechanizing the processes, translating small-scale laboratory data into engineering design data, and addressing the technical immaturity of second-generation (2G) lignocellulosic technology, which includes difficulties in feeding high-solids suspension to a high-pressure reactor.

4.7 Conclusions

Waste biomasses are widely available and inexpensive, making them the most abundant source of biofuel materials. To convert these biomasses into bioenergy, a combination of biochemical, mechanical, thermochemical, and pretreatment technologies is used. To produce bioethanol from waste lignocellulosic materials, an effective pretreatment method is necessary to break down the plant cell wall and release monosaccharide sugars for fermentation. In an ideal biorefinery setup, the substantial amount of lignin present in biomass would be utilized for producing biofuels and biochemicals using thermochemical methods. However, there are no affordable processing technologies available at present that can efficiently convert a significant portion of waste biomass into liquid or gaseous fuels. To achieve a sustainable and economical supply of bioenergy, it is crucial to focus on developing new, cost-effective conversion technologies, optimizing existing technologies for enhanced productivity with minimal energy consumption, studying the chemistry of the biomass to understand the complexity of the process, and minimizing the generation of by-products and process waste to reduce carbon emissions. Due to the decline in fossil fuel resources and the increasing demand for renewable energy sources, the future of biofuels as an alternative energy option in the fuel market looks very promising.

References

Ab Rani, M. A., Brant, A., Crowhurst, L., Dolan, A., Lui, M., Hassan, N. H., Hallett, J. P., Hunt, P. A., Niedermeyer, H., Perez-Arlandis, J. M., Schrems, M., Welton, T., & Wilding, R. (2011). Understanding the polarity of ionic liquids. *Physical Chemistry Chemical Physics: PCCP, 13*, 16831–16840.

Achinas, S., Achinas, V., & Euverink, G. J. W. (2017). A technological overview of biogas production from biowaste. *Engineering, 3*, 299–307.

Anukam, A., & Berghel, J. (2021). *Biomass pretreatment and characterization: A review. Biotechnological Applications of Biomass.* IntechOpen.

Armstrong, R., Wolfram, C., de Jong, K. P., Gross, R., Lewis, N. S., Boardman, B., Ragaukas, A. J., Ehrhardt-Martinez, K., Crabtree, G., & Ramana, M. V. (2016). The frontiers of energy. *Nat. Energy, 1*, 1.

Badiei, M., Asim, N., Jahim, J. M., & Sopian, K. (2014). Comparison of chemical pretreatment methods for cellulosic biomass. *APCBEE Procedia, 9*, 170–174.

Banerjee, G., Car, S., Liu, T., Williams, D. L., Meza, S. L., Walton, J. D., & Hodge, D. B. (2011). Scale-up and integration of alkaline hydrogen peroxide pretreatment, enzymatic hydrolysis, and ethanolic fermentation. *Biotechnology and Bioengineering, 109*, 922–931.

Bhatt, S. M., & Shilpa. (2015). Lignocellulosic feedstock conversion, inhibitor detoxification and cellulosic hydrolysis – A review. *Biofuels, 5*(6), 633–649.

Binod, P., Sindhu, R., Singhania, R. R., Vikram, S., Devi, L., Nagalakshmi, S., Kurien, N., Sukumaran, R. K., & Pandey, A. (2010). Bioethanol production from rice straw: An overview. *Bioresource Technology, 101*, 4767–4774.

Bundhoo, Z. M. A., & Mohee, R. (2018). Ultrasound-assisted biological conversion of biomass and waste materials to biofuels: A review. *Ultrasonics Sonochemistry, 40*(Part A), 298–313, 2018.

Chandel, H., Kumar, P., Chandel, A. K., & Verma, M. L. (2022). Biotechnological advances in biomass pretreatment for bio-renewable production through nanotechnological intervention. *Biomass Conversion and Biorefinery*.

Chang, V. S., & Holtzapple, M. T. (2000). Fundamental factors affecting biomass enzymatic reactivity. *Applied Biochemistry and Biotechnology, 84*, 5–37.

Conde-Mejia, C., Jiménez-Gutiérrez, A., & El-Halwagi, M. (2012). A comparison of pretreatment methods for bioethanol production from lignocellulosic materials. *Process Safety and Environmental Protection., 90*, 189–202.

Coughlan, M. P., & Ljungdahl, L. G. (1988). Comparative biochemistry of fungal and bacterial cellulolytic enzyme systems. *FEMS Symposium - Federation of European Microbiological Societies, 43*, 11–30.

Dai, Y., Spronsen, G. J., Witkamp, R., & Verpoorte, Y. H. (2013). Ionic liquids and deep eutectic solvents in natural products research: mixtures of solids as extraction solvents. *Journal of Natural Products, 76*, 2162–2173.

De Castro, A. M., de Albuquerque de Carvalho, M. L., Leite, S. G. F., & Pereira, N. (2010). Cellulases from Penicillium funiculosum: Production, properties and application to cellulose hydrolysis. *Journal of Industrial Microbiology & Biotechnology, 37*, 151–158.

Deka, D., Das, S. P., Sahoo, N., Das, D., Jawed, M., Goyal, D., & Goyal, A. (2013). Enhanced cellulase production from Bacillus subtilis by optimizing physical parameters for bioethanol production. *ISRN Biotechnology, 2013*965310.

Dong, K., Wang, Q., Lu, X., Zhou, Q., & Zhang, S. (2014). *Structures and interactions of ionic liquids.* Berlin: Springer, Chap. 1.

Dutta, S. K., & Chakraborty, S. (2015). Kinetic analysis of two-phase enzymatic hydrolysis of hemicellulose of xylan type. *Bioresource Technology, 198*, 642–650.

Dutta, S. K., & Chakraborty, S. (2016). Pore-scale dynamics of enzyme adsorption, swelling and reactive dissolution determine sugar yield in hemicelluloses hydrolysis for biofuel production. *Scientific Reports, 6*, 38173.

Elgharbawy, A. A., Alam, M. Z., Moniruzzaman, M., & Goto, M. (2016). Ionic liquid pretreatment as emerging approaches for enhanced enzymatic hydrolysis of lignocellulosic biomass. *Biochemical Engineering Journal, 109*, 252–267.

Every, H. A., Bishop, A. G., MacFarlane, D. R., Oradd, G., & Forsyth, M. (2004). Transport properties in a family of dialkylimidazolium ionic liquids. *Physical Chemistry Chemical Physics: PCCP, 6*, 1758–1765.

Foreman, P. K., Brown, D., Dankmeyer, L., Dean, R., Diener, S., Dunn-Coleman, N. S., Goedegebuur, F., Houfek, T. D., England, G. J., Kelley, A. S., Meerman, M. J., & Ward, M. (2003). Transcriptional regulation of biomass-degrading enzymes in the filamentous fungus Trichoderma reesei. *The Journal of Biological Chemistry, 278*, 31988–31997.

Gaikwad, A., & Chakraborty, S. (2014). Mixing and temperature effects on the kinetics of alkali metal catalyzed, ionic liquid based batch conversion of cellulose to fuel products. *Chemical Engineering Journal, 240*, 109–115.

Galbe, M., & Wallberg, O. (2019). Pretreatment for biorefineries: A review of common methods for efficient utilisation of lignocellulosic materials. *Biotechnology for Biofuels, 12*, 294.

Gao, Z., Mori, T., & Kondo, R. (2012). The pretreatment of corn stover with Gloeophyllum trabeum KU-41 for enzymatic hydrolysis. *Biotechnology for Biofuels, 5*(1), 28.

Gorke, J. T., Srienc, F., & Kazlauskas, R. J. (2008). Hydrolase-catalyzed biotransformations in deep eutectic solvents. *Chemical Communications., 10*, 1235–1237.

Gou, Z., Ma, N. L., Zhang, W., Lei, Z., Su, Y., Chunyu, S., & Sun, Y. (2020). Innovative hydrolysis of corn stover biowaste by modified magnetite laccase immobilized nanoparticles. *Environmental Research, 188*109829.

Gould, J. M. (1985). Studies on the mechanism of alkaline peroxide delignification of agricultural residues. *Biotechnology and Bioengineering, 27*, 225–231.

Greaves, T. L., & Drummond, C. J. (2015). Protic ionic liquids: Evolving structure-property relationships and expanding applications. *Chemical Reviews, 115*, 11379–11448.

Griffin, H. L., Sloneker, J. H., & Inglett, G. E. (1974). Cellulase production by Trichoderma viride on feedlot waste. *Applied and Environmental Microbiology, 27*, 1061–1066.

Groenestein, C. M., Monteny, G. J., Aarnink, A. J. A., & Metz, J. H. M. (2007). Effect of urinations on the ammonia emission from group-housing systems for sows with straw bedding: Model assessment. *Biosystems Engineering, 97*(1), 89–98.

Guan, W., Chang, N., Yang, L., Bu, X., Wei, J., & Liu, Q. (2017). Determination and prediction for the polarity of ionic liquids. *Journal of Chemical & Engineering Data, 62*, 2610–2616.

Harmsen, P., Huijgen, W., Bermudez, L., & Bakker, R. (2010). Literature review of physical and chemical pretreatment processes for lignocellulosic biomass. *Report, no. 1184*, 1–49.

Himmel, M. E., Ding, S.-Y., Johnson, D. K., Adney, W. S., Nimlos, M. R., Brady, J. W., & Foust, T. D. (2007). Biomass recalcitrance: Engineering plants and enzymes for biofuels production. *Science (New York, N.Y.), 315*, 804–807.

Hogan, F., Mormede, S., Clark, P., & Crane, M. (2004). Ultrasonic sludge treatment for enhanced anaerobic digestion. *Water Science and Technology: A Journal of the International Association on Water Pollution Research, 50*, 25–32.

Hohn, H.-P., & Sahm, H. (1983). Induction of cellulases in Trichoderma reesei. *Enzyme Technology*, 55–68.

Khanal, S. K., Grewell, D., Sung, S., & (Hans) van Leeuwen, J. (2007). Ultrasound applications in wastewater sludge pretreatment: A review. *Critical Reviews in Environmental Science and Technology, 37*, 277–313.

Kim, D. W., Yang, J. H., & Jeong, Y. K. (1988). Adsorption of cellulose from Trichoderma viride on microcrystalline cellulose. *Applied Microbiology and Biotechnology, 28*, 148–154.

Kolbeck, C., Lehmann, J., Lovelock, K. R. J., Cremer, T., Paape, N., Wasserscheid, P., Froba, A. P., Maier, F., & Steinruck, H. P. (2010). Density and surface tension of ionic liquids. *J The Journal of Physical Chemistry B, 114*, 17025–17036.

Kumar, A., Parikh, B., & Pravakar, M. (2016). Natural deep eutectic solvent mediated pretreatment of rice straw: Bioanalytical characterization of lignin extract and enzymatic hydrolysis of pretreated biomass residue. *Environmental Science and Pollution Research, 23*, 9265–9275.

Kumar, P., Barrett, D. M., Delwiche, M. J., & Stroeve, P. (2009). Methods for pretreatment of lignocellulosic biomass for efficient hydrolysis and biofuel production. *Industrial & Engineering Chemistry Research, 48*, 3713–3729.

Kumar, R., Singh, S., & Singh, O. V. (2008). Bioconversion of lignocellulosic biomass: Biochemical and molecular perspectives. *Journal of Industrial Microbiology & Biotechnology, 35*, 377–391.

Lachke, A. H., & Rayali, S. L. (2009). Bioethanol from lignocellulosic biomass. In A. Pandey (Ed.), *Handbook of plant-based biofuels* (pp. 121–138). Boca Raton: CRC Press.

Laser, M., Schulman, D., Allen, S. G., Lichwa, J., Antal, M. J., & Lynd, L. R. (2002). A comparison of liquid hot water and steam pretreatments of sugar cane bagasse for bioconversion to ethanol. *Bioresource Technology, 81*, 33–44.

Lenihan, P., Orozco, A., O'Neill, E., Ahmad, M. N. M., Rooney, D. W., & Walker, G. M. (2010). Dilute acid hydrolysis of lignocellulosic biomass. *Chem. Eng. J, 156*, 395–403.

Li, A., Antizar-Ladislao, B., & Khraisheh, M. (2007). Bioconversion of municipal solid waste to glucose for bio-ethanol production. *Bioprocess and Biosystems Engineering, 30*, 189–196.

Liu, X., Ai, N., Zhang, H., Lu, M., Ji, D., Yu, F., & Ji, J. (2012). Quantification of glucose, xylose, arabinose, furfural, and HMF in corncob hydrolysate by HPLC-PDA–ELSD. *Carbohydrate Research, 353*, 111–114.

Malherbe, S., & Cloete, T. E. (2003). Lignocelluloses biodegradation: Fundamentals and applications. *Reviews in Environmental Science and Bio/Technology, 1*, 105–114.

Mielenz, J. R. (2020). Small-scale approaches for evaluating biomass bioconversion for fuels and chemicals. In (second editionA. Dahiya (Ed.), *Bioenergy* (17Academic Press.

Mishima, Y., Giraldez, A. J., Takeda, Y., Fujiwara, T., Sakamoto, H., Schier, A. F., & Inoue, K. (2006). Differential regulation of germline mRNAs in Soma and germ cells by zebrafish miR-430. *Current. Biology, 16*, 2135–2142.

Mishra, S., Singh, P. K., Dash, S., & Pattnaik, R. (2018). Microbial pretreatment of lignocellulosic biomass for enhanced biomethanation and waste management. *3 Biotech, 8*, 458.

Miskat, M. I., Ahmed, A., Chowdhury, H., Chowdhury, T., Chowdhury, P., Sait, S. M., & Park, Y. K. (2020). Assessing the theoretical prospects of bioethanol production as biofuel from agricultural residues in Bangladesh: A review. *Sustainability, 12*, 8583.

Morais, A. R. C., da Costa Lopes, A. M., & Bogel-Łukasik, R. (2015). Carbon dioxide in biomass processing: Contributions to the green biorefinery concept. *Chemical Reviews, 115*, 3–27.

Mosier, N. (2005). Features of promising technologies for pretreatment of lignocellulosic biomass. *Bioresource Technology, 96*, 673–686.

Mrudula, S., & Murugammal, R. (2011). Production of cellulose by Aspergillus niger under submerged and solid state fermentation using coir waste as a substrate. *Brazilian Journal of Microbiology, 42*, 1119–1127.

Mtui, G. Y. S. (2009). Recent advances in pretreatment of lignocellulosic wastes and production of value added products. *African Journal of Biotechnology, 8*, 1398–1415.

Nanda, S., Mohammad, J., Reddy, S. N., Kozinski, J. A., & Dalai, A. K. (2013). Pathways of lignocellulosic biomass conversion to renewable fuels. *Biomass Conversion and Biorefinery, 4*, 157–191.

Naseeruddin, S., Yadav, K. S., Sateesh, L., Manikyam, A., Desai, S., & Rao, L. V. (2013). Selection of the best chemical pretreatment for lignocellulosic substrate Prosopis juliflora. *Bioresource Technology, 136*, 542–549.

Olson, D. G., Tripathi, S. A., Giannone, R. J., Lo, J., Caiazza, N. C., Hogsett, D. A., Hettich, R. L., Guss, A. M., Dubrovsky, G., & Lynd, L. R. (2010). Deletion of the Cel48S cellulase from Clostridium thermocellum. *PNAS, 107*, 17727–17732.

Pachauri, P., V, A., More, S., Sullia, S. B., & Deshmukh, S. (2017). Purification and characterization of cellulase from a novel isolate of Trichoderma longibrachiatum. *Biofuels*, 1–7.

Paul, S. K., & Chakraborty, S. (2018). Microwave-assisted ionic liquid-mediated rapid catalytic conversion of non-edible lignocellulosic Sunn hemp fibres to biofuels. *Bioresource Technology, 253*, 85–93.

Paul, S. K., & Chakraborty, S. (2019). Mixing effects on the kinetics of enzymatic hydrolysis of lignocellulosic Sunn hemp fibres for bioethanol production. *Chemical Engineering Journal, 377*120103.

Pilli, S., Bhunia, P., Yan, S., LeBlanc, R. J., Tyagi, R. D., & Surampalli, R. Y. (2011). Ultrasonic pretreatment of sludge: A review. *Ultrasonics Sonochemistry, 18*, 1–18.

Pinkert, A., Ang, K. L., Marsh, K. N., & Peng, S. (2011). Density, viscosity and electrical conductivity of protic alkanolammonium ionic liquids. *Physical Chemistry Chemical Physics: PCCP, 13*, 5136–5143.

Procentese, A., Raganati, F., Olivieri, G., Russo, M. E., Rehmann, L., & Marzocchella, A. (2018). Deep Eutectic Solvents pretreatment of agro-industrial food waste. *Biotechnology for Biofuels, 11*, 37.

Rezania, S., Din, M. F. M., Mohamad, S. E., Sohaili, J., Taib, S. M., Yusof, M. B. M., Kamyab, H., Darajeh, N., & Ahsan, A. (2017). Review on pretreatment methods and ethanol production from cellulosic water hyacinth. *BioResources, 12*(1), 2108–2124.

Robert, C., Chassard, C., Lawson, P. A., & Bernalier-Donadille, A. (2007). Bacteroides cellulosilyticus sp. nov., a cellulolytic bacterium from the human gut microbial community. *International Journal of Systematic and Evolutionary Microbiology, 57*(7), 1516–1520.

Rodrigues, A. C., Haven, M. Ø., Lindedam, J., Felby, C., & Gama, M. (2015). Celluclast and Cellic® CTec2: Saccharification/fermentation of wheat straw, solid−liquid partition and potential of enzyme recycling by alkaline washing. *Enzyme and Microbial Technology, 79*, 70−77.

Rogalinski, T., Ingram, T., & Brunner, G. (2008). Hydrolysis of lignocellulosic biomass in water under elevated temperatures and pressures. *The Journal of Supercritical Fluids, 47*, 54.

Roy, S., & Chakraborty, S. (2019). Comparative study of the effectiveness of protic and aprotic ionic liquids in microwave-irradiated catalytic conversion of lignocellulosic June grass to biofuel precursors. *Bioresource Technology Reports, 8*100338.

Roy, S., & Chakraborty, S. (2020). A kinetic framework for microwave-irradiated catalytic conversion of lignocelluloses to biofuel precursors by employing protic and aprotic ionic liquids. In P. Verma (Ed.), *Biorefineries: A step towards renewable and clean energy. Clean energy production technologies* (pp. 173−215). Singapore: Springer.

Roy, S., & Chakraborty, S. (2023). Regulatory effects of water in two-phase protic ionic liquid-mediated catalytic conversion of non-edible lignocelluloses to biofuel precursors. *Biomass Bioenergy, 168*106674.

Sánchez, Ó. J., & Cardona, C. A. (2008). Trends in biotechnological production of fuel ethanol from different feedstocks. *Bioresource Technology, 99*, 5270−5295.

Sarkar, N., Ghosh, S. K., Bannerjee, S., & Aikat, K. (2012). Bioethanol production from agricultural wastes: An overview. *Renewable Energy, 37*(1), 19−27.

Schacht, C., Zetzl, C., & Brunner, G. (2008). From plant materials to ethanol by means of supercritical fluid technology. *The Journal of Supercritical Fluids, 46*, 299.

Shamsudin, S., Md Shah, U. K., Zainudin, H., Abd-Aziz, S., Mustapa Kamal, S. M., Shirai, Y., & Hassan, M. A. (2012). Effect of steam pretreatment on oil palm empty fruit bunch for the production of sugars. *Biomass and Bioenergy, 36*, 280−288.

Sharma, S., Tsai, M. L., Sharma, V., Sun, P. P., Nargotra, P., Bajaj, B. K., Chen, C. W., & Dong, C. D. (2023). Environment friendly pretreatment approaches for the bioconversion of lignocellulosic biomass into biofuels and value-added products. *Environments, 10*, 6.

Sindhu, R., Binod, P., & Pandey, A. (2015). Biological pretreatment of lignocellulosic biomass—An overview. *Bioresource Technology, 199*, 76−82.

Suhara, H., Kodama, S., Kamei, I., Maekawa, N., & Meguro, S. (2012). Screening of selective lignin-degrading basidiomycetes and biological pretreatment for enzymatic hydrolysis of bamboo culms. *International Biodeterioration & Biodegradation, 75*, 176−180.

Sun, Y., & Cheng, J. (2002). Hydrolysis of lignocellulosic materials for ethanol production: a review. *Bioresource Technology, 83*, 1−11.

Tan, J., Li, Y., Tan, X., Wu, H., Li, H., & Yang, S. (2021). Advances in pretreatment of straw biomass for sugar production. *Front. Chem., 9*696030.

Teymouri, F., Alizadeh, H., Laureano-Pérez, L., Dale, B., & Sticklen, M. (2004). Effects of ammonia fiber explosion treatment on activity of endoglucanase from Acidothermus cellulolyticus in transgenic plant. *Applied Biochemistry and Biotechnology, 116*, 1183−1192.

Tutt, M., Kikas, T., & Olt, J. (2012). Comparison of different pretreatment methods on degradation of rye straw. *Engineering for Rural Development*, Jelgava, Latvia.

Vandenbossche, V., Brault, J., Vilarem, G., Hernández-Meléndez, O., Vivaldo-Lima, E., Hernández-Luna, M., Barzana, E., Duque, A., Manzanares, P., Ballesteros, M., Mata, J., Castellon, E., & Rigal, L. (2014). A new lignocellulosic biomass deconstruction process combining thermo-mechano chemical action and bio-catalytic enzymatic hydrolysis in a twin-screw extruder. *Industrial Crops and Products, 55*, 258−266.

Verma, M. L., Barrow, C. J., & Puri, M. (2013). Nanobiotechnology as a novel paradigm for enzyme immobilisation and stabilisation with potential applications in biodiesel production. *Applied Microbiology and Biotechnology, 97*, 23−39.

Weiss, N., Borjesson, J., Pedersen, L. S., & Meyer, A. S. (2013). Enzymatic lignocellulose hydrolysis: Improved cellulase productivity by insoluble solids recycling. *Biotechnology for Biofuels, 6*, 5.

Wood, T. M. (1985). Properties of cellulolytic enzyme systems. *Biochemical Society Transactions, 13*, 407−410.

Wood, T. M., & Mccrae, S. I. (1978). The cellulase of Trichoderma koningii. Purification and properties of some endoglucanase components with special reference to their action on cellulose when acting alone and in synergism with the cellobiohydrolase. *The Biochemical Journal, 171*, 61−72.

Wyman, C. E., Dale, B. E., Elander, R. T., Holtzapple, M., Ladisch, M. R., & Lee, Y. Y. (2005). Coordinated development of leading biomass pretreatment technologies. *Bioresource Technology, 96*, 1959−1966.

Yadav, M., Singh, A., Balan, V., Pareek, N., & Vivekanand, V. (2019). Biological treatment of lignocellulossic biomassy Chaetomium globosporum: Process derivation and improved biogas production. *International Journal of Biological Macromolecules, 128*, 176−183.

Yang, B., Boussaid, A., Mansfield, S. D., Gregg, D. J., & Saddler, J. N. (2002). Fast and efficient alkaline peroxide treatment to enhance the enzymatic digestibility of steam-exploded softwood substrates. *Biotechnology and Bioengineering, 77*, 678−684.

Zadeh, Z. E., Abdulkhani, A., Aboelazayem, O., & Saha, B. (2020). Recent insights into lignocellulosic biomass pyrolysis: A critical review on pretreatment, characterization, and products upgrading. *Processes, 8*, 799.

Zhang, S., Sun, N., He, X., Lu, X., & Zhang, X. (2006). Physical properties of ionic liquids: Database and evaluation. *Journal of Physical and Chemical Reference Data, 35*, 4.

Zhang, Y.-H. P., Ding, S.-Y., Mielenz, J. R., Cui, J.-B., Elander, R. T., Laser, M., Himmel, M. E., McMillan, J. R., & Lynd, L. R. (2007). Fractionating recalcitrant lignocellulose at modest reaction conditions. *Biotechnology and Bioengineering, 97*, 2.

Zhao, X., Cheng, K., & Liu, D. (2009). Organosolv pretreatment of lignocellulosic biomass for enzymatic hydrolysis. *Applied Microbiology and Biotechnology, 82*, 815−827.

CHAPTER

5

Design and operation of advanced waste biomass processing system

Nitin Kumar, Jaydev Kumar Mahato and Sunil Kumar Gupta

Department of Environmental Science & Engineering, Indian Institute of Technology (Indian School of Mines), Dhanbad, Jharkhand, India

5.1 Introduction

Municipal solid waste (MSW) is mainly generated from urban local bodies that include waste streams from domestic areas, institutions, and commercial establishments (Vyas et al., 2022). At present, around 1.9 billion tonnes/year of MSW is generated globally, and it is expected to rise up to 3.4 billion tonnes/year by 2050. Approximately 70% of the collected MSW ends up in landfills, whereas 20% is recycled, and the rest is used for energy recovery (Nanda & Berruti, 2021). The management of MSW is one of the most difficult challenges that the municipalities of various countries are facing all around the globe (Prajapati et al., 2021). Inadequate disposal of MSW contaminates the nearby environment and leads to greenhouse gas emissions (Dabe et al., 2019). Waste characterization is considered one of the essential factors in selecting an effective waste management system and deciding the most appropriate treatment technology (Phuong et al., 2021). The composition of the MSW varies across different locations of the municipalities. The MSW is typically composed of biodegradable and nonbiodegradable fractions from various organic and inorganic sources. MSW typically includes paper, plastics, glass, metal, electronic waste, inert materials, solid organic wastes (SOW) (e.g., food waste, yard waste, wood waste), etc. Compositional analysis revealed that approximately 40%–70% of the total MSW is composed of SOW. These wastes are generally rich in nutrients, minerals, proteins, sugars, etc., which can be used as a feedstock to recover resources such as compost, green fuel, and energy (Hagos et al., 2017). Hence, SOW being an unmanaged resource has become a hot topic for research all across the globe (Cerda et al., 2018).

The conventional methods, including landfilling, incineration, and composting, are traditionally used technologies for the management of SOW. Traditional biological conversion processes include composting, vermicomposting, and anaerobic digestion that produce organic fertilizer and biogas. Nowadays, advanced thermochemical conversion technologies such as pyrolysis, gasification, hydrothermal, and thermal digestion technologies are gaining attention for the recovery of nutrients, bio-oil, syn-gas, and char. Some next-generation biological conversion techniques such as fermentation are also being utilized to produce resources such as biohydrogen/ethanol and other biofuels. In contrast to conventional methods, these technologies have a higher potential in significantly reducing the GHG emissions and transforming SOW into valuable resources. However, several technical and practical challenges need to be resolved for the widespread application of these technologies (Wainaina et al., 2020). This chapter covered important constituents of the SOW and focused on the detailed aspects of the different technological options for their treatment. A strength, weakness, opportunity, and threat (SWOT) analysis of various technologies has also been performed in this chapter.

5.2 Global trend of waste generation and compositional variation in household organic waste

The waste generation trend of various countries is shown in Fig. 5.1. China generates highest amount of SOW (17.7 MT/year) followed by India (11 MT/year) and the USA (6.1 MT/year). The fraction of organics in the total MSW in developed countries such as USA (27%), Japan (31%), France (32%), and United Kingdom (31%) is much lower than developing countries such as India (48%) and China (59%). The food waste (FW) originating from households and commercial establishments is the major component of the SOW. The waste generated from households and residential societies mainly includes avoidable or unavoidable food residues from kitchens, spoiled food items, etc. Compositional analysis of the waste is the initial and essential step for deciding the most effective approach for its better management. The generation of household organic waste depends on consumption behavior, socioeconomic status, and lifestyle pattern, which varies in urban and rural environments. The compositional variation in the SOW generated from different parts of the world is presented in Table 5.1.

Developed countries generate relatively higher fraction of fruit and vegetable waste than developing ones. China generates more amount of meat-based waste than other countries. In terms of continents, the amount of cooked food waste (45%) generated by Asia was found to be higher than others. On average, the global composition of total waste is dominated by the accumulation of fruit and vegetable waste (44%) followed by cooked items (29%). The organic wastes generated from households are rich in organic matters (> 80%) and nutrients, have higher calorific value (> 1500 kcal/kg on dry weight basis), which make it suitable for recovering resources such as compost and energy.

5.3 Sources of solid organic waste (SOW)

5.3.1 Food waste

FW is the major constituent of MSW and is considered the most abundant biowaste in the world (Sommaggio et al., 2018). USEPA defines FW as "Uneaten food and food preparation wastes from residences, commercial and institutional establishments" (Thyberg & Tonjes, 2016). Approximately 1.3 billion tonnes/year of FW are produced globally, and its production is gradually rising day by day. FWs are extremely biodegradable in nature having a high moisture content (>80%). This is found rich in organic matter (volatile solids between 70% and 90%, on dry weight basis), carbohydrates, lipids, proteins, cellulose, hemicellulose, sugars, fatty acids, amino acids, etc., which could be converted into different forms (Dahiya et al., 2015). The composition of FW varies among different regions across the globe. This mainly includes cooked and uncooked food items, vegetables,

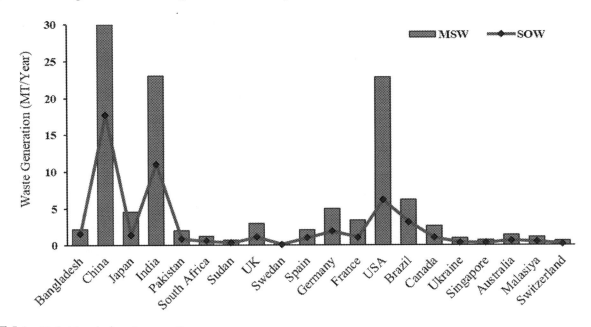

FIGURE 5.1 Global trend of waste generation.

TABLE 5.1 Compositional variation in household organic waste.

Origin	Composition (%)						References
	Fruits and vegetable	Cooked food	Meat and eggshells	Tea leaves/coffee grounds	Dairy products	Others	
India	52	27	08	06	02	05	Kumar and Gupta (2022)
China	43	32	20	–	–	05	Wang et al. (2017)
UK	63	17	07	09	01	03	Heaven et al. (2013)
Italy	60	25	07	–	–	08	Heaven et al. (2013)
North America	26	21	06	–	22	25	Chen, Chaudhary et al. (2020)
Europe	28	29	05	–	12	26	Chen, Chaudhary et al. (2020)
Asia	40	45	03	–	04	08	Chen, Chaudhary et al. (2020)
Africa	43	42.5	05	–	0.5	09	Chen, Chaudhary et al. (2020)
Global average	44	29	07	07	05	08	This study

(-): Not mentioned

fruits, bakery, dairy, and meat items, from households, food processing industries, hospitality sector, and commercial establishments (Parthiba Karthikeyan et al., 2018). Approximately around half of energy and materials could easily be recovered from FW using various technologies transforming waste into energy and resources (Chojnacka et al., 2019).

5.3.2 Garden/yard waste

The generation of garden waste (GW) has increased due to the growth of green spaces in urban areas. This mainly includes leaves, grass clippings, and other lignocellulosic residues generated from the pruning of trees and plants. Approximately 70%–80% of the GW is composed of grass clippings. GW represents a major component of the biodegradable municipal waste stream. The composition and physicochemical properties of GW depend on factors various such as topography, environment, and season of the locations (Boldrin & Christensen, 2010). In United Kingdom the fraction of GW (21%) is higher than FW (17%) of the total household waste with an average GW generation rate of 0.68 kg/household/day (Eades et al., 2020). The average moisture content in the GW varies between 10% and 30%, and 55%–60% of the dry matter constitutes of lignocellulosic material. Traditional GW disposal methods include landfilling and incineration, which lead to environmental issues such as air, water, and land pollution (Gabhane et al., 2012). Composting of GW is not recommended because it can't be completely degraded and generate low-quality products because the lignocellulose content in GW is recalcitrant to microbial degradation. However, GW could be used as an additive to improve the quality of compost products (Chen et al., 2019). GW could also be utilized for the production of biofuels (e.g., bioethanol, pellets), biochar, or energy recovery(Sofokleous et al., 2022).

5.3.3 Livestock waste

Livestock includes buffaloes, dairy animals, goats, pigs, sheep, horses, chickens, ducks, quails, rabbits, etc. Rumen and animal manure are also a part of livestock waste (Dhanya et al., 2020). Approximately 50% of the dairy industry's emissions come from manure. With an average of 10 tonnes of manure per year from one cow, the USA itself produces 10 million tonnes of animal manure annually. Proper management of manures would help in reducing the eutrophication and GHGs emissions. Livestock manures are produced from the digestion of animal feed and are found to be rich in lignocellulosic materials. Animal manure is found rich in nitrogen (up to 4%), and approximately 60%–70% of the dry matter constitutes of lignocellulose. This animal manure is now considered a new source of renewable energy as it has enormous potential for the production of biofuels (Jung et al., 2021). The conventional techniques could efficiently be utilized for producing biofuels from the manures (Chowdhury et al., 2020). Whereas animal by-products (feather meal, pig meal, poultry meal, bone meal, etc.) could be used in manufacturing fertilizers (Chojnacka et al., 2019).

5.3.4 Agricultural waste

Agricultural by-product includes wheat, corn stover, sugarcane bagasse, straw from rice, and other crops, shells of almond, pistachio, and nuts, stalks of sunflower, cotton, and soybean, husks of rice, maize, soybean, groundnut, etc. Agricultural residues are rich in lignocellulosic components, that is, lignin (10%–25%), cellulose (40%–60%), and hemicellulose (20%–40%) (Donar et al., 2016). Instead of utilizing it as a resource, open burning of the agricultural residues is commonly practiced to clear the agricultural field. This leads to the accumulation of inorganic salts and emissions of particulate matter, resulting in adverse health impacts. Pretreatment significantly improves the digestion efficiency of stubborn agricultural residues rich in lignocellulosic material (Kumar & Samadder, 2020). Anaerobic digestion does not completely recover the energy from agricultural residues. Due to the recalcitrant nature of the lignin, a significant amount of the organic matter remains left in the digestate (Antoniou et al., 2019). Hence, for complete harnessing the potential of agricultural residues, valorization of digestate via advanced processes such as pyrolysis, gasification, and hydrothermal treatment for the production of biochar and biofuels could prove to be an efficient option (Kapoor et al., 2020). Residues containing high sugar, such as potato peels, molasses, and sugarcane bagasse, are being utilized for the production of bioethanol (Rajaeifar et al., 2019). Several factors such as the availability of biomass, costs, and operating conditions should be considered for producing bioenergy using agricultural residues (Dhanya et al., 2020).

The global and Indian composition of the SOW is shown in Fig. 5.2. The global composition of the SOW is mainly dominated by FW (75%) followed by agricultural and livestock waste (15%). The characteristics of the waste assists in identifying the most suitable treatment option. The physicochemical characterization of the major fractions of SOW is shown in Table 5.2. The FW is found rich in plant essential nutrients, that is, N (1.5%–3.2%), P (0.2%–0.5%), and K (0.6%–2.0%), which make it suitable for recovering nutrients. Whereas GW and agricultural waste are found rich in lignocellulosic materials (>65%) making it recalcitrant to biodegradation; hence, they are more suitable for energy recovery.

5.4 Classification of technologies for processing of solid organic waste

Proper treatment of SOW using various technologies such as biological, thermal treatment, and sanitary landfilling is usually practiced in developed countries. These technologies demand funds, skilled manpower, and a large land area. Whereas, in developing countries, the open dumping into the landfills, open burning are still

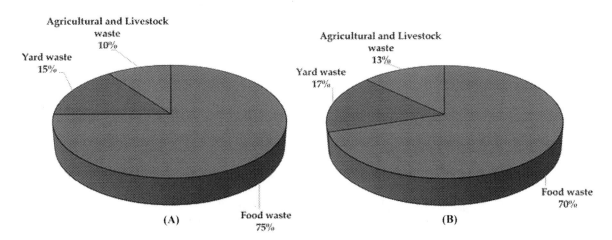

FIGURE 5.2 (A) Global, (B) Indian composition of solid organic waste.

5.4 Classification of technologies for processing of solid organic waste

TABLE 5.2 Physiochemical characteristic range of various types of organic wastes.

Properties (%)[a]	Substrate				References
	Food waste	Garden waste	Agricultural waste	Livestock waste	
Moisture content	60–90	10–30	5–15	50–80	Boldrin and Christensen (2010); Fisgativa et al. (2018); Huang et al. (2011); Perin et al. (2020); Quadar et al. (2022); Titiloye et al. (2013)
Volatile solids	75–90	75–85	60–80	65–85	
N	1.5–3.2	0.8–1.6	0.7–2.3	1.0–3.8	
P	0.2–0.5	0.06–0.2	0.1–0.3	0.4–1.7	
K	0.6–2.0	0.5–1.2	0.4–1.5	0.8–1.9	
Lignin	5–10	5–15	10–25	7–12	
Cellulose	10–20	30–55	40–60	10–42	
Hemicellulose	15–30	25–35	20–40	8–24	

[a]*Expressed in dry weight basis except moisture content.*

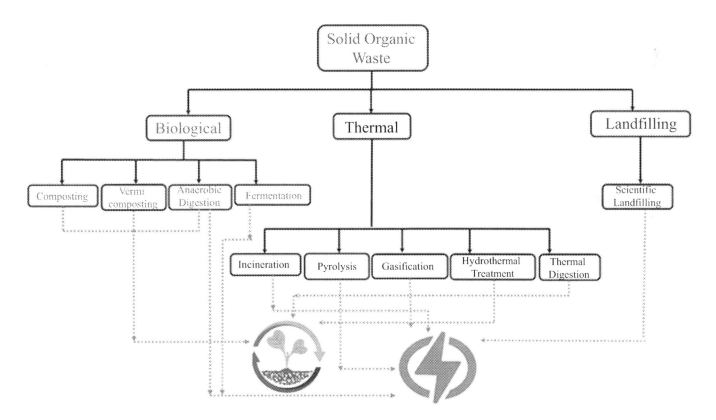

FIGURE 5.3 Schematics of various technologies for processing of SOW. *SOW*, Solid organic waste

mostly opted for waste disposal. Various techniques are available for converting SOW into useful resources while complying with all the regulatory norms for environmental protection. The waste management system aims to recover energy and resources, followed by the safe disposal of any residues. For any particular location, the waste processing technology that fulfills all the required criteria is finally selected for operation. Among the widely available techniques, the three most widely used technologies are (1) biological conversion (composting, vermicomposting, anaerobic digestion, and fermentation), (2) thermal conversion (incineration, pyrolysis, gasification, hydrothermal treatment, and thermal digestion), and (3) scientific landfilling. Various treatment techniques reviewed in this chapter are illustrated in Fig. 5.3.

5.4.1 Biological conversion techniques

5.4.1.1 Composting

Composting is a biological process that converts the SOW into a humidified substance suitable for soil application (Tiquia, 2010). During composting, microbes convert the organic matter into a stabilized end product. During composting of yard waste and animal manure, the temperature of piles reaches up to 70°C. The thermophilic reactions that occurred in composting make the end product devoid of pathogenic bacteria (Chojnacka et al., 2019). Moreover, the end product of composting is found rich in plant available nutrients such as nitrogen, phosphorus, and potassium. It efficiently reduces the mass of SOW and transforms it into a nutritious soil amendment suitable for improving the soil fertility (Awasthi et al., 2016). The currently employed composting technologies include aerated, windrows, and in-vessel systems. The main factors controlling composting processes include moisture content, mixing agent, particle size, temperature, pH, C/N ratio, aeration rate, and nutrients (Kumar et al., 2010). These parameter changes continuously along with the decomposition process and significantly affect the quality of the end product. The composting of various substrate and their critical findings are presented in Table 5.3. Li et al. (2021) revealed that an initial moisture content of 53% resulted in bio-augmentation of microbial agent during composting of chicken manure. Yang et al. (2019) found that mixing 10% matured compost along with the food waste during aerobic composting with intermittent aeration helps in reducing 69% GHGs emissions. While other additives such as biochar help in retaining the nutrients and thus improve the quality of the compost. Aeration combined with aerobic composting has proven effective in accelerating the rate of composting. (Kulcu & Yaldiz, 2004) found that an aeration rate of 0.4 L air min^{-1} kg$_{om}^{-1}$ resulted higher degradation of agricultural waste. Compared with landfill emissions, composting generates less amount of greenhouse gases and also improves the soil properties (Chen, Zhang, et al., 2020). The nutrient-rich fertilizers generated from composting process promote the circular economy concept by maintaining the soil organic matter. However, the process is time-intensive and requires careful management for a successful outcome.

5.4.1.2 Vermicomposting

Vermicomposting is the combined action of microbes and earthworms to achieve nonthermophilic decomposition and stabilization of SOW (Ali et al., 2015). The microbes assist in degrading the organic matter, while earthworms help in the conditioning of the substrate and accelerating the biochemical process. Earthworms could effectively utilize the organic matter having moisture content between 40% and 55%, pH in the range from 5 to 8,

TABLE 5.3 Composting of various substrate and their critical findings.

Sl. no.	Substrate used	Technology	Time period, days	Critical findings	References
1.	Food waste and matured compost	Aerobic composting with intermittent aeration	35	Mixing 10% matured compost reduced GHGs emissions by 69.2%.	Yang et al. (2019)
2.	Chicken manure	Aerobic bioreactor	45	Optimum moisture content (53%) resulted bio-augmentation of microbial agent.	Li et al. (2021)
3.	Mixed agricultural waste	Forced aeration reactors	21	Optimum aeration rate (0.4 L air min^{-1} kg$_{om}^{-1}$) resulted in highest degradation of organic matter	Kulcu and Yaldiz (2004)
4.	Green plant materials and biochar	Aerobic windrow composting	90	Addition of 4% biochar improves the nutrient retention, water holding capacity of the compost	Hagemann et al. (2018)
5.	Cow manure and sawdust	Aerobic reactor	63	Addition of 2% CaCN$_2$ assist in reaching the sanitary standard without any significant impact on composting process	Simujide et al. (2013)
6.	Food waste and wood chip	In-vessel forced aeration composting	30	An aeration rate of 0.15 m^3/m^3 min effectively maintained adequate oxygen level and temperature inside the vessel	Kim et al. (2008)

with a C/N ratio around 30. Although vermicomposting is considered superior to composting, this process still has less potential in inhibiting the pathogens, which is considered as the one of the major shortcomings of vermicomposting. Compared with conventional composting, where the temperature may reach up to 70°C, the optimum temperature in vermicomposting process is 35°C, which is considered suitable for earthworms. The earthworms are responsible for improving the bioavailability of nitrogen, phosphorus, and potassium, which ultimately enhances soil fertility. This increase in the bioavailability of nutrients is strongly attributed to the digestion of organics due to the action of gut enzymes in earthworms. The important factors influencing the vermicomposting process are stocking density, feeding rate, pH, temperature, moisture, and C:N ratio. The variation in compost quality due to variation in the substrate and earthworm species is presented in Table 5.4.

5.4.1.3 Anaerobic digestion

Anaerobic digestion (AD) is considered the most suitable biological treatment technique for the recovery of methane from SOW. The end product of AD is biogas (a mixture of CH_4 and CO_2 gases) along with digestate rich in plant essential nutrients (Rajak et al., 2020). In this process, organic matter is transformed into biogas by the microorganism in an anaerobic environment. The quality of the biogas generated from this process depends upon the initial composition of the substrate and process parameters. Biogas is a composition of various gases such as 60%–70% CH_4, 30%–40% CO_2, and a trace of other gases (such as NH_3, water vapor, and H_2S) (Kumar & Samadder, 2017). The anaerobic digestion process for various substrates and their methane yield are compared in Table 5.5. The digestate could effectively be used to improve the soil organic matter in the agricultural fields. It was estimated that 1 m^3 of biogas could generate 2.04 kWh of electricity (Murphy et al., 2004). The major factors that influence the effectiveness of AD include temperature, pH, total solids content, C/N ratio, particle size,

TABLE 5.4 Variation in compost quality due to variation in the substrate and earthworm species.

Sl. no.	Substrate	Time period, days	Earthworm species	Quality of end product			References
				N (%)	P (%)	K (%)	
1.	Cattle dung	90	*Eisenia fetida*	2.0	1.08	1.96	Bhat et al. (2013)
2.	Sorghum straw and cow dung	60	*Eudrilus eugeniei* and *Eisenia fetida*	0.94	0.8	0.81	Chander et al. (2018)
3.	Coconut husk and cattle dung	120	*Eisenia fetida*	2.7	1.08	2.73	Quadar et al. (2022)
4.	Household waste	90	*Perionyx excavatus* and *Perionyx sansibaricus*	2.03	0.63	0.96	Suthar and Singh (2008)
5.	Cow dung and kitchen waste	49	*Lumbricus rubellus*	2.01	0.29	0.99	Adi and Noor (2009)

TABLE 5.5 Variation in methane yield of anaerobic digestion processes for various types of SOW.

Sl. no.	Substrate	Type of reactor	Inoculum used	Methane yield	References
1.	Food waste and cattle manure	Semi-continuous reactor at mesophilic temperature	Activated Sludge	388 mL CH_4/gVS	Zhang et al. (2013)
2.	Kitchen waste	Biochemical methane potential (BMP) tests in 250 mL bottle at 35°C	Anaerobic sludge	520 mLCH_4/gVS	Liu et al. (2012)
3.	Wheat straw	BMP tests in 100 mL flask	Granular sludge	306 mLCH_4/g OM	Dumas et al. (2015)
4.	Grass crippling	Two-stage continuously stirred digester	Anaerobic sludge	350 L CH_4 kg^{-1} VS	Massanet-Nicolau et al. (2015)
5.	Rice straw	Anaerobic batch reactor	Digestate from biogas plant	292 mLCH_4/gVS	Dehghani et al. (2015)

etc. For effective functioning of an anaerobic digester, the optimal pH should be 7 and C/N ratio should be between 15 and 20. However, the major associated issue with AD is its longer treatment period and sudden bacterial inhibition (Pham et al., 2015).

5.4.1.4 Fermentation

The fermentation process is used to convert the organics containing lipids, proteins, and carbohydrates into useful products such as volatile fatty acids (acetic, propionic, and butyric acid) and alcohols (ethanol and butanol) (Sekoai et al., 2021). Before the fermentation process, sterilization of feedstock is usually done to avoid any microbial contamination. Factors influencing fermentation include the type of reactor, inoculum, the concentration of the substrate, temperature, pH, and metal ions (Ghimire et al., 2015). Biohydrogen is a sustainable form of energy that is produced from SOW through the fermentation process, which has higher a heating value (120MJ/kg). The pH level around 5.5 simulates the production of hydrogen in the temperature range of 35°C–40°C. The combined production of both VFAs and biohydrogen through the fermentation process is one of the best examples where biomolecules and energy are produced simultaneously. The dark fermentation process is widely used to produce fermented by-products (solvents and fatty acids). It presents an opportunity for combining various processes to recover higher energy (Turon et al., 2016). However, further studies are required to upscale this technique and lowering down the treatment cost; moreover, the overall economic viability of the process should be considered before starting any commercial operation (Pham et al., 2015). Comparison of different fermentation processes and their end product is presented in Table 5.6.

5.4.2 Thermal conversion techniques

5.4.2.1 Incineration

Incineration of organic waste is carried out at a temperature > 900°C, which not reduces its huge volume by 85% but also recovers energy from it. For an effective incineration with energy recovery, the average calorific value of waste should be between 1700 and 1900 kcal/kg. Incineration involves controlled combustion of organic waste, resulting into generation of heat energy that can be utilized for operating steam turbines to generate electricity (Pham et al., 2015). This entire practice works on a sequence of two steps, namely primary and secondary processes. The primary process involves drying, volatilization, fixed carbon combustion, and burn-out of char of the solids. At the same time, the secondary process includes the combustion of vapors, gases, and particulates driven off during the primary process (Begum et al., 2012). The incineration process yields water vapor (H_2O), carbon dioxide (CO_2), other gases, and a small portion of noncombustible residue that can be disposed of by secure landfilling in an environment-friendly manner. Generally, incineration of 1 tonne of MSW generates approximately 544 kWh of energy and produces 180 kg of solid residue (Zaman, 2010). The mineralization of

TABLE 5.6 Evaluation of fermentation processes and their end products.

Sl. no.	Substrate	Method	Experimental condition	End product	References
1.	Food waste and wheat straw	Dark Fermentation	Tests were carried out at 37°C for 14 days using waste activated sludge as inoculum.	Biohydrogen	Ghimire et al. (2015)
2.	Food waste	Anaerobic electro fermentation	Graphite was used as electrodes and applied with 0.6 V potential on working electrode. Consortium from sewage treatment plant was used as inoculum.	Biohythane and volatile fatty acids	Shanthi Sravan et al. (2018)
3.	Lemon peel	Enzymatic hydrolysis and fermentation	Hydrolysates from the hydrolysis process were placed in oven to inactivate enzymes, and fermentation was carried out at 37 °C	Bioethanol	Boluda-Aguilar and López-Gómez (2013)
4.	Organic municipal waste	Semicontinuous acidogenic fermentation	Mechanically stirred reactors was used for fermentation at mesophilic condition.	Volatile fatty acid	Cheah et al. (2019)
5.	Wheat bran	Three-stage enzymatic hydrolysis and fermentation	Fermentation reaction was initiated by *Sporolactobacillus inulinus* YBS1–5 inoculation, followed by incubation at 40°C	D-Lactate	Li et al. (2017)

organic substances into harmless end products is one of the major advantages of the incineration process. In contrast, the release of various toxic and noxious gases (dibenzofurans, chlorinated dibenzo-p-dioxins, greenhouse gases, etc.) discouraged its application (Chen, Zhang, et al., 2020).

5.4.2.2 Pyrolysis

It is an advanced thermochemical process that involves the thermal degradation of organic waste at a temperature range of 300°C–700°C under anoxic conditions (Kambo & Dutta, 2015). During the pyrolysis process, the quality of the end products, that is, pyro-gas (CO, CO_2, CH_4, and H_2), liquid (bio-oil), or carbon-rich solid (biochar), mainly depends on its operational parameters (process temperature, heating rate, and residence time). The pyro-gas and bio-oil have a heat value of around 15.7 MJ/m^3 and 30–40 MJ/kg, respectively (Cantrell et al., 2007). Moreover, carbon-rich solid products (biochar) have several applications and can also be used to improve soil fertility (Chojnacka et al., 2019). Sometimes, the high moisture content of biomass hinders the process by decreasing the operating temperature, so it is suggested to lower the moisture (10%–15%) prior to pyrolysis (Marzbali et al., 2021). The efficiency of the pyrolysis process could be enhanced by a pretreatment step termed torrefaction. This process is also known as mild pyrolysis, where the biomass is heated slowly in an oxygen-deficient condition at a maximum temperature of 300°C (Akdeniz, 2019). However, the end products of torrefaction cannot be considered biochar due to its intermediate properties of raw biomass and biochar. The energy produced by the pyrolysis is cleaner than the incineration process due to the reduced dioxins, sulfur oxide (SO_2), and nitrogen oxides (NO_2). It has several advantages over gasification or incineration, including short vapor residence time, high liquid yield production (425°C–600°C), production of char, and higher energy recovery (Marzbali et al., 2021).

5.4.2.3 Gasification

Gasification is another thermochemical process that converts organic waste into a combustible gaseous product (syngas) in a limited oxygen supply (Sheth & Babu, 2010). It changes the chemical structure of the biomass at a high temperature (600°C–900°C), and the gasification agent (other gaseous compound) assists in quickly converting the feedstock into gas with the help of different heterogeneous reactions. The primary gaseous product generated from this process includes H_2O, CO_2, H_2, CO, and CH_4; however, a small amount of biochar (<10%) is also produced during the process (Kambo & Dutta, 2015). Despite being a complicated process, gasification has several benefits, namely potential to reduce the release of lethal fumes (dioxins and detoxifying ashes), cost-effective, and lowering the release of heavy metals through evaporation (Al-Ghouti et al., 2021). Moreover, gasification is a more flexible waste conversion technique than incineration and pyrolysis, with a high level of public acceptance.

5.4.2.4 Hydrothermal treatment

Hydrothermal treatment is the best-suited technology for wet biomass (80%–90% moisture content) that could even operate in the presence of water or steam. It converts the organic material into value-added end products (solid, liquid, and biofuel) by the actions of water at elevated temperatures (>100°C) inside a closed chamber (Vlaskin et al., 2017). Based on operational temperature, hydrothermal treatment is classified into three techniques: (1) hydrothermal liquefaction (HTL), (2) hydrothermal carbonization (HTC), and (3) supercritical water gasification (SCWG). The HTC and HTL are usually performed at a temperature range of 180°C–250°C and 250°C–370°C and produce hydrochar and bio-oil as the end product, respectively. However, when the operating temperature is raised to >400°C to maximize the gas production (methane and hydrogen), the process is termed SCWG (Marzbali et al., 2021). Hydrothermal treatment possesses several advantages over traditional waste processing methods such as incineration, pyrolysis, and gasification. Due to the high moisture content of food waste, these techniques required a predrying step of feedstock to lower the moisture. Hence, the requirement for extra energy consumption and handling of wet waste is the major stumbling block. Whereas during hydrothermal waste processing, a predrying step is not required, and the energy consumption is nearly half compared with pyrolysis of feedstock with higher moisture content (Peterson et al., 2008).

5.4.2.5 Thermal digestion

Thermal digestion emerged as a novel technique for the rapid conversion of SOW into nutrient-rich organic fertilizer (Kumar & Gupta, 2021). In this method, SOW is shredded, and heated at a temperature range of

100°C–160°C, which digests the organics and converts into simpler form. Inside a thermal digester, the heated air is circulated toward the reaction chamber where the shredded SOW is uniformly agitated to improve the drying rate. The end product of the thermal digestion process is devoid of any pathogen and moisture; hence, it could easily be stored for longer duration without any change in its nutrient value. This method is quick, hygienic, easy to operate, and requires very small area for its operation, making it suitable for residential societies and commercial establishments. The variation in N, P, and K at different digestion temperature is shown in Fig. 5.4. Total N content reduced by 18%, whereas the availability of P and K increased by 14% and 20% respectively when the temperature was increased from 100°C to 160°C. The thermal digestion helps in degrading the high-molecular-weight organic N into low-molecular-weight organic N and creates a dry bond N structure, which converts the SOW into a slow-release fertilizer. At the same time, the organic P gets oxidized into inorganic phosphate form, while the physically bounded K gets converted into exchangeable K^+ form and thus enhances the bioavailability of P and K by converting it into plant available form (Kumar & Gupta, 2022).

The comparison of the operational parameters and the end product of various thermal conversion technologies is presented in Table 5.7.

5.4.3 Sanitary landfilling

Landfill is one of the oldest but still followed techniques by the developing or underdeveloped countries for the management of waste. Landfills break down and stabilize the disposed wastes over a period of time.

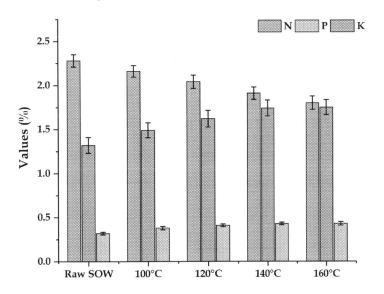

FIGURE 5.4 Variations in N, P, and K at different digestion temperature.

TABLE 5.7 Comparison between operational parameters of various thermal conversion technologies.

Technology	Parameters				End product
	Temperature (°C)	Pressure	Environment	Treatment duration (h)	
Incineration	900–1500	Atmospheric	Oxygen rich	4–8	Gaseous emissions, fly bottom ash
Pyrolysis	300–700	Atmospheric	Anoxic	2–4	Biochar, CO_2, CO, CH_4
Gasification	600–900	Atmospheric	Limited oxygen	1–2	Syngas and ash
Hydrothermal treatment	180–260	Autogenic (up to 45 MPa)	Oxygen rich	8–10	Hydrocar, biooil, CO_2, H_2
Thermal digestion	100–160	Atmospheric	Oxygen rich	4–6	Nutrient rich organic fertilizer

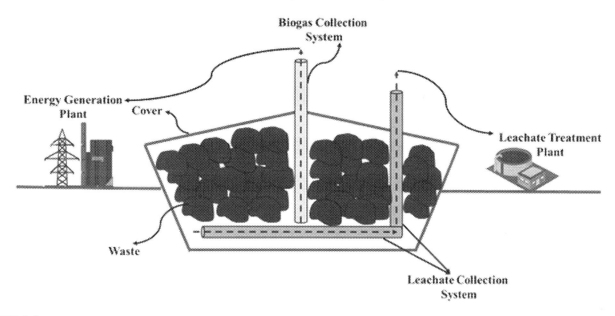

FIGURE 5.5 Schematic diagram of a sanitary landfill with biogas collection system.

Most of the landfills do not have energy recovery facilities. In most of the developing countries, unsanitary landfilling is a common practice for the disposal of waste that causes a serious environmental threat (Kumar & Samadder, 2017). Studies reported that unsanitary landfilling of waste causes more damage to the environment as compared with other waste management techniques (Zaman, 2010). The leachates generated from this landfill are a major pollutant that contaminates the nearby water bodies and groundwater. Sanitary landfilling is defined as the controlled and systematic disposal of waste to reduce the negative environmental impact through biogas recovery and leachate treatment Fig. 5.5. Developed nations have begun to discourage the disposal of waste in landfills by enforcing strict policies, reducing and recycling the waste. Landfilling should be the last option for waste disposal; if required, a maximum of 10%–15% of the total waste should go to landfills.

5.5 SWOT analysis of various technologies

The SWOT (strength, weakness, opportunity, and threats) analysis technique is widely used for the selection of suitable technology for the processing and disposal of SOW (Aich & Ghosh, 2016). Comparative assessment of the existing technologies in terms of strength, weakness, opportunity, and threats is given in Table 5.8. The biological treatment involves lower treatment cost; however, it requires regular monitoring, longer treatment time, larger environmental footprints, and also involves the chances of process failure due to bacteria inhibition (Pham et al., 2015). Incineration, pyrolysis, and gasification processes are well suited for SOW, having a high calorific value and low moisture content (Dhar et al., 2017). These processes require a predrying step and could only handle feedstock with lower moisture content. The requirement of an energy-intensive drying process for handling the SOW with high moisture is still a major challenge for these techniques. Whereas the hydrothermal treatment and thermal digestion process eliminate the requirement of the predrying process, and hence, it is suitable for biomass with high moisture content. These processes result in the production of disinfected, nutrient-rich end product. However, the requirement of extremely high energy for heating and difficulty in transforming it into commercial scale are major disadvantages of hydrothermal treatment. While the thermal digestion is a quick process relatively simpler in operation, higher scaling-up potential and doesn't possess any threats to the environment.

TABLE 5.8 SWOT analysis of various treatment technologies.

Technologies	Parameters			
	Strength (S)	Weakness (W)	Opportunity (O)	Threats (T)
Composting	Low-cost technology	Time intensive and requires stringent process monitoring; emission control is difficult	Recovery of organic fertilizer.	Potential threat to soil and water contamination; gaseous emission.
Vermicomposting	Simple and easy to start.	Requirement of vast land; frequent monitoring is required.	Recovery of nutrient-rich organic fertilizer	Gaseous emissions to the environment.
Anaerobic digestion	Energy and fertilizer recovery option	Chances of process failure are high; requires higher initial investment.	Recovery of biogas along with the fertilizer.	Emissions to the environment.
Fermentation	Mixed organic waste could be used	Pretreatment is required; weakness in transforming into commercial scale	Recovery of biofuels.	No such serious threat is identified.
Incineration	Reduces huge volume	Requires predrying step; higher investment cost; residue management is difficult	Energy recovery option	Emissions to the environment
Pyrolysis	Less treatment time; low amount of final residue	Energy intensive process; requirement of anoxic environment.	Generates biochar; opportunity for energy recovery	Potential threat to the atmosphere from emissions
Gasification	Low volume of final residue.	High initial investment; Full scale operations is not yet developed	Produces syngas suitable for various industrial applications	Emissions to the atmosphere.
Hydrothermal Treatment	Suitable for organic waste with high moisture	Higher energy consumption; difficulty in transforming into commercial scale	Recovery of various resources.	By-products may have negative environmental impact
Thermal digestion	Suitable for any type of organic waste; portable; ease in operation	Requirement of initial investment.	Recovery of nutrient rich organic fertilizer; easy to scale up for commercial use	No such threat is identified.
Sanitary landfilling	Natural process that could handle a huge volume of waste	Requirement of larger land area; reclamation of land takes long duration	Opportunity to recover biogas for energy generation	Potential threat to air, water and soil contamination.

5.6 Summary and conclusions

This chapter presented an overview of various technologies for the recycling of SOW. Traditional biological conversion processes such as composting, vermicomposting, and anaerobic digestion are widely used technologies for the production of organic fertilizer and biogas. Next-generation biological conversion processes such as fermentation could prove to be an efficient technology for the recovery of biohydrogen/ethanol and other biofuels. Nowadays, advanced thermochemical conversion technologies such as pyrolysis, gasification, hydrothermal technologies are gaining attention for the recovery of bio-oil, syngas, and char. However, there are still many practical constraints and technical issues that need to be addressed for the widespread application of these technologies. The thermal digestion of SOW is a quick and novel process for recovery and recycling of nutrients back to the nature. The selection of the process parameters, however, would vary and depends on various factors such as compositional variation, infrastructure, and socioeconomic aspects.

References

Adi, A. J., & Noor, Z. M. (2009). Waste recycling: Utilization of coffee grounds and kitchen waste in vermicomposting. *Bioresource Technology*, 100(2), 1027–1030. Available from https://doi.org/10.1016/j.biortech.2008.07.024.

Aich, A., & Ghosh, S. K. (2016). Application of swot analysis for the selection of technology for processing and disposal of MSW. *Procedia Environmental Sciences*, 35, 209–228. Available from https://doi.org/10.1016/j.proenv.2016.07.083.

References

Akdeniz, N. (2019). A systematic review of biochar use in animal waste composting. *Waste Management*, 88, 291–300. Available from https://doi.org/10.1016/j.wasman.2019.03.054, http://www.elsevier.com/locate/wasman.

Al-Ghouti, M. A., Khan, M., Nasser, M. S., Al-Saad, K., & Heng, O. E. (2021). Recent advances and applications of municipal solid wastes bottom and fly ashes: Insights into sustainable management and conservation of resources. *Environmental Technology and Innovation*, 21. Available from https://doi.org/10.1016/j.eti.2020.101267, http://www.journals.elsevier.com/environmental-technology-and-innovation/.

Ali, U., Sajid, N., Khalid, A., Riaz, L., Rabbani, M. M., Syed, J. H., & Malik, R. N. (2015). A review on vermicomposting of organic wastes. *Environmental Progress and Sustainable Energy*, 34(4), 1050–1062. Available from https://doi.org/10.1002/ep.12100, http://onlinelibrary.wiley.com/journal/10.1002/(ISSN)1944-7450.

Antoniou, N., Monlau, F., Sambusiti, C., Ficara, E., Barakat, A., & Zabaniotou, A. (2019). Contribution to Circular Economy options of mixed agricultural wastes management: Coupling anaerobic digestion with gasification for enhanced energy and material recovery. *Journal of Cleaner Production*, 209, 505–514. Available from https://doi.org/10.1016/j.jclepro.2018.10.055, https://www.journals.elsevier.com/journal-of-cleaner-production.

Awasthi, M. K., Pandey, A. K., Bundela, P. S., Wong, J. W. C., Li, R., & Zhang, Z. (2016). Co-composting of gelatin industry sludge combined with organic fraction of municipal solid waste and poultry waste employing zeolite mixed with enriched nitrifying bacterial consortium. *Bioresource Technology*, 213, 181–189. Available from https://doi.org/10.1016/j.biortech.2016.02.026, http://www.elsevier.com/locate/biortech.

Begum, S., Rasul, M. G., & Akbar, D. (2012). An investigation on thermo chemical conversions of solid waste for energy recovery. *World Academy of Science, Engineering and Technology*, 62, 624–630.

Bhat, S. A., Singh, J., & Vig, A. P. (2013). Vermiremediation of dyeing sludge from textile mill with the help of exotic earthworm Eisenia fetida Savigny. *Environmental Science and Pollution Research*, 20(9), 5975–5982. Available from https://doi.org/10.1007/s11356-013-1612-2, https://link.springer.com/journal/11356.

Boldrin, A., & Christensen, T. H. (2010). Seasonal generation and composition of garden waste in Aarhus (Denmark). *Waste Management*, 30(4), 551–557. Available from https://doi.org/10.1016/j.wasman.2009.11.031.

Boluda-Aguilar, M., & López-Gómez, A. (2013). Production of bioethanol by fermentation of lemon (Citrus limon L.) peel wastes pretreated with steam explosion. *Industrial Crops and Products*, 41(1), 188–197. Available from https://doi.org/10.1016/j.indcrop.2012.04.031.

Cantrell, K., Ro, K., Mahajan, D., Anjom, M., & Hunt, P. G. (2007). Role of thermochemical conversion in livestock waste-to-energy treatments: Obstacles and opportunities. *Industrial and Engineering Chemistry Research*, 46(26), 8918–8927. Available from https://doi.org/10.1021/ie0616895, 08885885 United States.

Cerda, A., Artola, A., Font, X., Barrena, R., Gea, T., & Sánchez, A. (2018). Composting of food wastes: Status and challenges. *Bioresource Technology*, 248, 57–67. Available from https://doi.org/10.1016/j.biortech.2017.06.133, http://www.elsevier.com/locate/biortech.

Chander, G., Wani, S. P., Gopalakrishnan, S., Mahapatra, A., Chaudhury, S., Pawar, C. S., Kaushal, M., & Rao, A. V. R. K. (2018). Microbial consortium culture and vermi-composting technologies for recycling on-farm wastes and food production. *International Journal of Recycling of Organic Waste in Agriculture*, 7(2), 99–108. Available from https://doi.org/10.1007/s40093-018-0195-9, http://www.springer.com/environment/pollution + and + remediation/journal/40093.

Cheah, Y. K., Vidal-Antich, C., Dosta, J., & Mata-Álvarez, J. (2019). Volatile fatty acid production from mesophilic acidogenic fermentation of organic fraction of municipal solid waste and food waste under acidic and alkaline pH. *Environmental Science and Pollution Research*, 26(35), 35509–35522. Available from https://doi.org/10.1007/s11356-019-05394-6, https://link.springer.com/journal/11356.

Chen, C., Chaudhary, A., & Mathys, A. (2020). Nutritional and environmental losses embedded in global food waste. *Resources, Conservation and Recycling*, 160104912. Available from https://doi.org/10.1016/j.resconrec.2020.104912.

Chen, M., Huang, Y., Liu, H., Xie, S., & Abbas, F. (2019). Impact of different nitrogen source on the compost quality and greenhouse gas emissions during composting of garden waste. *Process Safety and Environmental Protection*, 124, 326–335. Available from https://doi.org/10.1016/j.psep.2019.03.006, http://www.elsevier.com/wps/find/journaldescription.cws_home/713889/description#description.

Chen, T., Zhang, S., & Yuan, Z. (2020). Adoption of solid organic waste composting products: A critical review. *Journal of Cleaner Production*, 272122712. Available from https://doi.org/10.1016/j.jclepro.2020.122712.

Chojnacka, K., Gorazda, K., Witek-Krowiak, A., & Moustakas, K. (2019). Recovery of fertilizer nutrients from materials - Contradictions, mistakes and future trends. *Renewable and Sustainable Energy Reviews*, 110, 485–498. Available from https://doi.org/10.1016/j.rser.2019.04.063, https://www.journals.elsevier.com/renewable-and-sustainable-energy-reviews.

Chowdhury, T., Chowdhury, H., Hossain, N., Ahmed, A., Hossen, M. S., Chowdhury, P., Thirugnanasambandam, M., & Saidur, R. (2020). Latest advancements on livestock waste management and biogas production: Bangladesh's perspective. *Journal of Cleaner Production*, 272122818. Available from https://doi.org/10.1016/j.jclepro.2020.122818.

Dabe, S. J., Prasad, P. J., Vaidya, A. N., & Purohit, H. J. (2019). Technological pathways for bioenergy generation from municipal solid waste: Renewable energy option. *Environmental Progress and Sustainable Energy*, 38(2), 654–671. Available from https://doi.org/10.1002/ep.12981, http://onlinelibrary.wiley.com/journal/10.1002/(ISSN)1944-7450.

Dahiya, S., Sarkar, O., Swamy, Y. V., & Venkata Mohan, S. (2015). Acidogenic fermentation of food waste for volatile fatty acid production with co-generation of biohydrogen. *Bioresource Technology*, 182, 103–113. Available from https://doi.org/10.1016/j.biortech.2015.01.007, http://www.elsevier.com/locate/biortech.

Dehghani, M., Karimi, K., & Sadeghi, M. (2015). Pretreatment of rice straw for the improvement of biogas production. *Energy and Fuels*, 29(6), 3770–3775. Available from https://doi.org/10.1021/acs.energyfuels.5b00718, http://pubs.acs.org/journal/enfuem.

Dhanya, B. S., Mishra, A., Chandel, A. K., & Verma, M. L. (2020). Development of sustainable approaches for converting the organic waste to bioenergy. *Science of the Total Environment*, 723. Available from https://doi.org/10.1016/j.scitotenv.2020.138109, http://www.elsevier.com/locate/scitotenv.

Dhar, H., Kumar, S., & Kumar, R. (2017). A review on organic waste to energy systems in India. *Bioresource Technology*, 245, 1229–1237. Available from https://doi.org/10.1016/j.biortech.2017.08.159, http://www.elsevier.com/locate/biortech.

Donar, Y. O., Çağlar, E., & Sınağ, A. (2016). Preparation and characterization of agricultural waste biomass based hydrochars. *Fuel*, 183, 366–372. Available from https://doi.org/10.1016/j.fuel.2016.06.108.

Dumas, C., Silva Ghizzi Damasceno, G., Barakat, A., Carrère, H., Steyer, J.-P., & Rouau, X. (2015). Effects of grinding processes on anaerobic digestion of wheat straw. *Industrial Crops and Products*, 74, 450–456. Available from https://doi.org/10.1016/j.indcrop.2015.03.043.

Eades, P., Kusch-Brandt, S., Heaven, S., & Banks, C. J. (2020). Estimating the generation of garden waste in england and the differences between rural and urban areas. *Resources*, 9(1), 8. Available from https://doi.org/10.3390/resources9010008.

Fisgativa, H., Marcilhac, C., Girault, R., Daumer, M. L., Trémier, A., Dabert, P., & Béline, F. (2018). Physico-chemical, biochemical and nutritional characterisation of 42 organic wastes and residues from France. *Data in Brief*, 19, 1953–1962. Available from https://doi.org/10.1016/j.dib.2018.06.050.

Gabhane, J., William, S. P., Bidyadhar, R., Bhilawe, P., Anand, D., Vaidya, A. N., & Wate, S. R. (2012). Additives aided composting of green waste: Effects on organic matter degradation, compost maturity, and quality of the finished compost. *Bioresource Technology*, 114, 382–388. Available from https://doi.org/10.1016/j.biortech.2012.02.040.

Ghimire, A., Frunzo, L., Pirozzi, F., Trably, E., Escudie, R., Lens, P. N. L., & Esposito, G. (2015). A review on dark fermentative biohydrogen production from organic biomass: Process parameters and use of by-products. *Applied Energy*, 144, 73–95. Available from https://doi.org/10.1016/j.apenergy.2015.01.045, http://www.elsevier.com/inca/publications/store/4/0/5/8/9/1/index.htt.

Hagemann, N., Subdiaga, E., Orsetti, S., de la Rosa, J. M., Knicker, H., Schmidt, H.-P., Kappler, A., & Behrens, S. (2018). Effect of biochar amendment on compost organic matter composition following aerobic composting of manure. *Science of The Total Environment*, 613–614, 20–29. Available from https://doi.org/10.1016/j.scitotenv.2017.08.161.

Hagos, K., Zong, J., Li, D., Liu, C., & Lu, X. (2017). Anaerobic co-digestion process for biogas production: Progress, challenges and perspectives. *Renewable and Sustainable Energy Reviews*, 76, 1485–1496. Available from https://doi.org/10.1016/j.rser.2016.11.184.

HeavenS., ZhangY., ArnoldR., PaavolaT., VazF., & CavinatoC. (2013). *Compositional analysis of food waste from study sites in geographically distinct regions of Europe.*

Huang, G., Wang, X., & Han, L. (2011). Rapid estimation of nutrients in chicken manure during plant-field composting using physicochemical properties. *Bioresource Technology*, 102(2), 1455–1461. Available from https://doi.org/10.1016/j.biortech.2010.09.086.

Jung, S., Shetti, N. P., Reddy, K. R., Nadagouda, M. N., Park, Y. K., Aminabhavi, T. M., & Kwon, E. E. (2021). Synthesis of different biofuels from livestock waste materials and their potential as sustainable feedstocks – A review. *Energy Conversion and Management*, 236. Available from https://doi.org/10.1016/j.enconman.2021.114038, https://www.journals.elsevier.com/energy-conversion-and-management.

Kambo, H. S., & Dutta, A. (2015). A comparative review of biochar and hydrochar in terms of production, physico-chemical properties and applications. *Renewable and Sustainable Energy Reviews*, 45, 359–378. Available from https://doi.org/10.1016/j.rser.2015.01.050.

Kapoor, R., Ghosh, P., Kumar, M., Sengupta, S., Gupta, A., Kumar, S. S., Vijay, V., Kumar, V., Kumar Vijay, V., & Pant, D. (2020). Valorization of agricultural waste for biogas based circular economy in India: A research outlook. *Bioresource Technology*, 304. Available from https://doi.org/10.1016/j.biortech.2020.123036, http://www.elsevier.com/locate/biortech.

Kim, J. D., Park, J. S., In, B. H., Kim, D., & Namkoong, W. (2008). Evaluation of pilot-scale in-vessel composting for food waste treatment. *Journal of Hazardous Materials*, 154(1–3), 272–277. Available from https://doi.org/10.1016/j.jhazmat.2007.10.023.

Kulcu, R., & Yaldiz, O. (2004). Determination of aeration rate and kinetics of composting some agricultural wastes. *Bioresource Technology*, 93(1), 49–57. Available from https://doi.org/10.1016/j.biortech.2003.10.007, http://www.elsevier.com/locate/biortech.

Kumar, N., & Gupta, S. K. (2022). Exploring drying kinetics and fate of nutrients in thermal digestion of solid organic waste. *Science of the Total Environment*, 837. Available from https://doi.org/10.1016/j.scitotenv.2022.155804, http://www.elsevier.com/locate/scitotenv.

Kumar, N., & Gupta, S. K. (2021). Exploring the feasibility of thermal digestion process: A novel technique, for the rapid treatment and reuse of solid organic waste as organic fertilizer. *Journal of Cleaner Production*, 318. Available from https://doi.org/10.1016/j.jclepro.2021.128600, https://www.journals.elsevier.com/journal-of-cleaner-production.

Kumar, M., Ou, Y. L., & Lin, J. G. (2010). Co-composting of green waste and food waste at low C/N ratio. *Waste Management*, 30(4), 602–609. Available from https://doi.org/10.1016/j.wasman.2009.11.023.

Kumar, A., & Samadder, S. R. (2017). A review on technological options of waste to energy for effective management of municipal solid waste. *Waste Management*, 69, 407–422. Available from https://doi.org/10.1016/j.wasman.2017.08.046, http://www.elsevier.com/locate/wasman.

Kumar, A., & Samadder, S. R. (2020). Performance evaluation of anaerobic digestion technology for energy recovery from organic fraction of municipal solid waste: A review. *Energy*, 197117253. Available from https://doi.org/10.1016/j.energy.2020.117253.

Liu, X., Wang, W., Gao, X., Zhou, Y., & Shen, R. (2012). Effect of thermal pretreatment on the physical and chemical properties of municipal biomass waste. *Waste Management*, 32(2), 249–255. Available from https://doi.org/10.1016/j.wasman.2011.09.027.

Li, M. X., He, X. S., Tang, J., Li, X., Zhao, R., Tao, Y. Q., Wang, C., & Qiu, Z. P. (2021). Influence of moisture content on chicken manure stabilization during microbial agent-enhanced composting. *Chemosphere*, 264. Available from https://doi.org/10.1016/j.chemosphere.2020.128549, http://www.elsevier.com/locate/chemosphere.

Li, J., Sun, J., Wu, B., & He, B. (2017). Combined utilization of nutrients and sugar derived from wheat bran for D-Lactate fermentation by Sporolactobacillus inulinus YBS1–5. *Bioresource Technology*, 229, 33–38. Available from https://doi.org/10.1016/j.biortech.2016.12.101, http://www.elsevier.com/locate/biortech.

Marzbali, M. H., Kundu, S., Halder, P., Patel, S., Hakeem, I. G., Paz-Ferreiro, J., Madapusi, S., Surapaneni, A., & Shah, K. (2021). Wet organic waste treatment via hydrothermal processing: A critical review. *Chemosphere*, 279. Available from https://doi.org/10.1016/j.chemosphere.2021.130557, http://www.elsevier.com/locate/chemosphere.

Massanet-Nicolau, J., Dinsdale, R., Guwy, A., & Shipley, G. (2015). Utilising biohydrogen to increase methane production, energy yields and process efficiency via two stage anaerobic digestion of grass. *Bioresource Technology*, 189, 379–383. Available from https://doi.org/10.1016/j.biortech.2015.03.116, http://www.elsevier.com/locate/biortech.

Murphy, J. D., McKeogh, E., & Kiely, G. (2004). Technical/economic/environmental analysis of biogas utilisation. *Applied Energy*, 77(4), 407–427. Available from https://doi.org/10.1016/j.apenergy.2003.07.005, http://www.elsevier.com/inca/publications/store/4/0/5/8/9/1/index.htt.

Nanda, S., & Berruti, F. (2021). A technical review of bioenergy and resource recovery from municipal solid waste. *Journal of Hazardous Materials*, 403123970. Available from https://doi.org/10.1016/j.jhazmat.2020.123970.

Parthiba Karthikeyan, O., Trably, E., Mehariya, S., Bernet, N., Wong, J. W. C., & Carrere, H. (2018). Pretreatment of food waste for methane and hydrogen recovery: A review. *Bioresource Technology*, 249, 1025–1039. Available from https://doi.org/10.1016/j.biortech.2017.09.105, http://www.elsevier.com/locate/biortech.

Perin, J. K. H., Borth, P. L. B., Torrecilhas, A. R., Cunha., Kuroda, E. K., & Fernandes, F. (2020). Optimization of methane production parameters during anaerobic co-digestion of food waste and garden waste. *Journal of Cleaner Production*.

Peterson, A. A., Vogel, F., Lachance, R. P., Fröling, M., Antal, M. J., & Tester, J. W. (2008). Thermochemical biofuel production in hydrothermal media: A review of sub- and supercritical water technologies. *Energy and Environmental Science*, 1(1), 32–65. Available from https://doi.org/10.1039/b810100k, http://pubs.rsc.org/en/journals/journal/ee.

Pham, T. P. T., Kaushik, R., Parshetti, G. K., Mahmood, R., & Balasubramanian, R. (2015). Food waste-to-energy conversion technologies: Current status and future directions. *Waste Management*, 38(1), 399–408. Available from https://doi.org/10.1016/j.wasman.2014.12.004, http://www.elsevier.com/locate/wasman.

Phuong, N., Yabar, H., & Mizunoya, T. (2021). Characterization and analysis of household solid waste composition to identify the optimal waste management method: A case study in Hanoi City, Vietnam. *Earth*, 2(4), 1046–1058. Available from https://doi.org/10.3390/earth2040062.

Prajapati, P., Varjani, S., Singhania, R. R., Patel, A. K., Awasthi, M. K., Sindhu, R., Zhang, Z., Binod, P., Awasthi, S. K., & Chaturvedi, P. (2021). Critical review on technological advancements for effective waste management of municipal solid waste — Updates and way forward. *Environmental Technology & Innovation*, 23101749. Available from https://doi.org/10.1016/j.eti.2021.101749.

Quadar, J., Chowdhary, A. B., Dutta, R., Angmo, D., Rashid, F., Singh, S., Singh, J., & Vig, A. P. (2022). Characterization of vermicompost of coconut husk mixed with cattle dung: physicochemical properties, SEM, and FT-IR analysis. *Environmental Science and Pollution Research*. Available from https://doi.org/10.1007/s11356-022-21899-z, https://link.springer.com/journal/11356.

Rajaeifar, M. A., Sadeghzadeh Hemayati, S., Tabatabaei, M., Aghbashlo, M., & Mahmoudi, S. B. (2019). A review on beet sugar industry with a focus on implementation of waste-to-energy strategy for power supply. *Renewable and Sustainable Energy Reviews*, 103, 423–442. Available from https://doi.org/10.1016/j.rser.2018.12.056, https://www.journals.elsevier.com/renewable-and-sustainable-energy-reviews.

Rajak, R. C., Jacob, S., & Kim, B. S. (2020). A holistic zero waste biorefinery approach for macroalgal biomass utilization: A review. *Science of the Total Environment*, 716. Available from https://doi.org/10.1016/j.scitotenv.2020.137067, http://www.elsevier.com/locate/scitotenv.

Sekoai, P. T., Ghimire, A., Ezeokoli, O. T., Rao, S., Ngan, W. Y., Habimana, O., Yao, Y., Yang, P., Yiu Fung, A. H., Yoro, K. O., Daramola, M. O., & Hung, C. H. (2021). Valorization of volatile fatty acids from the dark fermentation waste Streams-A promising pathway for a biorefinery concept. *Renewable and Sustainable Energy Reviews*, 143. Available from https://doi.org/10.1016/j.rser.2021.110971, https://www.journals.elsevier.com/renewable-and-sustainable-energy-reviews.

Shanthi Sravan, J., Butti, S. K., Sarkar, O., Vamshi Krishna, K., & Venkata Mohan, S. (2018). Electrofermentation of food waste — Regulating acidogenesis towards enhanced volatile fatty acids production. *Chemical Engineering Journal*, 334, 1709–1718. Available from https://doi.org/10.1016/j.cej.2017.11.005, http://www.elsevier.com/inca/publications/store/6/0/1/2/7/3/index.htt.

Sheth, P. N., & Babu, B. V. (2010). Production of hydrogen energy through biomass (waste wood) gasification. *International Journal of Hydrogen Energy*, 35(19), 10803–10810. Available from https://doi.org/10.1016/j.ijhydene.2010.03.009.

Simujide, H., Aorigele, C., Wang, C. J., Lina, M., & Manda, B. (2013). Microbial activities during mesophilic composting of manure and effect of calcium cyanamide addition. *International Biodeterioration and Biodegradation*, 83, 139–144. Available from https://doi.org/10.1016/j.ibiod.2013.05.003.

Sofokleous, M., Christofi, A., Malamis, D., Mai, S., & Barampouti, E. M. (2022). Bioethanol and biogas production: an alternative valorisation pathway for green waste. *Chemosphere*, 296133970. Available from https://doi.org/10.1016/j.chemosphere.2022.133970.

Sommaggio, L. R. D., Mazzeo, D. E. C., Sant' Anna, D. D. A. E. S., Levy, C. E., & Marin-Morales, M. A. (2018). Ecotoxicological and microbiological assessment of sewage sludge associated with sugarcane bagasse. *Ecotoxicology and Environmental Safety*, 147, 550–557. Available from https://doi.org/10.1016/j.ecoenv.2017.09.009, http://www.elsevier.com/inca/publications/store/6/2/2/8/1/9/index.htt.

Suthar, S., & Singh, S. (2008). Vermicomposting of domestic waste by using two epigeic earthworms (Perionyx excavatus and Perionyx sansibaricus). *International Journal of Environmental Science and Technology*, 5(1), 99–106. Available from https://doi.org/10.1007/BF03326002, http://www.springerlink.com/content/1735-1472.

Thyberg, K. L., & Tonjes, D. J. (2016). Drivers of food waste and their implications for sustainable policy development. *Resources, Conservation and Recycling*, 106, 110–123. Available from https://doi.org/10.1016/j.resconrec.2015.11.016, http://www.elsevier.com/locate/resconrec.

Tiquia, S. M. (2010). Reduction of compost phytotoxicity during the process of decomposition. *Chemosphere*, 79(5), 506–512. Available from https://doi.org/10.1016/j.chemosphere.2010.02.040, http://www.elsevier.com/locate/chemosphere.

Titiloye, J. O., Abu Bakar, M. S., & Odetoye, T. E. (2013). Thermochemical characterisation of agricultural wastes from West Africa. *Industrial Crops and Products*, 47, 199–203. Available from https://doi.org/10.1016/j.indcrop.2013.03.011.

Turon, V., Trably, E., Fouilland, E., & Steyer, J. P. (2016). Potentialities of dark fermentation effluents as substrates for microalgae growth: A review. *Process Biochemistry*, 51(11), 1843–1854. Available from https://doi.org/10.1016/j.procbio.2016.03.018, http://www.elsevier.com/inca/publications/store/4/2/2/8/5/7.

Vlaskin, M. S., Kostyukevich, Y. I., Grigorenko, A. V., Kiseleva, E. A., Vladimirov, G. N., Yakovlev, P. V., & Nikolaev, E. N. (2017). Hydrothermal treatment of organic waste. *Russian Journal of Applied Chemistry*, 90(8), 1285–1292. Available from https://doi.org/10.1134/S1070427217080158, http://www.springerlink.com/content/1070-4272.

Vyas, S., Prajapati, P., Shah, A. V., & Varjani, S. (2022). Municipal solid waste management: Dynamics, risk assessment, ecological influence, advancements, constraints and perspectives. *Science of the Total Environment*, 814. Available from https://doi.org/10.1016/j.scitotenv.2021.152802, http://www.elsevier.com/locate/scitotenv.

Wainaina, S., Awasthi, M. K., Sarsaiya, S., Chen, H., Singh, E., Kumar, A., Ravindran, B., Awasthi, S. K., Liu, T., Duan, Y., Kumar, S., Zhang, Z., & Taherzadeh, M. J. (2020). Resource recovery and circular economy from organic solid waste using aerobic and anaerobic digestion technologies. *Bioresource Technology*, 301. Available from https://doi.org/10.1016/j.biortech.2020.122778, http://www.elsevier.com/locate/biortech.

Wang, L. E., Liu, G., Liu, X., Liu, Y., Gao, J., Zhou, B., Gao, S., & Cheng, S. (2017). The weight of unfinished plate: A survey based characterization of restaurant food waste in Chinese cities. *Waste Management, 66*, 3–12. Available from https://doi.org/10.1016/j.wasman.2017.04.007, http://www.elsevier.com/locate/wasman.

Yang, F., Li, Y., Han, Y., Qian, W., Li, G., & Luo, W. (2019). Performance of mature compost to control gaseous emissions in kitchen waste composting. *Science of the Total Environment, 657*, 262–269. Available from https://doi.org/10.1016/j.scitotenv.2018.12.030, http://www.elsevier.com/locate/scitotenv.

Zaman, A. U. (2010). Comparative study of municipal solid waste treatment technologies using life cycle assessment method. *International Journal of Environmental Science and Technology, 7*(2), 225–234. Available from https://doi.org/10.1007/BF03326132, http://www.ceers.org/ijest/issues/full/v7/n2/702002.pdf.

Zhang, C., Xiao, G., Peng, L., Su, H., & Tan, T. (2013). The anaerobic co-digestion of food waste and cattle manure. *Bioresource Technology, 129*, 170–176. Available from https://doi.org/10.1016/j.biortech.2012.10.138, http://www.elsevier.com/locate/biortech.

CHAPTER 6

Application of cutting-edge molecular biotechnological techniques in waste valorization

Poulomi Sarkar[1] and Angana Sarkar[2]

[1]Division of Plant Physiology, ICAR-Indian Agricultural Research Institute, New Delhi, India [2]Department of Biotechnology and Medical Engineering, National Institute of Technology Rourkela, Rourkela, Odisha, India

6.1 Introduction

Advancement in urbanization has revolutionized human civilization. The advent in engineering and technology has created devises that are now indispensable in our lives. However, the negative impacts of modernization are quite distressing. Depletion of natural resources and environmental pollution are the major drawbacks that are associated with increasing urbanization (Han et al., 2022). Development of sustainable as well as effective resources for energy and feedstock is a pressing need of this century (Usmani et al., 2021). Use of biological methods for waste bioprocessing is a cost-effective and environmentally benign alternative. Moreover, organic wastes are degraded using microbial metabolic pathways, which could lead to controlled growth conditions and production of valuable products from waste (Pandit et al., 2015).

Valorization of waste biomass, which contains rich resources of renewable energy and carbon sources, has become very important (Liu et al., 2018). Many physical, chemical, and biological methods of waste remediation have been implemented among which biological processes are the most preferred (Dash & Osborne, 2022). Recent advances in omics tools, metabolic engineering, synthetic biology, and several molecular techniques are helping us to designing nonmodel microorganisms for bioprocessing purposes in a systematic manner (Yan & Fong, 2017). The biological waste bioprocessing techniques work through either aerobic or anaerobic routes based on the innate microbial communities as well as the type of electron acceptors available (Purohit et al., 2016). The biological methods of waste valorization are eco-friendly and sustainable in nature. This chapter aims to describe the present scenarios in waste generation, challenges associated with their remediation, and recent advances in molecular biotechnological tools for waste-to-wealth transitions using bioprocessing.

6.2 Status of waste generation in India and world

The waste disposal methodologies have become difficult owing to the shift in nature of waste from more organic to synthetic inorganic compounds (Cheela et al., 2021; Prajapati et al., 2021). Formerly organic wastes could be safely laid into low-lying regions. Rapid industrialization has increased the genesis of solid waste from different sources such as glass, paper, plastic manufacturing industries, vegetable and fruit markets, domestic areas, healthcare units, construction sites, commercial centers, and trading zones. These anthropogenic processes make solid waste generation and management a worldwide problem (Gour & Singh, 2023). According to United Nations, the estimated urbanization of world population will reach 67.2% during 2050. As waste generation is directly proportional to population growth and gross domestic product utility, the rise in urbanization will hugely impact the waste genesis scenarios in future. Global solid waste generation is >17 billion tonnes per annum. The per capita municipal solid waste generation in India currently stands at 0.49 kg/day, which is less than most of the developed nations. However, it is projected that in future this waste generation statistics of developing nations will coincide with those in developed countries. In terms of plastic wastes, India generates 9.46 million tonnes per annum that comprises 43% of single-use plastic and 40% constitutes the uncollected waste plastics (Bansal et al., 2023). Global generation of agricultural waste is >140 billion metric tons (Ajayi & Lateef, 2023). Another problem of modernization is the growing demand for electronic and electrical equipment. The use of such devices is negatively impacting the environment by increasing the generation of E-wastes. Worldwide around 53.6 million tonnes of E-wastes were produced in 2019. India contributes 3.23 million tonnes of E-waste each year. Out of this, 9.3 million tonnes were properly collected, disposed, or recycled. However, 44.3 million tonnes of E-wastes remained undocumented (Forti et al., 2020; Sengupta et al., 2023). Similarly, the generation of agricultural and medical waste in India is also an alarming issue. Estimated >686 metric tonnes of lignocellulosic crop residues are produced in the agricultural sectors in India (Yadav et al., 2023). The biomedical waste generation in India was about 203,650 Mg in 2017 that has increased considerably in the postpandemic time (Kashyap & Ramaprasad, 2023). Food waste production is closely related to agricultural wastes. As per Food Waste Index Report of 2019, approximately 931 million tonnes of food wastes were generated from different household activities. Developed countries in Latin America and Europe produce 200 and 180 kg/capita/annum consumer waste, respectively. Similarly, developed and industrialized Asia and sub-Saharan Africa generate 155 and 150 kg/capita/annum consumer wastes. Overall, food waste that is contributed by high-income and developed nations is >307 g/capita/day (Capanoglu et al., 2022). If the available reports on food waste generation in different countries and continents are summarized, it can be seen 50,000–100,000 tonnes per annum vegetable oil waste is produced in United Kingdom, 4,000,000 tonnes per year tomato pomace are contributed by Europe, 57,000 tonnes per annum wheat straw waste are produced in the USA, 40,000–45,000 tonnes per annum cereal waste are generated in Europe, 700 tonnes per year orange peels and tomato pomace are produced in the USA and France, respectively (Capanoglu et al., 2022). However, developed nations have better waste management and resource recovery methods in use. However, waste management efficiency of developing nations is not as robust as developed countries. Currently, the waste management capacity in India is 62 million tonnes per annum. The waste collection efficiency of India is nearly 90%. However, only 27% of this collected waste gets treated, and the remaining untreated waste poses threat to human as well as environmental sustainability (Prajapati et al., 2021).

6.3 Sources and types of wastes

Based on different sources of waste generation, three broad categories of waste can be identified (Fig. 6.1).

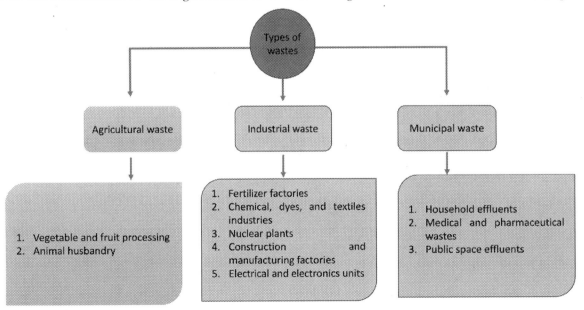

FIGURE 6.1 Types of waste categorized based on different sources of generation.

6.3.1 Agricultural and livestock waste

Over the last 50 years, agricultural processes have expanded more than three times producing >23.7 million tonnes food per day globally. However, this increase in productivity is associated with the generation of huge amounts of agricultural wastes. Agricultural activities are responsible for emission of 21% greenhouses gases (Duque-Acevedo et al., 2020). Agricultural wastes comprise residues, which are produced during growing and processing of food crops, fruits, vegetables, livestock rearing, dairy products, etc. The nature of the waste varies from gas, liquid, or solids. The different types of raw residues include sugar cane bagasse, corn stalks, cereal husks, drops and cuts from fruits or vegetable pruning, animal carcasses, manure, plant roots or tubers, meat, and animal by-products such as oils, fats, lards, shells, milk sludge, and whey (Capanoglu et al., 2022). Chemical pesticides or fertilizers mix with agricultural runoffs, thereby increasing the hazard parameters of agricultural wastes. As estimated, >998 million tonnes of agricultural wastes are produced annually (Obi et al., 2016). Considering published data, global agricultural residues such as crops, fruits, and oil seeds constitute 30%, 50%, and 20% of food waste, respectively. Similarly, households and food processing industries generate 42% and 38% wastes, respectively. The beverage industries add 26%, followed by dairy processing (21.3%), fruit or vegetable processing (14.8%), grain and starch industries (12.9%), meat processing industries (8%), oil processing mills (3.9%), and fisheries industry (0.4%) worldwide annually (Capanoglu et al., 2022). Huge amount of carbon is stored in agricultural lignocellulosic biomass waste. The major lignocellulosic wastes are categorized as hemicellulose, cellulose, and lignin (Koul et al., 2022). Persistence of these wastes is detrimental to the environment, human, and animal health (Obi et al., 2016).

6.3.2 Industrial waste

Industries are the backbone of country's economy. However, the biggest drawback of industrialization is unavoidable genesis of toxic industrial wastes, which increases the environmental hazards (Sarkar et al., 2017). Different industries such as petroleum, manufacturing, mining, power plants, and construction industries generate huge amounts of hazardous wastes. For example, petroleum processing plants produce oil sludge, production water, and wastewaters, which are endowed with recalcitrant hydrocarbons. Prolonged persistence of these hydrocarbons causes detrimental effects to health and ecosystems (Jagaba et al., 2022). Discharge of petroleum industry wastewater to local water bodies contaminates the aquatic ecosystems (Mokif et al., 2022). They are many health risks associated to petroleum hydrocarbons because they contain enormous amounts of carcinogenic, teratogenic, and neurotoxic fractions in them (Kuppusamy et al., 2020). Petroleum processing industries discharge substantial amounts of air pollutants such as hydrogen sulfide, sulfur oxide, carbon dioxide, carbon monoxide, particulate matters, and other toxic gases (Adebiyi, 2022). Industrial wastes produced from different metal mining facilities, petroleum refineries, and other manufacturing units such as fertilizers and dyes contain high amounts of toxic metal or heavy metals fractions, which are resistant to natural degradation (Jadhav & Hocheng, 2012). Similarly, menacing industrial waste outputs come from fertilizer, textiles, glass, and electrical material manufacturing industries. Such industries discharge massive quantity of chemical compounds such as phosphates, nitrates, fly ash, blast furnace slags, and obnoxious fumes into the nature (Cánovas et al., 2018; Dey et al., 2022). Apart from this, industrial wastes are also composed of fillers from concrete manufacturing factories or metal aluminosilicate-rich mine tailings (Dey et al., 2022). Many other types of industrial wastes that pollute the environment are red mud from aluminum refining industries, marble power dust, and chippings from manufacturing factories (Das et al., 2022). Mining operations for coal, oil, or gas generate liquid waste, drill cuttings, and formation waters, while the mine tailings contain gangue and water (Kalisz et al., 2022). Hazardous wastes are also discharged from nuclear power plants and industries. These wastes stand for potential sources of dangerous radioactive nuclides of different radioactive metals and compounds (Ojovan & Steinmetz, 2022). As per the International Atomic Enegry Agency (IAEA), radioactive wastes are classified as high-level wastes (need superior-level protection during handling and are capable of generating radiogenic heat), intermediate-level wastes (requiring high containment and greater depth disposal), low-level waste (having less amounts of long-lived radionuclides, so require robust containment methods), very short-lived waste (containing very short-lived radionuclides that do not require high-level containment or isolation method), and very low level wastes (contain radionuclides, which are prone to decay within few years) (IAEA R, 2014; IAEA; Ojovan & Steinmetz, 2022).

6.3.3 Municipal and electronic waste

One of the major sources of anthropogenic wastes comes in the form of municipal waste. A large variety of municipal wastes are accumulated into the nature. The most common items include plastics and rubber waste (polyethylene bags, polyvinyl chloride items, polystyrene items, zip lock bags, rubber bands, etc.), glass (storage containers, utensils, home décor, etc.), metals (metals cans, cookware, culinary equipment, etc.), textiles and leathers (discarded clothes and bags), food wastes (spoiled meats, vegetables, food leftovers, etc.), packaging materials (cardboard, newsprints, magazines, flyers, shredded papers, etc.), yard wastes (grass, leaves, trimmings, etc.), broken furniture, concrete, dry wall, inorganic materials, biomedical wastes (needles, synringes, pills, capsules, devices, etc.,), sanitary napkins, diapers, pet litter, cosmetic products, and used household materials (Jouhara et al., 2017; Nanda & Berruti, 2021).

A massive proportion of modern wastes contains humongous quantity of electronic wastes also called e-waste. Proper management of e-waste is a critical challenge worldwide. However, developing countries are gravely affected with e-waste accumulation (Osibanjo & Nnorom, 2007). The prime components of electronic wastes are electronic devices, thrashed computers, cell phones, used batteries, light bulbs, laptops, sound systems, damaged watches and clocks (Nanda & Berruti, 2021). The constant superseding of new smart devices has shortened the longevity of many devices, which rapidly becomes obsolete and transits into an electronic waste (Kiddee et al., 2013).

6.4 Waste remedial approaches and associated challenges

Management of huge amounts of anthropogenic wastes is a critical problem particularly within urban setup. The choice of waste treatment depends on nature and availability of the waste (Da Silva et al., 2012). The major waste treatment approaches that are prevalent can be broadly categorized as: (1) physical methods, (2) chemical methods, and (3) biological methods. The physical methods that are applied mostly to the treatment of petroleum hydrocarbon contaminated wastes are incineration, land farming, containment, and solidification (Ambaye et al., 2022). Similarly radioactive wastes are remediated using physical containment methods (Yanikomer et al., 2016). Recent advances in thermal plasma technology have also been employed for management of radioactive wastes (Prado et al., 2020). The mitigation of volatile chemical wastes is achieved by using techniques such as air extraction, sparging, and ignition (Ambaye et al., 2022). The treatment methods for heavy metal removal are adsorption, precipitation, ion exchange, electrolysis, solvent extraction, membrane separation, etc. (Maity et al., 2022). Generation of construction site wastes is another challenge in today's developing world. On-site sorting of construction wastes has been implemented as a method of physical technique for construction site solid waste management (Yuan et al., 2013). The available methods of chemical remediation are leaching, catalytic oxidation, fixation, and electrokinetic remediation (Koul & Taak, 2018). Based on broad classification of waste, the type of treatment processes are applied. However, the physical and chemical remedial processes are hazardous to humans as well as environment because these processes generate enormous amounts of obnoxious by-products (Sarkar et al., 2017). Considering a sustainable and environment-friendly approach of waste mitigation bioremediation is a suitable alternative. Bioremediation can be achieved by either promoting the growth of native waste inhabiting organisms or by adding exogenous metabolically superior organisms to the waste. Microbes play a quintessential role in bioremediation (Patel et al., 2022). With their vast metabolic repertoire, they can degrade almost any kind of recalcitrant waste. However, previously limited knowledge was available about microbial populations and their functional potential, which hampered the proper implementation of bioremediation processes. Interestingly, the recent progress in omics-based approaches such as metagenomics, metaproteomics, metatranscriptomics, and metabolomics is helping in bridging this gap (Gupta et al., 2020; Patil & Sarkar, 2022).

6.5 Waste bioprocessing as a valuable tool for resource recovery

Based on present waste generation scenarios, there is a pressing need of developing economic and eco-friendly remedial strategies for efficient management of waste (Banerjee & Arora, 2021). Report suggests that low- and middle-economy-based cities generate mostly biodegradable wastes, while high-income cities add more amount of nondegradable wastes (Banerjee & Arora, 2021; ISWA U, 2015) (ISWA). Improper disposal of waste not only contaminates the environment but also adds to the emission of several greenhouse gases, thereby affecting the environment and health. Moreover, wastes such as agricultural biomass, petroleum hydrocarbons, wastewater, and sewage are endowed with different carbon resources, which if not managed properly increases the carbon footprint on earth (Liu et al., 2018). Adoption of integrated waste remediation approach of reduction, reuse, recycling, and recovery of value-added product will be advantageous for reducing waste (Banerjee & Arora, 2021).

Waste-to-wealth transition using bioprocessing is feasible and sustainable tool, which is gaining impetus (Liu et al., 2018). Developing nations such as India mostly face the challenge of energy crises particularly, in rural areas. Hence, waste-to-energy transitions are becoming important. Further, resource recovery from waste also comes with its own benefits such as waste quantity reduction as high as 90%, minimization of land use, and most importantly reduction in environmental pollution (Prajapati et al., 2021). Recalcitrant wastes such as plastics can also be used as building materials in construction industries (Bansal et al., 2023). Agricultural residues find wide range of applications. Residues such as seeds, shells, or husks can act as rich resources of diverse compounds such as bioactive compounds (antioxidants, flavonoids, etc.), animal feed (single-cell proteins), renewable energy resources (briquettes, biochar, pyrolytic oil, etc.), biogas, biodiesel, different enzymes (cellulase, lipase, lignolytic enzymes), scavengers for wastewater treatment, organic acids, biofertilizers for plants, bioremediators, dyes, nanoparticles, and building material (Ajayi & Lateef, 2023; Capanoglu et al., 2022). The bioprocessing of waste for conversion to value-added alternatives involves several steps. Processes such as microbial anaerobic digestion, fermentation, and electrochemistry work together in valorization of wastes (Banerjee & Arora, 2021). Upcoming concepts for valorization also classify waste as biorefineries, which can be exploited for developing different valuable products (Ajayi & Lateef, 2023). Similarly, petroleum industry waste also offers many usages. Petroleum wastes act as resources for construction materials (cement clinker, mortar, building blocks, ceramics, bricks, etc.), road construction materials, asphalt production, adsorbent material for wasterwater treatment, generation of gas, metal catalyst synthesis, rubber tire manufacturing, carbon clay composite, etc. (Jagaba et al., 2022). Different strategies such as pyrolysis/calcination, hydrothermal synthesis, sol–gel chemical transformations, and ball milling are often used for conversion of industrial, electronic, or urban solid wates to functional materials (Chen et al., 2022). However, waste-to-wealth transitions are challenging to scale up. Therefore, extensive research is ongoing in this domain.

6.6 Recent advances in molecular biotechnology tools for waste valorization

Waste valorization methods are largely based on the nature of waste being treated. The recent advances in different molecular biotechnological tools (Fig. 6.2) have been described in detail in the following segment.

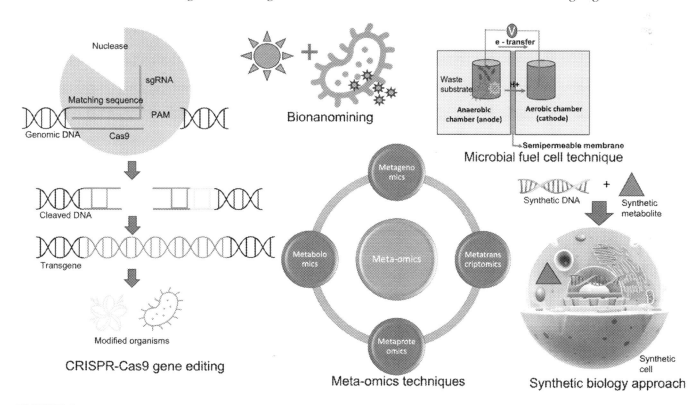

FIGURE 6.2 Molecular biotechnological techniques that are is applied to waste bioprocessing.

6.6.1 Synthetic biology approaches

Synthetic biology approaches rely on use of molecular biology tools and knowledge to synthesize, remodel, and fine-tune biological components such as cells, proteins, and genes for different beneficial applications (Rylott & Bruce, 2020). As per National Human Genome Research Institute, synthetic biology can be defined "as a field of science that involves redesigning organisms for useful purposes by engineering them to have new abilities" (Rylott & Bruce, 2020). Synthetic biology together with material science has helped in the formation of next-generation materials such as Engineered Living Materials (ELMs) (Kovelakuntla & Meyer, 2021). One of the promising examples of ELMs is the formation of aquaplastics, which are sustainable alternatives to petrochemical-based synthetic plastics. Duraj-Thatte et al. (2021) demonstrated the engineering of the *Escherichia coli* to produce bio-hydrogels, which eventually led to the formation of aquaplastics. Synthetic biology is also reshaping the use of extremophilic microbes in bioremediation of polycyclic aromatic hydrocarbons, wastewater systems, etc. The extremophilic microbes such as thermos-tolerant and halophilic bacteria are endowed with genomic repositories, which can be further modified for different bioprocessing approaches. For example, *Galdiera sulphuraria*, a thermo-tolerant microalgae together with heterotrophic bacteria, were experimented for remediation of phosphates and ammonium from wastewaters. These approaches also helped in the production of biofuels with minimal energy investments (Cheng et al., 2019; Rylott & Bruce, 2020). Synthetic biology has aided in the engineering of several nonmodel microbes such as *Geobacter sulfurreducens* and *Shewanella oneidensis* in the development of biosensors as well as fabrication of artificial organelles for bioremediation of metal or metalloid from environments (Rylott & Bruce, 2020).

6.6.2 Microbial fuel cell technology

Electroactive microbes have emerged as a renewable resource for energy and as greener alternatives for environmental management (Lovley & Holmes, 2022). These microbes can be exploited in the formation of microbial fuel cells for waste to wealth transition. Published literature suggests that microbial fuel cells (MFCs) were found as early as 1910 by Michael Cresse Potter at University of Durham, United Kingdom. However, the phenomenon of microbial fuel cells was described in around 1800 (Zamri et al., 2023). Studies have indicated the use of algal and bacterial strains in functioning of microbial fuel cells (Bond & Lovley, 2003; Koch & Harnisch, 2016). Microbes inhabiting natural environments and forming biofilms also participate in electron exchanges. These communities act as natural microbial fuel cells and are termed as "electromicrobiomes" (Chabert et al., 2015; Garbini et al., 2023). These electromicrobiomes, which are associated with fresh and marine water systems, soil/sediment interfaces or agricultural lands could provide in situ remediation solutions for heavy metals, pesticides, and hydrocarbon remediation. Over the decade, microbial fuel cells have been used in waste bioremediation and simultaneous electricity generation (Huang et al., 2011; Niu et al., 2023; Ren et al., 2014; Singh et al., 2019; Zhuang et al., 2012). These microbial batteries have helped in the biodegradation of organic pollutants (hydrocarbons) or heavy metals from wastewaters. Microbial fuel cells have also been tagged with plants in constructed wetlands for degradation of hazardous wastewater effluents (Garbini et al., 2023). Apart from wastewater treatment, researchers have also demonstrated sustainable functioning of MFCs in bioelectricity generation from formaldehydes (Idris et al., 2023). Similarly, another study has demonstrated the use of sediment-microbe-based fuel cells in generation of bioelectricity from Himalayan rock soils (Bhattacharya et al., 2023). Microbial fuel cells are becoming superior alternatives for waste bioprocessing due to their low operation cost, sustainability, and simplicity of functioning.

6.6.3 Omics-based and gene editing technologies

Biological waste bioprocessing techniques rely on use of complex microbial systems. Microbial conversion of hazardous waste to value-added products has attracted attention since ages. However, knowledge gap existed regarding the structure and function of different microbial communities. Metaomics-based approaches have helped in the understanding of this previously unexplored microbial life. Advent in gene or genome editing techniques has also augmented the fabrication of new methods for waste-to-wealth transitions. Over the past decades, several reports have highlighted the role of metagenomics in biological mitigation of environmental wastes (Chen et al., 2022; Patil & Sarkar, 2022; Sarkar et al., 2017). Use of metagenomics to elucidate the nature of microbial communities within wastewater treatment facilities, phosphorus or nitrate removal plants, heavy metal–contaminated groundwater, anaerobic oily sludge digesters, and sewage treatment units have been well documented (Chen et al., 2022; Christgen et al., 2015; Jadeja et al., 2019; Yadav et al., 2015). Medical or pharmaceutical waste bioprocessing has generated various antibiotic resistance microbes that got identified by metagenomics and culturomics methods (Nowrotek et al., 2019). Metagenomics and metatranscriptomic studies within anaerobic sludge digesters revealed microbial succession and anaerobic respiration processes, which were used in the development of full-scale digesters and wastewater treatment systems for bioremediation (Mei & Liu, 2022). The application of metaomics techniques can further be leveraged in assessing the soil health and food production by using the microbial communities as bioindicators (Djemiel et al., 2022). Currently, molecular biotechnological tools have become imperative to the development of sustainable and eco-friendly alternatives for waste bioprocessing as well as circular bioeconomy. Moreover, the discovery of CRISPR/Cas9 gene editing tool with its simple guided mechanisms is revolutionizing the waste bioresource scenarios (Hemalatha et al., 2023). Published resources suggest that CRISPR/Cas9 can assist in multifaceted way from enhancing food security to supporting bioconversion of waste (Gong et al., 2022; Hemalatha et al., 2023). VADER, a CRISPR-Cas9-based gene manipulation technique, was recently tested for its efficiency in antibiotic resistance gene degradation for successful wastewater bioremediation (Li et al., 2023). Genetic manipulations of different microbes have also helped in the bioconversion of plant-based lignocellulosic wastes to biofuels such as ethanol (Gong et al., 2022). Extensive research and development are ongoing on improving crop health, reducing postharvest loss, increasing yield, and imparting stress resilience using CRISPR/Cas9 gene editing (Hemalatha et al., 2023). Interestingly, studies have indicated successful manipulation of plant genomes using CRISPR/Cas9 in enhancement of synthesis of bioplastics (Dobrogojski et al., 2018). Apart from this, CRISPR technology could be further used for upgrading algal biorefinery toward carbon sequestration or carbon neutrality (Lee et al., 2023).

6.6.4 Microbe-based bionanomining

Mining industries contribute substantially to a country's economy. However, different mining activities produce enormous amounts of hazardous metal wastes. To combat this, bionanomining has evolved as a proficient biotechnological tool for generation of valuable nanoparticles from mining industry effluents (Wong-Pinto et al., 2020). The use of microbes for bionanomining using actual waste effluents is associated with several challenges (Wong-Pinto et al., 2020). However, different reports have highlighted the use of real waste effluents in bionanomining of metals. Brandao et al., have demonstrated the biosynthesis of copper nanoparticles (nps) in stirred tank reactors using *Rhodococcus erythropolis* ATCC 4277. Mine tailing was used a precursor for this production (Brandão et al., 2023). Similarly, copper NPs were produced using waste mine tailings and exploiting the metabolic potential of *Pseudomonas stutzeri* DSM 5190 (Wong-Pinto et al., 2021). Bionanomining also allows the formation of nanoparticles from solid wastes. Electronic wastes such as circuit boards were shown as promising sources for biomining of valuable metals (Yousef et al., 2018). Combined use of agricultural waste such as fruit extracts and *Rhodococcus opacus* also helped in the bionanomining of selenium and iron oxides (Maass et al., 2019; Wong-Pinto et al., 2020). Okanigbe et al., provided an insight into how bionanomining can be used as an energy-efficient alternative in gold mining operations (Okanigbe et al., 2022).

6.7 Future perspectives

For a sustainable future, reuse and recycle of waste are urgently needed. Proper bioprocessing of waste can help in extracting various value-added products. However, the scaling-up of such processes is often challenging. Moreover, chemical and physical processing of wastes generates toxic by- and end products. Thus biological techniques are most suitable for waste processing. To advance and upgrade these biotechnological processes, different molecular tools have been devised in the past decades. Still, the hunt for the best-suited approach is ongoing. Research is going on in increasing the efficiency and sustainability of field-scale applicability of molecular bioprocessing techniques. Metagenomics-enabled identification of yet uncultivable microbial assemblages will be helpful in coming times. Synthetic biology-based approaches of devising hybrid microbial communities and their field-scale applications will pave new avenues in waste biotransformation. In future, widespread manufacturing and use of bioplastics will help in eradicating the nuisance caused by synthetic plastics. Extensive research is required for harnessing the carbon sequestration processes for building a sustainable future and reducing the carbon footprints on earth. In line with diverse biotechnological research and development, a country's government legislation should be framed with a mission of zero waste generation. Overall greater reliance on biotechnological tools and products will help in maximum valorization of wastes for sustainable environment.

References

Adebiyi, F. M. (2022). Air quality and management in petroleum refining industry: A review. *Environmental Chemistry and Ecotoxicology*, 4, 89–96. Available from https://doi.org/10.1016/j.enceco.2022.02.001.

Ajayi, V. A., & Lateef, A. (2023). Biotechnological valorization of agrowastes for circular bioeconomy: Melon seed shell, groundnut shell and groundnut peel. *Cleaner and Circular Bioeconomy*, 4. Available from https://doi.org/10.1016/j.clcb.2023.100039.

Ambaye, T. G., Chebbi, A., Formicola, F., Prasad, S., Gomez, F. H., Franzetti, A., & Vaccari, M. (2022). Remediation of soil polluted with petroleum hydrocarbons and its reuse for agriculture: Recent progress, challenges, and perspectives. *Chemosphere*, 293. Available from https://doi.org/10.1016/j.chemosphere.2022.133572, http://www.elsevier.com/locate/chemosphere.

Banerjee, S., & Arora, A. (2021). Sustainable bioprocess technologies for urban waste valorization. *Case Studies in Chemical and Environmental Engineering*, 4. Available from https://doi.org/10.1016/j.cscee.2021.100166.

Bansal, S., Kushwah, S. S., Garg, A., & Sharma, K. (2023). Utilization of plastic waste in construction industry in India−A review. *Materials Today: Proceedings*, Apr 8.

Bhattacharya, R., Bose, D., Yadav, J., Sharma, B., Sangli, E., Patel, A., Mukherjee, A., & Ashutosh Singh, A. (2023). Bioremediation and bioelectricity from Himalayan rock soil in sediment-microbial fuel cell using carbon rich substrates. *Fuel*, 341. Available from https://doi.org/10.1016/j.fuel.2022.127019, http://www.journals.elsevier.com/fuel/.

Bond, D. R., & Lovley, D. R. (2003). Electricity production by Geobacter sulfurreducens attached to electrodes. *Applied and Environmental Microbiology*, 69(3), 1548–1555. Available from https://doi.org/10.1128/AEM.69.3.1548-1555.2003.

Brandão, I. Y., de Macedo, E. F., de Souza Silva, P. H., Batista, A. F., Petroni, S. L., Gonçalves, M., Conceição, K., de Sousa Trichês, E., Tada, D. B., & Maass, D. (2023). Bionanomining of copper-based nanoparticles using pre-processed mine tailings as the precursor. *Journal of Environmental Management*, 338117804, Jul 15.

Capanoglu, E., Nemli, E., & Tomas-Barberan, F. (2022). Novel approaches in the valorization of agricultural wastes and their applications. *Journal of Agricultural and Food Chemistry*, 70(23), 6787–6804, Feb 23.

Chabert, N., Ali, O. A., & Achouak, W. (2015). All ecosystems potentially host electrogenic bacteria. *Bioelectrochemistry (Amsterdam, Netherlands)*, 106, 88–96, Dec 1.

Cheela, V. R. S., Ranjan, V. P., Goel, S., John, M., & Dubey, B. (2021). Pathways to sustainable waste management in Indian Smart Cities. *Journal of Urban Management*, 10(4), 419–429. Available from https://doi.org/10.1016/j.jum.2021.05.002, https://www.journals.elsevier.com/journal-of-urban-management/.

Chen, Z., Wei, W., Chen, H., & Ni, B.-J. (2022). Recent advances in waste-derived functional materials for wastewater remediation. *Eco-Environment & Health*, 1(2), 86–104. Available from https://doi.org/10.1016/j.eehl.2022.05.001.

Cheng, F., Mallick, K., Henkanatte Gedara, S. M., Jarvis, J. M., Schaub, T., Jena, U., Nirmalakhandan, N., & Brewer, C. E. (2019). Hydrothermal liquefaction of Galdieria sulphuraria grown on municipal wastewater. *Bioresource Technology*, 292. Available from https://doi.org/10.1016/j.biortech.2019.121884, http://www.elsevier.com/locate/biortech.

Christgen, B., Yang, Y., Ahammad, S. Z., Li, B., Rodriquez, D. C., Zhang, T., & Graham, D. W. (2015). Metagenomics shows that low-energy anaerobic-aerobic treatment reactors reduce antibiotic resistance gene levels from domestic wastewater. *Environmental Science and Technology*, 49(4), 2577–2584. Available from https://doi.org/10.1021/es505521w, http://pubs.acs.org/journal/esthag.

Cánovas, C. R., Macías, F., Pérez-López, R., Basallote, M. D., & Millán-Becerro, R. (2018). Valorization of wastes from the fertilizer industry: Current status and future trends. *Journal of Cleaner Production*, 174, 678–690, Feb 10.

Das, O., Babu, K., Shanmugam, V., Sykam, K., Tebyetekerwa, M., Neisiany, R. E., Försth, M., Sas, G., Gonzalez-Libreros, J., Capezza, A. J., Hedenqvist, M. S., Berto, F., & Ramakrishna, S. (2022). Natural and industrial wastes for sustainable and renewable polymer composites. *Renewable and Sustainable Energy Reviews*, 158. Available from https://doi.org/10.1016/j.rser.2021.112054, https://www.journals.elsevier.com/renewable-and-sustainable-energy-reviews.

Da Silva, L. J., Alves, F. C., & de França, F. P. (2012). A review of the technological solutions for the treatment of oily sludges from petroleum refineries. *Waste Management & Research*, 30(10), 1016–1030, Oct.

Dash, D. M., & Osborne, W. J. (2022). A systematic review on the implementation of advanced and evolutionary biotechnological tools for efficient bioremediation of organophosphorus pesticides. *Chemosphere*, 137506, Dec 13.

Dey, Dhrutiman, Srinivas, Dodda, Panda, Biranchi, Suraneni, Prannoy, & Sitharam, T. G. (2022). Use of industrial waste materials for 3D printing of sustainable concrete: A review. *Journal of Cleaner Production* 130749. Available from https://doi.org/10.1016/j.jclepro.2022.130749, In this issue.

Djemiel, C., Dequiedt, S., Karimi, B., Cottin, A., Horrigue, W., Bailly, A., Boutaleb, A., Sadet-Bourgeteau, S., Maron, P. A., Chemidlin Prévost-Bouré, N., Ranjard, L., & Terrat, S. (2022). Potential of meta-omics to provide modern microbial indicators for monitoring soil quality and securing food production. *Frontiers in Microbiology*, 13. Available from https://doi.org/10.3389/fmicb.2022.889788, https://www.frontiersin.org/journals/microbiology#.

Dobrogojski, J., Spychalski, M., Luciński, R., & Borek, S. (2018). Transgenic plants as a source of polyhydroxyalkanoates. *Acta Physiologiae Plantarum*, 40(9). Available from https://doi.org/10.1007/s11738-018-2742-4, https://rd.springer.com/journal/11738.

Duque-Acevedo, M., Belmonte-Ureña, L. J., Cortés-García, F. J., & Camacho-Ferre, F. (2020). Agricultural waste: Review of the evolution, approaches and perspectives on alternative uses. *Global Ecology and Conservation*, 22. Available from https://doi.org/10.1016/j.gecco.2020.e00902, https://www.journals.elsevier.com/global-ecology-and-conservation.

Duraj-Thatte, A. M., Manjula-Basavanna, A., Courchesne, N. M. D., Cannici, G. I., Sánchez-Ferrer, A., Frank, B. P., Van'T Hag, L., Cotts, S. K., Fairbrother, D. H., Mezzenga, R., & Joshi, N. S. (2021). Water-processable, biodegradable and coatable aquaplastic from engineered biofilms. *Nature Chemical Biology, 17*(6), 732–738. Available from https://doi.org/10.1038/s41589-021-00773-y, http://www.nature.com/nchembio.

Forti V., Balde C.P., Kuehr R., & Bel G. The global E-waste monitor 2020: *Quantities, flows and the circular economy potential*.

Garbini, G. L., Barra Caracciolo, A., & Grenni, P. (2023). Electroactive bacteria in natural ecosystems and their applications in microbial fuel cells for bioremediation: A review. *Microorganisms., 11*(5), 1255, May 10.

Gong, C., Cao, L., Fang, D., Zhang, J., Kumar Awasthi, M., & Xue, D. (2022). Genetic manipulation strategies for ethanol production from bioconversion of lignocellulose waste. *Bioresource Technology, 352*. Available from https://doi.org/10.1016/j.biortech.2022.127105, http://www.elsevier.com/locate/biortech.

Gour, A. A., & Singh, S. K. (2023). Solid waste management in India: A state-of-the-art review. *Environmental Engineering Research, 28*(4). Available from https://doi.org/10.4491/eer.2022.249, https://www.eeer.org/journal/view.php?number = 1446.

Gupta, K., Biswas, R., & Sarkar, A. (2020). Advancement of omics: prospects for bioremediation of contaminated soils. *Microbial Bioremediation & Biodegradation*, 113–142.

Han, P., Teo, W. Z., & Yew, W. S. (2022). Biologically engineered microbes for bioremediation of electronic waste: Waypost, challenges and future directions. *Engineering Biology, 6*(1), 23–34.

Hemalatha, P., Abda, E. M., Shah, S., Venkatesa Prabhu, S., Jayakumar, M., Karmegam, N., Kim, W., & Govarthanan, M. (2023). Multi-faceted CRISPR-Cas9 strategy to reduce plant based food loss and waste for sustainable bio-economy – A review. *Journal of Environmental Management, 332*. Available from https://doi.org/10.1016/j.jenvman.2023.117382.

Huang, J., Yang, P., Guo, Y., & Zhang, K. (2011). Electricity generation during wastewater treatment: An approach using an AFB-MFC for alcohol distillery wastewater. *Desalination, 276*(1-3), 373–378. Available from https://doi.org/10.1016/j.desal.2011.03.077.

IAEA R, *Protection and safety of radiation sources: International basic safety standards, no GSR part 3*. International Atomic Energy Agency. (2014).

ISWA U, United Nations Environment Programme. International Solid Waste Association. 2015

Idris, M., Ibrahim, N., & Yaqoob, A. (2023). Sustainable microbial fuel cell functionalized with a bio-waste: A feasible route to formaldehyde bioremediation along with bioelectricity generation. *Chemical Engineering Journal, 455*.

Jadeja, N. B., Purohit, H. J., & Kapley, A. (2019). Decoding microbial community intelligence through metagenomics for efficient wastewater treatment. *Functional & Integrative Genomics, 19*, 839–851, Nov.

Jadhav, U. U., & Hocheng, H. (2012). A review of recovery of metals from industrial waste. *Journal of Achievements in Materials and Manufacturing Engineering, 54*(2), 159–167, Oct.

Jagaba, A. H., Kutty, S. R. M., Lawal, I. M., Aminu, N., Noor, A., Al-dhawi, B. N. S., Usman, A. K., Batari, A., Abubakar, S., Birniwa, A. H., Umaru, I., & Yakubu, A. S. (2022). Diverse sustainable materials for the treatment of petroleum sludge and remediation of contaminated sites: A review. *Cleaner Waste Systems, 2*. Available from https://doi.org/10.1016/j.clwas.2022.100010, http://www.elsevier.com/locate/issn/27729125.

Jouhara, H., Czajczyńska, D., Ghazal, H., Krzyżyńska, R., Anguilano, L., Reynolds, A. J., & Spencer, N. (2017). Municipal waste management systems for domestic use. *Energy, 139*, 485–506. Available from https://doi.org/10.1016/j.energy.2017.07.162, http://www.elsevier.com/inca/publications/store/4/8/3/.

Kalisz, S., Kibort, K., Mioduska, J., Lieder, M., & Małachowska, A. (2022). Waste management in the mining industry of metals ores, coal, oil and natural gas - A review. *Journal of Environmental Management, 304*. Available from https://doi.org/10.1016/j.jenvman.2021.114239.

Kashyap, S., & Ramaprasad, A. (2023). Geographical and temporal analysis of bio-medical waste management in India. *GeoJournal, 88*(4), 4269–4278. Available from https://doi.org/10.1007/s10708-023-10854-1.

Kiddee, P., Naidu, R., & Wong, M. H. (2013). Electronic waste management approaches: An overview. *Waste Management, 33*(5), 1237–1250, May 1.

Koch, C., & Harnisch, F. (2016). Is there a specific ecological niche for electroactive microorganisms. *ChemElectroChem, 3*(9), 1282–1295. Available from https://doi.org/10.1002/celc.201600079, http://onlinelibrary.wiley.com/journal/10.1002/(ISSN)2196-0216.

Koul, B., & Taak, P. (2018). *Chemical methods of soil remediation* (pp. 77–84). Springer Nature America, Inc. Available from https://doi.org/10.1007/978-981-13-2420-8_4.

Koul, Bhupendra, Yakoob, Mohammad, & Shah, Maulin P. (2022). Agricultural waste management strategies for environmental sustainability. *Environmental Research, 206*, 112285. Available from https://doi.org/10.1016/j.envres.2021.112285, In this issue.

Kovelakuntla, V., & Meyer, A. S. (2021). Rethinking sustainability through synthetic biology. *Nature Chemical Biology, 17*(6), 630–631. Available from https://doi.org/10.1038/s41589-021-00804-8, http://www.nature.com/nchembio.

Kuppusamy, S., Maddela, N. R., Megharaj, M., Venkateswarlu, K., Kuppusamy, S., Maddela, N. R., Megharaj, M., & Venkateswarlu, K. (2020). Impact of total petroleum hydrocarbons on human health. *Total Petroleum Hydrocarbons: Environmental Fate, Toxicity, and Remediation*, 139–165.

Lee, T. M., Lin, J. Y., Tsai, T. H., Yang, R. Y., & Ng, I. S. (2023). Clustered regularly interspaced short palindromic repeats (CRISPR) technology and genetic engineering strategies for microalgae towards carbon neutrality: A critical review. *Bioresource Technology, 368*. Available from https://doi.org/10.1016/j.biortech.2022.128350, http://www.elsevier.com/locate/biortech.

Li, X., Bao, N., Yan, Z., Yuan, X. Z., Wang, S. G., & Xia, P. F. (2023). Degradation of antibiotic resistance genes by VADER with CRISPR-Cas immunity. *Applied and Environmental Microbiology, 89*(4). Available from https://doi.org/10.1128/aem.00053-23, https://journals.asm.org/doi/10.1128/aem.00053-23.

Liu, Z., Si, B., Li, J., He, J., Zhang, C., Lu, Y., Zhang, Y., & Xing, X. H. (2018). Bioprocess engineering for biohythane production from low-grade waste biomass: Technical challenges towards scale up. *Current Opinion in Biotechnology, 50*, 25–31. Available from https://doi.org/10.1016/j.copbio.2017.08.014, http://www.elsevier.com/locate/copbio.

Lovley, D. R., & Holmes, D. E. (2022). Electromicrobiology: The ecophysiology of phylogenetically diverse electroactive microorganisms. *Nature Reviews. Microbiology, 20*(1), 5–19, Jan.

Maass, D., de Medeiros Machado, M., Rovaris, B. C., Bernardin, A. M., de Oliveira, D., & Hotza, D. (2019). Biomining of iron-containing nanoparticles from coal tailings. *Applied Microbiology and Biotechnology*. Available from https://doi.org/10.1007/s00253-019-10001-2, In this issue.

Maity, S., Bajirao Patil, P., SenSharma, S., & Sarkar, A. (2022). Bioremediation of heavy metals from the aqueous environment using Artocarpus heterophyllus (jackfruit) seed as a novel biosorbent. *Chemosphere, 307*. Available from https://doi.org/10.1016/j.chemosphere.2022.136115.

Mei, R., & Liu, W.-T. (2022). Meta-omics-supervised characterization of respiration activities associated with microbial immigrants in anaerobic sludge digesters. *Environmental Science & Technology, 56*(10), 6689–6698. Available from https://doi.org/10.1021/acs.est.2c01029.

Mokif, L. A., Jasim, H. K., & Abdulhusain, N. A. (2022). Petroleum and oily wastewater treatment methods: A mini review. *Materials Today: Proceedings, 49*, 2671–2674, Jan 1.

Nanda, S., & Berruti, F. (2021). Municipal solid waste management and landfilling technologies: A review. *Environmental Chemistry Letters, 19*(2), 1433–1456. Available from https://doi.org/10.1007/s10311-020-01100-y, http://springerlink.metapress.com/app/home/journal.asp?wasp = d86tgdwvtg0yvw9gvkwp&referrer = parent&backto = browsepublicationsresults,140,541.

Niu, T., Zhu, H., Shutes, B., Yu, J., He, C., Hou, S., Cui, H., & Yan, B. (2023). Wastewater treatment performance and gaseous emissions in MFC-CWs affected by influent C/N ratios. *Chemical Engineering Journal, 461*. Available from https://doi.org/10.1016/j.cej.2023.141876.

Nowrotek, M., Jałowiecki, Ł., Harnisz, M., & Płaza, G. A. (2019). Culturomics and metagenomics: In understanding of environmental resistome. *Frontiers of Environmental Science and Engineering, 13*(3). Available from https://doi.org/10.1007/s11783-019-1121-8, http://www.springerlink.com/content/2095-2201/.

Obi, F. O., Ugwuishiwu, B. O., & Nwakaire, J. N. (2016). Agricultural waste concept, generation, utilization and management. *Nigerian Journal of Technology, 35*(4), 957–964, Oct 11.

Ojovan, M. I., & Steinmetz, H. J. (2022). Approaches to disposal of nuclear waste. *Energies, 15*(20), 7804, Oct 21.

Okanigbe D.O., Popoola A.P., Malatji N., Lesufi T., & Sekgobela G. 2022 1 2022/01 Minerals, metals and materials series, 23671696 189-200 Springer Science and Business Media Deutschland GmbH South Africa Bionanomining: A Revised Insight into Processing of South Africa's Complex Gold Ores. Available from https://doi.org/10.1007/978-3-030-92662-5_19, http://www.springer.com/series/15240?detailsPage = titles.Part.F.

Osibanjo, O., & Nnorom, I. C. (2007). The challenge of electronic waste (e-waste) management in developing countries. *Waste Management & Research, 25*(6), 489–501, Dec.

Pandit, P. D., Gulhane, M. K., Khardenavis, A. A., & Vaidya, A. N. (2015). Technological advances for treating municipal waste. Microbial Factories: Biofuels. *Waste Treatment, 1*, 217–229.

Patel, A. K., Singhania, R. R., Albarico, F. P. J. B., Pandey, A., Chen, C. W., & Dong, C. D. (2022). Organic wastes bioremediation and its changing prospects. *Science of the Total Environment, 824*. Available from https://doi.org/10.1016/j.scitotenv.2022.153889, http://www.elsevier.com/locate/scitotenv.

Patil, P., & Sarkar, A. (2022). *Omics to field bioremediation: Current status, challenges, and future opportunities Omics for Environmental Engineering and Microbiology Systems* (pp. 1–17). India: CRC Press. Available from https://doi.org/10.1201/9781003247883-1, https://www.routledge.com/Omics-for-Environmental-Engineering-and-Microbiology-Systems/Kumar-Garg-Kumar-Biswas/p/book/9781032162836.

Prado, E. S. P., Miranda, F. S., de Araujo, L. G., Petraconi, G., & Baldan, M. R. (2020). Thermal plasma technology for radioactive waste treatment: a review. *Journal of Radioanalytical and Nuclear Chemistry, 325*(2), 331–342. Available from https://doi.org/10.1007/s10967-020-07269-4, http://www.wkap.nl/journalhome.htm/0236-5731.

Prajapati, K. K., Yadav, M., Singh, R. M., Parikh, P., Pareek, N., & Vivekanand, V. (2021). An overview of municipal solid waste management in Jaipur city, India - Current status, challenges and recommendations. *Renewable and Sustainable Energy Reviews, 152*. Available from https://doi.org/10.1016/j.rser.2021.111703, https://www.journals.elsevier.com/renewable-and-sustainable-energy-reviews.

Purohit, H. J., Kapley, A., Khardenavis, A., Qureshi, A., & Dafale, N. A. (2016). Insights in waste management bioprocesses using genomic tools. *Advances in Applied Microbiology, 97*, 121–170. Available from https://doi.org/10.1016/bs.aambs.2016.09.002, http://www.sciencedirect.com/science/journal/00652164.

Ren, L., Ahn, Y., & Logan, B. E. (2014). A two-stage microbial fuel cell and anaerobic fluidized bed membrane bioreactor (MFC-AFMBR) system for effective domestic wastewater treatment. *Environmental Science & Technology, 48*(7), 4199–4206, Apr 1.

Rylott, E. L., & Bruce, N. C. (2020). How synthetic biology can help bioremediation. *Current Opinion in Chemical Biology, 58*, 86–95. Available from https://doi.org/10.1016/j.cbpa.2020.07.004, http://www.elsevier.com/locate/cbi.

Sarkar, P., Roy, A., Pal, S., Mohapatra, B., Kazy, S. K., Maiti, M. K., & Sar, P. (2017). Enrichment and characterization of hydrocarbon-degrading bacteria from petroleum refinery waste as potent bioaugmentation agent for in situ bioremediation. *Bioresource Technology, 242*, 15–27. Available from https://doi.org/10.1016/j.biortech.2017.05.010, http://www.elsevier.com/locate/biortech.

Sengupta, D., Ilankoon, I. M., Kang, K. D., & Chong, M. N. (2023). Circular economy and household e-waste management in India. Part II: A case study on informal e-waste collectors (Kabadiwalas) in India. *Minerals Engineering, 200*, 108154, Sep 1.

Singh, H. M., Pathak, A. K., Chopra, K., Tyagi, V. V., Anand, S., & Kothari, R. (2019). Microbial fuel cells: A sustainable solution for bioelectricity generation and wastewater treatment. *Biofuels, 10*(1), 11–31. Available from https://doi.org/10.1080/17597269.2017.1413860, http://www.tandfonline.com/loi/tbfu20?open = 5&repitition = 0#vol_5.

Usmani, Z., Sharma, M., Awasthi, A. K., Sivakumar, N., Lukk, T., Pecoraro, L., Thakur, V. K., Roberts, D., Newbold, J., & Gupta, V. K. (2021). Bioprocessing of waste biomass for sustainable product development and minimizing environmental impact. *Bioresource Technology, 322*. Available from https://doi.org/10.1016/j.biortech.2020.124548, http://www.elsevier.com/locate/biortech.

Wong-Pinto, L. S., Menzies, A., & Ordóñez, J. I. (2020). Bionanomining: Biotechnological synthesis of metal nanoparticles from mining waste—opportunity for sustainable management of mining environmental liabilities. *Applied Microbiology and Biotechnology, 104*(5), 1859–1869, Mar.

Wong-Pinto, L. S., Mercado, A., Chong, G., Salazar, P., & Ordóñez, J. I. (2021). Biosynthesis of copper nanoparticles from copper tailings ore – An approach to the 'Bionanomining'. *Journal of Cleaner Production, 315*. Available from https://doi.org/10.1016/j.jclepro.2021.128107, https://www.journals.elsevier.com/journal-of-cleaner-production.

Yadav, R., Singh, S., & Singh, A. N. (2023). Cost-benefit analysis act as a tool for evaluation of agricultural waste to the economy: A synthesis. *Waste management and resource recycling in the developing world* (pp. 647–663). Elsevier, Jan 1.

Yadav, T. C., Pal, R. R., Shastri, S., Jadeja, N. B., & Kapley, A. (2015). Comparative metagenomics demonstrating different degradative capacity of activated biomass treating hydrocarbon contaminated wastewater. *Bioresource Technology, 188*, 24–32. Available from https://doi.org/10.1016/j.biortech.2015.01.141, http://www.elsevier.com/locate/biortech.

Yan, Q., & Fong, S. S. (2017). Challenges and advances for genetic engineering of non-model bacteria and uses in consolidated bioprocessing. *Frontiers in Microbiology*, 8. Available from https://doi.org/10.3389/fmicb.2017.02060, https://www.frontiersin.org/articles/10.3389/fmicb.2017.02060/full.

Yanikomer, N., Asal, S., Haciyakupoglu, S., & Akyil Erenturk, S. (2016). New solidification materials in nuclear waste management. *International Journal of Engineering Technologies, IJET*, 2(2). Available from https://doi.org/10.19072/ijet.54627.

Yousef, S., Tatariants, M., Makarevičius, V., Lukošiūtė, S. I., Bendikiene, R., & Denafas, G. (2018). A strategy for synthesis of copper nanoparticles from recovered metal of waste printed circuit boards. *Journal of Cleaner Production*, 185, 653–664. Available from https://doi.org/10.1016/j.jclepro.2018.03.036.

Yuan, H., Lu, W., & Hao, J. J. (2013). The evolution of construction waste sorting on-site. *Renewable and Sustainable Energy Reviews*, 20, 483–490, Apr 1.

Zamri M.L., Makhtar S.M., Sobri M.F., & Makhtar M.M. (2023). *Microbial fuel cell as new renewable energy for simultaneous waste bioremediation and energy recovery.* InIOP Conference Series: Earth and Environmental Science (Vol. 1135, No. 1, p. 012035). IOP Publishing.

Zhuang, L., Zheng, Y., Zhou, S., Yuan, Y., Yuan, H., & Chen, Y. (2012). Scalable microbial fuel cell (MFC) stack for continuous real wastewater treatment. *Bioresource Technology*, 106, 82–88. Available from https://doi.org/10.1016/j.biortech.2011.11.019.

CHAPTER

7

Microwave application in bioenergy production from waste biomass

Rejeti Venkata Srinadh and Neelancherry Remya

School of Infrastructure, Indian Institute of Technology Bhubaneswar, Bhubaneswar, Odisha, India

7.1 Introduction

With the rise in population and global industrialization, the need for energy has been substantially rising. One of the major sources of energy is fossil fuels such as coal, petroleum, and natural gas. Prolonged dependency on a limited quantity of nonrenewable resources leads to their near extinction. The extensive usage of nonrenewable sources causes a detrimental effect on the environment, not limited to an increase in Greenhouse Gas emissions. Therefore to provide novel energy sources that cause lesser damage to the environment, the energy providers are dwelling up for safer and more efficient sources and methods as well. Renewable sources of energy, including water, the solar, wind, and bioenergy, are becoming major energy suppliers in recent years. Bioenergy is obtained through proper treatment of biomass. Generally, waste biomass is used as biomass feedstock for bioenergy production (Williams et al., 2015). Several techniques, including physical, thermal, microbial, and chemical methods, are being used for converting biomass into bioenergy. There is a vital necessity in emphasizing the treatment of biomass, owing to the limitations on waste disposal that are majorly connected with cost and space. Due to the increasing expense of stringent pollution control/pollution palliation or compensating measures, air emissions are often the most significant environmental challenge for combustion systems. They act as a significant barrier to the feasibility of waste burning in a large number of places with poor quality of air. In contrast, the products of biomass pyrolysis cater to choices for mitigating greenhouse gas pollution and for mitigating biomass combustion's particle emissions, such as creating heat and power (Li et al., 2016).

Recent research has focused on the usage of microwave-assisted heating in the context of biomass processing and biofuel generation. Several biomass conversion methods have included microwave-assisted heating technology, a unique technique. Most research strives to enhance fuel properties and product yield while boosting overall efficacy and optimizing numerous aspects such as microwave power and duration, and different conversion methods are achieving varying degrees of success.

7.2 Need for bioenergy

Bioenergy is one of the various alternatives, accessible to contribute to addressing our energy requirements. It is a type of renewable energy obtained from biomass, and it may be harnessed to generate transportation fuels, heat, power, and goods. Renewable sources such as solar, wind, and water are applicable majorly for heat and power production. However, the biomass as a source has multidimensional applications, which are depicted in Fig. 7.1.

Bioenergy is most effective when it is generated from waste materials such as agricultural and farming by-products, fallen trees, and rubbish and waste that would otherwise rot in a landfill. These waste materials have the potential to provide important energy at a cheap cost, and utilizing them for energy minimizes the need for landfills and helps conserve our environment while providing another source of energy. Unlike fossil fuels,

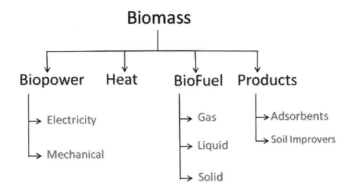

FIGURE 7.1 Derivatives from biomass.

bioenergy generation is termed carbon-neutral because they release the same amount of carbon dioxide into the atmosphere, which was taken up during the photosynthesis of respective biomass in the recent past (Arpia et al., 2021). Compared with fossil fuels, biofuels burn cleaner, leading to fewer greenhouse gas emissions, particulate matter emissions, and acid rain–causing chemicals such as sulfur. Bioenergy is theoretically renewable indefinitely until the amount of biomass utilized is less than the amount that can be grown back.

Feedstock supply and market demand are interconnected forces that depend on one another to stimulate bioeconomy progress. This reliance poses some practical challenges, but solutions such as the establishment of companion markets (for coproducts or marketable intermediate products) might help enhance the efficiency of biomass resources (Li et al., 2016).

Since different types of biomass contain a variety of components in different proportions, advanced technologies are also required to transform biomass more effectively and inexpensively for several end-use purposes. These needs provide jobs and encourage economic development in a wide range of industries, from scientific research to industrial operations, agriculture, and equipment design.

7.3 Biomass materials

The term "biomass" refers to material generated from living or recently living biological creatures. Biomass is carbon-based and consists of a variety of organic molecules comprising hydrogen, generally with oxygen and nitrogen atoms, as well as minor amounts of other atoms such as alkaline earth, alkali, and heavy metals. In the modern bioeconomy, agricultural by-products and forestry/wood resources are the principal sources of biomass. Agricultural biomass is mostly used to manufacture fuels and bio-based compounds. Woody biomass is often used to generate heat and electricity for the electrical, industrial, commercial, and residential areas. Animal dung is processed to generate power and heat for agricultural applications. The biologically active fraction of MSW and other waste materials is used to generate electricity and heat for a variety of industries.

7.3.1 Sources of biomass

These biomass feedstocks are the majority sources from Agriculture Sources, Forest, Waste, and Algae. The detailed sources of biomass are depicted in Fig. 7.2.

7.3.1.1 Forest source

The initial source of bioenergy is woody biomass obtained from forests. It is still the most common source of fuel for space heating and cooking in the world, especially in rural communities. Few public or private extended-rotation forests (i.e., the growth–harvest cycles lasting years) are or will be maintained specially to supply biomass for bioenergy. Instead, bioenergy from biomass is often a by-product of forest management efforts (e.g., fuel-hazard clearance) or commercial/industrial activities that emphasize greater-value raw materials such as saleable/merchantable timber. The bark is often burnt in mill fuel kilns or marketed as landscaping materials. It has a high potassium and silica content, which reduces its suitability as a feedstock despite having a higher energy density compared with wood chips (Bajpai, 2020). Wood chips, on the other hand, can be utilized as a

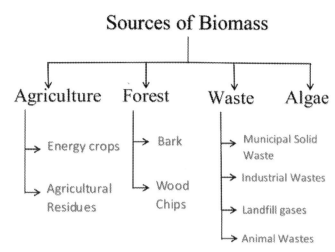

FIGURE 7.2 Different biomass sources.

solid fuel source (for burning) or can be improved and then densified as pellets. Wood chips are generally divided into three types: whole-tree chips, mill chips, and bole chips. Bole chips are made from low-grade or pulp logs, which are often sourced from managed forests. Whole-tree chips are made by cutting either the full low-grade tree or just the tops and branches removed from logs. While forest management produces the bulk of whole-tree chips, they are also generated through land clearance and land reformation operations that cater to roads, buildings, parking lots, and open spaces. Mill chips are clean wood chips that are generated by sawing debarked logs into lumber.

7.3.1.2 Agriculture

Sugars, lipids, starches, woody materials, and nonwoody cellulosic materials are all derived from agriculture. Agriculture-based biomass is derived from crops planted expressly for the production of bioenergy (i.e., specialized bioenergy crops) and agricultural leftovers. Agricultural residues are nonedible cellulosic components that stay back after the edible sections of crops have been harvested. Annual crops produced for sugars, starches, and oils and perennial herbaceous nonfood plants are cultivated for cellulose and are examples of dedicated bioenergy crops. Plant leaves and stems are examples of agricultural wastes. Some annual crops, such as maize, are designated as dedicated bioenergy crops because of their grain and cellulose wastes (Bajpai, 2020). The annual food crops such as soybean and palm are generally used as first-generation feedstock for the production of biofuels. Sugarcane and corn are the most commonly used biomass feedstock for bioethanol generation. Since the land available for agricultural purposes is limited, the major issue that has to be taken care of is the competition between the crops grown for bio-energy production and food supply. This competition may further lead either to a food shortage or to a rise in food prices.

The second-generation biofuels are produced majorly using perennial crops such as switchgrass and miscanthus as feedstock. Since they are not edible crops and also provide potential environmental advantages such as prevention of soil erosion and improvement of water quality, this biomass has achieved maximum attention for advanced biofuel production. Crop residues are also a significant source of cellulosic feedstock (Maryana et al., 2014). Crop residues are the substances that have remained in the fields after crop harvesting. For example, the leaves, stems, and stalks of maize left behind after grain harvest, known as stover, can be utilized to make bioethanol. Their usage can mitigate the effects of biofuel production on food security.

To prevent potential challenges with food and forage production on primary farming lands, government officials and experts are weighing the risks and benefits of periodic biomass harvesting from conservation lands, such as those reserved in agriculturally influenced terrain for soil and water conservation, improvement of water quality, wildlife habitat, or other nonagricultural purposes (Sarangi et al., 2018).

7.3.1.3 Waste

Organic material residues from industrial operations, agricultural solid and liquid wastes (such as manure), municipal solid wastes (MSW), and building wastes are all examples of waste-based biomass. Many industrial processes and manufacturing activities generate leftovers, trash, or by-products that can be utilized to generate bioenergy. Nonwoody wastes are mostly generated by waste paper, liquid by-products of paper manufacture (i.e., black liquor), and textile manufacturing. Used pallets, sawmill leftovers such as shavings and sawdust, cut-offs from furniture manufacture, and composite wood products incorporating nonwood resins, fillers, and adhesives are all major sources of woody waste materials. Conversion procedures for these wastes may be similar to those used for virgin wood (Zhang, 2017).

Agricultural wastes include the results of agroindustrial activities and livestock manure. Also, animal wastes are utilized as feedstock in anaerobic digesters, which eliminate potential pathogens while also producing biogas (i.e., methane). Biogas may be used to generate heat and electricity in place of propane, kerosene, and firewood. It can also be compressed and liquefied to be used as a transportation fuel.

Municipal solid waste (MSW), often known as garbage and urban solid waste, is primarily residential or domestic waste and is a significant source of biomass. Biodegradable garbage, such as food packaging and kitchen food waste; recyclable materials, such as paper, metals, and plastics; furniture and appliances, toys and clothing, and debris are all examples of municipal solid waste. The majority of MSW is directed to landfills, but in certain places, burnt for generating power. Gasification can be used to convert nonincinerated portions of waste to syngas. Syngas can be fired along with coal in boilers to generate power.

7.3.1.4 Algae

Algae is a very appealing alternative as a biofuel feedstock. Algae does not compete with food, land, or water. The high energy density and low-temperature fuel qualities of fuel generated from algae make it suited for use as fuel in jets, general vehicles, and domestic heating oil in colder climates. The production of algal fuel is also anticipated to be 100 times greater in comparison with other sources of biofuel. Furthermore, it assures a constant supply, captures waste carbon dioxide for biomass generation, manages farm nutrients runoff, and treats wastewater. The major disadvantage of algal biofuel production is that it is not cost-effective and also requires additional infrastructural development for large-scale fuel production (Ocreto et al., 2021). As a result, micro- and macroalgae are being investigated as commercially feasible feedstocks for third-generation biofuel production. Low-cost throughput feedstock for algae growth (e.g., carbon dioxide, wastewater containing algae, nutrients, cheap carbon source wastes such as grasses, bagasse, and corn stover) and the economic viability of commercial-scale systems can be influencing factors in the development of advanced biofuels.

7.3.2 Properties of biomass

7.3.2.1 Heating values

One of the most essential qualities of a fuel is its heating value, which is measured in MJ/kg or cal/g, and represents the total amount of energy accessible in the fuel. Heating value is mostly determined by the chemical constitutions of fuel and may be represented in two forms as Gross Calorific Value (GCV) or High Heating Value (HHV) and Net Calorific Value (NCV) or Low Heating Value (LHV). Net Calorific Value is equal to High Calorific Value after deducting the Latent heat of condensation of water vapor or steam. In general, the HHV is the best recommendation to be utilized for biomass combustion. Almost all biomass feedstock meant for combustion has an "as-received" (not oven dry) HHV of 15–19 MJ/kg, with most agricultural residues having an HHV of 15–17 MJ/kg and most woody materials having an HHV of 18–19 MJ/kg (Bakshi & Banerji, 2010). Woody biomass fuels have slightly higher HHV when compared with field crop biomass fuels because woody biomass has lower ash content compared with field crop biomass. Table 7.1 depicts the HHV of some common biomass fuels (Bakshi & Banerji, 2010; Maryana et al., 2014; Zhang, 2017).

7.3.2.2 Moisture content

Moisture content is one of the simplest biomass qualities to assess, which helps in determining a better biofuel. Because water has no energy value and much of the energy in the fuel is used up to heat and evaporate water, high-moisture biofuels burn more slowly and deliver less usable heat per unit mass. Fresh-green-wood, leafy crops contain approximately 50% water content. Extremely dry fuels, on the other hand, might generate dust issues, resulting in equipment damage and even explosion concerns. Moisture content is one of the major issues for using biomass as an alternative fuel to fossil fuels (Zhang, 2017). It is essential to decrease the moisture content of the biomass wastes by using drying techniques to increase their thermal efficiency and energy density.

TABLE 7.1 HHV of biomass fuels.

Biomass fuels (oven-dried)	Higher heating value (MJ/kg)
Birch	18.7–21.7
Bagasse	15.7–19.5
Wheat straw	17.5
Douglas fir	19.5–21.5
Hickory	18.8–22.4
Rice hulls	15.3
Switch grass	17.9–19.0
Willow	19.3–20.2
Rice straw	15.8
Maple wood	18.6–19.9
Corn stover	17.6–20.4

TABLE 7.2 Composition of various types of waste biomass.

Biomass	Proximate analysis (%)				Ultimate analysis (%)			
	Moisture content	Ash content	Fixed carbon	Volatile matter	C	H	N	O
Rice husk	0.84–0.94	20.87–20.91	16.41–16.47	61.20–62.40	40.52–41.12	5.22–5.28	0.37–0.39	53.18–53.58
Sugarcane	0.86–0.94	9.59–9.61	13.25–13.37	76.85–77.65	44.70–44.90	5.91–5.97	0.08–0.12	48.79–48.99
Sewage sludge	–	24.08–52.00	0.30–9.41	38.30–73.70	23.10–42.30	3.10–6.00	3.20–8.30	17.70–34.00
Douglas fir pellets	4.82	0.21	18.89	76.08	47.90	6.55	0.08	45.57
Rice straw	–	13.26–22.70	11.10–16.75	60.55–69.70	33.70–44.40	3.91–7.40	0.71–1.71	36.26–47.07
Palm shell kernel	0.38–0.42	3.49–3.51	22.35–22.45	73.10–74.30	48.18–48.38	5.36–5.42	0.10–0.14	46.08–46.28

7.3.2.3 Composition

In addition to heating value and moisture content, three other biomass attributes are important to assess the performance of biomass as a fuel: (1) ash content, (2) fixed carbon, and (3) volatile matter. The mass proportion of biomass constituted of noncombustible mineral material is referred to as ash content. Grasses, bark, and field crop wastes often contain far more ash than wood. Hence, the combustion systems need to have different configurations for different types of biomass. The major constituent elements of any biomass are carbon (C), hydrogen (H), nitrogen (N), and oxygen (O). Table 7.2 summarizes the proximate and elemental composition of various biomasses (Diaz et al., 2015; Mohd Fuad et al., 2019; Onokwai et al., 2022). There are some mineral components in biomass fuels, especially chlorine, potassium, and silica, which might lead to slagging and fouling issues developing at lower temperatures than expected. The presence of sulfur and chlorine in biomass leads to the corrosion of the combustion system, which further reduces its lifetime and increases the operational cost. There is also a risk of emission of toxic gases such as furans and dioxins during incineration due to the presence of excess chlorine in the biomass.

There are three major components of waste biomass materials to be considered for bioenergy production: sugars or starches, lipids, and cellulose or lignocellulose. Lipids are those components that are obtained from nonwoody plants and algae. These are rich in energy and are generally hydrophobic, such as fats and oils. Lipids-containing biomass is used as feedstock in the production of biofuels such as biodiesel. Sugar/starches represent the simplest carbohydrates that are present in the edible portion of the food crops. They are also termed as the first-generation feedstock (Maryana et al., 2014). The most prevalent first-generation biofuel feedstock are corn, wheat, and sugar cane. Cellulose and lignocellulose are the complex carbohydrates and noncarbohydrates that are present in the leaves and stems of plants. These plants do not have any food value and generally serve as second-generation feedstock for bioenergy production. Cellulosic/lignocellulosic feedstock is obtained from two types of plants: woody and nonwoody plants. Many plants include hemicellulose and lignin along with cellulose (Nomanbhay et al., 2013). Table 7.3 summarizes the major components in lignocellulosic biomass feedstock. Hemicellulose is a large and complex molecule of carbohydrate that aids in the cross-linking of fibers in the cell wall of plants. Lignin is a noncarbohydrate polymer found between cellulose and hemicellulose. When all three molecules are available in the same compound, it is called lignocellulose. Hemicellulose may be turned into fermentable sugar and ultimately into ethanol and certain other products. It is considered a byproduct because the conversion of lignin into useful compounds is challenging. As lignin transformation technologies advance, new markets for its utilization may arise.

7.3.2.4 Particle size and density

The particle size and density of biomass fuels are particularly significant because they influence the burning properties, namely the rate of drying and heating during the combustion process. The size of the fuel particle also determines the type of handling equipment required. The lower the size of the material, the greater shall be the heating efficiency leading to better energy production. Biomass generally has low bulk density, which is responsible for its low volume-based heating value. Thus, biomass requires a greater storage area and also requires more hauling trips, which result in increased transportation costs in comparison with denser materials. The size and bulk density of some biomass fuels are listed in Table 7.4 (Bajpai, 2020; Zhang, 2017). Therefore biomass is densified into condensed feedstock to increase its energy density, thereby reducing bioenergy conversion cost. Pelletization is one of the common methods to increase the bulk density of the biomass.

7.4 Conventional bioenergy production methods

7.4.1 Biochemical processes

There are many biochemical processes for the biomass conversion into energy products such as fermentation and anaerobic digestion. Fig. 7.3 explains the process of biochemical conversion of biomass into bioenergy products such as bioethanol and biogas. Microorganisms are employed in biomass fermentation to transform the feedstock into biofuels (such as bioethanol and biobutanol), combustible gases, materials, and chemicals. Another frequent biological technique for turning biomass or organic wastes into biogas (mostly methane and carbon dioxide) employing anaerobic bacteria is anaerobic digestion. In this process, the biogas generation efficiency is primarily determined by the pH, C/N ratio, and temperature. These processes are largely time-taking and majorly dependent on microorganisms.

TABLE 7.3 Major components in lignocellulosic biomass.

Biomass	Cellulose content (%)	Hemicellulose content (%)	Lignin content (%)	References
Oil-palm empty fruit bunch	22.5–25.3	24.5–27.8	24.0–26.6	Nomanbhay et al. (2013)
Rice husks	15.5–26.0	28.7–40.0	12.0–29.3	Diaz et al. (2015)
Sugarcane bagasse	42.0	25.0	20.0	Maryana et al. (2014)
Corn stover	32.7	20.9	25.4	Ravikumar et al. (2017)
Miscanthus	31.5–36.5	39.8–44.8	26–30	Zhu et al. (2015)

TABLE 7.4 Size and bulk density of biomass feedstock.

Biomass fuel	Size (m)	Density (kg/m^3)
Saw dust	0.0003–0.002	120
Chopped straw	0.005–0.025	60
Green-wood chips	0.025–0.075	180–190
Wood pellets	0.006–0.008	600
Biomass briquettes	0.025–0.010	600
Cord wood	0.3–0.5	230

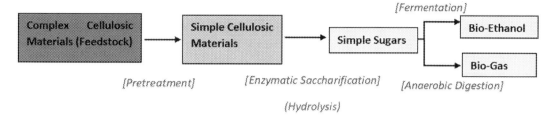

FIGURE 7.3 Biochemical processes of bioenergy production.

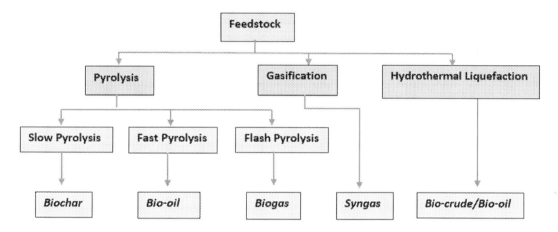

FIGURE 7.4 Thermochemical processes of bioenergy production.

Due to the improved process efficiency, better flexibility, faster conversion rate, product selectivity, and alternative market for by-products, the thermochemical conversion technologies are more commercially appealing. While biological conversion technologies are safe for the environment, the sensitivity of microorganisms to various factors (such as nutrients, sugar substrates, contaminants, inhibitors, and growth conditions), high-cost hydrolytic enzymes, and longer fermentation periods, make them relatively time-consuming processes (Diaz et al., 2015).

7.4.2 Thermochemical processes

7.4.2.1 Pyrolysis

Pyrolysis is the process where the thermal decomposition of biomass waste is done at temperature conditions of 400°C–800°C for the production of bioenergy products such as biochar, bio-oil, and biogas/py-gas. It is a thermochemical process where the polymeric molecules present in biomass (such as cellulose, hemicellulose, and lignin) break down into simpler compounds having lesser molecular weights, thus resulting in biofuel formation. The operating parameters including pyrolysis temperature, heating rate, the particle size of biomass, and sweeping gas flow rate influence the quality and quantity of the products obtained through pyrolysis. In this process, the temperature is the most critical operating parameter, and an intermediate temperature of 500°C–550°C maximizes the bio-oil output in the products. Pyrolysis can produce the most liquid biofuel at medium temperature, at a high heating rate, and at a lesser time. This bio-oil can be used both as a potential feedstock for the production of chemicals and as an energy source. Based on the heating rate provided, Pyrolysis is further classified into three types: Flash Pyrolysis, Slow Pyrolysis, and Fast Pyrolysis—majorly yielding biogas, biochar, and bio-oil, respectively. Char and gases are more likely to occur at lower and higher temperatures, respectively. Bio-oil production is maximized by using a moderate pyrolysis temperature, a high heating rate, and a short vapor residence time in the process. Sweeping gas flow rates have little impact on bio-oil production. It does, however, shorten the residence periods of volatiles, reducing secondary cracking and vapor repolymerization (Klinger et al., 2018). Furthermore, high bio-oil production is also dependent on a quick quenching of hot pyrolysis gases. It is the reactor design and operating parameters that determine the optimum pyrolysis product yield, which varies for various types of biomass. Fig. 7.4 describes various thermochemical processes involved in the bioenergy production.

7.4.2.2 Hydrothermal liquefaction

Various thermochemical conversion technologies have been developed through many years of research for different varieties of biomass waste, each with its own set of pros and cons. In this aspect, Hydrothermal Liquefaction (HTL) is the best procedure for converting wet feedstock to fuel energy (Dimitriadis & Bezergianni, 2017). In the process of HTL, biomass is first broken down into simpler fragments through hydrolysis, followed by dehydration, dehydrogenation, deoxygenation, and decarboxylation, which further degrades the biomass into smaller molecules. Repolymerization may be utilized to create certain complex molecules in the end. Hot compressed water acts as a solvent, catalyst, and reactant in the anoxic conditions at temperatures between 250°C and 375°C and pressures between 4 and 22 MPa in this HTL process. During this process, a range of acids, alcohols, aldehydes, esters, ketones, phenols, and other aromatic compounds are typically present in the biocrude result.

7.4.2.3 Gasification

Gasification is a method of converting solid feedstock, such as biomass and coal, using gasification agents, such as steam, oxygen, or air under substoichiometric conditions, into syngas, which is a combination of hydrogen and carbon monoxide that is easy to manage and can be further used to produce storable fuel and chemicals. There are various kinds of reactors available for gasification, namely fixed bed, fluidized bed, and entrained flow gasifiers. Process temperature, biomass feeding rate, and equivalence ratio (ER) are the most influential system performance and efficiency characteristics in such gasification systems. The syngas generated can be used for various purposes, as a feedstock for the chemical manufacturing, and utilized in energy generation. Decomposition of organic matter in a gasifier involves several different processes, including drying, pyrolysis, oxidation, and gasification. A gasifier's chemical operations are also described in detail in certain review papers. At roughly 100°C, the drying process begins. Due to the high temperature, steam is involved in the reaction between water and gas. During pyrolysis, which occurs when the temperature rises to 200°C–300°C, volatiles are liberated and char is formed. Carbon monoxide and carbon dioxide are formed as a result of the burning of volatiles and char. Exothermic reactions such as combustion and/or partial combustion provide heat for gasification reactions in the reduction zone. As a result, factors such as the nature of the fuel, operation temperature, reactor size, enthalpy, and the residence time are critical for achieving thermodynamic equilibrium (Ongen et al., 2016).

7.5 Microwave application in bioenergy production

Microwave pyrolysis is a highly efficient process where treatment of materials is done without any major emissions to the environment. It is generally applied for the second- and third-generation biomass feedstock. Compared with conventional heating methods, microwave-assisted organic reactions tend to proceed more quickly. Higher yields and selectivity can be attained with shorter response times. In addition, because microwaves travel through the vessel wall, there shall be no direct contact between the reactant feedstock and the energy source. To interact directly with the reaction mixture components, this vessel wall is often practically transparent to microwaves (Ocreto et al., 2021). Also, immediate beginning and stopping and the development of fast heating within the material are two of the many advantages of microwave heating over traditional heating. The comparison of microwave-assisted heating with conventional heating is described in Table 7.5. Furthermore, microwaves are used in nonthermal applications to measure the dielectric characteristics of several materials, including glass, rubber, paper, wood, and synthetic polymers (Ellison et al., 2018). Microwave nonthermal reactions may occur, resulting in a substantial increase in yield even under milder settings, besides the aforementioned benefits. These microwave nonthermal effects are connected to the reaction of the system to electromagnetic radiation and are unrelated to temperature fluctuation (Ellison et al., 2018). On the other hand, Lignocellulosic structures in the biomass can be disintegrated by the thermal effect of microwave irradiation on these microfibers. Microwaves are known to have physiochemical effects that speed up the breakdown of lignocellulosic biomass crystallization.

7.5.1 Microwave-assisted pretreatment of biomass

Pretreatment of lignocellulosic biomass is an essential step for a few thermochemical processes and biological conversion due to the structural complexity of lignin, cellulose, and hemicellulose, limited decomposability of lignin, and formation of unwanted products during the fermentation process (Hoang et al., 2021). Hence, an effective pretreatment technology depolymerizes the lignin, decreases cellulose crystallinity, and helps in extracting

TABLE 7.5 Comparison between conventional heating and microwave-assisted heating.

Conventional heating	Microwave-assisted heating
Heat transfer through conduction, convection, and radiation of heat.	Energy transfers directly to the feedstock material from the electromagnetic waves
It follows the heat transfer phenomena	It follows the energy conversion phenomena
Superficial heating, that is, heating at the surface takes place	Heat can be generated throughout the material since microwaves penetrate materials and deposit the energy
Conventional heating has a lower level of control	Microwave heating has a higher level of control and thus easy to operate.
The lesser yield of biofuel products due to less efficient heating rates with longer reaction time	More efficient heating rates in lesser residence time lead to higher product yields both in terms of quantity and quality.

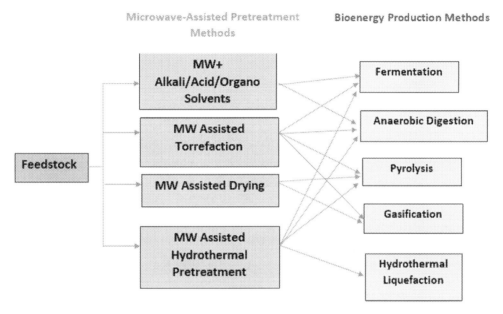

FIGURE 7.5 Microwave-assisted pretreatment for bioenergy production.

the fermentable sugars from the biomass waste feedstock. Microwave pretreatment of biomass is achieving popularity these days due to its faster processing, minimal cost of operation, volumetric heating, restricted side reactions, and little necessity for solvent. In contrast to conventional heating, microwaves deliver a homogeneous heating effect regardless of the biomass size. Microwaves are achieving popularity in biomass pretreatment and conversion as a result of their great heating efficiency, energy transfer, ease of operation, low cost, low maintenance, and less space requirement. Fig. 7.5 describes various MW-assisted pretreatment methods for different bioenergy production methods.

7.5.1.1 For biochemical process

Biomass wastes generally consist of strong lignocellulosic bonds and also have a complex structure of cell walls, which are the major drawbacks in using biomass directly as feedstock in biochemical processes of bioenergy production (especially fermentation process for bioethanol production). Hence, biomass-waste pretreatment is performed using conventional physiochemical processes such as treatment with alkali, acids, and organosolvents. The biomass wastes are microwave-assisted pretreated before being subjected to further biochemical processes of bioenergy production such as anaerobic digestion and bioethanol generation (fermentation). It results in enhanced enzymatic saccharification of biomass waste by reducing lignin and hemicellulose to a greater extent at a lesser residence time compared with conventional pretreatment. Microwave technology has the advantage of greatly reducing the time required for delignification and also utilizes significantly less electrical energy (Nomanbhay et al., 2013).

Anaerobic digestion (AD) pretreatment can help break down polymers into smaller molecules, which leads to hydrolysis. Pretreatment will dissolve the bacterial cell wall and aid in the flow of extracellular polymeric substances (EPSs) into the digestate that results in improved digestion and a shorter slurry mix retention time within the digester. MW-assisted heating can be combined alongside alkalis, H_2O_2, and acids as a pretreatment method prior to Anaerobic Digestion, which might enhance biogas generation, minimize sludge, reduce viscosity, and improve the soluble COD—to—total COD ratio (Arpia et al., 2021). Fig. 7.6 depicts the findings of various microwave-assisted pretreatment methods for different biomass wastes. Microwave-assisted alkali pretreatment of sugarcane bagasse improved the lignin removal to 90% compared with 59% through conventional alkali pretreatment (Maryana et al., 2014). Similarly, the improved delignification can also be observed in Miscanthus pretreatment with MW-assisted NaOH within half of the time taken by the conventional method (Zhu et al., 2015). There is a slight increment in the lignin removal efficiency in MW pretreatment of beechwood when compared with conventional pretreatment at similar conditions of temperature and time (Verma et al., 2011).

7.5.1.2 For thermochemical process

Drying, torrefaction, and hydrothermal treatment are the potential ways of waste biomass pretreatment since they are the lone thermochemical methods that have proved commercially and environmentally sustainable and are in use today.

The biomass waste consists majorly of moisture content, which gets transferred into bio-oil during the pyrolysis process, which is a major lacuna for energy applications of bio-oil. Hence, there is a necessity to maintain the moisture content in the biomass waste feed.

7.5.1.2.1 Microwave-assisted drying

Drying is one of the pretreatment techniques used for removing the excess water content from biomass waste before feeding it into the pyrolysis chamber. The conventional drying process involves the usage of furnaces and kilns demanding high costs, besides being time-consuming. Compared with conventional oven drying, microwave-assisted drying is a faster and more efficient technique as it not only removes excess moisture but also increases the surface area of the biofuel later obtained through pyrolysis of dried biomass waste. The faster release of moisture from biomass in microwave drying leads to the quicker removal of volatile compounds from the pretreated feedstock during the pyrolysis. This enhances the production of bio-oil and biochar from microwave-dried biomass wastes (X. Wang et al., 2008). The difference in the biofuel yield from various biomass waste feedstock from pyrolysis of conventionally dried and microwave-assisted drying as pretreatment of various biomass waste feedstock is depicted in Table 7.6. Microwave drying equipment has been utilized successfully in other fields; therefore there is minimal technical risk in using microwave energy for drying biomass materials

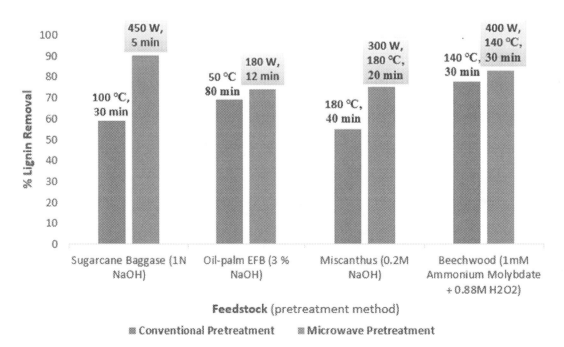

FIGURE 7.6 Comparison of microwave-assisted pretreatment of biomass with conventional pretreatment methods.

TABLE 7.6 Microwave-assisted drying as pretreatment for bioenergy production.

Biomass waste feedstock	Optimum operating parameter	Bioenergy production process	Salient output (%)			Remarks	References
			Yield	MW	OD		
Pinewood saw dust	600 W, 6 min	Pyrolysis at 500°C	Biochar	15.2	14.7	Dehydration was achieved approximately seven times faster with MW-drying compared to oven drying	X. Wang et al. (2008)
			Bio-oil	69.8	67.3		
			Biogas	15	18		
Peanut shell	600 W, 6 min		Biochar	30.5	30		
			Bio-oil	52.5	51.8		
			Biogas	17	18.2		
Maize stalk	600 W, 6 min		Biochar	29.8	29.4		
			Bio-oil	51.2	48.2		
			Biogas	19	22.4		
Kitchen waste	1500 W, 105.15°C, 24 h	Combustion	Combustion Characteristic Index by MW drying is 1.17×10^{-4} /%$min^{-1}K^{-2}$ Combustion Characteristic Index by oven drying is 8.9×10^{-5} /%$min^{-1}K^{-2}$			CCI by MW drying increased by approx. 36% compared to oven drying	Liu et al. (2016)
Rice straw	900 W, 1.5 min	Torrefaction at 250°C, 30 min	HHV_{MW} is 23 MJ/kg HHV_{OD} is 23.75 MJ/kg			Drying time reached 60 times faster while using MW-drying	Amer et al. (2019)

as pretreatment, which is more energy-efficient than conventional air drying. Due to the waste heat usage during the pyrolysis process, microwave drying in conjunction with conventional air drying should be an economically and technically viable pretreatment approach for the fast pyrolysis of biomass (Wang et al., 2008).

7.5.1.2.2 Microwave-assisted torrefaction

Torrefaction involves heating biomass waste feedstock at temperatures between 200°C and 300°C in an oxygen-free atmosphere. It produces biochar, a solid substance with combustion characteristics similar to coal. The calorific value, carbon content, and energy density of waste are considerably escalated as hemicelluloses are thermally reduced, and lignin and cellulose are largely dissolved during this process. MW-assisted torrefaction (MAT) pretreatment before enzymatic saccharification, pyrolysis, and gasification create a considerable variation in the quality of the product. For instance, when microwave-assisted catalytic torrefaction (MACT) was used as a pretreatment of Douglas fir sawdust pellets for pyrolysis, furfural output increased (Ren et al., 2012). During this process, the breakdown of hemicellulose content in the biomass takes place along with the modification of cellulose and lignin content. Hence, the torrefied biomass contains lesser hemicellulose in comparison with raw biomass, which relates to the reduction of furans in the pyrolysis products of torrefied biomass. Table 7.7 provides a summary of the various biomass materials used as feedstock, the optimum parameters of pretreatment, and significant results of bioenergy production using MW-assisted torrefied feedstock. MAT as a pretreatment shall be done at a relatively lower temperature and time, which helps in maintaining lignin and thereby improving phenol content in the end product of torrefied-biomass pyrolysis (Ren et al., 2013). Also, the pH value of end products from torrefied biomass increases with an increase in MAT pretreatment severity. This helps in making the pyrolyzed products safer to handle and store as they turn less corrosive with the increase in pH. However, the MAT pretreatment involves cross-linking, devolatization, and charring of biomass, which negatively impacts the bio-oil yield. Yet the quality of bio-oil is improved due to the increase in phenol and hydrocarbon content in the bio-oil due to lignin decomposition during MAT pretreatment (Ren et al., 2014). The MAT pretreatment followed by gasification leads to the production of the highest gaseous yields and reduction in tar yields in comparison with gasification of raw biomass, because of the adversity of the MAT to the secondary reactions for tar formation. Also, energy condensation and stronger carbonization in MAT lead to the increase in methane and carbon dioxide content of noncombustible gas produced through the gasification of torrefied biomass (Yan et al., 2021). Hence, MAT acts as a potential pretreatment method for gasification, as it results in the production of high-quality gas, generation of less tar, and lower energy consumption in comparison with gasification of raw biomass.

7.5.1.2.3 Microwave-assisted hydrothermal pretreatment (MAHT)

Hydrothermal pretreatment of lignocellulosic biomass waste could elevate the bioenergy production process performance by improving the production of bioethanol, hydrogen, and methane. This pretreatment of biomass feedstock at a temperature condition of 150°C–300°C has attracted interest in the manufacture of biofuels from lignocellulosic substrates because it eliminates the need for chemical additions and corrosion-resistant materials in hydrolysis reactors (López-Linares et al., 2019). Hydrothermal pretreatment of hemicellulose produces acetic acid that acts as a catalyst in hydrolysis, thereby degrading the biomass and increasing the sugar yield. Because of its crystalline and thermally resistant structure, cellulose dissolves only slightly in the hydrothermal pretreatment. When lignocellulosic biomass is hydrothermally pretreated, the activity of glucosidase, cellulose, and xylanase may be inhibited by the dissolved lignin. As a result, only reallocation of lignin occurs instead of its removal during the high temperature and pressure pretreatment (López-Linares et al., 2019). Thus this pretreatment technique helps in achieving an increase in acidogenic and methanogenic biodegradability, high organic-matter solubilization, and a further improvement in CH_4 production. This technique is majorly used during the anaerobic digestion (AD) process of bioenergy production. Hydrochar is produced by hydrothermal pretreatment of biomass feedstock at temperatures ranging from 150°C to 350°C. It is a high-value-added carbonaceous matter generally used as a catalyst, in energy storage, and soil amendments (Ahmed et al., 2019). Conventional heating has the drawback of a longer reaction period, which increases the chance of sugar decomposition. In recent decades, microwave heating has been developed in response to this constraint of traditional heating. In addition to intensifying the pretreatment process, microwave-assisted hydrothermal treatment methods can produce an eco-friendly technology with a greater pretreatment efficiency. Since water is particularly efficient at absorbing microwave energy, using microwave heating for conducting hydrothermal processes has emerged as a promising approach for the biomass valorization. Water has advantageous microwave absorption qualities because they possess high dielectric constant (δ') and loss tangent ($\tan \delta$). Water functions as both the catalyst and the reaction medium in the hydrothermal process. As a reaction medium, water under subcritical (100°C–374°C) or supercritical (over 374°C and pressure above 22.1 MPa)

TABLE 7.7 Microwave-assisted torrefaction for pretreatment of biomass.

Feed-stock	MAT optimum operating conditions	Bio-energy production method	Condition	Salient output (wt %)	Remarks	References
Corn stover	400 W, 275°C, 30 min	MW-Pyrolysis	700 W, 650°C, 15 min	Biochar- 38 Bio-oil- 21 Syngas- 41	The bio-oil yield was reduced from 34% in MW-pyrolysis of raw biomass to 21% in MW-pyrolysis of torrefied biomass.	Ren et al. (2014)
Douglas fir	650 W, 300°C, 20 min	MW-Pyrolysis	700 W, 480°C, 15 min	Biochar- 49.61 Bio-oil- 17.14 Syngas- 33.25	No hydrogen was detected in pyrolysis of raw biomass, whereas 20% (v/v) of Hydrogen was detected in MW-Pyrolysis of torrefied biomass.	Ren et al. (2013)
Herb residues	400 W, 225°C, 8.755 min	Steam Gasification	800°C	Biochar- 27.63 Bio-oil- 12.72 Syngas- 59.65	The gas yield was 0.86 Nm^3/kg which is almost two times greater than that of raw biomass gasification. Also, the LHV increased from 7.33 MJ/Nm^3 to 13.70 MJ/Nm^3.	(Yan et al., 2021)

conditions offers strong solubilization capabilities for dissolving organic molecules, thereby facilitating biomass conversion. MAHT maximizes hemicellulosic sugar and glucose recovery and is also used in increasing the yield of sugar (Sarker et al., 2021). The yield and physicochemical characteristics of the hydrochar obtained during the MAHT pretreated biomass are mostly dependent on the reaction temperature, catalyst (quantity and type), time, and solid-to-liquid ratio. In addition, various variables, such as biomass type, microwave power, and particle size, have a sizeable but less significant effect (Álvarez-Chávez et al., 2021). Table 7.8 depicts the effect of MAHT pretreatment on bioenergy production from various biomass feedstock.

7.5.2 Microwave-assisted bioenergy production

When compared with conventional/traditional systems of bioenergy production, microwave energy technologies may provide the following benefits: decrease in selective heating, lower waste volume, chemical reactivity, rapid heating, improved in-situ waste treatment capability, rapid and flexible processes, ease of control, overall cost-effectiveness, energy savings, and equipment portability. Fig. 7.7 depicts various microwave-assisted bioenergy production methods.

7.5.2.1 Microwave-assisted pyrolysis (MAP)

Pyrolysis is the most investigated thermochemical process used with MW-assisted heating to convert biomass feedstock into value-added products. MW-assisted pyrolysis (MAP) may be classified based on its operating temperature, heating rate, and residence duration. In comparison with traditional pyrolysis, MAP output yields and characteristics exhibited more significant variations. Higher product vapors are emitted by the devolatilization that happened during microwave-aided pyrolysis, consequently larger quantity of hydrocarbons in the vapors gets transformed into permanent gases (Shi et al., 2020). Moreover, MAP has been shown to provide better products because it minimizes the secondary reactions of the evolved gases. Due to high internal-heating rates and self-gasification between char and carbon dioxide, the solid fuel or biochar yield decreases in MAP in comparison with conventional pyrolysis (Shi et al., 2020). And also, the bio-oil obtained through MAP shows increased carbon content and calorific value accompanied by a fall in oxygen concentration. Table 7.9 summarizes the findings of the MAP process used for bioenergy production using different biomass waste feedstock.

The microwave process, design, and optimization heavily rely on the dielectric properties of pyrolysis feedstock materials. The capacity of a material to store and dissipate energy is described by the dielectric constant and the dielectric loss factor.

Water in the feedstock experiences dipole polarization, whereas the electric potentials within cell membranes of biological tissue are subject to ionic polarization in the case of organic materials such as biomass feedstock utilized in pyrolysis (Huang et al., 2010). Thus the water content and composition of the biomass material have a significant role in the coupling of microwave radiation with biomass. Carbon-rich structures with a high concentration of moisture and carbon quickly absorb microwaves while dry, unreacted biomass particles are nearly transparent to microwaves. Only a limited percentage of the microwave energy absorbed by pyrolysis feedstock such as dry lignocellulosic biomass is transferred to the biomass material and transformed to heat. To improve the heating efficiency of low-dielectric loss feedstock, the inclusion of a microwave absorber such as carbonaceous materials (such as activated carbon, pyrolytic char, and graphite), inorganic materials (catalyst, metal oxides, salts, etc.) is generally done to encourage catalysis and microwave absorption. For example, (Omar, 2010) evaluated the efficiency of microwave absorber addition and concentration for EFB pyrolysis. Five percent pyrolysis-derived char was applied to the samples, and the maximum temperature of 590°C was attained at the same time, whereas only 177°C was attained without using the absorber. Increasing the char content to 10% and 15% did not appear to increase the operating temperature. However, the rise in char content increased the output of hydrogen gas while decreasing the quantity of carbon monoxide and the lower heating value.

The feedstock characteristics such as moisture content, fixed carbon, ash content, volatile matter, elements present, particle size and density, calorific value have a direct influence on the MAP process of bioenergy production.

The MW power level has also been proven to alter product quantity and properties. As the heating rate rises with increased microwave power, more oxygen is released in the form of volatile chemicals with heavy molecular weight from the thermal degradation of proteins, water, cellulose, hemicelluloses, and lignin, hence boosting the bio-oil output. Increasing power levels also increase the biochar's caloric value and yield (Borges et al., 2014).

TABLE 7.8 Microwave-assisted hydrothermal pretreatment (MAHT) of biomass

Feedstock	Optimum operating parameter	Bioenergy production method	Salient outputs (%) Yield	Raw	MAHT	Remarks	References
Black spruce	160°C, 40 min, 1:13 (biomass: water)	Fast Pyrolysis at 550°C	Biochar Bio-oil syngas	50.9 16.1 33.0	44.2 20 35.8	There is no variation observed in the calorific value of biomass after MHT pretreatment	Álvarez-Chávez et al. (2021)
Bamboo sawdust	800 W, 230°C, 30 min	Pyrolysis at 550°C	Biochar	19.4	16	MHT pretreatment reduced the acetic acid content by 11.3% from raw BS, which is 3% greater than the conventional process	Dai et al. (2017)
Brewer's spent grain	1800 W, 192.7°C, 5.4 min	Enzymatic Hydrolysis (50°C, 48 h) + Fermentation (120 h)	Bio-Butanol yield = 46 kg/t BSG			The results from Enzymatic Hydrolysis of MHT pretreated biomass are four times higher compared to that of raw BSG	López-Linares et al. (2019)

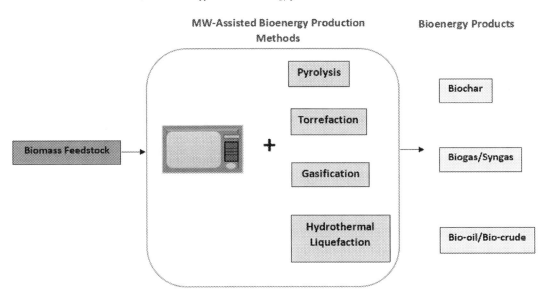

FIGURE 7.7 Microwave-assisted bioenergy production methods.

TABLE 7.9 Microwave-assisted pyrolysis of different biomass wastes.

Feedstock	Operating parameters	Salient output (yield %)		Remarks	References
Rice straw	300 W, 10 min, 400°C	Biochar Bio-oil Syngas	62.9 15.3 21.8	The HHV of syngas was found to be 10.56 MJ/Nm^3 which is greater than syngas HHV of 4–7 MJ/Nm^3 through gasification	Huang et al. (2010)
Corn cob	800 W, 10 min, 450°C	Biochar Bio-oil Syngas	32 42.1 25.9	The HHV of bio-oil obtained from corncob is 22.38 MJ/kg, which is approximately half of gasoline fuel HHV (42 MJ/kg)	Ravikumar et al. (2017)
Wood sawdust	750 W, 480°C, 2 g/min loading rate	Biochar Bio-oil Syngas	16 65 19	SiC is used as Microwave Absorber, and the HHV of Bio-oil was determined to be 20.38 MJ/kg	Borges et al. (2014)
Bamboo (reformed with coconut-derived activated carbon)	3 kW, 600°C, 15 min	Biochar Bio-oil Syngas	22.8 0.3 76.9	The gaseous yield has increased by 1.05 times and bio-oil decreased by 27 times due to feedstock reforming with Activated Carbon. Also, the LHV of syngas was found to be 24.5 MJ/kg	Shi et al. (2020)
Algae meal	250 W, 750°C, 60 min	Biochar Bio-oil Syngas	27.83 35.02 37.15	The HHV of bio-oil is 27.54 MJ/kg which is greater than that of biochar and syngas (24.23 MJ/kg and 17.24 MJ/kg respectively)	Ferrera-Lorenzo et al. (2014)

7.5.2.2 Microwave-assisted torrefaction (MAT)

As discussed earlier, Torrefaction is the thermochemical process of converting the feedstock into a bioenergy product in an inert atmosphere at a temperature range of 200°C–300°C. Conventionally, most research has accomplished torrefaction either utilizing a tube reactor or a fixed-bed reactor (Ahmad et al., 2018). Considering the torrefaction capability of microwave-assisted heating, there is a growth in research interest in using various residual wastes (such as agricultural residues, organic sewage sludge, municipal solid) and also the cotorrefaction of waste oils and other feedstock (Gronnow et al., 2013). Some process variables have a significant impact on torrefaction yield and product properties. The most influential parameter in the torrefied products' quality is found to be the MW power level (Ren et al., 2014), on which the heating rate, maximum temperature, and reaction time are dependent. Higher MW power levels are recommended (150–300 W) because elevated temperature conditions cause biomass secondary cracking or an onslaught of pyrolysis and negatively impact biochar quality. When compared with typical heating processes, MW torrefaction can produce energy yields of 70%–90%, but at higher MW power levels and for shorter periods (Gronnow et al., 2013). This process involves the degradation of hemicellulose by eliminating hydroxyl groups and generating nonpolar compounds, thereby increasing the hydrophobicity of the torrefied product. MAT helps in significantly improving the calorific values and fixed carbon of the terrified biomass compared with raw biomass. The obtained HHV values shall be similar to those of coals. The HHV of torrefied douglas fir sawdust pellets was found to be increased to 25.07 MJ/kg from 19.6 MJ/kg (raw biomass) using MAT (Ren et al., 2012). Table 7.10 summarizes the MAT process in increasing the HHV of biomass feedstock. This process also yields bio-oils mainly consisting of ketones/aldehydes, organic acids, and phenolics. The quantity of phenolic compounds increases with an increase in MAT time and temperature. Hence, the bio-oils produced shall have greater potential as fuel applications or as raw material for chemical production.

7.5.2.3 Microwave-assisted gasification (MAG)

Traditional/Conventional gasification involves the combustion of a portion of the biomass feedstock to produce sufficient thermal energy for the process. Hence, these are considered bulky and inflexible (Xie et al., 2014). The benefits given by microwave heating over the aforementioned traditional heating methods such as uniform and selective heating have prompted many investigations to employ the microwave heating mechanism for gasification. MAG optimizes biochar conversion to syngas at the appropriate carbon dioxide flow rate and increases specific gas production. Due to the superior heating mechanism of MAG, greater conversion efficiencies and improved fuel characteristics may be attained with lesser reaction times (Xiao et al., 2015). H_2, CO, CH_4, and CO_2 are major constituents of the syngas obtained from MAG. At elevated temperatures, the conversion of CO_2 to CO (Boudouard reaction, that is, $C + CO_2 \rightarrow 2CO$) takes place, which leads to an increase in CO yield in the released syngas (Ismail et al., 2015).

To avoid partial feedstock combustion and product gases dilution with combustion gases, an external energy source can be provided by thermal plasma, which can be generated through electric arc discharge or microwave. In the case of a microwave plasma gasification system, the plasma provides intense conditions of 3700°C–4700°C, which enables quick processing, reduced residence time, and hence volumetric process scaling down or shrinking. In addition, this intensity permits the use of feedstock that is resistant to thermal processing, such as municipal solid waste, which does not quickly decompose in traditional gasifiers. Contrary to the plasma formed by arc discharge, plasma generated by microwaves eliminates direct contact with electrodes, hence preventing electrode damage and can also operate under atmospheric conditions, whereas conventional gasifiers require high-strength structures for resisting high-pressure operations. In addition, microwave plasma has been demonstrated to accelerate the reformation and eradication of tar, a by-product of gasification that is normally seen as harmful. Lastly, the extra energy source provides a mechanism for compensating the heat losses, which become more limiting when the scale of typical non-plasma-gasification systems is decreased (Ismail et al., 2015). MAG is a potential conversion technique for achieving a sustainable bioeconomy, while being a relatively new technology. There is still a lot of research that is being done on microwave-assisted gasification processes. An extensive study on the catalysts under varied operating circumstances of the MAG system has shown that the

TABLE 7.10 Microwave-assisted torrefaction of various feedstocks.

Feedstock	Optimum operating conditions	Raw biomass HHV (MJ/kg)	Torrefied biomass HHV (MJ/kg)	References
Douglas fir	600 W, 300°C, 20 min	19.6	25.07	Ren et al. (2012)
Herb residue	600 W, 250°C, 12.5 min	13.53	17.1	Yan et al. (2021)
Oil-palm EFB	385 W, 300°C, 30 min	17.43	22.4	Ahmad et al. (2018)
Municipal solid waste	650 W, 250°C, 12 min	12.92	18.24	Siritheerasas et al. (2017)
Rice husk	350 W, 434°C, 10 min	17.4	21.9	Wang et al. (2012)
Sugarcane	250 W, 474°C, 15 min	17.8	27.8	Wang et al. (2012)

catalysts may improve the quality of syngas and gas production, prevent carbon deposition, and also reduce tar content (Xie et al., 2014). Some of the findings of Microwave-Assisted Gasification of certain biomass wastes are listed in Table 7.11.

7.5.2.4 Microwave-assisted hydrothermal liquefaction (MAHTL)

Microwave-assisted hydrothermal liquefaction (MAHTL) is lately developed as a viable alternative to traditional hydrothermal liquefaction because of the numerous benefits it provides. MAHTL is typically compared with regular HTL, although not in terms of biofuel generation, but the nature of the feedstock. Higher biocrude yield can be obtained from MAHTL of lipids over conventional HTL, but comparable biocrude yields shall be obtained from protein- and lignin-containing biomass (Jie Yang et al., 2020).

The catalysts' addition to feedstock for increasing the selectivity of the chemical in the biocrude generated is one of the recent developments in MAHTL. As the temperature, reaction time, and catalyst amount increase in MAHTL, the phenol formation increases and the concentration of aldehydes and carboxylic acids that are formed in the early stages decreases. (Remón et al., 2019) demonstrated that it is possible to convert 27% of the actual biomass into phenol-rich, that is, 47% bio-oil with greater HHV of 20 MJ/kg by using optimized conditions such as a temperature of 250°C, 80 bar pressure, and 0.25 g catalyst per gram of biomass for 1.9 h. The properties of that bio-oil indicate that it is also a sustainable source of phenolic-rich antioxidants and aromatics apart from being an excellent source of bioenergy (Remón et al., 2019). Similar kinds of MAHTL for bioenergy production from different biomass are summarized in Table 7.12.

7.6 Lab-scale to large-scale opportunities

Microwave-assisted bioenergy production is a revolutionary approach for effective in-situ biomass waste processing that utilizes microwave heating for the energy recovery from biomass and thereby repurposing those residuals into valuable products. Compared with traditional pyrolysis, this method yields oils with fewer dangerous constituents and a greater concentration of substances of industrial use. The fact that the gaseous portion provides larger quantities of hydrogen or syngas is similarly helpful and profitable. Microwave-assisted pyrolysis (MAP), on the other hand, gives an excellent opportunity to remove biomass wastes from current eco-damaging

TABLE 7.11 Microwave-assisted gasification for bioenergy production.

Feedstock	Optimum operating parameters	Salient output	Remarks	References
Pyrolytic rice straw biochar	550°C, 90 min	Biochar conversion effeciency = 49.01% Carbon conversion efficiency = 74.39% Gas yield- 1.32 m³/kg biochar	Over 90% vol. yields of CO + H_2 were obtained, similar to conventional heating at 800°C. The reaction time was reduced to 60 min from 90 min when 10% $Ca(OH)_2$ is used as catalyst	Xiao et al. (2015)
Corn stover with Ni/Al_2O_3 catalyst	750 W, 900°C, Catalyst: biomass = 1:3	Gas yield > 82% Tar content = 7% Char = 11%	SiC is used as Microwave Absorber. Maximum Syngas (>60% vol) and minimum tar content were obtained from MAG using Ni/Al_2O_3 as the catalyst. HHV of syngas produced was calculated to be 12.03 MJ/m³	Xie et al. (2014)
Oil-palm EFB biochar	800 W, 5 min, 10% A.C., 1500°C	Syngas yield = 1.55 m³/kg Gasification efficiency = 72.34% Unreacted carbon = 12%	Activated carbon is used as microwave absorber. The increase in temperature (>1000°C) occurred as early as the first minute.	Ismail et al. (2015)
			Since EFB is a better microwave absorber than OPS, the temperature is higher in EFB. HHV of syngas produced by MAG of EFB is 9.40 MJ/kg. HHV of syngas produced by MAG of OPS is 7.32 MJ/kg.	
Oil-palm shell (OPS) biochar	800 W, 5 min, 10% A.C., 920°C	Syngas yield = 1.98 m³/kg Gasification efficiency = 69.09% Unreacted carbon = 18%		Ismail et al. (2015)

TABLE 7.12 Microwave-assisted hydrothermal liquefaction of biomass waste.

Feedstock	Optimum operating parameters	Salient output (yield in wt%)	Remarks	References
Soyabean oil (lipid)	Solid ratio of 8.3 wt.% 270°C, 20 min	Biocrude yield = 103.8 Gas yield = 5 Solid residue yield = 0.4 Aqueous phase yield = -9.2	The negative aqueous phase yield tells that MAHTL used some compressed water for lipid hydrolysis. The bio-crude yield through MAHTL is 1.1 times greater compared to conventional HTL.	Yang et al. (2020)
Maple sawdust	Solid ratio of 8.3 wt% 270°C, 20 min	Bio-crude yield = 24 Gas yield = 10 Solid residue yield = 30 Aqueous phase yield = 36 HHV of biocrude = 24.3 MJ/kg ERE = 26.7%	The biocrude yield and its HHV through MAHTL are comparable with conventional HTL. The energy recovery efficiency of MAHTL is 1.05 times lower compared to the conventional process.	Yang et al. (2020)
Herbal crop (HEC)	Solid ratio of 10 wt% 1.2 kW, 240°C, 60 min	Biocrude yield = 15.1 Gas yield = 5.8 Solid residue yield = 51 Aqueous phase yield = 28.1 HHV of biocrude = 33.5 MJ/kg ERE = 26.1%	The HHV of biocrude oil obtained from MAHTL is significantly greater than the HHV of raw HEC (19.39 MJ/kg).	Zhuang et al. (2022)
Fungus chaff (FUC)	Solid ratio of 10 wt% 1.2 kW, 240°C, 60 min	Biocrude yield = 24.6 Gas yield = 6.5 Solid residue yield = 24 Aqueous phase yield = 44.9 HHV of biocrude = 34.1 MJ/kg ERE = 44.0%	A significant rise in HHV of biocrude oil through MAHTL is observed as compared with raw FUC HHV (19.06 MJ/kg).	Zhuang et al. (2022)
Spent coffee grounds (SCG)	Solid ratio of 8.3 wt% 270°C, 20 min	Biocrude yield = 30.1 Gas yield = 13.3 Solid residue yield = 28.6 Aqueous phase yield = 28 HHV of biocrude = 31.3 MJ/kg	The pH of the aqueous phase from MAHTL was found to be 3.9, similar to conventional HTL (pH = 4). Also, comparable biocrude yields were obtained through both processes.	Yang et al. (2022)

TABLE 7.13 Commercial applications of microwave-assisted heating.

Organization	Capacity	Country	Process involved	Product
Carbon-Scape	—	New-Zealand	Microwave-Assisted Torrefaction	Biochar and Bio-oil
Future Blends	Demonstration scale	UK	Microwave-Assisted Pyrolysis	Bio-oil
Scandinavian Biofuel Company	25,000 tonnes/yr	Norway	Microwave-Assisted Pyrolysis	Bio-gas and Biochar
SAIREM	50 kW, 27.12 MHz	France	Microwave-Assisted Drying	Timber Drying

disposal techniques such as landfilling and incineration, as well as a feasible approach for commercial extraction of useful products from biomass waste such as biochar. Yet, there are still questions regarding the parameters and operating circumstances that regulate this process, which may require additional optimization. The optimization of the process will enable the definition of acceptable process parameters, which will be advantageous for scaling up the microwave-assisted pyrolysis process. It was also revealed that microwave-assisted pyrolysis may be easily scaled and deployed to enhance distributed energy systems. In 1989, the United Kingdom built the first pyrolysis plant, which utilized microwave radiation to break down polymers in old tires. (Ravikumar et al., 2017). In the United States, the availability of patents for the microwave pyrolysis of wastes increased significantly during the 1990s. In recent years, several attempts have been made to transform microwaves into a viable method for biomass pyrolysis process treatment and the conversion of biomass into liquids and solids. Some of the commercial applications of Microwave-Assisted heating in biomass processing are listed in Table 7.13 (Zhang, 2017). Microwave-enhanced fast pyrolysis has also been advocated for commercial operations in addition to its advantages for studying basic pyrolysis reactions (Zhang, 2017).

Carbonscape of New Zealand is one of the most active firms in the field of microwave-assisted torrefaction producing biochar. They have graduated from the pilot-scale to the commercial level, and they manufacture charcoal and pyrolysis oils.

Initially, biochar was sold as a soil amendment, but its technology has already advanced to the point where it may be used as green coke in the steel industry and activated carbon. There is limited information available on the commercial usage of microwaves in the thermochemical processing of biomass wastes.

Microwave-assisted heating has been effectively used in numerous conversion processes of biomass wastes for increased ethanol and bioenergy production. However, a majority of the cases indicated that the increased product output was not sufficient enough in balancing the input energy from microwave devices, indicating negative energy efficiencies (Ravikumar et al., 2017). This efficiency is contingent on the microwave oven design being suitable for the requirements of the process, which might not be optimal when lab-scale studies are considered. On the other hand, the use of microwave heating to remediate and pyrolyze scrap tyres was shown to be superior to conventional heating in terms of both economic and environmental variables, despite the lack of material-specific information such as dielectric characteristics and compositional consequences (Appleton et al., 2005).

Exploring the relationship between the feedstock and electromagnetic field and optimizing several operational and mechanical parameters are the two main pillars that may lead to the successful implementation of MW heating for biofuel production on a large scale and thus be commercialized to replace the traditional/conventional method of bioenergy production for a more sustainable future.

References

Ahmad, M. I., Rizman, Z. I., Rasat, M. S. M., Alauddin, Z. A. Z., Soid, S. N. M., Aziz, M. S. A., Mohamed, M., Amini, M. H. M., & Amin, M. F. M. (2018). The effect of torrefaction on oil palm empty fruit bunch properties using microwave irradiation. *Journal of Fundamental and Applied Sciences*, 9(3S), 924. Available from https://doi.org/10.4314/jfas.v9i3s.67.

Ahmed, B., Aboudi, K., Tyagi, V. K., álvarez-Gallego, C. J., Fernández-Güelfo, L. A., Romero-García, L. I., & Kazmi, A. A. (2019). Improvement of anaerobic digestion of lignocellulosic biomass by hydrothermal pretreatment. *Applied Sciences (Switzerland)*, 9(18). Available from https://doi.org/10.3390/app9183853, https://res.mdpi.com/d_attachment/applsci/applsci-09-03853/article_deploy/applsci-09-03853.pdf.

Álvarez-Chávez, B. J., Godbout, S., & Raghavan, V. (2021). Optimization of microwave-assisted hydrothermal pretreatment and its effect on pyrolytic oil quality obtained by an auger reactor. *Biofuel Research Journal*, 8(1), 1316–1329. Available from https://doi.org/10.18331/BRJ2021.8.1.3, http://www.biofueljournal.com/.

Amer, M., Nour, M., Ahmed, M., Ookawara, S., Nada, S., & Elwardany, A. (2019). The effect of microwave drying pretreatment on dry torrefaction of agricultural biomasses. *Bioresource Technology*, 286121400. Available from https://doi.org/10.1016/j.biortech.2019.121400.

Appleton, T. J., Colder, R. I., Kingman, S. W., Lowndes, I. S., & Read, A. G. (2005). Microwave technology for energy-efficient processing of waste. *Applied Energy*, 81(1), 85–113. Available from https://doi.org/10.1016/j.apenergy.2004.07.002, http://www.elsevier.com/inca/publications/store/4/0/5/8/9/1/index.htt.

Arpia, A. A., Chen, W. H., Lam, S. S., Rousset, P., & de Luna, M. D. G. (2021). Sustainable biofuel and bioenergy production from biomass waste residues using microwave-assisted heating: A comprehensive review. *Chemical Engineering Journal*, 403. Available from https://doi.org/10.1016/j.cej.2020.126233, http://www.elsevier.com/inca/publications/store/6/0/1/2/7/3/index.htt.

Bajpai, P. (2020). *Biomass properties and characterization* (pp. 21–29). Elsevier BV. Available from 10.1016/b978-0-12-818400-4.00003-7.

Bakshi, B. R., & Banerji, B. (2010). *Bioenergy and biofuel from biowastes and biomass*. American Society of Civil Engineers.

Borges, F. C., Du, Z., Xie, Q., Trierweiler, J. O., Cheng, Y., Wan, Y., Liu, Y., Zhu, R., Lin, X., Chen, P., & Ruan, R. (2014). Fast microwave assisted pyrolysis of biomass using microwave absorbent. *Bioresource Technology*, 156, 267–274. Available from https://doi.org/10.1016/j.biortech.2014.01.038, http://www.elsevier.com/locate/biortech.

Dai, L., He, C., Wang, Y., Liu, Y., Yu, Z., Zhou, Y., Fan, L., Duan, D., & Ruan, R. (2017). Comparative study on microwave and conventional hydrothermal pretreatment of bamboo sawdust: Hydrochar properties and its pyrolysis behaviors. *Energy Conversion and Management*, 146, 1–7. Available from https://doi.org/10.1016/j.enconman.2017.05.007.

Diaz, A. B., Moretti, M. M. D. S., Bezerra-Bussoli, C., C.d.C. Carreira Nunes, A., Blandino, R., da Silva, E., & Gomes. (2015). Evaluation of microwave-assisted pretreatment of lignocellulosic biomass immersed in alkaline glycerol for fermentable sugars production. *Bioresource Technology*, 185, 316–323. Available from https://doi.org/10.1016/j.biortech.2015.02.112, http://www.elsevier.com/locate/biortech.

Dimitriadis, A., & Bezergianni, S. (2017). Hydrothermal liquefaction of various biomass and waste feedstocks for biocrude production: A state of the art review. *Renewable and Sustainable Energy Reviews*, 68, 113–125. Available from https://doi.org/10.1016/j.rser.2016.09.120.

Ellison, C., McKeown, M. S., Trabelsi, S., Marculescu, C., & Boldor, D. (2018). Dielectric characterization of bentonite clay at various moisture contents and with mixtures of biomass in the microwave spectrum. *Journal of Microwave Power and Electromagnetic Energy*, 52(1), 3–15. Available from https://doi.org/10.1080/08327823.2017.1421407, http://www.tandfonline.com/loi/tpee20#.VscCNbdf1Fo.

Ferrera-Lorenzo, N., Fuente, E., Bermúdez, J. M., Suárez-Ruiz, I., & Ruiz, B. (2014). Conventional and microwave pyrolysis of a macroalgae waste from the Agar-Agar industry. Prospects for bio-fuel production. *Bioresource Technology*, 151, 199–206. Available from https://doi.org/10.1016/j.biortech.2013.10.047, http://www.elsevier.com/locate/biortech.

Gronnow, M. J., Budarin, V. L., Mašek, O., Crombie, K. N., Brownsort, P. A., Shuttleworth, P. S., Hurst, P. R., & Clark, J. H. (2013). Torrefaction/biochar production by microwave and conventional slow pyrolysis - comparison of energy properties. *GCB Bioenergy*, 5(2), 144–152. Available from https://doi.org/10.1111/gcbb.12021, http://onlinelibrary.wiley.com/journal/10.1111/(ISSN)1757-1707.

References

Hoang, A. T., Nižetić, S., Ong, H. C., Mofijur, M., Ahmed, S. F., Ashok, B., Bui, V. T. V., & Chau, M. Q. (2021). Insight into the recent advances of microwave pretreatment technologies for the conversion of lignocellulosic biomass into sustainable biofuel. *Chemosphere*, 281. Available from https://doi.org/10.1016/j.chemosphere.2021.130878, http://www.elsevier.com/locate/chemosphere.

Huang, Y. F., Kuan, W. H., Lo, S. L., & Lin, C. F. (2010). Hydrogen-rich fuel gas from rice straw via microwave-induced pyrolysis. *Bioresource Technology*, 101(6), 1968–1973. Available from https://doi.org/10.1016/j.biortech.2009.09.073, http://www.sciencedirect.com.

Ismail, N., Ho, G. S., Amin, N. A. S., & Ani, F. N. (2015). Microwave plasma gasification of oil palm biochar. *Jurnal Teknologi*, 74(10), 7–13. Available from https://doi.org/10.11113/jt.v74.4827, http://www.jurnalteknologi.utm.my/index.php/jurnalteknologi/article/download/4827/3341.

Klinger, J. L., Westover, T. L., Emerson, R. M., Williams, C. L., Hernandez, S., Monson, G. D., & Ryan, J. C. (2018). Effect of biomass type, heating rate, and sample size on microwave-enhanced fast pyrolysis product yields and qualities. *Applied Energy*, 228, 535–545. Available from https://doi.org/10.1016/j.apenergy.2018.06.107, http://www.elsevier.com/inca/publications/store/4/0/5/8/9/1/index.htt.

Liu, H., Jiaqiang, E., Ma, X., & Xie, C. (2016). Influence of microwave drying on the combustion characteristics of food waste. *Drying Technology*, 34(12), 1397–1405. Available from https://doi.org/10.1080/07373937.2015.1118121, http://www.tandf.co.uk/journals/titles/07373937.asp.

Li, J., Dai, J., Liu, G., Zhang, H., Gao, Z., Fu, J., He, Y., & Huang, Y. (2016). Biochar from microwave pyrolysis of biomass: A review. *Biomass and Bioenergy*, 94, 228–244. Available from https://doi.org/10.1016/j.biombioe.2016.09.010, http://www.journals.elsevier.com/biomass-and-bioenergy/.

López-Linares, J. C., García-Cubero, M. T., Lucas, S., González-Benito, G., & Coca, M. (2019). Microwave assisted hydrothermal as greener pretreatment of brewer's spent grains for biobutanol production. *Chemical Engineering Journal*, 368, 1045–1055. Available from https://doi.org/10.1016/j.cej.2019.03.032, http://www.elsevier.com/inca/publications/store/6/0/1/2/7/3/index.htt.

Maryana, R., Ma'rifatun, D., Wheni, I. A. K. W. S., & Rizal, W. A. (2014). Alkaline pretreatment on sugarcane bagasse for bioethanol production, . *Energy Procedia* (47, pp. 250–254). Indonesia: Elsevier Ltd. 18766102. Available from http://www.sciencedirect.com/science/journal/18766102, 10.1016/j.egypro.2014.01.221.

Mohd Fuad, M. A. H., Hasan, M. F., & Ani, F. N. (2019). Microwave torrefaction for viable fuel production: A review on theory, affecting factors, potential and challenges. *Fuel*, 253, 512–526. Available from https://doi.org/10.1016/j.fuel.2019.04.151, http://www.journals.elsevier.com/fuel/.

Nomanbhay, S. M., Hussain, R., & Palanisamy, K. (2013). Microwave-Assisted alkaline pretreatment and microwave assisted enzymatic saccharification of oil palm empty fruit bunch fiber for enhanced fermentable sugar yield. *Journal of Sustainable Bioenergy Systems*, 03(01), 7–17. Available from https://doi.org/10.4236/jsbs.2013.31002.

Ocreto, J. B., Chen, W. H., Ubando, A. T., Park, Y. K., Sharma, A. K., Ashokkumar, V., Ok, Y. S., Kwon, E. E., Rollon, A. P., & De Luna, M. D. G. (2021). A critical review on second- and third-generation bioethanol production using microwaved-assisted heating (MAH) pretreatment. *Renewable and Sustainable Energy Reviews*, 152. Available from https://doi.org/10.1016/j.rser.2021.111679, https://www.journals.elsevier.com/renewable-and-sustainable-energy-reviews.

OmarR. (2010). Evaluation of microwave pyrolysis of oil palm empty fruit bunches..

Ongen, A., Ozcan, H. K., & Ozbas, E. E. (2016). Gasification of biomass and treatment sludge in a fixed bed gasifier. *International Journal of Hydrogen Energy*, 41(19), 8146–8153. Available from https://doi.org/10.1016/j.ijhydene.2015.11.159, http://www.journals.elsevier.com/international-journal-of-hydrogen-energy/.

Onokwai, A. O., Ajisegiri, E. S. A., Okokpujie, I. P., Ibikunle, R. A., Oki, M., & Dirisu, J. O. (2022). Characterization of lignocellulose biomass based on proximate, ultimate, structural composition, and thermal analysis. *Materials Today: Proceedings*, 65, 2156–2162. Available from https://doi.org/10.1016/j.matpr.2022.05.313, https://www.sciencedirect.com/journal/materials-today-proceedings.

Ravikumar, C., Senthil Kumar, P., Subhashni, S. K., Tejaswini, P. V., & Varshini, V. (2017). Microwave assisted fast pyrolysis of corn cob, corn stover, saw dust and rice straw: Experimental investigation on bio-oil yield and high heating values. *Sustainable Materials and Technologies*, 11, 19–27. Available from https://doi.org/10.1016/j.susmat.2016.12.003.

Remón, J., Randall, J., Budarin, V. L., & Clark, J. H. (2019). Production of bio-fuels and chemicals by microwave-assisted, catalytic, hydrothermal liquefaction (MAC-HTL) of a mixture of pine and spruce biomass. *Green Chemistry*, 21(2), 284–299. Available from https://doi.org/10.1039/c8gc03244k, http://pubs.rsc.org/en/journals/journal/gc.

Ren, S., Lei, H., Wang, L., Bu, Q., Chen, S., Wu, J., Julson, J., & Ruan, R. (2013). The effects of torrefaction on compositions of bio-oil and syngas from biomass pyrolysis by microwave heating. *Bioresource Technology*, 135, 659–664. Available from https://doi.org/10.1016/j.biortech.2012.06.091, http://www.elsevier.com/locate/biortech.

Ren, S., Lei, H., Wang, L., Bu, Q., Wei, Y., Liang, J., Liu, Y., Julson, J., Chen, S., Wu, J., & Ruan, R. (2012). Microwave torrefaction of douglas fir sawdust pellets. *Energy and Fuels*, 26(9), 5936–5943. Available from https://doi.org/10.1021/ef300633c.

Ren, S., Lei, H., Wang, L., Yadavalli, G., Liu, Y., & Julson, J. (2014). The integrated process of microwave torrefaction and pyrolysis of corn stover for biofuel production. *Journal of Analytical and Applied Pyrolysis*, 108, 248–253. Available from https://doi.org/10.1016/j.jaap.2014.04.008.

Sarangi, P. K., Nanda, S., & Mohanty, P. (2018). Recent advancements in biofuels and bioenergy utilization. *Recent advancements in biofuels and bioenergy utilization* (pp. 1–402). Singapore, India: Springer. Available from http://doi.org/10.1007/978-981-13-1307-3, 10.1007/978-981-13-1307-3.

Sarker, T. R., Pattnaik, F., Nanda, S., Dalai, A. K., Meda, V., & Naik, S. (2021). Hydrothermal pretreatment technologies for lignocellulosic biomass: A review of steam explosion and subcritical water hydrolysis. *Chemosphere*, 284. Available from https://doi.org/10.1016/j.chemosphere.2021.131372, http://www.elsevier.com/locate/chemosphere.

Shi, K., Yan, J., Menéndez, J. A., Luo, X., Yang, G., Chen, Y., Lester, E., & Wu, T. (2020). Production of H2-rich syngas from lignocellulosic biomass using microwave-assisted pyrolysis coupled with activated carbon enabled reforming. *Frontiers in Chemistry*, 8. Available from https://doi.org/10.3389/fchem.2020.00003, http://journal.frontiersin.org/journal/chemistry.

Siritheerasas, P., Waiyanate, P., Sekiguchi, H., & Kodama, S. (2017). Torrefaction of municipal solid waste (MSW) pellets using microwave irradiation with the assistance of the char of agricultural residues. *Energy Procedia* (138, pp. 668–673). Thailand: Elsevier Ltd. 18766102. Available from https://doi.org/10.1016/j.egypro.2017.10.190, http://www.sciencedirect.com/science/journal/18766102.

Verma, P., Watanabe, T., Honda, Y., & Watanabe, T. (2011). Microwave-assisted pretreatment of woody biomass with ammonium molybdate activated by H2O2. *Bioresource Technology*, 102(4), 3941–3945. Available from https://doi.org/10.1016/j.biortech.2010.11.058.

Wang, X., Chen, H., Luo, K., Shao, J., & Yang, H. (2008). The influence of microwave drying on biomass pyrolysis. *Energy and Fuels*, 22(1), 67–74. Available from https://doi.org/10.1021/ef700300m, 08870624 China.

Wang, M. J., Huang, Y. F., Chiueh, P. T., Kuan, W. H., & Lo, S. L. (2012). Microwave-induced torrefaction of rice husk and sugarcane residues. *Energy*, 37(1), 177–184. Available from https://doi.org/10.1016/j.energy.2011.11.053, http://www.elsevier.com/inca/publications/store/4/8/3/.

Williams, C. L., Dahiya, A., & Porter, P. (2015). *Introduction to bioenergy* (pp. 5–36). Elsevier BV. Available from 10.1016/b978-0-12-407909-0.00001-8.

Xiao, N., Luo, H., Wei, W., Tang, Z., Hu, B., Kong, L., & Sun, Y. (2015). Microwave-assisted gasification of rice straw pyrolytic biochar promoted by alkali and alkaline earth metals. *Journal of Analytical and Applied Pyrolysis*, 112, 173–179. Available from https://doi.org/10.1016/j.jaap.2015.02.001.

Xie, Q., Borges, F. C., Cheng, Y., Wan, Y., Li, Y., Lin, X., Liu, Y., Hussain, F., Chen, P., & Ruan, R. (2014). Fast microwave-assisted catalytic gasification of biomass for syngas production and tar removal. *Bioresource Technology*, 156, 291–296. Available from https://doi.org/10.1016/j.biortech.2014.01.057, http://www.elsevier.com/locate/biortech.

Yang, J., He, Q. S., Niu, H., Dalai, A., Corscadden, K., & Zhou, N. (2020). Microwave-assisted hydrothermal liquefaction of biomass model components and comparison with conventional heating. *Fuel*, 277118202. Available from https://doi.org/10.1016/j.fuel.2020.118202.

Yang, J., Niu, H., Corscadden, K., He, Q., & Zhou, N. (2022). MW-assisted hydrothermal liquefaction of spent coffee grounds. *Canadian Journal of Chemical Engineering*, 100(8), 1729–1738. Available from https://doi.org/10.1002/cjce.24270, http://onlinelibrary.wiley.com/journal/10.1002/(ISSN)1939-019X.

Yan, B., Jiao, L., Li, J., Zhu, X., Ahmed, S., & Chen, G. (2021). Investigation on microwave torrefaction: Parametric influence, TG-MS-FTIR analysis, and gasification performance. *Energy*, 220119794. Available from https://doi.org/10.1016/j.energy.2021.119794.

Zhang, W. (2017). Alternative energy sources for green chemistry. *Green Processing and Synthesis*, 6(1). Available from https://doi.org/10.1515/gps-2016-0201.

Zhuang, X., Liu, J., Wang, C., Zhang, Q., & Ma, L. (2022). Microwave-assisted hydrothermal liquefaction for biomass valorization: Insights into the fuel properties of biocrude and its liquefaction mechanism. *Fuel*, 317123462. Available from https://doi.org/10.1016/j.fuel.2022.123462.

Zhu, Z., Simister, R., Bird, S., McQueen-Mason, S. J., Gomez, L. D., & Macquarrie, D. J. (2015). Microwave assisted acid and alkali pretreatment of Miscanthus biomass for biorefineries. *AIMS Bioengineering*, 2(4), 449–468. Available from https://doi.org/10.3934/bioeng.2015.4.449.

CHAPTER

8

Bioprospecting of algal biomass for value generation from municipal waste

Renupama Bhoi[1], Shubham Agrawal[1], Jasmita[2] and Angana Sarkar[1]

[1]Department of Biotechnology and Medical Engineering, National Institute of Technology Rourkela, Rourkela, Odisha, India [2]Department of Biotechnology, College of Commerce, Arts, and Science Patna (Bihar) Patliputra University, Patna, Bihar, India

8.1 Introduction

Increasing population day by day also led to an increase in industrialization and urbanization worldwide, which also led to the increasing generation of more wastewater from many areas including manufacturing industries, food industries, agricultural industries, piggery industries, dairy industries, and domestic activities (Purba et al., 2022).

Large quantity of wastewater is discharged by household activities that are rich in nutrient media. Conventional sludge active system used to treat the wastewater removed the chemical oxygen demand, but several phosphate and nitrogen-containing compounds are still released by the system and lead to mixing with the lake, causing eutrophication. Therefore conventional activated sludge system releases a large amount of effluent, which is also rich in inorganic compound and also has a limitation as it requires a high amount of energy for its process (Abdelfattah et al., 2022).

Generally, nitrogen is present in the form of ammonia, nitrate, or nitrite in wastewater (Shengnan et al., 2022). Several heavy metals are also present in municipal wastewater such as arsenic, zinc, cadmium, chromium, copper, and mercury as per the Europian Commission on Environment 2022. To treat the nutrient-rich wastewater unnecessarily discharged into water systems such as lakes and rivers, it can be treated with microalgal cultivation as it is cost-effective and provides economic benefits leading to sustainable waste-to-energy management. Microalgae cultivation is affected by various physiochemical parameters such as pH (acidic or alkaline), temperature, salinity, light, and nutrient media for optimum biomass production. However, microalgae biomass has many applications in the biorefinery aspect that includes biofuel or biodiesel production, extraction of lipids, protein, carbohydrates, pigment, etc. (Goveas et al., 2022).

The cultivation of microalgae for wastewater treatment depends on the types of algal species and the ability to uptake nutrients from the wastewater for its growth (Chaleshtori et al., 2022). Growing microalgae in wastewater for bioremediation has various advantages over conventional methods, for example, the nutrient removal is more, Less or no toxic by-product from sludge, it's a cost-effective and eco-friendly method, and chances of the high growth rate of algal species due to the rich nutrient medium of wastewater, much value-added product can be extracted (Purba et al., 2022).

Microalgae, on the other hand, have a highly adaptable metabolic system. They can grow in photoautotrophic, mixotrophic, or heterotrophic conditions, as well as in other metabolic states. Furthermore, because urban wastewater is a good source of both nitrogen and phosphorous, it can also serve as a good source of microalgae cultivation. As a result, algal systems are thought to be effective approaches to treating a variety of wastewater sources. When algae are exposed to light, they begin to absorb nutrients from the effluent while also producing oxygen through photosynthesis (Chaleshtori et al., 2022).

Because of their versatility and tolerance to a wide range of nutrients and organic matter, microalgal cultivation on wastewater effluents has gained a lot of attention recently. The varied group of organisms known as microalgae can be found in a variety of native habitats. While the majority of microalgal species are photoautotrophs, it is known that some of them can also grow in heterotrophic or mixotrophic environments (Ghaffar, et al., 2022).

Photoautotrophic growth conditions are frequently found in microalgae. While some species can alternate between photoautotrophic and heterotrophic growth, both metabolic ways of life coexist simultaneously in a mixotrophic growth (Dragone, 2022). Microalgae can grow heterotrophically by using organic substrates to generate energy through aerobic metabolism in the absence of sunlight. It has been demonstrated that the cultivation of microalgae is effective for the commercial production of high-value chemical compounds, such as cosmetic products, drugs, and nutritional supplements, under anaerobic conditions (Rahman et al., 2022).

Microalgae use both inorganic and organic carbon sources during the process known as mixotrophic growth, which occurs in the presence of light (Salla et al., 2016). In the bioprocessing sector, fed-batch cultivation is typical. By using a semibatch method, it is possible to prevent the accumulation of toxic substances throughout cultivation as well as the limitation or suppression of substrate.

This oxygen is consumed by heterotrophic bacterial species to mineralize organic substances. Microalgae methods are very cost-effective and require less energy to initiate. They use CO_2 to minimize sludge formation and emissions of greenhouse gases (Ahmed et al., 2022). Among the most significant advantages of microalgae treatment methods is their ability to absorb hazardous heavy metals, provide oxygen, increase pH and the amount of dissolved oxygen (DO), and thus provide indirect decontamination.

8.2 Composition of municipal wastewater

Microalgae is a promising feedstock as it has several advantages for its higher rate of biomass production property, photosynthetic efficiency, and growth rate as compared with other feedstock. India is rich in different industries such as pharmaceutical, food, textile, and dairy (Ghosh, 2018). Releasing a higher amount of effluent that is rich in nutrients served as ideal nutrients for algal growth. Dairy industries release a huge amount of wastewater consisting of wasted milk, nutrient, washing detergent, fats, and nutrients, etc. Generally, dairy waste is composed of different micro- and macronutrients including potassium, sodium, calcium, magnesium, calcium, cobalt, iron, manganese, nickel, phosphorus, and sulfate, which are enriched by various nutrient sources leading to eutrophication (Kusmayadi et al., 2022). This enriched nutrient wastewater can be utilized before discharging into a lake or river for sustainable environmental management. Microalgae are in demand and receiving great interest in treating wastewater due to their various advantages. Microalgae require various feasible parameters for their cultivation such as light, temperature, carbon dioxide, water, and nutrient, and for that dairy, and wastewater can provide a suitable and sustainable medium for their cultivation. Pig farming discharges comprise large amounts of organic matter, nitrogen, and phosphorus; therefore several researchers have been using piggery sewage water for microalgae cultivation (Shengnan et al., 2022). A large amount of piggery wastewater produced has been a problem for water bodies. Several researchers have used piggery wastewater to cultivate microalgae (Luan, et al., 2019). However, it isn't suggested to use it in concentrated form because its darker color affects absorption and biomass yield, and it contains a high concentration of toxic ammonia for microalgae.

Nutrient media varies with which types of industries we are considering such as food and beverages producing industries, domestic wastewater, and municipal wastewater. Dairy wastewater is rich in potassium, sodium, calcium, magnesium, calcium, cobalt, iron, manganese, nickel, phosphorus, and sulfate (Dhandayuthapani et al., 2022).

Municipal wastewater is the combining discharged water from domestic household activities, schools, and colleges released effluent, effluent from hospitals, and water discharged from toilets and bathrooms, etc. (Arrojo et al., 2022). The quality of municipal wastewater is characterized by physical and chemical biological properties, which can be determined by using different parameters for analyzing its physicochemical composition (Neveux et al.,2016). The parameters include pH, temperature, total suspended solids (TSSs), TDSs (total dissolved solids), dissolved oxygen (DO), chemical oxygen demand (COD), electrical conductivity (EC), biological oxygen demand (BOD), turbidity, alkalinity, nitrate sulfate concentration, and metals such as zinc, chromium, copper, lead, iron, nickel, and cadmium are described in Table 8.1.

TABLE 8.1 Physicochemical parameters of municipal wastewater.

Sl. no	Parameters	Units	Municipal wastewater	References
	pH	–	8.10	Salama et al. (2017)
			7.6	Neveux et al. (2016)
			7.8	
			8.1	
			8.39	Kumar & Chopra (2012)
			7.37	Masindi et al. (2022)
	COD	mg/L	1420.54	Kumar & Chopra (2012)
			71	Masindi et al. (2022)
			31	Salama et al. (2017)
			12	Neveux et al. (2016)
			24	
			311.5	Sharma & Kansal (2011)
			260	Nawakar et al. (2019)
	BOD	mg/L	620.27	Kumar & Chopra (2012)
			230	Mishra et al. (2021)
			146.975	Sharma et al. (2011)
	DO	mg/L	2.42	Kumar & Chopra (2012)
			23.81	Neveux et al. (2016)
			0.89	
			2.05	
	EC	ds/m	2.84	Kumar & Chopra (2012)
			2.29	Neveux et al. (2016)
			2.16	
			2.20	
			2.29	Salama et al. (2017)
			2.4	Zhang et al. (2022)
	TDS	mg/L	1300	Nawarkar & Salkar (2019)
	TSS	mg/L	1824.42	Kumar & Chopra (2012)
			50	Neveux et al. (2016)
			2	
			22	
			605	Mishra et al. (2021)
			257.45	Sharma & Kansal (2011)
	Turbidity	NTU	20.86	Kumar & Chopra (2012)
			59.2	Sharma & Kansal (2011)
			60	Nawarkar & Salkar (2019)
	Alkalinity	mg/L	254.33	Kumar & Chopra (2012)

(Continued)

TABLE 8.1 (Continued)

Sl. no	Parameters	Units	Municipal wastewater	References
			813	Masindi et al. (2022)
			272	Salama et al. (2017)
	Free CO2	mg/L	122.99	Kumar & Chopra (2012)
	Sulfate	mg/L	336.19	Kumar & Chopra (2012)
			225.95	Sharma & Kansal (2011)
			40.4	Salama et al. (2017)
	Nitrate	mg/L	54.5	Sharma & Kansal (2011)
	Iron	mg/L	7.74	Kumar & Chopra (2012)
			0.526	Sharma & Kansal (2011)
	Chromium	mg/L	0.21	Kumar & Chopra (2012)
			0.036	Sharma & Kansal (2011)
	Copper	mg/L	0.78	Kumar & Chopra (2012)
			0.0258	Sharma & Kansal (2011)
	Lead		0.065	Sharma & Kansal (2011)
	Zinc	mg/L	3.24	Kumar & Chopra (2012)
			0.064	Sharma & Kansal (2011)
			0.009	Salama et al. (2017)
	Cadmium	mg/L	0.64	Kumar & Chopra (2012)
	Chlorine	mg/L	346.58	Kumar & Chopra (2012)
			164.675	Sharma & Kansal (2011)
	Nickel	mg/L	0.46	Kumar & Chopra (2012)

when the illuminance reached 10,000 lux or less. This indicates that higher intensity can raise the total amount of fatty acids in algal species and the amount of triacylglycerol, which is needed to produce biodiesel (Han et al., 2015). In the oxidation process of polyunsaturated fatty acids, long-term exposure intensity of light can often increase the rate of triacylglycerol and subsequently decrease the amount of polar lipids production. The most crucial factors affecting algal growth are light, pH, turbulence, salinity, and temperature, as well as the quantity and quality of nutrients. The composition of fatty acids and microalgal lipid yield both enhanced to their maximum levels of growth.

Microalgae photosynthesize or assimilate inorganic carbon for transformation into organic matter as all plants do. The energy that generates this reaction comes from light. The importance of light intensity is significant, but the necessities differ depending on the depth and cell concentration of the microalgal biomass to penetrate the culture at greater depths and cell densities, illuminance must be enhanced such as fluorescent tubes and natural light can both be used (Goswami et al., 2022).

Photoinhibition may be caused by excessively bright light sources, such as direct sunlight or small containers near artificial light. Additionally, it's important to prevent temperature rise brought about by both artificial and natural lighting. It is preferable to use fluorescent bulbs that transmit light in the blue or red-light parts of the spectrum because these are the parts of the spectrum of light that are most favorable to photosynthetic activity (Chen et al., 2022). Even though cultured algae generally grow under continuous illumination, the amount of artificial light should be at least 18 hours per day. The majority of cultured algal species have a pH range of 7–9, with 8.2–8.7 being the ideal range. Aeration is a significant factor because it prevents microalgal deposition by thoroughly mixing the culture (Wang et al., 2010). To properly cultivate algae that produce more organic matter, mixing is important as it ensures that every cell in the community is subjected to light and essential minerals in an adequate proportion. It also improves gaseous exchange between both the growth media and the air.

The majority of commonly cultivated microalgal species can withstand temperatures between 16°C and 27°C. Supplements such as organic, inorganic, micronutrients, macronutrients, and vitamins were necessary for algae culture (Kumar & Chopra, 2012; Salama et al., 2017).

8.3 Biorefinery approaches of microalgal cultivation in municipal wastewater

A physical, chemical, or biological process that separates, refines, or transforms biological resources from the kingdoms Monera, Protista, Plantae, Animalia, or Fungi that originate in an oceanic or terrestrial environment and their finished biomass or raw materials used for other byproduct production is referred to as biorefinery (Cherubini et al., 2009). The different value-added products that can be recovered from microalgal biomass have important applications in different industries (Nishshanka et al., 2021). The biorefinery approaches to recover various bioactive compounds after cultivation in wastewater simultaneously help to treat the sewage effluent and are described in the flow diagram (Fig. 8.1).

The Cascading Algal Biomethane-Biorefinery System (CABBS) means that upgrading biogas photosynthetically by using microalgae may address the economic and environmental limitations associated with the conventional biogas process. To produce bicarbonate, carbon dioxide is removed from biogas using a carbonate-rich solution. Bicarbonate-rich solutions with a moderately high pH are used as inorganic carbon sources for microalgal development. Photosynthetic biomethane improvements would be more sustainable if microalgae had high geomorphological productivity, rapid CO_2 fixation, and the ability to develop under a wide range of conditions (Bose et al., 2020).

Micosporine-like amino acids (MAAs) from microalgae can now be identified and quantified using liquid chromatography and electrospray ionization mass spectrometry. Following that, liquid chromatography in combination with tandem mass spectrometry was developed as a more advanced method for the fast and precise determination and quantification of MAA. Novel genome engineering techniques are appropriate for increasing

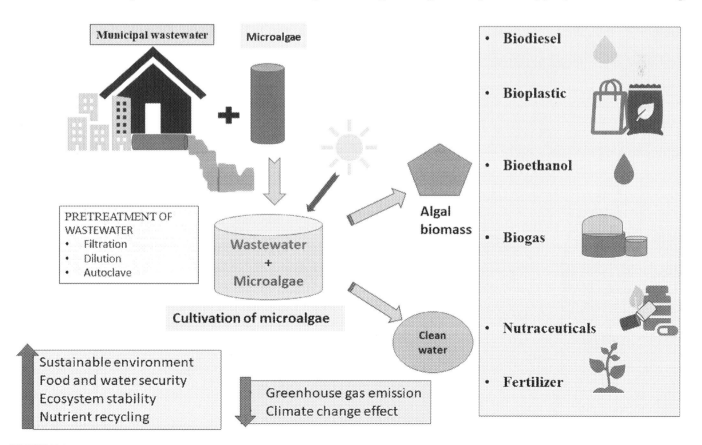

FIGURE 8.1 Flowchart of biorefinery approaches of microalgal municipality wastewater treatment.

biomass output. However, the primary drawbacks of these applications are low efficiency, cytotoxicity, and a lack of knowledge concerning the genetic information of microalgae (Raj et al., 2021).

Algal cell lysis can be followed by an extraction procedure using different downstream processing techniques, such as centrifugation and filtration, to recover the nanomaterials from the algal broth cell suspension, in addition to using live algal cells directly for the synthesis of nanoparticles (Mukherjee et al., 2021).

Nanoparticles can be used to extract lipids without killing the cells by milking microalgae in conjunction with a nanobiorefinery, significantly lowering the cost of lipid extraction (Kashyap & Kiran, 2021). Alga-based biochar has a good capacity for adsorbing organic and inorganic pollutants from textile effluents when compared with other activated carbon (Khan et al., 2022). Carotenoids can be used in the pharmaceutical and nutritional industries because of their antiinflammatory, antioxidant, and cardiovascular disease—reducing qualities, among other things (Ren et al., 2021). Through the exchange of metabolites or components such as dissolved organic matter, vitamins, iron, sulfur, nitrogen, and phosphorus, bacteria and microalgae interact symbiotically Microalgal cells can store carbohydrates that can be transformed into hydrogen, and solar energy can be directed to triglycerides to be converted to diesel fuel. The leftover microalgal biomass can be fermented by fermentation process or anaerobic bacteria to create ethanol or biomethane. Most of the signaling molecules, genes, and enzymes involved in the symbiotic relationship between microalgae and bacteria are still unknown. However, it is now understood that this symbiosis with microalgae can help improve biomass production and enrich biomass with valuable energy and chemical products, whereas bacteria have previously been considered contaminants in algal cultures (Yao et al., 2019). The advantages of these systems are described in Table 8.2. Bioenergy systems present significant opportunities for reducing greenhouse gas emissions because of their enormous potential to replace fossil fuels in energy generation. Increased use of biomass-based fuels can enhance the health of rural areas, keep the environment clean, generate new work opportunities, and encourage economic sustainability (Ferreira, 2017). The formation of an integrated closed-loop biorefinery that maximizes the use of biomass, decreases waste, and derives the most value-added product from the available resources ensures their economic viability and sustainability. The formation of an integrated closed-loop biorefinery that maximizes the use of biomass, minimizes waste, and derives the greatest economic advantage from the available resources helps to ensure their commercial feasibility (Conteratto et al., 2021).

8.4 Cultivation of microalgae in municipal wastewater

An efficient and realistic way to address the high costs of the cultivation of microalgae and lower environmental pollution is to combine the growth of algal species with wastewater treatment. By removing the inorganic nutrients such as nitrogen and phosphorus, which are necessary for the production of biomass, algae farming combined with sewage from urban areas produces high-value biomass, thereby also ensuring sewage treatment safety. These ingredients could have a serious impact on aquatic ecosystems, such as nutrient enrichment, which would affect aquatic organism activity and people's health (Luan et al., 2019).

Algae have demonstrated significant potential as a substitute source of bioenergy, which included biofuels, biomethane, biogas, heat, and electricity, protein-rich food and animal feed, dietary supplements, and industrial applications containing compounds. We are very familiar with the aquatic microalgae that live in freshwater, saline water, and brackish water (Bahman et al., 2022). Additionally, many of these microalgae have already grown quickly and produced a lot of biomasses in wastewater, and the process has proven that it can lower the concentration of pollutants in raw sewage. Utilizing algae to reduce the pollutant loads of household effluents is much more affordable and environmentally sustainable than using conventional or chemical methods since these have advantages of resource recovery, wastewater reprocessing, and biomass production (Mishra et al., 2021).

For the production of fatty acids, carbohydrates, vitamins, pigments, and biofuels such as bioethanol, biodiesel, and biogas, among other things, microalgae biomass consists of a variety of beneficial bioactive components (Guldhe et al., 2017). The method of cultivation affects the green algae biomass yield and biochemical composition. The biochemical structure is influenced by several physicochemical factors, including pH, temperature, light, and nutritional content such as carbon, phosphorus, nitrogen, and sulfur. These factors also play a significant role in high yield (Liu & Yildiz, 2018; Luan et al., 2019; Nur & Buma, 2019).

Municipal wastewater is a mixture of toilet, kitchen, bathroom, and other domestic effluents and is therefore rich in many nutrient compounds. Raw municipal wastewater undergoes many changes after microalgal treatment. The color is transformed from darkness to light. There won't be any obnoxious odor. Alkalinity drops and water hardness is reduced, and inorganic phosphate is recovered after wastewater treatment (Singh et al., 2022).

TABLE 8.2 List of different biorefinery approaches for advanced microalgal cultivation.

Sl. no.	Biorefinery approaches	Description	Advantages	Future perspective	Reference
1.	Cascading Algal Biomethane-Biorefinery System (CABBS)	Biogas production (Anaerobic digestion), biogas upgradation to biomethane, and digest treatment are done with the help of microalgae.	Waste products such as CO_2 and digestate from anaerobic digestion are used as substrates to cultivate algae for various useful products.	Maximizing the effectiveness of biogas upgrading and digestate treatment simultaneously has not been achieved yet but extensive research is going on to use microalgae as the advantages are numerous.	Bose et al. (2020)
2.	Micosporine-like amino acids (MAAs) from microalgae	MAAs are produced by algae as a part of their defense mechanism which protects them from UVB rays, salt stress, desiccation stress, etc. Microalgal species are a good source of MAAs.	MAAs have several useful properties such as antiinflammatory, wound healing agents, UV absorbing compounds, etc.	The mass production of MAAs has a huge opportunity when using molecular cloning methodologies and genome editing tools like TALEN, ZFN, and (CRISPR/Cas9).	Raj et al. (2021)
3.	Metal nanoparticles synthesis using algae	Many species of microalgae are proven efficient in promoting the production of Ag, Au, and Pt nanoparticles.	The biosynthesis of nanoparticles instead of chemical or physical processes is better for the environment.	The use of microalgae for nanoparticle synthesis is limited right now, but both micro- and macroalgae are frontrunners.	Mukherjee et al. (2021)
4.	Milking microalgae in conjugation with nano-biorefinery	Utilizing bioderived nanoparticles for milking the same or different species of microalgae for oil is possible.	Nanoparticles can be used to extract lipids without killing the cells, significantly reducing lipid extraction costs.	Nanoparticles will play an essential role in the future as they have antimicrobial, photocatalytic, photothermal, etc properties. Milking is one of the most important applications of nanoparticles on which intensive research is going on.	Kashyap and Kiran (2021)
5.	Sustainable microalgal biomass production in wastewater from the food industry for cheap biorefinery products	Microalgae are potential biocatalysts for the bioremediation of wastewater.	Microalgae can efficiently remove effluent from food processing that contains significant amounts of organic carbon, phosphorus, and nitrogen to generate various cheap value-added products.	Some challenges are still there for the cultivation of algae that needs to be solved such as the growth of algal feeders, protozoans, and other organisms stimulated by the cultivation of microalgae in wastewater because of their nonsterile habitats.	Ummalyma et al. (2022)
6.	Treatment of wastewater from the textile sector with microalgae-derived biochar	For the creation of biochar with volatile applications, such as fertilizer and adsorbent, microalgal biomass is ideal.	Compared to other carbonaceous materials, alga base biochar has a good capacity for adsorbing organic and inorganic pollutants from wastewater.	Algae biochar is very efficient in adsorbing heavy metals and dyes. However, very little research has been done till now. New research should be in the direction to use biochar made from lipid-extracted algae.	Khan et al. (2022)
7.	Dual-purpose microalgae-bacteria symbiosis system to treat wastewater and biodiesel production	In algal cultures bacteria have been regarded as pollutants, however, it is now realized that this symbiosis can help improve biomass production and to enrich biomass with valuable energy and chemical products.	Bacteria provide vitamins, nitrogen, phosphorus, and sulfur to algae which are important for optimum growth. Carbon source is supplied to bacteria from algae in the form of extracellular polymeric substance.	This system offers substantial savings over microalgae biomass production. The main challenge is the controlled integration of specific bacteria in the specific microalgal species for industrial use. More insight at the molecular level will be useful.	Yao et al. (2019)

(Continued)

TABLE 8.2 (Continued)

Sl. no.	Biorefinery approaches	Description	Advantages	Future perspective	Reference
8.	Carotenoids from microalgae	Different types of carotenoids can be extracted from algae	Carotenoids can be used in nutraceutical, and pharmaceutical industries as they have properties like they are antioxidant, antiinflammatory, reduce risk of cardiovascular diseases, anticancer activities, etc.	Salt stress method is used to increase carotenoid accumulation and can be combined with other treatments such as light induction, etc. chemical mutagens such as NMMG, EMS, etc or novel tools such as CRISPR, ZEN TALEN, etc. to enhance carotenoid accumulation are being researched.	Ren et al. (2021)

8.5 Challenges in microalgal cultivation in municipal wastewater

Even though microalgae have promising qualities, there are enormous obstacles that must be resolved before this path is used in a way that is both ecologically and commercially sustainable. The challenges are regarded as the biggest obstacles to overall microalgae cultivation. Due to the comparatively slow growth, microalgae grown in the effluent may be contaminated by fungi, zooplankton such as rotifers, protozoans such as amebas, flagellates, and other algal (Ummalyma et al., 2022). Microalgae can also be inhibited by abiotic wastewater contaminants. Poor growth, low biomass, and low lipid productivity can happen when there is a small amount of trace mineral elements in the sewage. An appropriate effluent composition must be assessed for the production of a particular algal strain because research has shown that some microalgae-based species may be inhibited by high nutrient concentrations (Mishra & Mohanty, 2019). Most domestic wastewater lacks organic carbon, which can prevent the growth of microalgal species and potentially have an impact on how the wastewater is treated. High oxygen content in the effluent can harm microalgae cells oxidatively and prevent photosynthetic activity. Toxic accumulation by microalgae biomass and seasonal changes also inhibit the growth of microalgae after a certain time (Song et al., 2022). The most frequent and difficult problem for many hydrogen production technologies is oxygen sensitivity. When sewage microalgae are used to produce biogas, several contaminants are frequently produced, including polluted coal, tar, carbon dioxides, carbon monoxide, methane, and other pollutants (Raj et al., 2021). Therefore some changes should be made to all processes to make sure that all these pollutants are reduced at the source for a sustainable, eco-friendly environment. Another major challenge is the production costs related to large-scale biohydrogen production from sewage microalgae. To encourage the adoption of this technology, these costs need to be decreased (Ahmed et al., 2022).

8.6 Future perceptive

Because of their capability to produce different kinds of products in biorefinery techniques by utilizing wastewater, microalgae are thought to be a promising source of raw material in the future. After cultivation, biorefinery techniques are required to utilize all microalgal products. Separating the various fractions while avoiding harming any or all of the product constituents is the main challenge. Proteins, lipids, and carbohydrates are the main building blocks of microalgal species, along with other substances such as pigments. The use of sewage water for microalgal growth coupled with the biorefinery concept in downstream processing could lower manufacturing costs in the microalgae sector.

The entire cell may eventually be used to produce food, fuel, fertilizer, and other products, and wastewater may eventually replace the costs of adding synthetic nutrients. Additionally, the selection of microalgal species according to their biochemical composition for effective wastewater treatment and biomass production plays a significant role in industrial applications.

Although simultaneous biogas upgrading and sludge treatment effectiveness optimization have not yet been accomplished, extensive research into the use of microalgae is ongoing due to its many benefits. Nanoparticles will play a vital role in the future because of their antimicrobial, photocatalytic, and photothermal properties, among others. One of the most significant uses of nanoparticles is milking, an area of active research (Mukherjee et al., 2021). The growth of protozoans, algal feeders, and other organisms stimulated by the cultivation of microalgae in wastewater because of their nonsterile habitats are a few challenges that still need to be resolved for the cultivation of algae.

To avoid biotic pollutants, sterilized wastewater should be taken into consideration for microalgal cultivation. To increase the production of biomass, immobilized or embedded microalgae are grown in wastewater streams. To achieve high productivity, it is necessary to supplement nutrients in wastewater when there is a low concentration of those nutrients. When some microalgae-based species are grown utilizing domestic sewage and no carbon source, CO_2 bubbling can improve algae growth. The development of cost-effective biomass harvesting methods from large-scale wastewater culture to increase its economic appeal. Future advancements in these areas may thoughtfully affect how commercially viable microalgal value-added products are. Future studies should concentrate on developing effective pretreatment techniques, determining the best cultural conditions for the species, and performing technoeconomic analyses and life cycle assessments to put their findings into practice (Ummalyma et al., 2022). Future studies should concentrate on locating the best wastewater sources, enhancing the growth environment and product accumulation, and simultaneously extracting the product or converting the biomass. To promote an ideal environment for microalgae growth, cost-effective and cost-effective pretreatment methods as well as integrated reactor systems must be developed. The toxicity and safety of the final product should be carried out following the guidelines established by the food safety authorities (Nishshanka et al., 2021).

References

Abdelfattah, A., Ali, S. S., Ramadan, H., El-Aswar, E. I., Eltawab, R., Ho, S.-H., Elsamahy, T., et al. (2022). Microalgae-based wastewater treatment: Mechanisms, challenges, recent advances, and future prospects. *Environmental Science and Ecotechnology*100205.

Ahmed, S. F., Mofijur, M., Nahrin, M., Chowdhury, S. N., Nuzhat, S., Alherek, M., Rafa, N., Chyuan Ong, H., Nghiem, L. D., & Mahlia, T. M. I. (2022). Biohydrogen production from wastewater-based microalgae: Progresses and challenges. *International Journal of Hydrogen Energy*, 47(88), 37321–37342. Available from https://doi.org/10.1016/j.ijhydene.2021.09.178.

Arrojo, M. Á., Regaldo, L., Orquín, J. C., Figueroa, F. L., & Abdala Díaz, R. T. (2022). Potential of the microalgae Chlorella fusca (Trebouxiophyceae, Chlorophyta) for biomass production and urban wastewater phycoremediation. *AMB Express*, 12(1). Available from https://doi.org/10.1186/s13568-022-01384-z.

Bahman, M., Aghanoori, M., Jalili, H., Bozorg, A., Danaee, S., Bidhendi, M. E., & Amrane, A. (2022). Effect of light intensity and wavelength on nitrogen and phosphate removal from municipal wastewater by microalgae under semi-batch cultivation. *Environmental Technology*, 43(9), 1352–1358. Available from https://doi.org/10.1080/09593330.2020.1829087.

Bose, A., O'Shea, R., Lin, R., & Murphy, J. D. (2020). A perspective on novel cascading algal biomethane biorefinery systems. *Bioresource Technology*, 304. Available from https://doi.org/10.1016/j.biortech.2020.123027.

Chaleshtori, S., Najafi, M., Shamskilani, A., Babaei, M., & Behrang. (2022). Municipal wastewater treatment and fouling in microalgal-activated sludge membrane bioreactor: Cultivation in raw and treated wastewater. *Journal of Water Process Engineering*, 49.

Chen, H., Fu, Q., Jiang, P., & Wu, C. (2022). Mitigation of photoinhibition in Isochrysis galbana by the construction of microalgal-bacterial consortia. *Journal of Applied Phycology*, 34(6), 2883–2894. Available from https://doi.org/10.1007/s10811-022-02742-x.

Cherubini, F., Jungmeier, G., Wellisch, M., Willke, T., Skiadas, I., Van Ree, R., & de Jong, E. (2009). Toward a common classification approach for biorefinery systems. *Biofuels, Bioproducts and Biorefining*, 3(5), 534–546. Available from https://doi.org/10.1002/bbb.172.

Conteratto, C., Dalzotto Artuzo, F., Inácio Benedetti Santos, O., & Talamini, E. (2021). Biorefinery: A comprehensive concept for the sociotechnical transition toward bioeconomy. *Renewable and Sustainable Energy Reviews*, 151. Available from https://doi.org/10.1016/j.rser.2021.111527.

Dhandayuthapani, K., Senthil Kumar, P., Yi Chia, W., Wayne Chew, K., Karthik, V., Selvarangaraj, H., Selvakumar, P., Sivashanmugam, P., & Loke Show, P. (2022). Bioethanol from hydrolysate of ultrasonic processed robust microalgal biomass cultivated in dairy wastewater under optimal strategy. *Energy*, 244. Available from https://doi.org/10.1016/j.energy.2021.122604.

Dragone, G. (2022). Challenges and opportunities to increase economic feasibility and sustainability of mixotrophic cultivation of green microalgae of the genus Chlorella. *Renewable and Sustainable Energy Reviews*, 160112284.

Ferreira, A. F. (2017). Biorefinery concept. *Biorefineries* (pp. 1–20). Cham: Springer.

Ghaffar, I., Deepanraj, B., Syam Sundar, L., Vo, D.-V. N., Saikumar, A., & Hussain, A. (2022). A review on the sustainable procurement of microalgal biomass from wastewaters for the production of biofuels. *Chemosphere*137094.

Ghosh, U. K. (2018). An approach for phycoremediation of different wastewaters and biodiesel production using microalgae. *Environmental Science and Pollution Research*, 25(19), 18673–18681.

Goswami, R. K., Agrawal, K., Upadhyaya, H. M., Gupta, V. K., & Verma, P. (2022). Microalgae conversion to alternative energy, operating environment and economic footprint: An influential approach towards energy conversion, and management. *Energy Conversion and Management*, 269. Available from https://doi.org/10.1016/j.enconman.2022.116118.

Goveas, L. C., Nayak, S., Vinayagam, R., Show, P. L., & Selvaraj, R. (2022). Microalgal remediation and valorisation of polluted wastewaters for zero-carbon circular bioeconomy. *Bioresource Technology*, 365. Available from https://doi.org/10.1016/j.biortech.2022.128169.

Guldhe, A., Kumari, S., Ramanna, L., Ramsundar, P., Singh, P., Rawat, I., & Bux, F. (2017). Prospects, recent advancements and challenges of different wastewater streams for microalgal cultivation. *Journal of Environmental Management*, 203, 299–315. Available from https://doi.org/10.1016/j.jenvman.2017.08.012.

Han, F., Pei, H., Hu, W., Song, M., Ma, G., & Pei, R. (2015). Optimization and lipid production enhancement of microalgae culture by efficiently changing the conditions along with the growth-state. *Energy Conversion and Management*, 90, 315–322. Available from https://doi.org/10.1016/j.enconman.2014.11.032.

Kashyap, M., & Kiran, B. (2021). Milking microalgae in conjugation with nano-biorefinery approach utilizing wastewater. *Journal of Environmental Management*, 293, 112864. Available from https://doi.org/10.1016/j.jenvman.2021.112864.

Khan, A. A., Gul, J., Naqvi, S. R., Ali, I., Farooq, W., Liaqat, R., AlMohamadi, H., Štěpanec, L., & Juchelková, D. (2022). Recent progress in microalgae-derived biochar for the treatment of textile industry wastewater. *Chemosphere*, 306. Available from https://doi.org/10.1016/j.chemosphere.2022.135565.

Kumar, V., & Chopra, A. K. (2012). Monitoring of physico-chemical and microbiological characteristics of municipal wastewater at treatment Plant, Haridwar City(Uttarakhand) India. *Journal of Environmental Science and Technology*, 5, 109–118.

Kusmayadi, A., Lu, P.-H., Huang, C.-Y., Leong, Y. K., Yen, H.-W., & Chang, J.-S. (2022). Integrating anaerobic digestion and microalgae cultivation for dairy wastewater treatment and potential biochemicals production from the harvested microalgal biomass. *Chemosphere*, 291. Available from https://doi.org/10.1016/j.chemosphere.2021.133057.

Liu, Yu, & Yildiz, I. (2018). The effect of salinity concentration on algal biomass production and nutrient removal from municipal wastewater by Dunaliella salina. *International Journal of Energy Research*, 42(9), 2997–3006.

Luan, D. S. L., Hoffmann, M. T., & Daniel, L. A. (2019). Microalgae cultivation for municipal and piggery wastewater treatment in Brazil. *Journal of Water Process Engineering*, 31100821.

Masindi, V., Fosso-Kankeu, E., Mamakoa, E., Nkambule, T. T., Mamba, B. B., Naushad, M., Pandey, S., et al. (2022). Emerging remediation potentiality of struvite developed from municipal wastewater for the treatment of acid mine drainage. *Environmental Research*, 210, 112944. Available from https://doi.org/10.1016/j.envres.2022.112944.

Mishra, H., Gaurav, G., Khandelwal, C., Dangayach, G. S., & Rao, P. N. (2021). Environmental assessment of an Indian municipal wastewater treatment plant in Rajasthan. *International Journal of Sustainable Engineering*, 14(5), 953–962. Available from https://doi.org/10.1080/19397038.2020.1862349.

Mishra, S., & Mohanty, K. (2019). Comprehensive characterization of microalgal isolates and lipid-extracted biomass as zero-waste bioenergy feedstock: An integrated bioremediation and biorefinery approach. *Bioresource Technology*, 273, 177–184. Available from https://doi.org/10.1016/j.biortech.2018.11.012.

Mukherjee, A., Sarkar, D., & Sasmal, S. (2021). A review of green synthesis of metal nanoparticles using algae. *Frontiers in Microbiology*, 12.

Nawarkar, C. J., & Salkar, V. D. (2019). Solar powered electrocoagulation system for municipal wastewater treatment. *Fuel*, 237, 222–226. Available from https://doi.org/10.1016/j.fuel.2018.09.140.

Neveux, N., Magnusson, M., Mata, L., Whelan, A., de Nys, R., & Paul, N. A. (2016). The treatment of municipal wastewater by the macroalga Oedogonium sp. and its potential for the production of biocrude. *Algal Research*, 13, 284–292. Available from https://doi.org/10.1016/j.algal.2015.12.010.

Nishshanka, G. K. S. H., Liyanaarachchi, V. C., Premaratne, M., Nimarshana, P. H. V., Ariyadasa, T. U., & Kornaros, M. (2021). Wastewater-based microalgal biorefineries for the production of astaxanthin and co-products: Current status, challenges and future perspectives. *Bioresource Technology*, 342. Available from https://doi.org/10.1016/j.biortech.2021.126018.

Nur, M. M. A., & Buma, A. G. J. (2019). Opportunities and challenges of microalgal cultivation on wastewater, with special focus on palm oil mill effluent and the production of high value compounds. *Waste and Biomass Valorization*, 10(8), 2079–2097.

Purba, L. D. A., Othman, F. S., Yuzir, A., Mohamad, S. E., Iwamoto, K., Abdullah, N., Shimizu, K., & Hermana, J. (2022). Enhanced cultivation and lipid production of isolated microalgae strains using municipal wastewater. *Environmental Technology & Innovation*, 27. Available from https://doi.org/10.1016/j.eti.2022.102444.

Rahman, Md. M., Hosano, N., & Hosano, H. (2022). Recovering microalgal bioresources: A review of cell disruption methods and extraction technologies. *Molecules (Basel, Switzerland)*, 27(9), 2786.

Raj, S., Kuniyil, A. M., Sreenikethanam, A., Gugulothu, P., Jeyakumar, R. B., & Bajhaiya, A. K. (2021). Microalgae as a source of mycosporine-like amino acids (MAAs); advances and future prospects. *International Journal of Environmental Research and Public Health*, 18(23). Available from https://doi.org/10.3390/ijerph182312402.

Ren, Y., Sun, H., Deng, J., Huang, J., & Chen, F. (2021). Carotenoid production from microalgae: Biosynthesis, salinity responses and novel biotechnologies. *Marine Drugs*, 19(12). Available from https://doi.org/10.3390/md19120713.

Salama, E.-S., Kurade, M. B., Abou-Shanab, R. A. I., El-Dalatony, M. M., Yang, I.-S., Min, B., & Jeon, B.-H. (2017). Recent progress in microalgal biomass production coupled with wastewater treatment for biofuel generation. *Renewable and Sustainable Energy Reviews*, 79, 1189–1211. Available from https://doi.org/10.1016/j.rser.2017.05.091.

Salla, A. C. V., Margarites, A. C., Seibel, F. I., Holz, L. C., Brião, V. B., Bertolin, T. E., Colla, L. M., & Al Vieira Costa, J. (2016). Increase in the carbohydrate content of the microalgae Spirulina in culture by nutrient starvation and the addition of residues of whey protein concentrate. *Bioresource Technology*, 209, 133–141.

Sharma, D., & Kansal, A. (2011). Water quality analysis of River Yamuna using water quality index in the national capital territory, India (2000–2009). *Applied Water Science*, 1, 147–157. Available from https://doi.org/10.1007/s13201-011-0011-4.

Shengnan, L., Qu, W., Chang, H., Li, J., & Ho, S.-H. (2022). Microalgae-driven swine wastewater biotreatment: Nutrient recovery, key microbial community and current challenges. *Journal of Hazardous Materials*129785.

Singh, D. V., Upadhyay, A. K., Singh, R., & Singh, D. P. (2022). Microalgal competence in urban wastewater management: Phycoremediation and lipid production. *International Journal of Phytoremediation*, 24(8), 831–841. Available from https://doi.org/10.1080/15226514.2021.1979463.

Song, Y., Wang, L., Qiang, X., Gu, W., Ma, Z., & Wang, G. (2022). The promising way to treat wastewater by microalgae: Approaches, mechanisms, applications and challenges. *Journal of Water Process Engineering*, 49. Available from https://doi.org/10.1016/j.jwpe.2022.103012.

Ummalyma, S. B., Sirohi, R., Udayan, A., Yadav, P., Raj, A., Sim, S. J., & Pandey, A. (2022). Sustainable microalgal biomass production in food industry wastewater for low-cost biorefinery products: a review. *Phytochemistry Reviews*. Available from https://doi.org/10.1007/s11101-022-09814-3.

Wang, L., Min, M., Li, Y., Chen, P., Chen, Y., Liu, Y., Wang, Y., & Ruan, R. (2010). Cultivation of green algae Chlorella sp. in different wastewaters from municipal wastewater treatment plant. *Applied Biochemistry and Biotechnology*, 162(4), 1174–1186. Available from https://doi.org/10.1007/s12010-009-8866-7.

Yao, S., Lyu, S., An, Y., Lu, J., Gjermansen, C., & Schramm, A. (2019). Microalgae-bacteria symbiosis in microalgal growth and biofuel production: A review. *Journal of Applied Microbiology*, 126(2), 359–368. Available from https://doi.org/10.1111/jam.14095.

Zhang, L. Q., Jiang, X., Rong, H.-W., Wei, C.-H., Luo, M., Ma, W.-C., & Ng, H.-Y. (2022). Exploring the carbon and nitrogen removal capacity of a membrane aerated biofilm reactor for low-strength municipal wastewater treatment. *Environmental Science: Water Research & Technology*, 8(2), 280–289. Available from https://doi.org/10.1039/D1EW00724F.

CHAPTER
9

Bioenergy from waste biomass

Subhrajit Roy and Saikat Chakraborty

Biological Systems Engineering, Plaksha University, Mohali, Punjab, India

9.1 Introduction

9.1.1 Role of bioenergy in combating climate change and the greenhouse effects

As a result of growing anthropogenic activity, there is a constant search for clean energy and fuel supply that would help maintain our ultimate survival by combating climate change and the greenhouse effect. Fossil fuels are the most widely used fuel in the world due to their high calorific value and ease of transportation, but unfortunately, they are also major sources of pollution (Prat et al., 2016). This has led researchers and scientists worldwide to become interested in developing new alternative energy sources.

The most efficient method of mitigating the severe effects of climate change is by utilizing renewable energy sources. In recent years, the capacity for renewable energy has been growing at a faster rate than that of nonrenewable sources (Harmsen et al., 2010). While solar and wind energy have made significant progress, they still cannot meet the world's energy demands on their own. Bioenergy appears to be the most promising alternative for filling the energy gap or at least reducing it. To meet the global commitment of limiting global warming to below 2°C, a combination of measures will be required, including expanding the use of bioenergy, implementing land-based initiatives such as reforestation, and utilizing carbon storage and capture techniques (Rogelj et al., 2016).

Our dependency on fossil fuels can be significantly decreased by using bioenergy as a source of power in both the industrial and transportation sectors (Miguel et al., 2016). It can effectively replace fossil fuels because it is portable and has no impact on the environment. However, to maintain sustainable development, the cost and competitiveness of producing bioenergy should be assessed considering social and environmental benefits in addition to economic gains, as well as the growth of local skills. Unfortunately, encouraging the development of bioenergy has had some unintended negative effects. Concerns including food poverty, water scarcity, biodiversity loss, land degradation, and desertification have gotten worse as a result of the conversion of tropical forests and other significant ecosystems into bioenergy-producing regions. According to the IPCC's Special Report on Climate Change and Land, scale and context are important things to consider when comparing the advantages and disadvantages of producing bioenergy (Rogelj et al., 2016).

9.1.2 Waste biomass as a potential source of bioenergy

This study investigates the potential for producing bioenergy from waste biomass and the obstacles and advancements in the technological processes involved. Global issues such as environmental concerns and food security are of great concern to us all. Creating bioenergy and biomaterials from waste biomass can help maintain a balance between energy and the environment while also replacing fossil fuels as the primary source of fuel, resulting in a cleaner and more environmentally friendly world. Over the past decade, bioenergy has received significant attention, with several countries undertaking extensive research initiatives to define and quantify green technology. Generating bioenergy from waste biomass has numerous benefits and can aid in mitigating environmental issues related to the use of fossil fuels, waste management, and urbanization.

The first step in producing bioenergy from waste biomass involves identifying and classifying the various waste biomass types based on their unique characteristics. Waste biomass classification primarily depends on its physical properties and chemical composition (Paul & Chakraborty, 2018). Physical properties such as crystallinity, porosity, viscosity, and particle size are crucial in categorizing waste biomass (Roy & Chakraborty, 2019). On the other hand, waste biomass's chemical composition determines its cellulose, hemicellulose, lignin, lipid, moisture, and ash content. These factors play a significant role in determining the type of bioenergy that can be generated from waste biomass and the most efficient bioenergy production technique. Additionally, bioenergy can be grouped into three categories, namely solid, liquid, and gaseous fuels. The majority of solid fuels comprise biochar and some value-added drop-in fuels, while liquid fuels include bioethanol, biodiesel, and furanic fuels. Biomethane and biohydrogen are examples of gaseous fuels (Ahmed et al., 2021; Cucchiella et al., 2018).

The subsequent sections will cover different approaches to pretreatment methods for waste biomass, including mechanical techniques such as drying, milling, and sonication, physicochemical methods such as acid, alkali, ionic liquid, steam, and others, as well as biological methods such as bacterial and enzymatic, and catalytic methods such as microwave and oil-bath (Paul & Chakraborty, 2019; Roy & Chakraborty, 2019). Additionally, we will discuss various conversion techniques such as thermochemical conversions (pyrolysis, combustion, and hydrothermal liquefaction) and biochemical conversions (anaerobic digestion, enzymatic, and fermentation) (Gaikwad & Chakraborty, 2014). With ongoing modifications and improvements in these processes, the production of bioenergy from waste biomass will continue to move closer to resolving the energy crisis, thus paving the way for a better, more sustainable future for humanity.

9.2 Waste biomass: sources, classification, and characterization

Waste biomasses are seen to be a great source for producing sustainable bioenergy because they are affordable and easily accessible in some areas, such Latin America and Asia (Bhatt & Bal, 2019). Environmental laws are being drafted to control the waste produced from numerous sources, such as agriculture, forestry, wood production, and municipal pruning. The prospect of using this waste continues to grow significantly. Particularly in emerging economies where the need for biofuels and chemicals is growing and waste biomass is more readily available than in wealthy economies, the chemical industry is becoming more and more interested in using renewable feedstocks. The chemical and physical characteristics of each type of biomass determine its worth. Reliance on imported petroleum oils can be reduced by using local waste biomass as a feedstock, increasing marketability, generating employment opportunities, and fostering rural development in developing nations. By 2030, it is anticipated that lignocellulosic waste would be used to make 25% of chemicals and 20% of transportation fuels (Clauser et al., 2021).

9.2.1 Various sources of waste biomass

Waste biomass for energy production is receiving a lot of attention as a result of growing global concerns about environmental pollution and the paucity of energy resources (Pishvaee et al., 2021). It is now known that environmentally hazardous waste biomass can be transformed into usable energy utilizing workable technology. Reducing reliance on fossil fuels and addressing environmental pollution issues are both achieved by using waste biomass as an energy source (Gao et al., 2018).

Biofuels can be made from waste products from the processing of wood, including sawdust, chips, and discarded logs. The waste and sawdust produced by paper and sawmills can be used to make ethanol and boiler fuel (Naik et al., 2010). Straw is an important source of biomass energy in China and is produced as a by-product of harvesting food crops. For producing biobutanol, maize stover can also be transformed into fermentable sugars. In addition to palm kernel press cake, a by-product of the extraction of palm oil, which can be used to make bioethanol via fermentation, sugarcane bagasse and leaves are viable candidates for the production of bioethanol in tropical nations (Li et al., 2007).

Virgin oils of the highest quality for use in food are made from precisely selected feedstocks, while waste oils such as leftover cooking oils can be utilized to make biodiesel at a lower cost. The material cost of producing biodiesel is reduced by 60%–90% when used in place of virgin oil (Zhang et al., 2003). Reusing waste oils also solves the issue of getting rid of a lot of frying oil that is no longer safe to eat because of its high free fatty acid concentration. The quality of old edible oils is comparable with that of fresh oils, and straightforward pretreatments such as heating and filtration can get rid of water and other impurities before transesterification (Talebian-Kiakalaieh et al., 2013).

Macroalgae (seaweeds) and microalgae are the two categories into which algae can be divided. While microalgae are microscopic unicellular algae found in waterbodies, macroalgae are giant multicellular algae that are found in ponds. Microalgae are a prospective source of oil because of their high lipid accumulation and quick growth rates, whereas macroalgae include a variety of medicinal chemicals (Lee et al., 2019). Microalgae do not contend with macroalgae for extensive freshwater resources or arable land. Spent microalgae biomass can be transformed into biofuels after oils or other valuable components have been extracted. After the goal products, such as oils or other useful compounds, have been removed from the microalgae biomass, discarded biomass from the microalgae, such as biomass leftovers and waste materials, can be converted into biofuels.

9.2.2 Classification of waste biomass

Forestry and agriculture make up the two primary groups of waste biomass. Fuelwood and industrial roundwood for paper, building, and furniture are examples of forestry biomass. The conversion of these resources into a variety of usable goods, such as food, animal feed, wood products, processed biomass, and biomass fuels, determines how these resources are transformed (Stoeglehner & Narodoslawsky, 2009). Biomass from agricultural waste is divided into residues, crops, and livestock products, all of which can be used as food, fuel, building materials, and feed. These biomass wastes are produced all over the world and are renewable (Usmani et al., 2021).

Unlike biomass that is purposefully farmed for energy use, biomass residues and waste are produced as by-products during the manufacture and consumption of desirable raw items. The types of biomass residues include primary, secondary, and tertiary (Li et al., 2007). While secondary residues are created during the processing of food, primary residues are created during the planting of food crops and forest products (Fig. 9.1). After a biomass-derived product has been used, tertiary residues are still available and come in the form of municipal solid waste, sewage sludge, and/or wastewater (Born et al., 2014). Utilizing these resources to produce the energy not only aids in lowering environmental pollution but also lessens reliance on fossil fuels. Among the numerous types of biomass residues and waste, microalgae biomass, waste cooking oils, and wood and agricultural residues have all demonstrated remarkable potential (Lee et al., 2019).

9.2.3 Characterization of waste biomass

Waste biomasses are a viable source of bioenergy precursors. However, producing biofuels from waste especially lignocellulosic biomasses is difficult. The primary components of waste lignocellulosic biomass are cellulose, lignin, and hemicellulose, whose compositions differ depending on the source and have a substantial impact on the bioenergy production process (Roy & Chakraborty, 2020). As a result, a compositional study of the

FIGURE 9.1 Types of waste biomass residues.

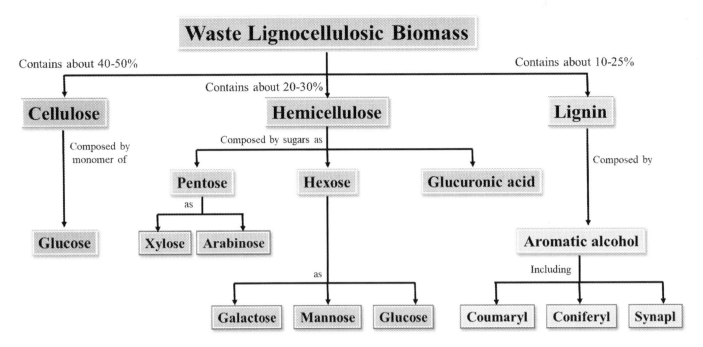

FIGURE 9.2 Composition of waste lignocellulosic biomass.

waste biomass is utmost important to determine the proportions of cellulose, lignin, hemicelluloses, and other components (Fig. 9.2). The synthesis of biofuel precursors from lignocellulosic waste is determined not only by the content of the biomass but also by its physical attributes, which include crystallinity, porosity, Degree of Polymerization (DP), and particle size (Paul & Chakraborty, 2018). High crystallinity, DP, and low surface area prevent the enzymes and the chemicals from accessing the crystalline cellulose polymer chains and rupturing the β-(1—4)-glycosidic bonds between their monomeric glucose units (Yang et al., 2013). Therefore crystallinity, DP, and porosity of waste lignocellulosic biomass are also measured using various techniques.

9.2.3.1 Determination of cellulose, hemicellulose, and lignin content

The waste biomass is crushed into a fine powder with an average particle size of 100 mesh and mixed with 10 mL of a 0.275 M H_2SO_4 solution and stirred for 24 hours at 150 rpm and 25°C. The solution mixture is then autoclaved at 118°C (Toribio-Cuaya et al., 2014) for 30 minutes. The flask is then removed from the autoclave, and the reactor is cooled by reducing the pressure. The supernatant is filtered using filter paper after the filtrate has been centrifuged at 3000 rpm for 15 minutes. A UV-vis spectrophotometer is used to measure the concentration of lignin that is soluble in acids.

The delignification procedure is carried out next utilizing an alkaline media and hydrogen peroxide as a catalyst. A 50 mL flask containing the oven-dried waste biomass is added to with 6 mL of 0.375 M NaOH and 90 μL of 12.8 M H_2O_2. The flask's temperature is held constant at 50°C, and stirring is done for 3 hours at a speed of 150 rpm. When the reaction is complete, the solid is recovered by filtration and removed the NaOH by washing. The solid is then dried for 10 hours at 80°C and weighed to determine its cellulose content (Paul & Chakraborty, 2018). To precipitate the alkaline lignin, the filtered liquid is acidified with CH_3COOH (50% v/v) and centrifuging the solution for 20 minutes at 3000 rpm. The precipitate is dried at 80°C for 8 hours after the supernatant has been removed. The alkaline lignin's weight is then calculated.

A 250 mL beaker containing 500 mg of waste lignocellulosic biomass is washed 8–10 times with 5 mL of hot water using filter papers. After that, 5 mL of anhydride C_2H_5OH is utilized to wash the biomass. To get rid of the sample's ethanol, the recovered solid is filtered and dried for 8 hours at 80°C. The fractions A and B, which make up the hemicelluloses, are extracted using an alkaline extraction method after the drying process (Roy & Chakraborty, 2019). The dried biomass is placed in a 50 mL flask, and 10 mL of a 1 M NaOH solution is then added to perform the alkaline extraction. The mixture is then agitated for 24 hours at 100 rpm at 25°C. Using 50% (v/v) acetic acid, the pH of the filtrate is reduced from 9 to 4. The resulting suspension is then agitated at ambient temperature for 15 minutes and centrifuged for 20 minutes at 3000 rpm. After draining the supernatant, the solid is dried for 8 hours at 90°C to recover the hemicellulose A. The product's weight is then measured. The supernatant obtained from the first part is concentrated during the evaporation procedure. To precipitate the hemicellulose B, the supernatant is gradually added to a CH_3OH solution in a 1:3 by volume ratio and centrifuged at 3000 rpm. By draining the CH_3OH supernatant and drying the solid at 90°C, hemicellulose B can be recovered. The weight of hemicellulose B is then measured. Hemicellulose fractions A and B together make up the total amount of hemicelluloses present in the waste lignocellulose material (Roy & Chakraborty, 2020).

9.2.3.2 Determination of degree of polymerization (DP)

The reducing end concentration method for measuring DP needs less time and a lower sample volume. As a result, we prefer to use the reducing end concentration method to determine the DP of cellulose in the waste lignocellulosic biomass (Dutta & Chakraborty, 2015). DP is calculated as the ratio of the phenol-sulfuric acid method-measured glycosyl monomer concentration and the modified 2, 2' bicinchoninate (BCA) method-measured reducing end concentration.

Solution A and Solution B are mixed in a 1:1 (by volume) ratio to form the BCA working solution. For solution A, distilled water is used to dissolve disodium 2,2-bicinchoninate, sodium carbonate, and sodium bicarbonate, whereas solution B uses $CuSO_4 0.5H_2O$ and L-serine. To prevent the precipitation of cellulose, 1 ML of the sample and 1 ML of the BCA working solution are added to a vial and incubated at 75°C for 30 minutes. The reaction mixture containing cellulose particles is placed into a centrifuge tube when the vials have cooled to room temperature. The centrifuge is then run for 5 minutes at 5000 rpm. The glucose concentration is then determined using a standard glucose calibration plot that has been plotted in the range of $0-50\ \mu M$, and the absorbance of the supernatant is measured at 560 nm (Paul & Chakraborty, 2019).

The reaction mixture containing cellulose particles is centrifuged for 5 minutes at 5000 rpm when the vials have cooled to room temperature. The supernatant's absorbance is then measured at 560 nm, and the amount of glucose is determined using a calibration plot for standard glucose that has been made with a range of $0-50\ \mu M$.

9.2.3.3 Determination of moisture and ash content

A muffle furnace is used to keep 1 g of waste lignocellulosic biomass at 550°C for up to 5 hours to completely burn the biomass. The biomass is weighted upon combustion. By dividing the initial dry waste lignocellulosic biomass weight by the ash weight, the ash percentage may be computed. One gram of lignocellulosic biomass is placed at 105°C for 5 hours in a hot air oven to measure its moisture content. The weight difference between the waste biomass and the dried biomass determines the moisture content (Paul & Chakraborty, 2018).

9.2.3.4 XRD, BET-BJH, and particle size analysis

Utilizing a diffractometer and Cu-K radiation at 40 kV and 30 mA with 2θ angle scanning from 10 degrees to 90 degrees, waste biomass's powder XRD patterns were obtained. The peak height approach is used to determine the relative concentration of crystalline material in each sample (Park et al., 2010; Thygesen et al., 2005). A BET surface area analyzer is used to calculate the surface area and the pore size. The samples are degassed for 8 hours at 110°C under high vacuum. The Barrett–Joyner–Halenda (BJH) method is used to calculate the specific surface area, whereas the Brunauer–Emmett–Teller (BET) method is used to calculate the pore surface area, pore volume, and average pore diameter (Dutta & Chakraborty, 2016). By estimating the average particle size and the particle size distribution using the Dynamic Light Scattering (DLS) approach and a Zetasizer nanoparticle analyzer with laser at a wavelength of 633 nm at 25°C, the particle size of the waste biomass is determined. The sample is placed in a cuvette, and the laser's scattering power is measured at a 90 degree angle with 80 seconds counting period (Tuoriniemi et al., 2014).

9.3 Sources and classification of bioenergy obtained from waste biomass

In recent times, many studies have been conducted to assess the future demand and supply of bioenergy from biomass waste. Overall, it seems that the world's potential for bioenergy is sufficient to meet global energy needs by 2050 (Reid et al., 2020). However, meeting this demand will require significant investment in equipment, infrastructure, and research and development. There is a growing global awareness of the need for sustainable use of biomass, as evidenced by research and policy initiatives in many countries (Usmani et al., 2021). Based on the sources and state of bioenergy derived from waste biomass, they can be broadly categorized into three types: solid, liquid, and gaseous fuels. Each of these categories will be thoroughly discussed in the following sections.

9.3.1 Solid fuels

Residues produced by agroindustrial activities are a significant type of solid waste, especially in countries where agriculture and agroindustries are major economic sectors. Wine and olive oil production, as well as rice processing, are lucrative agroindustries that generate substantial amounts of by-products and solid waste. Recently, there has been growing interest in using these waste materials as resources for both energy and high-value products, as part of a circular economy. Waste biomass from agroindustrial activities, as well as municipal waste, could be highly promising in this regard due to their composition and abundance (Pellera et al., 2021).

9.3.1.1 Biochar

Biochar production by pyrolysis is a promising way to repurpose solid waste biomass (Tag et al., 2016). Biochar is appealing due to its potential to reduce environmental and health risks from waste biomass and its various useful applications such as water contaminant removal, soil amendment for enhanced properties, and remediation of polluted areas. The versatility of biochar stems from its properties that are dependent on the pyrolysis conditions and the type of feedstock utilized (Pellera et al., 2021).

When characterizing biochar, various properties are analyzed to determine their potential applications. Proximate and ultimate composition provides information on the stability of biochar in soil, with higher fixed carbon, aromaticity, and hydrophobicity leading to greater carbon sequestration (Rehrah et al., 2016). Biochars produced at lower temperatures are more hydrophilic and increase soil water-holding capacity. Electrical conductivity of biochar indicates its impact on soil quality and fertility, while biochars with high surface functional groups and cation exchange capacity can improve nutrient exchange sites and prevent nutrient leaching in soils.

9.3.1.2 Drop-in fuels

Drop-in fuels are those that may be used directly as an alternative of typical fossil fuels in prevailing engines without requiring infrastructure or engine modifications. Gasification, pyrolysis, and hydrothermal liquefaction are a few of the processes that can turn waste biomass into drop-in fuels such as biofuels (Danquah et al., 2018). Through these procedures, the waste biomass is broken down into its constituent sugars, gases, and oils, which are subsequently transformed into a form that may be used as fuel. Drop-in fuels made of waste biomass have the ability to decrease greenhouse gas emissions, lessen reliance on fossil fuels, and keep waste out of landfills. Drop-in fuels must overcome obstacles including feedstock availability, economic viability, and technical scalability to be widely adopted and commercialized. To enhance existing biofuel pathways for drop-in fuel production, it is necessary to better understand the economic viability, technical scalability, and environmental effects of drop-in fuel production. To evaluate the economic and environmental implications of proposed processes, technoeconomic analyses (TEA) and life cycle analyses (LCA) are frequently utilized (Kargbo et al., 2021).

9.3.2 Liquid fuels

Depending on the variety of biomass employed in their production, liquid biofuels are categorized into three generations. Food-grade biomass that is abundant in sugars, carbohydrates, or lipids is the source of first-generation biofuels. However, this biomass must be used for food due to the growing human population, making it a scarce resource. Algae and cyanobacteria are examples of aquatic biomass that is used to make third-generation biofuels. Lignocellulosic biomass (LCB) or nonedible biomass is used to make second-generation biofuels (Fig. 9.3). The main emphasis is on 2G and 3G liquid biofuels when assessing the food versus fuel argument (Roy & Chakraborty, 2019). 2G biofuels require less land, water, and fertilizers than 1G and 3G biofuels. Large-scale biofuel production is possible due to their low production and commercialization costs as well as the fact that they do not compete with food.

9.3.2.1 Bioethanol

With a global production of 87.2 billion liters of bioethanol in 2013, it has gained popularity as a biofuel (Rossetti, 2016). The top two producers are Brazil and the United States, which make bioethanol mostly from sugar cane or maize, respectively. However, because these sources compete with the food and feed chain, this has raised questions regarding social sustainability. Second-generation biofuels have been created from waste biomass or energy crops that don't compete with agriculture to solve this problem. Despite this, first-generation bioethanol is being produced in large quantities in the USA and Europe. Compared with 7% a decade prior, 40% of the corn harvested in the USA in 2011 was utilized to produce bioethanol (Brunet et al., 2015).

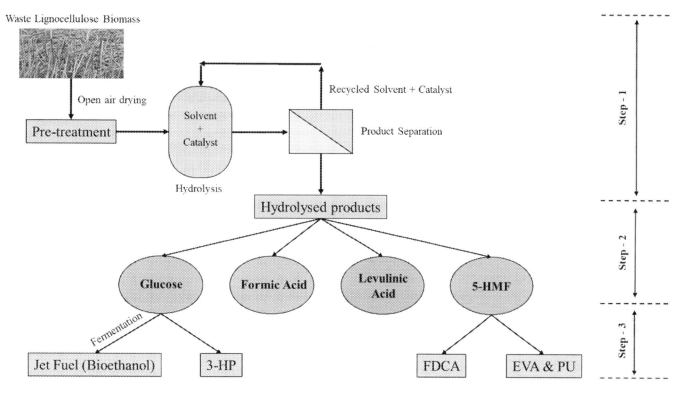

FIGURE 9.3 Synthesis of liquid biofuels from waste lignocellulosic biomass.

9.3.2.2 Biodiesel

A sustainable, eco-friendly liquid biofuel that can aid in lowering greenhouse gas emissions is biodiesel. Vegetable, nonedible, and waste oils were initially used to make it, but these feedstocks have disadvantages including requiring a lot of land and labor and being expensive (Nathan et al., 2019). In recent experiments, oleaginous bacteria have been employed to produce biodiesel from waste lignocellulosic substrates. Under stressful conditions, these microorganisms acquire more lipid, which is similar to vegetable oils. A sustainable biorefinery requires a variety of feedstocks, including vegetable oils, nonedible oils, microalgae, fungi, yeast, and bacteria (Subramanian et al., 2010). Pretreatment, saccharification, microbe-mediated fermentation, lipid extraction, transesterification, and biodiesel purification are all crucial phases in the synthesis of biodiesel from waste lignocellulosic substrates (Chintagunta et al., 2021).

9.3.2.3 Furanic fuels

One of the key research areas in the development of renewable energy is the feasible and effective conversion of waste biomass to platform chemicals. It is widely agreed that we must create a variety of green technologies for renewable energy (Li & Yang, 2014). From an atom-economy perspective, the conversion of waste biomass to platform chemicals may reserve all carbon atoms, making it the most effective method. Furanic fuels are regarded as the link between biomasses and important biochemicals and are excellent platform chemicals (Guo et al., 2020). The primary ingredient used to make furanic fuels is furfural. It is a chemical that may be produced from a variety of waste biomass feedstocks, including maize cobs, wheat straw, and sugarcane bagasse, and is renewable, nontoxic, and easily accessible. There are numerous ways to make furfural, including acid hydrolysis and steam explosion.

9.3.3 Gaseous fuels

Biogas is produced by the anaerobic digestion process, which is a proven renewable energy technique. Energy crops, municipal solid waste, industrial organic waste, agricultural residues (such as manure and crop residues), and other feedstocks can all be treated using this method. Biomethane can be created from properly processed biogas through purification. If appropriately upgraded, landfill gas can potentially be used as a source of biomethane. There are numerous upgrading technologies that can be used, including membrane separation, pressure swing adsorption, chemical scrubbing, and water scrubbing (Cucchiella et al., 2018).

9.3.3.1 Biomethane

Natural gas can be replaced with biomethane, a renewable energy source that also helps to reduce greenhouse gas emissions. It is also referred to as green gas and shares numerous properties with methane. Biomethane is adaptable and can be used to generate energy, be delivered through natural gas infrastructures, or be used as a vehicle fuel (Morero et al., 2017). To produce biomethane, advanced and effective technologies are required. Biomethane has many advantages, both for the environment and for enhancing economic performance.

9.3.3.2 Biohydrogen

The synthesis of valuable metabolites as coproducts has increased interest in microalgal biohydrogen production (Ahmed et al., 2021). However, issues with process engineering, limited microalgal productivity, high operating costs, and a lack of knowledge about strain capacity provide obstacles to the commercialization of microalgae-based biohydrogen generation. The biohydrogen output from microalgae can be increased by a variety of metabolic pathways; however, recent assessments lack a thorough economic analysis, making it difficult to assess the viability of the process. Biohydrogen is one of the most promising bioenergy products that can be produced from microalgal biomass. From microalgae, biohydrogen can be produced using two different methods, but combining these methods can increase output.

9.4 Catalytic conversion of waste biomass to bioenergy

After pretreatment of waste lignocellulosic biomass, hydrolysis is used to produce monomeric sugars by rupturing the β-1−4 glycosidic linkages in cellulose and hemicellulose polymer chains. Acidic, catalytic, and enzymatic hydrolysis techniques are available. Acid hydrolysis of waste biomass demands a high temperature and generates several undesirable by-products (Jeffries & Jin, 2004). By-product formation eventually diminishes product concentrations and raises process costs. The acids utilized in this technique are highly caustic, and during hydrolysis, the acids create hazardous components. On the other hand, catalytic hydrolysis is a safer and more environmentally friendly procedure than acid hydrolysis (Zhou et al., 2011). Noncorrosive catalysts are employed in the catalytic hydrolysis process. After the catalytic hydrolysis process, solvent recovery and recycling may be undertaken, making commercial production of biofuels cost-efficient (Guo et al., 2012).

9.4.1 Microwave-irradiated catalytic conversion

All facets of chemistry are now impacted by the diverse area of "green chemistry." Its main goal is to safeguard the environment by identifying novel chemical processes and pollution-reduction solutions. Technology and solvents are crucial in determining chemical process cost, safety, potential health risks, and environmental impact (Klein-Marcuschamer et al., 2011). As conventional chemical synthesis is typically slow, it cannot meet the rising demand for compound manufacturing. Chemical synthesis techniques that are quick and safe for the environment are highly desired.

These demands may be satisfied by microwave technology. Microwave technology can cut the reaction time needed to produce valuable molecules from days to minutes (Dabrowska et al., 2018). The interaction between molecules in a reaction mixture is the basis for microwave-irradiated technology, which is accelerated by the electromagnetic waves produced by the microwave. This technique depends on the polarity of the molecules and generally uses two basic reaction mechanisms: dipole rotation and ionic conduction of polar molecules (Thawarkar et al., 2015). The dipole rotation of the polar molecules and reactants in the electromagnetic field causes the microwave to irradiate a reaction mixture (Grewal et al., 2013).

9.4.2 Oil-bath-mediated catalytic conversion

Silicone oil or mineral oil with a high heat capacity is used in the oil-bath-heated catalytic conversion of waste lignocelluloses to biofuels. In small-batch reactors submerged in an oil bath, the reactions are conducted

(Paul & Chakraborty, 2018). A common heating technique is oil-bath heating, in which heat is transferred both outside (by convection and conduction) and internally (by conduction only) of the reactor. The processing time is constrained by the rate of heat transmission from the reactor surface into the material's body. As a result, the time needed for microwave-assisted conversion of waste lignocellulose is shorter than that needed for oil-bath-mediated catalytic conversion (Roy & Chakraborty, 2023).

9.5 Biochemical conversion of waste biomass to bioenergy

Apart from chemical processes, biological processes are also useful for the transformation of waste biomasses to biofuels. Waste lignocellulosic biomasses are extremely resistant to enzymatic attack because of their structural stability. Therefore, proper pretreatment is necessary to increase the accessibility of the biomass to the enzyme. The production of sugars from enzymatic hydrolysis can be improved by controlling the process conditions. The process parameters are dependent on the nature of the substrates and the enzymes (Gupta & Lee, 2009; Wood & Bhat, 1988). Enzymatic processes are preferred because of their mild reaction conditions (temperature 40°C–50°C and pH 4–5) (Dubois et al., 1956; Wilson & Walker, 1991).

9.5.1 Anaerobic digestion

Waste biomass can be converted into a variety of bioenergy products, including ethanol, butanol, methane, hydrogen, power, and biofuels. Anaerobic digestion is one such process; it entails microorganisms breaking down organic molecules without the use of oxygen. Under the appropriate conditions, the organic material goes through fermentation throughout this process and creates biogas (Mahjoub et al., 2020). The type, density, and content of the raw materials, the pH level, the absence of oxygen in the atmosphere, and the length of time the fluid spends in the tank are all significant variables that might impact the production of biogas (Rao et al., 2010). Using separate electricity and heat units at the same time or a combined heat and power (CHP) unit, the biogas can be used to produce both electricity and heat (Li et al., 2018).

Anaerobic digestion in CHP biogas power plants normally requires a number of processes in order to produce electricity. First, feasible raw materials are gathered and transported to an appropriate area where they are crushed and blended with liquid at a predetermined rate. These resources include organic waste, animal waste, and plant wastes. Before the mixture enters the digestive phase, where biogas production occurs, it is then transferred to a predigestion tank. Methane, which may be used as a fuel for generators to produce heat and electricity, makes up between 50% and 70% of the biogas produced (Piñas et al., 2018). The liquid that is left over after the biogas generation process is finished is moved to storage tanks where it can be used as fertilizer, either wet or dry (Bijarchiyan et al., 2020). The generated biogas is then sent to a gas purification system, where it is converted into energy that powers generators that provide heat and electricity. The generated electricity is subsequently connected to the power distribution network for usage by the general population.

9.5.2 Enzymatic conversion

After pretreating the waste lignocellulosic biomass to extract the cellulose from the lignin and hemicelluloses, enzyme hydrolysis is carried out. The enzymatic hydrolysis of cellulose generated from waste biomass is carried out using the cellulase enzyme from *Trichoderma viride* (Pal & Chakraborty, 2013). The most effective enzymes for enzymatic hydrolysis of cellulose are cellulases from fungi. *Trichoderma ressei, T. viride, Sporotrichum pulverulentum*, and *Trichoderma koningii* are only a few of the microorganisms from which the enzyme can be produced (Gaikwad & Chakraborty, 2013). A multicomponent enzyme system made up of 1,4-β-D-glucan cellobiohydrolase (exo I, II, and III), 1,4-β-D-glucanohydrolase (endo I, II, III, IV, V, and VII), and β-D-glucoside glucohydrolase (β-glucosidase) is produced by *T. viride*. Endoglucanase, exoglucanase, and β-glucosidase, among other enzymes, must work in sync to produce soluble sugars from cellulose. Exoglucanase reacts with both the nonreducing and reducing ends of cellulose to create cellobiose, while endoglucanase randomly breaks the β-(1–4) glycosidic link of cellulose. Cellobiose, on the other hand, reacts with β-glucosidase and splits the dimer into two monomeric glucose molecules.

During enzymatic hydrolysis, soluble enzymes are moved from the liquid phase's bulk to the solid–liquid interface by convection before they diffuse into the solids' pores as a result of capillary action, where the enzyme's Carbohydrate-Binding Domain (CBD) binds to the insoluble substrate at the pore scale (Dutta & Chakraborty, 2016). As a result, during hydrolysis, an enzyme–substrate complex is formed. The catalytic domain (CD) of the cellulase enzyme is where the glycosidic link in the cellulose chain is broken, resulting in short carbohydrate chains (Sun et al., 1998). After that, the soluble sugars spread from the pores into the liquid in its entirety (Lee & Fan, 1982). Following these solid-phase processes, the soluble sugars are hydrolyzed by enzymes in the liquid phase to create glucose monomers.

9.5.3 Fermentation

The process of fermenting involves employing yeasts, bacteria, or other microorganisms to turn glucose and other fermentable carbohydrates into ethanol. The production of bioethanol employs a variety of fermentation techniques. The most popular fermentation techniques include separate hydrolysis and fermentation (SHF), simultaneous saccharification and fermentation (SSF), simultaneous saccharification and cofermentation (SSCF), and separate hydrolysis and cofermentation (SHCF) (Azhar et al., 2017). Direct microbial fermentation (DMC) is another technique that is frequently used.

Direct microbial conversion (DMC) involves simultaneous fermentation and cellulose hydrolysis to produce glucose. The three main processes—production of enzymes, hydrolysis, and fermentation—are all combined in this process. To make this method cost-effective, advanced pretreatment techniques are applied (Christakopoulos et al., 1992). However, due to the high amounts of by-products produced by this sort of fermentation process, the amount of ethanol produced is low.

Separate hydrolysis and fermentation (SHF) is a procedure in which the two processes are carried out in different vessels. The biomass used in SHF is first hydrolyzed into hexose or pentose, and the sugars are subsequently fermented to ethanol. Both enzymatic hydrolysis and fermentation take place at temperatures that are ideal for each process (30°C for fermentation and 40°C for hydrolysis). To increase the process's performance and efficiency, it is carried out under ideal reaction circumstances (Philippidis et al., 1993).

During the simultaneous saccharification and fermentation process (SSF), hydrolysis and fermentation are carried out simultaneously in a single vessel (Deshpande et al., 1981). The benefit of this process over other processes is the decrease in enzyme inhibition, which raises the amount of glucose and cellobiose. As a result, this method produces ethanol quickly and with a high yield. However, the ideal conditions for hydrolysis and fermentation differ, making it challenging to monitor the event as it happens. As the ideal temperatures for hydrolysis and fermentation differ, thermotolerant yeast strains are utilized for SSF (Ghosh et al., 1984).

Simultaneous saccharification and cofermentation (SSCF) is the method of fermenting pentose and hexose sugar in the same reactor (Canilha et al., 2012). This approach can lower the cost of producing ethanol as the reaction can take place in the same reactor without wasting any of the soluble sugars generated during hydrolysis. This method produces ethanol in a short amount of time. In SSCF, ethanol yield can also be increased.

Hexose and pentose sugars are fermented together in the separate hydrolysis and cofermentation (SHCF) process after the substrate has undergone separate hydrolysis (Erdei et al., 2012). This form of fermentation can result in complete pentose sugar fermentation. Ethanol is also produced by consolidated bioprocessing (CBP) in a cell recycle batch fermenter (CRBF) (Kida et al., 1991). The price for producing inoculums is decreased by this method. Stable operation and long-term ethanol productivity are two additional benefits of this method (Vazirzadeh & Mohsenzadeh, 2012).

9.6 Thermochemical conversion of waste biomass to bioenergy

In order to achieve existing emission targets, it is imperative to reduce the discharge of greenhouse gases and other harmful compounds. Waste biomass can be used to produce bioenergy. The amount of renewable energy sources can be increased by using waste biomass as a source of energy. Combustion, pyrolysis, and hydrothermal liquefaction are the three most often employed thermochemical processes for converting biomass into energy or chemicals (Mlonka-Mędrala et al., 2021). Due to its low energy density, high moisture content, and impurity levels, waste biomass typically requires additional processing, such as thermochemical conversion. Its energy attributes are improved through this process to meet the requirements for direct combustion.

9.6.1 Pyrolysis

Using pyrolysis, waste biomass can be transformed into various useful products while preserving its stored energy. Biomass is broken down by the thermochemical process of pyrolysis in the absence of oxygen. Although more recent studies have recommended a wider temperature range of 250°C–900°C, the normal temperature range for the slow pyrolysis of waste biomass is between 300°C and 650°C (Mishra & Mohanty, 2020). Numerous intricate and varied chemical reactions take place instantly throughout the pyrolysis process. The thermal cracking reaction, in which organic and inorganic gases are produced during heating, is the main component of the process. Gases and biochar are among the initial products of pyrolysis. Carbon monoxide, carbon dioxide, hydrogen, methane, and higher hydrocarbons such as ethyne, ethylene, and ethane are among the final gaseous products (Aqsha et al., 2017). Tars and solid char are also among the liquid products. These products are made by further dissolving the condensable gas. The composition of the biomass, specifically its hydrogen-to-carbon (H/C) ratio, as well as process variables such as the heating rate, pressure, temperature, and residence time, has an impact on the yield of pyrolysis products. Pyrolysis is a complex process that depends on a variety of variables. Only a few studies have looked at straw as a feedstock, even if many have concentrated on the experimental investigation of waste biomass pyrolysis.

9.6.2 Combustion

Direct combustion, evaporation combustion, decomposition combustion, surface combustion, and smoldering combustion are a few examples of the different forms of combustion (Marks, 1992). When a sample with a low melting temperature evaporates under heat and combines with oxygen in the gas phase, this is known as evaporation combustion. In decomposition combustion, flames are created when gases (such as H_2, CO, C_mH_n, H_2O, and CO_2) from thermal decomposition interact with gaseous oxygen (McKendry, 2002). After these combustion processes, char or biochar typically persists and burns by surface combustion. Thermal degradation of reactive biomass samples at a temperature lower than the ignition point of the volatile component is known as smoldering combustion. Decomposition combustion and surface combustion are the two main types of combustion used in the industrial direct combustion of biomass.

Based on the type of combustion method used, different waste biomass products are burned in different ways. There are two types of combustion traits for waste biomass: macroscopic characteristics and microscopic characteristics (Goyal et al., 2008). The ultimate analysis, heating value, moisture content, particle size analysis, bulk density, and ash fusion temperature (AFT) are among the macroscopic characteristics of agricultural waste biomass (Suman et al., 2021). Thermal analysis, chemical kinetics, and mineral information are all part of the microscopic examination of agricultural waste biomass. For simplicity of investigation, the fuel combustion characteristics of biomass can be divided into physical, chemical, thermal, and mineral properties.

9.6.3 Hydrothermal liquefaction

An extremely adaptable process that can handle a variety of bio-based and waste feedstocks is hydrothermal liquefaction (HTL), also known as hydrous pyrolysis (Biller et al., 2012). Among these feedstocks include woody biomass, food and industrial waste, swine manure, algae, and waste from the forestry sector. The HTL process converts biomass into biocrude using a solvent and sometimes catalysts, when the temperature is less than 400°C (Zhang et al., 2008). Previous studies have mainly used small-scale batch-type reactors with slow heating rates and long residence times. However, for this technology to become more economically and chemically feasible, there is a need for further research on continuous reactors.

A solvent is present during the HTL process, which creates a highly reactive environment at high pressures and temperatures. As a result of breakdown and repolymerization events that take place during HTL, biocrude, dissolved chemicals, solid residue, and gas are produced (Bensaid et al., 2012). High pressure and temperature together reduce the solvent's dielectric constant and density, making it easier for hydrocarbons to dissolve in water. The solvent, which is often water, is helped to maintain a liquid state through the high pressure. In addition, as the polarity of water molecules decreases, the electron is distributed more evenly across the oxygen and hydrogen atoms, increasing the dissociation of water into H^+ and OH^- ions (Sanghi & Singh, 2012). As a result, heated compressed water can serve as an effective catalyst for processes involving acids or bases.

HTL has received a lot of attention as a potential technology for converting biomass in recent years. However, due to the many feedstock types, beginning states, reaction settings, and catalysts used, the information gathered up to this point is extremely fragmented, which makes it challenging to compare research findings (Dimitriadis & Bezergianni, 2017). Thus the goal is to compile, examine, and assess the experimental data on HTL that are now accessible. This section provides a concise summary of the state-of-the-art in HTL for the conversion of various bio-based and waste feedstocks, algae, and woody biomass.

9.7 Production of transportable bioenergy from waste biomass

The production of transportable bioenergy from waste biomass involves converting waste biomass into a form that can be easily transported and used as fuel. Methods include pelletization, pyrolysis, gasification, and anaerobic digestion. Pelletization compresses waste biomass into pellets for burning, pyrolysis heats waste biomass to produce bio-oil, biochar, and gas, gasification produces syngas, and anaerobic digestion breaks down waste biomass to produce biogas such as biohydrogen and biomethane. The production of transportable bioenergy from waste biomass has several advantages. It reduces the amount of waste that would otherwise be sent to landfills or incinerators, thereby reducing greenhouse gas emissions and air pollution. It also provides a source of renewable energy that can be used to displace fossil fuels, thereby reducing dependence on imported oil and promoting energy independence.

9.7.1 Solid fuel

Solid fuels from waste biomass have several advantages, including reducing waste and greenhouse gas emissions, promoting renewable energy sources, and providing a cost-effective alternative to conventional fuels. However, there are challenges in the production process, including the availability and quality of waste biomass feedstocks, as well as the costs and efficiency of the production processes. Despite these challenges, solid fuels from waste biomass have significant potential in various sectors such as residential heating, industrial processes, and power generation. With proper management and technological improvements, solid fuels from waste biomass can help promote a circular bioeconomy.

9.7.1.1 Biochar

Pyrolysis is the process of heating organic waste material without oxygen to create biochar. In order to improve soil quality and store carbon, the waste biomass—such as agricultural by-products, wood chips, or municipal garbage—must be transformed into a stable, carbon-rich substance. High surface area and a porous structure of biochar allow it to retain water and nutrients, encourage microbial activity in the soil, and improve plant growth.

Biochar production has several benefits, including reducing greenhouse gas emissions, improving soil quality, and managing waste biomass. Biochar can also be used in other applications, such as water filtration, animal feed supplements, and energy production. However, the production of biochar can have potential environmental impacts, such as air pollution, soil degradation, and water contamination. Therefore proper production and application methods are essential to ensure the benefits of biochar while minimizing its negative impacts.

Biochar pH plays a significant role in both environmental and agricultural applications. Biochars are usually alkaline, which can help neutralize acidic soils, immobilize metals in contaminated soils, and remove them from aqueous solutions. The point of zero charge (pH_{PZC}) is also important, as it indicates the pH at which the surface charge of biochar is zero (Pellera et al., 2021). Biochar with a higher pyrolysis temperature is more effective in sorbing organic contaminants due to its high surface area, microporosity, and hydrophobicity. Conversely, biochar produced at lower temperatures is better at sorbing inorganic and polar contaminants through interactions with surface functional groups and precipitation/coprecipitation. Comprehensive characterization is essential to ensure efficient and profitable solutions for both agronomic and environmental applications. However, there are few studies that comprehensively characterize biochar for multiple applications.

9.7.1.2 Drop-in fuels

Drop-in fuels offer several benefits over fossil fuels, such as sustainability and a significant reduction of 50%–100% in greenhouse gas emissions. However, the production of drop-in fuels is currently more expensive, approximately twice as much as fossil fuels, especially when second-generation feedstock is used. The primary cost factors in manufacturing drop-in fuels are the price of feedstock, syngas cleaning and conditioning, and bio-oil upgrading. Gasification is believed to be the most promising method for drop-in fuel production due to its ability to accommodate different feedstocks and create high-yield liquid fuel, alongside other commercially viable fuels. However, gasification still faces challenges such as syngas cleaning and upgrading, catalyst deactivation, and liquid fuel upgrading to fit into existing infrastructure. Comparing different studies on drop-in fuels can be difficult due to various factors such as feedstock type and price, plant design, operating conditions, product output, economic assumptions, functional unit, system boundary, and other factors.

No known pilot plants have used the upgraded technique to produce "drop-in" fuel. Given that second-generation biofuels have a better potential to cut GHG emissions than first-generation biofuels, waste biomass is a more sensible financial and environmental choice. Low yields are a result of insufficient microbial adaptation to lignocellulosic sugars and ongoing research on biochemical pathways to "drop-in" fuels.

9.7.2 Liquid fuel

Liquid biofuels can be produced from waste biomass through fermentation, transesterification, and gasification. Fermentation can produce bioethanol from waste biomass such as sugarcane bagasse, corn stover, and switchgrass. Transesterification converts vegetable oils or animal fats into biodiesel. Gasification can produce syngas, which can be turned into liquid biofuels. These biofuels are renewable, eco-friendly and produce less greenhouse gas emissions than fossil fuels. However, their production is costly and requires significant infrastructure and technological development.

9.7.2.1 Bioethanol

The use of waste biomass to produce various types of liquid biofuels can help to reduce the reliance on petroleum products and also lead to reduced greenhouse gas emissions in the transportation industry. Two primary methods for producing biofuels and energy from waste biomass are biochemical and thermochemical processes. The former utilizes chemical reactions with enzymes, mushrooms, and microorganisms and is best suited for biomass with a C/N ratio less than 30 and humidity at collection above 30% (Clauser et al., 2021). On the other hand, thermochemical processes are an alternative method. One popular liquid biofuel is bioethanol, which is produced commercially by fermenting sugars found in waste biomass. Bioethanol has numerous applications, such as pharmaceuticals, cosmetics, solvents, and chemicals. In the energy sector, it is typically used as an oxygenating agent in gasoline and can be made from renewable sources such as waste agricultural and forestry feedstocks.

The most efficient methods for producing lignocellulosic bioethanol include pretreating the biomass with diluted acid treatment, steam explosion, or similar thermomechanical processes. To maintain the financial sustainability of the production process, lignin is frequently collected and used as fuel in these processes. By dissolving lignin in an organic solvent, the "Organosolv" procedure, on the other hand, adopts a different strategy and enables it to be recovered in a pure state that may be utilized as a chemical or additive (Rossetti, 2016). Because this technique is more difficult and requires more money up front, its financial success hinges on finding a use for lignin other than as fuel. The production of three essential chemicals—ethylene, acetic acid, and ethylacetate from ethanol—has been the subject of a systematic framework provided in the literature for the design and evaluation of bio-based chemical processes. A methodical methodology has been suggested to minimize cost functions in a life-cycle cost analysis utilizing interval linear programming techniques in order to analyze various scenarios and strategies for using waste biomass.

9.7.2.2 Biodiesel

Transesterification is a method for turning waste biomass into biodiesel. In this method, leftover vegetable or animal fats are combined with sodium hydroxide or potassium hydroxide, a catalyst, and an alcohol such as methanol or ethanol. Biodiesel is the mixture of fatty acid methyl esters (FAMEs) that is the end result. Waste biomass can be utilized as feedstocks for the manufacture of biodiesel, including agricultural residues, forestry waste, food waste, and municipal solid waste. A sustainable alternative to fossil fuels that lowers greenhouse gas emissions and dependence on nonrenewable resources is the generation of biodiesel from waste biomass. This procedure, however, can be costly and necessitates major infrastructure and technological advancement. It might be difficult to ensure constant and dependable production because the quality of the waste biomass feedstock can affect the quantity and quality of the biodiesel generated.

Fatty acid methyl esters (FAMEs) must meet standards such as ASTM D-3751 and EN14214 and have physico-chemical qualities that are comparable with those of petrodiesel in order to be considered suitable as a biofuel.

The fatty acid profiles of the oleaginous oils utilized as feedstocks, which are comparable to edible and nonedible oils with significant fatty acids, similarly affect FAME characteristics (Chintagunta et al., 2021). Within specific bounds, the presence of unsaturated fatty acids is suitable for the synthesis of FAME. The iodine value (IV) and polyunsaturated fatty acids, which have an impact on the oxidative stability of FAMEs, dictate their unsaturation. The saponification value of FAMEs determines whether soap will form if there are unreacted fatty acids present.

FAMEs contain both saturated and unsaturated fatty acids, and a high percentage of saturated fatty acids can lengthen the shelf life of food by reducing auto-oxidation. It is advised to maintain an ideal balance of saturated and unsaturated fatty acids in FAMEs because unsaturated fatty acids have cold flow blocking qualities. By measuring IV, which must be within set parameters to prevent gum formation, one can estimate the amount of unsaturated fatty acids present. Better ignition and full combustion of FAMEs are indicated by high cetane values, which lower gaseous and particle emissions.

FAMEs have qualities that can extend their shelf life and improve engine performance, such as their high heating value and oxidative stability. FAMEs, however, can also have high kinematic viscosity and density, which can cause problems with combustion and emission. Cetane number and cold filter plugging point of FAMEs must be assessed in order to determine their appropriateness as transportation fuels due to the presence of unsaturation, soaps, particle pollutants, and sulfur.

9.7.2.3 Furanic fuels

Furanic fuels are biofuels created by converting waste biomass into a combination of furfural and other furan derivatives using a catalyst. Various biomass feedstocks are converted into furfural using acid hydrolysis or steam explosion. Methylfuran, dimethylfuran, and tetrahydrofurfuryl alcohol, which have properties comparable with those of regular petrol and diesel fuel and may be used in existing cars without modification, can all be made from furfural by additional refinement. Furanic fuels are a possible replacement for fossil fuels as they are renewable, nontoxic, and easily accessible.

High-value furanic chemicals including furandicarboxylic acid (FDCA), dimethylformamide (DMF), ethoxymethylfurfural (EMF), and methylfurfural (MF) are precursors to hydroxymethylfurfural (HMF). Some substances can replace chemicals and polymers obtained from fossil fuels and be used as the main component of new products (Yan & Chen, 2014).

When compared with petrol, DMF has a higher energy density and can be used as a biofuel. The primary method for producing DMF is the hydroxygenation of HMF (Kazi et al., 2011). Under atmospheric pressure in the presence of HCl, a bimetallic catalyst made of Pd-Au, Pd-Au/C, or Pd/C could catalyze the conversion of HMF to DMF. High yield (>85%) hydroxygenation of HMF to DMF was accomplished by these catalysts. A higher yield (of about 99%) of HMF to DMF conversion has recently been shown to be possible using copper and cobalt bimetallic catalysts on nanoparticles (Leshkov et al., 2007).

HMF can be used to create several important materials, including 2, 5-diformylfuran (DEF) for poly-Schiff bases, medicinal goods, and functional materials. Hollow Fe-Co bimetallic nanoparticles and graphene oxide can catalyze the oxidation of HMF to DEF (Chen et al., 2014; Zhang et al., 2015), while Au-Pd alloy nanoparticles and functionalized carbon nanotubes can catalyze the conversion of HMF to FDCA (Ouyang et al., 2019; Xia et al., 2018), another important component of green plastic. Both catalytic and biological processes can be used to produce FDCA (Sajid et al., 2018).

Another significant furan derivative that can be added to ordinary diesel is EMF. High boiling point (508K), high energy density (30.3 MJ/L), low toxicity, high cetane number, and low SO_2 emission rate are all characteristics of this fuel (Lewkowski, 2001). EMF is created through the etherification of HMF with ethanol in the presence of an acid catalyst. EMF can be made using a multistep process in which biomass and HCl are used to create 5-Chloromethylfurfural (CMF), which is then nucleophilically substituted into ethanol.

9.7.3 Gaseous fuel

Biohydrogen, biomethane, and carbon dioxide make up the majority of biogas, a renewable energy source that also contains trace amounts of other gases. The types of waste biomass used and the conditions during the digestion process can have an impact on the composition of biogas. Numerous uses for biogas exist, including the production of heat and power as well as fuel for vehicles. Utilizing biogas can lessen reliance on fossil fuels and greenhouse gas emissions.

9.7.3.1 Biomethane

Researchers are increasingly interested in biomethane due to its potential as a renewable energy source and its ability to contribute to circular economy models by utilizing waste materials. Biomethane can be particularly useful for countries heavily reliant on fossil fuels, as it can help reduce their dependence and contribute to low carbon fuel policies. Many EU countries are adopting gas grid decarbonization strategies and offering subsidies to promote the development of the biomethane market. Subsidies are viewed as essential to the growth of the biomethane sector globally, which can provide energy storage that is available anytime and anywhere.

The usage of the biogas—biomethane cycle can offer a carbon-negative substitute for the use of natural gas. Pure biomethane, at 5g CO_2 eq/km, greatly reduces the GHG emissions of vehicle powertrain systems as compared with petrol, diesel and compressed natural gas (Cucchiella et al., 2018). The cost of producing biomethane varies depending on its intended use, and its economic viability is assessed from a variety of angles. The distance between the site of substrate acquisition and the site of energy transformation and the technology employed and the volume of biogas processed all have an impact on the cost of upgrading. Transport expenses have a significant impact on the plant's profitability.

9.7.3.2 Biohydrogen

Production of biohydrogen by fermentation is a possible, environmentally benign replacement for fossil fuels. The two types of fermentation are photofermentation, which employs light as an energy source, and dark fermentation. The plastoquinone pool found in Photosystems I and II receives the electrons produced from the endogenous substrate by catabolism. Through the transfer of electrons to the equivalent level of [Fe]-H2ase and ferredoxin, the redox potential of these electrons can be increased.

Water is broken down into oxygen, energy, and a reducing agent by the biophotolysis process, which harnesses solar energy. By reducing protons with enzymes such as nitrogenase or hydrogenase, this reducing agent is employed to create hydrogen. In order to make ATP and decrease ferredoxin, which are involved in metabolic processes that result in the production of hydrogen, the electrons from the water splitting travel through the electron transport chain. Depending on where the electrons come from, the biophotolysis process can be divided into indirect and direct biophotolysis (Ahmed et al., 2021).

The microbial electrolytic cell (MEC) is a technique for producing sustainable biohydrogen from sources of renewable biomass while using less energy. MECs are flexible devices that can transform organic carbons or CO_2 into beneficial compounds. Microalgae that are electrochemically active produce CO_2 and electrons, which are subsequently transported from the anode to the cathode where they react with hydrogen atoms in the solution to form H^+. The cathode of the microbial cell must have an external voltage source for this electrolytic process to produce biohydrogen. In comparison to fermentation alone, the MEC is a hybrid system that produces biohydrogen more efficiently. Therefore when photosynthetic and nonphotosynthetic bacteria are inoculated in a biomass source, such wastewater, in an electrolytic cell, the efficiency of biohydrogen production is increased.

9.8 Bioeconomy and biorefinery

Petroleum, often known as fossil fuel, has long been the primary natural resource used for generating energy (such as transportation fuels) and synthetic goods (such as plastics and chemicals). They are limited resources, though, and they harm the environment by releasing greenhouse gases (GHGs), primarily carbon dioxide (CO_2), into the atmosphere, which contribute to climate change (Venkata-Mohan et al., 2016). Global awareness of these environmental issues has increased, and research on carbon mitigation and adaptation is receiving a lot of attention. A major step toward carbon management and GHG mitigation is switching from a petroleum refinery model to a waste biorefinery model. By moving away from the conventional linear economy model of take, make, and dispose, waste biorefinery seeks to develop a sustainable circular bioeconomy based on the concepts of recycling, reusing, remanufacturing, and sustaining.

Waste-to-treasure, also known as the use of waste materials, such as municipal solid and liquid waste in bioprocesses to produce valuable bioproducts and metabolites, has gained popularity over the years. This method results in renewable, sustainable, and degradable products. In a waste biorefinery, bioprocessing not only improves waste management but also addresses energy and environmental security issues (Dahiya et al., 2018). This economically and environmentally sustainable platform uses a cheap and environmentally friendly feedstock. Food waste, industrial wastes, forest and agricultural waste, lignocellulosic materials, wastewater, and sludge are only a few examples of the trash that has been successfully transformed into marketable bio-based goods (Leong et al., 2021).

Global environmental and food security issues have sparked initiatives to find solutions by numerous organizations. Food production is threatened by climate change caused by the combustion of fossil fuels, fast industrialization, and urbanization. According to an IPCC report from 2018, a 1.5°C increase in global warming might have a severe impact on sea levels and agricultural supply. Waste biorefinery offers a low carbon economy and sustainable waste disposal when combined with circular bioeconomy. An environmentally beneficial alternative to using fossil fuels as a production feedstock is the conversion of waste and side streams into useful bioproducts, such as biopolymers and biofuels (Mishra et al., 2019). This strategy not only offers a cheap means of disposing of waste but also lowers pollution and encourages a greener world. The purpose of studies is to increase the economic viability, efficacy, and efficiency of wastewater treatment, which can be used to create bio-based goods and bioenergy. Utilizing biofuels for a variety of needs can aid in managing carbon emissions and reducing greenhouse gas emissions.

The adoption of a waste biorefinery and circular bioeconomy approach can guarantee both energy and environmental security. Environmental security is crucial for ensuring food security on a global scale, as having enough food resources is essential for preventing world hunger and maintaining human health and well-being. By implementing circular bioeconomy practices, the world can achieve sustainable and prosperous living while mitigating the negative economic impacts of environmental, food, and energy concerns.

9.9 Technological upgradation and scale-up

The use of waste biomass for the generation of renewable energy is an efficient approach that also has an added benefit of cleaning the environment. However, there are drawbacks to using waste and biomass residues as a quick source of energy. Due to the high energy needed for pretreatment, purification, and maintenance, the cost-competitiveness of waste-to-bioenergy production is currently lower than that of fossil fuels. These issues also restrict the commercialization of waste-to-bioenergy technology (Lee et al., 2019). Process optimization is therefore being investigated at present to boost manufacturing yield and process effectiveness.

Utilizing waste biomass for the generation of bioenergy has the potential of reducing the quantity of waste dumped in landfills, which would benefit the environment. However, if the waste-to-bioenergy technology is not correctly built or run, this process could potentially lead to the release of hazardous by-products into the atmosphere, such as furans, polychlorinated dioxins, lead, mercury, and cadmium. There have been measures taken to stop the emission of dangerous substances, including regulating the temperature and air/fuel mixing ratios and avoiding "quench" zones in the boiler. To reduce dangerous gas emissions during waste-to-bioenergy operations, it is crucial to continuously develop efficient control solutions. Anaerobic digestion is a cleaner and better choice for converting high-moisture waste biomass with low thermal energy demand, and it should be taken into account when choosing a waste-to-bioenergy system. In order to maximize energy recovery efficiency in power generation and reduce environmental consequences, proper waste classification is also essential.

9.10 Conclusions

Bioenergy from waste biomass has become popular as a renewable and sustainable energy source. Recent technological advancements have allowed the conversion of waste biomass into bioenergy. Anaerobic digestion, gasification, and pyrolysis are some of the available technologies used to produce bioenergy. Anaerobic digestion produces biogas, gasification converts solid biomass into syngas, and pyrolysis breaks down biomass into bio-oil, char, and gas, which can be used as a substitute for fossil fuels. Despite the advantages of bioenergy from waste biomass, there are some issues that need to be resolved. These include the availability of feedstock, which varies in quantity and quality, making it challenging to produce consistent bioenergy output. Some of the technologies for producing bioenergy from waste biomass are still in development and require significant investments in research and development to become commercially viable. Environmental concerns, such as greenhouse gas emissions, land use changes, and water usage, need to be considered and managed. The cost of producing bioenergy from waste biomass can be high, making it challenging to compete with other energy sources.

The integration of waste biomass in biorefineries is an important strategy for the sustainable use of waste biomass. The biorefinery concept allows for the production of a range of value-added products from waste biomass, which has significant environmental and economic benefits.

References

Ahmed, S. F., Rafa, N., Mofijur, M., Badruddin, I. A., Inayat, A., Ali, M. S., Farrok, O., Yunus., & Khan, T. M. (2021). Biohydrogen production from biomass sources: Metabolic pathways and economic analysis. *Frontiers in Energy Research*, 9753878.

Aqsha, A., Tijani, M. M., Moghtaderi, B., & Mahinpey, N. (2017). Catalytic pyrolysis of straw biomasses (wheat, flax, oat and barley) and the comparison of their product yields. *Journal of Analytical and Applied Pyrolysis*, *125*, 201–208.

Azhar, S. H. M., Abdulla, R., Jambo, S. A., Marbawi, H., Gansau, J. A., Faik, A. A. M., & Rodrigues, F. (2017). Yeasts in sustainable bioethanol production: A review. *Biochemistry and Biophysics Reports*, *10*, 52–61.

Bensaid, S., Conti, R., & Fino, D. (2012). Direct liquefaction of lingo-cellulosic residues for liquid fuel production. *Fuel*, *94*, 324–332.

Bhatt, S. M., & Bal, J. S. (2019). Bioprocessing perspective in biorefineries. In N. Srivastava, M. Srivastava, P. K. Mishra, S. N. Upadhyay, P. W. Ramteke, & V. K. Gupta (Eds.), *Sustainable approaches for biofuels production technologies* (pp. 1–23). Cham, Switzerland: Springer.

Bijarchiyan, M., Sahebi, H., & Mirzamohammadi, S. (2020). A sustainable biomass network design model for bioenergy production by anaerobic digestion technology: Using agricultural residues and livestock manure. *Energy, Sustainability and Society*, *10*, 19.

Biller, P., Ross, A. B., Skill, S. C., Lea-Langton, A., Balasundaram, B., Hall, C., Riley, R., & Llewellyn, C. A. (2012). Nutrient recycling of aqueous phase for microalgae cultivation from the hydrothermal liquefaction process. *Algal Research*, *1*, 70–76.

Born, G.J., van den Minnen, J.G., van Olivier, J.G. J., & Ros, J.P. M. (2014). *Integrated analysis of global biomass flows in search of the sustainable potential for bioenergy production*. PBL Netherlands Environmental Assessment Agency. November 2014 PBL publication number: 1509.

Brunet, R., Boer, D., Guillén-Gosálbez, G., & Jiménez, L. (2015). Reducing the cost, environmental impact and energy consumption of biofuel processes through heat integration. *Chemical Engineering Research and Design*, *93*, 203–212.

Canilha, L., Chandel, A. K., dos Santos Milessi, T. S., Fernandes Antunes, F. A., da Costa Freitas, W. L., Almeida Felipe, M. G., & da Silva, S. S. (2012). Bioconversion of sugarcane biomass into ethanol: An overview about composition, pretreatment methods, detoxification of hydrolysates, enzymatic saccharification, and ethanol fermentation. *Journal of Biomedicine & Biotechnology*, *2012*, 1–15.

Chen, P. X., Tang, Y., Zhang, B., Liu, R., Marcone, M. F., Li, X., & Tsao, R. (2014). 5-Hydroxymethyl-2-furfural and derivatives formed during acid hydrolysis of conjugated and bound phenolics in plant foods and the effects on phenolic content and antioxidant capacity. *Journal of Agricultural and Food Chemistry*, *62*, 4754–4761.

Chintagunta, A. D., Zuccaro, G., Kumar, M., Kumar, S. P. J., Garlapati, V. K., Postemsky, P. D., Kumar, N. S. S., Chandel, A. K., & Simal-Gandara, J. (2021). Biodiesel production from lignocellulosic biomass using oleaginous microbes: Prospects for integrated biofuel production. *Frontiers in Microbiology*, 12658284.

Christakopoulos, P., Koullas, D. P., Kekos, D., Koukios, E. G., & Macris, B. J. (1992). Direct ethanol conversion of pretreated straw by Fusarium oxysporum. *Bioresource Technology*, *35*, 297–300.

Clauser, N. M., González, G., Mendieta, C. M., Kruyeniski, J., Area, M. C., & Vallejos, M. E. (2021). Biomass waste as sustainable raw material for energy and fuels. *Sustainability*, *13*, 794.

Cucchiella, F., D'Adamo, I., Gastaldi, M., & Miliacca, M. (2018). A profitability analysis of small-scale plants for biomethane injection into the gas grid. *Journal of Cleaner Production*, *184*, 179–187.

Dabrowska, S., Chudoba, T., Wojnarowicz, J., & Lojkowski, W. (2018). Current trends in the development of microwave reactors for the synthesis of nanomaterials in laboratories and industries: A review. *Crystals*, *8*, 379.

Dahiya, S., Kumar, A. N., Shanthi-Sravan, J., Chatterjee, S., Sarkar, O., & Mohan, S. V. (2018). Food waste biorefinery: Sustainable strategy for circular bioeconomy. *Bioresource Technology*, *248*, 2–12.

Danquah, J. A., Roberts, C. O., & Appiah, M. (2018). Elephant grass (Pennisetum purpureum): A potential source of biomass for power generation in Ghana. *Current Journal of Applied Science and Technology*, 1–12.

Deshpande, V. V., Sivaraman, H., & Rao, M. (1981). Simultaneous saccharification and fermentation of cellulose to ethanol using P. funiculosum cellulase and free or immobilized Saccharomyces uvarum cells. *Biotechnology & Bioengineering*, *25*, 1679–1684.

Dimitriadis, A., & Bezergianni, S. (2017). Hydrothermal liquefaction of various biomass and waste feedstocks for biocrude production: A state of the art review. *Renewable and Sustainable Energy Reviews*, *68*, 113–125.

Dubois, M., Gilles, K. A., Hamilton, J. K., Rebers, P. A., & Smith, F. (1956). Colorimetric method for determination of sugars and related substances. *Analytical Chemistry*, *28*, 350–356.

Dutta, S. K., & Chakraborty, S. (2015). Kinetic analysis of two-phase enzymatic hydrolysis of hemicellulose of xylan type. *Bioresource Technology*, *198*, 642–650.

Dutta, S. K., & Chakraborty, S. (2016). Pore-scale dynamics of enzyme adsorption, swelling and reactive dissolution determine sugar yield in hemicelluloses hydrolysis for biofuel production. *Scientific Reports*, *6*, 38173.

Erdei, B., Franko, B., Galbe, M., & Zacchi, G. (2012). Separate hydrolysis and co-fermentation for improved xylose utilization in integrated ethanol production from wheat meal and wheat straw. *Biotechnology for Biofuels*, *5*, 12.

Gaikwad, A., & Chakraborty, S. (2013). Mixing effects on the kinetics of enzymatic hydrolysis of avicel for batch production of cellulosic ethanol. *Industrial & Engineering Chemistry Research*, *52*, 3988–3999.

Gaikwad, A., & Chakraborty, S. (2014). Mixing and temperature effects on the kinetics of alkali metal catalyzed, ionic liquid based batch conversion of cellulose to fuel products. *Chemical Engineering Journal*, *240*, 109–115.

Gao, N., Zhang, L., & Wu, C. (2018). Biomass and wastes for bioenergy: Thermochemical conversion and biotechnologies. *BioMed Research International* 9638380.

Ghosh, T. K., Roychoudhury, P. K., & Ghosh, P. (1984). Simultaneous saccharification and fermentation (SSF) of lignocellulosics to ethanol under vacuum cycling and step feeding. *Biotechnology & Bioengineering*, *26*, 377–381.

Goyal, H. B., Seal, D., & Saxena, R. C. (2008). Biofuels from thermochemical conversion of renewable resources: A review. *Renewable and Sustainable Energy Reviews, 12*, 504−517.

Grewal, A. S., Kumar, K., Redhu, S., & Bhardwaj, S. (2013). Microwave assisted synthesis: A Green chemistry approach. *International Research Journal of Pharmaceutical and Applied Sciences, 3*(5), 278−285.

Guo, W., Heeres, H. J., & Yue, J. (2020). Continuous synthesis of 5-hydroxymethylfurfural from glucose using a combination of AlCl3 and HCl as catalyst in a biphasic slug flow capillary microreactor. *Chemical Engineering Journal*, 381122754.

Guo, H., Qi, X., Li, L., & Jr. Smith, R. L. (2012). Hydrolysis of cellulose over functionalized glucose-derived carbon catalyst in ionic liquid. *Bioresource Technology, 116*, 355−359.

Gupta, R., & Lee, Y. Y. (2009). Mechanism of cellulase reaction on pure cellulosic substrate. *Biotechnology and Bioengineering, 102*, 1570−1581.

Harmsen, P., Huijgen, W., Bermudez, L., & Bakker, R. (2010). Literature review of physical and chemical pretreatment processes for lignocellulosic biomass. *Report no, 1184*, 1−49.

Jeffries, T. W., & Jin, Y. S. (2004). Metabolic engineering for improved fermentation of pentoses by yeasts. *Applied Microbiology and Biotechnology, 63*, 495−509.

Kargbo, H., Harris, J. S., & Phan, A. N. (2021). Drop-in" fuel production from biomass: Critical review on techno-economic feasibility and sustainability. *Renewable and Sustainable Energy Reviews*, 135110168.

Kazi, F. K., Patel, A. D., Serrano-Ruiz, J. C., Dumesic, J. A., & Anex, R. P. (2011). Techno-economic analysis of dimethylfuran (DMF) and hydroxymethylfurfural (HMF) production from pure fructose in catalytic processes. *Chemical Engineering Journal, 169*, 329−338.

Kida, K., Morimura, S., Kume, K., Suruga, K., & Sonoda, Y. (1991). Repeated-batch ethanol fermentation by a flocculating yeast, Saccharomyces cerevisiae IR-2. *Journal of Bioscience and Bioengineering, 71*, 340−344.

Klein-Marcuschamer, D., Simmons, B. A., & Blanch, H. W. (2011). Techno-economic analysis of a lignocellulosic ethanol biorefinery with ionic liquid pre-treatment. *Biofuels, Bioproducts and Biorefining*.

Lee, Y.-H., & Fan, L. T. (1982). Kinetic studies of enzymatic hydrolysis of insoluble cellulose: Analysis of the initial rates. *Biotechnology and Bioengineering, 24*, 2383−2406.

Lee, S. Y., Sankaran, R., Chew, K. W., Tan, C. H., Krishnamoorthy, R., Chu, D. T., & Show, P. L. (2019). Waste to bioenergy: A review on the recent conversion technologies. *BMC Energy, 1*(4).

Leong, H. Y., Chang, C. K., Khoo, K. S., Chew, K. W., Chia, S. R., Lim, J. W., Chang, J. S., & Show, P. L. (2021). Waste biorefinery towards a sustainable circular bioeconomy: A solution to global issues. *Biotechnology for Biofuels, 14*(87).

Leshkov, Y. R., Barrett, C. J., Liu, Z. Y., & Dumesic, J. A. (2007). Production of dimethylfuran for liquid fuels from biomass-derived carbohydrates. *Nature Letters*, 447.

Lewkowski, J. (2001). Synthesis, chemistry and applications of 5-hydroxymethylfurfural and its derivatives. *ARKIVOC Free Online Journal of Organic Chemistry / Arkat-USA, Inc* (i), 17−54.

Li, A., Antizar-Ladislao, B., & Khraisheh, M. (2007). Bioconversion of municipal solid waste to glucose for bio-ethanol production. *Bioprocess and Biosystems Engineering, 30*, 189−196.

Li, K., Liu, R., Cui, S., Yu, Q., & Ma, R. (2018). Anaerobic co-digestion of animal manures with corn stover or apple pulp for enhanced biogas production. *Renewable Energy, 118*, 335−342.

Li, H., & Yang, S. (2014). Catalytic transformation of fructose and sucrose to HMF with proline-derived ionic liquids under mild conditions. *International Journal of Chemical Engineering, 978708*, 7.

Mahjoub, N., Sahebi, H., Mazdeh, M., & Teymouri, A. (2020). Optimal design of the second and third generation biofuel supply network by a multi-objective model. *Journal of Cleaner Production*120355.

Marks, J. (1992). Wood powder: An upgraded wood fuel. *Forest Products Journal, 42*, 52−56.

McKendry, P. (2002). Energy production from biomass (part 1): Overview of biomass. *Bioresource Technology, 83*(1), 37−46.

Miguel, V., Jose, L. G., Joaquina, L., & Juan, L. R. (2016). Biofuels 2020: Biorefineries based on lignocellulosic materials. *Microbial Biotechnology, 9*(5), 585−594.

Mishra, R. K., & Mohanty, K. (2020). Kinetic analysis and pyrolysis behaviour of waste biomass towards its bioenergy potential. *Bioresource Technology*, 311123480.

Mishra, S., Roy, M., & Mohanty, K. (2019). Microalgal bioenergy production under zero-waste biorefinery approach: Recent advances and future perspectives. *Bioresource Technology*, 292122008.

Mlonka-Medrala, A., Evangelopoulos, P., Sieradzka, M., Zajemska, M., & Magdziarz, A. (2021). Pyrolysis of agricultural waste biomass towards production of gas fuel and high-quality char: Experimental and numerical investigations. *Fuel*, 296120611.

Morero, B., Groppelli, E. S., & Campanella, E. A. (2017). Evaluation of biogas upgrading technologies using a response surface methodology for process simulation. *Journal of Cleaner Production, 141*, 978−988.

Naik, S. N., Goud, V. V., Rout, P. K., & Dalai, A. K. (2010). Production of first and second generation biofuels: A comprehensive review. *Renewable and Sustainable Energy Reviews, 14*, 578−597.

Nathan, R. M., Kelley, P., Klaski, R., Bosco, A., Moore, B., & Traviss, N. (2019). Characterization and comparison of oxidative potential of real-world biodiesel and petroleum diesel particulate matter emitted from a non-road heavy duty diesel engine. *The Science of the Total Environment, 655*, 908−914.

Ouyang, Q., Liu, J., Li, C., Zheng, L., Xiao, Y., Wu, S., & Zhang, B. (2019). *Polymer Chemistry, 10*, 5594.

Pal, R. K., & Chakraborty, S. (2013). A novel mixing strategy for maximizing yields of glucose and reducing sugar in enzymatic hydrolysis of cellulose. *Bioresource Technology, 148*, 611−614.

Park, S., Baker, J. O., Himmel, M. E., Parilla, P. A., & Johnson, D. K. (2010). Cellulose crystallinity index: Measurement techniques and their impact on interpreting cellulose performance. *Biotechnology for Biofuels, 3*(10).

Paul, S. K., & Chakraborty, S. (2018). Microwave-assisted ionic liquid-mediated rapid catalytic conversion of non-edible lignocellulosic Sunn hemp fibres to biofuels. *Bioresource Technology, 253*, 85−93.

Paul, S. K., & Chakraborty, S. (2019). Mixing effects on the kinetics of enzymatic hydrolysis of lignocellulosic Sunn hemp fibres for bioethanol production. *Chemical Engineering Journal, 377*120103.

Pellera, F. M., Regkouzas, P., Manolikaki, I., & Diamadopoulos, E. (2021). Biochar production from waste biomass: Characterization and evaluation for agronomic and environmental applications. *Detritus, 17*, 15–29.

Philippidis, G. P., Smith, T. K., & Wyman, C. E. (1993). Study of the enzymatic hydrolysis of cellulose for production of fuel ethanol by the simultaneous fermentation and fermentation process. *Biotechnology and Bioengineering, 41*, 846–853.

Pishvaee, M. S., Mohseni, S., & Bairamzadeh, S. (2021). *An overview of biomass feedstocks for biofuel production. Biomass to biofuel supply chain design and planning under uncertainty* (pp. 1–20). Amsterdam, The Netherlands: Elsevier.

Piñas, J. A. V., Venturini, O. J., Lora, E. E. S., & Roalcaba, O. D. C. (2018). Technical assessment of mono-digestion and codigestion systems for the production of biogas from anaerobic digestion in Brazil. *Renewable Energy, 117*, 447–458.

Prat, D., Wells, A., Hayler, J., Sneddon, H., McElroy, R. C., Abou-Shehadad, S., & Dunne, P. J. (2016). CHEM21 selection guide of classical- and less classical-solvents. *Green Chemistry an International Journal and Green Chemistry Resource GC, 18*, 288.

Rao, P. V., Baral, S. S., Dey, R., & Mutnuri, S. (2010). Biogas generation potential by anaerobic digestion for sustainable energy development in India. *Renewable and Sustainable Energy Reviews, 14*(7), 2086–2094.

Rehrah, D., Bansode, R. R., Hassan, O., & Ahmedna, M. (2016). Physico- chemical characterization of biochars from solid municipal waste for use in soil amendment. *Journal of Analytical and Applied Pyrolysis, 118*, 42–53.

Reid, W. V., Ali, M. K., & Field, C. B. (2020). The future of bioenergy. *Global Change Biology, 26*(1), 274–286.

Rogelj, J., den Elzen, M., Hohne, N., Fransen, T., Fekete, H., Winkler, H., Schaeffer, R., Sha, F., Riahi, K., & Meinshausen, M. (2016). Paris agreement climate proposals need a boost to keep warming well below 2°C. *Nature, 534*, 631–639.

Rossetti, I. (2016). Economic assessment of biorefinery processes: The case of bioethanol. *Ind. Chem., 2*(1).

Roy, S., & Chakraborty, S. (2019). Comparative study of the effectiveness of protic and aprotic ionic liquids in microwave-irradiated catalytic conversion of lignocellulosic June grass to biofuel precursors. *Bioresource Technology Reports, 8*100338.

Roy, S., & Chakraborty, S. (2020). A kinetic framework for microwave-irradiated catalytic conversion of lignocelluloses to biofuel precursors by employing protic and aprotic ionic liquids. In P. Verma (Ed.), *Biorefineries: A step towards renewable and clean energy. Clean energy production technologies* (pp. 173–215). Singapore: Springer.

Roy, S., & Chakraborty, S. (2023). Regulatory effects of water in two-phase protic ionic liquid-mediated catalytic conversion of non-edible lignocelluloses to biofuel precursors. *Biomass Bioenergy, 168*106674.

Sajid, M., Zhao, X., & Liu, D. (2018). Production of 2,5-furandicarboxylic acid (FDCA) from 5-hydroxymethylfurfural (HMF): Recent progress focusing on the chemical-catalytic routes. *Green Chemistry an International Journal and Green Chemistry Resource GC, 20*, 5427.

Sanghi, R., & Singh, V. (2012). *Green chemistry for environmental remediation*. John Wiley & Sons.

Stoeglehner, G., & Narodoslawsky, M. (2009). How sustainable are biofuels? Answers and further questions arising froman ecological footprint perspective. *Bioresource Technology, 100*(16), 3825–3830.

Subramanian, R., Dufreche, S., Zappi, M., & Bajpai, R. (2010). Microbial lipids from renewable resources: production and characterization. *Journal of Industrial Microbiology & Biotechnology, 37*, 1271–1287.

Suman, S., Mohan., Yadav, A., Tomar, N., & Bhushan, A. (2021). Combustion characteristics and behaviour of agricultural biomass: A short review. *Renewable Energy - Technologies and Applications*, IntechOpen.

Sun, J. L., Sakka, K., Karita, S., Kimura, T., & Ohmiya, K. (1998). Adsorption of Clostridium stercorarium xylanase A to insoluble xylan and the importance of the CBDs to xylan hydrolysis. *Journal of Bioscience and Bioengineering, 85*, 63–68.

Tag, A. T., Duman, G., Ucar, S., & Yanik, J. (2016). Effects of feedstock type and pyrolysis temperature on potential applications of biochar. *Journal of Analytical and Applied Pyrolysis, 120*, 200–206.

Talebian-Kiakalaieh, A., Amin, N. A. S., & Mazaheri, H. (2013). A review on novel processes of biodiesel production from waste cooking oil. *Applied Energy, 104*, 683–710.

Thawarkar, S., Khupse, N. D., & Kumar, A. (2015). Solvent-mediated molar conductivity of protic ionic liquids. *Physical Chemistry Chemical Physics PCCP, 17*, 475–482.

Thygesen, A., Oddershede, J., Lilot, H., Thomsen, A. B., & Stahl, K. (2005). On the determination of crystallinity and cellulose content in plant fibres. *Cellulose, 12*, 563–576.

Toribio-Cuaya, H., Pedraza-Segura, L., Macias-Bravo, S., Gonzalez-Garcia, I., Vasquez-Medrano, R., & Favela-Torres, E. (2014). Characterization of lignocellulosic biomass using five simple steps. *Journal of Chemical, Biological and Physical Sciences, 4*(5), 28–47.

Tuoriniemi, J., Johnsson, A.-C. J. H., Holmberg, J. P., Gustafsson, S., Gallego-Urrea, J. A., Olsson, E., Pettersson, J. B. C., & Hassellöv, M. (2014). Intermethod comparison of the particle size distributions of colloidal silica nanoparticles. *Science and Technology of Advanced Materials, 15*035009.

Usmani, Z., Sharma, M., Awasthi, A. K., Sivakumar, N., Lukk, T., Pecoraro, L., Thakur, V. K., Roberts, D., Newbold, J., & Gupta, V. K. (2021). Bioprocessing of waste biomass for sustainable product development and minimizing environmental impact. *Bioresource Technology, 322*124548.

Vazirzadeh, M., & Mohsenzadeh, M. (2012). Bioethanol production from white onion by yeast in repeated batch. *Iranian Journal of Science and Technology, 2012*, 477–480.

Venkata-Mohan, S., Modestra, J. A., Amulya, K., Butti, S. K., & Velvizhi, G. (2016). A circular bioeconomy with biobased products from CO2 sequestration. *Trends in Biotechnology, 34*(6), 506–519.

Wilson, D. B., & Walker, L. P. (1991). Enzymatic hydrolysis of cellulose: An overview. *Bioresource Technology, 36*, 3–14.

Wood, T. M., & Bhat, K. M. (1988). Methods for measuring cellulase activities. *Methods in Enzymology, 160*, 87–112.

Xia, H., Xu, S., Hu, H., An, J., & Li, C. (2018). Efficient conversion of 5-hydroxymethylfurfural to high-value chemicals by chemo- and biocatalysis. *RSC Advance, 8*, 30875.

Yang, H., Chen, Q., Wanga, K., & Sun, R.-C. (2013). Correlation between hemicelluloses-removal-induced hydrophilicity variation and the bioconversion efficiency of lignocelluloses. *Bioresource Technology, 147*, 539–544.

Yan, K., & Chen, A. (2014). Selective hydrogenation of furfural and levulinic acid to biofuels on the ecofriendly Cu–Fe catalyst. *Fuel*, *115*, 101–108.

Zhang, Y., Dubé, M., McLean, D., & Kates, M. (2003). Biodiesel production from waste cooking oil: Economic assessment and sensitivity analysis. *Bioresource Technology*, *90*, 229–240.

Zhang, B., Keitz, M., & Valentas, K. (2008). Thermal effects on hydrothermal biomass liquefaction. *Applied Biochemistry and Biotechnology*, *147*, 143–150.

Zhang, J., Li, J., Tang, Y., Lu, Lin, L., & Long, M. (2015). Advances in catalytic production of bio-based polyester monomer 2,5-furandicarboxylic acid derived from lignocellulosic biomass. *Carbohydrate Polymers*, *130*, 420–428.

Zhou, C. H., Xia, X., Lin, C. X., Tong, D. S., & Beltramini, J. (2011). Catalytic conversion of lignocellulosic biomass to fine chemicals and fuels. *Chemical Society Reviews*, *40*, 5588–5617.

CHAPTER

10

Lignocellulosic wastes: different dimensions to a sustainable 2 G bioethanol production

Subhodeep Banerjee[1,2], Subhara Dey[2], Anusha[3] and Rintu Banerjee[2,3]

[1]Advanced Technology Development Centre, Indian Institute of Technology Kharagpur, Kharagpur, West Bengal, India
[2]P K Sinha Centre for Bioenergy and Renewables, Indian Institute of Technology Kharagpur, Kharagpur, West Bengal, India
[3]Agricultural & Food Engineering Department, Indian Institute of Technology Kharagpur, Kharagpur, West Bengal, India

10.1 Introduction

The exponential growth in population around the world leads to rise in demand for energy, which was traditionally met by nonrenewable sources. However, with time people have become more self-reliant and shift toward renewable sources to meet the energy demands. Currently, 85% of the world's energy consumption is derived from nonrenewable sources. Oil, natural gas, and coal are among the resources that will inevitably deplete and also contribute to atmospheric pollution, which leads to economic dependency on oil-producing countries. As of 2021, renewable energy sources accounted for approximately 29% of the planet's total electricity production, according to the International Energy Agency (IEA) (IEA, 2021). The use of renewable resources, for example, biofuels is now being promoted globally due to their potential for reducing atmospheric CO_2 levels. The use of lignocellulosics, such as agricultural and forestry waste, to produce biofuels such as ethanol and butanol, has shown promise in meeting the rising energy demand (Raj et al., 2022). The modernization of biorefineries, enzymatic hydrolysis, and other biotechnological advancements in converting cellulosic components into fermentable sugars have significantly reduced the cost of biofuels. To make them commercially viable on a large scale and profitable for the industry, there are, however, a few obstacles that must be overcome (Viikari et al., 2012).

This chapter aims at highlighting the important factors responsible for bioethanol production process and economic viability at an industrial scale.

10.2 Diversity and potentiality of lignocellulosic residue

As the most ubiquitous, natural, and renewable feedstock, lignocellulosic biomass has emerged as a potential alternative to fossil fuels. Their renewability, physical and chemical characteristics provided them a great potential to be used in biotechnological applications. Nonetheless, burning is still a common method for disposing of many lignocellulosic materials, which pollute the environment by releasing hazardous gases and particulate matter leading to global warming. being burned; these enormous amounts of lignocellulosic materials can be converted into value added by-products in the form of chemicals, biofuels, cheap sources of energy (Iqbal et al., 2013).

A wide array of lignocellulosic has been identified, which is probable feedstock for the generation of bioethanol. Agricultural leftover (wheat straw, rice straw, bagasse, cobs, husks, etc.), industrial by-products (pulps, solid wastes, etc.), forest biomass and wastes (hardwoods, softwoods, wood chips, forest thinnings, residues, etc.),

TABLE 10.1 Lignocellulosic content and ethanol yield from lignocellulosic biomass.

Source of lignocellulosic biomass	Examples	Abundance	Lignocellulose content	Theoretical ethanol yield	References
Agricultural waste (straws, stalks, husks, cobs, shells, leaves, roots, peels, and nuts, etc.)	Wheat straw	354 million tons/year	37% cellulose, 26.5% hemicellulose, 14% lignin	460 L/MT	Hoang and Nghiem (2021)
	Rice straw	800–1000 million tons/year	43.4% cellulose, 27.9% hemicellulose, 17.2% lignin	517 L/MT	
	Sugarcane bagasse	540 million tons/year	41.6% cellulose, 25.1% Hemicellulose, 20.3% lignin	483 L/MT	
Industrial waste	Pulp and paper	400 million tonnes of paper and paperboard & 188 million tonnes of virgin pulp	79% cellulose, 18% hemicellulose, very small amounts of lignin	350–400 L/MT	Sharma & Saini (2020)
	Food processing	1.3 billion tons/year	45%*** 35%*** 15% Lignin	107.58 g/kg dry material	Matsakas et al., 2014
Forest biomass and waste	Hardwoods	-	46.2% cellulose, 29.2% hemicellulose, 22% lignin	546 L/MT	Hoang and Nghiem (2021)
	Softwoods	-	40%–45% cellulose, 7%–14% hemicellulose, 26%–34% lignin	493 L/MT	
Municipal solid waste	Waste paper	100 million tons/year	33%–49% cellulose, 9%–16% hemicellulose, 10%–14% lignin	150 g/kg	Shi et al. (2009)
Dedicated energy crops	*Miscanthus spp.*,	15–25 tons/ha	Cellulose -38%–40%, Hemicellulose- 18%–24% Lignin- 24%–25%	260 L/MT	Sharma & Saini (2020)
	Switchgrass spp.	14.2 tons/ha	32%–45% cellulose, 21%–31% hemicellulose, 12%–28% lignin	100–249 L/MT	
	Poplar	10 tons/ha/year	50% cellulose, 30% hemicellulose, 20% lignin	247 L/MT	
Weed plants	*Amaranthus viridis*	60–100 tons of biomass/ha/year	37.4% cellulose, 34.2% hemicellulose, 5.1% lignin	521 L/ton	Premjet (2018)

municipal solid wastes (food waste, waste newspaper, kraft paper, other refuse, etc.), dedicated energy crops (*Miscanthus* spp., *Switchgrass* spp., etc.), and weed plants (*Eichhornia crassipes, Lantana camara*, etc.) are the possible lignocellulosic biomass for bioethanol generation (Sharma & Saini, 2020). Table 10.1 represents the detailed characterization of lignocellulosic biomass.

10.3 Biochemistry of lignocellulose

The most prevalent renewable resource on the planet is lignocellulose. The key constituents of lignocellulosic biomass are the carbohydrate cellulose, hemicellulose, and the noncarbohydrate aromatic compound lignin.

Lignin attaches to cellulose fibers to strengthen and solidify the cell walls of plants. The components are explained concisely in the following points:

- **Cellulose**

    Cellulose is the main structural polysaccharide of the primary plant cell wall, and it comprises approximately 35%−50% of the dry weight of lignocellulose. Cellulose is a linear homopolysaccharide of D-glucopyranose joined together by β-(1, 4) glycosidic linkages. The long polymers in cellulose are held together by hydrophobic interactions and intra- and interchain hydrogen bonding, forming a highly organized crystalline structure. Cellulose obtained from nonfood energy crops can be degraded by the action of microbial cellulases to glucose monomers, with subsequent conversion to biofuels or other value-added chemicals (Yang, 2007).

- **Hemicellulose**

    The second-most prevalent amorphous polymer in lignocellulosic biomass is hemicellulose (20%−35%). Hemicelluloses are imbedded in the plant cell walls, and one of their main functions is to bind cellulose microfibrils to strengthen the cell wall. Unlike cellulose, hemicellulose has a random and amorphous structure, which is composed of several heteropolymers including pentose (-L-arabinose and -D-xylose), hexose (-D- glucose, -D-mannose, and -D-galacatose) sugars, as well as uronic acids such -D- glucuronic acid, -D-galacturonic acid, and 4-O-methyl-D-glucuronic acid. The glycosidic linkages β-(1, 4) and β-(1, 3) hold these monomers together. Hemicelluloses can be hydrolyzed by dilute acid or base as well as several microbial hemicellulases, a complex set of enzymes with components that remove the side chains and others that attack the backbone randomly to release oligosaccharides that are subsequently degraded to simple sugars (Yang, 2007).

- **Lignin**

    The third main component of lignocellulose is lignin, which accounts for 15%−30% of lignocellulose dry mass. Lignin is found in all vascular plants, representing the second most abundant carbon source after cellulos. Lignin is a naturally existing amorphous, interconnected, a three-dimensional polymer made of the monomeric units p-hydroxyphenyl (H), guaiacyl (G), and, syringyl (S) units formed by the ester bonding of the monolignols sinapyl, coniferyl, and p-coumaryl alcohols. Softwood lignin is composed solely of coniferyl alcohol, whereas hardwood lignin is primarily composed of coniferyl alcohol and sinapyl alcohol. Grass lignin contains three varieties of monomers: coniferyl, sinapyl, and p-coumaryl alcohol. These units are connected by a variety of interunit aryl- or alkyl-ether bonds (Zoghlami & Paës, 2019).

All three biopolymers are included in the lignocellulose superstructure, with the cellulose microfibrils being cross-linked with lignin aggregates by ester and ether bonds and embedded in a network of hemicellulose by hydrogen bonds along with van der Waals forces. The high recalcitrance constituting the lignocellulosic biomass is accounted for by rigid arrangement of lignocellulose, the amount of lignin, and the high cellulose crystallinity. Other than lignin and holocellulose, proteins, pectin, nitrogenous chemicals, and inorganic compounds also make up a small amount of biomass.

10.4 Pretreatment methods of biomass for 2 G bioethanol

The pretreatment of biomass presents the biggest processing hurdle in the production of biofuel. Three fundamental components of lignocellulosic biomass are hemicellulose, lignin, and cellulose. The solubilization and separation of one or more of these biomass components are referred to as pretreatment methods. It increases the solid biomass's susceptibility to additional chemical or biological processing. To reduce the amount of crystallized cellulose and increase the proportion of amorphous cellulose, the form of cellulose that is most accessible to enzyme attack, the matrix must be broken through pretreatment (Amin et al., 2017). Pretreatment is used to change the size, shape, and chemical makeup of biomass on a macroscopic and microscopic scale as well as at the submicroscopic level. Increasing the yields of monomeric sugars, it makes the lignocellulosic biomass accessible to rapid hydrolysis. An efficient pretreatment procedure is intended to (1) produce sugars directly or indirectly via hydrolysis, (2) prevent the loss and/or breakdown of sugars produced, (3) restrict the production of inhibitory compounds, (4) reduce energy needs, and (5) reduce costs.

10.5 Combined pretreatment

Individual pretreatment methods for various feedstocks make a significant contribution to the final bioethanol production. However, due to inherent flaws, each of these approaches is constrained. Therefore, combining two or more pretreatment strategies could decrease the degree of these processes' drawbacks and enhance the yield of the final product, along with reducing the time and chemicals required to qualify as a green process. For instance, combining chemical and microbiological pretreatments is seen as a cost-effective method that can reduce secondary pollution, shorten pretreatment periods, and utilize less potent chemicals (Meenakshisundaram et al., 2021).

Two-step pretreatment of lignocellulosic biomass generally shows improved performance of the second pretreatment. As a result, both the substrate properties and the effectiveness of biomass enzymatic hydrolysis will be significantly impacted by the order of the pretreatment procedures. Hazelnut shells were pretreated with diluted acid (DAP), liquid hot water (LHW), and alkaline pretreatment (AP). LHW demonstrated the best cellulose recovery, DAP generated the highest hemicellulose solubilization, and AP generated the highest lignin removal, among the single pretreatments. The highest lignin removal was accomplished by AP-LHW at 60.7%, whereas the best hemicellulose removal was shown by LHW-DAP at 93.8%. The highest cellulose recovery was achieved by DAP-LHW at 94.0%. Increased sugar recovery from lignocellulosic materials may also be achieved through partial lignin displacement and cellulose structural alteration (Hoşgün et al., 2021).

An advanced pretreatment method for bamboo biomass employed 2-methyl tetrahydrofuran, an environmentally friendly, reusable organic solvent derived from biomass, combined with microwave-assisted degradation. This accelerated hemicellulose removal, exposed the cellulosic fibers, and enhanced the subsequent hydrolysis process as compared with the individual treatments (Li et al., 2017). The use of 1-butyl-3-methylimidazolium chloride ([Bmim]Cl) as an ionic liquid (IL) along with microwave radiation to pretreat lignocellulosic rice straw (RS) was investigated. The results showed that microwave-[Bmim]Cl reduced lignin more efficiently than [Bmim]Cl alone (57.02 1.24%), and the enzymatic saccharification of microwave-[Bmim]Cl-regenerated cellulose produced the highest glucan and xylan yields, indicating that it is a potentially useful alternative pretreatment method (Sorn et al., 2019). Additionally, microwave radiation is used in conjunction with weak acids and high-pressure treatments, which cause the enhanced breakdown of hemicellulose and degradation of lignin, respectively (Mikulski & Kłosowski, 2020).

Understanding the contribution of white-rot fungi in the biological decay process aids in overcoming the drawbacks of conventional biotreatment, such as an extended residence period and inefficient delignification. Meenakshisundaram et al. reviewed several works on combined pretreatments involving biological methods. The initial phase in a biological-alkaline combination is preferred to be biological pretreatment since it aids in the delignification of lignocellulosic fibers and lowers the concentration of alkali required and the temperature of the process, minimizing heating expenses. It is also preferable to combine acid pretreatment with biological pretreatment for reducing the severity of the process and the resulting formation of inhibitory compounds. The production of glucose and ethanol from the fungal pretreatment followed by the dilute acid pretreatment was far more impressive, despite the fact that the degradation of fiber from both sequential pretreatments was comparable. Prior fungal pretreatment in the biological-organosolv process facilitated the elimination of hemicellulose and lignin, which in turn increased the accessibility of the solvent during the organosolv process, thereby increasing cellulosic digestibility and reducing the severity of the process. Due to the ethanol and solubilized lignin's role as free radical scavengers, which prevent lignin from recondensing, it has very little residual lignin and high glucan retention. The combination oxidative-biological treatment using H_2O_2 and a fungal preparation (*Pleurotus ostreatus*) increased the rate of delignification by twofold and lowered carbohydrate losses compared with the only biological treatment. The structural modifications brought about by the oxidative pretreatment allowed the fungal hyphae to rapidly penetrate the feedstock, consequently reducing the combined pretreatment time to 18 days as compared with 60 days for sole pretreatment (Meenakshisundaram et al., 2021).

As illustrated above, a variety of pretreatment techniques used in combination significantly improve carbohydrate recovery and aid in increasing the enzymatic degradability of lignocellulosics for the manufacture of bioethanol. However, to scale the process, it is required to assess the effects of these combined processes on the environment, cost-effectiveness, and energy balance.

10.6 Biological pretreatment method treat above combined

Similar to physical, chemical, and physicochemical processes, biological pretreatment is another alternative option that can be explored for lignocellulosic ethanol production, from the name itself it can be well understood that the process is being eco-friendly, slow and targets specific interactions. In order to decompose the hemicellulosic and lignin components of lignocellulosic biomass, microorganisms are normally employed in biological pretreatment. Typically, white-rot fungi, which generate ligninolytic enzymes such as laccase, lignin peroxidase, and manganese peroxidase, are the microorganisms employed for biological pretreatment. By dissolving the lignin and hemicellulose components of lignocellulosic biomass, these enzymes increase the accessibility of the cellulose for subsequent saccharification. The advantages of biological pretreatment are as follows:

1. **Environmentally Friendly**: Biological pretreatment is an eco-friendly process for pretreating lignocellulosic biomass since it uses microorganisms to degrade the hemicellulosic and lignin components instead of harsh chemicals that can be harmful to the environment.
2. **High Selectivity**: Biological pretreatment is highly selective since the ligninolytic enzymes produced by the microorganisms only target the hemicellulosic and lignin components of the lignocellulosic biomass, leaving the cellulose intact.
3. **Low Cost**: Biological pretreatment is a low-cost method for pretreating lignocellulosics since it uses microorganisms that are readily available in nature and does not require the use of expensive equipment.

Despite the advantages, there are few drawbacks of biological pretreatment, that is, long pretreatment time: Biological pretreatment requires a longer pretreatment time in comparison to other methods of pretreatment. Depending on the kind of lignocellulosic biomass utilized, the type of microorganisms used, and the environmental factors, the pretreatment period could range anywhere from a few weeks to a few months (Zhang et al., 2020). The other drawback is variation in pretreatment efficiency, that is, the efficiency of biological pretreatment can vary based on the type of microorganism used, the type of lignocellulosic biomass, and the environmental conditions. This can lead to variability in the quality and quantity of the fermentable sugars produced.

The production of fermentable biofuels from lignocellulosics requires the conversion of complex carbohydrates, such as hemicellulose and cellulose, into fermentable sugars through the process of saccharification. Biological saccharification involves the use of enzymes produced by certain microorganisms to hydrolyze complex carbohydrates into simple fermentable sugars. Some examples of microbes include fungi belonging to phylum ascomycetes, which are often referred to as sac-fungi. They are subject to the subkingdom Dikarya (dikaryon's presence). Some ascomycetes are saprophytes, while others are pathogens that cause diverse diseases in animals and plants. Some ascomycetes can be consumed as mushrooms. Some of them exist as lichens and mycorrhiza in a symbiotic relationship. These filamentous ascomycetes produce enzymes such as cellulase, xylanase, and laccase. These enzymes act upon the carbohydrate and noncarbohydrate portion of the lignocellulosic biomass and lead to the formation of monomeric sugars, which are to be used further by other microorganisms for the generation of biofuel and other products of added value (Ferreira et al., 2016). Other than ascomycetes, white-rot fungi, which belong to phylum basidiomycetes, have inherent ability to degrade lignocelluloses using extracellular enzymes such as peroxidase, laccase, and oxidase. These fungi can be cultivated on any carbon source, primarily agricultural waste, and utilized as substrate (Madadi et al., 2017).

10.7 Enzymatic pretreatment

Laccase, lignin peroxidase, and manganese peroxidase play crucial roles in the degradation of lignin found in lignocellulose, thereby enhancing the saccharification processes. Laccase, an oxidoreductase group of enzyme, is capable of oxidizing various phenolic compounds within lignin, breaking down its complex structure. Lignin peroxidase, another lignin-degrading enzyme, exhibits high redox potential, enabling it to efficiently depolymerize lignin by generating free radicals. Similarly, manganese peroxidase, relying on manganese ions, generates highly reactive intermediates that facilitate lignin degradation. Together, these enzymes contribute to lignin breakdown, which ultimately leads to enhanced saccharification. By promoting lignin degradation, laccase lignin peroxidase and manganese peroxidase pave the way for increased accessibility of cellulose, allowing for improved conversion into fermentable sugars, thus bolstering the saccharification processes. Rajak and Banerjee (2016) provided details of how they used laccase for successful pretreatment of lignocellulose biomass Kans gras in amicable time and environment, the delignification enhanced the saccharification efficiency as the pretreatment method was mild and did not generate any toxic chemicals that would hamper further fermentation process or the need for neutralization with corrosive chemicals.

10.8 Different enzymes used for the pretreatment and hydrolysis of lignocellulosic biomass

Several enzymes can be used for the pretreatment of lignocellulosic biomass, including:

- Cellulases: These enzymes break down cellulose into carbohydrate monomers, such as glucose, and then to further fermented into biofuels. Cellulases can be produced by various fungi, bacteria, and other microorganisms.
- Hemicellulases: These enzymes break down hemicellulose, A ribose sugar complex polymer that binds with lignin. Hemicellulases can be produced by fungi, bacteria, and other microorganisms.
- Laccases: These enzymes can oxidize and break down lignin, which is a complex polymer that makes up a significant portion of lignocellulosic biomass. Laccases are made by fungi and bacteria.
- Peroxidases: These enzymes can disintegrate lignin, but they require the presence of hydrogen peroxide. Peroxidases are produced by bacteria and fungi.
- Proteases: There are several types of proteinases that help to break down the proteinaceous moieties that may be present in lignocellulosic biomass depending upon the specificity, types of proteinases are selected for bioconversion. The major challenge that lies with this enzyme is that proteases being a proteinaceous group attack all other proteinanceous enzyme resulting into faster denaturation. The report on protease production is plenty available and majorly through microorganisms.
- Pectinases: These enzymes can break down pectin, which is a polysaccharide that can interfere with the breakdown of holocellulose. Pectinases are produced by fungi, bacteria, and other microorganisms.
- Lignin peroxidase (LiP): LiP acts on lignin by oxidizing it, which leads to the breaking of aryl-ether bonds in the lignin polymer. This oxidation process generates highly reactive intermediates, such as quinones and free radicals, which can react with other lignin molecules or with other components of the lignocellulosic matrix.
- Manganese peroxidase (MnP): It is produced by certain fungi that is capable of breaking down lignin in lignocellulosic biomass. MnP acts on lignin by oxidizing it, which leads to the disintegratingof aryl-ether bonds in the lignin polymer. This oxidation process generates highly reactive intermediates, such as quinones and free radicals, which can react with other lignin molecules or with other components of the lignocellulosic matrix (Vasić et al., 2021).

The choice of enzyme(s) used for pretreatment will depend on the specific composition and characterization of biomass being processed along with the formation of required end product. Different combinations of enzymes may also be used to achieve optimal results.

10.9 Saccharification

Saccharification of lignocellulosic biomass is the process of converting the complex structure of lignocellulosic biomass into simple carbohydrates such as glucose, xylose, and arabinose. This is a crucial stage in the production of biofuels and other value-added products from lignocellulosic biomass, as microorganisms can ferment the sugars to produce biofuels and other beneficial compounds.

Saccharification is a crucial step in the production of biofuels and other value-added products from lignocellulosic biomass because the monosaccharides obtained from saccharification can be fermented into biofuels, such as ethanol, butanol, or other chemicals, such as enzymes, organic acids, or other chemicals. The process of saccharification can be carried out using either chemical or biological means.

10.9.1 Chemical hydrolysis

The mechanism of acid or alkaline hydrolysis determines the chemical hydrolysis process (Sarkar et al., 2012). Frequently, acid pretreatment facilitates the hydrolysis of hemicellulose, the readily hydrolyzable carbohydrate component of lignocellulosic biomass. The solid fraction consists of lignin and cellulose, while the liquid fraction is composed of degradation products of hemicellulose, which includes pentoses (such as xylose, arabinose) and some hexoses (mannose, and galactose) and oligosaccharides, which result from the partial hydrolysis of hemicellulose and cellulose during pretreatment. The cellulose component can later be subjected to chemical or enzymatic hydrolysis. The chemical procedure involves the use of either concentrated or diluted acid hydrolysis. To accomplish this goal, a variety of inorganic acids, including phosphoric acid (H3PO4), nitric acid (HNO3), hydrochloric acid (HCl), sulfuric acid (H2SO4), and formic acid (CH_2O_2), were utilized. The hydrolysis of van der Waals forces, covalent bonds, hydrogen bonds, and other intermolecular interactions in sugar molecules is facilitated by the catalytic effect of H^+ ions present in the acid. The hydrolysis rate exhibits a direct proportionality to both the concentration of $H3O+$ ions and the temperature (Bensah and Mensah, 2013). The chemical hydrolysis of lignocellulosic biomass typically involves the following steps:

- Pretreatment: The biomass is first subjected to a pretreatment step to break down its structural components and make them more accessible to chemical hydrolysis. This can be achieved by using mechanical, thermal, or chemical methods, such as steam explosion, acid pretreatment, or alkaline pretreatment.
- Acid hydrolysis: The biomass is then subjected to acid hydrolysis, which utilizes an acid catalyst, such as sulfuric acid or hydrochloric acid, to break down the cellulose and hemicellulose into their constituent sugars, such as glucose, xylose, and arabinose. To maximize sugar yield, the acid hydrolysis reaction is typically conducted at high temperature and pressure.
- Neutralization: After the acid hydrolysis reaction, the resulting mixture is neutralized to remove any excess acid and adjust the pH to a more neutral level.
- Separation and purification: The resulting sugar solution is then separated from any remaining solid residues, such as lignin, and purified to remove any impurities that may interfere with subsequent fermentation or chemical reactions.

Chemical hydrolysis is a crucial step in the conversion of lignocellulosic biomass into useful products, and it is extensively employed in the production of biofuels and chemicals. To increase the efficiency and sustainability of lignocellulosic biomass conversion, however, researchers are actively investigating alternative techniques, such as enzymatic hydrolysis.

10.9.2 Biological hydrolysis

Biological hydrolysis of lignocellulosic biomass entails the use of microorganism-produced enzymes to break down the intricate structure of the biomass into simpler sugars that can be used as substrates for the production of biofuels, chemicals, and other value-added products (Keskin et al., 2019). This process is also known as enzymatic hydrolysis and is an alternative to chemical hydrolysis, which uses acids or bases to break down the biomass.

The biological hydrolysis process typically involves the following steps:

- Pretreatment: The lignocellulosic biomass is first subjected to a pretreatment step to deconstruct its structural components and render them more accessible to enzymatic hydrolysis. Physical, chemical, or biological processes, such as steam detonation, acid pretreatment, or microbial pretreatment, can be used to perform pretreatment.
- Enzymatic hydrolysis: The pretreated biomass is then subjected to enzymatic hydrolysis, which utilizes enzymes such as cellulases and hemicellulases to degrade cellulose and hemicellulose into their constituent sugars, such as hexose and pentose sugars. The enzymes are typically produced by microorganisms, such as fungi and bacteria, and can be obtained from commercial sources or produced in-house using rDNA technology.
- Fermentation: The resultant sugar solution is fermented using microorganisms, such as yeast or bacteria, to produce biofuels, such as ethanol, or other value-added products, such as organic acids, enzymes, or specific chemicals.

The biological hydrolysis process has several benefits over chemical hydrolysis, including lower energy consumption, reduced environmental impact, and higher selectivity and specificity. However, the process is still relatively expensive and requires further optimization to achieve commercial viability. Utilizing genetically engineered microorganisms, developing more efficient enzymes, and optimizing fermentation conditions are some of the methods that researchers are actively pursuing to improve the efficacy and cost-effectiveness of biological hydrolysis.

10.10 Fermentation

The simple sugars produced by the hydrolysis of cellulase and hemicellulase are now available for microorganisms to convert metabolically into ethanol and carbon dioxide. The fermentable sugars are consisting of both hexose and pentose sugars. However, microorganisms such as yeast and bacteria can easily metabolize hexose sugar but not pentose sugar, thus making it a challenge for industrial-scale commercialization and cost-effectiveness of the process. Baker's Yeast or *Saccharomyces cerevisiae* is the most widely and frequently used microorganism to date; it can utilize hexose sugars for ethanol production. Hence, the problem of utilization of pentose sugar can be solved by using hybrid yeast strains or coculture of two yeast strains. Genetic engineering plays a crucial role in developing hybrid 2 yeast strains by fusing the protoplast of *S. cerevisiae* and pentose sugar fermenting yeast strain such as *Pachysolen tannophilus, Candida shehatae,* and *Pichia stipites*.

- Separate Hydrolysis and Fermentation (SHF): SHF is a fermentation technique for bioethanol production in which the saccharification and fermentation steps are done separately. First, the lignocellulosic biomass is pretreated to remove lignin and hemicellulose, and then it is hydrolyzed to produce a sugar solution. The sugar solution is then fermented using yeast or bacteria to produce bioethanol (Malik et al., 2022).
- Simultaneous Saccharification and Fermentation (SSF): SSF is a fermentation technique for bioethanol production in which the saccharification and fermentation steps are performed simultaneously. The lignocellulosic biomass is pretreated and then hydrolyzed to produce monomeric sugars. The reducing sugars are then fermented using yeast or bacteria to produce bioethanol. SSF can be more efficient than SHF because the enzymes and microorganisms work together more effectively.
- Prehydrolysis and Simultaneous Saccharification and Fermentation (PSSF): PSSF method for bioethanol production combines the advantages of both SHF and SSF. The lignocellulosic biomass is first pretreated with steam or boiling water to eliminate lignin and hemicellulose. The biomass is then subjected to prehydrolysis to produce a mix of monomeric sugars. Finally, yeast or other microbes are used to ferment the reducing sugars to produce bioethanol.
- Simultaneous Saccharification and Co-Fermentation (SSCF): SSCF is a bioethanol fermentation technique that integrates the hydrolysis and anaerobic digestion of cellulose and hemicellulose in a single step. The lignocellulosic biomass is pretreated and then hydrolyzed to produce a sugar solution. The sugar solution is then fermented using a mixed culture of microorganisms that can ferment both cellulose and hemicellulose to produce bioethanol.
- Consolidated Bioprocessing (CBP): CBP is a fermentation technique for bioethanol production that combines the saccharification and fermentation steps into a single process using a single microorganism or enzyme system. The microorganism or enzyme system can convert lignocellulosic biomass into sugar, which can then be fermented into bioethanol (Periyasamy et al., 2023). This method is more productive and economical than distinct hydrolysis and fermentation procedures.

10.11 Pilot plant: case study

Since 2004, a demonstration plant in Canada has produced cellulosic ethanol, and in 2009, wheat straw production capability reached 1,464,978 L/year. The facility's annual capacity for producing bioethanol derived from sugarcane waste is 40 million liters.

The Sekab facility in Sweden began producing ethanol from waste. Today, the company manufactures ED95 fuel to be used in large vehicles, for example, buses and trucks, which is composed of 95% ethanol, an ignition enhancer along with acorrosion inhibitor. These modifications render the fuel compatible with diesel engines.

2003 saw the signing of a multimillion dollarcontract between the US Department of Energy and Abengoa Bioenergy New Technologies to develop pilot-scale processes in order to integrate lignocellulosic bioethanol production with cereal bioethanol manufacturing in order to obtain the most economically advantageous results overall. The initiative involved bioethanol production from residual starch from agro waste, cellulose, and hemicellulose, principally corn stover. The initial phase in this project, the pilot plant trial phase, was concluded with the successful demonstration of starch conversion and enhanced by-products production. Since September 2007, 1.5 metric tonne per day lignocellulosic biomass refinery pilot plant is being employed for process optimization, engineering data gathering for scale-up, and for the generation of substantial quantities of various coproducts for additional research and analysis. This technology involves hydrothermal pretreatment with dilute acid and enzymatic hydrolysis. Inbicon (Dong Energy), based in Kalundborg (Denmark), harvests approximately 300 metric tonnes per year of cellulosic alcohol produced from soft feedstock such as bagasse from the sugar industry, miscanthus plants, and fruit clusters that were removed. Poet (US) is involved in cellulosic ethanol research and development, and in 2008, the technology was expanded up to a pilot plant. Over $40 million has been spent in research for the establishment of an industrial cellulosic ethanol plant in Emmetsburg, Iowa, which began operations. Other projects related to Bioethanol pilot plant from lignocellulosics include Intalian Beta Renewables (60,000 t/year from straws and giant reed), US based Gevo (54,000 t/year via residues), Chinese Longlive Biotech (50,000 t/year from residues), and Ineos-Bio in Florida (US) (24,000 t/year from kitchen and yard waste). Several pilot and demonstration-scale plants have been established throughout the globe, such as Belgian Bio Base Europe Pilot Plant, the Biorefinery Demonstration Plant of Malaysia, and the Guangxi Academy of Sciences Pilot Plant. These bioethanol production facilities use various feedstocks, which involve agricultural residues, residues from forests, and municipal solid waste accomodating diverse technological intervention (Soccol et al., 2019).

Pilot plant studies are a necessity to validate the projects that are developed in lab scale. Often the lab-scale or bench-top experiments fail to perform when the quantity of the experimental setup increases due to many factors such as enhanced toxicity of the by-product and also the experimental and reactor designs. The studies involved give opportunity to other aspiring researchers to follow the same path and keep the hopes of commercialization alive.

10.12 Why bioethanol has not yet been commercialized in accordance with fermentation techniques? Major challenges that lie with commercialization of lignocellulosic bioethanol

Exponential growth of the world's economy has increased both energy consumption and concern regarding the buildup of greenhouse gases and their effects on climate change. In response to such an alarming situation, various countries have started developing renewable energy from different sources and have tried to commercialize them by blending or other different strategies to culminate polluting emissions into the environment, to decrease the prices of fuel, reducing country's reliance on current petroleum sources, boosting rural, and creating jobs in a sustainable manner and safe environment — a step toward better mankind.

The main factors that contribute to successful commercialization for 2 G bioethanol production include cheap raw materials, economic pretreatment techniques, in-house enzyme production with high and efficient titers, fermentation mode, high ethanol yield, downstream recovery of ethanol, and maximal by-product use (Chandel et al., 2010). The factors mentioned above play a crucial role in commercial production of 2 G ethanol. Till date, transfer of cellulosic ethanol technology from laboratory scale to a commercial scale is an expensive preposition. In this section, we will try to understand the main reasons responsible for unsuccessful commercialization of bioethanol, in accordance with fermentation techniques.

Second-generation bioethanol production via fermentation techniques has not been fully commercialized due to several demerits and limitations. Second-generation bioethanol is produced from lignocellulosic biomass, such as energy crops, forestry residues, and agricultural waste, which are abundant and low-cost feedstocks. However, the lignocellulosics contain intricate and recalcitrant structures of hemicellulose, cellulose, and lignin, which are difficult to break down and ferment into ethanol.

The high expense of pretreatment and hydrolysis procedures is a significant drawback of 2 G bioethanol production (Dey et al., 2020). To make cellulose and hemicellulose available for enzymatic hydrolysis, the lignocellulosic structure must be broken down during the pretreatment process. This procedure uses a lot of energy, produces inhibitors that prevent fermentation, and necessitates toxic chemicals. Utilizing certain enzymes, the enzymatic hydrolysis process transforms cellulose and hemicellulose into fermentable sugars. Enzymes are expensive, and several factors, including the biomass's structure and the pretreatment procedure, might affect their efficiency.

The low ethanol yield and productivity of second-generation bioethanol production are another drawback. The yield and productivity of ethanol are decreased by the lignocellulosics' complex structure and the presence of inhibitory substances (Dey et al., 2020). The effectiveness of the fermentation process, which is impacted by variables including the type of microbe utilized, the conditions of the fermentation, and the composition of the hydrolysate, also affects the yield of ethanol.

However, second-generation bioethanol production also has several merits. One of the key benefits is the utilization of low-cost and abundant feedstocks, which reduces the competition for food and land resources. The use of lignocellulosic biomass also reduces greenhouse gas emissions and enhances sustainability compared with first-generation bioethanol production. Second-generation bioethanol also provides an opportunity to use waste materials and convert them into a valuable product, reducing environmental pollution and waste disposal costs.

In conclusion, second-generation bioethanol production via fermentation techniques is a promising approach for sustainable biofuel production. However, several limitations, including high pretreatment and hydrolysis costs, low ethanol yield and productivity, and the need for efficient fermentation strategies, need to be addressed to fully commercialize the process.

10.13 Supply chain management of lignocellulosics bioethanol

One of the most promising approaches for manufacturing sustainable and renewable energy is the manufacture of bioethanol derived from agricultural waste. Using agricultural waste as a base for bioethanol production can cut waste disposal expenses greatly and with little adverse impact on the nature. However, efficient and effective supply chain management is crucial for the lignocellulosic bioethanol production industry's success.

The supply chain management in the lignocellulosic bioethanol production industry is complex and involves multiple stages, including feedstock collection, transportation, pretreatment, enzymatic saccharification, fermentation, separation, and product distribution. Each of these stages can have a noteworthy impact over the efficiency and profitability regarding the supply chain.

1. Biomass Collection and Transportation: The gathering and distribution of agriculture waste for bioethanol production can have a substantial impact on the supply chain's cost and viability. Efficient collection and transportation strategies can reduce costs and minimize environmental impacts. Studies have shown that centralized collection and transportation systems are more cost-effective than decentralized systems. However, decentralized systems can be more sustainable and environmentally friendly, as they reduce transportation distances and emissions. The use of advanced logistics and optimization tools can also improve the efficiency of feedstock collection and transportation (Yan et al., 2008).
2. Pretreatment and Enzymatic Hydrolysis: The pretreatment and enzymatic saccharification are critical for converting lignocellulosic residues into convertible carbohydrate monomers. These choices of pretreatment and enzymatic hydrolysis methods can significantly impact the efficiency and cost of the supply chain. Various pretreatment methods have been studied, including acid, alkaline, steam explosion, and microwave-assisted methods. The choices made for a suitable pretreatment method depend on various aspects, such as the nature of feedstock, the desired product quality, plus the cost of the process. The use of advanced enzymatic hydrolysis technologies, such as high solid loading and fed-batch processes, can also increase the efficacy of the supply chain.
3. Fermentation and Distillation: The fermentation and distillation stages involve the conversion of 13 fermentable sugars into bioethanol. The efficiency and cost of these stages can be influenced by various factors. The use of high-performance microorganisms and advanced fermentation technologies, for example, simultaneous saccharification and fermentation (SSF) combined with consolidated bioprocessing (CBP) improve the efficiency and profitability of the mode of supply. The usefulness regarding advanced distillation methods, such as membrane distillation and reactive distillation, can also reduce energy consumption and improve product quality.
4. Product Distribution: The distribution of bioethanol products can also impact the efficiency and cost of the supply chain. Efficient product distribution strategies can reduce transportation costs and improve customer satisfaction. The use of advanced logistics and distribution technologies, such as route optimization and real-time tracking, has the ability to foster new channels in supply chain.

One is transportation cost for raw material to the plant cite, and second is the distribution of the final product and by-products that are generating through the biorefinery concept. As India is an agriculture-based country, huge by-products are available that can be considered for bioethanol production. The environmental as well as economic impact is presented as the transportation costs per tonne per kilometer and carbon emissions per kilometer, which keeps on increasing with diameter of biomass traveled form central collection point. It is a major challenge and for that only the recommendation of scientist and technocrats that involved in 2 g ethanol production to go for establishment of decentralized factories that will not only take care of appropriate utilization of agricultural by-product but also help in solving the unemployment ratio of rural sector (Duc et al., 2021).

The inclusion of certain by-product or other end products such as biogas generation from lignocellulose waste after ethanol generation, biomanure from waste, extraction of any biochemical from the process such as glycerol or other organic acid can be integrated for better economy for bioethanol generation.

10.14 Challenges of 2 G bioethanol production: biorefinery concept

Biorefinery comes from the concept of utilizing biomass to conceive an array of commercially viable products and energy. Lignocellulosic waste is a favorable renewable feedstock for the sustainable generation of energy and value-added chemicals that are currently derived from fossil fuels, due to their refuse nature and availability, low cost, steady cost, adaptability, high carbohydrate content, and noncompetitiveness with food production. Nonetheless, lignocellulosic biomass is structurally and chemically heterogeneous and resistant to decomposition. In addition to hexose and pentose sugar, lignocellulosic biomass hydrolysis produces plant fiber, carbon dioxide, organic acids, and lignin residues as by-products (Bensah and Mensah, 2013). To make the process commercially viable, it is necessary to separate and utilize these compounds in a biorefinery or circular bioeconomy system for the generation of energy and additional value chemical molecules for the maximum utilization of lignocellulosic biomass.

A biorefinery or circular bioeconomy is a system that uses a near waste-free method to change biomass into greener fuels, power, and chemicals with added value. Similar to petroleum refineries, a biorefinery extracts multiple fractions from biomass based on their properties and fractions. As a result, biorefineries are able to produce value-added chemicals from the various waste fractions and intermediates generated from the processing of lignocellulosic feedstock.

The common challenges faced by 2 G bioethanol production are the lower yield of energy obtained from lignocellulosic material as compared with 1 G bioethanol crops such as sugarcane and starch-rich crops, the elevated production costs due to pretreatment processes to dismantle native rigid biopolymers, the poor efficiency of enzyme hydrolysis of resistant biomass, and the accumulation of residual sugars and other toxic compounds during the process.

Numerous conversion techniques, such as combustion, thermochemical, and microbial conversions, are integrated in a biorefinery infrastructure to create sustainable value-added products. Using processes such as gasification, pyrolysis, and hydrothermal liquefaction, biomass is converted into syngas, char and bio-oil. The syngas or bio-oil is then used as a feedstock for the creation of value-added products such as biofuels and chemicals. Burning biomass generates heat and power that can be used in the biorefinery or sent to the grid. In a biorefinery, waste heat from combustion is used to power other processes. Sugars and other organic compounds in biomass are converted by microorganisms into biofuels, biochemicals, and other products with added value. Fermentation, anaerobic digestion, and algae cultivation are all examples of microbial conversion technologies. By combining these various conversion technologies, the infrastructure of a biorefinery can maximize the use of biomass feedstocks and minimize waste by creating a closed-loop system in which waste generated in one process is utilized as feedstocks for another process. As a result, the environmental impact is reduced, and the economic feasibility of bio-based products is increased (Sharma & Saini, 2020).

The efficiency of the pretreatment process, which typically involves harsh solvothermal and pyrolytic processes assisted by toxic acids or alkalis that nonselectively degrade lignin into a heterogeneous product stream that is inappropriate to derive energy or chemical compounds and also constitutes accumulated toxic agents, determines the upgradation of lignocellulosic materials into fine chemicals or energy. The condensed lignin obtained in the process is of low quality and is used only for heat generation. Therefore, the focus of research for the valorization of lignocellulosic biomass has shifted to investigating a gentle, selective, and greener industrial method. In this context, a lignin-first approach that realizes high-yield and highly selective aromatic monomers, along with carbohydrate preservation, is regarded as one of the best prospective strategies. An effective "lignin-first" method is reductive catalytic fractionation, where the lignin derivatives are obtained in the liquid phase, assisting the direct hydrodeoxygenation process for the synthesis of coproducts with added value. This method shows promise in producing lignin derivatives of high quality that is utilized for a variety of purposes, including vanillin, carbon fiber, flavors and fragrances, pharmaceuticals, resins, and adhesives (Salapa et al., 2017). The following possible bio-based products and chemicals were identified by the US Department of Energy (USDOE) in 2004: malic, fumaric, and succinic acids, furan dicarboxylic acid, Aspartic acid, itaconic acid, levulinic acid, 3-hydroxybutyrolactone, glutamic acid, 3-hydroxypropionic acid, and 3-hydroxypropionic acid Propanediol, glycerol carbonate, epichlorohydrin, furans (furfural, 5-hydroxymethylfurfural, 2,5-FDCA), lactic acid, sorbitol, succinic acid/aldehyde/3-hydroxy propionic acid, and xylitol (Komesu et al., 2022).

Consequently, a comprehensive method would be preferable for evaluating the extent to which the pretreatment and hydrolysis factors affect the process's efficiency and yield of ethanol. A fundamental practical obstacle to producing and scaling up lignocellulosic ethanol is the fermentation of glucose and xylose sugar components. By discovering naturally occurring bacteria that can hydrolyze several types of sugars and are resilient to inhibitory compounds, this bottleneck is avoided. Alternately, a strain altered to utilize pentose, upregulate already-existing pathways (such as the pentose phosphate route), and downregulate by-product pathways that lead to the buildup of toxins can be used for strain enhancement in lignocellulosic valorization.

The bioconversion of lignocellulose to ethanol continues to be carried out on a small scale despite intensive research efforts and significant improvements in pretreatment, eliminating toxins, hydrolysis, and fermentation. To make the generation of 2 G ethanol economically and industrially practical, a variety of saccharification and ethanol fermentation integration technologies have been proposed. Cellulosic ethanol production uses fewer process units, which directly affects operating and capital costs, energy consumption, and process time (Zabed et al., 2016). The yield of ethanol and other chemicals with added value from lignocellulosics can be significantly increased through the use of efficient microbial strains with improved metabolic pathways, as well as optimized hydrolysis and pretreatment conditions.

10.15 Future scope

Our environment is rich in fermentative microorganisms, few of which are discovered by scientists, but still, there is a wide range of microbes that are left unexplored. The wild strains of microbes that are used for industrial production of 2 G ethanol are unable to meet the requirement of high ethanol yield from lignocellulosic biomass, for example, wild strains of fermentative yeast such as *Saccharomyces cerevisae*, which cannot ferment pentose sugars, whereas strains such as *C. shehatae*, *P. stipites*, and *P. tannophilus* can metabolize pentose into ethanol. Hence, there are a few drawbacks to the use of wild strains such as less heat resistance, less tolerance to pH change, and high ethanol stress. Therefore, to deal with the abovementioned drawbacks, scientists aimed at modification of genes to increase their resistance to the condition of fermentation, to improve their tolerance toward inhibitors generated during fermentation, and also resistance toward high sugar concentration (Ximenes et al., 2010).

In recent years, several techniques for genome modification have been developed. These techniques include the CRISPR/Cas system (Javed et al., 2019), the nuclease-based TALEN system, the zinc finger domain-based ZFN system, the meganuclease system, and the oligonucleotide-based YOGE system (Ulaganathan et al., 2017). Numerous studies have documented that genetic engineering of fermenting bacteria such as *Escherichia coli* and *Zymomonas mobilis* has increased ethanol yield. In other genetic modification techniques, microbes have been engineered in such a way that it can metabolize multiple sugars and enable them to produce ethanol via the glycolysis pathway. The genetic engineering studies are mainly dedicated to the simultaneous use of mixed sugars for ethanol production using strains such as *S. cerevisiae* (yeast), Gram-positive bacteria such as *Clostridium cellulolyticum* and *Lacticaseibacillus casei*, and Gram-negative bacteria such as *Z. mobilis*, *E. coli*, and *Klebsiella oxytoca*. Techniques used for obtaining recombinant strain include mutagenesis and the introduction of a heterologous metabolic pathway for pentose sugar utilization into a well-known conventional strain such as *S. cerevisiae*. The primary objective for developing recombinant genotypes is to provide efficient and cost-effective conversion of feedstock into bioethanol and to reduce processing capital expenditures. Constructing yeast strains capable of cofermenting pentoses and hexoses for the production of lignocellulosic ethanol involves the application of specific genetic engineering techniques. Although genetic engineering has contributed to progress in this field, there is still a lack of both reaction intermediates and efficient pentose transporters.

By 2025, biofuels are predicted to make up a sizeable fraction of the world's renewable energy mix, with bioethanol output rising by 40% over that time period, according to a report by the IEA (2020).

The formation of inhibitors, such as furfural and hydroxymethylfurfural (HMF), during the pretreatment and hydrolysis stages of bioethanol production from the lignocellulosic residue by chemical processes, is one of the biggest hurdles in this process. These inhibitors can inhibit the growth and fermentation of microorganisms, leading to reduced ethanol yield. Several strategies have been developed to overcome this challenge, including detoxification, strain improvement, and cofermentation with multiple microorganisms.

In the coming years, many crop residues could be used as substrates for ethanol production. However, the utilization of each biomass source poses a technological challenge and has to be tailored for each substrate, thus the major challenge lies with the scientific community to develop its process/product based on the availability of the agro residues, at the same time the biggest challenge lies with the policymaking process for uniform applicability throughout the globe. This will not only give us a suitable scheme for the appropriate utilization of biomass but also address the acute environmental pollution related to biomass burning and its proper handling.

10.16 Conclusion

Every nation must determine the most effective and cost-effective method to utilize its raw materials and residues to produce biofuels. Since there are multiple operational steps and high costs of complex conversion and separation, the whole process of second-generation biomass- to-product is typically regarded as unfeasible when only one feedstock and a single product are involved. To increase production capacity and ensure feed supply for each facility, the trend has been toward mixing biomass from various sources and multiple products, also known as a network supply chain. Moreover, the commercialization of lignocellulosic bioethanol has already begun in some countries. For example, the Brazilian sugarcane industry has started to produce bioethanol from sugarcane bagasse, a lignocellulosic waste generated during sugarcane processing. The United States also has several commercial-scale bioethanol plants that use corn stover and other lignocellulosic waste as feedstock. The future of lignocellulosic bioethanol production looks promising through a greener approach and will be of great importance to mankind (Table 10.2).

TABLE 10.2 Different pretreatment methods utilized to obtain high-reducing sugar.

Method type		Pretreatment type		
Conventional method	Physical treatment	Milling		
		Extrusion		
		Ultrasonication		
		Microwave		
	Chemical treatment	Acid		
		Alkali		
		Ionic		
		Organosolv		
		Ozonolysis		
		Metalsalts		
		Sulfite		
		oxidative		
		Pyrolysis		
	Physico-chemical treatment	Steam explosion		
		Liquid hot water (LHW)		
		Ammonia fiber explosion (AFEX)		
		Ammonia percolation (ARP)	recycle	
		Supercritical fluid (SCF)		
Advanced method	pretreatment	Combined biological and chemical/physicochemical pretreatment	Biological − Acid/Alkaline /Oxidative/ Organosolv/LHW/HWE/Autohydrolysis/Steam Explosion	
			Combined Dilute Acid & steam Explosion	
			Microwave-assisted liquid	Ionic
			Mechanical Size reduction and Electrostatic Separation	
			Alkaline pretreatment and autoclave	

References

Amin, F. R., Khalid, H., Zhang, H., Rahman, S. U., Zhang, R., Liu, G., & Chen, C. (2017). *Pretreatment methods of lignocellulosic biomass for anaerobic digestion* (7, pp. 1–12). AMB Express.

Bensah, E. C., & Mensah, M. (2013). Chemical pretreatment methods for the production of cellulosic ethanol: Technologies and innovations. *International Journal of Chemical Engineering*, 2013.

Chandel, A. K., Singh, O. V., Chandrasekhar, G., Rao, L. V., & Narasu, M. L. (2010). Key drivers influencing the commercialization of ethanol-based biorefineries. *Journal of Commercial Biotechnology*, 16, 239–257.

Dey, P., Pal, P., Kevin, J. D., & Das, D. B. (2020). Lignocellulosic bioethanol production: Prospects of emerging membrane technologies to improve the process– a critical review. *Reviews in Chemical Engineering*, 36(3), 333–367.

Duc, D. N., Meejaroen, P., & Nananukul, N. (2021). Multi-objective models for biomass supply chain planning with economic and carbon footprint consideration. *Energy Reports*, 7, 6833–6843.

Ferreira, J. A., Mahboubi, A., Lennartsson, P. R., & Taherzadeh, M. J. (2016). Waste biorefineries using filamentous ascomycetes fungi: Present status and future prospects. *Bioresource Technology*, 215, 334–345.

Hoang, T. D., & Nghiem, N. (2021). Recent developments and current status of commercial production of fuel ethanol. *Fermentation*, 7(4), 314.

Hoşgün, E. Z., Biran Ay, S., & Bozan, B. (2021). Effect of sequential pretreatment combinations on the composition and enzymatic hydrolysis of hazelnut shells. *Preparative Biochemistry & Biotechnology*, 51(6), 570–579.

IEA. (2021). Bioenergy. Available from https://www.iea.org/reports/bioenergy, License: CC by 4.0.

IEA. (2020). World Energy Outlook 2020. Available from https://www.iea.org/reports/world-energy-outlook-2020, License: CC BY 4.0.

Iqbal, et al. (2013). 'Biotech applications of biomass,'. *BioResources*, 8(2), 3157–3176.

Javed, M. R., Noman, M., Shahid, M., Ahmed, T., Khurshid, M., Rashid, M. H., & Khan, F. (2019). Current situation of biofuel production and its enhancement by CRISPR/Cas9-mediated genome engineering of microbial cells. *Microbiological Research*, 219, 1–11.

Keskin, T., Abubackar, H. N., Arslan, K., & Azbar, N. (2019). *Biohydrogen production from solid wastes. Biohydrogen* (pp. 321–346). Elsevier.

Komesu, A., Oliveira, J., Moreira, D. K. T., Khalid, H., Neto, J. M., & da Silva Martins, L. H. (2022). *Biorefinery approach for production of some high-value chemicals. Advanced biofuel technologies* (pp. 409–429). Elsevier.

Li, H., Yelle, D. J., Li, C., Yang, M., Ke, J., Zhang, R., & Mo, J. (2017). Lignocellulose pretreatment in a fungus-cultivating termite. *Proceedings of the National Academy of Sciences*, 114(18), 4709–4714.

Madadi, M., Tu, Y., & Abbas, A. (2017). Recent status on enzymatic saccharification of lignocellulosic biomass for bioethanol production. *Electronic Journal of Biology*, 13(2), 135–143.

Malik, K., Sharma, P., Yang, Y., Zhang, P., Zhang, L., Xing, X., & Li, X. (2022). Lignocellulosic biomass for bioethanol: Insight into the advanced pretreatment and fermentation approaches. *Industrial Crops and Products*, 188, 115569.

Matsakas, L., Kekos, D., Loizidou, M., et al. (2014). Utilization of household food waste for the production of ethanol at high dry material content. *Biotechnology for Biofuels*, 7, 4. Available from https://doi.org/10.1186/1754-6834-7-4.

Meenakshisundaram, S., Fayeulle, A., Leonard, E., Ceballos, C., & Pauss, A. (2021). Fiber degradation and carbohydrate production by combined biological and chemical/physicochemical pretreatment methods of lignocellulosic biomass–a review. *Bioresource Technology*, 331, 125053.

Mikulski, D., & Kłosowski, G. (2020). Microwave-assisted dilute acid pretreatment in bioethanol production from wheat and rye stillages. *Biomass and Bioenergy*, 136, 105528.

Periyasamy, S., Isabel, J. B., Kavitha, S., Karthik, V., Mohamed, B. A., Gizaw, D. G., & Aminabhavi, T. M. (2023). Recent advances in consolidated bioprocessing for conversion of lignocellulosic biomass into bioethanol–a review. *Chemical Engineering Journal*, 453, 139783.

Premjet, S. (2018). *Potential of weed biomass for bioethanol production. Fuel ethanol production from sugarcane* (pp. 83–98). IntechOpen.

Raj, T., Chandrasekhar, K., Kumar, A. N., Banu, J. R., Yoon, J. J., Bhatia, S. K., & Kim, S. H. (2022). Recent advances in commercial biorefineries for lignocellulosic ethanol production: Current status, challenges and future perspectives. *Bioresource Technology*, 344, 126292.

Rajak, R. C., & Banerjee, R. (2016). Enzyme mediated biomass pretreatment and hydrolysis: A biotechnological venture towards bioethanol production. *Rsc Advances*, 6(66), 61301–61311.

Salapa, I., Katsimpouras, C., Topakas, E., & Sidiras, D. (2017). Organosolv pretreatment of wheat straw for efficient ethanol production using various solvents. *Biomass and Bioenergy*, 100, 10–16.

Sarkar, N., Ghosh, S. K., Bannerjee, S., & Aikat, K. (2012). Bioethanol production from agricultural wastes: An overview. *Renewable Energy*, 37(1), 19–27.

Sharma, D., & Saini, A. (2020). Lignocellulosic ethanol production from a biorefinery perspective. Singapore: Springer.

Shi, A. Z., Koh, L. P., & Tan, H. T. (2009). The biofuel potential of municipal solid waste. *GCB Bioenergy*, 1(5), 317–320.

Soccol, C. R., Faraco, V., Karp, S. G., Vandenberghe, L. P., Thomaz-Soccol, V., Woiciechowski, A. L., & Pandey, A. (2019). *Lignocellulosic bioethanol: Current status and future perspectives. Biofuels: Alternative feedstocks and conversion processes for the production of liquid and gaseous biofuels* (pp. 331–354). Elsevier.

Sorn, V., Chang, K. L., Phitsuwan, P., Ratanakhanokchai, K., & Dong, C. D. (2019). Effect of microwave-assisted ionic liquid/acidic ionic liquid pretreatment on the morphology, structure, and enhanced delignification of rice straw. *Bioresource Technology*, 293, 121929.

Ulaganathan, K., Goud, S., Reddy, M., & Kayalvili, U. (2017). Genome engineering for breaking barriers in lignocellulosic bioethanol production. *Renewable and Sustainable Energy Reviews*, 74, 1080–1107.

Vasić, K., Knez, Ž., & Leitgeb, M. (2021). Bioethanol production by enzymatic hydrolysis from different lignocellulosic sources. *Molecules (Basel, Switzerland)*, 26(3), 753.

Viikari, L., Vehmaanperä, J., & Koivula, A. (2012). Lignocellulosic ethanol: From science to industry. *Biomass and Bioenergy*, 46, 13–24.

Ximenes, E., Kim, Y., Mosier, N., Dien, B., & Ladisch, M. (2010). Inhibition of cellulases by phenols. *Enzyme and Microbial. Technology*, 46(3–4), 170–176.

Yan, N., Zhao, C., Dyson, P. J., Wang, C., Liu, L. T., & Kou, Y. (2008). Selective degradation of wood lignin over noble-metal catalysts in a two-step process. *ChemSusChem: Chemistry & Sustainability Energy & Materials, 1*(7), 626–629.

Yang, S. T. (2007). *Bioprocessing—from biotechnology to biorefinery. Bioprocessing for value-added products from renewable resources* (pp. 1–24). Elsevier.

Zabed, H., Sahu, J. N., Boyce, A. N., & Faruq, G. (2016). Fuel ethanol production from lignocellulosic biomass: An overview on feedstocks and technological approaches. *Renewable and Sustainable Energy Reviews, 66,* 751–774.

Zhang, J., Zhou, H., Liu, D., & Zhao, X. (2020). *Pretreatment of lignocellulosic biomass for efficient enzymatic saccharification of cellulose. Lignocellulosic biomass to liquid biofuels* (pp. 17–65). Academic Press.

Zoghlami, A., & Paës, G. (2019). Lignocellulosic biomass: Understanding recalcitrance and predicting hydrolysis. *Frontiers in Chemistry, 7,* 874.

CHAPTER

11

Biodiesel from lipid-rich wastes: prospects and challenges in commercialization

Sarveshwaran Saravanabhupathy, Swagata Dutta and Rintu Banerjee

Agricultural & Food Engineering Department, Indian Institute of Technology, Kharagpur, Kharagpur, West Bengal, India

Abbreviations

WCO	Waste cooking oil
FAME	Fatty acid methyl ester
UCO	Used Cooking Oil
CAGR	Compound annual growth rate
IEA	International Energy Agency
ASTM	American Standard Test Method
EN	European
ROI	Return on investment
PBT	PayBack time
NPV	Net present value
PC	Production capacity
FFA	Free fatty acid

11.1 Introduction

Energy is one of the crucial components for preserving a nation's economic development. The need for energy has significantly increased due to rapid industrialization, population growth, expanding urbanization, and economic expansion; fossil fuels have been a major factor in meeting this demand. However, the worries about the depletion of nonrenewable fossil fuel reserves and their availability in the near future, as well as the serious warnings about greenhouse gas emissions caused by excessive fossil fuel use that would cause a dramatic change in the global climate, have prompted researchers to look for alternative, sustainable, and renewable energy sources. In this context, biodiesel has lately gained importance as a supplement or alternative for conventional diesel fuel. Many countries have shown interest in biodiesel production due to its renewability, technicality, and eco-friendliness. The transport sector alone consumes 28% of global energy needs, with biofuel accounting for 3% of that energy (Dewangan et al., 2018).

From chemical viewpoint, biodiesel is made up of monoalkyl fatty acid esters having long C-chains and is produced by transesterifying bio-derived fats with an alcohol having short C-chain. As the biodiesel's characteristics are comparable with that of commercial diesel, it can be used in the contemporary engines (Balat & Balat, 2010). Moreover, biodiesel combustion emits lesser amount of environment-damaging substances compared with commercial fuels, thus reducing environmental risks significantly. It also overcomes the dependency of imported petroleum, offering rural economic sustainability using local resources. It thus helps in circumventing the energy needs while mitigating environmental damages. The primary advantages of using biodiesel are its ease of combustion, biodegradability, and renewability. Therefore, using biodiesel might be a great way to help the transportation industry, which primarily relies on diesel fuel, while also lowering hazardous gas emissions and preserving a healthy environment.

Various feedstocks have been explored for the biodiesel production; however, commercial viability of biodiesel is dependent on its large-scale production using low-cost feedstocks. Hence, finding the inexpensive feedstock is critical for lowering the production costs. The principal feedstocks for biodiesel manufacturing can be classified into different groups such as (1) Agro/Vegetable source, which includes edible vegetable oil sources such as palm, olive, ground nut, soybean, sunflower (1G Biodiesel) along with nonedible vegetable oil sources such as *Jatropha*, rubber seed, karanja, safflower, and cottonseed, which represent 2G biodiesel feedstocks (Udayakumar et al., 2022); (2) Animal source, which includes slaughter house animal fats such as beef tallow, fleshing oil, pork lard, veal, mutton fat, lamb meat; poultry farm animal fats/poultry fats such as chicken fat, yellow grease, duck tallow, feather meal; and wastes from fish oil (Foroutan et al., 2021). (3) Microbial sources; microalgae, which represent 3G biodiesel feedstocks; (4) Industrial source; (5) Other lipid-rich wastes such as waste frying oil, waste acidified oil, and waste water; (6) Alcohols such as methanol, ethanol, propanol, butanol, and pentanol, as transesterification agents and can also act as a feedstock for microbial oil production (Verma et al., 2016). Out of all the above discussed feedstocks, lipid-rich feedstock has the ability to fulfill the world's energy demands due to the limited price and favorable quality of biodiesel.

Currently, biofuels are more costly than fuels made from petroleum, but this may change. Unfortunately, enhanced production economics are insufficient for environmentally sustainable output or the viability of it on a large scale. For the production of biofuels to be feasible, a low-cost point supply of raw materials along with the other necessary resources is required. Variations in price indicate an uncertain business environment for biodiesel. A switch to biofuels should be primarily motivated by their possibly substantially lower carbon impact than petroleum-based energy. Hence, in this chapter, more emphasis has been given to the commercial aspects of biodiesel, and major focus has been given on the present biodiesel market statistics and difficulties in commercializing biodiesel production from diverse lipid-rich feedstocks with some successful commercial case studies.

11.2 Biodiesel market and Indian scenario

Different generations of biodiesel have different level of commercial reach worldwide, the location and feedstock availability being the major deciding factors. According to the International Energy Agency (IEA), 89% of the feedstocks utilized in the manufacturing of biodiesel globally in 2019 were from vegetable oils (IEA, 2021). Argentina, Brazil, Germany, and the United States produced the most 1G biodiesel in 2019. The usage of 1G biodiesel has been criticized for leading to food poverty and deforestation (Masi et al., 2021).

The use of 2G biodiesel has been advocated as a strategy to use waste and by-products from different industries and to lessen competition for food crops. According to the IEA, spent cooking oil accounted for 50% of the feedstocks used in the manufacture of biodiesel in 2019, accounting for 2% of the world's total output (IEA, 2021). On the other hand, producing 2G biodiesel is having problems such as high feedstock production costs, water requirement for obtaining a quality oil seeds, and the requirement for energy-efficient oil extraction methodology. Whereas the third-generation biodiesel solves many of the discussed problems, its commercial reach requires more research.

The size of the global biodiesel market was estimated to be US$ 92.3 billion in 2021. It is anticipated that the global biodiesel market would rise from $189.7 billion in 2030, with a compound annual growth rate (CAGR) of 8.33% between 2022 and 2030. The market size in 2022 was roughly US$ 99.99 billion and will increase (US$ 189.7 billion by 2030). Within the context of the expanding global biodiesel market, the market overview for the period of 2023–2028 has indicated that the Indian biodiesel market attained a value of US$ 383.4 million in the year 2022 and will continue to witness further growth in the future (Biodiesel Market, 2022–2030). As a consequence, there is an increasing demand for biodiesel as a substitute for traditional fossil fuels in various sectors such as transportation and power production (Balat & Balat, 2010). The primary factors that industrialists consider when establishing themselves in any sector are financial investment and technological advancements. The presence of multiple vendors results in significant market variability. Furthermore, the extensive research and development involved in selecting feedstocks for product manufacturing, coupled with the demand and supply imbalance resulting from inadequate production efficiency, are associated with significant advantages for fresh players in the industry (Lopez & Laan, 2008).

The biodiesel sector in India has faced significant challenges in the past 2 years due to the closure of factories related to the COVID-19 pandemic and high operating costs. Efforts to establish a reliable delivery system for used cooking oil (UCO) are currently underway. The current demand for biodiesel remains insufficient, with a

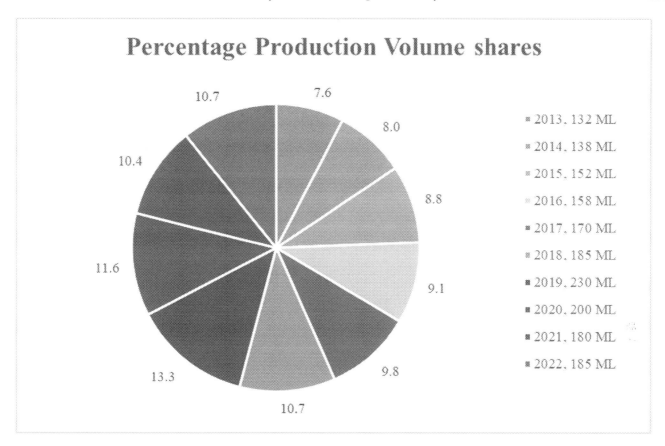

FIGURE 11.1 Production volume of biodiesel India 2013–2022 represented in percentage shares by year wise. *Source: In the data legends the actual volume in million litre (ML) is expressed (Statista, June 2022).*

significant portion of its distribution occurring in unorganized regions lacking substantial support from India's oil marketing companies. Furthermore, the expansion of the market has been impeded due to a scarcity of superior feedstock resources.

From Fig. 11.1, it is observed that 185 million liter s of biodiesel were expected to be produced in 2022 in India. The anticipated national blending rate for the year 2022 was 0.07%. Biodiesel is generated through the utilization of animal fats procured from the local area, along with minimal quantities of nonconsumable oils and UCO. Additionally, imported palm stearin and palm acid oil are also incorporated into the production process. The countries of Malaysia and Indonesia, which are major producers of palm oil, require a mandatory minimum national biodiesel blend ranging from B10 to B30. India has also set a target of achieving a B5 biodiesel mix for on-road use by the year 2030 (Statista Production volume of biodiesel in India from 2013 to 2022). But the utilization of biodiesel remains limited owing to constraints such as import restrictions, high feedstock expenses, an unstructured and uncommitted supply chain, and limited feedstock accessibility. The prevalence and proliferation of fake, low-cost oil alternatives that are advertised as biodiesel impede the efficacy of regulatory and supervisory measures. The restricted availability of alternative feedstocks constrains domestic production and technological investments in India, thereby posing a hindrance to the expansion of the biodiesel industry. Hence, large-scale biodiesel production needs various alternative potential lipid-rich feedstock sources, some of which has been discussed below.

11.3 Potential of lipid-rich feedstocks for large-scale biodiesel production

Various feedstocks have been widely explored globally for biodiesel production. Table 11.1 presents some of the lipid-rich feedstocks that have been reported in different literatures. The principal feedstocks for biodiesel manufacturing can be classified into different groups, which are discussed below:

TABLE 11.1 A brief table for some sources of lipid-rich feedstocks.

Lipid-rich feedstocks	Sources	Biodiesel yield	References
Vegetable-based edible oil source	Palm + cottonseed oil	96.67%	Razzaq et al. (2022)
Vegetable-based nonedible oil source	Jatropha oil	–	Pramanik (2003)
	Sunflower + Mahua oil	–	Udayakumar et al. (2022)
	Rubber seed oil	94%	Rahman et al. (2022)
Microalgal source	Microalgae	86%	Rahman et al. (2022)
Animal source	Beef tallow animal fat residues	71%	Olubunmi et al. (2022)
	Mutton fat residues	93%	Bhatti et al. (2008)
	Chicken fat residues	99%	Foroutan et al. (2021)
		90%	Odetoye et al. (2021)
	Chicken and pig fat residues	80%	Cunha et al. (2013)
Others	Grease trap waste	96.7%	Tran et al. (2018)
	Waste oil	78%–98%	Talebian-Kiakalaieh et al. (2013), Wang et al. (2007)
	Waste water	14%–33%	Balasubramanian et al. (2018) Patino et al. (2021)

11.3.1 Agro/vegetable source

11.3.1.1 Edible vegetable oil sources

Lipid-rich edible oil sources utilized for the production of first-generation biodiesel account for the major percentage of biodiesel produced in the world. Although use of feedstock varies with its availability in different countries, the different feedstocks that are used worldwide include soybean, palm, canola, coconut, sunflower, sesame, mustard, rapeseed, hazel nut, peanut, olive, ground nut, rice bran oil, etc. (Chakraborty et al., 2014). These edible oils have been used as single oil, like biodiesel production from soybean oil using whole immobilized cell as catalyst (Sharma et al., 2021) or immobilized fungal cells as biocatalyst (Rial et al., 2021). They have also been used as blending of oils, like blending of palm oil with cotton seed oil used in a study to optimize the yield of the biodiesel (Razzaq et al., 2022).

But using these edible oils as biodiesel creates a continuous competition between their use as food and their use as fuel. Also the lands meant for growing food crops are being used for fuel. Hence, there is a possibility that massive production of biodiesel from edible vegetable oil sources might result into shortage of global food. Environmentalists have already started linking the detrimental effects such as deforestation and ecological destruction with the biodiesel derived from edible oil sources. This further resulted in increase of the price of edible vegetable oils. To combat such disastrous circumstances, use of nonedible agro/vegetable feedstock came into the picture.

11.3.1.2 Nonedible vegetable oil sources

To reduce dependency on edible agro-based lipid-rich feedstocks, lipid-rich nonedible vegetable oils were explored for biodiesel. These feedstocks are not suitable for human use and can be grown easily at a low cost in wastelands (Banković-Ilić et al., 2012). This approach thus proved to be a better alternative, mainly for the developing world, as it reduced the usage of expensive palatable food crops for biodiesel and also minimized the land competition. A large number of lipid-rich nonedible plants are already found in abundance in nature such as jatropha, jojoba, mahua, rubber seed, karanja, safflower, tobacco, cottonseed, linseed, *Pongamia*, *Calophyllum*, and reetha. These represent the 2G biodiesel feedstocks. In a study, jatropha oil blended with commercial diesel in a ratio 1:1 or 2:3 produced best-quality alternative diesel for contemporary engines (Pramanik, 2003). In another study, crude sunflower oil and mahua oil were mixed equally as a potential feedstock for biodiesel production (Udayakumar et al., 2022). In another experiment, biodiesel with a higher yield of 94% was obtained when rubber seed oil was used with a dual catalyst CaO and Zn (Rahman et al., 2022).

11.3.2 Microalgal source

Another prospective feedstock for biodiesel production is microalgae. Using microalgae as a green energy source for producing 3G biodiesel following acid and alkali-based esterification provided a biodiesel yield of 86% (Rahman et al., 2022). Time independency, rapid and ease of cultivation in any type of land, eco-friendliness, and other advantages of cultivating microalgae makes microalgal fuel a potential replacement for fossil fuels.

11.3.3 Animal source

Animal residue is another source that can act as a cheap feedstock for biodiesel generation. Huge amount of animal waste residues are generated from slaughter houses that cause pollution, especially the fats need to be treated or recycled. Such fats include beef fat from cattle, sheep based mutton fat, pig pork-derived lard and poultry fat, blood, skin, offal, and trims after rendering process; and wastes from fish oil. Although these animal fats are from various sources, they contain some common saturated C16–C18 fatty acids such as palmitic, stearic acid and unsaturated fatty acids such as oleic acid as monounsaturated fatty acid; and linoleic and arachidonic acids as polyunsaturated fatty acids. Among these, most possess a large percentage of saturated fatty acids, which requires pretreatment but produces more stable biodiesel having high cetane numbers.

Beef tallow animal fat was utilized for biodiesel synthesis using CaO as catalyst to achieve a yield of 71% at 60°C for 1.5 h (Olubunmi et al., 2022). In another study, methyl ester yield as high as 93% was obtained from waste mutton fat after 1-day treatment with acid at 60°C (Bhatti et al., 2008). Among the animal residues for biodiesel making, poultry fat comprises high amount of free fatty acids (FFAs). Hence, pretreatment is required for removal of these FFAs and to achieve high biodiesel yield. A pretreated poultry waste containing waste chicken fat and egg shell as catalyst generated biodiesel with a yield of about 90% (Odetoye et al., 2021). Methyl ester yield as high as 99% was obtained from chicken fat after 1-day treatment with acid at 50°C (Bhatti et al., 2008). Poultry wastes are also used in combination with other animal residues. For example, a biodiesel yield of 80% was achieved during ethanolysis of a pretreated mixture of chicken and pig fat residues at 30°C using mild alkaline treatment (Cunha et al., 2013). Using heterogeneous waste glass catalyst for producing biodiesel from waste chicken fat under alkaline condition showed 99% yield when optimized by response-surface method and artificial neural network (Foroutan et al., 2021).

Thus huge amount of generated animal wastes all around the globe on a regular basis can serve as an excellent feedstock source for producing cost-effective biodiesel on a commercial scale.

11.3.4 Other lipid-rich sources

11.3.4.1 Industrial waste

A high lipid-containing feedstock for cost-effective biodiesel production is grease trap waste. Grease trap waste is captured using interceptors used in the sewage pipes of several restaurants and food processing plants to prevent drain clogging. Although its collection cost is low, but the highly expensive pretreatment decreases its feasibility from becoming a commercial biodiesel. Hence, various production processes are being adopted to bring down the cost of pretreatment such as using ethanol alone or using ethanol with acetone. A study conducted using ethanol alone as well as with acetone showed production cost of US$1337.5 per tons. Not only the amount of lipid extraction solvent and the extraction time impacts the biodiesel production, but also the quality of the feedstock is a major contributor to the richness of the produced biodiesel. Moreover, different extraction solvents or nonsolvent extraction techniques and enzymatic biodiesel synthesis might be applied for further improvement of the biodiesel production (Tran et al., 2018).

11.3.4.2 Waste water

Waste water acts as another potential resource for biodiesel production as a lipid-rich feedstock. Sewage sludge, milk processing industrial sludge, etc., have been utilized as lipid-rich feedstocks for the making of biodiesel using different homogeneous and heterogeneous catalysts. An initial pretreatment step can further enhance the biodiesel production from these feedstocks. Sewage sludge with heterogeneous catalysts can be used to produce biodiesel. In a recent study, for producing biodiesel from sewage sludge, a mixture of acidic ion exchange resins was selected as heterogeneous catalyst with ultrasound pretreatment, which resulted in a yield of about 33% (Patiño et al., 2021). The fatty acids present were esterified, and the biodiesel was extracted by solvent extraction method. At pilot plant scale, dried milk processing sludge produced biodiesel with a yield of about 14% (Balasubramanian et al., 2018).

11.3.4.3 Wasteoil

Any edible oil containing animal or vegetable materials that has been used in food preparation but is no longer safe for human consumption is referred to as waste edible oil. Waste frying oil can be obtained from households, restaurants, and food centers. A large amount of waste cooking oil (WCO) is generated worldwide, but its disposal seems difficult as it obstructs the drains and sewers and causes odor and vermin issues leading to water pollution and causing wildlife concerns. Dumping it in a municipal solid waste landfill or a municipal sewage treatment facility is also illegal and causes difficulties. Its use in animal feeds has also been blocked due to its harmful effect on animals. Although the property of this waste oil differs from that of fresh oil due to repeated use, yet being a lipid-rich source, this can be utilized as a biodiesel feedstock. Studies on using WCO for biodiesel production are rapidly increasing due to the increase in food consumption with increased human population. Using this waste oil can bring down the production cost of biodiesel to about 60%–90% (Talebian-Kiakalaieh et al., 2013). Using a heterogeneous catalyst (ferric sulfate), 97.22% of FFA in the WCO was converted to methyl esters at a temperature of 95°C, when esterification followed transesterification process (Wang et al., 2007).

11.4 Different techniques for large-scale production of biodiesel

Biodiesel can be produced on a large scale using a variety of techniques. Some of the most common methods are:

11.4.1 Transesterification

Lipids can be directly converted to biodiesel through the chemical way of transesterification, which encompasses the reaction of an oil/fats in the presence of alcohol to yield esters and glycerol. The rate of reaction and product yield are improved with the application of a catalyst. Excess alcohol is utilized to move the reaction in equilibrium toward product side since the nature of the reaction is reversible.

$$\text{Glycerides}(\text{Lipid rich substrate}) + \text{Alcohol}_{(\text{excess})} \rightarrow \text{Esters} + \text{Glycerin}$$

The transesterification reaction must be adequately complete in order to achieve fuel quality requirements, such as the European Standard EN 14214, stating that for biodiesel fatty acid methyl or ethyl esters content >96.5% m/m. Transesterification's primary goal is to achieve triacyl glycerol's viscosity considerably comparable with that of conventional petroleum-derived diesel (Pullen & Saeed, 2015).

Transesterification can be achieved either to be catalyzed or noncatalyzed. Homogeneous and heterogeneous reactions can be acid or alkali-catalyzed, and enzyme-catalyzed procedures are the widely explored options. Supercritical methanol transesterification, which operates at greater pressure and temperature, is often the noncatalyzed alternative. Each process has pros and cons depending on the properties of the feedstock.

1. Catalyzed transesterification
   a. Base-catalyzed transesterification

   This is the most widely used method for producing biodiesel on a large scale. It involves the reaction of animal fats or vegetable oil with an alcohol, usually ethanol or methanol, in the presence of an alkaline catalyst (NaOH or KOH). The resulting product is biodiesel and glycerin, which can be separated through a process called glycerolysis. The alkoxide anion ($R'O^-$) produced by the base catalyst acts as a driving force of the reaction.

   Saponification and catalyst loss occur during base-catalyzed transesterification when the FFA level of the feedstock is high, as FFA gets converted to soap by the base catalyst; while less catalyst loss occurs for feedstocks containing less amount of FFA forming less water and soap. Alkali catalyst usage for feedstocks with FFA levels > 5% is stated to interfere with soap formation along with water, which prevents the separation of glycerol and esters and leads to emulsion development during washing. As a result, in order to prepare the feedstocks for base-catalyzed transesterification with high FFA concentration, acid-catalyzed esterification to alkyl esters is essential (Pullen & Saeed, 2015).

b. Acid-catalyzed transesterification

Acid-catalyzed transesterification, one of the several biodiesel manufacturing methods, facilitates the use of feedstock with a greater FFA concentration (cheaper feedstock). The downstream separation and purification procedures are made easier by the absence of soap production when acid is used as a catalyst. This technology can use feedstock with higher FFA content, more than 5%. (Lam et al., 2010; Miao et al., 2009). In practice of biodiesel generation utilizing animal fats and used oils, acid-catalyzed transesterification process is one of the suitable ways. This is primarily because no extra pretreatment step is necessary to minimize FFA, separation and purification of the product may be carried out using straightforward process steps since no soap is formed (Kulkarni et al., 2006; Zhang et al., 2003). Sulfuric acid is the most explored of the several potential forms of acid catalysts for the generation of biodiesel. It is capable of catalyzing the reaction under atmospheric pressure and at a moderate temperature range, ranging from 55°C to 88°C (Gebremariam & Marchetti, 2018).

c. Enzyme-catalyzed transesterification

The traditional method demands feedstock criteria, such as low water and FFA levels, to avoid the production of soap. The development of an alternative enzymatic approach that can function at relatively low temperatures and normal pressure, lowering energy consumption and making the process more environmentally friendly, has received increasing attention as a result of these disadvantages. Here, enzymes act as catalysts to speed up the reaction between the oil or fat and an alcohol, typically methanol or ethanol, to form fatty acid methyl esters (FAMEs). The enzyme used in this process is typically lipase that can break down fats and oils into their constituent fatty acids and alcohols. Lipase can be sourced from a variety of organisms, including bacteria, fungi, and plants. It also produces a higher biodiesel yield and can be used with a wider range of feedstocks (Hama & Kondo, 2013).

2. Noncatalyzed transesterification

Supercritical methanol process: This method involves the use of supercritical methanol, which is methanol heated to high temperatures and pressures. The high pressure and temperature create a supercritical state, which makes the methanol highly reactive and able to convert vegetable oil or animal fat into biodiesel in a single step. This process is faster and more efficient than transesterification, but it requires more energy. As there is no consideration of a catalyst in the process, this technique can result in the lowest material cost. However, because of its greater pressure and temperature necessities, this process results in expensive utility, thus making the method costly. (Tomás-Pejó et al., 2021).

11.4.2 Pyrolysis

This method involves the thermal decomposition of vegetable oil or animal fat in lack of oxygen to produce a mixture of liquid and gaseous products, which can then be separated and purified to obtain biodiesel. Pyrolysis is more energy-intensive than transesterification, but it can be used with a wider range of feedstocks. This technique utilizes heat along with nitrogen or air sparging to chemically alter the fat molecules. To identify a bio-based fuel suitable for diesel engines, many studies have been conducted on the pyrolysis of triglycerides. Waste cooking and frying oil and waste polyolefins were subjected to microwave copyrolysis to convert them into biofuel. The conversion of triglycerides to paraffins/oleins-like end products has been studied in relation to some parameters such as temperature and types of catalysts as used in petrodiesel production (Lam et al., 2017; Mahari et al., 2018).

11.4.3 Microemulsion

Emulsion is the dispersion of oil, surfactant, and water, which are thermodynamically unstable but remains in conjunction with an amphiphilic molecule known as cosurfactant. This method involves the use of a surfactant to create a stable emulsion of vegetable oil or animal fat and alcohol, which can then be reacted to produce biodiesel. Microemulsion can be used with a wide range of feedstocks and is less energy-intensive than other methods, but it requires more processing steps and can be more expensive. Using waste oils as emulsion fuel enhances combustion, particularly for extremely viscous gasoline. Water decreases the flash point, which results in reduced NOx generation. It also encourages puffing and microexplosion events, which prevents the creation of particulates (Kerihuel, 2007).

Each of these techniques has pros and cons of its own, and the choice of method will depend on factors such as the feedstock, the scale of production, and the desired properties of the biodiesel.

11.5 Scale-up of biodiesel production and technoeconomic analysis

Scale-up studies and technoeconomic analysis are both critical steps in the commercialization of biodiesel production. Scale-up studies involve optimizing the process of biodiesel production at the laboratory or pilot scale to ensure that it can be successfully implemented on a larger commercial scale. This includes evaluating various parameters such as feedstock selection, reaction conditions, and catalyst types to determine the optimal conditions for maximum yield and quality of biodiesel. Technoeconomic analysis is used to evaluate the feasibility of commercial-scale biodiesel production by estimating the cost of production and potential revenue streams. It includes evaluating the principal and operational costs associated with the production process, as well as factors such as feedstock costs, market demand, and potential revenue from by-products or coproducts.

11.5.1 Biodiesel and its scaling-up

Majority of the studies on biodiesel have been conducted on a lab or pilot size; therefore production of biodiesel on a large volume for industrial usage has not begun yet. However, researchers have studied/ modeled the scale-up of biodiesel production utilizing various waste from a lab scale to a commercial scale, some of the examples are discussed below.

Olkiewicz has used experimental data and performed computational scale-up to assess the commercial feasibility of biodiesel made from metropolitan and rural wastewater sludge. The volume of solvent used and the extraction time were the operational factors that had the largest impact on cost. The modified extraction method had a biodiesel break-even rate of 1232 $/tonne, which was reasonably comparable with the price of fossil diesel. Thus the gate price suggested by the author for their biodiesel was less expensive than mineral diesel as well as microalgae-based biodiesel. The price of biodiesel was reduced by eliminating the costly phase of sludge drying in the proposed method for producing biodiesel from waste sludge (modeling the process performance with computational tools) (Olkiewicz et al., 2016).

In another study, feasibility of producing biodiesel from WCO was determined from lab to pilot scale. The parameters for biodiesel production in lab scale were used as a factorial design of experimental procedure for pilot plant, which resulted in biodiesel yield of about 97% and residual losses up to 3%. Taking into account the costs of the machinery, power, labor, and energy, Mohammadshirazi et al. estimated the production overall cost to be 1.2 US$/liter, producing a net profit of roughly 52% and an income of 2.5 US$/liter. Government subsidies may also further lower the cost because WCO is less expensive than petrodiesel (Mohammadshirazi et al., 2014).

Grease trap waste, a typical example of high lipid content, has been regarded as a potential biodiesel feedstock that is both affordable and efficient due to its relatively low collecting costs. In a study by Tran et al., the economic viability of generating biodiesel utilizing trap grease waste was examined in order to identify feasible biodiesel manufacturers in Adelaide, South Australia. Based on the experimental findings, two distinct production routes—one where only ethanol acts as a cosolvent for esterification and another where acetone and ethanol as cosolvents for esterification were used, hasvebeen modlled using Aspen Plus V8.8 having a capacity of approximately 4400 tonnes per year (Tran et al., 2018).

Like in the abovementioned example, in several research studies, software has been used to simulate the large-scale commercialization of biodiesel. Although it is highly feasible to increase the cultivation of microbes using lipid-rich wastes (such as piggery and grease trap water) for the removal of nutrients and the production of biodiesel, the constraints of this technique could be impacted by several unpredictable factors and therefore should be further investigated, which will be covered in the upcoming section of this chapter.

Some of the examples for worldwide large-scale production of biodiesel include the following, Enzymatic transesterification with a capacity of 30 million gallons per year was originally introduced by Blue Sun Energy Ltd., the pioneer company to effectively commercialize the production of enzyme-catalyzed biodiesel. Novozymes' Callera Trans L enzyme is used in the manufacturing process to create biodiesel. Additionally, in collaboration with Novozymes and Tactical Fabrication LLC, Viesel Fuel LLC constructed an enzymatic biodiesel manufacturing facility using a stirred tank continuous reactor with a capacity of 5 million gallons per year. Genuine Bio-Fuel Inc., a pioneer in Florida's biodiesel production and technology, uses continuous-flow ultrasonic advanced technologies from Hielscher to produce biodiesel utilizing a chemically assisted process from a variety of feedstocks (Liow et al., 2022).

11.5.2 Technoeconomic analysis

A technoeconomic analysis of industrial biodiesel entails figuring out if it is technically and financially possible to generate the fuel on commercial scale. A range of sustainable feedstocks, including edible vegetable oils, livestock lipids, used cooking oil, microorganisms, and wastewater grease, can be used to make biodiesel in a large scale. However, the feedstocks used in present-day industrial biodiesel manufacturing still largely consist of refined vegetable oils and nonedible oils. In a study comparing the advantages of waste-to-energy technologies such as incineration, anaerobic digestion, and biodiesel, it was discovered that the production of biodiesel using scum sludge was the most cost-effective option, adding between US$491,949 and US$610,624 to the extra value annually. However, because the composition of waste varies depending on its source and mode of processing, a specific economical agenda is not relevant to all feedstocks (Anderson et al., 2016).

The generation of biodiesel from kernel palm oil was simulated using ASPEN Base Case Simulation (BCS) software using experimental laboratory data. Multiobjective optimal revenue scenarios and technological-economic forecasting techniques were investigated. The Historical Data Design of Design Expert successfully developed prognostic models for forecasting and optimizing technoeconomic factors, such as return on investment (ROI), net present value (NPV), payback time (PBT), and production capacity (PC). The batch size, production rate, and production batch number for the 3,000,000 kg/year BCS findings were 13,482.48 kg, 5.72 kg/min, and 223 batches/year, respectively. At 39.9%, $1,001,320, 2.48 years, and 3,000,690 kg were the ideal ROI, NPV, PBT, and PC values respectively as predicted feasible by the model (Oke et al., 2022).

From the above literature review, the authors have come to this conclusion that various factors such as the type of feedstock, catalyst, reaction conditions, and reactor design are playing vital role in scaling-up studies. When FFA is very high in the feedstocks, acids should be used as catalyst instead of base or acid pretreatment should be adopted to maintain the same production rate of biodiesel. Other problems compromising the production rate may be addressed with serious considerations as was stated above when there are other technical limitations. The use of advanced technologies such as computational modeling and simulation can also aid in predicting and improving the biodiesel productivity. Overall, a holistic approach is necessary for scaling up biodiesel production and meeting the growing demand for sustainable energy sources.

11.6 Commercialization of biodiesel

The growing need for cleaner and more sustainable energy sources has given the marketing of biodiesel a boost in recent years. Government initiatives to promote the use of biodiesel as well as the rising need for renewable energy sources have been the main drivers of this rise. Fig. 11.2 comprises a summary of many

Aspects of Biodiesel commercialization

Research: Research on alternative energy sources is top on the agenda. Various researchers have overtime sought to redefine the trajectory of research on biodiesel.

Strategy: International Governments have opted for strategic and radical policy adaptation measures towards scaling up biodiesel production and use.

Subsidy: Governmental subsidies and interventions are crucial at several phases, including manufacturing, marketing, by-product chain, and purchasing, to promote the supply and commercialization of biodiesel.

Goal: The goal of biodiesel is to minimize the carbon footprint of transportation and other industries while simultaneously boosting energy security, enhancing the environment, and reducing air pollution.

Vision: To promote biodiesel as a sustainable renewable fuel with the goal of making it a popular low-carbon fuel option with superior performance and emission characteristics.

FIGURE 11.2 Aspects of biodiesel commercialization.

factors to be taken into account for the commercialization of biodiesel in a national level. To make a commercial reach of biodiesel toward the end user, coordinated activities are required at multiple stages of the defined levels.

For example, Brazil has successfully commercialized biodiesel and is currently the third-largest producer of biodiesel in the world, after Europe and United States. Brazilian policymakers established measures such as mandated blending standards and tax breaks for biodiesel producers to promote the production and consumption of the fuel. Vegetable oils are consumed by Brazilian households at a rate of around 7 kg annually. (de Orçamentos Familiares, 2010). Considering the biodiesel's density (0.877 kg/L at 25°C) and a 94% yield through most commonly used alkaline-catalyzed production process in industries, it is now feasible to make sure that the availability of WCO can supply up to 13% of the country's yearly need for biodiesel (Santos et al., 2006). Merely the amount of oil produced by homes was counted; businesses and industries were not included in this consideration. This fact seems to raise the possibility that spent cooking oil may boost biodiesel demand. Technically, the procedures for making biodiesel from used oil are similar to those for regular transesterification processes employing acid, alkaline, and enzymatic catalysts. Each catalyst has benefits and drawbacks depending on the undesired components (particularly FFA and water) in the spent oil. In order to meet the quality requirements for the transesterification process, a feedstock pretreatment step may be necessary, taking into consideration the quality of the oil.

Extensive commercialization of biodiesel is a crucial piece of a bigger effort to reduce our reliance on fossil fuels. One such example includes RKA Petroleum Cos from Michigan, which is a distributor, retailer, transporter, and terminal operator of petroleum and is now offering preblended biodiesel at its terminal facility, thereby enhancing the accessibility of a cleaner and renewable source in a profitable manner. Preblending is more advantageous than splash-blending method due to improved quality, decreased risk and cost (https://www.rka.com).

Another significant supplier of biodiesel fuel and feedstock in the United States, Canada, and European Union is Targray. Their biodiesel initiatives assist fleet operators, refineries, dealers of fuel, and distributors in increasing their profitability while fostering a better sustainable economy for upcoming generations. Though commercialization aspects are been broadly explored, there are several drawbacks associated with it, some of them are discussed in the below section (https://www.targray.com/biofuels/feedstock).

11.7 Challenges and major bottlenecks of biodiesel commercialization

First-generation biodiesel production from edible feedstocks has several shortcomings, such as elevated raw materials cost, reliance of food product for fuel need, fertile agricultural land confiscation, deforestation, and adaptation of plants to changing environmental conditions made 1G biodiesel from edible plants eventually unsuccessful (Pikula et al., 2020). Second-generation biodiesel provides solutions to expansion of feedstock options, energy security, countryside financial growth, greenhouse gas mitigation along with the new infrastructure for harvesting, transporting, storing, and refining biomass (Goh et al., 2020). For 3G biodiesel, algal biomass is considered as a potential feedstock for biodiesel synthesis. In commercial-scale production, there are several shortcomings for production and handling the algal biomass that include upstream processing, where algal selection and growing techniques, nutrient supplies, energy requirements for running photobioreactors, algal footprint, water reusability and sensitivity to external environment pose a major challenge, whereas, in downstream processing, the process of harvesting algal cells and its drying for maximum lipid extraction gives a new challenge for maximum biodiesel recovery (Lam et al., 2017).

For engineers, advancing the practice of biodiesel making from a laboratory to an industrial scale might offer a number of difficulties and encounters a number of challenges, including reactor design—designing a suitable reactor to the required scale without compromising the yield of the product, and to ensure proper distribution of catalyst and heat uniformly to the entire reacting vessel through agitation and baffle aids are crucial. Feedstock variability—the variability in the feedstocks is based on seasonal or regional parameters, hence opting high-lipid-yielding feedstocks is vital for optimum recovery of lipid vis-à-vis biodiesel. Catalyst selection—selection of a suitable catalyst having higher efficiency and easy recovery is crucial; also the nature of catalyst should not have harmful interaction with the reaction vessel and thus always corrosion-free metal-made reactors are recommended. Separation and refinement of the product—biodiesel separation and purification from glycerol byproduct and unreacted feedstock can be challenging in large scale. Engineers must develop efficient and cost-effective separation and purification techniques that can remove impurities and ensure good-quality biodiesel.

Energy efficiency—biodiesel production on large scale can consume significant energy and efficient process designing can reduce the energy dependence. Other bottlenecks for the commercialization of biodiesel production include geographical variation, system stability and productivity, cost-operating costs, and policies implications (Zhu et al., 2013).

Another major bottleneck is the economic disadvantage of biodiesel (in comparison to commercial fuels) being a significant barrier to its commercialization. Although the cost of biodiesel has considerably decreased these days, it remains more expensive than commercial diesel, raising concerns about the economic feasibility of biodiesel production (Goh et al., 2020). It is well known that the cost of production is quite unresolved due to the dependence of fuel costs on the price of biomass and crude oil, neither of its price tends to fluctuate. Therefore the most crucial consideration to make is the feedstock choice before choosing the biodiesel in a commercial production. A poor choice of feedstock may be the cause of the problem of an excessive budget needed for feedstock. In any particular production line, the feedstock cost shouldn't be higher than 50% of the whole production cost.

Achieving the fuel criteria is another crucial factor for commercial-scale production of biodiesel. While designing a fuel system for a compression ignition engine that runs on diesel, biodiesel, or biodiesel blends, the physiochemical characteristics of the fuel are vital. A critical feature in defining the quality of biodiesel fuel is the cetane number. It has a higher cetane rating than diesel fuel and contains no aromatics and 10%−12% oxygen by weight. Due to the higher fuel quality of biodiesel, vehicles utilizing it release less CO, particulate pollution, and HC than engines using petroleum-based mineral diesel. The specification for biodiesel was developed using the American Standard Test Method (ASTM) and the European (EN) standard. France, Germany, Italy, and other countries have developed their own biodiesel standards. In spite of this, an Indian standard with the denomination IS: 15607 exists. India accepted the international technical standard IS 15607:2005 and 2016 for biodiesel as a blending stock (Jain & Sharma, 2010).

The implementation of the law/legislation is one of the major initiatives to support the enforcement of biodiesel. Taxes, regulations, and government subsidies are also encompassed. In particular situations, the enforcement requires specific rules and regulations in order to systematically enhance the economics of biodiesel. Failing in this aspect is a major bottleneck and a drawback in the large-scale commercialized production of biodiesel worldwide. According to Ong et al., ethanol once held a bigger part of the worldwide biofuel industry. However, biodiesel has risen to the top of the list as the fuel that consumers are most interested in purchasing. Additionally, this circumstance demonstrated how the regulations that were implemented (taxes, mandates, and incentives) successfully fueled the expansion of biodiesel, particularly in the European Unions and Asia (Ong et al., 2012). Many national governments are still maintaining a backward position in this domain.

Through the continuous research activities by the researchers worldwide and through governmental support, the abovementioned bottlenecks can be eliminated in the upcoming years to make biodiesel an easy-to-access low-cost sustainable alternative.

11.8 Conclusion

Even though fossil fuels are affordable and widely accessible, we still need to find their substitutes or create technologies that can reduce the negative effects of consuming them. Key fuel properties of the resultant fuel are quite similar to those of conventional diesel fuel made from petroleum, making the two compatible and blending them in any ratio possible. In the current situation, WCO is a cost-effective option for producing biodiesel due to its availability and cheap price. This oil contains several unwanted substances that are generated during frying, such as polymers, FFA, and many other chemicals, which are extremely problematic during the transesterification step. It is impractical to pretreat the WCO in order to get rid of these unwanted chemicals, and the choice of a transesterification technique should be made based on the water and FFA content of the WCO. By examining their economic effects, numerous different methods used in large-scale biodiesel production were examined for their efficacy in this chapter. In addition, this chapter also discussed several lipid-rich waste feedstocks and also discussed few techniques widely used to convert lipid-rich wastes to biodiesel, with a focus on scale-up and commercialization. Simulated technoeconomic feasibility study was also discussed to support the commercial viability of biodiesel as an alternative fuel in the future. Also, this chapter outlined the market demand of biodiesel industry in both Indian and world scenarios and the bottlenecks, governmental support needed for making the product economically viable in commercial level has been thoroughly discussed in this chapter.

References

Anderson, E., Addy, M., Ma, H., Chen, P., & Ruan, R. (2016). Economic screening of renewable energy technologies: Incineration, anaerobic digestion, and biodiesel as applied to waste water scum. *Bioresource Technology*, 222, 202–209. Available from https://doi.org/10.1016/j.biortech.2016.09.076.

Balasubramanian, R., Sircar, A., Sivakumar, P., & Anbarasu, K. (2018). Production of biodiesel from dairy wastewater sludge: A laboratory and pilot scale study. *Egyptian Journal of Petroleum*, 27(4), 939–943. Available from https://doi.org/10.1016/j.ejpe.2018.02.002.

Balat, M., & Balat, H. (2010). Progress in biodiesel processing. *Applied Energy*, 87(6), 1815–1835. Available from https://doi.org/10.1016/j.apenergy.2010.01.012.

Banković-Ilić, I. B., Stamenković, O. S., & Veljković, V. B. (2012). Biodiesel production from non-edible plant oils. *Renewable and Sustainable Energy Reviews*, 16(6), 3621–3647. Available from https://doi.org/10.1016/j.rser.2012.03.002.

Bhatti, H. N., Hanif, M. A., Qasim, M., & Ata-ur-Rehman. (2008). Biodiesel production from waste tallow. *Fuel*, 87(13–14), 2961–2966. Available from https://doi.org/10.1016/j.fuel.2008.04.016.

Biodiesel Market (By Feedstock: Vegetable Oil, Animal Fats; By Application: Fuel, Power Generation, Others; By Production Process: Alcohol Trans-Esterification, Hydro-Heating) - Global Industry Analysis, Size, Share, Growth, Trends, Regional Outlook, and Forecast 2022–2030. Available Online: https://www.precedenceresearch.com/biodiesel-market Accessed on March 2023.

Chakraborty, R., Gupta, A. K., & Chowdhury, R. (2014). Conversion of slaughterhouse and poultry farm animal fats and wastes to biodiesel: Parametric sensitivity and fuel quality assessment. *Renewable and Sustainable Energy Reviews*, 29, 120–134. Available from https://doi.org/10.1016/j.rser.2013.08.082.

Cunha, A., Feddern, V., De Prá, M. C., Higarashi, M. M., De Abreu, P. G., & Coldebella, A. (2013). Synthesis and characterization of ethylic biodiesel from animal fat wastes. *Fuel*, 105, 228–234. Available from https://doi.org/10.1016/j.fuel.2012.06.020.

de Orçamentos Familiares, I. P. (2010). Familiares 2008–2009: aquisição alimentar e domiciliar per capita. *São Paulo: IBGE*.

Dewangan, A., Yadav, A. K., & Mallick, A. (2018). Current scenario of biodiesel development in India: prospects and challenges. *Energy Sources, Part A: Recovery, Utilization and Environmental Effects*, 40(20), 2494–2501. Available from https://doi.org/10.1080/15567036.2018.1502849, http://www.tandf.co.uk/journals/titles/15567036.asp.

Foroutan, R., Mohammadi, R., & Ramavandi, B. (2021). Waste glass catalyst for biodiesel production from waste chicken fat: Optimization by RSM and ANNs and toxicity assessment. *Fuel*, 291. Available from https://doi.org/10.1016/j.fuel.2021.120151.

Gebremariam, S. N., & Marchetti, J. M. (2018). Biodiesel production through sulfuric acid catalyzed transesterification of acidic oil: Techno economic feasibility of different process alternatives. *Energy Conversion and Management*, 174, 639–648. Available from https://doi.org/10.1016/j.enconman.2018.08.078.

Goh, B. H. H., Chong, C. T., Ge, Y., Ong, H. C., Ng, J. H., Tian, B., Ashokkumar, V., Lim, S., Seljak, T., & Józsa, V. (2020). Progress in utilisation of waste cooking oil for sustainable biodiesel and biojet fuel production. *Energy Conversion and Management*, 223. Available from https://doi.org/10.1016/j.enconman.2020.113296, https://www.journals.elsevier.com/energy-conversion-and-management.

Hama, S., & Kondo, A. (2013). Enzymatic biodiesel production: An overview of potential feedstocks and process development. *Bioresource Technology*, 135, 386–395. Available from https://doi.org/10.1016/j.biortech.2012.08.014.

IEA (2021). Available online: https://www.iea.org/reports/india-energy-outlook-2021. Accessed on March 2023.

Jain, S., & Sharma, M. P. (2010). Prospects of biodiesel from Jatropha in India: A review. *Renewable and Sustainable Energy Reviews*, 14(2), 763–771. Available from https://doi.org/10.1016/j.rser.2009.10.005.

KerihuelA. (2007). *Etude de la formulation et de la combustion de carburantsalternatifs à base de graissesanimales pour moteur à allumagespontané* (Doctoral dissertation).

Kulkarni, M. G., Gopinath, R., Meher, L. C., & Dalai, A. K. (2006). Solid acid catalyzed biodiesel production by simultaneous esterification and transesterification. *Green Chemistry*, 8(12), 1056–1062. Available from https://doi.org/10.1039/b605713f.

Lam, M. K., Lee, K. T., & Mohamed, A. R. (2010). Homogeneous, heterogeneous and enzymatic catalysis for transesterification of high free fatty acid oil (waste cooking oil) to biodiesel: A review. *Biotechnology Advances*, 28(4), 500–518. Available from https://doi.org/10.1016/j.biotechadv.2010.03.002.

Lam, S. S., Mahari, W. A. W., Jusoh, A., Chong, C. T., Lee, C. L., & Chase, H. A. (2017). Pyrolysis using microwave absorbents as reaction bed: an improved approach to transform used frying oil into biofuel product with desirable properties. *Journal of Cleaner Production*, 147, 263–272. Available from https://doi.org/10.1016/j.jclepro.2017.01.085.

Liow, M. Y., Gourich, W., Chang, M. Y., Loh, J. M., Chan, E. S., & Song, C. P. (2022). Towards rapid and sustainable synthesis of biodiesel: A review of effective parameters and scale-up potential of intensification technologies for enzymatic biodiesel production. *Journal of Industrial and Engineering Chemistry*, 114, 1–18. Available from https://doi.org/10.1016/j.jiec.2022.07.002, http://www.sciencedirect.com/science/journal/1226086X.

Lopez, G., & Laan, T. (2008). *Biofuels-at what cost? Government Support for Biodiesel in Malaysia*.

Mahari, W. A. W., Chong, C. T., Lam, W. H., Anuar, T. N. S. T., Ma, N. L., Ibrahim, M. D., & Lam, S. S. (2018). Microwave co-pyrolysis of waste polyolefins and waste cooking oil: Influence of N2 atmosphere versus vacuum environment. *Energy Conversion and Management*, 171, 1292–1301. Available from https://doi.org/10.1016/j.enconman.2018.06.073.

Masi, M., Oddo, E., Rulli, M. C., & Seabra, J. E. (2021). *Sustainable Development Solutions Network and Fondazione Eni Enrico Mattei Roadmap to 2050 The Land-Water-Energy Nexus of Biofuels*, 1–164.

Miao, X., Li, R., & Yao, H. (2009). Effective acid-catalyzed transesterification for biodiesel production. *Energy Conversion and Management*, 50(10), 2680–2684. Available from https://doi.org/10.1016/j.enconman.2009.06.021.

Mohammadshirazi, A., Akram, A., Rafiee, S., & Bagheri Kalhor, E. (2014). Energy and cost analyses of biodiesel production from waste cooking oil. *Renewable and Sustainable Energy Reviews*, 33, 44–49. Available from https://doi.org/10.1016/j.rser.2014.01.067.

Odetoye, T. E., Agu, J. O., & Ajala, E. O. (2021). Biodiesel production from poultry wastes: Waste chicken fat and eggshell. *Journal of Environmental Chemical Engineering*, 9(4). Available from https://doi.org/10.1016/j.jece.2021.105654.

Oke, E. O., Okolo, B. I., Adeyi, O., Adeyi, J. A., Ude, C. J., Osoh, K., Otolorin, J., Nzeribe, I., Darlinton, N., & Oladunni, S. (2022). Process Design, techno-economic modelling, and uncertainty analysis of biodiesel production from palm kernel oil. *Bioenergy Research*, 15(2), 1355–1369. Available from https://doi.org/10.1007/s12155-021-10315-y, http://www.springer.com/life + sci/plant + sciences/journal/12155.

Olkiewicz, M., Torres, C. M., Jiménez, L., Font, J., & Bengoa, C. (2016). Scale-up and economic analysis of biodiesel production from municipal primary sewage sludge. *Bioresource Technology, 214*, 122–131. Available from https://doi.org/10.1016/j.biortech.2016.04.098, http://www.elsevier.com/locate/biortech.

Olubunmi, B. E., Alade, A. F., Ebhodaghe, S. O., & Oladapo, O. T. (2022). Optimization and kinetic study of biodiesel production from beef tallow using calcium oxide as a heterogeneous and recyclable catalyst. *Energy Conversion and Management: X, 14*. Available from https://doi.org/10.1016/j.ecmx.2022.100221.

Ong, H. C., Mahlia, T. M. I., & Masjuki, H. H. (2012). A review on energy pattern and policy for transportation sector in Malaysia. *Renewable and Sustainable Energy Reviews, 16*(1), 532–542. Available from https://doi.org/10.1016/j.rser.2011.08.019.

Patiño, Y., Faba, L., Díaz, E., & Ordóñez, S. (2021). Biodiesel production from wastewater sludge using exchange resins as heterogeneous acid catalyst: Catalyst selection and sludge pre-treatments. *Journal of Water Process Engineering, 44*. Available from https://doi.org/10.1016/j.jwpe.2021.102335.

Pikula, K., Zakharenko, A., Stratidakis, A., Razgonova, M., Nosyrev, A., Mezhuev, Y., Tsatsakis, A., & Golokhvast, K. (2020). The advances and limitations in biodiesel production: feedstocks, oil extraction methods, production, and environmental life cycle assessment. *Green Chemistry Letters and Reviews, 13*(4), 11–30. Available from https://doi.org/10.1080/17518253.2020.1829099, https://www.tandfonline.com/loi/tgcl20.

Pramanik, K. (2003). Properties and use of jatropha curcas oil and diesel fuel blends in compression ignition engine. *Renewable Energy, 28*(2), 239–248. Available from https://doi.org/10.1016/s0960-1481(02)00027-7.

Pullen, J., & Saeed, K. (2015). Investigation of the factors affecting the progress of base-catalyzed transesterification of rapeseed oil to biodiesel FAME. *Fuel Processing Technology, 130*(C), 127–135. Available from https://doi.org/10.1016/j.fuproc.2014.09.013.

Rahman, W. U., Khan, A. M., Anwer, A. H., Hasan, U., Karmakar, B., & Halder, G. (2022). Parametric optimization of calcined and Zn-doped waste egg-shell catalyzed biodiesel synthesis from Hevea brasiliensis oil. *Energy Nexus, 6*. Available from https://doi.org/10.1016/j.nexus.2022.100073, https://www.journals.elsevier.com/energy-nexus.

Razzaq, L., Mujtaba Abbas, M., Miran, S., Asghar, S., Nawaz, S., Soudagar, M. E. M., Shaukat, N., Veza, I., Khalil, S., Abdelrahman, A., & Kalam, M. A. (2022). Response surface methodology and artificial neural networks-based yield optimization of biodiesel sourced from mixture of palm and cotton seed oil. *Sustainability, 14*(10). Available from https://doi.org/10.3390/su14106130.

Rial, R. C., de Freitas, O. N., Cavalheiro, L. F., Nazário, C. E. D., & Viana, L. H. (2021). Biodiesel production using Candida rugosa as biocatalytic lipase immobilized on p-nitrobenzyl cellulose xanthate (NBXCel). *Biofuels, Bioproducts and Biorefining, 15*(6), 1789–1801. Available from https://doi.org/10.1002/bbb.2278, http://onlinelibrary.wiley.com/journal/10.1002/(ISSN)1932-1031.

Santos, R. M., Maia, I. D. S., Santos, N. A. C. D., Amaral, B. A., Castro, V. D., & Carvalho, J. R. M. (2006). *Biodiesel of soybean-experimental practice of transesterification to organic chemical lessons; Biodiesel de soja-pratica experimental de transesterificacao para as aulas de quimica organica*.

Sharma, A., Melo, J. S., Prakash, R., & Tejo Prakash, N. (2021). Lab-scale production of biodiesel from soybean acid oil using immobilized whole cells as catalyst. *Biocatalysis and Biotransformation, 39*(6), 443–454. Available from https://doi.org/10.1080/10242422.2021.1964486, http://www.tandfonline.com/loi/ibab20.

Statista Production volume of biodiesel in India from 2013 to 2022. Available online: https://www.statista.com/statistics/1051902/india-biodiesel-production-volume/. Accessed on March 2023.

Talebian-Kiakalaieh, A., Amin, N. A. S., & Mazaheri, H. (2013). A review on novel processes of biodiesel production from waste cooking oil. *Applied Energy, 104*, 683–710. Available from https://doi.org/10.1016/j.apenergy.2012.11.061, http://www.elsevier.com/inca/publications/store/4/0/5/8/9/1/index.htt.

Tomás-Pejó, E., Morales-Palomo, S., & González-Fernández, C. (2021). Microbial lipids from organic wastes: Outlook and challenges. *Bioresource Technology, 323*. Available from https://doi.org/10.1016/j.biortech.2020.124612.

Tran, N. N., Tišma, M., Budžaki, S., McMurchie, E. J., Gonzalez, O. M. M., Hessel, V., & Ngothai, Y. (2018). Scale-up and economic analysis of biodiesel production from recycled grease trap waste. *Applied Energy, 229*, 142–150. Available from https://doi.org/10.1016/j.apenergy.2018.07.106, http://www.elsevier.com/inca/publications/store/4/0/5/8/9/1/index.htt.

Udayakumar, M., Sivaganesan, S., & Sivamani, S. (2022). Process optimization of KOH catalyzed biodiesel production from crude sunflowermahua oil. *Biofuels, 13*(8), 1031–1039. Available from https://doi.org/10.1080/17597269.2022.2071068.

Verma, P., Sharma, M. P., & Dwivedi, G. (2016). Impact of alcohol on biodiesel production and properties. *Renewable and Sustainable Energy Reviews, 56*, 319–333. Available from https://doi.org/10.1016/j.rser.2015.11.048.

Wang, Y., Pengzhan Liu, S. O., & Zhang, Z. (2007). Preparation of biodiesel from waste cooking oil via two-step catalyzed process. *Energy Conversion and Management, 48*(1), 184–188. Available from https://doi.org/10.1016/j.enconman.2006.04.016.

Zhang, Y., Dubé, M. A., McLean, D. D., & Kates, M. (2003). Biodiesel production from waste cooking oil: 1. *Process design and technological assessment. Bioresource Technology, 89*(1), 1–16. Available from https://doi.org/10.1016/S0960-8524(03)00040-3, http://www.elsevier.com/locate/biortech.

Zhu, L., Wang, Z., Takala, J., Hiltunen, E., Qin, L., Xu, Z., Qin, X., & Yuan, Z. (2013). Scale-up potential of cultivating Chlorella zofingiensis in piggery wastewater for biodiesel production. *Bioresource Technology, 137*, 318–325. Available from https://doi.org/10.1016/j.biortech.2013.03.144.

CHAPTER 12

Biogas from organic wastes

Joseph Sekhar Santhappan[1], Rajalingam Arumuganainar[1], Godwin Glivin[2] and V.M. Jaganathan[2]

[1]Engineering Department, College of Engineering and Technology, University of Technology and Applied Sciences, Shinas, Oman [2]Department of Energy and Environment, National Institute of Technology, Tiruchirappalli, Tamil Nadu, India

12.1 Introduction

12.1.1 Biogas worldwide development worldwide

Worldwide, because of population and industries growth, electricity demands are increased. To solve this problem, researchers have started to identify many other novel methods to meet increased electricity demands by using many integrated methods. To solve the economic problems of using fossil fuels, every country is focusing to use nonfossil fuel to generate electrical energy. In 2030, the target of 40% of electric energy will be expected to derive from nonfossil fuel (Liu et al., 2020). The Stated Policies Scenario (STEPS) and Announced Pledges Scenario (APS) have lower energy demand growth than the WEO2021 due to high energy prices and a pessimistic economic outlook. Market uncertainty and high prices are reducing consumer spending and corporate productivity. Hence, energy consumption rises more slowly in both the STEPS and APS, and the fuel mix changes significantly from previous estimates (Fig. 12.1). The STEPS projected trend on energy consumption in Exajoules (EJ), up to 2030 shows that the world is concerned about climate change, rising fuel prices, and cost-effective clean energy technology (International Energy Agency, 2022).

Renewable energy sources (RESs) are beneficial to the fight against global warming because of their relatively low carbon emissions. Global warming is caused by the use of fossil fuels (Glivin, Kalaiselvan, et al., 2021). However, biogas is the one of the important cost-effective and eco-friendly RESs. The opportunity to study the

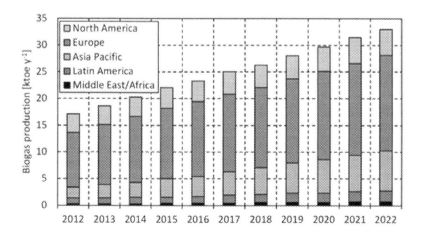

FIGURE 12.1 Biogas production: world status 2012–2022 (PIKE Research, 2012).

numerous services that biogas and biomethane can offer in various nations is made possible by this. These services can vary greatly depending on local conditions and legislative priorities. In a great number of underdeveloped countries, the use of solid biomass could be replaced by biogas as a local source of heat, energy, and clean cooking fuel in the near future. If the "digestate" that is left over from biodigesters is used as fertilizer, there is the potential for enhanced agricultural output as well as a reduction in the amount of land cleared for agriculture, which are both cobenefits of this practise.

The increasing amounts of organic waste created by modern economies and cultures have the potential to be utilized in the production of clean energy sources, which has a wide range of potential benefits for the development of sustainable communities. Despite the fact that biogas and electricity are two separate products with their own specific applications, they are both derived from a wide array of organic feedstocks that have significant unrealized potential (International Energy Agency, 2020). In addition, the use of biogas presents an opportunity for rural communities and enterprises to participate in the rapidly evolving energy sector. There are many different methods for producing biogas, which is utilized on a massive scale all around the world. Every part of the globe possesses a sizeable potential to produce biogas, and it is projected that the availability of environmentally friendly feedstocks to produce biogas will rise by 40% by the year 2040. In the Asia-Pacific area, natural gas imports and use have expanded rapidly (International Energy Agency, 2020). Africa, Europe, South and North America provide significant prospects. It is anticipated that, over the next two decades, there will be a rapid increase in overall potential as a result of an increase in the availability of diverse feedstocks within a larger global economy, as well as an improvement in waste management and collection programs within a great number of developing countries (Ferrer et al., 2014).

In locations where it is challenging to connect to national grids or when there is a high demand for heat that cannot be satisfied by renewable energy, biogas is a sustainable solution for communities to meet their energy needs. Local factors affect its use and level of competition. In disadvantaged countries, 200 million additional people will have access to clean cooking by the year 2040 owing to biogas. The production of biogas, which turns organic waste into a renewable energy source, offers a glimpse into a future in which resources are continuously used and recycled, the rising demand for energy services can be met, and wider environmental benefits can also be realized (International Energy Agency, 2020). Anaerobic digestion will be employed to produce biogas from all the biowaste that has been methodically gathered from restaurants, residential areas, schools, and other sources (Glivin & Sekhar, 2016).

For anaerobic digestion, a wide range of biomasses from all over the world can be used, such as crop residues, animal dung, used fats, fatty wastes, used oils, used grains, grasses, distillery waste, dairy waste, chicken manure, cattle manure, pig manure, and other agricultural leftovers. Food waste is waste that comes from restaurants, homes, and food processing facilities. Waste from industrial sources, such as breweries, distilleries, and paper mills, is included in this category. Sewage sludge is the leftover solid substance from sewage treatment. Crops planted expressly for the generation of energy are referred to as energy crops. Examples include sugarcane, corn, and switchgrass. Gas produced by the breakdown of organic material in landfills is known as landfill gas. Organic municipal solid waste is the portion of home waste that is made up of organic materials. Cost, availability, and energy content are only a few of the factors that affect the choice of biomass for anaerobic digestion (Glivin & Sekhar, 2020a; Glivin, Premalatha, et al., 2021; Kalaiselvan et al., 2022). Agricultural and food wastes are two of the most commonly used biomasses for anaerobic digestion because of their abundance and relatively high energy content. It's important to keep in mind that using biomass for anaerobic digestion might affect the ecosystem in both positive and negative ways. On the other hand, the generation of biogas can also result in an increase in greenhouse gas emissions, and the use of energy crops may change how land is used and directly conflict with food production. Therefore, the responsible management of biomass sources and the meticulous examination of its environmental impacts are essential to the long-term viability of anaerobic digestion as a technology. Researchers from all over the world are therefore interested in creating biogas from biowaste to create power in order to meet the enormous demand for electricity and other heating needs in the future.

12.1.2 Anaerobic digestion process

Anaerobic organic waste digestion (OWD) is a biological process that takes place in the absence of oxygen. This process turns organic materials (feedstocks) a combustible gas methane, besides few other gases including CO_2. Fermentation is an integral part of the digestive process. Fig. 12.2 is a schematic representation of the anaerobic digester that is used to generate biogas from organic wastes.

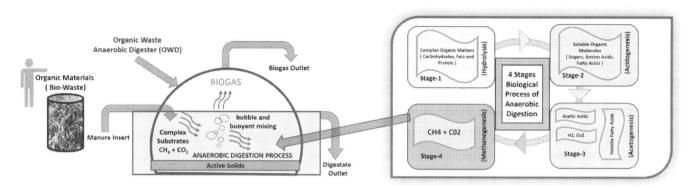

FIGURE 12.2 Organic Waste Anaerobic Digester process.

The anaerobic digestion process looks like the above diagram. Organic substances, including carbs, proteins, lipids, and oils, are what kick off the cycle in the first place. Sugars, amino acids, and fatty acids are also converted as H_2, CO_2, and organic acids are. After that, the water and any solids that are still present in the digester effluents are extracted. The biogas that is produced by the digestor is then used to get rid of CH_4, CO_2, and H_2S (Neves et al., 2018). Particularly when it comes to the farming of cattle, making use of an anaerobic digester might be beneficial. In terms of waste treatment, it lessens the volume, mass, organic content, and biodegradability of the waste. This lessening of the waste's volume, mass, organic content, and biodegradability improves the efficiency with which the residual material can be used as an amendment to soil and as a fertilizer. Less volatile organic compounds and methane are formed, and the digester also gets rid of viruses in the trash, which are other benefits for the environment. Anaerobic digestion is a synergistic process that incorporates anaerobic bacteria. The progression and the many different product kinds that correspond to each phase are shown in Fig. 12.2. Previous studies detail the primary reactions that take place throughout each of the four stages (Neves et al., 2018).

The process of acetogenesis, which is triggered by acetogenic bacteria attacking the intermediates of acidogenesis, results in the production of acetic acid, carbon dioxide, and hydrogen gas as by-products. A wide variety of microbes, such as the propionate decomposer Syntrophobacter wolinii, are essential to the process of acetogenesis. Clostridium species, Peptococcus anaerobes, Lactobacillus, and Actinomyces are examples of organisms that can produce acid. Methanogenesis is helped along by a wide variety of bacterial species, including Methanosarcina, Methanobacterium, and Methanobacillus, among others. Any form of organic material can be put into an anaerobic digester and broken down into its component parts. Inorganic materials such as lignin, peptidoglycan, and membrane-associated proteins are some examples of substances that do not undergo biodegradation. Volatile solids (VSs) are the only substances that have the capacity to be both nonbiodegradable and biodegradable at the same time. The majority of the biogas is composed of carbon dioxide and other trace gases, with CH4 accounting for approximately $50\% - 60\%$ of the mixture on average. The digestate is rich in fiber and can be utilized in the production of composite boards, compost, and animal bedding due to the presence of essential nutrients (Safferman et al., 2012).

The anaerobic digestion process will be determined by a variety of factors. The quantity and yield of biogas vary due to many factors. For instance, Fig. 12.3 shows the biogas yield from some common organic wastes (Neves et al., 2018). It is dependent on a variety of elements, including the moisture level of each feed material, the availability of carbon and nutrients, the biological degradability, the methane potential, and more. As can be observed in Fig. 12.3, fats are the most prolific producers of biogas and contain the largest proportions of volatile substances. It takes longer to digest solid feedstocks since they are more difficult to break down than soluble feedstocks. Because it has a high ratio of carbon to nitrogen, lignocellulosic biomass, for example, must have additional nitrogen sources added to it. The microbial population of the digester will also influence how well it performs its function. To accomplish this, adequate quantities of methanogens and microorganisms involved in fermentation need to be preserved.

It is essential to have a strong understanding of the appropriate retention times, mixing, and total and volatile solids (TS and VS) in the diets. There are operational considerations that must be made regarding the quantity and nature of the feedstocks that are added to the digester. The bacteria, feed, and nutrients must stay in close touch with one another at all times in order for the mixing process to serve its goal. The breakdown of volatile solids will be helped along by mixing, which will result in an increase in the amount of biogas produced. Yet, mixing requires additional energy; therefore it is necessary to strike a balance between the two. Both mechanical mixing and the bubbling of gas are ways of mixing that can be used in this system.

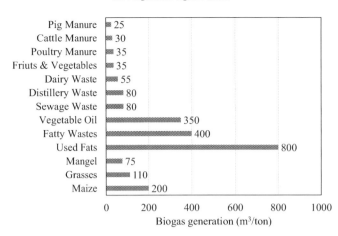

FIGURE 12.3 Biogas yields from a few biomasses (Neves et al., 2018).

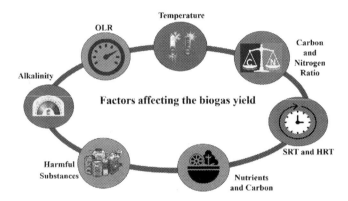

FIGURE 12.4 The important parameters controlling biogas yield.

Environmental conditions including the temperature, pH, and concentrations of salt, cationic ions, volatile fatty acids, and ammonia in the reactor influence the gas generation. Different methanogens act differently at different temperatures. For thermophile methanogens to make the most biogas, the digester needs to be running between 40°C and 70°C. Additionally, methanogens favor neutral pH levels (6.8–8.2). When too much organic material that can be digested, a dangerous substance, or a quick change in temperature is added, volatile fatty acids (VFAs) can build up and stop digestion from making gas. Oxygen, antibiotics, cleaning products, inorganic acids, alkaline and alkaline earth salts, heavy metals, sulfides, and ammonia are all toxic. When the methanogen bacteria can't keep up, the digester turns acidic, which is also known as "sour." There are, however, a number of designs including complete mixing, plug flow, batch sequencing, and fixed film that can control the temperature of different reactors.

12.1.3 Factors affecting the biogas yield

To make the technique more effective, it is necessary to adjust the critical parameters of the anaerobic digestion process, as well as the growth kinetics and environmental conditions. The following provides a concise explanation of these parameters as well as their optimal ranges, as published in a variety of different literatures. The process of anaerobic digestion is affected by a variety of factors, some of which are alkalinity, temperature, the durations of hydraulic and solid retention, acidic substances whose pH values are determined, the availability of nutrients and carbon, the rate of organic loading (OLR), hazardous compounds, and others illustrated in Fig. 12.4.

Temperature has an effect not only on the metabolic activity of the microbes but also on the rate of digestion. The temperatures at which anaerobic digestion can take place can vary widely from case to case, and its range is classified as psychrophilic (living at temperatures lower than 20°C), mesophilic (living at temperatures between

30°C and 42°C), and thermophilic (living at temperatures between 43°C and 55°C) (Al Seadi et al., 2008). In actuality, the majority of anaerobic digestion systems are designed to perform in temperatures ranging from 30°C to 38°C, while some are designed to function in temperatures ranging from 50°C to 57°C (Metcalf & Eddy, 2021). In general, mesophilic digestion is more stable, less likely to be affected by the toxicity of ammonia nitrogen, and requires less process heat than thermophilic digestion. This may allow for greater loadings with shorter hydraulic retention durations. The thermophilic process has several advantages over the mesophilic and psychrophilic processes, despite the fact that both of those processes have some disadvantages as well. As a result, many current anaerobic digestion plants have chosen temperatures that are conducive to the thermophilic process (Al Seadi et al., 2008). The performance of tubular anaerobic digesters at temperatures of 55°C (psychrophilic), 20°C (thermophilic), and 35°C (mesophilic) was observed by experimental studies. It shows that the thermophilic settings might create 144% and 41% more biogas than the other two conditions, respectively. Benefits of the thermophilic process include kill pathogens effectively with shorter retention times, a more rapid and effective digesting process, greater substrate breakdown and utilization, and increased capability for solid − liquid separation. The thermophilic method has drawbacks, including a greater degree of imbalance, a requirement for more energy due to the high temperature, and a greater danger of ammonia inhibition (Browne et al., 2014). During digestion, solids and liquids are typically kept for a period of time that is referred to as the solids retention time (SRT) and the hydraulic retention time (HRT), respectively. These factors have a direct influence on the anaerobic reactions (hydrolysis, fermentation, and methanogenesis), as well as the size of the anaerobic reactors. For high-rate digestion, SRT values might range anywhere from 10 to 20 days (Browne et al., 2014). The C/N ratio can be improved by mixing substrates that have low C/N ratios with those that have high C/N ratios in the appropriate quantities. One of the primary motivations for carrying out AD through the use of codigestion in clinical settings is the enhancement of the process's inherent stability (Zhang et al., 2006).

The pH level is one of the primary operational factors that greatly influences the digestion process. The alkalinity regulates pH by absorbing acidity produced during acidogenesis, a crucial phase in the anaerobic digestion process (Olvera & Lopez-lopez, 2012). The total alkalinity of a well-established digester normally ranges between 2000 and 5000 mg/L and is proportional to the solid feed content of the digester (Zhang et al., 2006). If the pH is outside of this range, the mesophilic digestion process is greatly hindered. Between 6.5 and 8 is the optimum pH range for mesophilic digestion (Al Seadi et al., 2008). The optimal pH range for maximizing biogas production in AD is 6.8 − 7.2. The methanogenesis bacteria in the AD process prefer a pH of around 7.0 and are quite sensitive to pH fluctuations. Acidogenesis bacteria survive a pH range of 4.0−8.5 and are far less pH sensitive than other bacteria. The methane yield decreases with increase in OLR (Aramrueang et al., 2016). The methane production is inhibited even when pH remains constant. In addition to a high methane output, adjusting the pH value to its optimal level increases biodegradability (Montañés et al., 2014). By reducing the size of the input material from 0.84 to 0.3 mm using bead milling (pretreatment), the methane output may be increased by 28% compared with disposal treatment (Izumi et al., 2010).

A wide variety of chemical and inorganic poisons and inhibitors can cause an anaerobic digester to malfunction or fail. In anaerobic digesters, hazardous chemicals such as ammonia, hydrogen sulfide, light metal ions, and heavy metal ions are frequently detected (Chen et al., 2008). The following is a list of digestion inhibitors detailed in Sustainable Energy Authority of Ireland (SEAI) materials. Hydrogen sulfide (H_2S) is produced by the digestive process and is also present in organic debris. Moreover, sulfuric acid and hydrogen sulfide are very corrosive compounds that could damage the digester's components. High quantities of nitrogen and ammonium (NH_4) can occasionally lead to a high ammonia concentration, a cellular poison. In this instance, nitrogen-poor substrates, such as chicken manure or pig slurry, should be blended or diluted with nitrogen-rich substrates. Carbon, nitrogen, phosphorus, and sulfur are important nutrients for the survival and growth of organisms participating in the anaerobic digestion process. Some nutrients and trace elements must be taken in adequate quantities to avoid anaerobic digestion from becoming inhibited and unstable. Additionally recommended are acetate levels of 0.02 mg/g for iron, 0.04 mg/g for cobalt, 0.03 mg/g for nickel, and 0.02 mg/g for zinc (Al Seadi et al., 2008).

12.1.4 Importance of codigestion in biogas production

The codigestion process is one of the essential strategies to boost the production of biogas, and it should not be overlooked. In this part of the article, the viability of the codigestion method is tested by examining and analyzing the results of a number of research that have been published in the academic journals that are listed further down. Researchers (Martí-Herrero et al., 2015) evaluated the production of biogas in a psychrophilic environment (one that has a chilly climate) by using dung from cows, llamas, and sheep. The experiment was carried out in four low-cost tubular digesters, and the dung from cows, llamas, cows and sheep, as well as llamas and sheep, were used as fertilizers. In contrast to the cow manure substrate, the cow — sheep codigestion demonstrates a promising performance with a 100% increase in the production of biogas. In addition, research has shown that the production of biogas from llama — sheep manure is around 50% lower than that of llama manure substrate. After a period of 100 days, the codigestion process stabilized in terms of the creation of biogas, the amount of methane produced, and the pH level. The research scientists came to the conclusion that additional research is required to perfect the process of producing biogas from llama and sheep excrement.

The codigesting of food scraps and cattle dung has the potential to boost the production of both biogas and methane (Zhang et al., 2013). The modes of testing that were used were batching and semicontinuous testing. The production of methane rose by 41.1% during the batch testing, and by 55.2% during the semicontinuous tests. A higher biodegradability as well as a higher carbon to nitrogen ratio would be the explanation for the improved biogas generation. Several different research studies investigated the potential benefits of codigesting food scraps with cow manure and sewage sludge. The codigestion resulted in the production of biogas with an HRT of 17 days. The HRT measures the amount of time it takes for the gas to be produced. According to the data, optimal codigestion occurs when the ratio of animal fat to vegetable fat is somewhere around 28%. Codigestion allows for the production of a carbon-to-nitrogen (C/N) ratio that is regulated and falls within the range of 20—25 L. When codigested with 50% of the VS, it was found that the bacteria and archaea that are essential to the metabolic pathway that is responsible for the transformation of organic matter into methane did not experience any changes. This illustrates that the level of inclusion and the type of substrate may serve as valuable inputs for selecting agricultural by-products and managing pig slurry. A few studies came to the conclusion that substrates high in starch might be codigested with material high in nitrogen in order to achieve efficient digestion. Karanja seed cake that had been stripped of its oil was the subject of an experiment that (Barik & Murugan, 2015) carried out. The best mixing ratio for this industrial waste from karanja biodiesel manufacturers and cow dung was discovered to be 25:75, which also produced an environmentally benign fertilizer and produced a methane output of 73%. Furthermore, the codigestion procedure helps to maintain the digester for greater performance and increased biogas production. This approach may help to obtain high methane yield by revealing the varied compositions of the biowaste.

12.1.5 Motivation and objectives of the study

The use of organic waste from university is more beneficial because many other energy sources, such as LPG, coal, and kerosene, may be conserved or used less frequently for cooking and other thermal applications. Although there have been numerous studies available to produce biogas from various type of garbage, studies on the use of wastes from institutions are limited. The food waste from hostels and canteens collected at the institution level may have different characteristics. Due to the fluctuations in student enrollment over the course of a year, the supply of biowaste on academic campuses is not consistent. As a result, the quantity and quality of biogas produced will change throughout the year, necessitating specific research to determine the viability of biogas usage in these conditions (Glivin & Sekhar 2019, 2020a; Glivin, Premalatha, et al., 2021; Kalaiselvan et al., 2022). The biowastes such as rice waste, rice trash mixed with vegetables, meat, and fish, and vegetable waste as well as cow dung from the campus's cattle farm can be used together and separately to generate biogas. To use biogas for heating and power generation, it must contain above 50% methane, and it must be ensured to confidently use this conversion technology. As a result, the current study's main objective is to focus the possibility of biogas production from educational institution and to forecast the technical methods used for production by installing biogas plan in some institution in India. Therefore, the wastes from educational institutions in and around the study region were collected besides finding their potential through a survey. To study the physical and chemical characteristics, standard method was used. Furthermore, four anaerobic digesters were used in the experimental study to find the variation in the quality of biogas over a year and the impact of nonuniform loading of anaerobic digesters.

12.2 Materials and methods

12.2.1 Feedstock survey

The raw resources are sourced from a university campus in the southern region of India. In addition to cow dung, the most prevalent form of biowaste on campus is uneaten rice, vegetables, meat, fruits, and so on. To select the biowaste that would be used in the study, a survey was carried out using the appropriate questionnaire to discover the particulars of the biowaste, which included the quantity, content, and kind of the biowaste. The shape of the questionnaire was also contingent on the kind of institution being surveyed, the academic schedules being followed, the number people inside and outside the campus, the sources for producing biowaste, the conventional fuels being used for cooking, and so on. Details regarding the gender of students and staff, as well as information regarding the sources from which biowaste is produced, how it is used, and how it is disposed of, are some of the details that should be provided. The reliability of the data was validated with the assistance of the relevant authorities from a few different institutions. The step-by-step process that was used in this investigation is mentioned in Fig. 12.5.

The energy sources for biogas (BG) that have been identified include leaves, cotton waste, paper waste, sewage sludge, and food waste. For instance, Fig. 12.7 provides the significant data calculated for educational university. With a projected population of 100% literate and a total population of 1000 − 2500, we chose one of the institutions in the southern region of India. The university's livestock population density ranges from 0 to 15, and the amount of biowaste that is available each day is between 100 and 700 kg, while the predicted daily generation of dung from the animals is between 0 and 70 kg. They were able to cook 12 − 15 kg of food each day using conventional fuel (LPG). Daily surveys on the amount of food waste in university were done for different academic schedules. This study considers the types of meal menus that different institutions employ the most frequently. Included is the food waste that is produced before and after eating. This biowaste is wasted and dumped in vast quantities. Fig. 12.6 shows the average biowaste for each month as estimated and plotted. Data on the availability of feedstocks are daily gathered from the canteens and dormitories.

The data were collected over the period of one year, spread out throughout the four distinct seasons. According to the graph (Fig. 12.6), the amount of biowaste that can be accessed daily ranges from 394 kg during the months of January, February, March, April, and May to 190 kg during the months of November and December. This indicates

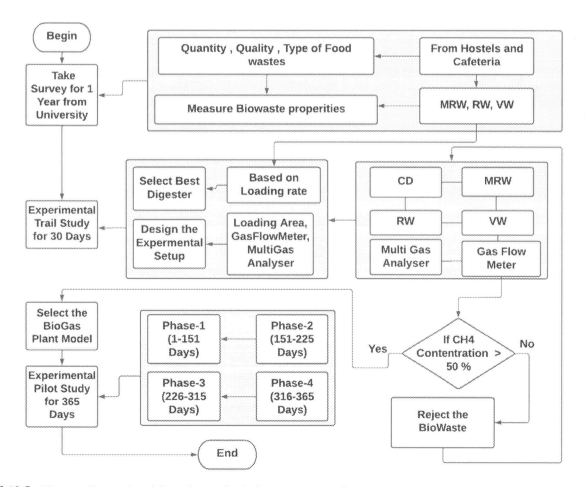

FIGURE 12.5 The overall procedure followed to study the biogas generation from organic wastes.

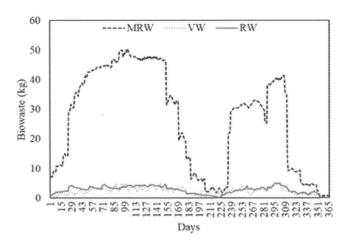

FIGURE 12.6 Biowaste availability per day throughout a year.

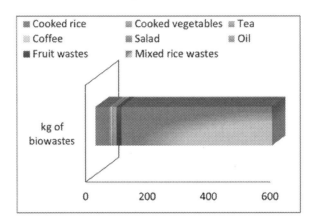

FIGURE 12.7 Sample data for biowastes generated in an institution.

that the amount of biowaste that can be accessed daily during the months of August, September, and October is significantly higher than the amount that can be accessed during the remaining months. Fig. 12.6 illustrates the typical amount of biowaste, and Fig. 12.7 shows the percentage of biowaste that is generated at one of the universities (Glivin, Vairavan, et al., 2021). The hostels were the locations where the samples of biowaste were collected before the food wastes were disposed of. Each type of food waste was placed in its own individual bucket for collection. Instructions on how to properly dispose of any uneaten food were provided to both the staff and the pupils. Even though it was recognized that certain wastes, such as waste from fruit, meat, and fish, were less common than others, the quantity of these wastes in the general trash was examined at least twice a month to search for any significant fluctuations. The variation was not significant in they were taken together as mixed rice waste (MRW). The other two groups are rice waste (RW) and vegetable waste (VW).

The wastes have been categorized into three types as RW, MRW, and VW, and the respective quantities are 5–50, 70–400, and 5–50 kg. The quantity was evaluated so that the correct size of the biodigester could be determined. The MRW comprises food scraps that have been discarded after being consumed, including rice scraps, fish scraps, vegetable scraps, and meat scraps, among others. The digesting process made use of the collected biowaste in the form of feedstock. All these different types of biowaste, apart from waste from raw vegetables and rice, are blended with MRW because there are so few other options.

12.2.2 Evaluation of biowaste characteristics

These variables, pH, VS, and TS, were subjected to experimental testing utilizing the criteria that are presented in the following paragraphs since they are critical in regulating the generation of biogas. In compliance with the regulations set forth by the APHA, the following method was utilized to determine the TS of the feed: As shown in Figs. 12.8 and 12.9, 50 g of each type of biomass was put into porcelain jars that were first preweighed before being placed in an oven heated to 60°C for 24 hours and then heated to 103 degrees Celsius for 3 hours (Glivin & Sekhar, 2020a, 2020b; Glivin, Kalaiselvan, et al., 2021). Weighed along with the container were the dry samples. Following drying, a weighing balance with a 0.001 g precision was used to quantify the combined weight of the samples and the container. Each sample's TS % was computed using:

$$TS = \left(\frac{W_d}{W_w}\right) \cdot 100 \tag{12.1}$$

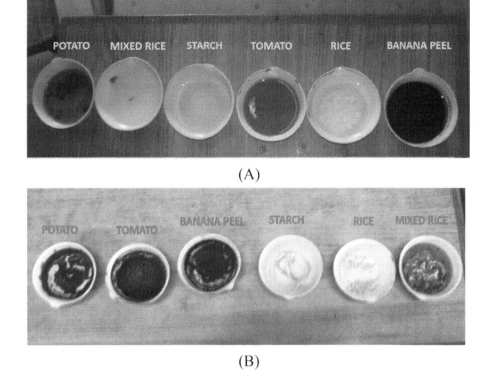

FIGURE 12.8 (A) Biowaste samples before heating. (B) Hot air oven and samples used to find the total solids.

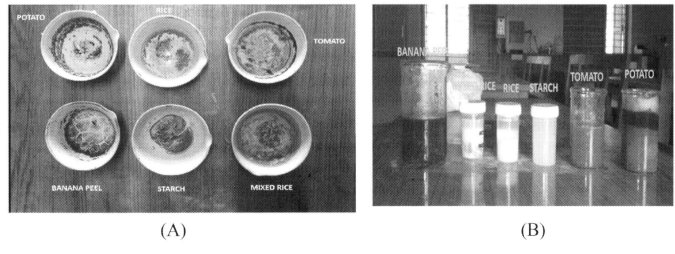

FIGURE 12.9 (A) Biowaste samples to find the volatile solids. (B) Samples used to measure pH values.

where W_d and W_w are the relative weights of the dry and wet samples

The VS of feed materials was determined using the accepted procedure. The dried samples were then ignited entirely inside the muffle furnace as illustrated in Fig. 12.9A after being further dried at 550°C for around an hour. The desiccator samples were weighed after cooling, and Eq. (12.2) was used to determine the samples' specific surface area (VS).

$$VS = \left[\frac{(W_d - W_a)}{W_a}\right].100 \tag{12.2}$$

In this equation, W stands for weight, and d and a stand for the samples before and after the test, respectively. The list of samples used to determine the PH values can be shown in Fig. 12.9B.

The pH of the biowaste was checked a minimum of one time in a day with an accuracy of 0.05%. The chemical properties of the four types of biowaste used in this study were looked at. Eqs. (12.1) and (12.2) were used to figure out what TS and VS are. When the results of the tests were compared with what was already known, it was clear that they were correct.

12.2.3 Experimental setup

Fig. 12.10 is a schematic diagram of the apparatus used in the experiment, which consists of three floating biogas plants made using fiberglass-reinforced polyester (FRP) in the manner of drums. The capacities of these plants are as follows: 2, 1, and 0.25 m³, respectively. The FRP biogas plant has many different parts, such as a digester tank, a floating drum, a water jacket, a central guide, a drain stopper, an intake pipe, and an exit pipe. The biogas is stored in a floating drum. The water jacket stops both the floating drum and the smell of the digester. It also stops biogas from leaking out of the digester. The galvanized steel (GS) guide pipe in the middle of the floating drum makes it so that it floats straight up and down. After the digestion process is done, the outlet pipes will be used to get rid of the leftover digestate. Both the pipes going in and out are made of PVC. The drain plug lets the filled substrate be taken out of the digester tank so it can be cleaned. The four digesters that make up the study are called OWD1, OWD2, OWD3, and OWD4. The plants chosen are easy to fill and empty at any

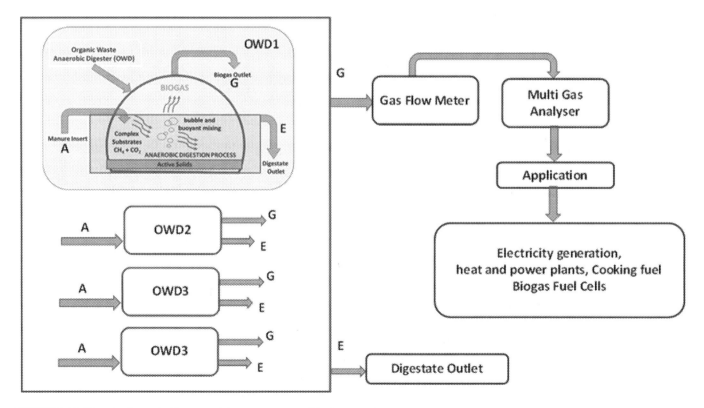

FIGURE 12.10 A schematic diagram of the apparatus used in the experiment.

time (Pareek & Nagarsheth, 2015). It is also easier to move these biogas plants from one place to another. FRP digesters are better than other fixed-type biogas plants in the following ways: A pressure gage is attached to the top of the floating drum, and resistance temperature detector (RTD) sensors are placed in the substrate (Glivin & Sekhar, 2016). To measure the amount and quality of the gas, a multigas analyzer and a thermal gas flow meter that are set up for biogas are used. Many bypass lines are attached before each instrument to keep people from suffocating. A cooking stove is given so that the biogas can be burned safely.

The digester tank, gas holder, water jacket, inlet, and output pipes, drain plugs, and guide rod make up the biodigester. A water jacket surrounds the digester tank to keep ambient oxygen from coming into contact with the inoculum (Lehtomäki et al., 2007). The gas holder, also known as the floating drum, rises and falls in response to the production of biogas. A central guide made of GS is installed in the middle of the digester tank to prevent getting stuck between the gas holder and the water jacket. A drain tube is also installed in the water jacket in addition to the outflow pipe (Estoppey, 2011). Table 12.1 lists the four digester components along with a description of each one.

12.2.4 Experimental procedure

A slurry has been made using the cow dung that is located within the campus so that methanogenic bacteria can be produced. At the beginning, cow manure and water were combined in a proportion of one to one, and the resulting slurry was put through anaerobic digestion for a period of 50 days in four small-scale digesters. CD was utilized for the creation of the slurry since its anaerobic digestion will result in the production of methanogen bacteria at a faster rate than that of any other type of biowaste. Over a period of 50 days, the methanogenic bacteria that were produced were capable of digesting any type of biowaste (Singh et al., 1985). In addition, this was ensured by conducting tests to evaluate the quality of the biogas that was produced from the slurry. Following a period of 50 days, the digesters had the appropriate quantity of CD, RW, MRW, and VW slurries added to them according to the preparation instructions. During this time, cooking stoves were used to discharge the gas generated. Following verification that bacteria were no longer producing biogas from cow dung and were starved for food, campus-collected biowastes such as RW, MRW, and VW were gradually added to each digester.

Each anaerobic digester contained an entrance for the feedstock and an exit for the digested feedstock. For biogas storage, a floating cylinder with a water seal was provided. The gas discharge from the digesters was connected to a 0.5% full-scale accuracy thermal gas flow meter and a 0.3% accuracy multigas analyzer (NUCON) (FS). A pH electrode and a thermometer were placed inside the digester to measure the pH and temperature of the substrate. In addition, the average temperature, pH, and quality were calculated after being measured at least four times per day. The annual average atmospheric temperature in the research area ranged from 19°C to 34°C,

TABLE 12.1 Design features of the biogas plants.

Sl. No.	Capacity of the biogas plant	OWD1 (2 m^3)	OWD2 (1 m^3)	OWD3 + OWD4 (0.25 m^3)
1	Type of the biogas plant	FRP	FRP	FRP
2	Loading rate (kg)	20–50	10–25	1–6
3	Feeding type	Manual	Manual	Manual
4	Inner diameter of the digester (mm)	1100	770	450
5	Length of the digester (mm)	1100	1170	650
6	Inner diameter of the floating drum (mm)	1200	820	550
7	Length of the floating drum (mm)	950	970	550
8	Inner diameter of the water jacket (mm)	1250	870	600
9	Length of the water jacket (mm)	950	970	550
10	Central guide thickness (mm)	38.1	38.1	31.75
11	Size of the inlet pipe (mm)	76.2	76.2	76.2
12	Size of the outlet pipe (mm)	101.6	101.6	76.2

but more than 90% of the data indicated a temperature above 28°C. The organic loading rate (OLR) was not constant throughout the duration of the test because biowaste was not always available. Due to the inability of the digesters to process all the waste, a constant component of the waste supply was taken for loading in accordance with the digesters' capacity. The production of methane was observed, and its daily quantity and quality were evaluated. Moreover, measurements of the temperature, pH, and quality were made at least four times each day, and the average of each was calculated. In the region where the survey was conducted, the annual temperature range was from 19°C to 34°C; however, more than 90% of the readings indicated that the temperature was higher than 28°C (Glivin & Sekhar, 2019; Glivin et al., 2022). After ensuring that the bacteria were starving and that biogas production from bovine dung had ceased, the biowaste was progressively added to each digester as digested in OWD1, OWD2, OWD3, and OWD4 for 365 days with variable loading in accordance with the waste availability. The mixing ratio for all feedstocks with water was maintained as 1:1.

12.2.4.1 Experimental procedure: trail study

All four digesters were initially filled with a 1:1 mixture of cow dung and water in order to cultivate methanogenic bacteria over a 55-day period. The wastes of four sorts that were collected from the educational institution were gradually loaded for 30 days with a constant OLR, once it was confirmed that CD had completely digested. Using a multigas analyzer and a thermal gas flow meter, respectively, the quantity and quality of methane produced each day were determined. During the digestion process, the feedstock's quality, pH, and temperature were also measured at regular intervals, and an average was determined. The temperature was recorded as being between 29°C and 34°C throughout the trial investigation.

12.2.4.2 Experimental procedure: pilot study

According to the experimental plan, the identical digesters were planned to utilize for the pilot study, which lasted 365 days, after the trial research. Unfortunately, the loading rate changed daily depending on the amount of biowaste that was available. Only 10% of each waste was taken daily and used to fill the digesters because the amount of biowastes was greater than the permitted OLR of the chosen digesters. As a result, the pilot study has considered the influence of biogas generation that is not uniform, and Fig. 12.5 depicts the loading pattern used for this study. Over the course of 365 days, changes in the pH, temperature, TS, VS, yield, and chemical makeup of the biogas were tracked. The research area's annual average atmospheric temperature ranged from 25°C to 36°C; however, more than 90% of the data indicated that the temperature was higher than 28°C. As a result, mesophilic was thought to be a state of digestion. The sample production biogas quantity and quality are recorded for 1 year, and it shown in Table 12.2, together with representative readings of RW during various phases in a 0.25 m^3 digester. Considering the academic calendars, the entire test period was divided into four sections as mentioned in Table 12.2.

12.2.5 Biogas quantity measurement

Thermometers and pH electrodes are submerged inside the digester tank so that variations in temperature and pH can be monitored as they occur inside the digesters during the digesting process. Each floating drum comes equipped with a set of control valves that connect it to a manifold. The outlet from the manifold is connected to a multigas analyzer called NUCON that has an accuracy of 0.5% through a thermal gas flow meter that has an accuracy of 0.1% and does mass flow measurements of liquids. The molecules of the gas carry the heat from the sensors and cool them when the biogas flows through the flow sensor. With the help of proper heating techniques, this heat has been offset. In this device, the flow stream is passed over a heated sensor.

TABLE 12.2 Sample reading for RW (rice waste) for different phases in 0.25 m^3 digester.

Phase	Month	pH	Total solids (%)	Volatile solids (%)	Organic loading rate (kg)	CH$_4$ (%)	Yield (m^3)	Temperature (°C)
I	January–May	6.1–6.7	9.0–31.0	90–95	0.12–4.5	51–53.9	0.007–0.14	29–32
II	June–August				0.7–3	51–53.5	0.01–0.11	29–34
III	August to November				2–4.5	51–53.9	0.09–0.16	29–33
IV	December				1–3	52–53.9	0.02–0.05	29–30

A temperature sensor is built into the design and is used for temperature compensation since convective heat transfer depends on the temperature difference between the heated sensor and the fluid. Even though the temperature differential (ΔT) between the heated RTD and the reference sensor will fluctuate, the circuit quickly makes up for the energy loss by heating the flow sensor in order to manage the overheated temperature.

The mass flow rate of the biogas is shown by the amount of electricity required to maintain this temperature. The specifications of a gas flow meter are with accuracy $\pm 0.1\%$, repeatability $<4\%$ of full scale, resolution 0.2% of full scale, permissible pressure fluctuation of 75,000–120,000 Pa, and a working temperature of 5°C–40°C. The manufacturers calibrated the gas flow meter used in this study in accordance with their standards. To prevent reading inaccuracies, manufacturers do a routine inspection once every 3 months. After quantity analysis, the biogas produced by the thermal gas flow meter is also permitted to pass through a multigas analyzer, where it is possible to track the compositions of CH_4, CO_2, O_2, and CO with various detectors. Since the work's intended use did not center on operating engines, where it plays a big role, the H2S composition is not relevant in the overall yield and was not measured in this experiment. The device offers a continuous online digital display and is easy to use. Also, it is a portable instrument calibrated to measure the aforementioned gases. The suppliers' assistance was used to measure the accuracy once every 6 months. The instrumentation panel, which controls the flow of biogas from digesters to the tools that test the quality and quantity, is the last. The instrumentation panel's parts include a multigas analyzer, a thermal gas flow meter, a gas chamber, bypass lines, and control valves. Before the gas is sent to the cooking burner, a number of by-pass lines controlled by a number of control valves are connected to prevent the needless use of the equipment.

12.2.6 Uncertainty analysis

An uncertainty analysis is required so that the correctness of the experiments may be demonstrated. Experiment errors can be caused by many factors such as selection of an instrument, the quality of that instrument, its calibration, the environment, the observational methods that were utilized, the results, and the test design. Based on a comprehensive characterization of the uncertainties in the primary experimental measurements, an analysis of the uncertainties in a number of different experimental outcomes has been carried out. If the amount R being measured is affected by an independent variable such as x_1, x_2, and x_3, the uncertainty in individual terms was calculated as per the procedure discussed in previous studies (Holman, 2002):

$$R = R(x_1, \ x_2, \ x_3, \ldots x_n) \tag{12.3}$$

The experimental observations made in this study include those of temperature, flow rate, gas composition, weight, pressure, and pH. Table 12.3 lists each of these parameters' respective uncertainty. The manufacturer's specifications and calibration data are used to determine the accuracy for each instrument.

12.3 Results and discussion

The results of research that was conducted both theoretically and practically on the effectiveness of anaerobic digesters at educational institutions with varying rates of organic loading are presented in this chapter. Properties of the collected biowaste are utilized so that theoretical investigations can be carried out. Fig. 12.11 shows the average TS, VS, and pH of the selected samples.

TABLE 12.3 Uncertainties in measurement of parameters during biogas generation.

Sl. No.	Instrument	Range	Accuracy range
1	Flow meter	0–10 m^3/s	$\pm 0.1\%$
2	RTD	0°C–100°C	± 0.1°C
3	Weighing balance	0–320 g	± 0.01 g
4	pH electrode	0–15	$\pm 0.1\%$
5	Pressure gage	0–2 bar	$\pm 0.01\%$
6	Multigas analyzer	0%–100%	$\pm 0.5\%$

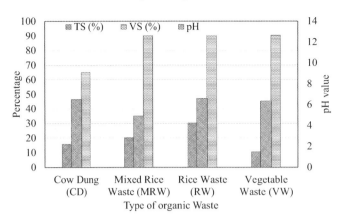

FIGURE 12.11 Characterization of feedstock.

12.3.1 Physical characterization of feedstocks

Animal waste, food waste, agricultural by-products, and energy crops are just a few examples of the various organic resources that can be converted into biogas, a renewable energy source. The physical properties of these feedstocks have a significant impact on how efficiently and effectively biogas is produced. Particle size, moisture content, bulk density, and the presence of volatile substances are only a few of the physical characteristics of the feedstocks used to produce biogas. These variables control the feedstocks' homogeneity and flowability, which in turn influence how the material is mixed, pumped, and fed into the biogas reactor. Particle size is a crucial factor that influences how well the feedstock is digested. Smaller particle sizes boost the amount of microbial activity-available surface area, resulting in quicker and more effective digestion. Larger particle sizes, on the other hand, may clog the reactor and result in poor mixing, which will reduce the yield of biogas. Another critical factor that influences the production of biogas is moisture content. High moisture feedstocks are typically more viscous and challenging to handle, whereas dry feedstocks may be excessively dusty and problematic for air infiltration. Depending on the feedstock and the type of biogas reactor being utilized, the ideal moisture content for producing biogas will vary. A measure of the mass of the feedstock per unit volume is bulk density. Low-bulk-density feedstocks may settle and cause blockages in the reactor, whereas high-bulk-density feedstocks require more energy to mix and pump. The kind of reactor and the feeding mechanism determine the appropriate bulk density for the production of biogas. The amount of organic matter in the feedstock that can be transformed into biogas is gauged by its volatile solids content. Since they include more easily digestible organic matter, feedstocks with a high volatile solid's concentration are often favored for the production of biogas. Yet, a high-volatile solids content might also result in a biogas reactor with high quantities of ammonia and other unfavorable substances. In conclusion, a crucial component of producing biogas is dependent upon the physical characterization of feedstocks. Biogas producers can enhance the effectiveness and efficiency of their biogas production processes by comprehending and optimizing the physical properties of feedstocks.

12.3.2 Chemical characterization of feedstocks

The chemical makeup of the feedstock can have a big impact on the quality of the biogas produced as well as the biogas production process. Chemical characterization includes examining the feedstock's physical characteristics, biological content, and elemental makeup. The amount of carbon, nitrogen, hydrogen, and oxygen in the feedstock is included in the elemental composition. These components serve as the fundamental building blocks for methane, the primary component of biogas, making them necessary for its generation. The concentration of organic matter, carbohydrates, proteins, and lipids is part of the feedstock's biochemical makeup. The ideal feedstock-to-water ratio, pH, and temperature for biogas generation can all be determined with the help of chemical characterization of the feedstocks. Also, it aids in locating any potential impediments to the creation of biogas present in the feedstock, such as elevated concentrations of ammonia, sulfur, or heavy metals. The chemical characterization of the feedstock also aids in estimating the feedstock's potential energy yield, which can be used to determine the economic viability of producing biogas. In general, chemical characterization of the feedstocks for the production of biogas is an essential step in streamlining the biogas production process and increasing the energy yield from biowaste materials. pH, TS, and volatile solids (VSs) are three crucial chemical characteristics that determine the quality of biowastes (Brandt et al., 2014), and they were tested in accordance with the accepted practices, and the results are shown in Fig. 12.12.

12.3.3 Experimental findings of trial and pilot studies

The outcomes of the 30-day trial research are given in Fig. 12.12. It shows the average methane composition in CD, MRW, RW, and VW as determined by experimental studies. The practical research reveals that the methane composition for CD on the 30th day is 59.69%. So, the efficacy of the current strategy is established. The same method has been used to anticipate the methane generation from MRW, RW, and VW, and the results are shown in Fig. 12.12B-C. The methane production from the cow dung serves as evidence that this approach is reliable. Based on the anaerobic digestion of MRW in the pilot study carried out for 365 days, the volume of the biogas is validated. The actual biogas yield over a year for the tested waste (MRW) is shown in Fig. 12.13.

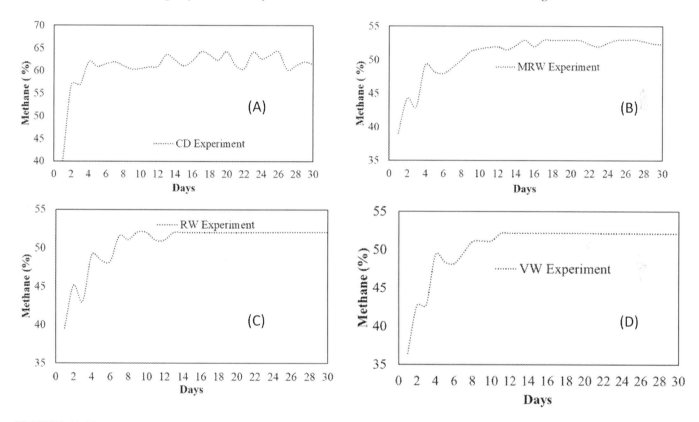

FIGURE 12.12 Composition of methane in the biogas from (A) CD, (B) RW, (C) MRW, and (D) VW. *MRW*, mixed rice waste; *RW*, rice waste; *VW*, vegetable waste.

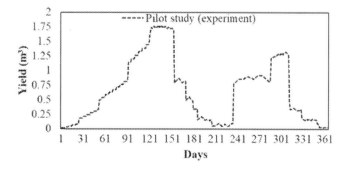

FIGURE 12.13 Biogas yield over a year for MRW. *MRW*, mixed rice waste.

12.3.4 Effect of nonuniform loading rate in a pilot study

To determine the viability of digester performance in terms of methane composition and biogas yield, the 365-day nonuniform creation of biowaste in a school is analyzed. The study period was grouped into four phases to incorporate the effect of academic schedules. The loading rate of RW produced from educational institutions throughout a 365-day period is depicted in Fig. 12.14A. RW availability during phase I, which lasts for the first 150 days, is shown to range between 7 and 46 kg. Since the digester can only hold 10% of the biowaste, or 0.07–4.6 kg, was added. Due to the digesters' size limitations, a similar process has been used to load the other digesters as well. The biogas yield measured in accordance with the loading arrangement is depicted in Fig. 12.14B. The graph demonstrates that at this period, the maximum yield is 0.18 m^3, with an average yield of 0.16 m^3. Due to exams and breaks, phase II is seen as having a nonacademic timetable. The average loading during this interval was between 1 and 3.2 kg. As a result, the biogas yields gradually decreased in accordance with the loading rate, reaching a level of 0.05 m^3 on the 180th day and a minimum of 0.01 m^3 on the 228th day. The

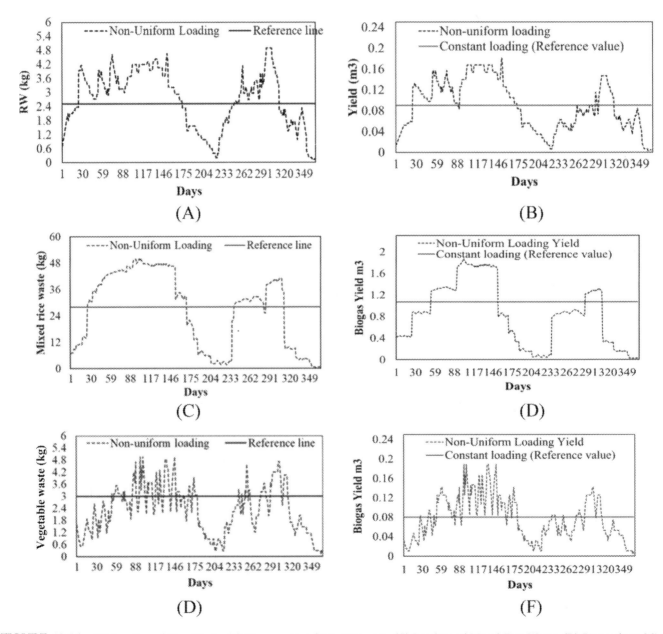

FIGURE 12.14 (A) Loading of Rice Waste. (B) Biogas output from rice waste. (C) Loading of Mixed Rice Waste. (D) Biogas from Mixed Rice Waste. (E) Loading of Vegetable Waste. (F) Biogas from Vegetable Waste.

growth in the student population during phase III led to a significant availability of biowaste. Like phase I, an increase in biowastes loading rate resulted in a higher yield of biogas. A loading rate of $1-4.9$ kg on average and a biogas yield of $0.03-0.15$ m^3 on average are noted. Like phase II, phase IV saw a decline in student enrollment due to extracurricular activities, with an average loading rate of $1-2$ kg and a biogas yield of $0.01-0.08$ m^3. As indicated in the figure, the yield in this instance was projected to be 0.09 m^3, using 2.5 kg as the uniform loading of RW.

Fig. 12.14C and D illustrate the biogas loading and yield patter of MRW over a year. Initially recorded at 0.4 m^3, the MRW biogas yield increased to a maximum of 1.8 m^3 on day 145. All biowastes displayed the initial lag; however, it was also noted that the biogas generation steadily rose to a constant level. The availability of MRW was higher than RW during phase II, which is the nonacademic timetable, but it was lower than it was during phase I. According to Fig. 12.14C, the loading rate change from 0.7 to 32 kg could give a biogas yield of 0.3 m^3 on the 180th day and a minimum of 0.1 m^3 on the 225th. Phase III is comparable to phase I in that it sees an average loading rate of $7-38$ kg and an average biogas yield of $0.8-1.3$ m^3. The student population declines in phase IV due to nonacademic schedules, like phase II. With the loading rate of $0.7-10$ kg, the biogas yield during phase IV ranged from 0.1 to 0.3 m^3.

The uniform and nonuniform loading rates of VW are depicted in Fig. 12.14E. In contrast to the real highest and minimum loading, which ranged between 1 and 5 kg, the uniform loading considered in this research is 2.5 kg. As demonstrated in Fig. 12.14F, the biogas yield produced from VW varies during phase I from 0.01 to 0.18 m^3. The figure shows the biogas yield of 0.03 m^3 on the 180th day and 0.01 m^3 on the 225th day with a loading rate between 0.5 and 2 kg, indicating that the availability of VW is significantly lower during academic schedules. Due to nonacademic schedules, student population declines in phase IV with an average loading rate of $1-5$ kg. The same case's biogas production was discovered to range from 0.01 to 0.09 m^3.

Proper digestion and a high yield of biogas are caused by the methanogen bacteria that were created during phase I with a constant loading rate. Phase II's limited yield potential was brought on by a lower loading rate, which prevented methanogen bacteria from undergoing anaerobic digestion. This affects the phase III biogas yields. The biogas yield was not similar to phase I even though the loading rate was similar to phase I; this is because in phase I, cow manure was loaded initially for the purpose of producing methanogen bacteria (Zhu et al., 2014). The activity of the anaerobic digestion process decreases due to the lower loading rate, and as a result, methanogen bacteria are insufficient (Feng et al., 2015). Phase IV saw a decrease in loading rate and a similar decline in biogas yield to phase II. This demonstrates how the loading rate's nonuniformity may affect the biogas yield. However, not all phases of biogas have a considerable difference in methane composition.

12.3.5 Sensitivity analysis

The sample uncertainties calculated as per the procedure discussed in the experimentation section are presented in Table 12.4. The error values are not significant, the experimental reliability is confirmed.

TABLE 12.4 The uncertainties in parameters.

Sl. No	Parameter	Uncertainty calculation
1	Temperature (28°C)	$\frac{\partial T_{Temp}}{T} = \pm \left(\frac{0.1}{28}\right) = \pm 0.36\%$
2	Velocity	$\frac{\partial V}{V} = \pm 0.65\%$
3	Weight (0.5 kg)	$\frac{\partial W_P}{W_P} = \pm \left(\frac{0.01}{0.5}\right) = \pm 2\%$
4	pH	$\frac{\partial W_{pH}}{W_{pH}} = \pm \left(\frac{0.1}{7}\right) = \pm 1.42\%$
5	Pressure	$\frac{\partial W_{Pressure}}{W_{Pressure}} = \pm \left(\frac{0.01}{0.5}\right) = \pm 2\%$
6	Flow rate	$\frac{\partial F_{Quantity}}{F_{Quantity}} = \sqrt{\left(\frac{0.1}{28}\right)^2 + (0.0065)^2 + \left(\frac{0.01}{0.5}\right)^2 + \left(\frac{0.1}{7}\right)^2 + \left(\frac{0.01}{0.5}\right)^2} = 2.85\%$
7	Methane content	$\frac{\partial M_{Quality}}{M_{Quality}} = \sqrt{\left(\frac{0.5}{52}\right)^2} = 0.96\%$

12.4 Conclusion

The food wastes are collected from educational institutions throughout the year, and they are classified as rice waste, mixed rice waste, and vegetable waste. The total solids, volatile solids, and moisture content of the biowaste are all found to be within ranges that are good for anaerobic digestion. The pH of the substrate was evaluated, and although the biowastes' pH before loading was less than 5, it was found to be between 6.5 and 7.5 inside the digester. This indicates that the digesting process inside the selected digester has a conducive environment for the optimum biogas yield. All the feedstocks used in the trial study could generate biogas with a methane content above 50%. This demonstrates the technical viability of employing food wastes collected from the selected institutions for useful energy generation. In all four phases during a year, the biogas quantity differed based on the population. Whereas the methane quality was observed to be almost the same in all phases. According to experimental results, methane was present in biogas in the amounts of 61.13%, 52.98%, 52.01%, and 51.11% from four categories of biowastes. The range of the biogas yield was from 0.007 to 0.08 m^3. As a result, biogas produced from food wastes could be used for a variety of purposes because it contains a reasonable amount of methane. Also, the university where the study was being done could use biogas produced from food waste for applications such as electricity generation, heating, hydrogen generation, cooking, and fuel cells.

References

Al Seadi, T., Rutz, D., Prassl, H., Köttner, M., Finsterwalder, T., Volk, S., & Janssen, R. (2008). Biogas Handbook. University of Southern Denmark Esbjerg, Niels Bohrs Vej 9-10, DK-6700 Esbjerg, Denmark. Available from https://www.lemvigbiogas.com/BiogasHandbook.pdf.

Aramrueang, N., Rapport, J., & Zhang, R. (2016). Effects of hydraulic retention time and organic loading rate on performance and stability of anaerobic digestion of *Spirulina platensis*. *Biosystems Engineering, 147*, 174 –182. Available from https://doi.org/10.1016/j.biosystemseng.2016.04.006.

Barik, D., & Murugan, S. (2015). Assessment of sustainable biogas production from de-oiled seed cake of karanja-an organic industrial waste from biodiesel industries. *Fuel, 148*, 25–31. Available from https://doi.org/10.1016/j.fuel.2015.01.072.

Brandt, A. J., del Pino, G. A., & Burns J. H. (2014) Experimental protocol for manipulating plant-induced soil heterogeneity. Available from https://doi.org/10.3791/51580.

Browne, J. D., Gilkinson, S. R., & Frost, J. P. (2014). The effects of storage time and temperature on biogas production from dairy cow slurry. *Biosystems Engineering, 129*, 48–56. Available from https://doi.org/10.1016/j.biosystemseng.2014.09.008.

Chen, Y., Cheng, J. J., & Creamer, K. S. (2008). Inhibition of anaerobic digestion process : A review. *Bioresource Technology, 99*, 4044–4064. Available from https://doi.org/10.1016/j.biortech.2007.01.057.

Estoppey, N., Zurbrügg, C., & Vögeli, Y. (2011). Digesting faeces at household level - Experience from a "Model Tourism Village" in south India. In: *Sustainable Sanitation Practice* (pp. 4–9), 9. Eco San Club.

Feng, R., Li, J., Dong, T., & Li, X. (2015). Performance of a novel household solar heating thermostatic biogas system. *Applied Thermal Engineering*. Available from https://doi.org/10.1016/j.applthermaleng.2015.12.003.

Ferrer, P., Cambra-López, M., Cerisuelo, A., et al. (2014). The use of agricultural substrates to improve methane yield in anaerobic co-digestion with pig slurry: Effect of substrate type and inclusion level. *Waste Management (New York, N.Y.), 34*, 196–203. Available from https://doi.org/10.1016/j.wasman.2013.10.010.

Glivin, G., Kalaiselvan, N., Mariappan, V., et al. (2021). Conversion of biowaste to biogas: A review of current status on techno-economic challenges, policies, technologies and mitigation to environmental impacts. *Fuel, 302*121153. Available from https://doi.org/10.1016/j.fuel.2021.121153.

Glivin, G., Mariappan, V., Premalatha, M., et al. (2022). Comparative study of biogas production with cow dung and kitchen waste in Fiber-Reinforced Plastic (FRP) biodigesters. *Materials Today: Proceedings, 52*, 2264–2267.

Glivin, G., Premalatha, M., Murugan, P. C., et al. (2021). Conversion of biowaste to biogas: A review of current status on techno-economic challenges, policies, technologies and mitigation to environmental impacts. *Fuel, 302*121153.

Glivin, G., & Sekhar, S. J. (2016). Experimental and analytical studies on the utilization of biowastes available in an educational institution in india. *Sustain, 8*. Available from https://doi.org/10.3390/su8111128.

Glivin, G., & Sekhar, S. J. (2019). Studies on the feasibility of producing biogas from rice waste. *Romanian Biotechnological Letters*, Doi 10.

Glivin, G., & Sekhar, S. J. (2020a). Waste potential, barriers and economic benefits of implementing different models of biogas plants in a few Indian educational institutions. *BioEnergy Research, 13*, 668–682.

Glivin, G., & Sekhar, S. J. (2020b). Simulation of anaerobic digesters for the non-uniform loading of biowaste generated from an educational institution. *Latin American Applied Research - An international journal, 50*, 33–40.

Glivin, G., Vairavan, M., Manickam, P., & Santhappan, J. S. (2021). Techno economic studies on the effective utilization of non-uniform biowaste generation for biogas production. *Anaerobic Digestion in Built Environments*, 81.

Holman, J. P. (2002). *Experimental methods for engineers* (Seventh Ed.). McGraw-Hill.

International Energy Agency. (2020). Outlook for biogas and biomethane. Prospects for organic growth. IEA Publ 1–93.

International Energy Agency. (2022). International Energy Agency (IEA) World Energy Outlook 2022. https://wwwieaorg/reports/world-energy-outlook-2022/executive-summary 524.

Izumi, K., Okishio, Y., Nagao, N., et al. (2010). *International Biodeterioration & Biodegradation, 64*, 601–608. Available from https://doi.org/10.1016/j.ibiod.2010.06.013.

Kalaiselvan, N., Glivin, G., Bakthavatsalam, A. K., et al. (2022). A waste to energy technology for Enrichment of biomethane generation: A review on operating parameters, types of biodigesters, solar assisted heating systems, socio economic benefits and challenges. *Chemosphere*133486.

Lehtomäki, A., Huttunen, S., & Rintala, J. A. (2007). Laboratory investigations on co-digestion of energy crops and crop residues with cow manure for methane production: Effect of crop to manure ratio. *Resources, Conservation and Recycling, 51*, 591–609. Available from https://doi.org/10.1016/j.resconrec.2006.11.004.

Liu, B., Liu, S., Guo, S., & Zhang, S. (2020). Economic study of a large-scale renewable hydrogen application utilizing surplus renewable energy and natural gas pipeline transportation in China. *International Journal of Hydrogen Energy, 45*, 1385–1398. Available from https://doi.org/10.1016/j.ijhydene.2019.11.056.

Martí-Herrero, J., Alvarez, R., Cespedes, R., et al. (2015). Cow, sheep and llama manure at psychrophilic anaerobic co-digestion with low cost tubular digesters in cold climate and high altitude. *Bioresource Technology, 181*, 238–246. Available from https://doi.org/10.1016/j.biortech.2015.01.063.

Metcalf & Eddy. (2021). Wastewater engineering treatment and reuse. 4th Edition.

Montañés, R., Pérez, M., & Solera, R. (2014). Anaerobic mesophilic co-digestion of sewage sludge and sugar beet pulp lixiviation in batch reactors: Effect of pH control. *Chemical Engineering Journal, 255*, 492–499. Available from https://doi.org/10.1016/j.cej.2014.06.074.

Neves, N. G., Berni, M., Dragone, G., Mussatto, S. I., & Forster-Carneiro, T. (2018), Anaerobic digestion process: technological aspects and recent developments. *International Journal of Environmental Science and Technology, 15*, 2033–2046. Available from https://doi.org/10.1007/s13762-018-1682-2.

Olvera, R., & Lopez-lopez, A. (2012). Biogas production from anaerobic treatment of agro-industrial wastewater. IntechOpen.

Pareek, R., & Nagarsheth, H. J. (2015). Preparation parametric comparison and performance evaluation of FRP and HDPE type biogas digesters. *Journal of Clean Energy Technologies*. Available from https://doi.org/10.18178/jocet.2016.4.5.312.

PIKE Research. (2012). Methane recovery and utilization in landfills and anaerobic digesters: Municipal solid waste, agricultural, industrial, and wastewater market report on analysis and forecasts. Renew biogas 87.

Safferman, S. I., Kirk, D. M., Faivor, L. L., & Haan, W. W. (2012). *Anaerobic digestion processes* (pp. 103–136). John Wiley & Sons, Ltd.

Singh, N., Wakil, S. J., & Stoops, J. K. (1985). The development and application of a novel chromophoric substrate for investigation of the mechanism of yeast fatty acid synthase. *Biochemical and Biophysical Research Communications, 131*, 786–792.

Zhang, C., Xiao, G., Peng, L., et al. (2013). The anaerobic co-digestion of food waste and cattle manure. *Bioresource Technology, 129*, 170–176. Available from https://doi.org/10.1016/j.biortech.2012.10.138.

Zhang, R., El-mashad, H. M., Hartman, K., et al. (2006). Characterization of food waste as feedstock for anaerobic digestion. *Bioresource Technology, 98*, 929–935. Available from https://doi.org/10.1016/j.biortech.2006.02.039.

Zhu, G., Li, J., & Jha, A. K. (2014). Anaerobic treatment of organic waste for methane production under psychrophilic conditions. *International Journal of Agriculture and Biology, 16*, 1025–1030.

CHAPTER

13

Hydrogen from waste and biowaste materials: production, separation, purification, and use

Prem Kumar Seelam[1], Putrakumar Balla[2] and Simona Liguori[3]

[1]Sustainable Chemistry Research Unit, Faculty of Technology, University of Oulu, Oulu, Finland [2]Department of Chemical Engineering and Applied Chemistry, Chungnam National University, Daejeon, Republic of Korea [3]Department of Chemical & Biomolecular Engineering, Clarkson University, Potsdam, NY, United States

13.1 Introduction: H_2 from biomass-derived biowastes

13.1.1 Potential biowaste sources

Hydrogen economy is more reality than hype in many countries. Undoubtedly, hydrogen is foreseen as one of the key energy carriers and the solution to the climate change, energy security, and air pollution control. At the moment, it is serious concern to reduce the carbon emissions and mitigate the air pollution. Low- and zero-carbon technologies such as hydrogen can be promising and an important transition to green and sustainable energy. Hydrogen is the most abundant element on the earth but not available in free form. Hydrogen possesses highest specific energy (120 kJ/g) among all conventional fuels. Nevertheless, hydrogen can be produced from wide variety of feedstocks and raw materials. Sustainable hydrogen production is the most important to achieve for the complete deployment of H_2 economy in energy and fuel sectors. As seen from Fig. 13.1, an exponential growth on research articles published on hydrogen production using biomass-based biowastes and other potential wastes was foreseen. Today, hydrogen can be generated from any ubiquitous biomass source. However, there huge challenges exist in many areas of hydrogen economy, that is, production, separation, purification, distribution, and use. A complete H_2 network must be established, and efficient gas infrastructure and storage facilities must be built.

Today, hydrogen is categorized into different colors based on the production method and CO_2 capture technology. Black H_2: hydrogen from coal gasification, and it has highest CO_2 emissions, and this method is most harmful to environment. Gray H_2: it is most widely and currently the major H_2 is produced via steam reforming of natural gas with CO_2 emissions. Blue H_2: it is the traditional steam reforming of natural process with CO_2 capture and storage technology (CCS). This technology least produces CO_2 but with extra cost on CCS. Turquoise H_2: utilizing the natural gas or biogas in thermocatalytic decomposition or simply methane pyrolysis to H_2 and carbon without CO_2 emissions, this technology is promising and foreseen as a short- to long-term solution if the process is cost-effective via carbon products. Finally, green H_2: using renewable energies such as solar, wind, geothermal, biomass in the H_2 production, and this method does not produce any direct emissions (according to life cycle); for example, water electrolysis, water splitting via photolysis and electro photolysis of water and other H_2 sources. Nevertheless, each method has its own pros and cons. The green and turquoise H_2 looks promising and the share of the renewables in H_2 is increasing at faster rate. Utilization of biowastes and other potential wastes can be promising in green hydrogen production. The whole process works under closed carbon cycle if H_2 from biomass is implemented. There are many technologies and methods to produce H_2 from biowastes via thermochemical, photochemical, electrophotochemical, and biological routes.

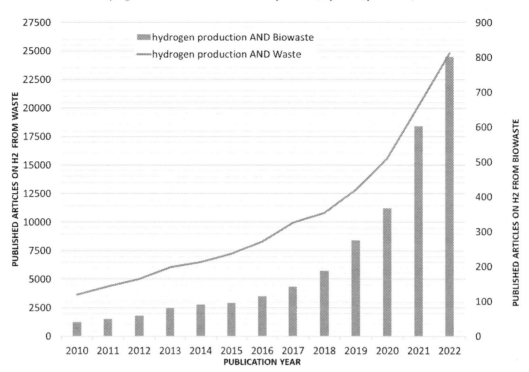

FIGURE 13.1 List of published articles (2010–22) in Sciencedirect database with key words search: hydrogen production AND biowaste and hydrogen production AND waste.

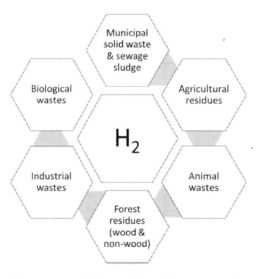

FIGURE 13.2 Hydrogen generation from different types of biomass-derived biowastes and other potential waste feedstocks.

Biomass is a renewable material and can be derived from any form of living entity such as plants, microorganisms, or woody materials (Fig. 13.2). Typically, biomass is ubiquitous and abundant on earth and can be presented in any form such as woody, nonwoody, plants, or microorganisms (e.g., algae and fungi). The composition of different biomass varies depends on type, origin, and it consists of main components, that is, lipids, fats, carbohydrates, and water. Biomass-derived wastes are generated in millions of tons per day, and there is a huge potential to utilize this biowastes in many forms, for example, fuels, energy, chemicals, and heat. Herein exist many technoeconomic challenges in utilizing biowaste in decentralized networks, and this solely depends on geographical location and high potential to integrate the renewable technologies with biowaste utilization in

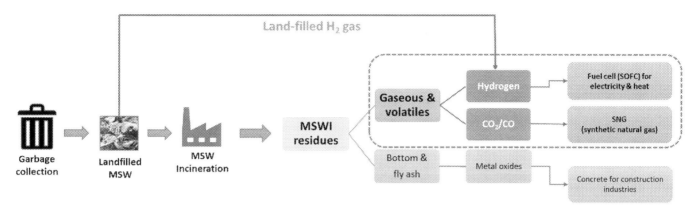

FIGURE 13.3 Schematic presentation of separation and purification of H_2 and other gaseous compounds from MSWs.

most efficient manner. In this chapter, insights into the critical research and development on producing hydrogen (H_2) from biomass-derived biowaste through various conversion and processing technologies are discussed.

Global challenge in tackling the food waste and municipal solid waste (MSW) problems and providing sustainable solutions is the key. In Finland, 300 000 tons of bottom ash (20–30 wt.%) from incenerated MSW (MSWI) was generated. MSWI residues are potential H_2 sources and high in H_2 vol.% besides other compounds (e.g., 5–8 L_{H_2} per $Kg_{b.ash}$). MSWI residues in aqueous phase result in H_2 generation via following equations (13.1–13.2). Significant amounts of H_2 (e.g., 2–3 m^3/h LEL) and carbon oxides are produced in abiotic conditions over days (land-filled and incineration). Clean hydrogen production from MSWI in alkaline conditions will be studied. In addition, processing H_2 with CO_2 to produce synthetic natural gas (SNG) and feeding to a gas grid for clean electricity can be viable solution (end application). There is huge scope of abatement and utilization of abundant MSW. As the MSW residues are potential source for producing wide variety of by-products, for example, H_2 and oxides. MSWI waste-to-energy conversion (WtE) technique can be utilized in potential moderate amount of clean energy. The MSW consists of varied composition of C, H, N, O, S (ultimate analysis) and the significant amount of moisture, volatiles, carbons, ash, and high amount of calorific value 3000–2000 kcal/kg wasted (proximate analysis). The hydrogen composition varied from 5 to 10 wt.% and separating and purifying the H_2 are a challenge due to coexistent of other hetero atoms such as N, S, O, and C (Fig. 13.3) (Liu et al., 2023; Saffarzadeh et al., 2016).

$$2Al + 6H_2O = 2Al(OH)_3 + 3H_2 \qquad (13.1)$$

$$2Al + 6H_2O + 2Ca(OH)_2 = 2CaAl(OH)_5 + 3H_2 \qquad (13.2)$$

Hydrogen economy is more reality than hype in many countries. Undoubtedly, hydrogen is foreseen as one of the key energy carriers and the solution to the climate change, energy security, and air pollution control. At the moment, it is serious concern to reduce the carbon emissions and mitigate the air pollution. Low- and zero-carbon technologies such as hydrogen can be promising and an important transition to green and sustainable energy. Hydrogen is the most abundant element on the earth but not available in free form. Nevertheless, hydrogen can be produced from wide variety of feedstocks and raw materials. Sustainable hydrogen production is most important to achieve the complete deployment of H2 economy in energy and fuel sectors.

13.2 Hydrogen production processes from biowastes and industrial waste gases

Utilization of biowastes for hydrogen production provides inexpensive energy generation with simultaneous waste treatment. This chapter summarizes biohydrogen production from some waste materials. The major criteria for the selection of waste materials to be used in biohydrogen production are the availability, cost, carbohydrate content, and biodegradability. Simple sugars such as glucose, sucrose, and lactose are readily biodegradable and preferred substrates for hydrogen production. However, pure carbohydrate sources are expensive raw materials for hydrogen production. The production of hydrogen from biomass as feedstocks has a variety of positive effects on the environment and the economy, and it has the potential to significantly reduce the amount of fuel that is required at this time. Major waste materials that can be used for hydrogen gas production may be summarized as follows.

13.2.1 Chemical and thermal

Hydrogen production from various organic biomasses via thermochemical process is considered as a promising and economical viable technique. The advantages of this process are higher product yield and flexibility with current available facilities than other hydrogen production methods. Nowadays, large-scale hydrogen production from various biomass is the challenging process. Many researchers have documented that the clean H_2 energy production is possible using different types of biomasses via thermochemical conversion. Most commonly, biomass with high organic content is preferred as potential feedstock, for example, municipal solids waste, lignocellulosic biomass, algal waste, herbaceous vegetation, solid biomass, marine-based biomass, farming, and other tree residues can be used for the production of H_2. The selection of organic biomass for thermochemical conversion mainly depends on the high amount of carbohydrates and lipid content, proteins, lipids, hemicellulose, cellulose, and lignin. which is more responsible for recovery of valuable products such as H_2, biooil, and biochar. The above listed biomasses are the major source (feedstock) and available at free cost for H_2 production worldwide. About 50–70 wt% of H_2 can be recovered using biomass via thermochemical conversion. The four major types of thermochemical conversion technologies used for H_2 production are (1) pyrolysis, (2) gasification, (3) steam reforming techniques, (4) sorption enhanced steam reforming, and (5) sorption-enhanced chemical looping. Among various thermochemical biomass conversion platforms, gasification is the oldest and commercially applied throughout the world for pyrolysis of the waste biomass. Biomass gasification process includes various sequential steps such as drying, pyrolysis, and partial oxidation as well as reduction reaction. Volatile materials that are produced during pyrolysis step are oxidized and reacted with char to generate H_2, CO, CO_2, CH_4, and hydrocarbons. The hydrocarbons and char generated during this process can react with gasifying agents to produce H_2. In this process, the sources of organic biomass can be transformed into ethanol, bio-oils, and liquid biofuels. The liquid fuel in turn reacts with steam at increased temperature in the presence of catalysts to generate gas that chiefly consists of H_2, CO, and CO_2. The syngas produced from biomass gasification and freely available natural gas undergo copyrolysis to produce H_2 clean energy for domestic and industrial use. The applied heat energy, retention, rate of heating, pressure, catalysts, and reaction operating conditions of the pyrolysis reactor alter the composition of finally obtained H_2 during pyrolysis (Fig. 13.4).

13.2.1.1 Hydrogen from industrial emissions and side-stream gases

Volatile organic compounds (VOCs) belong to the most significant air pollutants. These air pollutants are of major concern to the environment and have significant effects on the earth climate change and human health (Ioannis, 2020; Moore, 2009). Nonmethane VOCs (NMVOCs) contribute to the tropospheric ozone formation and smog in the cities, and certain NMVOCs such as 1,3 butadiene, benzene, methyl mercaptan, and chloromethanes

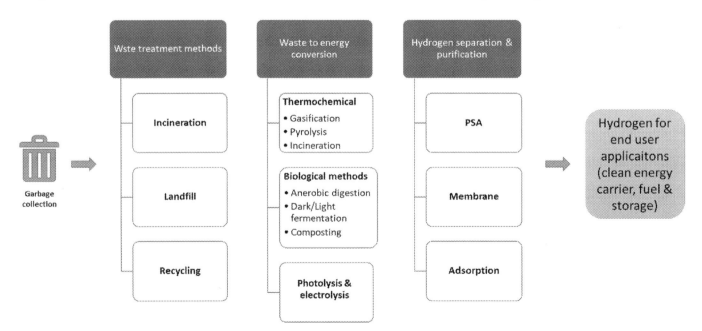

FIGURE 13.4 Schematic of the main processes involved in a thermochemical conversion route, including gas separation.

are hazardous and carcinogenic to human health (Nattaporn, 2022). VOCs are present naturally in the atmosphere as a part of the carbon cycle. Over the years, the global VOC emissions (excluding methane) have significantly been reduced, but during the recent years, the emissions have started to increase again. There is an urgent need for new approaches to abate the VOCs even too much lower levels due to strict regulations (Yunlong, 2021). For example, in the EU countries, emissions of VOCs are explicitly regulated (Arja, 2016; Christos, 2022). VOCs are chemicals having such a vapor pressure, which makes them evaporate easily at room temperature or at specific temperatures of their use. The main sources for nonmethane VOCs (NMVOCs) emissions are combustion activities, solvents, and products use, for example, paints as well as coatings, chemical and the process industries (Arja, 2016; Christos, 2022; Rajesh, 2023). The VOC emissions are typically a mixture of several condensable vapor compounds with varied concentrations. The annual anthropogenic NMVOC emissions were ~142 Tg equivalent of carbon and ~20.1 Tg of H_2 in 2010 (Li, 2019; Yang, 2019). Many technological challenges have to be solved to achieve low-carbon society, which in Finland would mean the reduction of around 80% of carbon emissions by 2050 (Lin, 2022). Despite their polluting nature, even the VOCs containing flue gas streams could be valuable sources in the production of cleaner energy and fuels. Fig. 13.5 proposes a new approach to overcome the low-carbon society challenge: converting the abundant industrially emitted VOCs into valuable energy source (i.e., hydrogen or syngas). This solution provides environmental benefits with VOC emissions abatement and economic benefits with new value creation from the emissions. This would enhance the overall energy efficiency of the processes and would improve their environmental performance.

Traditionally, the abatement of VOCs is done via thermal oxidation of VOCs and will eventually increase the CO_2 emissions. Production of H_2 can be potentially done efficiently and in an environmentally friendly way from the VOCs with high H_2 yield and selectivity. Oxygenated VOCs such as alcohols, aldehydes can be reformed at relatively low temperatures (below 550°C), whereas aromatics and aliphatic compounds are reformed at high temperatures (above 600°C). Most of the studies are reported for individual model compounds such as methanol, acetone, toluene, or ethanol (Ikuo, 2017; Moreno, 2018; Veerapandian, 2021). The amount of H_2 produced from VOCs reforming is unpredictable due to varied VOCs concentration. However, it is still sufficient to run an SOFC to generate heat and electricity. The conceptual design of the whole process of utilizing the VOCs in hydrogen and syngas generation via steam and dry reforming technologies can be applied (Fig. 13.5). However, there are big challenges in catalysts and process development to separate and purify the H_2 rich stream.

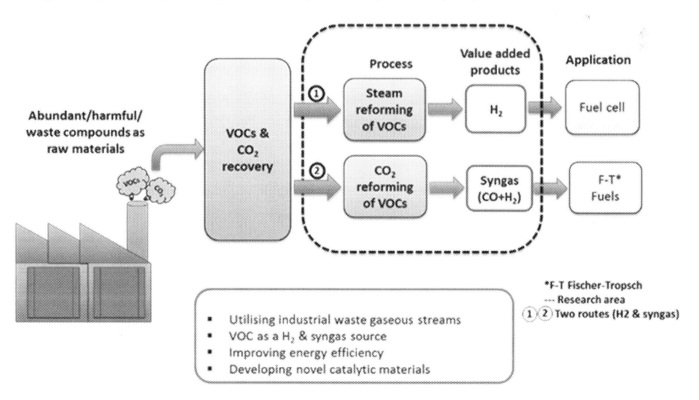

FIGURE 13.5 Conceptual overall process in VOCs utilization for H_2 and syngas production. Two innovative routes: that is, 1. steam and 2. dry reforming of VOCs. Most critical research and development areas are pointed in dot lines.

13.2.2 Biological processes

In accordance with sustainable development and waste minimization issues, biohydrogen gas production from renewable sources, also known as "green technology," has received considerable attention in recent years. Biological production of hydrogen gas has significant advantages over chemical methods. Biological hydrogen production is a viable alternative to the aforementioned methods for hydrogen gas production. The major biological processes utilized for hydrogen gas production are biophotolysis of water by algae, dark and photofermentation of organic materials, usually carbohydrates by bacteria. Carbohydrate-rich, nitrogen-deficient solid wastes such as cellulose and starch containing agricultural and food industry wastes and some food industry wastewaters such as cheese whey, olive mill, and baker's yeast industry wastewaters can be used for hydrogen production by using suitable bioprocess technologies. Complex nature of these wastes may adversely affect the biodegradability. Biohydrogen production can be realized by anaerobic (fermented) and photosynthetic microorganisms using carbohydrate-rich and nontoxic raw materials. Sequential dark and photofermentation process is a rather new approach for biohydrogen production. Under anaerobic conditions, starch containing solid wastes is easier to process for carbohydrate and hydrogen gas formation. Starch can be hydrolyzed to glucose and maltose by acid or enzymatic hydrolysis followed by conversion of carbohydrates to organic acids and then to hydrogen gas. Photosynthetic processes include algae, which use CO_2 and H_2O for hydrogen gas production. Some photoheterotrophic bacteria utilize organic acids such as acetic, lactic, and butyric acids to produce H_2 and CO_2. The advantages of the later method are higher H_2 gas production and utilization of waste materials for the production. However, the rate of H_2 production is low, and the technology for this process needs further development. Fig. 13.6 depicts a schematic diagram for biohydrogen production from cellulose and starch containing agricultural residues.

In biophotolysis, the hydrogen ion is catalyzed either by nitrogenase or hydrogenase enzyme to produce hydrogen. It is further classified into direct biophotolysis and indirect biophotolysis. Green algae, cyanobacterium, etc., harvest solar energy for the water-splitting process to produce O_2 and reduce an electron carrier ferredoxin in the chloroplasts of these organisms. The oxygen sensitivity of hydrogenase enzyme restricts the process,

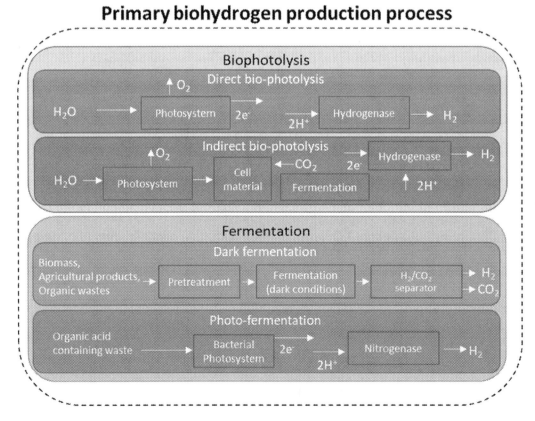

FIGURE 13.6 Schematic representation of the primary biological routes for H_2 production (Nikolaidis & Poullikkas, 2017). *Modified from Nikolaidis, P., & Poullikkas, A. (2017). A comparative overview of hydrogen production processes.* Renewable and Sustainable Energy Reviews, 67, 597–611.

so the yield of H_2 during this process is not much significant. This direct biophotolysis is like the photosynthesis method present in plants and algae. It is based on the ability to break water molecule into hydrogen and oxygen through photosynthesis. This method is generally found in cyanobacteria and algae. These microorganisms produce protons and electrons by using sunlight, and the enzyme hydrogenase converts $H+$ into H_2. Ni hydrogenase and Fe hydrogenase are extremely sensitive to oxygen, so only under special conditions, hydrogen production is possible by this method. Indirect biophotolysis involves two-stage process for the biohydrogen production. The reactions comprise separation of hydrogen and oxygen to overcome the problems of sensitivity of hydrogenase enzyme as observed in direct biophotolysis. By utilizing the sunlight in light-dependent reaction, splitting of water molecule takes place to produce protons and oxygen, along with simultaneous carbon dioxide fixation into carbohydrates. These substrates are used as a carbon source in the second stage for hydrogen production. Indirect photolysis is dependent on the photon conversion efficiency of the algal culture, and 10%–13% light utilization is realistic and achievable. In cyanobacteria such as Nostoc, Anabaena, and Calothrix, their thick filaments with heterocyst containing nitrogenase are protected from oxygen.

The electrohydrogenesis method is Microbial Electrolytic Cell, a type of bioelectrochemical system driven by the externally applied voltage across the electrodes. Here, hydrogen production takes place in an anaerobic system where bacterial oxidation of the feed at the anode and reduction of electrons at the cathode result into hydrogen production. This method is considered as a new cohort of sustainable biohydrogen production.

13.2.3 Photochemical processes

A promising technique of hydrogen evolution is photo reforming of biomass since it uses renewable biomass substrate, foreseeably unlimited solar energy inputs, and might make use of waste products from currently used industrial biomass processes. In addition to yielding useful carbon-free energy in the form of molecular hydrogen, photocatalytic biomass conversion opens a pathway for the creation of biomass products with commercial value. The only energy inputs required for this photocatalytic conversion are efficient, renewable reaction materials (biomass) and endless sunlight.

Molecular hydrogen can be produced from biomass through photocatalytic reforming, more research is required to find active photocatalysts made of inexpensive, nontoxic materials that can sustain high rates of H_2 production without the need for pretreatment of the biomass source. The electronic band structures and energies of photocatalysts are crucial in the photophysical events of electron/hole transport and charge carrier injection to produce H2 from biomass. The basic processes of photocatalytic hydrogen production involve the following steps. First, photons with enough energy are absorbed to activate semiconductor photocatalysts to generate electron–hole pairs. Second, the photogenerated electrons and holes migrate to the surface of the photocatalyst, during which process surface and bulk recombination happen. Third, the electrons on the photocatalyst's surface trigger a proton reduction reaction for hydrogen production (Fig. 13.7).

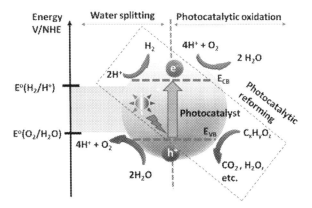

FIGURE 13.7 Schematic illustration of water splitting and photocatalytic oxidation of biomass over photocatalyst nanoparticle under light illumination (Huang et al., 2020; Kondarides et al., 2008). *Adapted from Huang, C.-W., Nguyen, B.-S., & Wu, J. C. S. et al. (2020). A current perspective for photocatalysis towards the hydrogen production from biomass-derived organic substances and water.* International Journal of Hydrogen Energy, *45, 18144–18159; Kondarides, D. I., Daskalaki, V. M., Patsoura, A. et al. (2008). Hydrogen production by photo-induced reforming of biomass components and derivatives at ambient conditions.* Catalysis Letters, *122, 26–32. This is an open-access article distributed under the terms of the Creative Commons CC BY license, which permits unrestricted use, distribution, and reproduction in any medium, provided the original work is properly cited.*

Titanium dioxide (TiO_2) has been known for its photoactive properties and is used for water splitting and hydrogen production. However, the utilization of TiO_2 as the photocatalyst has several problems due to its large bandgap of 3.2 eV, low electron mobility, and UV range absorbance. Therefore, most studies on photocatalyst for hydrogen production used TiO_2 doped with metal and nonmetal cocatalyst, such as Pt, Pd, Au, and NiOx. The combination of Pt/TiO_2 shows a promising result of light-induced water splitting and biomass oxidation with the cellulose, resulting in a relatively high hydrogen and glucose yield. Other types of photocatalyst are also studied and developed as alternatives to the Pt with higher cost and limited availability. However, most studies on hydrogen production by biomass photo-induced reforming require biomass pretreatment before feeding into the photocatalyst process. The biomass feedstock shall be processed to either monomeric substrates such as alcohols or saccharides such as glucose and cellulose. Meanwhile, the photocatalytic reaction from raw biomass is limited by the complex structure and chemical compound of biomass, which requires a pretreatment process to provide optimum reforming process by a specific photocatalyst type. There are reports on photocatalytic reforming of raw biomass to produce hydrogen, showing the maximum production rate of oat 5.31 mmol/hg_{cat} with the Pt/TiO_2 photo catalyst. Despite many challenges due to the complex chemical structure of biomass and limited photocatalytic reaction performance, the photo-induced reforming process is an alternative process to steam reforming in producing hydrogen from a renewable feedstock of biomass by utilizing the renewable energy source of solar. In order to increase photocatalytic performance under visible or natural sunlight and the use of mild conditions to enhance selectivity toward value-added products along with H_2 productivity, recent attention has shifted to designing narrow bandgap materials and/or metal-free photocatalysts for H_2 production from biomass (Fig. 13.8).

13.2.4 Electrochemical processes

Electrochemical hydrogen production coupled with renewable energies is undoubtedly an unavoidable alternative to thermoneutral methods based on fossil resources. Electrochemical hydrogen production coupled with renewable energies is undoubtedly an unavoidable alternative to thermos neutral methods based on fossil resources. Hydrogen electrochemical production through proton reduction in acidic media and water reduction in alkaline media at the cathode and oxidation of oxygenated organic compounds from biomass at the anode of an electrolysis cell are equivalent to an electroreforming reaction. In this context, thermodynamic data clearly speak for electroreforming of oxygenated organic compounds instead of water electrolysis. Oxygenated organic compounds produced as by-products/wastes by the biofuel industry, for example, can be electroreformed into hydrogen, which can help to make this industry more profitable and virtuous from the green chemistry point of view. However, hydrogen evolution rates remain very low compared with water electrolysis. Larger amounts of platinum group metals—based catalysts in anodes, bigger system dimensions, or higher metal loadings will increase the capital expenditure. Replacement of platinum group metals by less costly Ni-based catalysts involves higher cell voltages, increasing the operational expenditure. The electroreforming concept becomes very attractive in terms of sustainability, energy-saving, low hydrogen production cost, and so on, and more and more research groups get involved in insuring the future development of this technology.

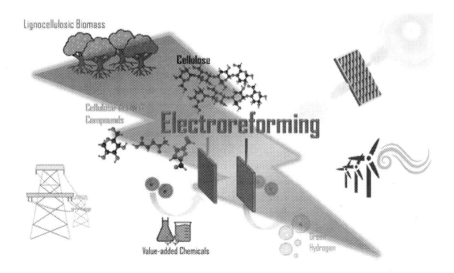

FIGURE 13.8 Schematic diagram of biomass electroreforming to green hydrogen production (Adapted from Lai et al., 2021). *Lai, Z. I., Lee, L. Q., Li, H. (2021). Electroreforming of biomass for value-added products. Micromachines, 12, 1405;* © *2021 by the authors. Licensee MDPI, Basel, Switzerland. This article is an open-access article distributed under the terms and conditions of the Creative Commons Attribution (CC BY) license.*

13.3 Hydrogen separation and purification

Most of the outlet streams coming from biowastes and containing hydrogen mixed with various coproducts need to be treated before using hydrogen as an energy vector. The treatment process is composed of the separation step, which is usually used as first-stage operation to concentrate hydrogen, followed by the purification step to upgrade the product to the final use. Several technologies are available for hydrogen treatment, which can be divided into three main categories: physical, chemical, and selective diffusion methods (Grashoff et al., 1983). Usually, each separation method is selected accordingly to (1) the purity requirements for any specific end use, (2) cost, and (3) hydrogen recovery (Speight, 2015). Table 13.1 compares several hydrogen separation technologies, which will be discussed in the following.

13.3.1 Pressure swing adsorption

The pressure swing adsorption (PSA) technology uses adsorbent material to remove impurities from the hydrogen-containing stream, and the separation effect is based on the binding strength between the gas and the adsorbents. A PSA unit usually consists of a set of columns, between 4 and 12, in order to perform the steps for pressure swing: a cycle of pressurization–depressurization. This hydrogen purification method is commonly used in the process of natural gas steam reforming, and it enables high pure hydrogen, between 98% and 99.999% with a maximum recovery of 85% (Sircar & Golden, 2000). These values are feasible due to the high operating pressure and the elevated column numbers. Higher hydrogen recovery can be achieved by using PSA units with more than 12 columns, which allows for increasing energy efficiency as well (Grande, 2012). However, the presence of carbon monoxide, nitrogen, and argon, which are difficult to remove due to the low affinity with the adsorbents (Besancon et al., 2009; Golmakani et al., 2017), does not allow to obtain both high hydrogen purity and recovery. In addition, nowadays, the main challenge is to scale-down the PSA to separate hydrogen produced in portable technologies. Therefore, less columns and lower pressure, from ~25 bar to < 8 bar can be employed, but this causes a low recovery, ~70% (Hellas, 2015). For this reason, several studies focused on the role of adsorbent and how to increase their

TABLE 13.1 Comparison of hydrogen separation technologies (Liguori et al., 2020).

Technique	Description	T (°C)	H_2-feed	H_2 purity (%)	H_2 recovery(%)	Drawbacks	Benefits
PSA	Selective adsorption of compounds from a gas stream	80–180	Any H_2 rich gas	99.99	70–85	Around 20% of H_2 is lost in the operation; Energy intensive	High H_2 purity >99.99%
Cryogenic distillation	Partial condensation of gas mixtures at low temperatures	−180	Petrochemical and refinery off-gases	90–98	95	Relatively low H_2 purity; Energy intensive	Better economies of scale
Metal hydride separation	Reversible reaction of hydrogen with metals to form hydrides	> RT	Ammonia purge gas	99	75–95	Relatively low H_2 recovery	High H_2 purity
Dense Pd-alloy membrane	Selective diffusion of hydrogen through a Pd-based alloy membrane	300–450	Any H_2 containing stream	99.999	Up to 99	High cost; Low mechanical resistance	High purity H_2 stream; H_2 recovery up to 99%
Polymeric membrane	Differential rate of diffusion of gases through a permeable membrane	25–100	Refinery off-gases and ammonia purge gas	92–98	>85	Relatively low H_2 purity	Low cost; Very high H_2 purity >99.999%
Solid polymer electrolyte cell	Electrolytic passage of hydrogen ions across a solid polymer membrane	RT	H_2 produced by thermochemical cycles	99.8	95	Sluggish anode reaction coupled with the inefficient cathode reaction (low overall cell performance)	High H_2 purity ≈99.8%; High H_2 recovery ≈95%

Source: Adapted from Liguori, S., Kian, K., Buggy, N., Anzelmo, B. H., Wilcox, J. (2020). Opportunities and challenges of low-carbon hydrogen via metallic membranes. Progress in Energy and Combustion Science, 80, 100851; Creative Commons. This is an open access article distributed under the terms of the Creative Commons CC-BY license, which permits unrestricted use, distribution, and reproduction in any medium, provided the original work is properly cited.

performance. For example, Ribeiro et al. (2008) packed two PSA columns with activated carbon and zeolite, respectively. The authors reported a hydrogen purity of 99.9958% with a maximum recovery of 52.11%. The operating pressure was 7 bar. More recently, a simulation investigation (Delgado, 2014) showed higher hydrogen purity and recovery, 99.993% and 90.3% respectively, by operating two columns at 16 bar. The columns were packed with activated carbon and zeolite 13x. However, the study reported contamination of the hydrogen stream with CO, ~63ppm, which does not allow to feed the hydrogen stream directly to a PEM fuel cell. In another simulation study (Agueda et al., 2015), 99.999% of hydrogen purity and 93% of hydrogen recovery were reached by using a PSA packed with MOF (metal organic framework, UTSA-16), but no mention was made for CO content in the stream.

It is evident that the use of zeolite, activated carbons, and MOF as adsorbents for hydrogen separation and purification in PSA is a doable path toward scaling-down this technology. Nevertheless, more studies are needed to investigate the CO separation owing to the highly favorable CO_2 adsorption on the adsorbents, which partially inhibits the CO removal.

13.3.2 Cryogenic distillation

The cryogenic distillation method exploits the difference in boiling point temperature of the species for the separation, and it can be used to purify hydrogen at large scale enabling a hydrogen purity of 90%–98% and recovery of 95%. However, due to the very low operating temperature, it is considered an energy-intensive process, which limits its use for hydrogen separation (Hinchliffe & Porter, 2000). It is industrially used though to purify industrial gases, such as nitrogen, oxygen, argon, helium, and hydrocarbon streams rich in $C4^+$, propane, ethane, etc. (Bernardo & Jansen, 2015)

13.3.3 Metal hydrides

Metal hydrides (MHs) can selectively absorb hydrogen allowing it to reach 99% hydrogen purity and relatively high hydrogen recovery between 75% and 95% at low operating conditions, and they are raising attention due to their potential use as hydrogen storage. Indeed, their benefit is that only hydrogen can be absorbed from a stream containing other gases at certain pressure and temperature (Lototskyy et al., 2011). This is due to the hydriding reaction, where (1) hydrogen molecules are adsorbed on the metal surface, (2) they are decomposed in atomic hydrogen, (3) which diffuses through the metal lattice, (4) to form solid solution of MHx, and (5) be converted to MHy (metal hydride) (Chen et al., 2014). The hydriding reaction is reversible, and it takes place by cooling or pressurization of the stream, while heating or depressurization is needed for the reversible reaction.

To obtain effective hydrogen purification, it is important to keep the MH poison-free. Indeed, MHs are susceptible to some impurities such as CO, H_2S, O_2, H_2O, and CO_2. Particularly, water and oxygen can progressively deteriorate the MH absorption and desorption characteristics, while CO can negatively affect the dehydrogenation MH properties. To prevent and avoid the poisoning effect from these gases, the MHs surface can be modified. The surface modification has—hence—the potential to protect the surface by forming a coating layer to block the permeation of impurities and increase the catalytic activity of the surface allowing for more hydrogen absorption. On this route, Lototsky et al. (2011) coated Pd by fluorination and electroless deposition of single or mixed metal on the MH. The results showed that the hydrogen absorption rate increased after the surface modification showing performance 100 times higher than unmodified MH. In another study (Klochko et al., 2013), the authors observed highly efficient hydrogen purification in the presence of impurities such as CO_2 and CO. Miura et al. (2012) developed a new system for hydrogen separation from gas reformer called COA-MIB (CO Adsorption Metal Hydride Intermediate buffer). The system removed CO from the stream prior to be in contact with the MH. They were able to reach 83% hydrogen recovery from the exhaust stream of reformer. Au et al. (1996) investigated the separation, purification, and transportation of hydrogen from an ammonia synthetic plant, and four MH reactors formed by $MlNi_{5-x}Al_x$ were used. The authors obtained 99.999% hydrogen purity with 70% of H_2 recovery. In addition, they showed that by using MH reactors to provide hydrogen from the exhaust stream of ammonia plant instead of using water electrolysis for float glass factory has the potential to reduce 60% of hydrogen production cost and save 97% of power consumption. Dunikov et al. (2012) analyzed AB_5 MH with composition of $Mm_{0.8}La_{0.2}Ni_{4.1}Fe_{0.8}Al_{0.1}$ for hydrogen separation and purification. The results showed that presence of N_2 as impurities decreases the hydrogen partial pressure causing a slowdown in the hydrogen sorption reaction. Saitou and Sugiyama (1995) investigated hydrogen purification with sintered MH pellets containing $FeTi_{0.95}Mm_{0.08}$ alloys. They reached 99.999% of hydrogen purity and 90% of product yield with a gas mixture including several impurities such as O_2, N_2, CH_4, Ar, CO, and CO_2. Kim and Lee (2015) investigated hydrogen separation, purification capacity, and impurities resistant of La, Nd-rich Mm-based AB_5 MH. The authors reported that the MH was able to separate hydrogen from gas mixture and reach ultrahigh purity in each cycle. However, a decrease trend of hydrogen concentration by 6.8% and 10.7% compared its initial value was observed after 220 and 660 adsorb and desorb cycles, respectively. The study showed that MH had high sensitivity toward CO presence, but it was able to recover after vacuum heat treatments. The authors reported the order of gases, which negatively affecting the hydrogen purification, as $CH_4 > CO > O_2 > N_2 > CO_2$. In summary, the MHs have a significant potential for hydrogen separation, purification, and storage. However, further studies are needed in order to enhance their hydrogen capacity, mechanical and chemical properties, and optimize the reactor design.

13.3.4 Dense Metal Membranes

Dense metal membranes are commonly used for high-purity hydrogen separation due to their unique characteristic of being completely selective toward hydrogen permeation. There are a few types of dense metal hydrogen perm-selective membranes: "Non-Pd," Pd, and Pd-alloy membranes. The "Non-Pd"-based membranes make use of cheaper materials than Pd, such as vanadium, tantalum, and niobium, which present higher hydrogen permeabilities compared with pure Pd. Nevertheless, these materials need a catalytic layer (usually Pd is used) on both sides to avoid their oxidation and facilitate the dissociation and recombination of hydrogen atoms. This type of membranes is not suitable for high operating temperatures since migration of metal elements between the core metal and the catalytic layer may prevail causing oxidation of the core metal with a corresponding decline in hydrogen permeation. Pure Pd membranes have shown great potential for their application to H_2 separation at high temperature. However, they are affected by problems such as the embrittlement, deposition of carbonaceous impurities, and poisoning by CO and H_2S, which often result in a mechanical failure of the membrane and in low chemical stability. It has been reported that the use of Pd-alloy membranes (mainly Pd-Ag Pd-Cu and Pd-Au) decrease these issues and even increase the permeability of H_2 compared with pure Pd. The Pd-Ag alloy system is the most widely studied. It is a stable alloy with face-centered cubic (FCC) structure, which reduces the temperature for metallic hydride formation, which causes hydrogen embrittlement and affecting membrane permeability. However, although considerably better than pure palladium, Pd-Ag still has relatively low strength, expands significantly on hydrogen absorption, and can experience grain coarsening during extended periods at high temperatures. The Pd-Cu membranes have also been widely studied due to the suppression of hydrogen embrittlement at low temperatures, reduction of membrane cost due to the low price of copper, and its most important property, the great resistance to sulfur poisoning compared with pure Pd and Pd-Ag. However, long-term stability may be an issue with Pd-Cu membranes due to an unstable FCC/BCC[1] mixed phase occurring in the system around the high permeability Pd-Cu (40wt.%) composition. Alloying Pd with Au has been also shown to improve sulfur tolerance and CO poisoning effect, while also increasing hydrogen permeability. Nevertheless, the membrane is affected by pinhole formation and short-term stability. Although these binary alloys show some bottlenecks, a combination of them in a ternary alloy can be considered for their application to the downstream process due to their ability to transport hydrogen at high temperature and be chemical- and mechanical-resistant. A typical schematic of a dense Pd-membrane system is shown in Fig. 13.9.

13.3.5 Polymeric Membranes and others

Commercial polymeric membranes have been successfully used to recover hydrogen from ammonia purge gas streams since 1979 (Baker, 2012) and currently compete with cryogenic separation and PSA on industrial scale to recover hydrogen from ammonia plants, refineries, and petrochemical manufacture (Edlund, 2009a; Paglieri & Way, 2002a, 2002b; Koros & Mahajan, 2000; Perry et al., 2006). Usually, polymeric membranes reach hydrogen recovery up to 95%, as reported in Table 13.1. However, they can reach up to 97%, such as the polyaramid membranes (Dupont) (Ekiner & Vassilatos, 1990) and Prism (Air products) (Sanders et al., 2013). The use of new polymeric membranes can have benefits compared with conventional technologies, such as lower carbon-footprint, energy consumption, reduced maintenance requirement and cost. However, the main limitation for their use for hydrogen separation is the trade-off between selectivity and permeability. Fig. 13.10 shows the hydrogen permeability versus the hydrogen/methane selectivity for several polymeric membranes. It is evident that the limit can be overcome by the adoption of advanced membrane structure with improved properties. Indeed, recently several strategies have been

---

[1] BCC: Body-Centered Cubic.

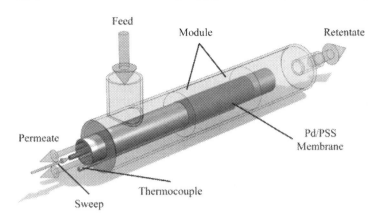

FIGURE 13.9 Schematic of a dense metallic membrane for hydrogen separation. *Adapted from Liguori, S., Kian, K., Buggy, N., Anzelmo, B. H., Wilcox, J. (2020). Opportunities and challenges of low-carbon hydrogen via metallic membranes.* Progress in Energy and Combustion Science, *80, 100851.*

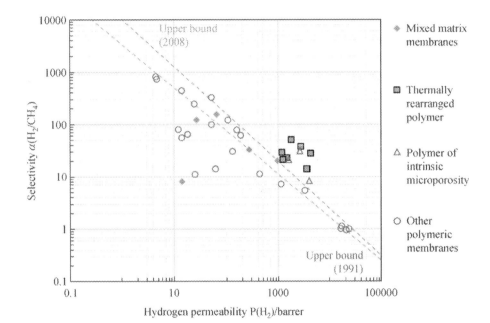

FIGURE 13.10 W. Robeson's upper bound for hydrogen/methane selectivity vs hydrogen permeability for several polymeric membranes (Lu et al., 2020). *Adapted from Lu, H. T., Li, W., Miandoab, E. S., Kanehashi, S., & Hu, G. (2020). The opportunity of membrane technology for hydrogen purification in the power to hydrogen (P₂H) roadmap: A review.* Frontiers of Chemical Science and Engineering, *15(3), 464–482 with permission.*

investigated with the aim to crossover the upper bound, such as thermally rearranged polymer membranes (Han, Lee et al., 2010; Han, Misdan et al., 2010; Kim & Lee, 2015; Park et al., 2007; Yeong et al., 2012).

In addition, the introduction of carbon nanotube, ZIF, UiO-66 into the polymeric membrane, the so-called mixed matrix membrane (MMM), has shown the potential to improve the hydrogen permeability without decreasing the selectivity (Kim et al., 2019; Safak Boroglu & Yumru, 2017; Weng et al., 2009).

Other membranes, such as zeolite, silica, MOF, and ceramic, have shown the potential to be used for hydrogen separation and purification although studies on polymeric and dense membranes have been predominant. It can be noticed from the exisiting data that those metallic membranes show the highest hydrogen/methane selectivity, but at the relatively high temperature range and their stability can limit their use to only a few applications, while polymeric membranes show a trade-off between permeability and selectivity although they are commercially used. The other type of membranes needed in the future and extensive to evaluate their feasibility for hydrogen separation and purification is highly critical.

13.4 Hydrogen distribution and end use

For a complete assessment of the hydrogen economy, it is critical to evaluate the hydrogen end use and distribution. As reported by Mansilla et al., 2018; the end use of hydrogen falls into three sectors: industry, building, and transportation.

13.4.1 End use

Hydrogen is mainly used today as feedstock in industrial applications, such as in the synthesis of ammonia and methanol, and in petroleum refining. Currently, 26.4 million tons of hydrogen is used in the refining for hydrotreating, desulphurization, and hydrocracking, followed by the ammonia synthesis with a demand of 22.8 million tons per year (Mansilla et al., 2018). Other applications involve the production of methanol, semiconductors, metallurgy, glass and food industry, and fuel for space rockets, which require 18% of the global hydrogen needs. The success of hydrogen in the transportation sector will depend on the development and commercialization of competitive fuel cells although the issue remains for maritime and aeronautic mobility. Hydrogen can also be used to generate electricity for stationary applications, such as households or buildings though combined heat and power (CHP). In this case, it is important to evaluate the application in-grid or off-grid. Indeed, in the latter case, hydrogen can be a good alternative to other fuels because of the high price of energy in-grid in remote areas.

13.4.2 Distribution

The choice for a large-scale hydrogen production and distribution network is usually the preferred option. Indeed, producing large amounts of hydrogen in a single plant enables low capital investments per unit of fuel produced. However, the major disadvantage of this approach is that the building of large-scale fuel production plants relies significantly on hydrogen storage.

13.5 Conclusions and summary

Biowaste and other potential waste can be utilized efficiently in the production of valuable H2. The processing of biowaste in an eco-friendly concept to convert abundant waste to useful energy carrier, that is, H_2 is a sustainble concept. Hydrogen is ubiquitous and exists in many forms from waste to industrial off-gases (e.g., from refineries, power plants, steel plants). Globally, millions of tons of wastes are generated from human and industrial activities in the form of food, agro-, forestry, wood, MSW, sewage sludge, and industrial waste gases. There are potential challenges that exist in utilizing the waste to hydrogen energy technology. Waste conversion, separation, and purification face technical challenges in thermal and other conversion or processing steps, whereas separation and purification using membrane technology achieved excellent progress. Anyhow commercialization is near to these coming years due to decarbonization of energy industries is thriving to implement the climate goals and reduce the dependency over fossil based fuels and energy sources. Hydrogen is foreseen as the solution for many problems, and it is already proved technology; for example, in steel industries, H_2 is replaced the fossil fuel coke reducer.

References

Agueda, V. I., et al. (2015). Adsorption and diffusion of H_2, N_2, CO, CH_4 and CO_2 in UTSA-16 metal-organic framework extrudates. *Chemical Engineering Science*, *124*, 159–169. Available from https://doi.org/10.1016/j.ces.2014.08.039.

Au, M., Chen, C., Ye, Z., Fang, T., Wu, J., Wang, O. (1996). *The recovery, purification, storage and transport of hydrogen separated from industrial purge gas by means of mobile hydride containers*. vol. 21.

Arja, A., et al. (2016). Reducing burden of disease from residential indoor air exposures in Europe (HEALTHVENT project). *Environmental Health*, *15*, 62, In this issue. Available from https://doi.org/10.1186/s12940-016-0101-8.

Baker, R. W. (2012). *Membrane technology and applications* (3rd edn.). Oxford: Wiley-Blackwell.

Bernardo, P., & Jansen, J. C. (2015). Polymeric membranes for the purification of hydrogen. In V. Subramani, A. Basile, & T. N. Veziroğlu (Eds.), *Compend Hydrog Energy* (pp. 419–443). Elsevier. Available from http://doi.org/10.1016/b978-1-78242-361-4.00014-5.

Besancon, B. M., et al. (2009). Hydrogen quality from decarbonized fossil fuels to fuel cells. *International Journal of Hydrogen Energy*, *34*(5), 2350–2360. Available from https://doi.org/10.1016/j.ijhydene.2008.12.071.

Chen, X. Y., Wei, L. X., Deng, L., Yang, F. S., & Zhang, Z. X. (2014). A review on the metal hydride based hydrogen purification and separation technology. *Applied Mechanics and Materials*, *448–453*, 3027–3036. Available from https://doi.org/10.4028/http://www.scientific.net/AMM.448-453.3027, Trans Tech Publications Ltd.

Christos, H. H., et al. (2022). Chemicals in European residences – Part I: a review of emissions, concentrations and health effects of volatile organic compounds (VOCs). *Science of the Total Environment*, *839*(156201). Available from https://doi.org/10.1016/j.scitotenv.2022.156201.

Delgado, J. A. (2014). Adsorption and diffusion of H_2, CO, CH_4, and CO_2 in BPL activated carbon and 13X zeolite: evaluation of performance in pressure swing adsorption hydrogen purification by simulation. *Industrial & Engineering Chemistry Research*, *53*(40), 15414–15426. Available from https://doi.org/10.1021/ie403744u.

Dunikov, D., Borzenko, V., & Malyshenko, S. (2012). Influence of impurities on hydrogen absorption in a metal hydride reactor. *International Journal of Hydrogen Energy, 37*, 13843–13848. Available from https://doi.org/10.1016/j.ijhydene.2012.04.078.

Edlund, D. (2009a). Hydrogen membrane technologies and fuel processing application. In K. Liu, C. Song, V. Subramani (Eds.) *Hydrogen membrane technologies and application in fuel processing* (pp. 357–384).

Ekiner, O., & Vassilatos, G. (1990). Polyaramide hollow fibers for hydrogen/methane separation—Spinning and properties. *Journal of Membrane Science, 53*(3), 259–273.

Golmakani, A., Fatemi, S., & Tamnanloo, J. (2017). Investigating PSA, VSA, and TSA methods in SMR unit of refineries for hydrogen production with fuel cell specification. *Separation and Purification Technology, 176*, 73–91. Available from https://doi.org/10.1016/j.seppur.2016.11.030.

Grande, C. A. (2012). Advances in pressure swing adsorption for gas separation. *ISRN Chemical Engineering, 2012*, 982934. Available from https://doi.org/10.5402/2012/982934.

Grashoff, G. J., Pilkington, C. E., & Corti, C. W. (1983). The purification of hydrogen. *Platinum Metals Review, 27*(4), 157–169.

Han, S. H., Lee, J. E., Lee, K. J., Park, H. B., & Lee, Y. M. (2010). Highly gas permeable and microporous polybenzimidazole membrane by thermal rearrangement. *Journal of Membrane Science, 357*(1-2), 143–151.

Han, S. H., Misdan, N., Kim, S., Doherty, C. M., Hill, A. J., & Lee, Y. M. (2010). Thermally rearranged (TR) polybenzoxazole: Effects of diverse imidization routes on physical properties and gas transport behaviors. *Macromolecules, 43*(18), 7657–7667.

Hellas, Ff.RaT.. (2015). Final report summary–HY2SEPS-2 (hybrid membrane–pressure swing adsorption (PSA) hydrogen purification systems). European Commission: Brussels, Beglium: http://cordis.europa.eu/result/rcn/169285_en.html.

Hinchliffe, A. B., & Porter, K. E. (2000). A comparison of membrane separation and distillation. *Chemical Engineering Research and Design, 78*, 255–268. Available from https://doi.org/10.1205/026387600527121.

Huang, C.-W., Nguyen, B.-S., Wu, J. C. S., et al. (2020). A current perspective for photocatalysis towards the hydrogen production from biomass-derived organic substances and water. *International Journal of Hydrogen Energy, 45*, 18144–18159.

Ikuo, U., et al. (2017). Desorption behavior of various volatile organic compounds from activated carbon in supercritical carbon dioxide: measurement and kinetic modeling. *The Journal of Supercritical Fluids, 121*, 41–51, In this issue.

Ioannis, M., et al. (2020). Environmental and health impacts of air pollution: a review. *Frontiers in Public Health, 8*, 1–13. Available from https://doi.org/10.3389/fpubh.2020.00014.

Kim, E., Kim, H., Kim, D., Kim, J., & Lee, P. (2019). Preparation of mixed matrix membranes containing ZIF-8 and UiO-66 for multicomponent light gas separation. *Crystals, 9*(1), 15.

Kim, S., & Lee, Y. M. (2015). Rigid and microporous polymers for gas separation membranes. *Progress in Polymer Science, 43*, 1–32.

Klochko, Y., Linkov, V., & Pollet, B. G. (2013). Application of surface-modified metal hydrides for hydrogen separation from gas mixtures containing carbon dioxide and monoxide. *Journal of Alloys and Compounds, 580*. Available from https://doi.org/10.1016/j.jallcom.2013.02.096.

Kondarides, D. I., Daskalaki, V. M., Patsoura, A., et al. (2008). Hydrogen production by photo-induced reforming of biomass components and derivatives at ambient conditions. *Catalysis Letters, 122*, 26–32, Return to ref 9 in article.

Koros, W. J., & Mahajan, R. (2000). *Journal of Membrane Science, 175*, 181–196.

Lai, Z. I., Lee, L. Q., & Li, H. (2021). Electroreforming of biomass for value-added products. *Micromachines, 12*, 1405. Available from https://doi.org/10.3390/mi12111405.

Li, M., et al. (2019). Persistent growth of anthropogenic non-methane volatile organic compound (NMVOC) emissions in China during 1990–2017: drivers, speciation and ozone formation potential. *Atmospheric Chemistry and Physics, 19*, 8897–8913, In this issue. Available from https://doi.org/10.5194/acp-19-8897-2019.

Liguori, S., Kian, K., Buggy, N., Anzelmo, B. H., & Wilcox, J. (2020). Opportunities and challenges of low-carbon hydrogen via metallic membranes. *Progress in Energy and Combustion Science, 80*, 100851. Available from https://doi.org/10.1016/J.PECS.2020.100851.

Lin, C., et al. (2022). Strategies to achieve a carbon neutral society: a review. *Environmental Chemistry Letters, 20*, 2277–2310, In this issue. Available from https://doi.org/10.1007/s10311-022-01435-8.

Liu, B., Han, B., Liang, X., & Liu, Y. (2023). Hydrogen production from municipal solid waste: Potential prediction and environmental impact analysis. *International Journal of Hydrogen Energy*.

Lototsky, M. v, Williams, M., Yartys, V. A., Klochko, Y. v, & Linkov, V. M. (2011). Surface-modified advanced hydrogen storage alloys for hydrogen separation and purification. *Journal of Alloys and Compounds, 509*. Available from https://doi.org/10.1016/j.jallcom.2010.09.206.

Lototskyy, M., Williams, M., Klochko, Y., Modibane, K., & Linkov, V. (2011). Hydrogen separation from CO_2-and CO-containing gases using surface-modified metal hydrides. *Journal of Alloys and Compounds, 509*(2), S555–S561. Available from https://doi.org/10.1016/j.jallcom.2010.09.206.

Lu, H. T., Li, W., Miandoab, E. S., Kanehashi, S., & Hu, G. (2020). The opportunity of membrane technology for hydrogen purification in the power to hydrogen (P_2H) roadmap: A review. *Frontiers of Chemical Science and Engineering, 15*(3), 464–482. Available from https://doi.org/10.1007/s11705-020-1983-0.

Mansilla,C., Bourasseau,C., Cany,C., Guinot,B., Le Duigou,A., Lucchese,P. (2018). Chapter 7 – Hydrogen applications: Overview of the key economic issues and perspectives. In Catherine A. Pantel (Ed.), *Hydrogen supply chains* (pp. 271–292). Academic Press. https://doi.org/10.1016/B978-0-12-811197-0.00007-5.

Miura, S., Fujisawa, A., & Ishida, M. (2012). A hydrogen purification and storage system using metal hydride. *International Journal of Hydrogen Energy, vol. 37*, 2794–2799. Available from https://doi.org/10.1016/j.ijhydene.2011.03.150.

Moore, F. (2009). Climate change and air pollution: exploring the synergies and potential for mitigation in industrializing countries. *Sustainability, 1*, 43–54, In this issue. Available from https://doi.org/10.3390/su1010043.

Moreno, F. (2018). Ir-based bimetallic catalysts for hydrogen production through glycerol aqueous-phase reforming. *Catalysts, 8*, 613, In this issue. Available from https://doi.org/10.3390/catal8120613.

Nattaporn, P., et al. (2022). Characteristics and impact of VOCs on ozone formation potential in a petrochemical industrial area, Thailand. *Atomsphere, 13*(5), 732, In this issue. Available from https://doi.org/10.3390/atmos13050732.

Nikolaidis, P., & Poullikkas, A. (2017). A comparative overview of hydrogen production processes. *Renewable and Sustainable Energy Reviews, 67*, 597–611. Available from https://doi.org/10.1016/J.RSER.2016.09.044.

Paglieri, S. N., & Way, J. D. (2002a). *Separation & Purification Reviews, 31*, 1–171.

Paglieri, S. N., & Way, J. D. (2002b). Innovations in palladium membrane research. *Separation and Purification Methods, 31*(1), 1–169. Available from https://doi.org/10.1081/SPM-120006115.

Park, H. B., Jung, C. H., Lee, Y. M., Hill, A. J., Pas, S. J., Mudie, S. T., Van Wagner, E., Freeman, B. D., & Cookson, D. J. (2007). Polymers with cavities tuned for fast selective transport of small molecules and ions. *Science (New York, N.Y.), 318*(5848), 254–258.

Perry, J. D., Nagai, K., & Koros, W. J. (2006). *MRS Bulletin, 31*, 745–749.

Rajesh, R., et al. (2023). Low-temperature total oxidation of propane using silver-decorated MnO_2 nanorods. *ACS Applied Nano Materials, 6*(13), 12258, In this issue. Available from https://doi.org/10.1021/acsanm.3c01952.

Ribeiro, A. M., et al. (2008). A parametric study of layered bed PSA for hydrogen purification. *Chemical Engineering Science, 63*(21), 5258–5273. Available from https://doi.org/10.1016/j.ces.2008.07.017, [Crossref], [CAS], Google Scholar.

Safak Boroglu, M., & Yumru, A. B. (2017). Gas separation performance of 6FDA-DAM-ZIF-11 mixed-matrix membranes for H_2/CH_4 and CO_2/CH_4 separation. *Separation and Purification Technology, 173*, 269–279.

Saffarzadeh, A., Arumugam, N., & Shimaoka, T. (2016). Aluminum and aluminum alloys in municipal solid waste incineration (MSWI) bottom ash: a potential source for the production of hydrogen gas. *International Journal of Hydrogen Energy, 41*(2), 820–831.

Saitou, T., & Sugiyama, K. (1995). Hydrogen purification with metal hydride sintered pellets using pressure swing adsorption method. *Journal of Alloys and Compounds, 231*, 865–870. Available from https://doi.org/10.1016/0925-8388(95)01774-7.

Sanders, D. F., Smith, Z. P., Guo, R., Robeson, L. M., McGrath, J. E., Paul, D. R., & Freeman, B. D. (2013). Energy-efficient polymeric gas separation membranes for a sustainable future: A review. *Polymer, 54*(18), 4729–4761.

Sircar, S., & Golden, T. (2000). Purification of hydrogen by pressure swing adsorption. *Separation Science and Technology, 35*(5), 667–687. Available from https://doi.org/10.1081/SS-100100183.

Speight, J. G. (2015). 6 - gasification processes for syngas and hydrogen production. *Gasification for Synthetic Fuel Production, 119*, In this issue. Available from https://doi.org/10.1016/B978-0-85709-802-3.00006-0.

Veerapandian, P., et al. (2021). Fractionation of lignin using organic solvents: a combined experimental and theoretical study. *International Journal of Biological Macromolecules, 168*(7), 792–805, In this issue. Available from https://doi.org/10.1016/j.ijbiomac.2020.11.139.

Weng, T. H., Tseng, H. H., & Wey, M. Y. (2009). Preparation and characterization of multi-walled carbon nanotube/PBNPI nanocomposite membrane for H_2/CH_4 separation. *International Journal of Hydrogen Energy, 34*(20), 8707–8715.

Yang, C., et al. (2019). Abatement of various types of VOCs by adsorption/catalytic oxidation: a review. *Chemical Engineering Journal, 370*, 1128–1153, In preparation. Available from https://doi.org/10.1016/j.cej.2019.03.232.

Yeong, Y. F., Wang, H., Pallathadka Pramoda, K., & Chung, T. S. (2012). Thermal induced structural rearrangement of cardo-copolybenzoxazolemembranes for enhanced gas transport properties. *Journal of Membrane Science, 397*, 51–65.

Yunlong, G., et al. (2021). Recent advances in VOC elimination by catalytic oxidation technology onto various nanoparticles catalysts: a critical review. *Applied Catalysis B: Environmental, 218*, 119447, In this issue. Available from https://doi.org/10.1016/j.apcatb.2020.119447.

CHAPTER

14

A biorefinery route to treat waste water through extremophilic enzymes: an innovative approach to generate value from waste

Tuhin Subhra Biswas, Thamizvani K., Kasturi Bidkar, Kavya Singh, Chandukishore T. and Ashish A. Prabhu

Bioprocess Development Laboratory, Department of Biotechnology, National Institute of Technology, Warangal, Telangana, India

14.1 Bioprospecting of novel and industrially relevant enzymes

14.1.1 Introduction

Enzymes, also referred to as the biocatalysts, play a vital role in the industrial sector due to its various favorable properties. Large scale production of enzymes with lower production cost has been the prime focus for various industries (Berini et al., 2017). Generally, one of the classic approaches to production involves identification and isolation of potent enzymes from microbes present in unexploited environmental niches with extreme conditions, and then utilize its properties for the production of high value products. Bioprospecting for these extremophilic microbes has led to discovery of enzymes with high tolerance to various extreme unnatural conditions. An alternate approach would be to cultivate inherent microorganisms from waste resources and utilize it for enzyme production as well as for efficient waste management. The rise in global population has led to increase in waste disposed and the rise in demand for fossil fuels. Thus, current researches on bio refinery approaches, also referred to as the waste to energy technology, comprehend numerous techniques which utilize waste products and effluents to produce high value products such as biofuels using different microorganisms. One of the major drawbacks to the above approaches is that many microorganisms are difficult to cultivate in laboratory conditions as it's difficult to replicate their exact environmental niche.

To overcome the above drawback, there has been a new active area of research known as the Metagenomics which does not require any isolation or cultivation of microorganisms. This is based on direct isolation of genomic DNA from samples and identification of novel enzymes with the help of genomic libraries constructed. Function and sequence-based screening are the two major approaches used for screening of enzymes (Madhavan et al., 2017). The pioneer of this technology was Grant et al. (2006), and their study was based on metatranscriptomics which was utilized for the search of novel eukaryotic enzymes. This shows that the development of microbial enzymes and recombinant DNA technology has played a vital role in meeting the current demands for greener alternatives of enzyme and energy production. This chapter deals with types of enzymes, sources of novel enzymes, immobilization techniques, identification of novel enzymes from extremophiles and the metagenomics approach.

14.1.2 Types of industrially relevant enzymes

14.1.2.1 Cellulase/hemicellulase

Cellulose is a polysaccharide containing D-glucose subunits connected by β-1, 4 linkages. Cellulase, also known as the cellulose degrading enzymes, is classified into endo-cellulase, exo-cellulase, and β-glucosidases.

The commonly found microbes which produce this enzyme include *Cellulomonas sp., Pseudomonas sp., Escherichia coli, Bacillus sp.*, and *Serratia marscens* (Tatta et al., 2022). Applications of cellulase include cotton softening, food mashing, and wastewater treatment. Hemicelluloses are polysaccharides with high molecular mass and complex structure, which also form 20%–40% of the lignocellulose composition. Hemicellulase also referred to as the hemicellulose degrading enzymes are usually in multiple forms. Large polymers are converted into smaller oligosaccharides, disaccharides, and monosaccharaides using this enzyme (Amoozegar et al., 2019). These enzymes are also known to produce biofuel from lignocellulosic biomasses. The basic principle behind this process is that these enzymes are used to hydrolyze agriculture-based waste products which later acts as the substrate for bioethanol fermentation by *Saccharomyces cerevisiae* (Matsuzawa et al., 2015). Studies reveal that numerous hydrolytic enzymes from halophiles are being utilized in biomass degradation for consequent production of biofuels. Moreover, active research is going on identifying novel cellulases from extremophiles which degrade lignocellulose without any pre-treatment (Amoozegar et al., 2019).

14.1.2.2 Laccases

Laccases are enzymes that oxidize phenolic and nonphenolic lignin-related compounds and pollutants. Their major application is in bioremediation which includes industrial effluent treatment; they detoxify effluents from various industries such as paper and pulp, textile, and petrochemical. Lignin, being the second most abundant raw material on earth is also one of the serious contaminants in industrial effluents because of its rigid structure and poor biodegradation. Studies reveal that laccases secreted by halo-tolerant microorganisms help in lignin degradation since waste streams of industries contain high amounts of salt. Studies also show that lignolytic fungi and their enzymes have the capability to detoxify biofuel feedstock. Moreover, laccases also play an important role in food processing and food quality improvement (Upadhyay et al., 2016).

14.1.2.3 Lipases

The hydrolytic enzyme which plays a pivotal role in biodiesel production is lipase. Application of this enzyme includes production of numerous products ranging from fruit juices, pharmaceuticals and baked foods. Fats, oils, and related compounds are the major targets of this enzyme. Recently, the major application of lipases is its use in removal of pitch which is the hydrophobic component of wood consisting of triglycerides and waxes. The main producers of lipases are bacteria and fungi. A research had shown that lipase isolated from the halophillic bacterial strain *Idiomarina sp.* W_{33}, was utilized for the production of biodiesel from Jatropha oil. Reports also reveal that biodiesel produced from free lipase enzymes extracted from this strain had lower yield compared to the immobilized lipases since free enzymes get aggregated in the low water media and lower the surface area available (Amoozegar et al., 2019).

14.1.2.4 Amylase

Starch is hydrolyzed with the help of the enzyme amylase to form glucose, glucose sirups and high fructose sirups in industries. The glucose product obtained will act as the substrate and can be fermented further to different forms of amylase available, including bioethanol. α-amylase, β-amylase, glucoamylase, and glucose isomerase. A research study revealed that the strain *Nesterenkonia sp* produced α-amylase, which was capable of producing ethanol and butanol directly from glucose. In general, the enzyme glucoamylase is utilized for saccharification of raw material and the subsequent ethanol production is carried out with the help of *S. cerevisiae* (Amoozegar et al., 2019).

14.1.2.5 Pectinase

Pectin is an important component of the plant cell wall and the enzyme pectinase is a depolymerizing agent. Paper and textile industries utilize these enzymes to develop cleaner processes and reduce production of waste. Pectinase from microbial origin are considered to have higher efficiency and are known for its application in textile processing and cotton fiber bio scouring. Research studies reveal that alkaline pectinase are used for treatment of pectic waste waters from various industries (Sanchez & Demain, 2011).

14.1.2.6 Proteases

Proteases are one of the important naturally occurring enzyme in almost all living cells and play a major role in maintaining the physiological functions. They are the most in demand industrial enzyme which covers 65% of the total global market. This enzyme has a wide array of applications ranging from detergents, food, textile to pharmaceuticals. The primary microbial producer of protease is *Bacillus subtilis*. However, fungal proteases have also gained a lot of attention in the past decade due to low media cost requirements. Majorly, agricultural waste products and effluents are used as a substrate for the fermentation process required for the production of this enzyme (Tatta et al., 2022).

14.1.2.7 Yeast enzymes

A recent trend has led to the use of enzymes produced by yeast due to its various advantages over other conventional model microorganisms. One of such advantage is that these diverse microorganisms have led to the use of waste water, industrial waste rather than the conventionally used agro industrial waste for fermentative conversions.

The application of yeast enzymes in food industry includes invertase, which is produced from *Kluyveromyces fragilis*, *Saccharomyces carlsbergensis*, and *S. cerevisiae* had been used for manufacturing candy and jam. Hydrolysis of lactose from milk or whey was carried out using the lactase enzyme produced from *K. fragilis* or *Kluyveromyces lactis*. Crystallization of beet sugar was carried out using α-galactosidase obtained from *S. carlsbergensis* (Sanchez & Demain, 2011).

The application of yeast enzymes in bioethanol production has been a developing technology in recent years. Hexose utilizing strains such as *S. cerevisiae*, Xylose fermenting strains such as *Pichia stipites*, *Candida tropicallis*, and *Rhodotorula* have been extensively studied for the production of ethanol. Bioethanol produced from fermentation of lignocellulosic feed stock and municipal waste using these strains are considered the second generation biofuel. In general, the *Saccharomyces* strains are referred to as the workhorses of fermentation industry due to their various advantages such as high ethanol, pH tolerance, and low oxygen requirement (Nandal et al., 2020).

Therefore, different types of industrially relevant enzymes mentioned above are mostly obtained from microbial, fungal, or yeast sources because of the ease in production processes. Recently, there has been an emerging technology which utilizes algal strains for the production of biofuels using wastewater.

14.1.3 Production of biofuel from waste water using algal strains

Microalgae are currently used for the production of third generation biofuel feedstock, due to their high lipid accumulating capability, high growth rate, and least requirement of sophisticated technologies. The major advantage is that they require minimal freshwater for growth, which in turn makes it potential for waste water treatment.

Nutrient-rich environment acts as a substrate for the growth of microalgae, which gives them the ability to accumulate nutrients and metals in waste water, and thus an efficient low-cost water treatment strategy. In addition, the waste water grown algae can be utilized for the production of biofuel which can also be referred to as the third generation biofuels. *Chlorella* and *Scenedesmus* genus are the most widely studied strains for algal growth in municipal sewage water (Pittman et al., 2011).

A research revealed that *Chlorella sorokiniana* grown in untreated dairy waste water can be utilized for the production of biofuels. According to the findings, it was proven that high biomass production can be obtained using this strain (Hamidian & Zamani, 2022).

The drawback of using algal strains is that there is a need for efficient algal harvesting procedures and further research in algal life cycle analysis. Therefore, in future, wastewater-grown microalgae can be the potential source for the production of biofuels (Fig. 14.1).

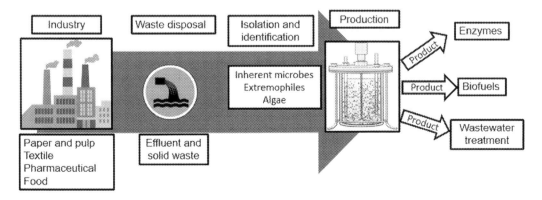

FIGURE 14.1 Ways to generate waste to value from industrial effluents.

14.2 Extremophilic enzymes

Biotechnology holds significant importance in various aspects, including industrial applications and our daily lives. The function of enzymes as a form of biocatalysts is now well established, and the enzymatic properties, including its affinity for the substrate, tolerance limits, temperature stability, and selectivity, can be enhanced through genetic modification. Studies on new generation of enzymes (biocatalysts), which are highly stable to different adverse conditions, can be replaced by chemical modification. Extremophiles are microorganisms capable of thriving in extreme temperatures, such as temperatures ranging from $-2°C$ to $15°C$ in cold environments and $60°C-120°C$ in hot environments. They can also tolerate high ionic strength (2–5 M NaCl) and extreme pH (<4.0 or >9.0). Extremozymes are also capable of catalyzing their respective reactions in several conditions including non-aqueous environments, high-pressure conditions, acidic and alkaline pH levels and, even at temperatures as high as 150°C or below the freezing point of water. A pictorial representation of classification of extremophilic enzymes based on temperature and functional groups is given in Fig. 14.2 (Saumaya et al., 2007).

14.2.1 Halophilic enzymes

Halophiles are a type of extremophiles capable of regulating their growth and development even when salt concentrations are very high. Most of the halophiles are classified into the domain Archaea, but some alga like *Dunaliella salina* and fungus like *Wallemia ichthyophaga* belong to this group. Slight halophiles can grow in salt concentrations ranging from 0.3 to 0.8 M, moderate halophiles in concentrations from 0.8 to 3.4 M, and extreme

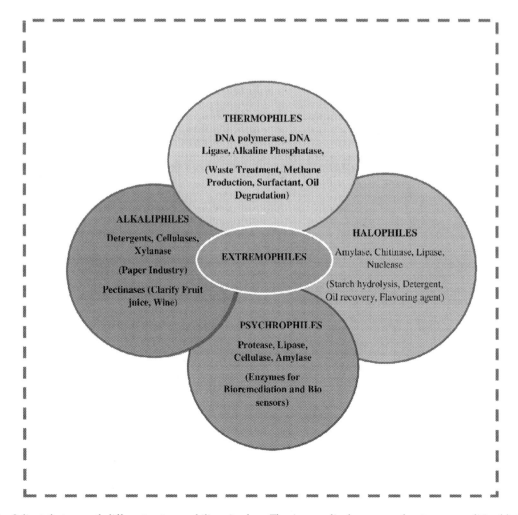

FIGURE 14.2 Salient features of different extremophilic microbes. The image displays several extreme condition(s) they can survive (Saumaya et al., 2007).

halophiles in concentrations from 3.5 to 5.2 M (Laye & Dassarma, 2018). After genomic and functional analyses, it is observed that the enzymes present in halophilies (Archaea, bacteria) are mainly negatively charged due to high content of acidic residue; their hydrophobic nature is also altered, ultimately enhancing their solubility and enabling them to perform effectively in low-water environments. As a result, hydrogen bonds are formed in between negatively charged residues and it is critical to maintain a stable hydration shell for water molecules (Jin, 2019). Other important factors for enzyme function in high concentration of salinity include increases in ion-pair networks. When water activity is low, it can easily mimic the conditions in a mixture comprising water-organic solvent. These make halophilic enzymes very stable, so these have the potential to act as industrial biocatalysts. Halophilic proteases and amylases have been introduced in detergent formulations (Abdel-Hamed, 2016). Halophiles are an effective source for several extremozymes such as amylases, proteases, cellulases, alcohol dehydrogenases, and lipases. For example, nuclease can be isolated from Micrococcus varians and is used commercially for the production of 5'-guanylic acid (5'-GMP), which is a flavoring agent.

14.2.2 Thermophilic enzymes

Extremophiles that survive under extremely high-temperature conditions (hot springs or geothermal vents) are known as *extreme thermophiles*. The term *thermophiles* are generally used to refer to bacteria and Archea (thermophilic prokaryotes) that can survive in habitats with high temperatures. Examples of bacterial thermophiles' scientific names are *Thermotoga maritima, Thermus aquaticus*, and *Thermus thermophilus*. *Methano pyruskandleri* is a thermophile Archaea that thrives at 250°F. Thermophiles have several thermostable proteins which help their cell membranes resist denaturation at high temperatures, and some may also resist proteolysis (Sysoev, 2021). Disulfide bridges(S-S) play an important role to increase thermal stability as they decrease the total entropy of unfolded protein. Other factors, such as compact barrel folding, short length of loops and helices, bonds between salt bridges, and interactions between inner hydrophobic amino acids, also help maintain the stable conformation of thermophilic enzymes. Notably, DNA polymerases extracted from thermophilic organisms provided the foundation for the discovery of **Polymerase chain reaction (PCR)**, which plays a key role in medical diagnostics and research (Kumar, 2019).

Thermophiles can play a significant biotechnological as well as an ecological role:

- Thermophiles are a significant source of biotechnological tools. Furthermore, they are considered as a holy grail as they are the source of *extremozymes* and *molecular chaperons*.
- *Thermococcus litoralis* and *Thermus aquaticus*, which are hyperthermophilic bacteria, are widely used for their DNA polymerases. Other extremozymes are *Proteases, Xylanases, Amylases, Glucoamylases, Glucose Isomerase Pullulanases, Aminotransferases*, and *Glutamate Synthetase (GS)*.

Molecular chaperones are proteins that are efficiently expressed under several extreme stress conditions such as temperature. *Sulfolobus shibate* and *Sulfolobus solfataricus* are used as the source of molecular chaperons (Jorquera et al., 2019):

- Thermophilic enzymes from *Bacillus stearothermophilus* are frequently used for the manufacturing of biological washing powder.
- Thermophiles are involved in the degradation of pollutants.
- Thermophiles carry out *bioremediation* of the oil-contaminated desert soil.
- Thermophiles help in the immobilization of heavy metals in the soil.
- Thermophilic enzymes are useful for polysaccharide processing and biofuel production.

14.2.3 Psychrophilic enzymes

Psychrophilic microorganisms are able to colonize in cold environments such as deep sea, mountain, glacier, and polar regions. According to their optimal growth temperature, they are categorized as psychrotolerant or psychrotroph. The most common example of psychrophiles is under the category of Gram-negative α-, β-, and γ-proteo bacteria (*Pseudomonas* sp and *Vibrio* sp). *Arthrobacter* sp. and *Micrococcus* sp. are under Gram-positive bacteria. In general, if growth temperature is low, it supports for high content of polyunsaturated fatty acids having methyl-branches, and a shorter length of acyl-chain. These studies assured that there is a high proportion of *cis*-unsaturated double-bonds and anterior branched fatty acids on the surface (Huston, 2008). This altered composition might play a key role in increasing the fluidity of the membrane. Typically, these steric constraints can alter the packing order and reduce the number of effective interactions in the cell membrane. The presence of

Cold-Acclimation Proteins (Caps) in psychrophiles is also helpful for beer stability in extreme conditions. **Antifreeze proteins (AFPs)** exhibit the property to bind efficiently with the ice crystals as they contain a large complementary surface. As a result, they create a thermal hysteresis, lowering the temperature, which allows organisms to grow (De Santi, 2016).

Psychrophilic enzymes tend to possess various key features which include: decreased core hydrophobicity, significantly increased surface hydrophobicity, a lower arginine/lysine ratio, reduced secondary structure content, a higher amount of glycine and less proline residues in loops, a small number of weak metal-binding sites, a limited number of disulfide bonds, and a decrease in electrostatic interactions such as hydrogen-bonds, cationic-anionic interactions, and aromatic–aromatic interactions (Jin, 2019). Cold-adapted enzymes have several applications in industries such as household detergents, molecular biology, baking, pharmaceutical engineering, molecular biology, textiles, and cosmetics.

14.2.4 Acidophilic enzymes

Acidophiles are those microorganisms that can continue their growth at an optimum pH < 2 (Baker-Austin and Dopson, 2007). There are some mechanisms by which acidophiles can easily maintain the cytoplasmic pH that is close to neutrality so that they can protect all acid-labile cellular components by several processes such as pumping of protons, also by decreaseing the permeability of the cellular membrane. They are particularly helpful in preventing the entry of protons into the cytosol. In most acid-stable proteins (for example, pepsin and the **soxF** protein extracted from Sulfolobus *acidocaldarius*), it is found an overabundance of several acidic residues is found. These residues readily minimize the destabilization at low pH and effectively induce stability by increasing positive charges (Hedrich and Schippers, no date). Other mechanisms such as solvent accessibility of acidic residues can be minimized when these bind on metal cofactors. When pH are less than 2, most of the surface amino acids will become protonated. As a result, less negative charge will be deposited on the surface and the subjected protein is stabilized (Dopson & Johnson, 2012).

Acidophilic **Glucoamylase** is important for starch processing as it reduces the time and cost of production of several oligosaccharides from raw starch. Amylolytic enzymes such as trehalase, which is isolated from acidophiles such as *S. solfataricus*, are used in medicine as preservatives and also as stabilizers (Schiraldi et al., 2002). Acidophilic enzymes are also useful in bioremediation and they can efficiently remove upto 90% of heavy metals

14.2.5 Alkaliphilic enzymes

Alkaliphiles are microbes thate can grow optimally around a pH of 10. So it is clear that these can capable of survival in alkaline environment easily. These enzymes have high alkaline isoelectric points (pI) with respect to their neutrophilic counterparts, due to large proportion of Arginine and Histidine residues. Alkaliphilic bacteria have cellular mechanisms that involve the activation of both symport and antiport systems. Electrolytic antiporters are helpful to generate an electrochemical gradient due to $Na+$ and $H+$ ions and the symport system helps to enable the uptake of $Na+$ ions and solutes (Abdel-Hamed, 2016).

Different enzymes such as **alkaline lipases, alkaline proteases, amylases, and cellulases** are readily stable in alkaline medium, present in detergents and used in laundry, can resist the denaturation property of surfactants and also effect of chelators such as EDTA. Alkaliphilic **proteases** have applications in leather processing. Alkaline lipases are mainly used to remove natural fat from skin and improve color (Hasan et al., 2006).

14.2.6 Radiophilic enzymes

Radiophiles are extremophiles that can resist the radiation. These can continue their growth when oxidative stress is high and mainly in radioactive environments having UV, gamma, and X-ray radiation. Direct exposure to intense or prolonged UV radiation may lead to mutagenic DNA dimers which can, in turn, increase the risk of various types of cancers in humans. Radiophiles produce several primary and secondary metabolites, some of which are used in the manufacturing of anticancer drugs and antioxidants. Natural pigments such as Bacterioruberin which is produced from *Halobacterium* & *Rubrobacter* and radioressitant carotenoid Deinoxanthin produced from *Deinococcus radiodurans* pose effective antioxidant property which are used as therapeutic agents for the treatment of cancer diseases (Dumorné, 2017).

14.2.6.1 Conclusions and future prospects

Extremophilic enzymes have been successfully introduced as a potential source for novel and suitable enzymes due to their high stability and performance even when extreme conditions prevail. Notwithstanding their advantages, the main limitation of extremozymes is that they are very less in amount. Particularly, extremophilic enzymes having large amount of biological applications as these can resist extreme conditions such as high and low temperature, organic solvents, pH, and radiation. The extremozymes have a large number of economical roles in agriculture, food industry, pharmaceutical, beverages, and detergent, textile, paper, leather, and biomining industries (Dumorné, 2017). Extremophilic enzymes has given new route for the expansion of industries in the field of biotechnology with increasing in demand for novel biocatalysts for various applications. The extremophiles are suitable sources that might be better applications in several biotechnological applications toward the expansion of a bio-based economy and industrial approaches (Kumar, 2019). But it is very difficult to mimic extreme environment in the lab for proper cultivation of those extremophiles. Moreover, insufficient knowledge of media compositions and long incubation time can hamper the process. According to current scenario of globalization of market, novel extremozymes play a significant role and definitely it will help to expand the market as much as possible.

14.3 Enzyme modification using immobilization

In terms of endurance and resilience to environmental changes, immobilized enzymes outlive unbound enzymes in reaction mixture. Furthermore, the diversity of immobilized enzyme systems can facile retrieval of both enzymes and products, numerous enzymes reutilize continual enzymatic reactions, quick reaction stoppage, and a broader diversity of bioreactor designs (Satish et al., 2018).

Enzyme usage has progressively increased in various industries, notably in the last two decades. Enzyme utilization in the food sector may be found in baking, starch conversion, dairy products, and beverage processing (wine, beer, vegetable, and fruit juices) (Rodrigues et al., 2013). Because of their effect on the end product, enzymes play a particular role in the textile industry. When a faultless final product is required, enzymes have become a required processing technique in fields such as pulp and paper manufacturing and detergents. Because of enzyme specificity, which is important in biosensors, their application in increasingly complex industries, such as biosensors, is rapidly improving. Many other critical industries, for example, healthcare, medicines, and chemical manufacturing are increasingly relying on nature's remarkable catalysts. In recent decades, enzymes have been extensively applied in biofuels such as biodiesel as well as ethanol. Waste treatment in general, and notably solid waste treatment and waste water purification, represents one of the most important enzyme uses in modern life. A significant problem in commercial biocatalyst is the production of reliable, long-lasting, and, most importantly, hydrophobic biocatalysts (Homaei et al., 2013).

Enzymes that have been immobilized have shown to be incredibly helpful in a variety of industrial applications. They offer various advantages over solution-based enzymes, including cheaper cost, greater stability, and the ability to rapidly disengage itself out from reaction, permitting for pure product isolation P. However, there are particular logistical challenges regarding its use in industry applications. Because enzymes are typically costly, isolating and purifying them costs several times of magnitude far beyond conventional catalysts. They are very vulnerable to denaturing conditions when they are not in their natural surroundings due to the protein makeup. They may act as inhibitors due to their sensitivity to process parameters including temperature, pH and a little amount of chemicals. Indeed, numerous immobilized enzymes, such as glucose isomerase or penicillin G acylase, had already found widespread commercial applications, and the immobilization of more enzymes has sparked the scientific community's interest (Fopase et al., 2020).

14.3.1 Overview of immobilization

Unlike typical heterogeneous chemical catalysts, the majority of enzymes in universal catalysis systems function diluted in water, leading in product contamination and limiting active enzyme removals from the majority of reaction mixtures. An immobilization method is one of the most effective ways to overcome these limits. Weak or covalent linkages can be used to bind a physical or chemical support. Physical bonding is inadequate in industrial settings and cannot keep the enzyme attached to the carrier (Pablos et al., 2021).

As a support, a synthesized polymer, an inorganic polymer including some zeolite or silica, or even a biopolymer might be employed. Enzyme entrapment is achieved by immersing it in a polymeric matrix, such as an organic polymeric matrix or silica sol-gel, or maybe a membranous structure, such as a hollow fiber or some microcapsule. Entrapment requires the creation of the polymeric matrix as in presence of the enzyme. Enzyme granules or particles were cross-linked with a bi-functional reagent to create carrier free macroparticles in the final category (Datta et al., 2013).

14.3.2 Modes of immobilization

Many diverse approaches have been developed, which are either innovative combinations of previously published methods or are occasionally specific to a particular support or enzyme. Regrettably, no one approach or support is suitable for the most of enzymes or processes. This is due to the fact that enzymes have diverse chemical properties and compositions, different classifications of substrates and outputs, and varied applications. Any strategy, however, may have a variety of benefits and drawbacks (Rafeeq et al., 2021).

Immobilization has traditionally relied on six techniques: (1) ionic binding, (2) affinity binding, (3) covalent binding, (4) cross linking, (5) entrapment, and (6) encapsulation.

14.3.2.1 Ionic bonding

In this technology, enzyme molecules adhere to substrate by electrostatic forces of attraction on interfaces via ionic binding. Electrostatic ionic interactions are a straightforward way for immobilizing dielectric catalysts as well as catalysts which can ionize immobilization conditions. pH influences creation of ionic charges on enzyme's surface. Isoelectric point changes as the pH changes (pI). At pH3, the surface enzyme molecules gain a net positive charge and link to a negatively charged surface. Enzymes which are negatively charged attached to a silicon material requires surface functionalization, such as silica support modification with amine groups (Fopase et al., 2020). Lei and fellow researchers immobilized organophosphorus hydrolase (OPH) with functionality -NH2 and -COOH groups on SBA-15 at pH 7.5. The mesoporous carrier having 2% -COOH groups had the greatest protein loadings (4.7% w/w) and activity (4182 units mg-1 support). The carrier with 20% -NH2 groups, on the other hand, had very little loading. The observed discrepancy might be attributed to the influence of surface charges. Electrostatic forces predominate in enzyme complexes with pure silica SBA-15, whereas weak hydrophobic forces prevail in associations with organofunctionalized SBA-15 (Amalia et al., 2021).

14.3.2.2 Affinity binding

In order to create high affinity linkages between the enzyme structure as well as a solid support generated with the proper affinity ligand, affinity tags located outside of the native protein structure that are far from the active region may be used. Because of the variety of affinity systems, there are several options for enzyme immobilization. Using this approach, competitively aminated silica beads with activated graphite-cellulose are connected to Histidine (His) as well as sugar-tagged enzymes. These approaches also have broad scope and can potentially be employed in any situation that necessitates immobilized biocatalysts, seeking to make upcoming enzyme innovations more sensitive, simple, recyclable, and affordable (Cargnin et al., 2020).

It is an appropriate approach for immobilizing labile materials and when the enzyme is too costly to employ in significant quantities during the immobilization process. When powerful inhibitors must be monitored, reversible immobilization is also helpful. An excess of enzyme is utilized in a normal analytical setup, and inhibitory effects are easily compensated for by resting enzyme molecules. As a result, inhibitor sensitivity is low. The inhibitory impact is considerably easier to assess when only a little quantity of enzyme is employed. The affinity immobilization method offers the added benefit of stabilizing the attached molecule. Because monoclonal antibodies are becoming more widely available, affinity immobilization may become even more desired than it is today. Affinity immobilization has the potential to broaden the use of immobilization technology to more labile and sensitive biological structures.

Certain alterations in confirmation have been seen following enzyme immobilization. The structure of enzyme subunits changes after immobilization, according to FTIR study of immobiliszd glucosidase enzyme on porous silicon via adsorption. In another example of physical adsorption of Laccase enzyme, circular dichroism (CD) with fluorescence spectroscopy revealed that perhaps the secondary structure of the enzyme altered to varied degrees. The -sheet ratio dropped considerably, the alpha helix ratio practically disappeared, and the turn and random coil ratios increased greatly, indicating that the enzyme had suffered considerable secondary structural damage (Yao et al., 2022).

14.3.2.3 Covalent binding

Covalent bonding is most popular approach for immobilizing homogeneous catalysts. The development of the covalent connections between functional groups and surface is referred to as covalent bonding. Covalent attachment is only advised if it improves the characteristics of the enzyme. It is not envisaged that the covalent bonds that bind the enzyme to a support surface would interfere with enzyme action. By preserving the enzyme's structural structure, the covalent approach improves enzyme stability. It is more resistant to high temperatures because of its increased structural stability. The enzyme is leached from the surface as little as possible due to the increased binding strength. Enzymes that have been covalently immobilized have a good long-term storage stability. It was revealed that a structurally diverse sample of SBA-15 had neither a high PGA loading nor exhibited quick coupling kinetics in between PGA nor the oxirane groups. The reason for this is because PGA attachment involves surface silanol groups (Morellon-Sterling et al., 2022). The primary goal of covalent binding is to keep the immobilized enzyme in place, which lowers enzyme leaching. It needs to be noted that the harsh conditions utilized for covalent attachment may cause the enzyme's structure to change, resulting in diminished enzymatic activity. Furthermore, interactions between the enzyme's active sites and a substrate may result in a reduction in functioning. The difficulties of the immobilization method, as well as economic considerations, are drawbacks of employing covalent bonding immobilization.

14.3.2.4 Cross linking

Cross-linking is the covalent connection of enzyme molecules that results in the development of three-dimensional enzyme complexes. The main disadvantages of this approach seem to be the difficulty in managing composite shape, substrate accessibility to aggregate cores, and the lack of mechanical strength of the cross-linked enzyme (Morellon-Sterling et al., 2022). These difficulties may be solved by combining such technology with a wide range of different enzyme immobilization approaches. For example, an enzyme might be successfully adsorbed inside a three-dimensional matrix of linked traps with diameters several orders of magnitude larger than the enzyme's, accompanied by cross-linking. The size of the cage controls the number of enzyme aggregates, and the diffusion restriction for substrate towards the center of an enzyme aggregation can be lowered.

14.3.2.5 Entrapment

The act of physically confining an enzyme in a medium that only allows the passage of the substrate and not the enzyme is known as entrapment. The following are the disadvantages of the entrapment method:

- The failure to regulate the restricting environment of the enzyme may impose significant diffusing obstacles on substrate as well as product transit, leading in reaction slowing;
- Continuous degradation of activity owing to enzyme leaching can occur unless the distribution of pore sizes of the entrapment medium is not tight;
- In the past, polymers used for enzyme trapping typically degraded.

Solgels can be used to entrap enzymes depending on the enzyme and reaction circumstances. When alkyltrialkoxysilanes (e.g., methyltrimethoxysilane) were added to an ambigel synthesis mixture, there was no matrix contraction and no holes for the interface of the sol-hydrophobic gel. To maintain the sol-gel structure intact, chemicals such as polyethylene glycol, polyvinyl alcohol (PVA), and even albumin may be utilized. In addition to shrinking the gel matrix, the chemicals also aid to lower internal tension. The inclusion of PVA to the fabrication of sol-gel for lipase B immobilization affects the surface attributes of surface area and hardness in testing (Kujawa et al., 2021).

14.3.2.6 Encapsulation

The method of encapsulation involves enclosing the catalyst inside the porous medium with a pore opening that is narrower than the pore space's diameter. This same tiny porous aperture not only keeps the encapsulated catalyst from being lost in the reaction fluid, but it also creates a high mass transfer barrier between reactant and product. The immobilized catalase displayed high re-utility (70% activity preserved after 25 batch procedures), pH resilience, and proteolysis tolerance. Encapsulation is the only technique of immobilizing a catalyst which does not involve any contact between the catalyst and the support; nonetheless, the dimension of the pore holes in the matrix should be relatively small than the immobilized catalyst's kinetic size (Zhao et al., 2006).

14.4 Metagenomics: a source of enzyme discovery: introduction

Microbial enzymes are used in practically every industry, from the biochemical, pharmacological, including agricultural sectors to detergent, fabric, suede, wood and paper production (Adrio and Demain, 2008). In 2015, industrial automation was used for the first money. According to global enzyme, available at https://www.gminsights.com/industry-analysis/enzymes-market, this is anticipated to rise to 9.4 billion USD in 2020 and 14.9 billion USD in 2025. These advent of non-culturable technologies based on gathering genetic sample from environmental resources (metagenomics, metatranscriptomics as well as metaproteomics) throughout the last 20 years has highlighted the importance of non-culturable bacteria as a supplier of novel enzymes. Even though recent advancements in metaproteomics (Wilmes et al., 2015) have revealed the existence for new lipases (Sukul et al., 2017) as well as cellulases (Speda et al., 2017), metagenomics is mainly considered to be an excellent procedure for recognizing inventive catalysts from eubacteria eDNA (Lorenz & Eck, 2005).

The word "metagenomics" refers to the study of a habitat's genomic complement by extracting and copying DNA from a collection of microorganisms directly (Handelsman et al., 1998). Glycosyl hydrolases (GHs), phosphatase, lipase and esterase are the most frequent ones studied in metagenomic research for biotechnological processes; a few other pharmaceutically important enzymatic types such as amylases, cellulase and phytase have also been discovered by metagenome (Coughlan et al., 2015; DeCastro et al., 2016). Table 14.1 also shows some examples of enzymes derived from metagenomics studies that have been commercialized. Table 14.2 shows instances of enzymes from metagenomics that have been patented in the recent 10 years.

14.4.1 Protease

Novel proteases, among the most commonly utilized industrial groups of enzymes in cleaning, suede, food industry, polymer, medicine production, brewing and sewage management have been identified through five metagenome research. Serine proteases were detected underneath in Yucatan which are two in number which further correspond a unique, unclassified suborder (Apolinar-Hernandez et al., 2016) and one protease which is serine isolated within stimulated tannery mud offers great potential due to its significantly increased stabilization in both organic as well as anionic solvents (Devi et al., 2016).

14.4.2 Phosphatases

New phytases which are five in general (histidine acid phosphatases having uses in agribusiness as well as reproduction as a livestock feed addition) were eventually found and categorized. (Tan et al., 2014) discovered two new, peculiar patterned phytases in plant fertilized areas. Moreover, the same scientists narrated the characterization of a phytase with an extremely extended half-life at an elevated heat from a fungus metagenome (Tan et al., 2016). Ultimately, the discovery of a new psychrophilic alkaline phosphatase was done by the analysis of increased expression of genes through tidal portions (Lee et al., 2015).

TABLE 14.1 Production of biofuels from different sources.

Sr. no	Enzyme	Source	Applications	References
Microbial source				
1.	α-amylase	*Nesterenkonia sp.* strain F	Bioethanol and butanol	Amoozegar et al. (2019)
2.	Cellulase	*Gracilibacillus sp.* strain SK1	Bioethanol	
3.	Cellulase	*Bacillus methylotrophicus* RYC0110	Bioethanol	
4.	Lipase	*Idiomarina sp.* W33	Bio-diesel production	
Fungal source				
1.	Laccase	*Digitatispora marina, Halocyphina villosa,* and *Nia vibrissae*	Lignin degradation	Upadhyay et al. (2016)
2.	Laccase	*Ulomyces chlamydosporum, Emericella nidulans,* and *Aspergillus phoenicis*	Decolorizing agent	

TABLE 14.2 List of extremozymes regarding their sources, characteristics and applications.

Types	Growth characteristics	Geographical location/environment	Enzymes	Applications
Acidophile	Optimum pH for growth at or below 3–4	Acid mine drainage, volcanic springs	Proteases Cellulases Oxidases	• As animal feed, improvement of digestion. • Removes the hemicellulosic material from Feed component. • Desulfurization of coal.
Alkaliphile	Optimal growth at pH values above 10	Soda Lakes.	Proteases, Cellulases Pectinases Amylase, glucoamylase Alkaline phosphatase	• Detergents, food, f bread making, fruit juice processing. • Bleaching of Pulp, papers and degumming. • Starch processing, Desizing of Starch. • Detergents, Recombinant DNA technology.
Thermophile	Can thrive at temperatures 60_oC; humidity	Hot Spring, Yellowstone National Park,	(Amylases, Glucosidases, Cellulases) Proteases and lipases Alcohol dehydrogenase DNA polymerases	• Diagnostics, hydrolysis of Polysaccharide. • Baking, brewing, production of amino acid, detergents. • Chemical synthesis. • Genetic engineering, diagnostics.
Hyperthermophile	Optimum growth at 80_oC or higher	Submarine Hydrothermal vents,	No information	No information.
Psychrophile	Optimum growth at 10_oC or lower.	Ice, snow, and Arctic Ocean	Proteases Amylase Dehydrogenases Cellulases	• Detergents, food applications • Detergents and bakery • Biosensors. • Detergents feed and textiles.
Halophile	Requiring at high salt concentration 1.5 (M) for growth	Salt Lakes, Marine and deep sea.	Proteases Dehydrogenases	• Peptides synthesis, fruit juice processing. • Catalysis in organic media, helps in chemical synthesis • (Phosphorylation)
Radioresistant	Can tolerate high levels of ionizing radiation.	Sunlight, high UV radiation	–	• Pharmacology, Anti- cancer agents.
Xerophile	Capable of growth when water activity is low; can resist to high desiccation	Desert, rock, surfaces	–	–

14.4.3 Oxidoreductases

Oxidoreductases are a diverse collection of enzymes with plethora of utility throughout the pharmacological, food sectors and in phytoremediation. Five dioxygenases originating from soil having bioremediation ability (Chemerys et al., 2014; dos Santos et al., 2015) as well as first metagenomics-originated d-amino acid oxidase having potential utility within antibiotic production pathway of 7-aminocephalosporanic acid out of cephalosporin C (Ou et al., 2015) were newly found oxidoreductases. Multi-copper oxidases (MCOs), having wide range applications to both on phenolic as well as non-phenolic targets have attracted interest because of its crucial function involved in the breakdown of lignocellulose biomasses, fall into this enzymatic category. Ausec et al. (2017) recently published a paper upon separation as well as characterization of initial metagenomics-derived acidobacterial MCO having greater heat tolerance as well as salt tolerance.

14.4.3.1 Others

Until 2014, seven metagenome research papers had described the expression as well as characterization of certain genuinely valuable enzymes along with a heat-stable and bulky metal-resistance nitrilase out of a marine that can be utilized to produce cyanocarboxylic acid (Sonbol et al., 2016) as well as β-agarase present within marshes land for utility in the skincare, pharmacological and related sectors (Mai et al., 2016). According to Ferrandi et al. (2017), co-expression in metagenomics-derived amine-transferases using chaperones that can withstand freezing throughout the unusual *E. coli* northern region induces the expression of RIL cells that enhances production of an enzyme upto many milligram/litre. Following this, one of the enzymes was found to be the most heat-stable natural amine transferase yet identified (Tables 14.3 and 14.4).

TABLE 14.3 Illustrations of metagenomics-derived enzymes supplied from BASF enzymes LLC (https://www.bsaf.com).

Commercialized label	Enzyme class	Uses
Luminase	Xylanase	Within the paper industry, pulp bio bleaching is performed.
Fuelzyme	α-Amylase	Production of fuel and alcohol
Pyrolase 160, Pyrolase 200	Cellulase	Recycling of gas as well as secondary oil.
Phyzyme XP	Phytase	Food supplements for animals.

TABLE 14.4 Patented metagenome-derived enzymes and their indicated applicability since the last 10 years.

Enzyme	Origin	Number of patent	Filling date of patent	Recommended uses
Peptidase	Compost	CN103409443 A	2013	Food industry, scientific studies, and the management of protein byproducts.
Alkaline protease	mud	CN103409443 A	2013	liquid detergents as a supplement
Alkaline phosphatase	na	US8647854 B2	2009	Genetic cloning
Alginate lyase	microbiome	CN102971426 B	2011	Seaweed treatment
DNA polymerase	Water	WO2012173905 A	2011	DNA polymerization
l-Methionine γ-lyase	Sediment	CN101962651 B	2010	Cancer therapy
Muramidase (Lysozyme)	Soil	CN101892252 B	2010	Antibiosis

Immobilization principle	Advantage	Disadvantage	Application
Ionic binding	Easy, High binding capacity, Highly stable.	KM value has risen somewhat	Suitable for big enzymes, amine oxidases, and peroxidase
Affinity binding	Possibility of immobilization in situ	Poor selectivity	Proteins capture during purification
Covalent binding	Stable to hydrolysis at pH 7, Easy to separate, Mostly used in PBR or FBR.	Enzyme conformation can be altered due to harsh conditions used; Reduction in the enzymatic activity. Interaction of binding site of enzyme and support may cause complete loss of enzyme activity	Immobilization of antibodies, oxidases and Proteases
Cross linking	Potent; Durable	Costly; easily degrades enzyme activity	Controlled enzyme replacement treatment
Entrapment	Maintain the surface chemistry; Mechanically and thermally very stable	Substrate to the enzyme diffusion is restricted; Controlling the enzyme confining environment reaction retardation is difficult and long response times	Applicable for most antibodies and enzymes, development of biosensors
Encapsulation	Highly reproducible; mechanically stable	Shear force inactivates enzyme	Medical enzyme replacement treatment, biomedical areas

14.5 Conclusion

The chapter mainly discusses about utilization of large amount of wastewater generated from both industrial and household process for production of extremophilic enzymes. Enzymes play a vital role in various industrial sectors such as food and beverage industries, pharma, and leather industries, but due to the harsh conditions at the industrial level, they limit their function which turns out into an economic burden. Extremophilic enzymes are one such group of biological catalyst nowadays used in the industrial process for the bioproduction of value-added products. The attempt to valorize the wastewater in the production of these extremophilic enzymes could solve the problem of water pollution and reduce the production cost of these enzymes. This chapter also discussed about advantages of physical and chemical modifications such as immobilization and their effect on efficacy of enzymatic action. Due to its unique features, the demand for these enzymes are also high in various industrial sectors. This challenge for efficient and economical production of these enzymes could be addressed by various metabolic engineering approaches with biorefinery approach for industrial scale operation.

References

Abdel-Hamed, A. R., et al. (2016). 'Biochemical characterization of a halophilic, alkalithermophilic protease from Alkalibacillus sp. NM-Da2'. *Extremophiles: Life Under Extreme Conditions*, 20(6), 885–894. Available from https://doi.org/10.1007/S00792-016-0879-X.

Amalia, S., Angga, S. C., Iftitah, E. D., Septiana, D., Anggraeny, B. O. D., Warsito., Hasanah, A. N., & Sabarudin, A. (2021). Immobilization of trypsin onto porous methacrylate-based monolith for flow-through protein digestion and its potential application to chiral separation using liquid chromatography. *Heliyon*, 7(8). Available from https://doi.org/10.1016/j.heliyon.2021.e07707.

Amoozegar, M. A., Safarpour, A., Noghabi, K. A., Bakhtiary, T., & Ventosa, A. (2019). Halophiles and their vast potential in biofuel production. *Frontiers in Microbiology*, 10, 1895.

Apolinar-Hernandez, M. M., Pena-Ramirez, Y. J., Perez-Rueda, E., Canto-Canche, B. B., De Los Santos-Briones, C., & O'Connor-Sanchez, A. (2016). Identification and in silico characterization of two novel genes encoding peptidases S8 found by functional screening in a metagenomic library of Yucatan underground water. *Gene*, 593(1), 154–161.

Ausec, L., Berini, F., Casciello, C., Cretoiu, M. S., van Elsas, J. D., Marinelli, F., et al. (2017). The first acidobacterial laccase-like multicopper oxidase revealed by metagenomics shows high salt and thermo-tolerance. *Applied Microbiology and Biotechnology*, 101(15), 6261–6276.

Berini, F., Casciello, C., Marcone, G. L., & Marinelli, F. (2017). Metagenomics: novel enzymes from non-culturable microbes. *FEMS Microbiology Letters*, 364(21), fnx211.

Cargnin, M. A., de Souza, A. G., de Lima, G. F., Gasparin, B. C., Rosa, D., dos, S., & Paulino, A. T. (2020). Pinus residue/pectin-based composite hydrogels for the immobilization of β-D-galactosidase. *International Journal of Biological Macromolecules*, 149, 773–782. Available from https://doi.org/10.1016/j.ijbiomac.2020.01.280.

Chemerys, A., Pelletier, E., Cruaud, C., Martin, F., Violet, F., & Jouanneau, Y. (2014). Characterization of novel polycyclic aromatic hydrocarbon dioxygenases from the bacterial metagenomic DNA of a contaminated soil. *Applied and Environmental Microbiology*, 80(21), 6591–6600.

Coughlan, L. M., Cotter, P. D., Hill, C., & Alvarez-Ordonez, A. (2015). Biotechnological applications of functional metagenomics in the food and pharmaceutical industries. *Frontiers in Microbiology*, 6, 672.

Datta, S., Christena, L. R., & Rajaram, Y. R. S. (2013). Enzyme immobilization: an overview on techniques and support materials. *3 Biotech*, 3(1), 1–9. Available from https://doi.org/10.1007/s13205-012-0071-7.

DeCastro, M. E., Rodriguez-Belmonte, E., & Gonzalez-Siso, M. I. (2016). Metagenomics of thermophiles with a focus on discovery of novel thermozymes. *Frontiers in Microbiology*, 7, 1521.

De Santi, C., et al. (2016). Characterization of a cold-active and salt tolerant esterase identified by functional screening of Arctic metagenomic libraries. *BMC Biochemistry*, 17(1), 1–13. Available from https://doi.org/10.1186/S12858-016-0057-X/FIGURES/7.

Devi, S. G., Fathima, A. A., Sanitha, M., Iyappan, S., Curtis, W. R., & Ramya, M. (2016). Expression and characterization of alkaline protease from the metagenomic library of tannery activated sludge. *Journal of Bioscience and Bioengineering*, 122(6), 694–700.

Dopson, M., & Johnson, D. B. (2012). *Minireview biodiversity, metabolism and applications of acidophilic sulfur-metabolizing microorganismse mi_2749 2620.2631*. Available from https://doi.org/10.1111/j.1462-2920.2012.02749.x.

dos Santos, D. F., Istvan, P., Noronha, E. F., Quirino, B. F., & Kruger, R. H. (2015). New dioxygenase from metagenomic library from Brazilian soil: insights into antibiotic resistance and bioremediation. *Biotechnology Letters*, 37(9), 1809–1817.

Dumorné, K., et al. (2017). Extremozymes: A potential source for industrial applications. *Journal of Microbiology and Biotechnology*, 27(4), 649–659. Available from https://doi.org/10.4014/jmb.1611.11006.

Ferrandi, E. E., Previdi, A., Bassanini, I., Riva, S., Peng, X., & Monti, D. (2017). Novel thermostable amine transferases from hot spring metagenomes. *Applied Microbiology and Biotechnology*, 101(12), 4963–4979.

Fopase, R., Paramasivam, S., Kale, P., & Paramasivan, B. (2020). Strategies, challenges and opportunities of enzyme immobilization on porous silicon for biosensing applications. *Journal of Environmental Chemical Engineering*, 8(5). Available from https://doi.org/10.1016/j.jece.2020.104266.

Grant, S., Grant, W. D., Cowan, D. A., Jones, B. E., Ma, Y., Ventosa, A., & Heaphy, S. (2006). Identification of eukaryotic open reading frames in metagenomic cDNA libraries made from environmental samples. *Applied and Environmental Microbiology*, 72(1), 135–143.

Hamidian, N., & Zamani, H. (2022). Potential of Chlorella sorokiniana cultivated in dairy wastewater for bioenergy and biodiesel production. *BioEnergy Research*, 15(1), 334–345.

Handelsman, Jo, Rondon, Michelle R., Brady, Sean F., Clardy, Jon, & Goodman, Robert M. (1998). Molecular biological access to the chemistry of unknown soil microbes: A new frontier for natural products. *Chemistry & Biology*, 5(10), R245–R249.

Homaei, A. A., Sariri, R., Vianello, F., & Stevanato, R. (2013). Enzyme immobilization: An update. *Journal of Chemical Biology*, 6(4), 185–205. Available from https://doi.org/10.1007/s12154-013-0102-9, Springer Verlag.

Huston, A. L. (2008). Biotechnological aspects of cold-adapted enzymes. *Psychrophiles: From Biodiversity to Biotechnology*, 347–363. Available from https://doi.org/10.1007/978-3-540-74335-4_20/COVER.

Jin, M., et al. (2019). Properties and applications of extremozymes from deep-sea extremophilic microorganisms: A mini review. *Marine Drugs*, 17(12). Available from https://doi.org/10.3390/md17120656.

Jorquera, M. A., Graether, S. P., & Maruyama, F. (2019). Editorial: Bioprospecting and biotechnology of extremophiles. *Frontiers in Bioengineering and Biotechnology*, 7(AUG), 204. Available from https://doi.org/10.3389/FBIOE.2019.00204/BIBTEX.

Kujawa, J., Głodek, M., Li, G., Al-Gharabli, S., Knozowska, K., & Kujawski, W. (2021). Highly effective enzymes immobilization on ceramics: Requirements for supports and enzymes. *Science of the Total Environment*, 801. Available from https://doi.org/10.1016/j.scitotenv.2021.149647, Elsevier B.V.

Kumar, S., et al. (2019). Thermozymes: Adaptive strategies and tools for their biotechnological applications. *Bioresource Technology*, 278, 372–382. Available from https://doi.org/10.1016/j.biortech.2019.01.088.

Laye, V. J., & Dassarma, S. (2018). An Antarctic Extreme halophile and its polyextremophilic enzyme: Effects of perchlorate salts. *Astrobiology*, 18(4), 412. Available from https://doi.org/10.1089/AST.2017.1766.

Lee, D. H., Choi, S. L., Rha, E., Kim, S. J., Yeom, S. J., Moon, J. H., et al. (2015). A novel psychrophilic alkaline phosphatase from the metagenome of tidal flat sediments. *BMC Biotechnology*, 15, 1.

Lorenz, P., & Eck, J. (2005). Metagenomics and industrial applications. *Nature Reviews. Microbiology*, 3(6), 510–516.

Madhavan, A., Sindhu, R., Parameswaran, B., Sukumaran, R. K., & Pandey, A. (2017). Metagenome analysis: a powerful tool for enzyme bioprospecting. *Applied Biochemistry and Biotechnology*, 183(2), 636–651.

Mai, Z., Su, H., & Zhang, S. (2016). Isolation and characterization of a glycosyl hydrolase Family 16 beta-Agarase from a Mangrove Soil Metagenomic Library. *International Journal of Molecular Sciences*, 17(8).

Matsuzawa, T., Kaneko, S., & Yaoi, K. (2015). Screening, identification, and characterization of a GH43 family β-xylosidase/α-arabinofuranosidase from a compost microbial metagenome. *Applied Microbiology and Biotechnology*, 99(21), 8943–8954.

Morellon-Sterling, R., Tavano, O., Bolivar, J. M., Berenguer-Murcia, Á., Vela-Gutiérrez, G., Sabir, J. S. M., Tacias-Pascacio, V. G., & Fernandez-Lafuente, R. (2022). A review on the immobilization of pepsin: A Lys-poor enzyme that is unstable at alkaline pH values. *International Journal of Biological Macromolecules*, 210, 682–702. Available from https://doi.org/10.1016/j.ijbiomac.2022.04.224, Elsevier B.V.

Nandal, P., Sharma, S., & Arora, A. (2020). Bioprospecting non-conventional yeasts for ethanol production from rice straw hydrolysate and their inhibitor tolerance. *Renewable Energy*, 147, 1694–1703.

Pablos, C., Govaert, M., Angarano, V., Smet, C., Marugán, J., & van Impe, J. F. M. (2021). Photocatalytic inactivation of dual- and mono-species biofilms by immobilized TiO_2. *Journal of Photochemistry and Photobiology B: Biology*, 221. Available from https://doi.org/10.1016/j.jphotobiol.2021.112253.

Pittman, J. K., Dean, A. P., & Osundeko, O. (2011). The potential of sustainable algal biofuel production using wastewater resources. *Bioresource Technology*, 102(1), 17–25.

Rafeeq, H., Hussain, A., Tarar, M. H. A., Afsheen, N., Bilal, M., & Iqbal, H. M. N. (2021). Expanding the bio-catalysis scope and applied perspectives of nanocarrier immobilized asparaginases. *3 Biotech*, 11(10). Available from https://doi.org/10.1007/s13205-021-02999-y, Springer Science and Business Media Deutschland GmbH.

Rodrigues, R. C., Ortiz, C., Berenguer-Murcia, Á., Torres, R., & Fernández-Lafuente, R. (2013). Modifying enzyme activity and selectivity by immobilization. *Chemical Society Reviews*, 42(15), 6290–6307. Available from https://doi.org/10.1039/c2cs35231a.

Sanchez, S., & Demain, A. L. (2011). Enzymes and bioconversions of industrial, pharmaceutical, and biotechnological significance. *Organic Process Research & Development*, 15(1), 224–230.

Satish, K., Neeraj., Viraj, K. M., & Santosh, Kr. K. (2018). Biodegradation of phenol by free and immobilized Candida tropicalis NPD1401. *African Journal of Biotechnology*, 17(3), 57–64. Available from https://doi.org/10.5897/ajb2017.15906.

Saumaya, G., Giraddi, R. S., & Patil, R. H. (2007). *Utility of Vermiwash for the Management of Thrips and Mites on Chilli (Capsicum annuum L.) Amended with Soil Organics* *, 20(3), 657–659.

Sonbol, S. A., Ferreira, A. J., & Siam, R. (2016). Red Sea Atlantis II brine pool nitrilase with unique thermostability profile and heavy metal tolerance. *BMC Biotechnology*, 16, 14.

Speda, J., Jonsson, B. H., Carlsson, U., & Karlsson, M. (2017). Metaproteomics-guided selection of targeted enzymes for bioprospecting of mixed microbial communities. *Biotechnology for Biofuels*, 10, 128.

Sukul, P., Schakermann, S., Bandow, J. E., Kusnezowa, A., Nowrousian, M., & Leichert, L. I. (2017). Simple discovery of bacterial biocatalysts from environmental samples through functional metaproteomics. *Microbiome.*, 5(1), 28.

Sysoev, M., et al. (2021). Bioprospecting of novel extremozymes from prokaryotes—The advent of culture-independent methods. *Frontiers in Microbiology*, 12. Available from https://doi.org/10.3389/FMICB.2021.630013.

Tan, H., Mooij, M. J., Barret, M., Hegarty, P. M., Harington, C., Dobson, A. D., et al. (2014). Identification of novel phytase genes from an agricultural soil-derived metagenome. *Journal of Microbiology and Biotechnology*, 24(1), 113–118.

Tan, H., Wu, X., Xie, L., Huang, Z., Peng, W., & Gan, B. (2016). A novel phytase derived from an acidic peat-soil microbiome showing high stability under acidic plus pepsin conditions. *Journal of Molecular Microbiology and Biotechnology*, 26(4), 291–301.

Tatta, E. R., Imchen, M., Moopantakath, J., & Kumavath, R. (2022). Bioprospecting of microbial enzymes: current trends in industry and healthcare. *Applied Microbiology and Biotechnology*, 106, 1813–1835.

Upadhyay, P., Shrivastava, R., & Agrawal, P. K. (2016). Bioprospecting and biotechnological applications of fungal laccase. *3 Biotech*, 6(1), 1–12.

Wilmes, P., Heintz-Buschart, A., & Bond, P. L. (2015). A decade of metaproteomics: where we stand and what the future holds. *Proteomics*, 15(20), 3409–3417.

Yao, L. W., Ahmed Khan, F. S., Mubarak, N. M., Karri, R. R., Khalid, M., Walvekar, R., Abdullah, E. C., Mazari, S. A., Ahmad, A., & Dehghani, M. H. (2022). Insight into immobilization efficiency of Lipase enzyme as a biocatalyst on the graphene oxide for adsorption of Azo dyes from industrial wastewater effluent. *Journal of Molecular Liquids*, 354. Available from https://doi.org/10.1016/j.molliq.2022.118849.

Zhao, X. S., Bao, X. Y., Guo, W., & Lee, F. Y. (2006). Immobilizing catalysts on porous materials. *Materials Today*, 9(3), 32–39. Available from https://doi.org/10.1016/S1369-7021(06)71388-8.

CHAPTER

15

Waste as a substrate for the production of organic acids and solvents

Kawinharsun Dhodduraj[*], *Durga Ashok Burande*[*], *Nivedhitha Ulaganathan and Ashish A. Prabhu*

Bioprocess Development Laboratory, Department of Biotechnology, National Institute of Technology Warangal, Warangal, Telangana, India

15.1 Introduction

An explosive population, (Global Itaconic Acid Market | Size | Share | Growth | 2022—2027 | - The Cowboy Channel (No Date). Available At, 2022/2022) aided by recent advances in technology, has resulted in increased production of daily goods/commodities. This, in turn, has led to the generation of significant amount of waste. Wastes produced in different sectors account for both toxic and nontoxic wastes. We can consider these nontoxic wastes as a key ingredient in the production of value-added products such as organic solvents, acids, and more. Over the past 5 years, the increasing prices of petroleum products including gasoline, diesel, and crude oil is a clear indicator that fossil fuel reserves are running low (Prices:Petroleum Planning & Analysis Cell, 2019). In addition, this gives a clear sign that renewable production should take over the commercial production market. Every person is accountable for producing more than a ton of waste annually, with global production reaching about 11 billion tons (Ferdous, 2021). According to a UN report, 931 million tons of food was wasted globally in 2019, which would have been sufficient to circumnavigate the world seven times (Estimated food wasted globally in 2019-UN report, 2021). Seemingly, the newly trending renewable bio-based production can help us valorize organic waste into value-added products. Primarily, the types of waste that can be used for the production of organic acids and solvents include agro waste, food waste, bakery waste, and sewage waste. These wastes contain significant amounts of carbon content that support microbial activity (Global Itaconic Acid Market | Size | Share | Growth | 2022—2027 | - The Cowboy Channel (No Date). Available At, 2022/2022), thereby reducing the need for fossil fuels as a substrate for production. Generally, bio-based production involves a suitable pretreatment of the substrates to make it feasible for microbial consumption. Thereafter, a sequence of steps with fixed conditions comprise the total upstream and downstream process, which can be scaled up for industrial production.

In this chapter, we will discuss a detailed microbial metabolic pathway that is applicable for the production of various products from different substrate sources. However, our discussion is restricted to three organic acids, namely, citric, succinic, and itaconic acid, and three organic solvents, namely, ethanol, butanol, and 2,3-butanediol.

15.2 Organic acids

Organic acids are chemical compounds that contain carbon, hydrogen, and oxygen as their prominent atoms, have at least one carboxylic acid group, and can exist in both saturated and unsaturated forms. These acids are of low

[*] Authors with equal contribution

molecular weight having an ability to form complexes with cationic metals and bind with other functional groups such as ketones, amides, peptides, and more. Over 130 acids have been discovered and can mostly be grouped into two categories based on their microbial production pathway, viz., (1) accumulation through glycolysis pathway followed by tricarboxylic acid (TCA) cycle, which is considered as long biosynthetic pathway (citric acid, succinic acid, itaconic acid, lactic acid, etc.). (2) Accumulation through oxidation of glucose, which consists of a very few enzymatic reactions and hence considered as short pathway or biotransformation (Gluconic acid, Kojic acid, etc.) (Mattey, 2008; Jones, 1998).

15.2.1 Citric acid

15.2.1.1 Physicochemical properties and brief history

2-hydroxy-propane-1,2,3-tricarboxylic acid ($C_6H_8O_7 \cdot H_2O$) with a molecular weight of 210.14 g/mol is commonly known as citric acid (CA). It contains three carboxylic groups, which account for three different pK_a values (3.1, 4.7, and 6.4) (Papagianni, 2007). The name *citric* is originally derived from the Latin word *citrus*, which refers to the trees from the *Citrus* genus that includes lemon trees. It is solid at room temperature and instantly dissolves in water resulting in a colorless, odorless, and slightly hygroscopic solution. CA (anhydrous) has a boiling point of 310°C and a melting point of 307.4°F (Show, 2015).

Synthesis of CA can be achieved from both natural (lemon, orange) and synthetic sources (microbial fermentation, chemical reaction-based production). Swedish chemist, Karl Wilhelm Scheele, was the pioneer of citric acid extraction from lemons in 1784. In England, around 1826, Scheele's method showed a monopoly on commercial-scale production. However, in 1893, Carl Wehmer showed that *Penicillium* mold could also produce CA with excess sugar as substrate. Due to its vulnerability to contamination, this fungus method of production did not succeed at the industrial level. In 1916, food chemist James Currie and microbiologist Charles Thom discovered a novel method of citric acid production from a fungus *Aspergillus niger* which generally grows in the garbage, soil, and decaying matter. Pfizer Inc. in Brooklyn commercialized this method of CA production in an industrial level. A high sugar concentration and an acidic medium with pH ranging from 2.5 to 3.5 not only increases CA production but also suppresses byproduct formation (Show, 2015), as well as the production of many byproducts. Presently, the production and isolation of CA in fermentation plants are in two forms: monohydrate ($C_6H_8O_7 \cdot H_2O$) and anhydrous ($C_6H_8O_7$) as a white powder (after crystallization and discoloration process).

15.2.1.2 Metabolic pathway of CA

The metabolic pathway of citric acid initiates with the conversion of glucose to pyruvate (glycolysis) followed by the TCA cycle (Tricarboxylic acid cycle). Pyruvate molecules are oxidized to form Acetyl-CoA and carboxylated to form oxaloacetate by pyruvate dehydrogenase and pyruvate carboxylase, respectively. Moreover, the action of the citrate synthase enzyme on oxaloacetate, Acetyl-CoA along with water molecule, results in the formation of citrate (Fig. 15.1). A single mole of glucose gives one mole of citric acid, one mole of ATP, and three moles of NADH. The maximum theoretical yield of CA is 1.067 g/g (Tong, 2019).

15.2.1.3 Biosynthetic production of CA

Recently, the production of organic acids from waste biomass is a trending area of research. CA can be produced from different forms of raw organic wastes such as sugarcane bagasse, cocoa pods, fruit peels, rice husk, and more. Using waste as a potential substrate for producing value-added products ensure indirect benefits including waste management, reduced pollution, and lower greenhouse gas emission. The typical production with *A. niger* uses molasses as a primary source of substrate. Many yeast, fungi, and bacteria are capable of producing citric acid like *Aspergillus wentii*, *Aspergillus carbonarius*, *Aspergillus awamori*, *Saccharomyces lypolytica*,

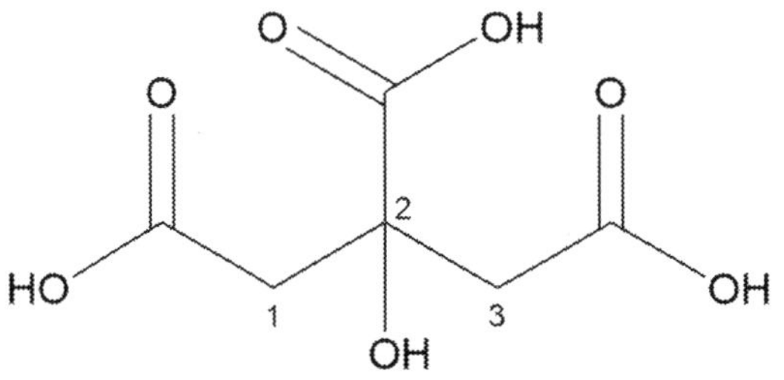

FIGURE 15.1 Citric acid structure.

Candida tropicalis, Yarrowia lipolytica, Cryptococcus citroformans, Bacillus licheniformis, and *Corynebacterium spp* (Grewal & Kalra, 1995; Vandenberghe, 1999; Moeller, 2013). However, a low yield renders most microorganisms commercially unacceptable in industrial production.

The production of CA through submerged microbial fermentation using *A. niger* is popular on a commercial scale. The synthesis of CA is affected by several fermentation broth factors, including the source of carbon and nitrogen, the pH, the amount of aeration, the presence of trace elements, the presence of lower alcohols, and more. For example, the addition of lower alcohols like methanol to the fermentation media containing crude carbohydrate sources showed enhancement in the production (Grewal & Kalra, 1995; Moyer, 1953). Typically, the waste biomass is pretreated with appropriate chemicals to make it feasible for fermentation. The whole process of production can be grouped into three categories: preparation of media and inoculation, fermentation, and downstream process (recovery) (Show, 2015) (Table 15.1).

15.2.1.4 Global market with applications

The global market for citric acid reached about 2.39 million tons per annum in 2020 and at a compounded annual growth rate (CAGR) of 4% (Global Itaconic Acid Market | Size | Share | Growth| 2022–2027 | - The Cowboy Channel (No Date). Available At, 2022/2022). The projected market of CA is 2.91 million tons per annum (approx.) by 2026 (*Citric Acid | Industry Report*, 2022). Around 59% of the production and 74% of exports worldwide are accounted for by China, which makes it the largest CA producer. Prominent areas of CA usage include detergents, dishwasher cleaners, cross-linkers, disinfectants, environmental remediation, extracting agent, preservatives, and more. Moreover, it gives a tangy acidic flavor to water, which makes it a popular flavoring agent in the food industry (Ciriminna, 2017).

15.2.2 Succinic acid

15.2.2.1 Physicochemical properties and brief history

Butanedioic acid ($C_4H_6O_4$) is a diprotic acid with a molar mass of 118.09 g/mol and is commonly known as succinic acid (SA). SA is also known as "amber acid" as it was initially extracted from amber through dry crystallization. It contains two carboxylic groups (Fig. 15.2), which account for two pK_a values (4.21 and 5.64) (Comuzzo & Battistutta, 2018). SA is solid at room temperature and is soluble in water (relatively low solubility). SA is an odorless and colorless crystal with a boiling point of 235°C and a melting point ranging from 185°C to 187°C (Saxena, 2017). SA is an intermediate product of the tricarboxylic acid (TCA) cycle or the citric acid cycle and also an end product of anaerobic fermentative metabolism (Song & Lee, 2006).

On a report by the US Department of Energy (US DOE), SA is one of the top 12 high-value-based organic acids (Werpy & Petersen, 2004/2004). Until the mid-16th century, people have been using amber (resin) as an antibiotic and healing agent without recognizing that its top layer is circumscribed by succinic acid. In 1546, a German mineralogist named Georgius Agricola extracted SA from amber by dry distillation technique and named it succinic acid, which was derived from the Latin word *"succinum,"* meaning amber (Smyth, 2007). For centuries, SA production has completely relied on petroleum products. This not only imparts a significant diminution of naturally forming fossil fuels but also causes greater effect to the environment through pollution (Venkateswar Rao, 2016). Utilization of organic waste materials as a substrate for the microbial production of succinic acid will be a game-changing replacement for fossil-fuel based biorefinery (Chu & Majumdar, 2012).

TABLE 15.1 Citric acid production from microbial cell factories using renewable waste biomass.

S. no	Waste (substrate)	Microorganism	Titer (g/L)	Yield (g/g)	References
1.	Acorn	*Aspergillus niger AA120*	130.8	—	Zhang (2019)
2.	Pomegranate peel (non-dried)	*A. niger B60*	—	0.351	Roukas and Kotzekidou (2020)
3.	Biodiesel industry waste	*Yarrowia lipolytica A-101–1.22*	—	0.6	Rymowicz (2010)
4.	Mango peel Sweet orange peel	*A. niger*	7.52 11.01	—	Abbas (2016)
5.	Extraction wastewater from citric acid producing industry	*A. niger*	126.6	—	Xu (2014)

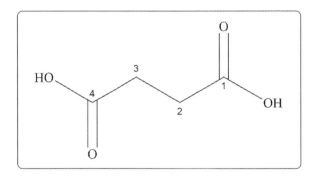

FIGURE 15.2 Succinic acid structure.

15.2.2.2 Metabolic pathway

As we already discussed, SA is an intermediate product of the tricarboxylic acid (TCA) cycle; microbes follow three different paths of accumulation, namely, oxidative, reductive, and glyoxylate pathways of TCA (Fig. 15.3). Succinate acts as a proton acceptor under anaerobic conditions; hence, minimal oxygen routes it through the reductive pathway of the TCA cycle, whereas in aerobic conditions, it routes through the oxidative mode of SA production. Moreover, microbes will alternatively switch their accumulation to the glyoxylate pathway in aerobic conditions, when acetate or ethanol is the prominent carbon source. Herein microbes produce 1 mole of SA by utilizing 2 moles of Acetyl-CoA (Ensign, 2006; Dittrich et al., 2009). In the reductive pathway of the TCA cycle, it takes 2 moles of NADH to produce 1 mole of SA where the Phosphoenolpyruvate (PEP) is converted to SA via 3 intermediates in order - oxaloacetate, malate and fumarate. In the reductive pathway, the maximum theoretical yield is 2 moles of SA from 1 mole of glucose, but due to the limitations, NADH hinders the possible production, resulting in the accumulation of a single mole of SA per mole of glucose. Although in aerobic conditions, Acetyl-CoA is considered as the key substrate where organisms like yeast are in a very active state for oxidative TCA pathway. The accumulation of SA in the oxidative pathway follows a series of intermediate formation including citrate, isocitrate, and succinate. The accumulated succinate is subsequently converted into fumarate by succinate dehydrogenase (SDH). Hence, deletion of *sdhA* is necessary to produce a substantial amount of SA by the oxidative route (Cheng, 2013).

15.2.2.3 Biosynthetic production of SA

SA is produced by diverse variety of microorganisms, which are mostly found in the rumen because the rumen area allows succinate to act as a propionate precursor, which is an energy source in the biosynthetic pathway of animals (Zeikus et al., 1999). Moreover, this explains the reason that propionate accumulating bacteria, such as. *Escherichia coli*, and *Pectinatus sp.*, also tend to produce succinate as one of their metabolic intermediates. In addition, rumen bacteria such as *Actinobacillus succinogenes, Ruminococcus flavefaciens, Bacteroides amylophilus, Cytophaga succinicans, Succinivibrio dextrinosolvens, Prevotella ruminicola, Wolinella succinogenes*, and more can also accumulate succinate (Bryant & Small, 1956; Bryant, 1958; Scheifinger & Wolin, 1973; Van Der Werf, 1997; Guettler et al., 1999; Chatterjee, 2001; Dittrich et al., 2009).

Even some of the robust yeasts, such as *Yarrowia lipolytica*, which may not able to consume certain carbon sources to produce a significant amount of products, can be genetically engineered to make a recombinant strain. These strains have been used to produce SA in an adequate amount as demonstrated by Prabhu (2020). The production yield depends on the microbe and the different fermentation strategies used such as batch, repeat-batch, fed-batch, and double-batch continuous fermentation (Table 15.2).

15.2.2.4 Global market with applications

The global market for SA reached about 50,000 metric tons in 2016 and is projected to reach about 100,000 metric tons by 2025 (Chinthapalli, 2018; Babaei, 2019). It was valued at USD 222.9 million in 2021 (Market (Global Itaconic Acid Market | Size | Share | Growth| 2022–2027 | - The Cowboy Channel (No Date). Available At, 2022/2022. Available At, 2022/2022)2022–2030, 2022). SA is considered a potential precursor chemical for the synthesis of 1,4-

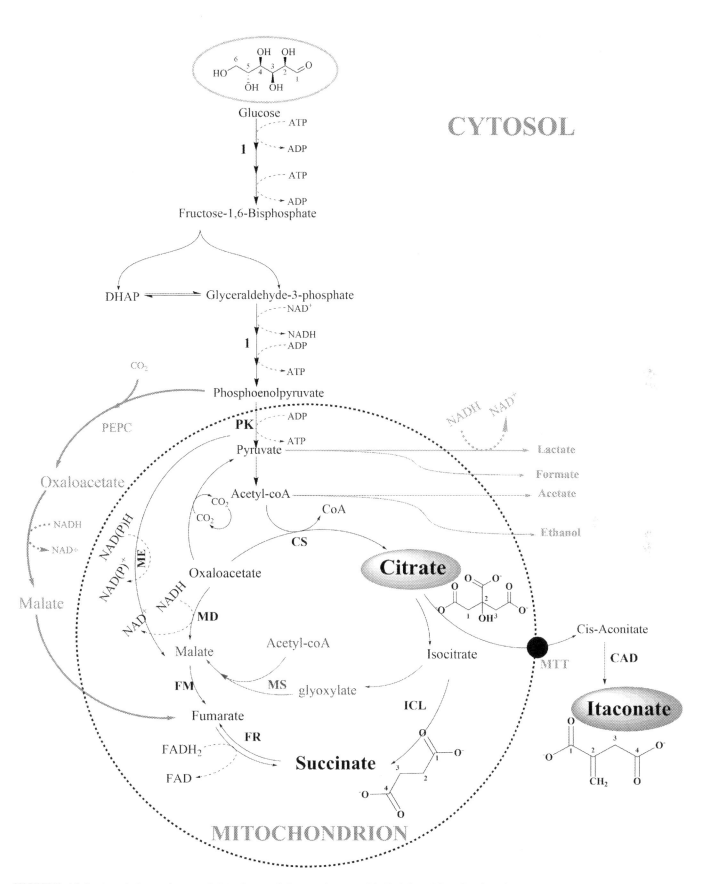

FIGURE 15.3 Metabolic pathway of CA, SA, and IA production. (1) Embden–Meyerh of pathway enzymes; *PK*, Pyruvate kinase; *CS*, Citrate synthase; *MD*, Maltate dehydrogenase; *FM*, Fumarase; *FR*, Fumarate reductase; *ICL*, Isocitrate lyase; *ME*, Malic enzyme; *Ms*, Maltate synthase; *PEPC*, Phosphoenolpyruvate carboxylase; *MTT*, Mitochondrial tricarboxylic transporter; *CAD*, cis-Aconitate decarboxylase.

TABLE 15.2 Succinic acid production from microbial cell factories using renewable waste biomass.

S. no	Waste (substrate)	Microorganism	Titer (g/L)	Yield (g/g)	References
1.	Oil palm trunk	*Actinobacillus succinogenes* 130Z	10.62	0.47	Bukhari (2020)
2.	Fruit and vegetable hydrolysate	*Yarrowia lipolytica* PSA02004	43.1	0.46	Li (2018)
3.	Bakery hydrolysate	*A. succinogenes*	31.7	—	Zhang (2013)
4.	Crop stalk waste	*A. succinogenes* BE-1	15.8	1.2	Li (2010)
5.	Mixed food	*Y. lipolytica* PGC202	71.6	0.61	Li (2019)
6.	Organic fraction of municipal solid waste	*A. succinogenes*	21.2	0.47	Stylianou (2020)

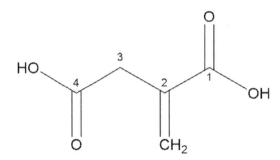

FIGURE 15.4 Itaconic acid structure.

butanediol, tetrahydrofuran, γ-butyrolactone, Adipic acid, biodegradable polymers, plant regulators, and more (Global Itaconic Acid Market | Size | Share | Growth | 2022—2027 | - The Cowboy Channel (No Date). Available At, 2022/2022). SA is used primarily in food industries, detergents and surfactants, green solvents, pharmaceutical industries, and more. In addition, SA is reacted with esters to give dimethylsuccinate, which are particularly used as coolants. These abundant applications result in a market value of $400,000,000 per year (Song & Lee, 2006; Zeikus et al., 1999). All the individual applications can be grouped into four main existing markets, viz., (1) Detergent or surfactant, (2) Ion chelation (corrosion prevention), (3) Food industry as an acidulant, preservative, and antimicrobial, (4) Pharmaceutical industry (Zeikus et al., 1999).

15.2.3 Itaconic acid

15.2.3.1 Physicochemical properties and brief history

2-Methylidenebutanedioc acid ($C_5H_6O_4$) is an unsaturated diprotic acid with a molecular weight of 130.10 g/mol and is commonly known as itaconic acid (IA). IA is otherwise known as 2-mythylidenesuccinic acid. It contains two carboxylic groups (Fig. 15.4) that account for two different pKa values (3.83, 5.5) (Klement & Büchs, 2013). In 1837, Baup extracted itaconic acid as a byproduct in the pyrolytic distillation of citric acid. IA exists as a white crystalline powder, which is hygroscopic and dissolves in water with a maximum solubility of 80.1 g/L at 20°C and in various alcohols (Yang, 2012; Klement & Büchs, 2013). IA has a boiling point of 268°C and a melting point of about 167°C—168°C with density of 1.632 g/cc (20°C) (Kobayashi, 1978/1978).

On a report by US DOE, IA is placed among the top 12 value-added chemicals produced from biomass (Werpy & Petersen, 2004/2004). Due to its numerous applications along with self-polymerization property, IA production has grown enormously accounting for more than 80,000 tons per year (Werpy & Petersen, 2004/2004). Kinoshita discovered a green IA accumulating fungal strain *Aspergillus itaconicus* from dried plums in 1932 (Willke & Vorlop, 2001). In 1955, Pfizer Co. Inc. initiated the production of IA commercially in their Brooklyn plant by submerged fermentation using an *A. terreus*-related strain which was identified by Calam et al. (1939). Since then, research has been growing to make it economically viable and cheaper. Nowadays, there has been an abrupt increase in the waste production from different sectors, resulting in numerous problems; however, they are being utilized as raw materials for the microbial production of value-added products.

15.2.3.2 Metabolic pathway of IA

IA metabolic pathway is described in the context of the strain *A. terreus*, which is considered as the prominent producer of IA and a model organism for analysis and construction of biosynthetic pathway of IA. As we discussed, IA was initially extracted as a by-product of CA pyrolytic distillation. The metabolic pathway initiates with the conversion of glucose to pyruvate through glycolysis in the cytosol, followed by dehydrogenation to give acetyl-CoA. The action of citrate synthase enzyme on acetyl-CoA and oxaloacetate produces citrate as the product in mitochondria. However, the dehydration of citrate takes place to form cis-aconitate. Hereafter, putative mitochondrial tricarboxylic transporter (MTTA) facilitates transportation of cis-aconitate to the cytosol. Wherein, cis-aconitate is converted to itaconate by cis-aconitate dehydrogenase (CAD) enzyme (Fig. 15.1) (Li, 2011; Zhao, 2018).

15.2.3.3 Biosynthetic production of IA

After the commercialization of IA production around 1960s, a plethora of studies have come across a variety of microbes that can accumulate IA other than *A. terreus*. However, every report concluded that no other strains would surpass the IA production of the *A. terreus*, which could be scaled up to the commercial sector for industrial production. Some microorganisms, including fungi *Ustilago maydis* and *Ustilago vetiveriae*, as well as yeasts such as *Candida* species, *Rhodotorula* species, *Pseudozyma antarctica*, and certain genetically engineered species such as *Saccharomyces cerevisiae* and *E. coli*, accumulate IA (Horgan & Murphy, 2011; Levinson et al., 2006; Carstensen, 2013; Blazeck, 2014; Harder et al., 2016; Zambanini, 2017).

The industrial sector has undergone an evolutionary shift in the economy as well as production process due to the utilization of complicated substrates (organic wastes) for the creation of value-added goods like organic acids. We are transitioning from submerged fermentation to solid-state fermentation because of the usage of solid wastes as a substrate. The IA production on the industrial level encompasses multiple sequential steps, viz., fermentation, filtration, crystallization, decolorization, and dying (Okabe, 2009; Zhao, 2018) (Table 15.3).

15.2.3.4 Global market with applications

The global market for itaconic acid reached about 40,000 tons per annum with an estimation of $2 per kilogram of IA. The global market value reached $98.8 million in 2020 and ng at a compounded annual (Global Itaconic Acid Market | Size | Share | Growth | 2022–2027 | - The Cowboy Channel (No Date). Available At, 2022/2022) growth rate (CAGR) of 2.8%; the projected market of CA is $119.9 by 2027 (Global Itaconic Acid Market; Zhao, 2018). IA is abundantly used in the production of polymers with umpteen different properties such as heat resistance, wet abrasion resistance, among others. It is also utilized in emulsion paints because of its excellent adhesive ability. IA is used in the production of surfactants, detergents, cleaners, artificial jewelry, and deodorants. It also serves as a co-monomer in thermoplastic production (Willke & Vorlop, 2001; El-Imam & Du, 2014). The applications of IA are extended to the production of bio-based methyl methacrylate, which serves as a monomer for poly methyl methacrylate (PMMA) that is used in plastic industries as well as in healthcare sector for knee fixations, joint implants, dental implants, etc. (Leggat et al., 2009; Le Nôtre, 2014).

15.3 Organic solvent

15.3.1 A Brief overview

Solvents are substances with the ability to dissolve a solute inside them and create a solution as a result (Stoye, 2000/2000). These solutions are utilized in industry for the preparation of nail polish remover, paint,

TABLE 15.3 Itaconic acid production from microbial cell factories using renewable waste biomass.

S. no	Waste (substrate)	Microorganism	Titer (g/L)	Yield (g/g)	References
1.	Jatropha cake	*Aspergillus terreus*	48.7	0.1	El-Imam (2013)
2.	Market refuse fruits (apple and banana)	*A. terreus* SKR10 mutant N54	30	0.36	Reddy and Singh (2002)
3.	Beech wood	*A. terreus*	69	0.357	Regestein (2018)
4.	Non-food resources (wheat bran and corn cobs)	*A. terreus*	–	0.05	Jiménez-Quero (2020)
5.	Corn stover hydrolysate	*A. terreus* M69	33.6	0.56	Liu (2020)

detergents, and perfumes. Depending on the presence of carbon atoms, solvents can be generally categorized as organic (for example, ethanol and benzene) or inorganic (for example, water and liquid ammonia).

Organic solvents contain different types of functional groups, such as ketone, alcohol, acid, and ether, and are composed of carbon, hydrogen, and oxygen atoms. Furthermore, the polarity of these functional groups determines the dielectric constant of the solvent. By virtue of the "like dissolves like" principle (Zhuang, 2021), which states that polar substances can only dissolve in polar solvents and vice versa, the solvent's polarity is determined and it provides us with an indication of which solutes can be dissolved in it. Accordingly, we can decide our solvent of interest based on our needs.

Physical characteristics of a few solvents are shown in (Table 15.4).

(Acetone CH3COCH3 PubChem; 1-Butanol | C4H9OH - PubChem; Ethanol | CH3CH2OH - PubChem; 2,3-Butanediol | C4H10O2 - PubChem)

In the next discussion, we will go into more detail regarding the production of solvents from waste, the types of waste that can be used, the microbes that are used for the fermentation process, the yield of production, and applications of the solvents (Table 15.5). In Fig. 15.5, we have provided an illustrative representation of solvent production, highlighting the key stages and processes involved in the synthesis of various solvents.

15.3.2 Butane-2,3-diol: glycol

2,3-Butanediol (BDO) is a colorless liquid with the chemical formula $(CH_3CHOH)_2$ (2,3-Butanediol | C4H10O2 - PubChem). It exists in three stereoisomeric forms (a chiral pair (levo, dextro) and meso isomer) (Boutron, 1992) you can find the different spatial arrangements of atoms in these forms in Fig. 15.6.

BDO is a significant chemical which is used to kickstart the synthesis of a wide variety of chemical derivatives. It is typically made from petrochemical fuels and oils (Köpke, 2011). However, considering the unsustainable nature of using nonrenewable natural resources like fossil fuels, we can focus on their creation, through microbial fermentation of a variety of wastes. During World War II, research in this field began with the use of sugarcane molasses for its production (Dai, 2015). In recent research, scientists have discovered a novel method for generating BDO on the surface of Mars utilizing cyanobacteria and *E. coli*. The resultant BDO can be used as rocket fuel (Rocket fuel).

TABLE 15.4 Physical properties of organic solvents.

Solvent	Molar mass (g/mol)	Boiling point (°C)	Dielectric constant	Density (g/mL)	Dipole moment (D)
Acetone	58.08	56.1	21	0.786	2.8
n-Butanol	74.12	117.7	18	0.810	1.63
Ethanol	46.07	78.2	24.55	0.789	1.69
2,3 Butanediol	90.12	177	28.8	0.987	4.12

TABLE 15.5 Solvent production from microbial cell factories using waste biomass.

Solvent	Feedstock	Microbe	Yield (g/g)	References
2,3 Butanediol	Fruit waste	*Enterobacter ludwigii FMCC204*	0.4	Liakou (2018)
	Rice straw	*Klebsiella sp. Zmd30*	0.6	Wong (2012)
	Bread waste	*E. ludwigi*	0.48	Narisetty (2022)
Ethanol	Textile waste	*Saccharomyces cervisiae*	0.4	Jeihanipour and Taherzadeh (2009)
	Potato peel waste	*S. cervisiae*	0.46	Arapoglou et al. (n.d)
	Municipal solid waste	*S. cervisiae*	–	Li et al. (2007)
Butanol	Starch industry waste water	*Clostridium beijerinckii NRRL B-466*	0.27	Maiti (2016)
	Agricultural waste	Bacterial microflora	0.52 mol/mol	Cheng (2012)
	Cauliflower waste	*Clostridium acetobutylicum NRRL B527*	–	Khedkar (2017)

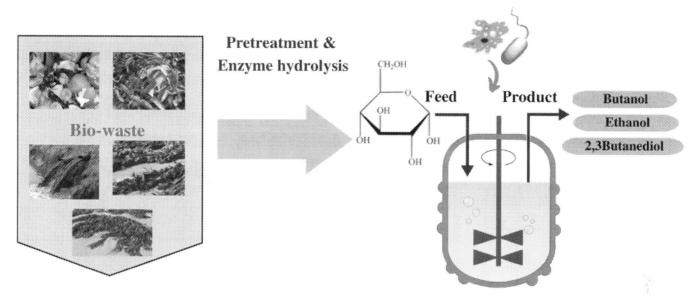

FIGURE 15.5 Illustrative representation of solvent production.

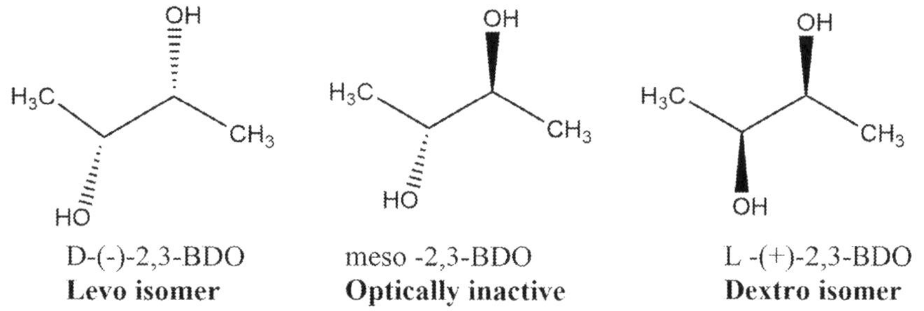

FIGURE 15.6 Stereoisomers of butane-2,3-diol.

The cost of producing BDO is currently $3.23/kg, and employing trash as a raw material will result in a cost-effective bioprocess (Tinôco, 2021).

15.3.2.1 Biorefinery-based BDO production

15.3.2.1.1 Vegetables and fruits

In 2014, there were approximately 35 million tons of fruits and 289.8 million tons of vegetables produced globally, (Liakou, 2018). However, until the customer consumes them, roughly 56% of these end up in the garbage. Agricultural production accounts for 20% of the total waste stream, followed by postharvest handling and storage (5%), processing and packaging (2%), distribution (10%), and consumption (19%) (Gustafsson et al., 2013).

The bacterial strain *Enterobacter ludwigii FMCC 204* was used to ferment waste which is abundant in glucose, fructose, and sucrose, to produce a sizable amount of BDO (Liakou, 2018). It gave a yield of 0.4 g/g for fruits, but it was less for vegetables as the C5 and C6 sugar molecules are not freely available and they need to be pretreated first (Chandel, 2012).

15.3.2.1.2 Farmland waste

The main agricultural waste products include starch, molasses, whey permeate, and lignocellulose materials derived from wood, corn cobs, bagasse, or rice straw. These materials are produced in significant numbers each year (Perego, 2000). Since liquidation produces a lot of methane, it is best to use it as a feedstock for the manufacturing of BDO because it can fix the problem.

Due to its refractory character, the lignocellulosic material needs to be pretreated with an acid or enzyme before fermentative destruction of sugar can occur. Subsequently, *Klebsiella sp.Zmd30* is incubated with this reasonably priced hydrolyzed waste (Wong, 2012).

It has been noted that trace elements are essential for the synthesis of BDO. Co^{2+} contributed in achieving high BDO yield (Petrov & Petrova, 2009). The productivity of hydrolyzed rice straw material was reported as 2.08 g/L/h, with a yield of 62% (Wong, 2012).

15.3.2.1.3 Bread waste

According to research studies, 33.3% of the food produced worldwide ends up in landfills, causing the world economy to suffer a massive loss of $750 billion (Kumar & Longhurst, 2018). Eliminating this waste could result in savings of up to £15–20 billion and a reduction of 27 million tons of "greenhouse gas" emissions.

More than 10% of all made bread is wasted worldwide, making bread the most wasted food in Europe. Additionally, bread can be utilized to produce BDO because it is a clean supply of fermentable sugar. Instead of using it directly, bread waste is first subjected to saccharification through hydrolysis, and then BDO is produced using *E. ludwigi* with a maximum yield of 0.48 g/g. Enzyme-hydrolyzed bread trash produces 138.8 g/L BDO (Narisetty, 2022).

15.3.2.2 Metabolic pathway

As the waste material is not made entirely of glucose molecules, it must first go through an enzyme or acid hydrolysis process to yield monosaccharides. It then undergoes anaerobic fermentation to create BDO through the phosphortransferase system, the sugar enters the microbial cell system which uses the glycolysis pathway to convert the glucose to pyruvate while also producing NADH and ATP molecules (Syu, 2012). Following that, pyruvate condenses to produce alpha-acetolactate using NADH. The next step is the anaerobic α-acetolactate decarboxylase-mediated conversion of alpha-acetolactate to R-acetoin. In contrast, alpha-acetolactate spontaneously decarboxylates in an aerobic environment to produce diacetyl, which the enzyme diacetyl reductase then converts to S-acetoin. In the last stage, BDO dehydrogenase reduces acetoin (R-acetoin formed in anaerobic condition, S-acetoin formed in aerobic conditions) to BDO. To renew NADH and maintain a steady oxidation–reduction state, the generated BDO can be reversibly transformed to acetoin (Maina, 2022) (Fig. 15.7).

While the creation of BDO in bacteria is considered a process for storing carbon and energy, at high concentrations of BDO, it can even hinder this process (Global Itaconic Acid Market | Size | Share | Growth| 2022–2027 | - The Cowboy Channel (No Date). Available At, 2022/2022), after which it will switch back to producing acetoin as a defense (Okonkwo et al., 2017).

15.3.2.3 Applications

BDO is a highly prized chemical that is a vital component of many different products used in the food, cosmetics, pharmaceutical, and plastic sectors. It serves as an additive in printing inks, fumigants, and perfumes. It can be used as liquid fuel as well, and most importantly BDO serves as the precursor material for the manufacture of 1,3-butadiene,diacetyl,methyl ethyl ketone (Chiao & Sun, 2007).

15.3.3 Ethanol

Ethyl alcohol, with the molecular formula C_2H_6O, is a flammable, volatile, colorless liquid that imparts characteristic aroma and taste to wine. Often abbreviated as EtOH (Ethanol | CH3CH2OH - PubChem).

As can be seen from the structure Fig. 15.8, that the molecule has the functional group -OH because it is an alcohol. Due to its ability to dissolve both hydrophobic and hydrophilic molecules and lower boiling point, ethanol is a unique solvent (Ethanol Extraction in Cannabis).

For centuries, ethanol has been manufactured through fermentation of sugars. Michael Faraday created it synthetically in 1825. He discovered that sulfuric acid could absorb significant amounts of coal gas, and Henry Hennell responded by stating that sufovinic acid (ethyl hydrogen sulfate) would form and then breakdown into ethanol (Hennell & Brande, 1826). It has also been created for industrial use by utilizing petrochemical methods such as ethylene hydration (Richard L. Myers), but considering the prevailing fuel shortage situation, the urge for fermentation techniques is obvious. Therefore, it is manufactured using waste resources that can yield sugar for fermentation rather than using main food materials.

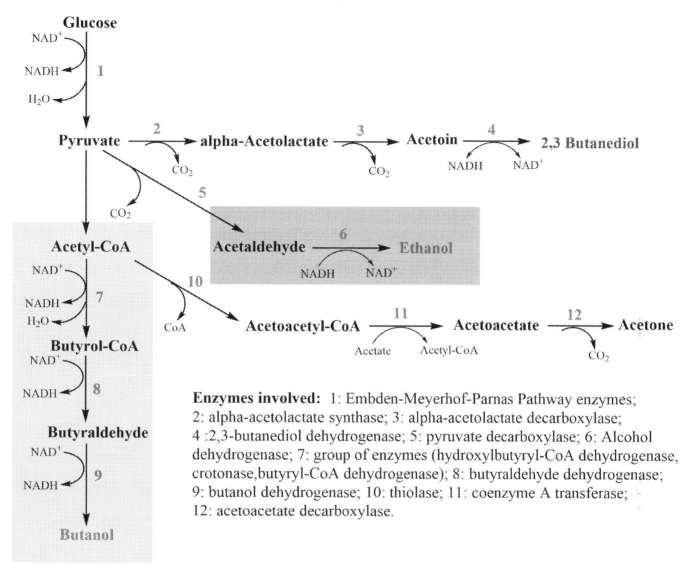

FIGURE 15.7 Metabolic pathway for the production of butanol, ethanol and 2,3-butanediol.

FIGURE 15.8 Structure of ethanol.

On average, 73% of the ethanol produced is used to make fuel, 17% is used to make beverages, and 10% is used for industrial purposes (GlobalPetrolPrices.com). Ethanol's current (2022) market price is $1.48 per liter, and in 2019, it was predicted to be a part of the $89.1 billion global market (Global Itaconic Acid Market | Size | Share | Growth | 2022–2027 | - The Cowboy Channel (No Date). Available At, 2022/2022). Furthermore, its high value reflects its strong demand (Ethanol Market Size, (Global Itaconic Acid Market | Size | Share | Growth | 2022–2027 | - The Cowboy Channel (No Date). Available At, 2022/2022) & Trends Report).

15.3.3.1 Biorefinery-based ethanol production
15.3.3.1.1 Cotton-based waste

Used textiles, along with excess cotton, constitute more than one-third of all textile waste generated. The textile industry disposes of this waste in landfills or incinerators (Miranda, 2007). Cotton, derived from Gossypium seed hairs, is typicaly composed of 88%–96% cellulose. Prior to use, it undergoes a bleaching process to enhance its cellulose content to 99% (Miranda, 2007). Despite the fact that sugar and starchy materials are typically used to make ethanol, the importance of employing lignocellulosic material as a feedstock cannot be overstated given its high availability and low cost (Lynd, 2005). It undergoes processing to get rid of lignin and hemicellulose and to reduce cellulose's crystallinity before being used (Galbe & Zacchi, 2002). Followed by simultaneous saccharification and fermentation (SSF) of the cleaned-up waste to produce ethanol. 99.1% of cotton was reported to be digested to glucose during hydrolysis at a temperature of 0°C and 12% NaOH, which when fermented with *S. cerevisiae* yielded 0.4 g/g of product (Jeihanipour & Taherzadeh, 2009).

15.3.3.1.2 Potato peel waste (PPW)

The potato processing industry, including manufacturers of French fries, chips, and puree, peel the potato in order to get crispiness to the product, which produces PPW, a waste with no value (ZMP-Marktbilanz / EconBiz). About 90 kg of waste is produced for every Mg of influent potatoes, of which 50 kg constitute potato peel, 30 kg is starch, and 10 kg comprises of inert material (Arapoglou et al.). It contains a sufficient amount of cellulose and starch, which can be utilized as a feedstock for ethanol synthesis. Enzymes are used in this case for pretreatment hydrolysis rather than acid because they are more efficient and less hazardous (in terms of water use, by-product formation, and energy consumption), and works better in milder environments. It has been shown that PPW that has been enzyme-hydrolyzed releases more reducing sugar than PPW that has been acid-hydrolyzed (Barnett, 1999). The production of ethanol was 7.58 g/L, and the yield matched 91.6% of theoretical value, or 0.46 g/g (Arapoglou et al.).

15.3.3.1.3 Municipal solid waste (MS)

European legislation has established targets to limit the volume of solid waste dumped in landfills (EUR-Lex — 32003L0030 - EN - EUR-Lex). As a solution to this, bio-ethanol production can be employed. Paper, kitchen scraps, yard trimmings, and textiles are the typical MSW fractions that will be used to produce ethanol (Li et al., 2007). Furthermore, we require substantial glucose yields for the generation of ethanol (McMillan, 1994). To improve the hydrolysis of lignocellulosic material, the waste is first processed and using cellulase isolated from *Trichoderma reesei*, enzymatic hydrolysis is carried out. It has been found that when only one type of substrate is utilized, a high glucose yield of between 70% and 89% is achieved. Consequently, the high glucose yield suggests a similarly high ethanol yield.

15.3.3.2 Metabolic pathway

The yeast *S. cerevisiae* typically produces ethanol, and while producing using waste, it must first undergo pretreatment with an acid or an alkali. It has been found that pretreatment with phosphoric acid increased the synthesis of ethanol from textiles by up to 66% (Jeihanipour & Taherzadeh, 2009). Followed by an enzymatic hydrolysis process lasting up to 4 days or longer depending on the waste's composition, the liberated glucose molecules go through fermentation to make ethanol, either simultaneously (SSF) or in batches.

During the fermentation process, glucose first enters glycolysis and gives pyruvate as a product which in anaerobic condition is converted to acetaldehyde by pyruvate decarboxylase enzyme, subsequently giving ethanol due to the catalytic activity of alcohol dehydrogenase by utilizing NADH (Fig. 15.7). It has been found that several factors such as temperature, pH, aeration rate, salt concentration, and carbohydrate concentration affect the production of bioethanol (Lugani, 2020).

For the reaction mentioned above, the chemical equation is:

$$C_6H_{12}O_6 \rightarrow 2CH_3CH_2OH + 2CO_2$$

At a temperature of 35°C–40°C, the above reaction occurs in the cytosol. It has been discovered that the ethanol concentration is constrained by ethanol's toxicity to yeast. Most tolerant strains are able to endure up to approximately 18% ethanol by volume, and so, the ethanol is concentrated using distillation.

15.3.3.3 Applications

Ethanol is favored as a solvent due to its excellent solubility in both polar and nonpolar molecules. Thanks to its antiseptic properties, it can be used as a sanitizer, a wiping agent, a fuel (biofuel), a medical solvent, and in the production of inks, perfumes, alcoholic beverages, fires, and other products (Arapoglou et al.).

15.3.4 Butanol

Butanol, a linear alcohol with a chemical formula of C_4H_9OH, exists in four structural isomeric forms: n-butanol (or) 1-butanol, sec-butanol (or) 2-butanol, isobutanol (or) 2-methyl propan-1-ol, tert-butanol (or) 2-methyl propanol (Fig. 15.9) (1-Butanol | C4H9OH - PubChem) and is represented using bond line structure as.

It is primarily used as a solvent and is one of the important intermediate chemicals. It is significant as a biofuel as well (Ni & Sun, 2009). It was discovered by Pasteur in 1861 (Qureshi, 2009). Though butanol production is one of the oldest fermentative processes, petrochemical production using propene took over it rapidly. However the depletion of fossil fuels demands the use of renewable sources again (Maiti, 2016). Since the feedstock cost accounts for about 60% of the total production cost, utilization of inexpensive materials such as waste carbon sources can be used as a substrate (Qureshi & Blaschek, 2000). Butanol is currently produced at 5–6 million tons per year with total value of $7–8.4 billion in the worldwide market (Lee, 2008).

15.3.4.1 Production using

15.3.4.1.1 Agro-industrial waste

According to a report given by the Federation of Indian Chambers of Commerce and Industry (FICCI) in year 2011, about 60%–70% of agro waste is discharged into the environment without any processing (Methane (Initiative, 2011/2011)). To manage this waste accumulation, it can be used as a feedstock in butanol production.

Agro-industrial waste, abundant in the food and beverage processing industry, is inexpensive and micronutrient-rich, making it an ideal carbon source for bioprocessing (Maiti, 2016). Wastes from the food and beverage processing industry can be classified as suspended brewery liquid waste, starch industry waste water, and apple pomace ultra-filtration sludge. This is used for ABE (Acetone-Butanol-Ethanol) fermentation using *Clostridium beijerinckii* NRRL B-466. Maiti et al. (Maiti, 2016) in their studies found that waste water from starch industry gives the maximum yield of butanol 0.27 g/g along with 0.46 g/g of ABE.

15.3.4.1.2 Agricultural waste

Asian nations have the potential to produce bioenergy from crop waste since they practice large-scale farming. In Taiwan, rice straw is the main agricultural waste, and it is used to produce bioenergy (Capello, 2009).

The waste is first treated using NaOH, $Ca(OH)_2$, or ammonia for breaking the cellulosic structure followed by enzyme hydrolysis to get glucose and then, the obtained glucose is converted to value-added product by the microbe during fermentation (Taherzadeh & Karimi, 2008). The production done using SHF (Separate Hydrolysis and Fermentation) process gave the maximum butanol concentration of 2.29 g/L with a yield of 0.52 mol butanol/mol reducing sugar (Cheng, 2012).

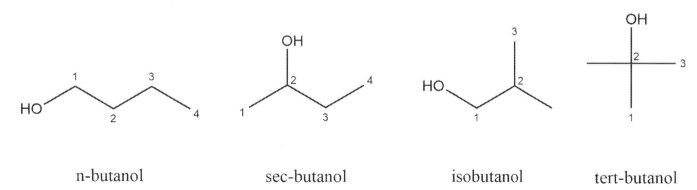

FIGURE 15.9 Isomers of butanol.

15.3.4.1.3 Cauliflower waste

For the efficient yet economic production of fuels, we need to explore more on second-generation feedstock instead of using primary food as a carbon source, considering the growing population and global food deficit (Narisetty, 2022). One such carbon source is cauliflower waste.

India ranks second in the world for cauliflower production, and it produces almost 35% of the total global market which is 7.89 million tons per year. During the harvesting, processing it generates 45%−60%(w/w) cauliflower waste (Dhillon et al., 2007). It is made up of 16.6% cellulose, 14.9% crude protein, 8.4% hemicellulose, 17% total sugar, 6.25% phenolics, 14% ash, and 10%−20% minerals (Oberoi, 2007).

As it has good amount of carbohydrates, it can be used as a feedstock for ABE fermentation; however, it is recommended to do pretreatment and drying before hydrolysis of waste. It has been reported that waste dried at 80°C gave maximum total sugar yield of 26.05 g/L (Khedkar, 2017) and the highest yield obtained for butanol stands at 0.17 g/g (Khedkar, 2017).

15.3.4.2 Metabolic pathway

Butanol is produced usually by *Clostridium sp.* during ABE fermentation (Sunwoo, 2018). It shows biphasic pattern of acidogenesis and solventogenesis consecutively. 1st phase is acidogenesis, which occurs duing exponential phase of bacterial growth (Jones & Woods, 1986) and 2nd phase, solventogenesis starts at the end of exponential phase and continues in stationary phase as well; during this phase, acid produced during acidogenesis is converted into solvents by complex metabolic pathway as shown in Fig. 15.7.

It was found that pH has a considerable impact on ABE fermentation. Rate of *Clostridium acetobutylicum XY16* increased rapidly when pH was controlled at 5.5 and for solvent production pH levels are maintained in the range 4.7−4.9 (Guo, 2012).

For getting good yield, it is recommended to do pretreatment before performing the hydrolysis. Additionally, having optimal amount of glucose in the feed before the start of fermentation gave the best yield of butanol (Maiti, 2016).

15.3.4.3 Applications

Butanol has wide application as a solvent in industries. It is a potential fuel because of its high energy content, less volatility, and less corrosive properties (Ni & Sun, 2009). It has the capability to easily replace ethanol because of its good ability to blend with gasoline and hence can be used in vehicles without modifying the engine (Biobutanol − A Replacement for Bioethanol? | AIChE).

References

Abbas, N., et al. (2016). Citric acid production from Aspergillus niger using mango (Mangifera indica L.) and sweet orange (Citrus sinensis) peels as substrate. *International Journal of Scientific and Engineering Research*, 7(2), 868−872. Available from https://doi.org/10.14299/IJSER.2016.02.001.

Acetone | CH3COCH3 - PubChem (no date). Available at: https://pubchem.ncbi.nlm.nih.gov/compound/180 Accessed 07.07.22.

Arapoglou, D. et al. (no date) Ethanol production from potato peel waste (PPW), Elsevier [Preprint]. Available at: https://www.sciencedirect.com/science/article/pii/S0956053X10002448 Accessed 30.06.22.

Babaei, M., et al. (2019). Engineering oleaginous yeast as the host for fermentative succinic acid production from glucose. *Frontiers in Bioengineering and Biotechnology*, 7, 361. Available from https://doi.org/10.3389/FBIOE.2019.00361/BIBTEX.

Barnett, C., et al. (1999). 'Pullulan production by Aureobasidium pullulans growing on hydrolysed potato starch waste'. *Carbohydrate Polymers*, 38(3), 203−209. Available from https://doi.org/10.1016/S0144-8617(98)00092-7.

Biobutanol − A Replacement for Bioethanol? | AIChE (no date). Available at: https://www.aiche.org/resources/publications/cep/2008/august/biobutanol-replacement-bioethanol Accessed 13.07.22.

Blazeck, J., et al. (2014). Metabolic engineering of Saccharomyces cerevisiae for itaconic acid production. *Applied Microbiology and Biotechnology*, 98(19), 8155−8164. Available from https://doi.org/10.1007/S00253-014-5895-0/FIGURES/6.

Boutron, P. (1992). Cryoprotection of red blood cells by a 2,3-butanediol containing mainly the levo and dextro isomers. *Cryobiology*, 29(3), 347−358. Available from https://doi.org/10.1016/0011-2240(92)90036-2.

Bryant, M. P., et al. (1958). Bacteroides ruminicola n. sp. and Succinimonas amylolytica; the new genus and species; species of succinic acid-producing anaerobic bacteria of the bovine rumen. *Journal of Bacteriology*, 76(1), 15−23. Available from https://doi.org/10.1128/JB.76.1.15-23.1958.

Bryant, M. P., & Small, N. (1956). Characteristics of two new genera of anaerobic curved rods isolated from the rumen of cattle. *Journal of Bacteriology*, 72(1), 22−26. Available from https://doi.org/10.1128/jb.72.1.22-26.1956.

Bukhari, N. A., et al. (2020). Organic acid pretreatment of oil palm trunk biomass for succinic acid production. *Waste and Biomass Valorization*, 11(10), 5549−5559. Available from https://doi.org/10.1007/S12649-020-00953-2/FIGURES/6.

Calam, C. T., Oxford, A. E., & Raistrick, H. (1939). Studies in the biochemistry of micro-organismsItaconic acid, a metabolic product of a strain of Aspergillus terreus Thom. *Biochemical Journal*, 33(9), 1488−1495. Available from https://doi.org/10.1042/BJ0331488.

References

Capello, C., et al. (2009). A comprehensive environmental assessment of petrochemical solvent production. *International Journal of Life Cycle Assessment*, 14(5), 467–479. Available from https://doi.org/10.1007/S11367-009-0094-4.

Carstensen, F., et al. (2013). Continuous production and recovery of itaconic acid in a membrane bioreactor. *Bioresource Technology*, 137, 179–187. Available from https://doi.org/10.1016/J.BIORTECH.2013.030.044.

Chandel, A. K., et al. (2012). Dilute acid hydrolysis of agro-residues for the depolymerization of hemicellulose: State-of-the-art. *D-Xylitol: Fermentative Production, Application and Commercialization*, 39–61. Available from https://doi.org/10.1007/978-3-642-31887-0_2/COVER/.

Chatterjee, R., et al. (2001). Mutation of the ptsG gene results in increased production of succinate in fermentation of glucose by Escherichia coli. *Applied and Environmental Microbiology*, 67(1), 148–154. Available from https://doi.org/10.1128/AEM.67.1.148-154.2001.

Cheng, C. L., et al. (2012). Biobutanol production from agricultural waste by an acclimated mixed bacterial microflora. *Applied Energy*, 100, 3–9. Available from https://doi.org/10.1016/J.APENERGY.2012.050.042.

Cheng, K. K., et al. (2013). Improved succinate production by metabolic engineering. *BioMed Research International*, 2013. Available from https://doi.org/10.1155/2013/538790.

Chiao, Jui-shen, & Sun, Zhi-hao (2007). History of the acetone-butanol-ethanol fermentation industry in China: Development of continuous production technology. *Microbial Physiology*, 13(1-3), 12–14. Available from https://doi.org/10.1159/000103592.

Chinthapalli, R. et al. (2018) Succinic acid: New bio-based building block with a huge market and environmental potential? – Preview | Renewable Carbon Publications. Available at: https://renewable-carbon.eu/publications/product/succinic-acid-new-bio-based-building-block-with-a-huge-market-and-environmental-potential-preview/ Accessed 08.07.22.

Chu, S., & Majumdar, A. (2012). Opportunities and challenges for a sustainable energy future. *Nature*, 488(7411), 294–303. Available from https://doi.org/10.1038/nature11475.

Ciriminna, R., et al. (2017). Citric acid: Emerging applications of key biotechnology industrial product. *Chemistry Central Journal*, 11(1), 1–9. Available from https://doi.org/10.1186/S13065-017-0251-Y.

Citric Acid Market Size, (Global Itaconic Acid Market | Size | Share | Growth| 2022-2027 | - The Cowboy Channel (No Date). Available At, 2022/2022)(2022 – 27) | Industry Report (2022). Available at: https://www.mordorintelligence.com/industry-reports/citric-acid-market Accessed 03.07.22.

Comuzzo, P., & Battistutta, F. (2018). *Acidification and pH control in red wines red wine technology* (pp. 17–34). Italy: Elsevier. Available from https://doi.org/10.1016/B978-0-12-814399-5.00002-5, https://www.sciencedirect.com/book/9780128143995.

Dai, J. Y., et al. (2015). Enhanced production of 2,3-butanediol from sugarcane molasses. *Applied Biochemistry and Biotechnology*, 175(6), 3014–3024. Available from https://doi.org/10.1007/S12010-015-1481-X/FIGURES/2.

Dhillon, G. S., Bansal, S., & Oberoi, H. S. (2007). Cauliflower waste incorporation into cane molasses improves ethanol production using Saccharomyces cerevisiae MTCC 178. *Indian Journal of Microbiology*, 47(4), 353–357. Available from https://doi.org/10.1007/S12088-007-0063-1.

Dittrich, C. R., Bennett, G. N., & San, K. Y. (2009). Metabolic engineering of the anaerobic central metabolic pathway in Escherichia coli for the simultaneous anaerobic production of isoamyl acetate and succinic acid. *Biotechnology Progress*, 25(5), 1304–1309. Available from https://doi.org/10.1002/BTPR0.222.

EL-IMAM, A. M. A., et al. (2013). Production of Itaconic acid from Jatropha curcas seed cake by Aspergillus terreus. *Notulae Scientia Biologicae*, 5(1), 57–61. Available from https://doi.org/10.15835/NSB518355.

EUR-Lex - 32003L0030 - EN - EUR-Lex (no date). Available at: https://eur-lex.europa.eu/legal-content/en/ALL/?uri=CELEX%3A32003L0030 Accessed 10.07.22.

El-Imam, A. A., & Du, C. (2014). Fermentative Itaconic acid production. *Journal of Biodiversity, Bioprospecting and Development*, 1(2), 1000119. Available from https://doi.org/10.4172/2376-0214.1000119.

Ensign, S. A. (2006). Revisiting the glyoxylate cycle: Alternate pathways for microbial acetate assimilation. *Molecular Microbiology*, 61(2), 274–276. Available from https://doi.org/10.1111/j.1365-2958.2006.05247.x.

Estimated 931 mn tonnes of food wasted globally in 2019; India's(Global Itaconic Acid Market | Size | Share | Growth| 2022-2027 | - The Cowboy Channel (No Date). Available At, 2022/2022) 68 mn: UN report - The Economic Times (2021). Available at: https://economictimes.indiatimes.com/news/economy/indicators/estimated-931-mn-tonnes-of-food-wasted-globally-in-2019-indias-share-68-mn-un-report/articleshow/81345719.cms Accessed 11.07.22.

Ethanol Market Size,(Global Itaconic Acid Market | Size | Share | Growth| 2022-2027 | - The Cowboy Channel (No Date). Available At, 2022/2022)& Trends Report, 2020-2027 (no date). Available at: https://www.grandviewresearch.com/industry-analysis/ethanol-market Accessed 11.07.22.

Ethanol prices around the world, 04-Jul-2022 | GlobalPetrolPrices.com (no date). Available at: https://www.globalpetrolprices.com/ethanol_prices/ Accessed 11.07.22.

Ethanol | CH3CH2OH - PubChem (no date). Available at: https://pubchem.ncbi.nlm.nih.gov/compound/702 Accessed 07.07.22.

Ferdous, W., et al. (2021). Recycling of landfill wastes (tyres, plastics and glass) in construction – A review on global waste generation, performance, application and future opportunities. *Resources, Conservation and Recycling*, 173, 105745. Available from https://doi.org/10.1016/J.RESCONREC.2021.105745.

Galbe, M., & Zacchi, G. (2002). A review of the production of ethanol from softwood. *Applied Microbiology and Biotechnology*, 59(6), 618–628. Available from https://doi.org/10.1007/s00253-002-1058-9.

Global Itaconic Acid Market | Size | Share | Growth| 2022-2027 | - The Cowboy Channel (no date). Available at. 2022.

Grewal, H. S., & Kalra, K. L. (1995). Fungal production of citric acid. *Biotechnology Advances*, 13(2), 209–234. Available from https://doi.org/10.1016/0734-9750(95)00002-8.

Guettler, M. V., Rumler, D., & Jain, M. K. (1999). Actinobacillus succinogenes sp. nov., a novel succinic-acid-producing strain from the bovine lumen. *International Journal of Systematic Bacteriology*, 49(1), 207–216. Available from https://doi.org/10.1099/00207713-49-1-207/CITE/REFWORKS.

Guo, T., et al. (2012). Enhancement of butanol production and reducing power using a two-stage controlled-pH strategy in batch culture of Clostridium acetobutylicum XY16. *World Journal of Microbiology and Biotechnology*, 28(7), 2551–2558. Available from https://doi.org/10.1007/S11274-012-1063-9/FIGURES/6.

Gustafsson, J., Cederberg, C., Sonesson, U.G., & Emanuelsson, A. (2013). The methodology of the FAO study: Global Food Losses and Food Waste - extent, causes and prevention"- FAO, 2011. Available from https://www.diva-portal.org/smash/get/diva2:944159/FULLTEXT01.pdf.

Harder, B. J., Bettenbrock, K., & Klamt, S. (2016). Model-based metabolic engineering enables high yield itaconic acid production by Escherichia coli. *Metabolic Engineering, 38*, 29–37. Available from https://doi.org/10.1016/J.YMBEN.2016.050.008.

Hennell, H., & Brande, W. T. (1826). XVII. On the mutual action of sulphuric acid and alcohol, with observations on the composition and properties of the resulting compound. *Philosophical Transactions of the Royal Society of London, 116*, 240–249. Available from https://doi.org/10.1098/RSTL.1826.0021.

Horgan, K. A., & Murphy, R. A. (2011). *Pharmaceutical and chemical commodities from fungi fungi: Biology and applications: Second edition* (pp. 147–178). Ireland: Wiley. Available from https://doi.org/10.1002/9781119976950.ch6, http://onlinelibrary.wiley.com/book/10.1002/9781119976950.

Initiative. (2011). *Resource assessment for livestock and agro-industrial wastes-India prepared for: The Global Methane Initiative*.

Jeihanipour, A., & Taherzadeh, M. J. (2009). Ethanol production from cotton-based waste textiles. *Bioresource Technology, 100*(2), 1007–1010. Available from https://doi.org/10.1016/j.biortech.2008.070.020.

Jiménez-Quero, A., et al. (2020). Optimized bioproduction of itaconic and fumaric acids based on solid-state fermentation of lignocellulosic biomass. *Molecules (Basel, Switzerland), 25*(5), 1070. Available from https://doi.org/10.3390/MOLECULES25051070.

Jones, D. L. (1998). Organic acids in the rhizosphere – A critical review. *Plant and Soil, 205*(1), 25–44. Available from https://doi.org/10.1023/A:1004356007312.

Jones, D. T., & Woods, D. R. (1986). Acetone-butanol fermentation revisited. *Microbiological Reviews, 50*(4), 484–524. Available from https://doi.org/10.1128/MR.50.4.484-524.1986.

Khedkar, M. A., et al. (2017). Cauliflower waste utilization for sustainable biobutanol production: revelation of drying kinetics and bioprocess development. *Bioprocess and Biosystems Engineering, 40*(10), 1493–1506. Available from https://doi.org/10.1007/S00449-017-1806-Y/TABLES/6.

Klement, T., & Büchs, J. (2013). Itaconic acid - A biotechnological process in change. *Bioresource Technology, 135*, 422–431. Available from https://doi.org/10.1016/j.biortech.2012.110.141, http://www.elsevier.com/locate/biortech.

Kobayashi, T. (1978). Production of Itaconic acid from wood waste. *Process Biochemistry, 13*(5), 15–22.

Köpke, M., et al. (2011). 2,3-butanediol production by acetogenic bacteria, an alternative route to chemical synthesis, using industrial waste gas. *Applied and Environmental Microbiology, 77*(15), 5467–5475. Available from https://doi.org/10.1128/AEM.00355-11.

Kumar, V., & Longhurst, P. (2018). Recycling of food waste into chemical building blocks. *Current Opinion in Green and Sustainable Chemistry, 13*, 118–122. Available from https://doi.org/10.1016/j.cogsc.2018.050.012, http://www.journals.elsevier.com/current-opinion-in-green-and-sustainable-chemistry.

Le Nôtre, J., et al. (2014). Synthesis of bio-based methacrylic acid by decarboxylation of itaconic acid and citric acid catalyzed by solid transition-metal catalysts. *ChemSusChem, 7*(9), 2712–2720. Available from https://doi.org/10.1002/CSSC.201402117.

Lee, S. Y., et al. (2008). Fermentative butanol production by clostridia. *Biotechnology and Bioengineering, 101*(2), 209–228. Available from https://doi.org/10.1002/BIT.22003.

Leggat, P. A., Smith, D. R., & Kedjarune, U. (2009). Surgical applications of methyl methacrylate: A review of toxicity. *Archives of Environmental and Occupational Health, 64*(3), 207–212. Available from https://doi.org/10.1080/19338240903241291.

Levinson, W. E., Kurtzman, C. P., & Kuo, T. M. (2006). Production of itaconic acid by Pseudozyma antarctica NRRL Y-7808 under nitrogen-limited (*Global Itaconic Acid Market | Size | Share | Growth| 2022-2027 | - The Cowboy Channel (No Date). Available At*, 2022/2022) conditions. *Enzyme and Microbial Technology, 39*(4), 824–827. Available from https://doi.org/10.1016/J.ENZMICTEC.2006.010.005.

Li, A., Antizar-Ladislao, B., & Khraisheh, M. (2007). Bioconversion of municipal solid waste to glucose for bio-ethanol production. *Bioprocess and Biosystems Engineering, 30*(3), 189–196. Available from https://doi.org/10.1007/S00449-007-0114-3.

Li, A., et al. (2011). A clone-based transcriptomics approach for the identification of genes relevant for itaconic acid production in Aspergillus. *Fungal Genetics and Biology, 48*(6), 602–611. Available from https://doi.org/10.1016/J.FGB.2011.010.013.

Li, C., et al. (2018). Hydrolysis of fruit and vegetable waste for efficient succinic acid production with engineered Yarrowia lipolytica. *Journal of Cleaner Production, 179*, 151–159. Available from https://doi.org/10.1016/J.JCLEPRO.2018.010.081.

Li, C., et al. (2019). Bio-refinery of waste streams for green and efficient succinic acid production by engineered Yarrowia lipolytica without pH control. *Chemical Engineering Journal, 371*, 804–812. Available from https://doi.org/10.1016/J.CEJ.2019.040.092.

Li, Q., et al. (2010). Efficient conversion of crop stalk wastes into succinic acid production by Actinobacillus succinogenes. *Bioresource Technology, 101*(9), 3292–3294. Available from https://doi.org/10.1016/J.BIORTECH.2009.120.064.

Liakou, V., et al. (2018). Valorisation of fruit and vegetable waste from open markets for the production of 2,3-butanediol. *Food and Bioproducts Processing, 108*, 27–36. Available from https://doi.org/10.1016/J.FBP.2017.100.004.

Liu, Y., et al. (2020). Itaconic acid fermentation using activated charcoal-treated corn stover hydrolysate and process evaluation based on Aspen plus model. *Biomass Conversion and Biorefinery, 10*(2), 463–470. Available from https://doi.org/10.1007/S13399-019-00423-3/FIGURES/4.

Lugani, Y., et al. (2020). Recent advances in bioethanol production from lignocelluloses: a comprehensive review with a focus on enzyme engineering and designer biocatalysts. *Biofuel Research Journal, 7*(4), 1267–1295. Available from https://doi.org/10.18331/BRJ2020.70.4.5.

Lynd, L. R., et al. (2005). Consolidated bioprocessing of cellulosic biomass: An update. *Current Opinion in Biotechnology, 16*(5), 577–583. Available from https://doi.org/10.1016/J.COPBIO.2005.080.009.

Maina, S., et al. (2022). Prospects on bio-based 2,3-butanediol and acetoin production: Recent progress and advances. *Biotechnology Advances, 54*, 107783. Available from https://doi.org/10.1016/J.BIOTECHADV.2021.107783.

Maiti, S., et al. (2016). Agro-industrial wastes as feedstock for sustainable bio-production of butanol by Clostridium beijerinckii. *Food and Bioproducts Processing, 98*, 217–226. Available from https://doi.org/10.1016/J.FBP.2016.010.002.

Mattey, Michael (2008). The production of organic acids. *Critical Reviews in Biotechnology, 12*(1-2), 87–132. Available from https://doi.org/10.3109/07388559209069189.

McMillan, J. D. (1994). *Conversion of hemicellulose hydrolyzates to ethanol*, 411–437. Available from https://doi.org/10.1021/BK-1994-0566.CH021.

Miranda, R. et al. (2007) 'Pyrolysis of textile wastes', 100th Annual Conference and Exhibition of the Air and Waste Management Association 2007, ACE 2007, 3, 1780–1794. doi:10.1016/J.JAAP.2007.03.008.

Moeller, L., et al. (2013). Citric acid production from sucrose by recombinant Yarrowia lipolytica using semicontinuous fermentation. *Engineering in Life Sciences*, 13(2), 163–171. Available from https://doi.org/10.1002/ELSC.201200046.

Moyer, Andrew J. (1953). Effect of alcohols on the mycological production of citric acid in surface and submerged culture. *Applied Microbiology*, 1(1), 1–7. Available from https://doi.org/10.1128/am.1.1.1-7.1953.

Narisetty, V., et al. (2022). Fermentative production of 2,3-butanediol using bread waste – A green approach for sustainable management of food waste. *Bioresource Technology*, 358, 127381. Available from https://doi.org/10.1016/J.BIORTECH.2022.127381.

Ni, Y., & Sun, Z. (2009). Recent progress on industrial fermentative production of acetone-butanol-ethanol by Clostridium acetobutylicum in China. *Applied Microbiology and Biotechnology*, 83(3), 415–423. Available from https://doi.org/10.1007/s00253-009-2003-y.

Oberoi, H. S., et al. (2007). Effects of different drying methods of cauliflower waste on drying time, colour retention and glucoamylase production by Aspergillus niger NCIM 1054. *International Journal of Food Science & technology*, 42(2), 228–234. Available from https://doi.org/10.1111/J.1365-2621.2006.01331.X.

Okabe, M., et al. (2009). 'Biotechnological production of itaconic acid and its biosynthesis in Aspergillus terreus. *Applied Microbiology and Biotechnology*, 84(4), 597–606. Available from https://doi.org/10.1007/S00253-009-2132-3/TABLES/5.

Okonkwo, C. C., Ujor, V., & Ezeji, T. C. (2017). Investigation of relationship between 2,3-butanediol toxicity and production during(Global Itaconic Acid Market | Size | Share | Growth| 2022-2027 | - The Cowboy Channel (No Date). Available At, 2022/2022) of Paenibacillus polymyxa. *New Biotechnology*, 34, 23–31. Available from https://doi.org/10.1016/J.NBT.2016.100.006.

Papagianni, Maria (2007). Advances in citric acid fermentation by Aspergillus niger: Biochemical aspects, membrane transport and modeling. *Biotechnology Advances*, 25(3), 244–263. Available from https://doi.org/10.1016/j.biotechadv.2007.010.002.

Perego, P., et al. (2000). 2,3-Butanediol production by Enterobacter aerogenes: selection of the optimal conditions and application to food industry residues. *Bioprocess Engineering*, 23(6), 613–620. Available from https://doi.org/10.1007/S004490000210.

Petrov, K., & Petrova, P. (2009). High production of 2,3-butanediol from glycerol by Klebsiella pneumoniae G31. *Applied Microbiology and Biotechnology*, 84(4), 659–665. Available from https://doi.org/10.1007/s00253-009-2004-x.

Prabhu, A. A., et al. (2020). Bioproduction of succinic acid from xylose by engineered Yarrowia lipolytica without pH control. *Biotechnology for Biofuels*, 13(1), 1–15. Available from https://doi.org/10.1186/S13068-020-01747-3/TABLES/1.

Prices:Petroleum Planning & Analysis Cell (2019). Available at: https://www.ppac.gov.in/content/149_1_pricespetroleum.aspx Accessed 12.07.22.

Qureshi, N. (2009). *Solvent production* (pp. 512–528). Elsevier BV. Available from 10.1016/b978-012373944-5.00160-7.

Qureshi, N., & Blaschek, H. P. (2000). Economics of butanol fermentation using hyper-butanol producing Clostridium beijerinckii BA101. *Food and Bioproducts Processing*, 78(3), 139–144. Available from https://doi.org/10.1205/096030800532888.

Reddy, C. S. K., & Singh, R. P. (2002). Enhanced production of itaconic acid from corn starch and market refuse fruits by genetically manipulated Aspergillus terreus SKR10. *Bioresource Technology*, 85(1), 69–71. Available from https://doi.org/10.1016/s0960-8524(02)00075-5.

Regestein, L., et al. (2018). From beech wood to itaconic acid: Case study on biorefinery process integration. *Biotechnology for Biofuels*, 11(1), 1–11. Available from https://doi.org/10.1186/S13068-018-1273-Y/FIGURES/6.

Rocket fuel made on Mars could propel astronauts back to Earth (no date). Available at: https://www.dpaonthenet.net/article/188369/ Accessed 09.07.22.

Roukas, T., & Kotzekidou, P. (2020). Pomegranate peel waste: a new substrate for citric acid production by Aspergillus niger in solid-state fermentation under non-aseptic conditions. *Environmental Science and Pollution Research*, 27(12), 13105–13113. Available from https://doi.org/10.1007/s11356-020-07928-9, https://link.springer.com/journal/11356.

Rymowicz, W., et al. (2010). Citric acid production from glycerol-containing waste of biodiesel industry by Yarrowia lipolytica in batch, repeated batch, and cell recycle regimes. *Applied Microbiology and Biotechnology*, 87(3), 971–979. Available from https://doi.org/10.1007/S00253-010-2561-Z/TABLES/5.

Saxena, R. K., et al. (2017). Production and applications of succinic acid. *Current Developments in Biotechnology and Bioengineering: Production, Isolation and Purification of Industrial Products*, 601–630. Available from https://doi.org/10.1016/B978-0-444-63662-1.00027-0.

Scheifinger, C. C., & Wolin, M. J. (1973). Propionate formation from cellulose and soluble sugars by combined cultures of Bacteroides succinogenes and Selenomonas ruminantium. *Applied Microbiology*, 26(5), 789–795. Available from https://doi.org/10.1128/aem.26.5.789-795.1973.

Show, P. L., et al. (2015). Overview of citric acid production from Aspergillus niger, 8(3), 271–283. Available from https://doi.org/10.1080/21553769.2015.1033653, https://doi.org/10.1080/21553769.2015.1033653.

Smyth, H. F., et al. (2007). Range-finding toxicity data: List VI, 23(2), 95–107. Available from https://doi.org/10.1080/00028896209343211, http://doi.org/10.1080/00028896209343211.

Song, H., & Lee, S. Y. (2006). Production of succinic acid by bacterial fermentation. *Enzyme and Microbial Technology*, 39(3), 352–361. Available from https://doi.org/10.1016/j.enzmictec.2005.110.043.

Stoye, D. (2000). *Ullmann's encyclopedia of industrial chemistry, Ullmann's encyclopedia of industrial chemistry*. Wiley-VCH. Available from 10.1002/14356007.a24_437.

Stylianou, E., et al. (2020). Evaluation of organic fractions of municipal solid waste as renewable feedstock for succinic acid production. *Biotechnology for Biofuels*, 13(1), 1–16. Available from https://doi.org/10.1186/S13068-020-01708-W/FIGURES/9.

Succinic Acid Market(Global Itaconic Acid Market | Size | Share | Growth| 2022-2027 | - The Cowboy Channel (No Date). Available At, 2022/2022)&(Global Itaconic Acid Market | Size | Share | Growth| 2022-2027 | - The Cowboy Channel (No Date). Available At, 2022/2022)Report, 2022-2030 (no date). Available at: https://www.grandviewresearch.com/industry-analysis/succinic-acid-market Accessed 10.07.22.

Sunwoo, I. Y., et al. (2018). Acetone–butanol–ethanol production from waste seaweed collected from Gwangalli Beach, Busan, Korea, based on pH-controlled and sequential fermentation using two strains. *Applied Biochemistry and Biotechnology*, 185(4), 1075–1087. Available from https://doi.org/10.1007/S12010-018-2711-9/TABLES/3.

Syu, M.-J. (2012) 'Mini-review. Biological production of 2,3-butanediol'. doi:10.1007/s002530000486.

Taherzadeh, M. J., & Karimi, K. (2008). Pretreatment of lignocellulosic wastes to improve ethanol and biogas production: A review. *International Journal of Molecular Sciences*, 9(9), 1621–1651. Available from https://doi.org/10.3390/ijms9091621Sweden, http://www.mdpi.org/ijms/papers/i9091621.pdf.

The 100 Most Important Chemical Compounds: A Reference Guide - Richard L. Myers - Google Books (no date). Available at: https://books.google.co.in/books?id = 0AnJU-hralEC&pg = PA122&redir_esc = y#v = onepage&q&f = false Accessed 10.07.22.

Tinôco, D., et al. (2021). Bioprocess development for 2,3-butanediol production by Paenibacillus strains. *ChemBioEng Reviews*, 8(1), 44–62. Available from https://doi.org/10.1002/CBEN.202000022.

Tong, Z., et al. (2019). Systems metabolic engineering for citric acid production by Aspergillus niger in the post-genomic era. *Microbial Cell Factories*, 18(1), 1–15. Available from https://doi.org/10.1186/S12934-019-1064-6/FIGURES/3.

Vandenberghe, L. P. S., et al. (1999). Microbial production of citric acid. *Brazilian Archives of Biology and Technology*, 42(3), 263–276. Available from https://doi.org/10.1590/S1516-89131999000300001.

Van Der Werf, M. J., et al. (1997). Environmental and physiological factors affecting the succinate product ratio during carbohydrate fermentation by Actinobacillus sp. 130Z. *Archives of Microbiology*, 167(6), 332–342. Available from https://doi.org/10.1007/S002030050452.

Venkateswar Rao, L., et al. (2016). Bioconversion of lignocellulosic biomass to xylitol: An overview. *Bioresource Technology*, 213, 299–310. Available from https://doi.org/10.1016/J.BIORTECH.2016.040.092.

WerpyT., & PetersenG. (2004). *Top value added chemicals from biomass: Volume I — Results of screening for potential candidates from sugars and synthesis gas*. Us Nrel, p. Medium: ED., doi: 10.2172/15008859.

Willke, T., & Vorlop, K. D. (2001). Biotechnological production of itaconic acid. *Applied Microbiology and Biotechnology*, 56(3-4), 289–295. Available from https://doi.org/10.1007/s002530100685.

Wong, C. L., et al. (2012). Producing 2,3-butanediol from agricultural waste using an indigenous Klebsiella sp. Zmd30 strain. *Biochemical Engineering Journal*, 69, 32–40. Available from https://doi.org/10.1016/J.BEJ.2012.080.006.

Xu, J., et al. (2014). Production of citric acid using its extraction wastewater treated by anaerobic digestion and ion exchange in an integrated citric acid-methane fermentation process. *Bioprocess and Biosystems Engineering*, 37(8), 1659–1668. Available from https://doi.org/10.1007/S00449-014-1138-0.

Yang, W., et al. (2012). Solubility of itaconic acid in different organic solvents: Experimental measurement and thermodynamic modeling. *Fluid Phase Equilibria*, 314, 180–184. Available from https://doi.org/10.1016/J.FLUID.2011.090.027.

Your Guide to Ethanol Extraction in Cannabis - Cannabis Business Times (no date). Available at: https://www.cannabisbusinesstimes.com/article/your-guide-to-ethanol-extraction/ Accessed 10.07.22.

Zambanini, T., et al. (2017). Efficient itaconic acid production from glycerol with Ustilago vetiveriae TZ1. *Biotechnology for Biofuels*, 10(1), 1–15. Available from https://doi.org/10.1186/S13068-017-0809-X/TABLES/3.

ZMP-Marktbilanz / Zentrale Markt- und Preisberichtstelle GmbH / Forst und Holz - EconBiz (no date). Available at: https://www.econbiz.de/Record/zmp-marktbilanz-zentrale-markt-und-preisberichtstelle-gmbh-forst-und-holz/10004550378 Accessed 10.07.22.

Zeikus, J. G., Jain, M. K., & Elankovan, P. (1999). Biotechnology of succinic acid production and markets for derived industrial products. *Applied Microbiology and Biotechnology*, 51(5), 545–552. Available from https://doi.org/10.1007/s002530051431.

Zhang, A. Y. Z., et al. (2013). Valorisation of bakery waste for succinic acid production. *Green Chemistry*, 15(3), 690–695. Available from https://doi.org/10.1039/C2GC36518A.

Zhang, N., et al. (2019). Citric acid production from acorn starch by tannin tolerance mutant Aspergillus niger AA120. *Applied Biochemistry and Biotechnology*, 188(1), 1–11. Available from https://doi.org/10.1007/s12010-018-2902-4.

Zhao, M., et al. (2018). Itaconic acid production in microorganisms. *Biotechnology Letters*, 455–464. Available from https://doi.org/10.1007/s10529-017-2500-5, Springer Netherlands.

Zhuang, B., et al. (2021). Like dissolves like: A first-principles theory for predicting liquid miscibility and mixture dielectric constant. *Science Advances*, 7(7), 7275–7287. Available from https://doi.org/10.1126/SCIADV.ABE7275/SUPPL_FILE/ABE7275_SM.PDF.

1-Butanol | C4H9OH - PubChem (no date). Available at: https://pubchem.ncbi.nlm.nih.gov/compound/263 Accessed 07.07.22.

2,3-Butanediol | C4H10O2 - PubChem (no date). Available at: https://pubchem.ncbi.nlm.nih.gov/compound/262 Accessed 07..07.22.

CHAPTER

16

Pigments and paints from wastes

Kumari Guddi[1], G. Vijay Chithra[1], R. Bhavani[1], Sambit Naik[2] and Angana Sarkar[1]

[1]Department of Biotechnology and Medical Engineering, National Institute of Technology Rourkela, Rourkela, Odisha, India [2]Fakir Mohan University, Balasore, Odisha, India

16.1 Introduction

The term "agroindustrial residues" refers to a broad range of agricultural and food-related wastes. Agricultural wastes or agrowastes are derived from various types of agricultural activities. These wastes are fruit and vegetable peels, whey, fruit seeds, bagasse, molasses, etc. Canada produces the largest amount of trash about 1.33 billion waste per year (Dey et al., 2021), which contains all agricultural as well as industrial wastes. As per the Food and Agricultural Organization (FAO), 250 million tons of agricultural waste are produced annually around the world during the processing of various plant products, primarily grains, vegetables, and fruits (Heredia-Guerrero et al., 2017). The agrowastes have an antimicrobial, antioxidant, and dyeing nature. These wastes consist of the agricultural remnant that helps to prevent soil infertility. Some industries such as dairy, food, and cereal are responsible for the production of agrowastes. At the time of food processing, the industries release wastes that are hazardous to the environment. In the dairy industry, whey is the residual water that is extracted from milk in the process of cheese manufacturing. In the fruit industry, waste is produced from unpleasant fruits. In the vegetable industry, the remnant of vegetables such as peels and seeds are considered waste. In the cereal industry, a by-product, that is, corn steep liquor, is obtained from the industry of corn wet milling. The main examples are straw from rice manufacturing and molasses from sugarcane manufacturing. The agrowastes contain a huge amount of biomaterial that can be used in different industries. Agroindustrial residues are used to grow bacteria, which helps in the extraction of microbial pigments such as violacein, riboflavin, melanin, carotenoid, prodigiosin, phycocyanin, and azaphilones phycoerythrobilin (Ramesh et al., 2022). For example, anthocyanin pigment is extracted from agrowastes and used as an indicator of time temperature to know the freshness of fish. Anthocyanin is used as a color indicator, which is responsible for color changing and is also extracted from basic foods such as cabbage and potato (Ramadhan & Handayani, 2021). Different methods are used to extract the pigments from agrowastes such as solvent extraction, microwave-assisted extraction, enzyme-assisted extraction, ultrasound-assisted extraction, and supercritical fluid extraction. From the leaves of *Tectona grandis*, the red pigment is extracted using acetone as solvent. Similarly, from the wood of *Areca catechu*, with the help of methanol and column chromatography, a brownish color was extracted (Nathan et al., 2020). The pigment laccase was extracted from the fungus *Aspergillus oryzae*, which is cultivated on agrowaste (Majumdar & Bhowal, 2022). The utilization of agricultural waste might be a profitable method of decreasing the production cost because synthetic culture medium is often expensive (Panesar et al., 2015). If we can utilize this waste, then we will get a lot of raw materials, which can be used in making different commercial products such as paints and pigments. It can also help in preventing environmental pollution, which is a big problem in the world. This waste can be used as a good source of nutrition for the cultivation of bacteria, which may be further used for the extraction of pigments. These pigments are used to make food color, dye, and fabric coloring material. The pigments are natural coloring agents and are much safer than synthetic dyes. Synthetic dyes are made from different chemicals, which are harmful to human beings as well as nature. If we extract these pigments from waste, then we can obtain pigments and paints besides reducing pollution. These pigments can be used in the textile industries as well as in the food industries. By using this trash, difficulties with the disposal and environmental contamination (including harm to aquatic life, tainted surface and ground waters, deteriorated soil quality, and foul-smelling natural waterways) are eliminated (Panesar et al., 2015). In this book chapter, various sources of agrowastes, methods for extraction of pigments from agrowaste, applications of these pigments, and production of pigments from microbes using agrowaste as a cost-effective substrate have been discussed in detail.

16.2 Different sources of agrowastes

There are several sources of agrowaste such as markets, agricultural fields, homes, and industries (Fig. 16.1). Flowers are a common source of agrowaste because they are used in marriage grounds, temples, hotels, and various other ceremonies. The main source of this garbage is the temple, masjid, gurdwara, and flower market. The main component of this garbage is the flowers such as roses, marigolds, and lotus (Elango & Govindasamy, 2018). When compared with the degradation of kitchen waste, floral waste degradation is a very slow process (Jadhav et al., 2013/2013). Therefore proper and environmentally friendly procedures for treating floral waste are needed. Studies have shown that it is possible to manage and use flower waste. The sources of fruit and vegetable waste are fruit or vegetable markets, the agricultural field, and homes. This waste contains fruit and vegetable peels, seeds as well as rejected fruits and vegetables. These peels are used in various industries. Orange peels are used in pharmaceutical and cosmetic industries and are also used to make biochar (Selvarajoo et al., 2022). The sources of cereal waste are the agricultural field and the cereal industry where the cereals are processed. A by-product of the cereal industry is corn-steep liquor. Researchers are interested in using corn-steep liquor because of its rich nutritional content to produce microbial pigments. It is shown that corn-steep liquor may replace various salts and yeast extracts for the synthesis of a red pigment by *Monascusruber* in solid-state fermentation (Hamano & Kilikian, 2006).

16.3 Different forms of agrowaste

Agrowaste is the waste that is generated from agricultural activity or agroindustry. Agrowastes are generally biodegradable. However, in some cases, it bears a capacity for severe pollution. These types of waste are full of bionutrients, and if they can be recyclable, then many types of components can be extracted. Different forms of agrowastes are flowers, fruits, vegetables, and cereals.

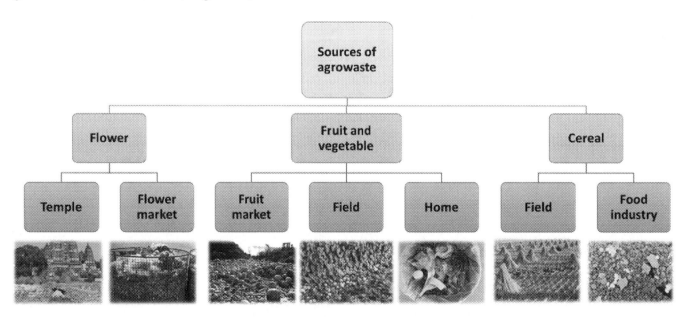

FIGURE 16.1 Different sources of agrowaste in the environment.

16.3.1 Flower waste

The flowers, which are regularly offered by people in temples, are never used and become waste. The direct disposal of this flower waste into lakes, rivers, and oceans affects not only the water quality but also the aquatic life that exists there. Flower garbage is the main source of pigment extraction as well as perfume making. Flower waste is used in composting and vermicomposting by 8.33% of farmers. This type of waste can be used in biofuel production, briquette, or biogas production if it doesn't contain any antibiotic or inhibitory compound (Frankowski et al., 2020). The pigments extracted from flowers are used in dyeing fabrics in the textile industry. In the case of synthetic dyeing, regular single step is followed, but in the natural dyeing process, two steps are involved, that is, identification and application (Amutha & Annapoorani, 2019). Different types of flower garbage such as petals, stems, and leaves are produced from different sources. Flowers and leaves have pigments that are used in many industries to extract these pigments and apply them as fabrics in industries and also used to make incense sticks, rose water, etc. (Kumar et al., 2020). About 2 million temples are there in India from which daily flower waste is generated. The famous cities with temples from which a large amount of flower garbage is generated are Bhubaneswar, puri, Varanasi, Haridwar, Siridi, etc. (Kumar et al., 2020).

16.3.2 Fruit and vegetable waste

Fruit and vegetable garbage is mainly used for vermicompost, which is full of nutrients and extraction of pigments such as carotenoid. About 50% of total waste is the remnant of fruit and vegetable such as fruit peel, cereals, and seeds. Pigments from these wastes can be used as food colorants due to low toxicity (Gupta et al., 2019). The fruit and vegetable garbage consists of peels, seeds, and fruits, which are rich in nutrients as well as pigments. Carotenoids can be extracted from the peels of tomato, anthocyanin can be extracted from grapes, and betacyanin and betaxanthin can be extracted from the vegetable Malabar spinach (Gupta et al., 2019).

16.3.3 Cereal waste

Cereal waste is the waste that is generated from the processing of cereals in the food industry. This can be the straw and bran in the rice processing industry and the molasses in the sugarcane industry. It can be disposed of through cattle feed or composting. Cereals cultivated in the field by farmers are full of nutrients and carbohydrates. The garbage of cereals consists of the remnant of the cereals used in industry, the extracts such as straw in the field. In the cereal industry with the help of enzymatic action, the fermentation process is done in which the endosperm of grains is distinguished (Verni et al., 2019). The main source of cereal garbage is the bread or food processing industry. The wastes from cereal are nontoxic and can be in solid form or liquid form. The solid form comes from the baking industry, which is the residue of corn extracts, and the liquid waste is from the rice industry, which is formed from the boiling of rice, rice mill wastewater, etc. (Hassan et al., 2021).

16.4 Types of pigments produced from agrowastes

Agrowastes are majorly grouped into floral wastes, fruit wastes, vegetable wastes, and cereals. Most of these wastes are rarely recycled, which creates environment-related issues Thus they can be processed further to get benefits out of it. All these components contribute to the source of pigment extraction using various physical, chemical, and biological methodologies. The various kinds of pigments found in these wastes are namely anthocyanins, carotenoids, betalains, lutein, zeaxanthin, etc. (Carrillo et al., 2022). Anthocyanins are a group of secondary metabolites pigments that are rich in fruits such as grape skin and blueberry (Świeca et al., 2014) and in certain flowers such as *Hibiscus sabdariffa* and *Bougainvillea spectabilis* (Sankaralingam et al., 2022). These pigments are used as food colorants and also in the textile industries for dyeing clothes and fabrics. Carotenoids are the organic fat-soluble pigments that impart yellow, orange, and reddish colors to fruits and vegetables. Maize, carrots, pumpkin, Broomcorn millet (Li et al., 2021), and African marigolds (Waghmode et al., 2018) are enhanced with these carotenoid pigments. These pigments are used as the active ingredient in food coloration, antioxidant and antimutagenic activity against prostate cancer (Goula et al., 2017). Carthamin and carthamidin are the yellow-red pigments in the safflower that are used as a staining agent and additive in beverages (Waghmode et al., 2018; Grewal et al., 2022). Beetroot and dragon fruit peel are primarily enriched with betalains, which include betacyanins (violet) and betaxanthins (yellow). These are composed of nitrogen-based water-soluble

TABLE 16.1 Different pigments that can be produced by agro-wastes.

S. No.	Type of source	Source	Pigment	Pigment color	Applications	References
1.	Fruits and vegetables	Purple yam peel (*Dioscorea alata* L.)	Cyanidin-3-glucoside	Red to blue	Used in bakery products as a color enhancer	Sukmawati and Maryanti (2022)
		Dragon fruit peel (*Selenicereus undatus*)	Betacyanin	Red	Natural food colorant.	Sukmawati and Maryanti (2022)
		Waste Pomegranate peel	Carotenoids Lutein and beta carotene	Red	Antioxidant, antimutagenic, and antiatherosclerotic against osteoarthritis and prostate cancer	Goula et al. (2017)
		Grape skin	Anthocyanin	Red	Food coloring agent	Świeca et al. (2014)
		Beetroot (*Beta vulgaris* L.)	Betalains	Red-purple	Formulation of functional food and beverages	Nirmal et al. (2021)
2.	Flowers	Safflower	Carthamin carthamidin	Yellow Red	Stain, additives in beverages and cosmetics, printing, and food colorants for ice creams	Waghmode et al. (2018)
		Hibiscus sabdariffa	Anthocyanin	Red	Dyeing clothes and used in the cosmetics industries	Sankaralingam et al. (2022)
		African marigold (*Tagetes erecta* L.)	Carotenoids and lutein	Yellow to orange-red	The active ingredient in the food industry and textile coloration	Waghmode et al. (2018)
3.	Cereals	Triticum durum wheat	Lutein and zeaxanthin	Yellow	Food colorants in pasta and semolina	Lachman et al. (2017)
		Broomcorn millet	Carotenoid	Yellow	Color enhancers in food and beverages	Li et al. (2021)

betalamic acid. These pigments are associated with various biological activities such as antiinflammatory, DNA damage repair mechanism, and free radicals scavenging potential (Carrillo et al., 2022). The various types of pigments and their applications are represented in Table 16.1.

16.5 Different methods of pigment production

16.5.1 Physical methods

16.5.1.1 Supercritical fluid extraction

Fluids (solvents) are used in the supercritical fluid extraction (SFE) process at temperatures and pressures that are above or close to the respective critical values to extract the pigments of interest. One of the most widely utilized solvents is carbon dioxide (CO_2) as it has a moderate critical temperature of 31°C and is simple to separate from the extract by lowering pressure. The manufacture of pharmaceuticals and the food and biofuel industries both heavily rely on robust technology (Salamatin, 2020).

16.5.1.2 CO_2 depressurization

With a low critical pressure of 7.38 MPa and a temperature of 31.1°C, the extraction process is conducted in a closed system with no oxygen. Conceivably, by altering cell membranes and eliminating vital cell components and membranes, SC-CO2 extraction may disturb intracellular electrolyte balance. The mixing of ethanol and water, which are often used as cosolvents, could improve the solubility of polar molecules such as anthocyanins and phenolic compounds as well as the solvating power of carbon dioxide (Idham et al., 2022).

16.5.1.3 Moderate electric field (MEF)

The voltage applied during MEF extraction damages the cell membrane and alters the permeability of the cell membrane, resulting in cell rupture. By electroporating tissues, MEF increases the extraction yield by allowing cells to be permeable to foreign substances. The reason is that the electric field strength or voltage gradient plays a vital part in the MEF extraction process (Wani et al., 2021).

16.5.1.4 Ultrasonication

Two theories underlie ultrasonication: the acoustic cavitation action and the sponge or piston effect (compression-decompression) (compression-collapse). Ultrasonic waves in the medium are compressed and expanded as a result of the piston or sponge effect. On the other hand, cavitation is an effect that results in high temperatures and pressures (50–100 MPa) (Kutlu et al., 2022).

16.6 Chemical methods

16.6.1 Soxhlet extraction

A plant sample in a thimble is repeatedly percolated with the condensed vapor of solvent during the extraction process until the extraction is finished (or the solvent is no longer able to solubilize the sample's interesting chemicals), which is indicated when the solvent has turned colorless. Soxhlet extraction is straightforward and simple to use, but it uses a lot of solvents, takes a long time, and damages pigment because of heat (Ngamwonglumlert et al., 2017).

16.6.2 Solvent extraction

The traditional method for extracting the pigments from microalgae and cyanobacteria involves solvent extraction that may or may not be followed by thermal treatment (heat or cold). In order to allow the solute to interact with the solvent until an equilibrium is attained, the two phases (solvent and biomass) must be mixed prior to the solvent extraction process. Chemical affinities are what cause the transmission from one phase to the next. The term "extract" refers to the mixture of solvent and solute that is produced at the end of the procedure. For the extraction of carotenoids, however, solvent extraction usually requires substantial amounts of organic solvents (Pagels et al., 2021).

16.7 Biological methods

Biological methods of extraction are the extraction or synthesis of pigments by enzyme-catalyzed reactions or microorganisms in large-capacity fermenters. A variety of microorganisms can be utilized to produce a wide range of pigments when provided with optimum fermentation conditions.

16.8 Production of pigments from agrowastes without using microorganisms

Several agrowaste sources along with their pigment extraction methodologies are mentioned in Table 16.2. Beta carotene pigment is extracted from *Cucurbita pepo Duch* and carrot peel by maceration and solvent extraction methods with the help of ethanol and methanol, respectively (Hemalatha & Kailasam, 2022/2022; Kausar et al., 2022). High levels of beta carotene were extracted by the method of maceration by using 95% ethanol and later separation of the pigment into a separate layer with the help of petroleum ether and additional ethanol. *Bougainvillea spectabilis* wastes are highly rich in anthocyanin pigment, extracted by the method of maceration. The pigment extracted was used for the indication of formalin food preservatives and boron food additives (Rusita & Hastuti, 2022). Organic solvent extraction was mostly preferred for the anthocyanin pigment isolation from cranberry pomace and purple barley waste (Fahmideh et al., 2022/2022). Pressurized ethanol-based (100%) extraction has led to the maximum yield of anthocyanins from the cranberry pomace (Wani et al., 2021). Ultrasound-assisted extraction is one of the best suitable methods for the extraction of anthocyanin from purple corn with an ultrasound power of 400 W (Chen et al., 2018). Red-violet pigment from the *Hylocereus polyrhizus* was extracted by an aqueous two-phase extraction method. Here the upper phase consists of ethanol with betacyanin pigment content and the other lower layer consists of undesired substances (Polturak & Aharoni, 2018). Amaranthus is rich in antioxidant properties and free-radicals scavenging activities due to the presence of betalains (Kantifedaki et al., 2018).

TABLE 16.2 Pigments obtained directly from agrowaste.

S. no.	Raw material	Pigment name	Color	Method	References
1.	*Cucurbita pepo* Duch.	Beta carotene	Yellow	Maceration (95% ethanol)	Hemalatha and Kailasam (2022/2022)
2.	*Bougainvillea spectabilis*	Anthocyanin	Violet	Maceration method (24 h)	Rusita and Hastuti (2022)
3.	*Hylocereus polyrhizus*	Betacyanins	Red-violet	Aqueous two-phase extraction (ammonium sulfate)	Polturak and Aharoni (2018)
4.	Yellow pitahaya (*Selenicereus megalanthus*)	Betalains	Yellow reddish color	Solvent extraction	Otálora et al. (2023)
5.	Purple corn (*Zea mays* L.)	Anthocyanins	Purple	Ultrasound-assisted method	Chen et al. (2018)
6.	Cranberry pomace	Anthocyanins	Red	Solvent extraction (pressurized ethanol)	Wani et al. (2021)
7.	*Opuntia* fruit peel	Betalains	Red-purple	Ultrasound-assisted extraction	Melgar et al. (2019)
8.	Purple barley (*Hordeum vulgari*)	Anthocyanins	Purple	Organic solvent extraction (methanol)	Ed Nignpense et al. (2022)
9.	Carrot peel	Beta carotene	Orange	Solvent extraction	Kausar et al. (2022)
10.	Amaranth	Beta cyanin	Red	Solvent extraction (methanol)	Kantifedaki et al. (2018)

16.9 Production of pigment from microorganisms using agrowastes as substrate

Prodigiosin is an alkaloid available from microbial sources. It exhibits various medicinal properties such as antifungal, antibacterial, and algicidal activities. This red pigment can be synthesized using solid-state fermentation (SSF) employing several bacterial strains such *Serratia marcescens* UCP 1549 and BWL 1001 with sugar cane bagasse and soybean oil as substrates at an ideal temperature of 28°C (Dos Santos et al., 2021; Liu et al., 2021). The class of phytonutrients known as carotenoids, sometimes known as "plant chemicals," is present in the cells of a diverse range of plants, algae, and bacteria. They support plants' ability to utilize light energy for photosynthesis. Additionally, they perform a crucial antioxidant role by neutralizing free radicals. These compounds absorb the blue-green region of the spectrum in visible light and hence appear in yellow to red color. They can be produced by common environmental yeast *Rhodotorula* species. Using onion peel and high-test molasses as substrates, carotenoids were produced at the yield of 710.33 mg/g and 2.45 mg/L with the help of the strains *Rhodotorula mucilaginosa* and *Rhodotorula glutinis* 32, respectively. The former was observed to produce three forms of carotenoids, namely torularhodin, β-carotene, and torulene (Sharma & Ghoshal, 2020; Galal & Ahmed, 2020). β-carotene being one of the widely recognized carotenoids in various elements of nature, having a characteristic orange color, can be produced by the cocci-shaped bacterium *Planococcus sp.* TRC1 when fermentation is carried out for 48 hours with an initial moisture content of 80% by providing paper mill sludge as a substrate (Majumdar et al., 2020). The vitamin A precursor can also be produced by *Blakeslea trispora* MTCC 88, a mold belonging to Zygomycota phylum of fungi by utilizing fruits and vegetable peels as substrates (Kaur et al., 2019). Under ideal fermentation circumstances, the fermentation of oil palm fronds and Bengal gramme husk with the fungi *Monascus purpureus* FTC5356 and *Talaromyces purpureogenus* CFRM02 respectively results in the production of red pigments that can be used as an alternative to toxic synthetic sources. Citrinin is a mycotoxin formed from polyketides that crystallize producing lemon-yellow color. *Penicillium citrinum* was the first citrinin producer to be reported. *Monascuspurpureus* EBY3 is described to produce this toxin during 12-day solid-state fermentation in dark at a pH of 27°C employing rice as the substrate (Marič et al., 2019). Violacein is a bis-indole pigment that occurs naturally having an eye-catching purple color. As stated by *Chromobacterium vaccinii* DSM 25150 consumes pulverized wheat bran with particle sizes ranging from 0.8 to 2 nm as a substrate in minimal media M9 during solid-state fermentation in order to produce violacein (Cassarini et al., 2021). Various microorganisms that have been studied for the production of pigment using agrowastes have been depicted in Table 16.3.

TABLE 16.3 Microorganisms used for pigment production using agrowastes as substrates.

Raw materials	Pigment	Color of the pigment	Microbes used	Optimum condition	Pigment yield	References
Wheat bran (in ground form of particle sizes 0.8–2 mm)	Violacein	Violet	*Chromobacterium vaccinii* DSM 25150	Temperature: 30°C Agitation: 100 rpm pH: 7.2 Media composition: Minimal salt medium M9 (NH4Cl 1 g/L, Na2HPO4.2H2O 6 g/L, KH2PO4 3 g/L, NaCl 0.5 g/L)	1.47 mg/L	Cassarini et al. (2021)
Sugarcane bagasse	Prodigiosin	Red	*Serratia marcescens* UCP 1549	Temperature: 28°C Fermentation time: 120 h Media composition: KH2PO4 3 g/L, K2HPO4 7 g/L, MgSO4.7H2O 0.2 g/L, (NH42SO4 1 g/L and 5% WSO	1.80 g/kg	Dos Santos et al. (2021)
Soybean oil	Prodigiosin	Red	*S. marcescens* BWL1001	Temperature: 28°C PH: 5.0 Medium composition: 100 g/L soybean oil, 10 g/L peptone, 10 g/L NaCl, 0.5 g/L MgSO4·7H2O, 0.3 g/L K2HPO4	27.65 g/L	Liu et al. (2021)
Rice	Citrinin	Yellow	*Monascus purpureus* EBY3	pH: 6.7 Fermentation period: 20 days Temperature: 27°C in the dark	1.1 mg/g	Marič et al. (2019)
Onion peel	Carotenoid (torularhodin, β-carotene, and torulene)	Yellow to red	*Rhodotorula mucilaginosa*	pH 6.1 Temperature: 25.8°C Agitation: 119.6 rpm	710.33 mg/g	Sharma and Ghoshal (2020)
Paper mill sludge	β-carotene	Orange	*Planococcus sp.* TRC1	Fermentation time: 48hrs Temperature: 30°C Ph: 7.0 10% inoculum Initial moisture content: 80%	31.05 mg/g	Majumdar et al. (2020)
High test molasses	Carotenoids	Yellow to red	*Rhodotorula glutinis* 32	1 ml of standard inoculum Agitation: 150 rpm Incubation period: 4 days Temperature:30°C.	2.45 mg/L	Galal and Ahmed (2020)
Oil palm frond	Red pigment	Red	*Monascus purpureus* FTC5356	Initial Moisture Content: 50% (w/w) pH: 6.0 Media composition: 2% (w/w) peptone, 100% (w/w) petiole Inoculum size: 108 spores/ml	3.68 AU/g dry matter	Said and Hamid (2019)
Bengal gram husk	Red pigment	Red	*Talaromyces purpureogenus* CFRM02	Temperature: 30°C (illuminated 12 h day and night condition) Incubation period: 10 days Agitation: 110 rpm Inoculum size: ~1.93 × 10⁶ spore/mL	0.565 AU/mL	Pandit et al. (2019)
Fruits and vegetable waste (orange, carrot, and papaya peels)	β-carotene	Orange	*Blakeslea trispora* MTCC 88	Temperature: 28°C pH: 6.2–6.5 Fermentation period: 96 h	0.127 mg/mL	Kaur et al. (2019)

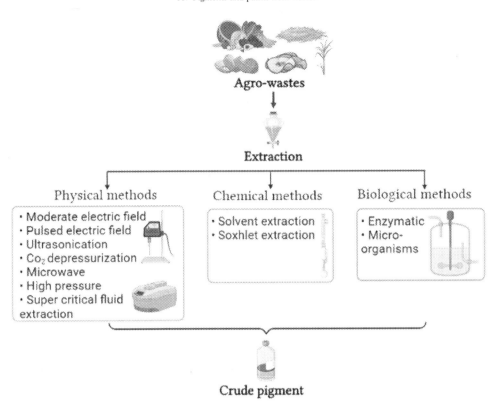

FIGURE 16.2 Process of pigment production from agrowastes.

16.10 Process of pigment production from agrowaste

Agrowastes can be utilized to extract pigments through physical, chemical, and biological methods (Fig. 16.2). Bioreactors are used to carry out fermentation reactions employing a variety of microorganisms, such as bacteria and yeast, for the large-scale manufacture of colorants and pigments. The resulting crude pigment is then subjected to additional downstream processing to produce a purified pigment that can then be packaged and marketed commercially (Wani et al., 2021).

16.11 Commercially available pigments from agrowaste and drawbacks of using agrowastes for pigments production

There is a report on the commercial use of *Tagetes erecta* waste flowers for dyeing cotton, wool, and silk (Vankar et al., 2009). According to a report, xanthophylls from Mahua flower extract are used in the food industry to make jams, biscuits, and jellies because of their nutritional value, which includes vitamins, enzymes, sugars, and crude pigments (Bhattacharya et al., 2012). The use of floral wastes in the creation of herbal incense sticks has been discovered. Incense sticks are made from flowers such as marigolds, and rosewater is made from roses.

There are several drawbacks to using agrowastes for pigment production. It requires more capital investment and regulatory approval for the extraction of value-added pigments from agrowastes. At the commercial scale, it is very important to consider the economic and marketing difficulties that can deprive the process of pigment extraction from agrowastes using microbes. The agrowastes are mostly composed of lignocellulosic compounds that require pretreatment of the raw materials before being processed with the microorganisms and thus make the initial process more costly. Some of the natural pigments such as betalains and anthocyanins (Constantin & Istrati, 2022) do not have the tendency to retain their color at high temperatures and pH, thus they are more prone to degradation. Large-scale production of bacterial pigments is not cost-effective, and the pigment's color varies from batch-to-batch production (Usman et al., 2017). The source of marine microorganisms is still being unexplored for the production of pigments from agrowastes, thus proper research could be done to get better pigments. Some of the agrowastes (Sanna & Fadda, 2022) and poultry wastes are also not been completely studied for pigment extraction (Ramesh et al., 2022).

16.12 Challenges and future aspects

Massive amounts of residue are produced during the processing of crops, postharvest operations, and by-products of industrial processes. If these residues are not treated, they act as environmental pollutants. The need for environmental preservation, sustainable agricultural productivity, and global food security has caused a rapid shift in how the world views agrowaste. Producing microbial pigments from agrowaste products may not only lower process costs but also aid in environmental management. The stability of natural pigments, which are occasionally problematic for the finished product, is a key strategy. To find new pigments with high stability for use in various industries, more research into various production processes is required. Finally, more agrowaste residue used in industries is required to implement a circular economy, which is an environmentally friendly strategy, as well as to reduce production costs.

References

Amutha, K., & Annapoorani, S. G. (2019). Natural dye extraction from agro-waste and its application on textiles. *Asian Dyers*, 16(1), 35–39.

Bhattacharya, Amrik, Saini, Vandana, & Gupta, Anshu (2012). Novel application of Mahua (Madhuca sp.) flowers for augmented protease production from Aeromonas sp. S1. *Natural Product Communications*, 7(10). Available from https://doi.org/10.1177/1934578x1200701028.

Carrillo, C., Nieto, G., Martinez-Zamora, L., Ros, G., Kamiloglu, S., Munekata, P. E., Pateiro, M., Lorenzo, J. M., Fernandez-Lopez, J., Viuda-Martos, M., & Perez-Alvarez, J. A. (2022). Novel approaches for the recovery of natural pigments with potential health effects. *Journal of Agricultural and Food Chemistry*, 70(23), 6864–6883.

Cassarini, M., Besaury, L., & Rémond, C. (2021). Valorisation of wheat bran to produce natural pigments using selected microorganisms. *Journal of Biotechnology*, 339, 81–92. Available from https://doi.org/10.1016/j.jbiotec.2021.08.003, http://www.elsevier.com/locate/jbiotec.

Chen, L., Yang, M., Mou, H., & Kong, Q. (2018). Ultrasound-assisted extraction and characterization of anthocyanins from purple corn bran. *Journal of Food Processing and Preservation*, 42(1), e13377.

Constantin, O. E., & Istrati, D. I. (2022). Extraction, quantification and characterization techniques for anthocyanin compounds in various food matrices—a review. *Horticulturae*, 8(11). Available from https://doi.org/10.3390/horticulturae8111084, http://www.mdpi.com/journal/horticulturae.

Dey, Tamanna, Bhattacharjee, Tarashree, Nag, Piyali, Ritika., Ghati, Amit, & Kuila, Arindam (2021). Valorization of agro-waste into value added products for sustainable development. *Bioresource Technology Reports*, 16. Available from https://doi.org/10.1016/j.biteb.2021.100834.

Dos Santos, R. A., Rodríguez, D. M., da Silva, L. A. R., de Almeida, S. M., de Campos-Takaki, G. M., & de Lima, M. A. B. (2021). Enhanced production of prodigiosin by Serratia marcescens UCP 1549 using agrosubstrates in solid-state fermentation. *Archives of Microbiology*, 203(7), 4091–4100.

Ed Nignpense, B., Latif, S., Francis, N., Blanchard, C., & Santhakumar, A. B. (2022). Bioaccessibility and antioxidant activity of polyphenols from pigmented barley and wheat. *Foods*, 11(22), 3697.

Elango, G., & Govindasamy, R. (2018). Analysis and utilization of temple waste flowers in Coimbatore District. *Environmental Science and Pollution Research*, 25, 10688–10700.

Fahmideh, L., Mazarie, A., Madadi, S., & Pahlevan. (2022). Comparing the antioxidant enzymes, osmotic regulators and photosynthetic pigments activities of two barley cultivars in Sistan region under salinity-stress conditions. *Environmental Stresses in Crop Sciences*, 485–499.

Frankowski, Jakub, Zaborowicz, Maciej, Dach, Jacek, Czekała, Wojciech, & Przybył, Jacek (2020). Biological waste management in the case of a pandemic emergency and other natural disasters. Determination of bioenergy production from floricultural waste and modeling of methane production using deep neural modeling methods. *Energies*, 13(11). Available from https://doi.org/10.3390/en13113014.

Galal, Gehan F., & Ahmed, Rania F. (2020). Using of some agro-industrial wastes for improving carotenoids production from yeast Rhodotorula glutinis 32 and bacteria Erwinia uredovora DSMZ 30080. *Microbiology Research Journal International*, 15–25. Available from https://doi.org/10.9734/mrji/2020/v30i130186.

Goula, A. M., Ververi, M., Adamopoulou, A., & Kaderides, K. (2017). Green ultrasound-assisted extraction of carotenoids from pomegranate wastes using vegetable oils. *Ultrasonics Sonochemistry*, 34, 821–830.

Grewal, J., Wołącewicz, M., Pyter, W., Joshi, N., Drewniak, L., & Pranaw, K. (2022). Colorful treasure from agro-industrial wastes: A sustainable chassis for microbial pigment production. *Frontiers in Microbiology*, 13. Available from https://doi.org/10.3389/fmicb.2022.832918, https://www.frontiersin.org/journals/microbiology#.

Gupta, N., Poddar, K., Sarkar, D., Kumari, N., Padhan, B., & Sarkar, A. (2019). Fruit waste management by pigment production and utilization of residual as bioadsorbent. *Journal of Environmental Management*, 244, 138–143. Available from https://doi.org/10.1016/j.jenvman.2019.05.055, http://www.elsevier.com/inca/publications/store/6/2/2/8/7/1/index.htt.

Hamano, P. S., & Kilikian, B. V. (2006). Production of red pigments by Monascus ruber in culture media containing corn steep liquor. *Brazilian Journal of Chemical Engineering*, 23(4), 443–449. Available from https://doi.org/10.1590/s0104-66322006000400002.

Hassan, G., Shabbir, M. A., Ahmad, F., Pasha, I., Aslam, N., Ahmad, T., Rehman, A., Manzoor, M. F., Inam-Ur-Raheem, M., & Aadil, R. M. (2021). Cereal processing waste, an environmental impact and value addition perspectives: A comprehensive treatise. *Food Chemistry*, 363. Available from https://doi.org/10.1016/j.foodchem.2021.130352, http://www.elsevier.com/locate/foodchem.

Hemalatha, T., & Kailasam, S. P. (2022). *Carotenoid screening in selected flowers*.

Heredia-Guerrero, J. A., Heredia, A., Domínguez, E., Cingolani, R., Bayer, I. S., Athanassiou, A., & Benítez, J. J. (2017). Cutin from agro-waste as a raw material for the production of bioplastics. *Journal of Experimental Botany*, 68(19), 5401–5410. Available from https://doi.org/10.1093/jxb/erx272, http://jxb.oxfordjournals.org/.

Idham, Z., Putra, N. R., Aziz, A. H. A., Zaini, A. S., Rasidek, N. A. M., Mili, N., & Yunus, M. A. C. (2022). Improvement of extraction and stability of anthocyanins, the natural red pigment from roselle calyces using supercritical carbon dioxide extraction. *Journal of CO_2 Utilization*, 56. Available from https://doi.org/10.1016/j.jcou.2021.101839, http://www.journals.elsevier.com/journal-of-co2-utilization/.

Jadhav, A. R., Chitanand, M. P., & Shete, H. G. (2013). Flower waste degradation using microbial consortium. *IOSR Journal of Agriculture and Veterinary Science*, 3(5), 1–63.

Kantifedaki, A., Kachrimanidou, V., Mallouchos, A., Papanikolaou, S., & Koutinas, A. A. (2018). Orange processing waste valorisation for the production of bio-based pigments using the fungal strains Monascus purpureus and Penicillium purpurogenum. *Journal of Cleaner Production*, 185, 882–890. Available from https://doi.org/10.1016/j.jclepro.2018.03.032.

Kaur, P., Ghoshal, G., & Jain, A. (2019). Bio-utilization of fruits and vegetables waste to produce β-carotene in solid-state fermentation: Characterization and antioxidant activity. *Process Biochemistry*, 76, 155–164. Available from https://doi.org/10.1016/j.procbio.2018.10.007, http://www.elsevier.com/inca/publications/store/4/2/2/8/5/7.

Kausar, S., Aziz, R. B., Waseem, M., Ahmad, M., Shafiq, H., Asim, M., Zia, U., Afzal, S., Xi, W., Hameed, M., & Shoukat, M. U. (2022). Carotenoid metabolism, regulation in tomato (Solanum lycopersicum) and health benefits: An updated review. *Sarhad Journal of Agriculture*, 38(5), 12–25.

Kumar, V., Kumari, S., & Kumar, P. (2020). Management and sustainable energy production using flower waste generated from temples. *Agriculture and Environmental Science Academy*, 154–165. Available from https://doi.org/10.26832/aesa-2020-edcrs-011.

Kutlu, N., Pandiselvam, R., Kamiloglu, A., Saka, I., Sruthi, N. U., Kothakota, A., Socol, C. T., & Maerescu, C. M. (2022). Impact of ultrasonication applications on color profile of foods. *Ultrasonics Sonochemistry*, 89. Available from https://doi.org/10.1016/j.ultsonch.2022.106109, http://www.elsevier.com/inca/publications/store/5/2/5/4/5/1.

Lachman, J., Martinek, P., Kotíková, Z., Orsák, M., & Šulc, M. (2017). Genetics and chemistry of pigments in wheat grain—A review. *Journal of Cereal Science*, 74, 145–154.

Li, M., Wang, Z., Chen, L. Q., Wang, J. J., Li, H. Y., Han, Y. H., & Zhang, B. (2021). The relationship between the photosynthetic pigments, carotenoids and yield of broomcorn millet (Panicum miliaceum; Poaceae). *Appl. Ecol. Environ. Res*, 19, 191–203.

Liu, Weijie, Yang, Jing, Tian, Yanning, Zhou, Xuge, Wang, Shiwei, Zhu, Jingrong, Sun, Di, & Liu, Cong (2021). An in situ extractive fermentation strategy for enhancing prodigiosin production from Serratia marcescens BWL1001 and its application to inhibiting the growth of Microcystis aeruginosa. *Biochemical Engineering Journal*, 166. Available from https://doi.org/10.1016/j.bej.2020.107836.

Majumdar, S., & Bhowal, J. (2022). Studies on production and evaluation of biopigment and synthetic dye decolorization capacity of laccase produced by A. oryzae cultivated on agro-waste. *Bioprocess and Biosystems Engineering*, 45(1), 45–60.

Majumdar, S., Mandal, T., & Dasgupta Mandal, D. (2020). Production kinetics of β-carotene from Planococcus sp. TRC1 with concomitant bioconversion of industrial solid waste into crystalline cellulose rich biomass. *Process Biochemistry*, 92, 202–213. Available from https://doi.org/10.1016/j.procbio.2020.01.012, http://www.elsevier.com/inca/publications/store/4/2/2/8/5/7.

Marič, A., Skočaj, M., Likar, M., Sepčić, K., Cigić, I. K., Grundner, M., & Gregori, A. (2019). Comparison of lovastatin, citrinin and pigment production of different Monascus purpureus strains grown on rice and millet. *Journal of Food Science and Technology*, 56, 3364–3373.

Melgar, B., Dias, M. I., Barros, L., Ferreira, I. C. F. R., Rodriguez-Lopez, A. D., & Garcia-Castello, E. M. (2019). Ultrasound and microwave assisted extraction of Opuntia fruit peels biocompounds: Optimization and comparison using RSM-CCD. *Molecules (Basel, Switzerland)*, 24(19). Available from https://doi.org/10.3390/molecules24193618, https://www.mdpi.com/1420-3049/24/19/3618/pdf.

Nathan, V. K., Rani, M. E., Rathinsamy, G., & Dhiraviam, K. N. (2020). Fabricating bio-active packing material made from alkali-steam exploded agro-waste using natural colorants. *International Journal of Environmental Science and Technology*, 17(1), 195–206. Available from https://doi.org/10.1007/s13762-019-02387-3.

Ngamwonglumlert, L., Devahastin, S., & Chiewchan, N. (2017). Natural colorants: Pigment stability and extraction yield enhancement via utilization of appropriate pretreatment and extraction methods. *Critical Reviews in Food Science and Nutrition*, 57(15), 3243–3259.

Nirmal, Nilesh Prakash, Mereddy, Ram, & Maqsood, Sajid (2021). Recent developments in emerging technologies for beetroot pigment extraction and its food applications. *Food Chemistry*, 356. Available from https://doi.org/10.1016/j.foodchem.2021.129611.

Otálora, M. C., Wilches-Torres, A., & Gómez Castaño, J. A. (2023). Mucilage from yellow pitahaya (Selenicereus megalanthus) fruit peel: Extraction, proximal analysis, and molecular characterization. *Molecules (Basel, Switzerland)*, 28(2), 786.

Pagels, Fernando, Pereira, Ricardo N., Vicente, António A., & Guedes, A. Catarina (2021). Extraction of pigments from microalgae and cyanobacteria—A review on current methodologies. *Applied Sciences*, 11(11). Available from https://doi.org/10.3390/app11115187.

Pandit, S. G., Mekala Ramesh, K. P., Puttananjaiah, M. H., & Dhale, M. A. (2019). Cicer arietinum (Bengal gram) husk as alternative for Talaromyces purpureogenus CFRM02 pigment production: Bioactivities and identification. *LWT*, 116. Available from https://doi.org/10.1016/j.lwt.2019.108499, https://www.journals.elsevier.com/lwt.

Panesar, R., Kaur, S., & Panesar, P. S. (2015). Production of microbial pigments utilizing agro-industrial waste: A review. *Current Opinion in Food Science*, 1(1), 70–76. Available from https://doi.org/10.1016/j.cofs.2014.12.002, http://www.journals.elsevier.com/current-opinion-in-food-science/.

Polturak, G., & Aharoni, A. (2018). "La Vie en Rose": Biosynthesis, sources, and applications of betalain pigments. *Molecular Plant*, 11(1), 7–22. Available from https://doi.org/10.1016/j.molp.2017.10.008.

Ramadhan, M. O., & Handayani, M. N. (2021). Anthocyanins from agro-waste as time-temperature indicator to monitor freshness of fish products. *ASEAN Journal of Science and Engineering*, 1(2), 67–72.

Ramesh, Chatragadda, Prasastha, V. R., Venkatachalam, Mekala, & Dufossé, Laurent (2022). Natural substrates and culture conditions to produce pigments from potential microbes in submerged fermentation. *Fermentation*, 8(9). Available from https://doi.org/10.3390/fermentation8090460.

Rusita, Youstiana Dwi, & Hastuti, Rini Tri (2022). Utilization of Bougainvillea Spectabilis flower extract and red Ashoka extract as formaline and Boraxs Teskit. *Atlantis Press SARL*, 11–19. Available from https://doi.org/10.2991/978-94-6463-018-3_3.

Said, F. M., & Hamid, N. F. (2019). Natural red colorant via solid-state fermentation of oil palm frond by Monascus purpureus FTC 5356: Effect of operating factors. *Journal of Engineering Science and Technology*, *14*(5), 2576–2589. Available from http://jestec.taylors.edu.my/Vol%2014%20issue%205%20October%202019/14_5_10.pdf.

Salamatin, A. A. (2020). Supercritical fluid extraction of the seed fatty oil: Sensitivity to the solute axial dispersion. *Industrial and Engineering Chemistry Research*, *59*(40), 18126–18138. Available from https://doi.org/10.1021/acs.iecr.0c03329, http://pubs.acs.org/journal/iecred.

Sankaralingam, B., Balan, L., Chandrasekaran, S., & Muthu Selvam, A. (2022). Anthocyanin: A natural dye extracted from Hibiscus sabdariffa (L.) for textile and dye industries. *Applied Biochemistry and Biotechnology*, 1–16.

Sanna, Daniele, & Fadda, Angela (2022). Waste from food and agro-food industries as pigment sources: Recovery techniques, stability and food applications. *Nutraceuticals*, *2*(4), 365–383. Available from https://doi.org/10.3390/nutraceuticals2040028.

Selvarajoo, A., Wong, Y. L., Khoo, K. S., Chen, W. H., & Show, P. L. (2022). Biochar production via pyrolysis of citrus peel fruit waste as a potential usage as solid biofuel. *Chemosphere*, *294*133671.

Sharma, R., & Ghoshal, G. (2020). Optimization of carotenoids production by Rhodotorula mucilaginosa (MTCC-1403) using agro-industrial waste in bioreactor: A statistical approach. *Biotechnology Reports*, *25*e00407.

Sukmawati, D., & Maryanti, R. (2022). Development of education and economic circulation in supporting local potential as community empowerment efforts amid the Covid-19 pandemic. *Indonesian Journal of Multidiciplinary Research*, *1*(2), 235–250.

Świeca, M., Sęczyk, Ł., Gawlik-Dziki, U., & Dziki, D. (2014). Bread enriched with quinoa leaves–The influence of protein–phenolics interactions on the nutritional and antioxidant quality. *Food Chemistry*, *162*, 54–62.

Usman, H. M., Abdulkadir, N., Gani, M., & Maiturare, H. M. (2017). Bacterial pigments and its significance. *MOJ Bioequiv Availab*, *4*(3), 00073.

Vankar, P. S., Shanker, R., & Wijayapala, S. (2009). Utilization of temple waste flower -Tagetus erecta for dyeing of cotton, wool and silk on industrial scale. *Journal of Textile and Apparel, Technology and Management*, *6*(1). Available from http://www.bioresourcesjournal.com/index.php/JTATM/article/download/471/333.

Verni, M., Rizzello, C. G., & Coda, R. (2019). Fermentation biotechnology applied to cereal industry by-products: Nutritional and functional insights. *Frontiers in nutrition*, 642.

Waghmode, M. S., Gunjal, A. B., Nawani, N. N., & Patil, N. N. (2018). Management of floral waste by conversion to value-added products and their other applications. *Waste and Biomass Valorization*, *9*, 33–43.

Wani, F. A., Rashid, R., Jabeen, A., Brochier, B., Yadav, S., Aijaz, T., Makroo, H. A., & Dar, B. N. (2021). Valorisation of food wastes to produce natural pigments using non-thermal novel extraction methods: a review. *International Journal of Food Science and Technology*, *56*(10), 4823–4833. Available from https://doi.org/10.1111/ijfs.15267, http://onlinelibrary.wiley.com/journal/10.1111/(ISSN)1365-2621.

CHAPTER 17

Plastic waste as a novel substrate for industrial biotechnology

Rajlakshmi, Priyadharshini Jayaseelan and Rintu Banerjee

Microbial Biotechnology and Downstream Processing Laboratory, Agricultural & Food Engineering Department, Indian Institute of Technology Kharagpur, Kharagpur, West Bengal, India

17.1 Introduction

Plastics have become an inexorable part of our day-to-day contemporary life, contributing to our daily requirements because of its versatile nature. They are synthetic polymers obtained from the polymerization of monomers. The chain structure produces several subcategories of polymeric materials including several types of products differing in their monomeric unit or additives, further differentiating into thermoset (crosslinked materials which cannot be molded upon heating) and thermoplastics (crosslinked material which can be molded upon heating) (PlasticsEurope, 2017/2017). The low cost, lightweight, great performance, and higher processability rate of plastics attract their use. This leads to the gathering of plastic waste becoming a serious concern globally. The annual production of plastic is approximately 8300 million tons, out of which 6300 million tons are dumped as waste and approximately 40% of this plastic obtained is meant for packaging. These 24 million ton of plastics used for packaging are meant for immediate discarding, which has an apparent hazardous impact on the environment. This single-use plastics for storage, sterile packaging, disposable medical parts, and transportation contributes a significant part to the waste. The recycling of post-consumed plastic is often difficult because these are mixed plastics with the unknown composition of the polymer and contains several contaminants such as inorganic compounds (ink, salt, etc.) and organic contaminants (food waste, agricultural waste, etc.). A major portion of these wastes is either discarded in landfill or oceans or incinerated in power plants polluting the surroundings and also resulting in low or no value. Most of the plastic wastes get dumped into the oceans either by industrial disposal, improper discarding (trashes on beaches), direct dumping or wind transfer from ships/boats, offshore platforms, losses during transportation, etc. (da Costa et al., 2017). This ultimately leads to inevitable damage to the ecosystem and biota. It has been predicted that the plastic wastes that are disposed of in oceans will overshoot the marine fish population by 2050, as approximately 8 million tons of plastics are dumped directly into oceans (Uekert et al., 2018).

The conventional approaches for plastic waste utilization are incineration, landfills, recycling, etc. These processes have several limitations like low-quality polymer, costly process, etc. The nonselective discarding of plastic waste is also one of the major concerns. Microplastics, having a diameter <5 mm originating from the fragmentation of plastic used in personal care products and industries, littering on beaches etc., are a fresh substrate for the hydrophobic contaminant adhesion, unique bacterial colonization, deposition of insects' eggs, etc. Researchers found that these microplastics are been ingested by the deposit-feeding marine worm which reduces their feeding activity, increasing the longer gut residence time causing inflammation (Wright et al., 2013). On the other hand, biodegradable plastics contribute to a more sustainable process by utilizing renewable resources, dropping the CO_2 emission during production, and providing lower or no negative effect on the environment. These polymers include thermoplastic starch (TPS), polylactic acid (PLA), poly (butylene adipate-co-terephthalate) (PBAT), and polyhydroxyalkanoate (PHA). However, the polymer's origin does not compulsorily influence its end-life fate. In fact, the bio-based origin does not signify biodegradability. For example, several technologies have been used to make polyethylene terephthalate (PET) and polyethylene (PE) from bio-based sources, but they are nonbiodegradable and thus the source does not address the end-life pollution encounters (Narancic & O'Connor, 2017).

The complete utilization of all side streams is obligatory to reduce the waste and eventually the formation of CO_2. Initially, the action plan of the Environment Protection Agency (EPA) and European Union (EU) for plastic waste management was reuse, prevention, recycling, and landfill disposal as the last resources (European Commission, 2013). However, the post-consumed plastics ended up in improper and unmanaged disposal in the environment, making the action plan insufficient. Thus, to avoid the problems caused by conventional processes, researchers have tried utilizing plastic as a substrate to obtain novel products with optimum quality and product efficiency. A suitable biotechnological approaches are tried to modify the plastic waste and use plastic as feedstock for better plastic waste management. This will eventually help to contemplate the circular economy. Several processes, such as thermochemical depolymerization, potential product formation, and other techniques, have been successfully tried to get the best possible result. Thus, this chapter focuses on the contribution of biotechnology to get an advanced understanding of the pathway to utilize plastic waste and obtain plastic waste-based novel products. In addition, the challenges and recent progress in this field are highlighted.

17.2 Plastics and their classification

Plastics are broadly classified into single-use and multiple-use plastics. As the name suggests, single-use plastics are used for once, and they are also known as disposable plastics. Some of the polymers which are used for the manufacturing of single-use plastics are low-density polyethylene (LDPE), high-density polyethylene (HDPE), polystyrene (PS), polypropylene (PP), and most commonly PET. Single-use plastics are generally seen in the industrial domains for packaging purposes, accounting for 36% of the total plastic usage (Geyer et al., 2017). A rough estimate is that 400 million tons of single-use plastics are produced every year.

Single-use plastics are disastrous to the environment because of the following reasons (Borg et al., 2022):

- The usage time of these plastics is very minimal but the measure of time they retain in the environment before degradation is humongous.
- They are one of the leading causes of litter, which is unhealthy for the environment and human health as well as considerably difficult to clear the mess.
- These kinds of plastics are fragile and might end up as microplastics which when entered into the food chain causes adverse effects on human and animal health.
- They are difficult to recycle which is why they often end up dumped in landfills.

Multi-use plastics as the name suggests are produced from used plastic. These are known for their prolonged working lifetime and recyclability. The plastic after its usage can be converted to another desired product after appropriate treatment of conversion. Multiple-use plastics, when recovered for reuse, either go for closed-loop or end-loop recycling. In closed-loop recycling, they are made into identical applications if the recovered plastic is homogenous or into any low values product as an end application, if it is mixed with other polymers.

Plastics can also be further classified based on the type of the monomer, their arrangement, and the nature of the functional group or the side chain. The physical and chemical properties of plastic are influenced by these factors. Based on the thermal property of the plastic, they can broadly be classified into thermoplastics and thermosetting plastics. The thermoplastic's chemical property is not altered when it is subjected to heating; rather it can be deformed into the required shape multiple times. Some examples of thermoplastics are PE, PS, polyvinyl chloride (PVC), and polytetrafluoroethylene (PTFE). The thermosets can only once be melted and shaped. Upon heating, they undergo an irreversible chemical reaction, which makes them incapable of reshaping and retaining their original properties. Based on the physical and chemical properties of plastic there are seven types of synthetic plastics such are PETE or PET, HDPE, PVC, LDPE, PP, PVC, PS, and miscellaneous plastics. Table 17.1 represents the compilation of the polymer's symbol based on degradability code, properties, and application with advantages and disadvantages.

TABLE 17.1 Broad classification of plastic polymers and their properties.

Plastic polymer	Production of the polymer	Properties	Disadvantages	Application	References
Polyethylene Terephthalate	It is formed by the esterification between terephthalic acid (TPA) and ethylene glycol or sometimes the trans-esterification between ethylene glycol and dimethyl terephthalate (DMT).	High hardness, good Strength, stiffness, excellent insulating property, low gas permeability, maximum dimensional stability, safe for food (<60°C) and beverage packing, suitable for the transparent application.	At higher temperatures, it is vulnerable to diluted acids, bases, and ketones, aromatic and chlorinated hydrocarbons.	Bottles for beverages, food packing, artificial silk, non-woven fabrics, disposable medical garments, engineering plastics	Ji, 2013; Koshti et al. (2018)
High-Density Polyethylene (HDPE)	• It is formed by ethylene monomer. Ethanae is subjected to steam cracking in which the intramolecular bonds are broken to form ethylene. ethylene. • The ethylene is then polymerized in presence of a chemical catalyst (Ziegler-Natta catalyst, metallocene catalyst and chromium/silica catalyst).	Resistant to temperature, moisture, scratching, and dents, cannot be easily leached, UV-resistant, or densely packed.	Highly flammable, stress cracking, poor weathering, has no resistance against oxidizing acids, chlorinated hydrocarbons.	It is used for making house wrap, flexible HDPE pipes, monobloc chairs, bottle crates, toys, milk can, and wire insulation.	Benham and McDaniel (2000/2000)
Polyvinyl chloride	• The polymerization of vinyl chloride monomer forms the PVC. • It is produced by the chlorination of ethylene and pyrolysis of the resulting compound ethylene dichloride (EDC).	Durability, flame retardancy, electrical property, abrasion-resistance and chemical resistance.	Low heat stability, plasticizer migration, instability of properties over time.	Car seat belts, safety equipment, switches, window frames, sidings, ports, roofing, life jackets, etc.	Ameer et al. (2013)
Low-density polyethylene	• The ethylene in gaseous form is polymerized in the presence of oxide initiators under high temperature and pressure. • The polymer formed is non-crystalline because its long branches hinder the tight packing of molecules.	Flexibility, chemical resistance, and waterproofing capabilities, low material weight, tensile strength, and high recyclability.	High thermal expansion, fails under mechanical and thermal stress.	Plastic wrap and film, flexible packing material, etc.	Jordan et al. (2016)
Polypropylene	• PP is prepared by the additional polymerization of propylene in the presence of a catalyst under high temperature and high-energy radiation. • The properties of PP are influenced by the co-polymer components and molecular weight distribution.	PP has great temperature resistance, high glass-transition point, low water absorption, high resistance to flexing stress, good electrical resistance, high impact strength, non-toxic.	High thermal expansion coefficient, susceptible to UV radiation, not easy to paint, flammability, low resistance to chlorinated solvents.	It is used to make bags, jugs, trays, pipes, tubes, and other clinical and laboratory plastic equipment which can be autoclaved	Maddah (2016/2016)

(Continued)

TABLE 17.1 (Continued)

Plastic polymer	Production of the polymer	Properties	Disadvantages	Application	References
Polystyrene	Styrene a clear hydrocarbon liquid, is obtained when ethylene reacts with benzene in the presence of aluminum chloride to yield ethylbenzene. • In the presence of free-radical initiators, it is polymerized to give polystyrene.	Low-cost production, good insulation and cushioning property, lightweight,	Susceptible to UV degradation, it is flammable, brittle and vulnerable to organic chemicals.	Used for making foam, CDs, DVD cases, meat poultry trays, tissue culture trays, test tubes, Petri dishes, etc.	Merrington (2011)
Miscellaneous plastic: • The plastics which doesn't fall into the above-mentioned types are included here. • Examples: polycarbonate, Acrylic or Polymethyl Methacrylate (PMMA), acrylonitrile butadiene styrene, fiberglass, and nylon.	Polycarbonate: The polymerization reaction between bisphenol A (BPA) and carbonyl dichloride ($COCl_2$) Occurs in the presence of methyl chloride solvent.	Despite being impact-resistant, extremely strong and tough, polycarbonate is easy to cut, mold, thermo-formed or cold-formed and has inherent design flexibility.	The presence of bisphenol during leaching harms human health and the environment.	It is used to make a broad range of products like sunglasses, riot gear, DVDs, CDs, and greenhouses.	Kumar ans Singh (2020)
	Polymethyl methacrylate (PMMA): It is a transparent thermoplastic, which is formed by the polymerization of methyl methacrylate	PMMA is known for its lightweight, it is transparent like glass. The shatter-resistant property of PMMA makes it a better alternative to glass.	It has poor wear, abrasion, heat (can bear only up to 80°C) and impact resistance.	It is prominently used in making optical devices, acrylic mirrors, plexiglass, and also it finds application in the makeup industry as it is approved by, the food and drug administration (FDA) and Cosmetic Ingredient Review (CIR) as a safe ingredient.	
	Acrylonitrile Butadiene Styrene (ABS): It is an amorphous thermoplastic which is produced through emulsion or bulk polymerization of three monomers: (acrylonitrile, butadiene and styrene)	ABS is an opaque ivory material, its characteristic property includes impact resistance, great insulating property, good abrasion property, mechanical and dimensional stability, and strain resistance.	Some of the disadvantages of ABS include poor resistance to chemicals and chlorinating agents. It is tolerant to aliphatic chemicals but not to aromatic or chlorinated hydrocarbons. It is also negatively influenced by UV light.	It finds its application in household appliances like vacuum cleaners, and food processors. Apart from that, ABS finds its application in 3-D printing technology, In fusion technology modeling (FTM) ABS is melted and spun into long threads.	

17.3 Problems associated with conventional plastic usage and its disposals

Plastic wastes are mostly dumped directly into landfills releasing several toxic gases, such as furans, cyclic chlorinated hydrocarbons, greenhouse gases (GHGs), and dioxins, into the surrounding (Lebreton et al., 2018). The statistic for waste generation varies among different countries. Plastic being a small component of waste by weight but are larger component by volume. Thus, the spatial and temporal evaluations can be confounded, and data on quantities of waste recycled can be directed as per the categorization of waste. In many localities, the areas of landfilling are running out and because of the longevity of the plastics, disposal of landfilling is just stockpiling toward the problems for the future (Barnes et al., 2009).

Several evidences are found where conventional plastic waste and its disposal has presented significant source of contaminants that are spread not only to the terrestrial but also to the aquatic environment (Teuten et al., 2009). Plastic waste in addition to the physical problem has the potential to pass the toxic substances to the food chain. Plastic debris spreads in the form of fragments, microplastics, and pellets that contain organic contaminants such as alkyl phenols, polycyclic aromatic hydrocarbons (PAHs), and polybrominated diphenyl ethers (PDBE) at concentrations extending from ng/g to µg/g. These compounds get assimilated to the plastic in the course of manufacturing while some get adsorbed to plastic debris from the surrounding.

Plasticizers and other additives leaching out from landfills also are one of the major concerns. Many additives are used in substantial quantities in a wide range of products. These chemicals are not only toxic but there exist several controversies about the additives leaching out of plastic products. The major issue lies in the quantities and types of additives present in plastics to uptake and get accumulated by living organisms. Additives like BPA, brominated flame retardants (BFRs), and anti-microbial agents such as phthalate plasticizers are of major concern, as they are found in products like food packaging, medical devices, cosmetics, computers, flooring materials, etc. Phthalates are found to leach out of the product as they are not chemically bound to the plastic matrix. They are of major concern because of their wide usage and high productivity (Wagner & Oehlmann, 2009). In addition to the dependence on finite plastic production, sources and concerns related to additives usage and effect, the current scenario of plastic consumption has posed a great challenge in global waste management.

17.4 Biodegradable plastics: limitations and its market studies

The fact that the prefix "bio" does not mean or make it more environment-friendly. This may give a glance that the bioplastic are obtained from plant source labeled under the "biodegradable" category. While plastics obtained from petroleum sources show wide applications and benefits, they are categorized under non-biodegradable.

The biodegradability of the plastic depends on the chemical structure and not on the source from where it is derived. Currently, plastics derived from biological sources including both animal and plants source are mostly manufactured from the starch source on the industrial scale. Bio-based polymers including bio-PET, bio-PE, and bio-PP exhibit similar mechanical features like fossil-based PET, PE, and PP, but they are non-biodegradable as no hydrolysable bonds like esters and amides are present that can be converted into well-defined oligomers or monomers. One of the important factors regarding degradability of the plastic is also the degradation condition, which is presumed to be similar for all bioplastics. However, the reality implies that various bioplastics require specific conditions to degrade and may take a longer time for complete biodegradation.

Petrochemical plastics often contain solely saturated carbon-carbon units that cannot be degraded to particular smaller unit or the oligomer/monomer. It requires a particular introduction of double bond or oxygen via specific treatment or UV irradiation for further degradation by microbial or enzymatic processes. PS is one of the examples that provide degradation only in the sulfonated form by particular fungi.

The global scenario of the bio-based plastic market indicates top players such as Nature Works LLC, Novamont s. p.A., Total Corbia PLA, BASF SE and Rodenburg Biopolymers that account for 34% of the global market (Goel et al., 2021). The current research studies are based on converting conventional plastics into biodegradable products by various techniques, such as oxo-degradation technology, enzyme-mediated technology, among others. In oxo-degradation technology, the high molecular weight polymers are mixed with certain additives that degrade the polymer under ultraviolet light (UV) into oligomers/monomers/dimers. Those plastics degrading under this condition are considered as oxo-degradable plastic. However, many ground reports have provided evidence that the oxo-degraded plastic gets degraded to microplastic which is further degraded into very minute invisible residues that are even more difficult to handle and also the entire process is highly time-consuming. This makes the process highly unsuitable for long-term usage and recycling. The other process is enzyme-based process that involves enzymatic breakdown utilizing living organisms through bio-fragmentation. Several companies like M/s Repsol (Partnership with enzyme technology to develop biodegradable polyolefin's), M/s Earth Nurture, M/s PEP Licensing, M/s BNT Force Biodegradable Polymers Private Limited, M/s Biosphere Plastic LLC, M/s Enzymoplast are involved in biodegradable additives depending on enzymatic technologies (Goel et al., 2021). They claim the reusability, recycling, and degradation of the plastic, but a detailed analysis of the reports indicates that only partial degradation is achieved. In some cases, data on the percentage degradation are manipulated and various companies offering enzymatic technologies are not accredited or certified. Thus, the reliability of such research studies create vicious data, so much more data with proper credibility are required to completely ascertain the enzyme-based technique for the degradation of plastics. Perhaps, above all the limitations, bioplastic still remains the forefront candidate for the replacement of fossil-based plastics.

17.5 Strategies for management of plastic waste (reduction, replacement and reuse)

The dynamic properties of plastic and its versatility in almost every industry have made plastic an inevitable artefact. Ever since 1950s, the plastic industry exponentially boomed so does plastic waste accumulation (Lear et al., 2021). Conventionally plastic waste is managed by dumping them in landfills. Two approaches such as "dry-tomb" and "wet-cell" are in practice. In the wet-cell landfill approach, there is a possibility that the dumped plastic would leachate and pollute the environment. On the other hand, the dry-fill approach has the limitation of land scarcity. The landfills are a huge threat to the groundwater, as the migration of toxic elements from the plastic would contaminate it (Bassi, 2017). Recycling plastic is carried out to overcome the disadvantages posed by the landfill method of waste management.

Recycling of plastic can be done mechanically, thermo-chemically, and by biotechnological interventions. Mechanical recycling is thermoplastic-specific. Incineration, pyrolysis, hydrogenation, and gasification are some of the approaches under thermo-chemical methods that are applied to produce energy out of plastic waste. The sorting of plastic waste before treatment is a time-consuming and labor-expensive factor. Apart from that, during incineration, certain harmful gases like polychlorinated biphenyls and dioxins are also produced. The biotechnological intervention has come up with novel ways of handling plastic waste safely, economically, and eco-friendly. The biotechnological strategies for handling plastic waste are broadly brought under reduction, modification, replacement, and conversion methods and represented in Fig. 17.1. In the subsequent sections, different strategies on reduction, replaecement, and reuse have been discussed in detail.

FIGURE 17.1 Biotechnological strategies for plastic waste management.

17.6 Reduction strategies

This reduction strategy involves the application of biological systems such as micro-organisms, mealworms, enzymes and gut bacteria under engineered conditions that aid the degradation of plastic waste. Plastics are naturally non-degradable or extremely degradable by the haunt micro-organism. Identification of specific micro-organisms and optimizing the growth condition suitable for the micro-organism to use plastic as its carbon source would aid in the degradation of plastic.

17.7 Micro-organisms in plastic reduction

The major constrain of microbial plastic reduction is due to its hydrophobic nature. The reduction of plastic by microorganisms is done through metabolic or enzymatic actions. The action of micro-organisms on plastic is influenced by various parameters such as hydrophobicity, hydrophilic property, crystallinity, chemical structure, molecular weight, surface area, and the functional groups present in the plastic. The synthetic polymers involved in the plastic's structure are of utmost stability. Some examples of such plastics are PE, PET, PP, PS, PVC, etc. The crystalline stable structure of these plastics makes it hard for the micro-organism to act upon its carbon thereby degrading it. Synthetic plastics have a high molecular weight which inversely influences their degradability potential. The hydrophobic nature of plastic makes it less available for microorganisms to seep in and fragment it. Whereas, bio-based plastic such as polyhydroxy butyrate (PHB), polyhydroxy hexanoate (PHH), polyhydroxy valerate (PHV), and polyhydroxy octanoate (PHC) are exceptions. These bio-based polymers are completely degradable (Brandon et al., 2018).

It has been reported that various bacteria, fungi, algae and their consortia have the potential enzymes to degrade synthetic plastics by acting on the backbone of the polymer. The major steps involved in the degradation of plastic by these micro-organisms are bio-deterioration, bio-fragmentation, and bio-assimilation followed by mineralization.

Bio-deterioration is a natural phenomenon in which superficial degradation of the polymer takes place in physical, mechanical, or chemical ways through microbial interventions. Mostly, this process is not a solo act, but rather a combined synergistic multistage process which affects and weakens the polymer structure. The nature and properties of the polymer as well as biotic and abiotic factors in the environment play a major role in biofilm formation which in turn contributes to deterioration (Krueger et al., 2015). The abiotic factors involved are temperature, oxygen level, light, and salinity (considering the ocean). The densely growing microbial communities on the surface of any material (living or static) which encircles themselves with a coating of extracellular polymeric substances (EPS) is called a biofilm. The biofilm aids in the attachment of other microbial communities to carry out further deterioration. Apart from being a matrix for microbial colonization, the biofilm also protects the microbial community from harsh environmental changes. The attached microorganisms seep into the pores of plastic, and as they grow, the pH in the environment changes which thereby causes the pores to enlarge upon degradation. This leads to cracks, which enlarge as the time passes and the strength of the polymer gradually decreases resulting in the collapse of the material.

The cleavage of the polymer through catalytic action through the free radical secreted by microbial colonization is known as bio-fragmentation. The polymer is broken down into monomers, dimers, and oligomers. The bacteria which have the oxygenases possess the potential to fragment the plastics. The enzyme oxygenase might be mono-oxygenases or dioxygenases (Ru et al., 2020). They carry out fragmentation by adding an oxygen molecule to the carbon chain of the polymer; this results in the formation of alcohol and some peroxyl radicals. Further various enzymes like lipase, esterases, and endopeptidases (in the case of the amide group in the polymer which depend upon the nature of the polymer) further modify the carboxylic group.

After biofragmentation, assimilation and mineralization of the oligomer occur, the end product obtained is dimers or monomers. Depending upon the size of polymer fragments, they either enter the cell or stay out, if it is not small enough to pass through the cell membrane. Those fragments which cannot enter the cell undergo anaerobic mineralization and the assimilated polymer fragments undergo aerobic breakdown inside the cell. The microbial system utilizes the polymer as a carbon source through the catabolic pathway to meet its energy as well as growth requirements. The secondary metabolites produced by these bacteria are further used by other bacteria for extensive degradation. During assimilation, some oxidized products like CO_2, CH_4, N_2, and H_2O are let out (Skariyachan et al., 2022). Mineralization of polymers can be either aerobic or anaerobic. The aerobic mineralization would yield the same as the products of microbial assimilation.

17.8 Enzymes in plastic reduction

The enzymatic breakdown of plastic occurs through catalytic reactions that involve the cleavage of chemical bonds between the polymers, mostly in the presence of water. Several enzymes have been reported to have the capability to degrade plastics. Though the efficiency is less, genetic modification has proved to improve the efficiency of degradation. Chen et al. (2022) studied the modified *Termobifida fusca* cutinase activity that enhanced the PET degradation by 90% exceeding almost 30% above the native-type enzyme. The biotechnological intervention proposes a novel modification strategy that involves the addition of positive charge amino acid and the remodeling of the binding site which directly enhances the efficiency of the enzyme during the degradation process. The PETase isolated from *Ideonella sakaiensis* 201-F6 shows the potential to consume the carbon from PET thereby degrading the plastic. The *I. sakaiensis* has two enzymes which convert PET to ethylene glycol and terephthalic acids. The PETase converts PET into Mono (2-hydroxyethyl) terephthalic acid (MHET). Further MHETase converts MHET to ethyl glycol and terephthalic acid through hydrolysis. These enzymes were expected to match *Thermobifida fusca cutinase*, an engineered double mutant (S238F/W159H), that showed surprisingly greater activity than the wild strain references. The obtained engineered enzyme also influences the crystallinity about 3.7 times more in only 96 hours span (Zhao et al., 2022).

17.9 Mealworm in plastic reduction

Another bio-degradation system that is faster than microorganisms in terms of plastic degradation is the mealworm. Recently, it has been scientifically reported that the gut of arthropods more specifically larvae of *Tenebrio molitor* generally known as mealworms has the capability to degrade plastics efficiently. Some of the plastic-degrading worms include *Tenebrio molitor* larvae, *Achroia grisella* larvae, *Lumbricus terrestris* (Earthworm), *Tenebrio obscurus* larvae, and *Zophobas atratus* larvae. A study conducted by Yang et al. (2015) revealed that mealworm's gut micro-biome has a significant role in plastic degradation. The mealworms were fed with gentamicin which inhibited the gut microbes. It was observed that those Gentamicin fed mealworms lost the ability to convert polystyrene and mineralize it into CO_2. It was identified that *Exigubacterium sp.* Strain YT2 in the gut of the mealworm also plays a significant role in this process. The biofilm formed over polystyrene over 28 days aided in creating cavities and pits that changed the surface property and increased the formation of polar groups. Another study conducted by Brandon et al. (2018) reported that mealworms degraded polyethylene at a similar rate to that of polystyrene.

The gut bacterial investigation of mealworms has shown that *Citobacter sp.* and *Kosakonia sp.* are associated with polyethylene and polystyrene degradation, respectively. It has been observed that plastic is depolymerized and oxidized inside the gut of mealworms. These oxidized products from the mealworm after degradation can be used as a substrate for the production of bio-degradable polymers such as PHAs. It was found that the larva of the wax moth (*Galleria mellonella*) which belongs to the Lepidoptera family does quick biodegradation of polyethylene. Bombelli et al. (2017) reported that a wax worm has the capability to make 2.2 ± 1.2 holes per hour. In 12 hours, 100 worms are enough to degrade a plastic bag and impact 92 mg of mass loss. In short, the wax worms can cause up to 13% weight loss over a period of 14 hours corresponding to 0.23 mg/cm h. Thus, it can be stated that mealworms have a faster plastic degradation rate than that of microbial isolates and the rich microbial community in the gut of the worms are responsible for such activity.

17.10 Replacement of conventional plastics using biotechnology

Plastics can be classified as degradable and non-biodegradable based on the degree of bio-degradability. The replacement of fossil-based synthetic plastics with unconventional bio-degradable plastic is an excellent approach to handle the plastic waste. The bio-based plastics are obtained from starch, plant-based resin, soy-based blend, microbial-based composites and poly (L-lactic acid). From an environmental point of view, these bio-plastics are completely preferable as their synthesis is green as well as can be directly degraded by the micro-organism in a short period when compared with synthetic polymers.

17.11 Starch-based polymer

The starch-based polymers take a huge share in the bio-based polymer as they are securable and easily degradable and could blend into other polymers easily. Starch is composed of two microstructures, namely, amylose and amylopectin. The amylose part of the glucose is a linear chain with amorphous nature; in contrast, the amylopectin of the glucose is a non-linear branched polymer with crystalline nature. The amylose has a low molecular weight resulting in a resemblance with the synthetic polymers, whereas the amylopectin with its branched structure has a high molecular weight making it viscous and less mobile. The amylose: amylopectin ratio can be tailored as desired (Bassi, 2017). By thermal processing, it has been proved that reduction in amylopectin proportion from starch makes it suitable for replacing plastics. The starch as such cannot be used to make plastics as it is heat liable. In order to combat this disadvantage and to induce better mechanical and thermal properties, the crystallinity should be reduced or removed. This is done by subjecting starch to some sequential steps like gelatinization followed by dehydration and the addition of some plasticisers. During the gelatinization process, irreversible damage to the crystallinity of the starch occurs as the starch undergoes swelling followed by melting of starch crystals leading to molecular solubilization. The gelatinized starch is obtained in the granular form (Jiang et al., 2020). These granules are further dehydrated and glycerol or water is added as a plasticiser. The plasticiser is added to hamper the re-crystallization of the starch. The processed starch is used in two ways to produce a bio-degradable polymer. The starch-filled polymer and starch-based polymer are the two types in use. The starch blend polymers are better surrogates in terms of their properties. In this method, the processed starch is extruded or subjected to injection molding with other polymers either inorganic natural fillers or synthetic polymers. Poly (butylene glycol adipate) is one such example where the blending of starch and polyurethane was carried out through microdispersion and the formed films were extruded. In the case of natural fillers starch reinforced with cellulose fiber is a good example (Jiang et al., 2020).

17.12 Polylactic acid-based polymer

Polylactic acid polymers are bio-based aliphatic semicrystalline non-toxic polyesters that are biodegradable and biocompatible. It is formed by the direct condensation of lactic acid monomer, and the formed oligomer undergoes ring-opening polymerization. The production of PLAs is sustainable as it involves renewable resources like dairy waste, agricultural residues, food waste, etc., which on fermentation, would produce lactic acid monomers. PLA's biodegradability makes it stand out from other polyester polymers. PLA as such is not accessible to microbial degradation though it is formed of lactic acid monomers. It undergoes degradation upon continual contact with water, which initiates hydrolysis followed by microbial degradation. PLA being non-toxic is also used in making diverse biomedical applications like screws, scaffolds, and drug carriers, as the degradation time can be controlled (Freeland et al., 2022). Besides the mentioned advantages, PLA is a thermosetting plastic, which implies that it can heat up to its melting point and then be cooled to reshape into desired secondary use without any change in its properties. Considering the recycling aspects, it can be subjected to chemical recycling. The process involves the depolymerization of PLA-based products in the presence of a suitable catalyst, followed by repolymerization to obtain reprocessed PLA-based plastics. The catalysts generally engaged in this process are montmorillonite K10 and some aliphatic diamines like 1,2-diamine ethane and 1,6 diamine-hexane (Naser et al., 2021). Thus, it is an excellent alternative to conventional plastics.

17.13 Modification of plastics for waste management

Modification of plastics is done to change their physical and chemical properties. The polymer is amalgamated with bio-degradable segments, such as itaconic acid, lactic acid, benzophenone, and nodox, where these additives inversely affect the crystallinity, chemical stability, thermal and UV-resistance, and hydrophobicity. These additive polymers also reduce the functionality as well as stability of the plastic but improve the degradation efficiency as the mentioned parameters are responsible for the dawdling of plastic degradation. The reduction in the hydrophobicity increases the exposure of polymers to the microbes and the susceptibility to photo-degradation and thermal instability considerably aiding in the breakdown of the polymer into simpler forms for facile microbial consumption or attack. These plastics with additives are referred to as oxo-degradable or oxo-biodegradable plastics depending upon the degradation that takes place after the oxidation reaction leading to fragmentation of the plastics.

The additive materials can be synthesized through microbial fermentation. Biotechnological arbitration is significant in this process as it provides economical and green monomer additives. Since the addition of the monomers influences the functional properties, the modification process for every plastic should be standardized. Factors like nature, polarity, quantity, and time play a significant role in deciding the physical property as well as degradability of the formed polymers during the addition of these additive monomers. Most of the additive monomers are added to the polymer in the melt phase while extrusion is carried out (Bassi, 2017).

The non-polar polymers need special attention in terms of additives to make them polar as well as to maintain the crystallinity for their strength. The additives like mono and dimethyl esters of itaconic acid were grafted into polypropylene in order to increase its polarity and reduce its molecular weight. The additive monomer was added to polypropylene during the melt phase. The standardized conditions for monomer addition were found to be 190°C for 6 minutes at 75 rpm. The addition of dimethyl itaconate acid and monomethyl itaconate could be done up to 1.5% and 1.6%, respectively. It was observed through differential scanning calorimetric measurements that the increase in the grafting percentage decreased the molecular weight of the polymer. Some of the other monomers grafted into polypropylene are maleic anhydride, carboxylic acid derivate that unsaturated, functional group containing vinyl or acrylic monomers (Krivoguz et al., 2006).

It was observed that the addition of some hydrophilic polymers such as polystyrene sulfonic acid (PSSA) into hydrophobic, mechanically, and thermally stable polymers such as poly-ether ketone, poly-ethylene-co-tetrafluoroethylene (ETFE) has positively correlated for polymer degradation. The addition of PSSA as a graft has lowered the thermal resistance as the grafted polymer cannot withhold beyond 85°C–95°C. The blended PSSA detached from the ETFE at higher temperatures. In an aqueous solution like propanol, the PSSA-grafted PEEK films have undergone extreme weight loss as they swell. It was concluded that the detachment of hydrophilic grafts from the hydrophobic polymer is reasoned by the stress-induced swelling at the interfacial boundary resulting in weight loss and eventual degradation (Enomoto et al., 2011). Some of the degradable polymers are made by graft polymerization by blending cellulose and poly (L-lactide) PLLA. The change in the crystalline structure resulted in degradability progressively. The lower the degree of crystallization, the higher the plastic degradability (Dai et al., 2013).

17.14 Conversion strategies using biotechnology

The conversion strategy is applied to transform plastic waste into a value-added product. Generally, the depolymerization of plastic waste is followed by the application of either genetically modified bacteria or some special bacterial isolates in order to produce a biopolymer or upgraded chemical. The depolymerization of plastic is done as mentioned before either through photo-degradation or thermo-chemical degradation or by mechanical methods. It was observed that the depolymerized polymer from pyrolysis is used to produce PHA by the intervention of *Pseudomonas aeruginosa* PAO-1. The PHA produced was nearly 25% of its dry cell weight (Bassi, 2017). The thermal means of depolymerization leads to the conversion of subjected plastic waste to crude oil under high pressure and extreme temperature in the presence of H_2O by hydrous pyrolysis. As the long chain of polymer breaks into monomers, the gaseous by-products such as carbon dioxide or methane fail to turn into oil in the process posing a disadvantage. Apart from the liquefaction of plastic waste under extreme pressure and temperature, it is also used for making H_2 which has a wide demand in the pharmaceutical, chemical, and agricultural industries. The photo-reforming of plastic is carried out to convert PET waste to H_2. The conversion strategies are further explained in detail in the upcoming section.

17.15 Renewal process for utilization of multiple use plastics waste

Biotechnology provides several ways and techniques to eradicate plastic waste in a controlled and sustainable manner. Microbial biotechnology delivers an opportunity to obstinate the plastic waste disaster and reform it toward utilization as a feedstock for the formation of a completely new value-added product. This will introduce a more efficient way to extemporize the resource and reduce the hazardous impact of plastics on the environment and living organisms, thus contributing to the entry of plastic into the circular economy model (European Commission, 2018). This model helps to reduce product inefficiency, overconsumption, and wastage of resources.

The current advancements and technologies show encouraging possibilities for the designing of pathways for the synthesis of new biodegradable materials and novel pathways for the sustainable degradation of plastics.

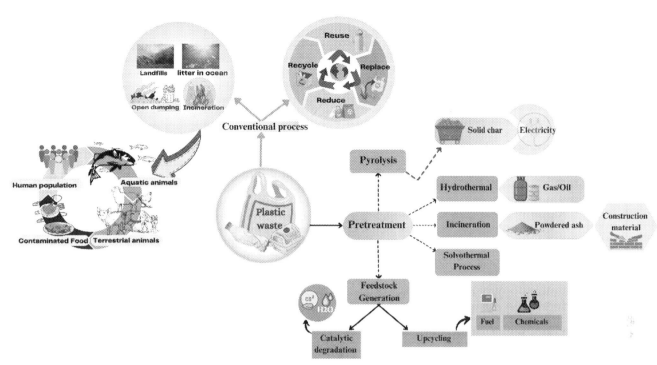

FIGURE 17.2 Conventional vs alternative methods for plastic waste utilization as novel substrate for biotechnological application.

Fig. 17.2 depicts the conventional versus alternative methods for plastic waste utilization as a novel substrate for biotechnological applications.

17.16 Pre-treatment of plastic waste by thermochemical depolymerization of plastic waste

Plastic polymers are derived from petrochemicals and fossil fuels. The decomposition of these polymers through the thermochemical route can generate green chemicals and fuels. The process can produce clean alternative fuels or useful mono- or oligomer plastics from several types of wastes including e-wastes by thermochemical conversion. This thermochemical conversion process includes pyrolysis, incineration, solvothermal process, hydrothermal process, among others.

17.17 Pyrolysis

Pyrolysis is an anaerobic thermochemical conversion process of biomass into liquid hydrocarbon fuels such as bio-char, bio-oil, and producer gas. It is a flexible and reliable approach to serve twofold purposes by supplementing the day-by-day growing energy demand from the petroleum industry and reducing plastic waste. Pyrolysis optimizes the processing of the plastic feed mixture to obtain high yield of certain products with desirable and attractive applications such as thermal cracking of plastic waste to provide oil for furnaces, generators, agricultural pumps, boilers, etc. (Miandad et al., 2019). In the initial step of the process, high temperature leads to the stepwise dehydration, fragmentation, and depolymerization of the biomass and organic products into the volatile compound. The volatile compound further quenches upon condensation to form bio-crude. The crude bio-oil quality majorly depends upon this quenching process and the residence time of the vapor. Several other process parameters like heating rate, temperature, physiochemical properties of the feedstocks (moisture content, particle size, elemental composition), and reactor type also play a role in determining the bio-oil, bio-char, and gas yield (Nanda et al., 2013).

The classification of pyrolysis depends upon two major factors, that is, vapor residence time and the heating rate. The heating can be done in flash, fast, intermediate, and slow pyrolysis processes. Flash pyrolysis utilizes a swift heating rate of more than 1000°C/s, a higher temperature ranging from 800°C to 1000°C and a vapor time of 0.5 seconds (Maschio et al., 1992). The fast pyrolysis possesses a fast heating rate of 10°C–200°C/s, a short vapor time of 30–1500 seconds, and a high-temperature range of 400°C–500°C, resulting in more yield of bio-oil than permanent gases and char (Bridgwater et al., 1999). However, intermediate pyrolysis prefers a heating rate of 2°C–10°C/s, vapor time of 10–20 seconds and a high-temperature range of 500°C–600°C, whereas slow pyrolysis utilizes a slow heating rate of 0.1°C–1°C/s, longer vapor duration of 10–100 minutes and moderate and wider temperature range of 300°C–700°C, producing more bio-char than the bio-oil (Ahmad et al., 2014) The condensation rate of the volatile components reduces due to the longer vapor residence time, favoring the secondary polymerization reaction producing tars and biochar. High temperature suppresses the bio-char formation because of the cracking reaction.

Based on the type of plastic feed characteristics, plastic waste can be considered as post-consumer and post-industrial utilization. Post-industrial plastics are clean and of better quality with specific defined composition. However, post-consumer plastics are generally contaminated containing other materials such as biomass, metals, and organic waste, making them not suitable for mechanical recycling and requiring additional processes such as cleaning, sorting, and separation. Here, the pyrolysis process can efficiently tolerate a considerably high amount of contaminants thereby providing dual benefit of reducing the additional steps of sorting and separation and valorization of plastic waste into valuable molecules and several oligomers.

Polymers like PMMA, PS, and PET show distinct chemical reactivities due to the presence of labile chemical bonds; thus, they are chemically treated by successive repolymerization to get virgin polymers or depolymerized to acquire monomers. The thermal cracking of plastics generates oil containing numerous compounds such as nitrogenous compounds (pyridines, nitriles, indole, etc.), oxygenated compounds (ketones, phenols, esters, etc.), hydrocarbons (olefns, parafns, naptha, isoparafns aromatic compounds, etc.), gases, light oil (boiling range 250°C–350°C), and heavy oil (boiling range >350°C) unlike biomass-derived oil (Toraman et al., 2014). Naptha generated from polystyrene consists of benzene, toluene, xylene, α-methyl styrene, izoproylbenzene and styrene. Gases consist of ethane, propane, butane, ethylene, propene, etc., with some amount of carbon dioxide (CO_2), hydrogen (H_2), and carbon monoxide (CO). However, light oil are hydrocarbons like paraffns, olefns, etc. and heavy oil contains high molecular weight compounds, aromatic compounds, etc. (Adrados et al., 2012). Pyrolysis of radioactive plastic waste materials primarily involves the immobilization of radioactive particles by volatilization of polymeric materials.

Co-pyrolysis is also an encouraging method to enhance the efficacy of plastic waste. It is a process where the synergistic effect of a mixture of two or more plastics and biomass are utilized as feedstock. The co-pyrolysis of PE and sugarcane bagasse is one of the examples where a higher yield of liquid, hydrocarbon and alcohol was observed (Hassan et al., 2020). The synergy between plastic waste and biomass is affected by several factors such as pyrolysis duration, heating rate, temperature, and catalyst. The refinement and the separation process of the product formed by co-pyrolysis need to be improved to obtain a high value-added product.

The factors affecting the pyrolysis selectivity and reactivity also depend upon the types of polymers. Reactor configurations such as fixed bed, vacuum moving bed, circulating fluidized bed, transported bed, mechanically fluidized bed, rotating cone, and ablative reactor are been utilized for the pyrolysis process. However, fluidized bed reactors are commonly used for enhanced bio-oil production with high thermal cracking because of their high heat and mass transfer rate.

17.18 Hydrothermal treatment

Hydrothermal treatment is an advanced tertiary treatment technology of plastic wastes that involves the utilization of subcritical/supercritical liquid or water as reaction media/solvent carried out inside an autoclave reactor for the degradation of plastic (Wang et al., 2019). It has a great potential to treat and convert small organic compounds like BFRs, contaminated e-plastic wastes, etc. (Zhan et al., 2020). It also provides flexibility to process different plastic waste regardless of the size, color, physical property, purity, etc. (Zhan et al., 2020) efficiently removed Br (90.13%) and Sb (85.6%) simultaneously utilizing an alkaline sulfide system via hydrothermal treatment from Sb_2O_3 and BFR-embedded e-plastic wastes.

Hydrothermal treatment takes place by dehydration followed by polymerization and carbonization process. The carbonized material obtained from this procedure (hydrochar) has several potential applications like supercapacitor, solid fuel, fuel cell, and sorbent formation. The hydrochar formed generally has a cross-linked aromatic polymer with organic functional group and surface porosity. However, these features can be altered by changing the hydrothermal carbonization reaction changing the fuel property and reactivity of the hydrochar. The hydrothermal treatment with superficial water/solvent generally functions at a lower temperature (180°C–350°C) and saturated pressure (2–10 MPa) compared to pyrolysis for the decomposition of plastic wastes to produce low molecular weight hydrocarbon products. It produces high amount of solid incineration of plastic products with enhanced combustion features but a very less amount of gaseous product. It also generates liquid oils as chemical feedstock and the separation of oil and char is easy to perform (Shen, 2016). Hydrothermal products can be utilized as efficient and clean solid fuels or chemical feedstock, thus creating environment-friendly utilization of plastic waste resources. This process also provides the benefit of processing wet plastic wastes, eliminating the requisite of energy-demanding watering step.

However high pressure, safety issues, and higher installation costs prevent the utilization of hydrothermal treatment process. The process is limited to a bench-scale system with an energy-intensive industrial process. In addition to this, the separation procedure is difficult to handle at the industrial scale.

17.19 Incineration

Incineration is a process to reduce waste landfilling and to recover the chemical energy from the waste in order to convert it into electricity. Incineration of plastic leads to the reduction of solid volume by 90%–99% (Arvanitoyannis, 2012).

The incineration of e-waste of plastic from computer motherboards, keyboards, casings, etc. is tested by several researchers. The e-waste plastic compounds contain the organic compounds related to flame retardants like active elements such as Cl, Br, etc. as well as Polybrominated biphenyls (PBB), PDBE, Tetrabromobisphenol A (TBBPA), etc. incineration of these waste in a well-controlled chamber has the ability to produce electricity with proper check of the toxic emission to the environment (Vehlow et al., 2000).

17.20 Solvothermal process

Solvothermal or solvolysis is a process where the individual plastics is chemically treated at 80°C–280°C using sub- and supercritical solvent molecule and depolymerized into monomers (Payne et al., 2019). The solvothermal process has several benefits such as lower critical temperature and pressure of organic solvent leading to reaction at optimal condition; solvent reuse and regeneration; and less energy exhaustive process. The process consists of alcoholysis, glycolysis, hydrolysis, aminolysis, and ammonolysis.

A non-oxygenated solvent such as toluene is used to degrade plastic wastes. Toulene being a nonpolar solvent has a lower critical temperature (318°C) and critical pressure of 598 psi. Saha et al. (2022) performed a study on mixed plastic and concluded that the production of lower hydrocarbon increased and also the process showed synergistic effect.

This process has been widely used to degrade plastic waste but it also has several disadvantages, such as the glycolysis of PET needs a longer reaction time and high temperature. The utilization of oxygenated solvent increases the risk of oxygen contamination in the process and also increase the cost of purification and separation. This reduction in the purity of the product because of problematic catalyst separation hinders the direct use of the product.

17.21 Plastic waste as feedstock for potential product formation

Using plastic waste as feedstock can be a significant achievement in the field of industrial biotechnology. These plastic wastes create a huge environmental impact, especially the ones that are fossil-fuel-derived. The world, on an average, produces over 275 million tons of waste every year (Jambeck et al., 2015). Therefore, taking the advantage of this fact, industrially compatible technology and processes capable of consuming these plastic wastes as feedstock can cause a monumental change in the current manufacturing arena by pushing it towards a more sustainable model of exploiting waste. In this avenue, thermochemical depolymerization of plastics holds some advantage in processing the plastic waste but is often undervalued compared to liquid/gaseous fuel production or re-polymerization of polymeric precursors. Combining thermochemical depolymerization of the plastic wastes with suitable microbial fermentation processes can be carried out to yield the respective monomer precursors and/or produce syngas (synthesis gas). This exciting approach has been reported by several researchers for plastic recycling (Drzyzga et al., 2015).

The first approach of using depolymerization techniques to yield monomeric precursors from these plastic wastes can be developed using biotechnological routes such as microbial degradation. Wei and co-workers have reported that PET can be subjected to bacterial depolymerization by incorporating bacterial/fungal-derived enzymatic hydrolysis (Wei et al., 2014). Similar types of hydrolytic enzymes are already being produced in large quantities (proteases and lipases) at relatively low costs for a wide range of applications such as biorefineries and detergent additives. Hence, these existing protocols can be used in situ for degrading the plastic wastes into monomeric precursor molecules by suitable microbial organisms at higher kinetic rates. Researchers can also aim to improve the degradation kinetics even further by incorporating suitable genetic engineering on these microbial agents and promoting the use of hydrolytic enzymes that can be extracted from nature. These next-generation microbial agents will be able to work at industrial scales for the conversion of plastic wastes into novel feedstock materials.

While devising innovative process routes for treating plastic wastes for feedstock production, exploring other efficient pathways is vital besides hydrolysis. Producing good-quality plastic monomers from the depolymerization route often suffers issues like inconsistent yields. For example, Kim et al. (2013) reported the bioengineering of xylose metabolizing microbes for producing lignocellulose feedstock from plastic wastes. Even though direct degradation of PET by *Pseudomonas* is not reported, still it is able to use PET as a carbon source for producing PHA. Furthermore, another byproduct formed by the degradation of PET, that is, ethylene glycol has also been reported as a feed material for certain *Pseudomonas* strains as well (Mückschel et al., 2012).

17.22 Synergistic approach for plastic waste conversion

The fundamental basis of a green-circular economy is to make sure that the majority of the waste is minimized and reutilized in a viable manner. Generating alternate feedstocks from plastic wastes is necessary as much as making a proper use of these recycled polymer precursors. As discussed in the previous section, that showed how enzymatic hydrolysis can be used to convert non-biodegradable plastic wastes into usable products such as PHA which, in turn, offers a scope for circular economy. The produced monomers can also be used as feedstock for plastics that need chiral building blocks. If successfully carried out, then the use of enzymes for processing can create a pathway for the complete biological recycling of plastic wastes in the form of circular economy. Hence, integrating hydrolytic enzymes in the form of unique microbial reactor platform can easily capture carbon trapped in the plastic wastes into valuable products by using microorganisms.

For example, some types of anaerobic microbes like Acetogens use WLP (Wood-Ljungdhal pathway) for the synthesis of acetyl coenzyme A which is an active constituent for aerobic respiration in biological systems (Youssef et al., 2019). This WLP is the only reported route that uses carbon dioxide fixation for carrying out energy conversion in the most energy-efficient manner possible till date. As a model example, *Acetobacterium woodii* constitutes a group of acetogens which exclusively convert carbon dioxide into acetate anions with a conversion efficiency of more than 94% (Claassens et al., 2019). These acetate moieties can be exposed to aerobic conditions where they can be microbially metabolized into biomass or generate energy. Using a combination of these two pathways also with the respiration process allows the production of a wide range of products that are commercially viable. Sakimoto et al. (2016) reported the use of CdS nanoparticles decorated *Moorella thermoacetica* that can carry out photosynthetic WLP leading to the formation of acetate anions for futuristic carbon capture and sequestration.

Another route for achieving a carbon dioxide-mediated circular economy is to implement microbes that can partially metabolize carbon dioxide into biomass or other usable bioproducts as well via less-efficient autotrophic metabolic routes or carboxylation reactions. For example, chemolithoautotrophic bacteria use gaseous substances and produce organic acids and some forms of short-chained alcohols by using natural as well as bioengineered pathways (Takors et al., 2018). Since these microbes are sensitive toward oxygen, they avoid the use of ATP-intensive metabolic routes during bioremediation. Hence, these micro-organisms can undergo mixotrophic growth modes that alleviate the use of ATP-intensive metabolic processes. However, chemolithotrophic bacteria that are aerobic in nature are able to deal with these ATP-limiting pathways more efficiently. For example, Krieg et al. (2018) bioengineered *Cupreavidue necator* to become aerobic in nature and demonstrated proof of the above concept.

Using a combined chemo-biological approach for utilizing carbon dioxide fixation into usable bioproducts can also be used as an alternative. Producing high-energy moieties by the co-polymerization of carbon dioxide is a quick route to direct the fixation of carbon dioxide into usable products. However, synthetically it is difficult to carry out carbon chain elongation but microbial cells are able to produce it to a certain extent. Hence, supplementing these microbial routes with some chemical carbon dioxide reduction pathways will boost the yield and energy efficiency of the fixation techniques. For example, Liebal et al. (2018/2018) computationally analyzed the performance of cyanobacteria, acetogens, methylotrophs, and synthetic CO_2 fixation pathways in *Saccharomyces cerevisiae* in terms of production rates, product yields, and the optimization potential. In an attempt, Li et al. (2012) successfully engineered a litho-autotrophic microorganism, *Ralstonia eutropha* H16, to produce higher-chain alcohols in an electro-bioreactor using carbon dioxide as the sole carbon source and electricity as the sole energy input.

The amalgamation of microbiology and pyrolysis for the conversion of non-degradable plastic provides an unconventional pathway to link the biological and technological portion for the valorization of plastic waste (Blank et al., 2020). Furthermore, alternative processes like the solvent pre-treatment method can be studied to degrade the plastic waste in order to reduce energy expenditure, etc. The dumping of landfills can be completely evaded by incineration leading to energy recovery.

According to Kusenberg et al. (2022), plastic waste pyrolysis oil can be implicated as steam cracking feedstock upon removal of contaminants. However, contaminants like sodium, silicon and sulfur do not require any additional treatment.

The composite formation of biopolymers can lead to the formation of new materials with novel and improved features due to the additive advantage of blending of polymers. However, the compatibility and stability of the blended polymers is a challenge that has to be addressed and studied further in detail.

17.23 Ecology and environment: modified plastic usage and its future

While human activities continue to alter the surrounding habitat, the major concern remains to analyze the impact of the actions taken to emigrate to the surrounding niche. The assessment of the biodegradability and the impact on the living being of the modified products from the plastic substrate is a necessary measure to assure the suitability of the product for both human consumption and environmental effect. The physical and chemical behaviors including morphology, molecular weight, stability, crystalline nature, copolymer, functional group and environmental factor including the effect of pH, temperature, hydrophobicity, salinity, and ultraviolet radiation need to be assessed for the novel substrate and the end product to omit any hazardous impact.

The conversion of plastic waste into other valuable products provides a radical streamlining of the markets, resources, customers, and producers. It is estimated that around 4.3 trillion $USD will be managed upon the conversion of current plastic waste into some new products. Thus, the 6R principle including recycle, redesign, remanufacture, recovered, reuse, and reduce is being acknowledged by many countries to acquire smarter waste management and a sustainable circular economy model (Jiang et al., 2020). Nonetheless, the plastic waste upcycling is still in its infancy stage, and the rigorous study and usage of the upcycling approaches, including nanocomposite fabrication, vetrimerization, additive manufacturing, etc., for producing value-added products will emerge out as a promising alternative to recycling plastic waste.

17.24 Conclusion

Plastic waste upcycling has recently gained attention in order to troubleshoot and find better possibilities to confiscate the conventional process of treating these wastes. The waste conversion strategy, reaction mechanisms, product value/ type, and polymer designs are widely studied in recent days by several researchers. Biotechnology can provide several methodologies for the optimum utilization of plastic as feedstock to gain resource efficiency through the valorization of waste streams, reducing the burden on ecosystem resources.

Parallell research studies based on the replacement of plastic with bio-based kind are also been analyzed and produced such as ethylene glycol, and terephthalic acid. In real terms, it still requires intense research to completely replace plastic with bio-based products to escape from the problem of plastic waste generation.

The use of plastic waste as feedstock has only a few industrial implementations until now. Several challenges persist in this procedure like specifics of the degradation pathway, its combination with the microbial process and elucidation of gap between the lab scales to the industrial scale. Researchers are very positive about the biotechnological strategy for switching plastic waste to a high-value products with optimum rate, titer and yield. It is necessary to properly understand the factors (reaction temperature, processing time, plastic composition and catalyst activity etc.) that are hindering the process. An advance and cost-effective catalyst with optimum activity needs to be used to improve the conversion process. Furthermore, additional research is required on upcycling of thermosets and tactics to shrink the growing speed of worldwide plastic demand and spread awareness about its reuse.

References

Adrados, A., de Marco, I., Caballero, B. M., López, A., Laresgoiti, M. F., & Torres, A. (2012). Pyrolysis of plastic packaging waste: A comparison of plastic residuals from material recovery facilities with simulated plastic waste. *Waste Management*, 32(5), 826–832. Available from https://doi.org/10.1016/j.wasman.2011.06.016.

Ahmad, R., Hamidin, N., Ali, U. F. M., & Abidin, C. Z. A. (2014). Characterization of bio-oil from palm kernel shell pyrolysis. *Journal Of Mechanical Engineering And Sciences*, 7(1), 1134–1140. Available from https://doi.org/10.15282/jmes.7.2014.12.0110.

Ameer, Ameer A., Abdallh, Mustafa S., Ahmed, Ahmed A., & Yousif, Emad A. (2013). Synthesis and characterization of polyvinyl chloride chemically modified by amines. *Open Journal of Polymer Chemistry*, 03(01), 11–15. Available from https://doi.org/10.4236/ojpchem.2013.31003.

Arvanitoyannis, I. S. (2012). *Waste management for polymers in food packaging industries.*

Barnes, D. K. A., Galgani, F., Thompson, R. C., & Barlaz, M. (2009). Accumulation and fragmentation of plastic debris in global environments. *Philosophical Transactions of the Royal Society B: Biological Sciences*, 364(1526), 1985–1998. Available from https://doi.org/10.1098/rstb.2008.0205, http://rstb.royalsocietypublishing.org/content/364/1526/1985.full.pdf.

Bassi, A. (2017). *Biotechnology for the management of plastic wastes* (pp. 293–310). Elsevier BV. Available from 10.1016/b978-0-444-63664-5.00013-7.

Benham, E., & McDaniel, M. (2000). Polyethylene, high density. *Kirk-Othmer Encyclopedia of Chemical Technology.*

Blank, L. M., Narancic, T., Mampel, J., Tiso, T., & O'Connor, K. (2020). Biotechnological upcycling of plastic waste and other non-conventional feedstocks in a circular economy. *Current Opinion in Biotechnology*, 62, 212–219.

Bombelli, P., Howe, C. J., & Bertocchini, F. (2017). Polyethylene bio-degradation by caterpillars of the wax moth Galleria mellonella. *Current Biology*, 27(8), R292–R293. Available from https://doi.org/10.1016/j.cub.2017.02.060, http://www.elsevier.com/journals/current-biology/0960-9822.

Borg, Kim, Lennox, Alyse, Kaufman, Stefan, Tull, Fraser, Prime, Renee, Rogers, Luke, & Dunstan, Emily (2022). Curbing plastic consumption: A review of single-use plastic behaviour change interventions. *Journal of Cleaner Production*, 344. Available from https://doi.org/10.1016/j.jclepro.2022.131077.

Brandon, A. M., Gao, S. H., Tian, R., Ning, D., Yang, S. S., Zhou, J., Wu, W. M., & Criddle, C. S. (2018). Biodegradation of polyethylene and plastic mixtures in mealworms (larvae of Tenebrio molitor) and effects on the gut microbiome. *Environmental Science and Technology*, 52(11), 6526–6533. Available from https://doi.org/10.1021/acs.est.8b02301, http://pubs.acs.org/journal/esthag.

Bridgwater, A. V., Meier, D., & Radlein, D. (1999). An overview of fast pyrolysis of biomass. *Organic Geochemistry*, 30(12), 1479–1493.

Chen, X. Q., Guo, Z. Y., Wang, L., Yan, Z. F., Jin, C. X., Huang, Q. S., Kong, D. M., Rao, D. M., & Wu, J. (2022). Directional-path modification strategy enhances PET hydrolase catalysis of plastic degradation. *Journal of Hazardous Materials*, 433. Available from https://doi.org/10.1016/j.jhazmat.2022.128816, http://www.elsevier.com/locate/jhazmat.

Claassens, N. J., Cotton, C. A. R., Kopljar, D., & Bar-Even, A. (2019). Making quantitative sense of electromicrobial production. *Nature Catalysis*, 2(5), 437–447. Available from https://www.nature.com/natcatal/.

da Costa, J. P., Duarte, A. C., & Rocha-Santos, T. A. (2017). *Microplastics—occurrence, fate and behaviour in the environment, . Comprehensive analytical chemistry* (Vol. 75, pp. 1–24). Elsevier.

Dai, Lin, Li, Dan, & He, Jing (2013). Degradation of graft polymer and blend based on cellulose and poly(L-lactide). *Journal of Applied Polymer Science*, 130(4), 2257–2264. Available from https://doi.org/10.1002/app.39451.

Drzyzga, O., Revelles, O., Durante-Rodríguez, G., Díaz, E., García, J. L., & Prieto, A. (2015). New challenges for syngas fermentation: towards production of biopolymers. *Journal of Chemical Technology & Biotechnology*, 90(10), 1735–1751.

Enomoto, K., Takahashi, S., Iwase, T., Yamashita, T., & Maekawa, Y. (2011). Degradation manner of polymer grafts chemically attached on thermally stable polymer films: Swelling-induced detachment of hydrophilic grafts from hydrophobic polymer substrates in aqueous media. *Journal of Materials Chemistry*, 21(25), 9343–9349. Available from https://doi.org/10.1039/c0jm04084c.

European Commission (2013) Impact assessment for a proposal for a Directive of the European Parliament and of the Council amending Directive 94/62/EC on packaging and packaging waste to reduce the consumption of lightweight plastic carrier bags.

European Commission, A. (2018). A European strategy for plastics in a circular economy. *Brussels*, 28(1), 1–17.

Freeland, B., McCarthy, E., Balakrishnan, R., Fahy, S., Boland, A., Rochfort, K. D., Dabros, M., Marti, R., Kelleher, S. M., & Gaughran, J. (2022). A review of polylactic acid as a replacement material for single-use laboratory components. *Materials*, 15(9). Available from https://doi.org/10.3390/ma15092989, https://www.mdpi.com/1996-1944/15/9/2989/pdf.

Geyer, R., Jambeck, J. R., & Law, K. L. (2017). Production, use, and fate of all plastics ever made. *Science Advances*, 3(7). Available from https://doi.org/10.1126/sciadv.1700782, http://advances.sciencemag.org/.

Goel, V., Luthra, P., Kapur, G. S., & Ramakumar, S. S. V. (2021). Biodegradable/bio-plastics: Myths and realities. *Journal of Polymers and the Environment*, 29(10), 3079–3104. Available from https://doi.org/10.1007/s10924-021-02099-1, http://www.kluweronline.com/issn/1566-2543/.

Hassan, H., Hameed, B. H., & Lim, J. K. (2020). Co-pyrolysis of sugarcane bagasse and waste high-density polyethylene: Synergistic effect and product distributions. *Energy*, 191. Available from https://doi.org/10.1016/j.energy.2019.116545.

Jambeck, J. R., Geyer, R., Wilcox, C., Siegler, T. R., Perryman, M., Andrady, A., Narayan, R., & Law, K. L. (2015). Plastic waste inputs from land into the ocean. *Science (New York, N.Y.)*, 347(6223), 768–771. Available from https://doi.org/10.1126/science.1260352, http://www.sciencemag.org/content/347/6223/768.full.pdf.

Ji, Li. Na (2013). Study on preparation process and properties of polyethylene terephthalate (PET. *Applied Mechanics and Materials*, 312, 406–410. Available from https://doi.org/10.4028/http://www.scientific.net/amm.312.406.

Jiang, Tianyu, Duan, Qingfei, Zhu, Jian, Liu, Hongsheng, & Yu, Long (2020). Starch-based biodegradable materials: Challenges and opportunities. *Advanced Industrial and Engineering Polymer Research*, 3(1), 8−18. Available from https://doi.org/10.1016/j.aiepr.2019.11.003.

Jordan, J. L., Casem, D. T., Bradley, J. M., Dwivedi, A. K., Brown, E. N., & Jordan, C. W. (2016). Mechanical properties of low density polyethylene. *Journal of Dynamic Behavior of Materials*, 2(4), 411−420. Available from https://doi.org/10.1007/s40870-016-0076-0, http://www.springer.com/materials/special + types/journal/40870.

Kim, S. R., Park, Y. C., Jin, Y. S., & Seo, J. H. (2013). Strain engineering of Saccharomyces cerevisiae for enhanced xylose metabolism. *Biotechnology Advances*, 31(6), 851−861. Available from https://doi.org/10.1016/j.biotechadv.2013.03.004, http://www.elsevier.com/inca/publications/store/5/2/5/4/5/5/index.htt.

Koshti, Rupali, Mehta, Linchon, & Samarth, Nikesh (2018). Biological recycling of polyethylene terephthalate: A mini-review. *Journal of Polymers and the Environment*, 26(8), 3520−3529. Available from https://doi.org/10.1007/s10924-018-1214-7.

Krieg, Thomas, Sydow, Anne, Faust, Sonja, Huth, Ina, & Holtmann, Dirk (2018). CO_2 to Terpenes: autotrophic and electroautotrophic α-humulene production with Cupriavidus necator. *Angewandte Chemie International Edition*, 57(7), 1879−1882. Available from https://doi.org/10.1002/anie.201711302.

Krivoguz, Y. M., Pesetskii, S. S., Jurkowski, B., & Tomczyk, T. (2006). Structure and properties of polypropylene/low-density polyethylene blends grafted with itaconic acid in the course of reactive extrusion. *Journal of Applied Polymer Science*, 102(2), 1746−1754. Available from https://doi.org/10.1002/app.23998.

Krueger, M. C., Harms, H., & Schlosser, D. (2015). Prospects for microbiological solutions to environmental pollution with plastics. *Applied Microbiology and Biotechnology*, 99(21), 8857−8874. Available from https://doi.org/10.1007/s00253-015-6879-4, https://link.springer.de/link/service/journals/00253/index.htm.

Kumar, Ranvijay, & Singh, Rupinder (2020). *Application of nano porous materials for energy conservation and storage*, 1−5. Available from https://doi.org/10.1016/b978-0-12-803581-8.11278-0.

Kusenberg, M., Eschenbacher, A., Djokic, M. R., Zayoud, A., Ragaert, K., De Meester, S., & Van Geem, K. M. (2022). Opportunities and challenges for the application of post-consumer plastic waste pyrolysis oils as steam cracker feedstocks: To decontaminate or not to decontaminate. *Waste Management*, 138, 83−115. Available from https://doi.org/10.1016/j.wasman.2021.11.009, http://www.elsevier.com/locate/wasman.

Lear, G., Kingsbury, J. M., Franchini, S., Gambarini, V., Maday, S. D. M., Wallbank, J. A., Weaver, L., & Pantos, O. (2021). Plastics and the microbiome: impacts and solutions. *Environmental Microbiome*, 16(1). Available from https://doi.org/10.1186/s40793-020-00371-w.

Lebreton, L., Slat, B., Ferrari, F., Sainte-Rose, B., Aitken, J., Marthouse, R., Hajbane, S., Cunsolo, S., Schwarz, A., Levivier, A., Noble, K., Debeljak, P., Maral, H., Schoeneich-Argent, R., Brambini, R., & Reisser, J. (2018). Evidence that the Great Pacific Garbage Patch is rapidly accumulating plastic. *Scientific Reports*, 8(1). Available from https://doi.org/10.1038/s41598-018-22939-w.

Li, H., Opgenorth, P. H., Wernick, D. G., Rogers, S., Wu, T. Y., Higashide, W., & Liao, J. C. (2012). Integrated electromicrobial conversion of CO2 to higher alcohols. *Science (New York, N.Y.)*, 335(6076), 1596, -1596.

Liebal, U. W., Blank, L. M., & Ebert, B. E. (2018). CO2 to succinic acid−Estimating the potential of biocatalytic routes. *Metabolic Engineering Communications*, 7.

Maddah, H. A. (2016). Polypropylene as a promising plastic: A review. *American Journal of Polymer Science*, 6(1), 1−11.

Maschio, G., Koufopanos, C., & Lucchesi, A. (1992). Pyrolysis, a promising route for biomass utilization. *Bioresource Technology*, 42(3), 219−231. Available from https://doi.org/10.1016/0960-8524(92)90025-s.

Merrington, A. (2011). Michigan Molecular Institute, 1910 W. St. Andrews Road, Midland, MI 48640, USA. *Applied Plastics Engineering Handbook: Processing and Materials*, 177.

Miandad, R., Rehan, M., Barakat, M. A., Aburiazaiza, A. S., Khan, H., Ismail, I. M. I., Dhavamani, J., Gardy, J., Hassanpour, A., & Nizami, A. S. (2019). Catalytic pyrolysis of plastic waste: Moving toward pyrolysis based biorefineries. *Frontiers in Energy Research*, 7. Available from https://doi.org/10.3389/fenrg.2019.00027 , https://www.frontiersin.org/articles/10.3389/fenrg.2019.00027/full.

Mückschel, Björn, Simon, Oliver, Klebensberger, Janosch, Graf, Nadja, Rosche, Bettina, Altenbuchner, Josef, Pfannstiel, Jens, Huber, Armin, & Hauer, Bernhard (2012). Ethylene glycol metabolism by Pseudomonas putida. *Applied and Environmental Microbiology*, 78(24), 8531−8539. Available from https://doi.org/10.1128/aem.02062-12.

Nanda, S., Mohanty, P., Pant, K. K., Naik, S., Kozinski, J. A., & Dalai, A. K. (2013). Characterization of North American lignocellulosic biomass and biochars in terms of their candidacy for alternate renewable fuels. *Bioenergy. Research; A Journal of Science and its Applications*, 6, 663−677.

Narancic, T., & O'Connor, K. E. (2017). Microbial biotechnology addressing the plastic waste disaster. *Microbial Biotechnology*, 10(5), 1232−1235.

Naser, A. Z., Deiab, I., & Darras, B. M. (2021). Poly (lactic acid)(PLA) and polyhydroxyalkanoates (PHAs), green alternatives to petroleum-based plastics: A review. *RSC Advances*, 11(28), 17151−17196.

Payne, J., McKeown, P., & Jones, M. D. (2019). A circular economy approach to plastic waste. *Polymer Degradation and Stability*, 165, 170−181. Available from https://doi.org/10.1016/j.polymdegradstab.2019.05.014.

PlasticsEurope, Plastics: The facts. PlasticEurope (2017).

Ru, Jiakang, Huo, Yixin, & Yang, Yu (2020). Microbial degradation and valorization of plastic wastes. *Frontiers in Microbiology*, 11. Available from https://doi.org/10.3389/fmicb.2020.00442.

Saha, Nepu, Banivaheb, Soudeh, & Toufiq Reza, M. (2022). Towards solvothermal upcycling of mixed plastic wastes: Depolymerization pathways of waste plastics in sub- and supercritical toluene. *Energy Conversion and Management: X*, 13. Available from https://doi.org/10.1016/j.ecmx.2021.100158.

Sakimoto, K. K., Wong, A. B., & Yang, P. (2016). Self-photosensitization of nonphotosynthetic bacteria for solar-to-chemical production. *Science (New York, N.Y.)*, 351(6268), 74−77. Available from https://doi.org/10.1126/science.aad3317, http://www.sciencemag.org/content/351/6268/74.full.pdf.

Shen, Yafei (2016). Dechlorination of poly(vinyl chloride) wastes via hydrothermal carbonization with lignin for clean solid fuel production. *Industrial & Engineering Chemistry Research*, 55(44), 11638−11644. Available from https://doi.org/10.1021/acs.iecr.6b03365.

Skariyachan, S., Taskeen, N., Kishore, A. P., & Krishna, B. V. (2022). Recent advances in plastic degradation − From microbial consortia-based methods to data sciences and computational biology driven approaches. *Journal of Hazardous Materials*, 426. Available from https://doi.org/10.1016/j.jhazmat.2021.128086, http://www.elsevier.com/locate/jhazmat.

Takors, Ralf, Kopf, Michael, Mampel, Joerg, Bluemke, Wilfried, Blombach, Bastian, Eikmanns, Bernhard, Bengelsdorf, Frank R., Weuster-Botz, Dirk, & Dürre, Peter (2018). Using gas mixtures of CO, CO_2 and H_2 as microbial substrates: The do's and don'ts of successful technology transfer from laboratory to production scale. *Microbial Biotechnology*, 11(4), 606–625. Available from https://doi.org/10.1111/1751-7915.13270.

Teuten, E. L., Saquing, J. M., Knappe, D. R. U., Barlaz, M. A., Jonsson, S., Björn, A., Rowland, S. J., Thompson, R. C., Galloway, T. S., Yamashita, R., Ochi, D., Watanuki, Y., Moore, C., Viet, P. H., Tana, T. S., Prudente, M., Boonyatumanond, R., Zakaria, M. P., Akkhavong, K., ... Takada, H. (2009). Transport and release of chemicals from plastics to the environment and to wildlife. *Philosophical Transactions of the Royal Society B: Biological Sciences*, 364(1526), 2027–2045. Available from https://doi.org/10.1098/rstb.2008.0284, http://rstb.royalsocietypublishing.org/content/364/1526/2027.full.pdf.

Toraman, H. E., Dijkmans, T., Djokic, M. R., Van Geem, K. M., & Marin, G. B. (2014). Detailed compositional characterization of plastic waste pyrolysis oil by comprehensive two-dimensional gas-chromatography coupled to multiple detectors. *Journal of Chromatography. A*, 1359, 237–246. Available from https://doi.org/10.1016/j.chroma.2014.07.017, http://www.elsevier.com/locate/chroma.

Uekert, T., Kuehnel, M. F., Wakerley, D. W., & Reisner, E. (2018). Plastic waste as a feedstock for solar-driven H_2 generation. *Energy and Environmental Science*, 11(10), 2853–2857. Available from https://doi.org/10.1039/c8ee01408f, http://pubs.rsc.org/en/journals/journal/ee.

Vehlow, J., Bergfeldt, B., Jay, K., Seifert, H., Wanke, T., & Mark, F. E. (2000). Thermal treatment of electrical and electronic waste plastics. *Waste Management and Research*, 18(2), 131–140. Available from https://doi.org/10.1034/j.1399-3070.2000.00107.x.

Wagner, Martin, & Oehlmann, J. örg (2009). Endocrine disruptors in bottled mineral water: Total estrogenic burden and migration from plastic bottles. *Environmental Science and Pollution Research*, 16(3), 278–286. Available from https://doi.org/10.1007/s11356-009-0107-7.

Wang, X., Sun, M., & Wang, Q. (2019). Liquefaction characteristics of waste electronic plastics in supercritical isopropyl alcohol. *Ekoloji.*, 28(107), 3235–3247. Available from http://ekolojidergisi.com/download/liquefaction-characteristics-of-waste-electronic-plastics-in-supercritical-isopropyl-alcohol-5963.pdf.

Wei, R., Oeser, T., & Zimmermann, W. (2014). Synthetic polyester-hydrolyzing enzymes from thermophilic actinomycetes. *Advances in Applied Microbiology*, 89, 267–305.

Wright, S. L., Rowe, D., Thompson, R. C., & Galloway, T. S. (2013). Microplastic ingestion decreases energy reserves in marine worms. *Current Biology*, 23(23), R1031–R1033. Available from https://doi.org/10.1016/j.cub.2013.10.068, http://www.elsevier.com/journals/current-biology/0960-9822.

Yang, Y., Yang, J., Wu, W. M., Zhao, J., Song, Y., Gao, L., Yang, R., & Jiang, L. (2015). Biodegradation and mineralization of polystyrene by plastic-eating mealworms: Part 1. Chemical and physical characterization and isotopic tests. *Environmental Science and Technology*, 49(20), 12080–12086. Available from https://doi.org/10.1021/acs.est.5b02661, http://pubs.acs.org/journal/esthag.

Youssef, N. H., Farag, I. F., Rudy, S., Mulliner, A., Walker, K., Caldwell, F., Miller, M., Hoff, W., & Elshahed, M. (2019). The Wood–Ljungdahl pathway as a key component of metabolic versatility in candidate phylum Bipolaricaulota (Acetothermia, OP1). *Environmental Microbiology Reports*, 11(4), 538–547. Available from https://doi.org/10.1111/1758-2229.12753, http://onlinelibrary.wiley.com/journal/10.1111/(ISSN)1758-2229.

Zhan, Lu, Zhao, Xuyuan, Ahmad, Zahoor, & Xu, Zhenming (2020). Leaching behavior of Sb and Br from E-waste flame retardant plastics. *Chemosphere*, 245. Available from https://doi.org/10.1016/j.chemosphere.2019.125684.

Zhao, X., Boruah, B., Chin, K. F., Đokić, M., Modak, J. M., & Soo, H. S. (2022). Upcycling to sustainably reuse plastics. *Advanced Materials*, 34(25). Available from https://doi.org/10.1002/adma.202100843, http://onlinelibrary.wiley.com/journal/10.1002/(ISSN)1521-4095.

CHAPTER

18

Catalytic conversion of biomass-based wastes: upgrading and valorization to value-added intermediates and platform molecules

Putrakumar Balla[1], Satya Kamal Chirauri[2], Srinivasarao Ginjupalli[3], Rajenidran Rajesh[4], Prathap Challa[4], Sungtak Kim[1] and Prem Kumar Seelam[5]

[1]Department of Chemical Engineering and Applied Chemistry, Chungnam National University, Daejeon, Republic of Korea [2]Micron Technology Operations LLP, Hyderabad, Telangana, India [3]Department of Applied Science, University of Technology and Applied Science, Muscat, Sultanate of Oman [4]Energy & Environmental Engineering Department, CSIR-Indian Institute of Chemical Technology, Hyderabad, Telangana, India [5]Sustainable Chemistry Research Unit, Faculty of Technology, University of Oulu, Oulu, Finland

18.1 Introduction

The ubiquitous fossil-based resources had dominated over the years in manufacturing and producing wide spectrum of products such as fuels, plastics, polymers, intermediates, and chemicals has significantly influenced today's civilization. However, due to the limited nature of fossil feedstocks, more attention has been focused on discovering alternative non-conventional renewable energy sources such as biomass (Tong et al., 2019; Ventura et al., 2020). Other documented negative climatic consequences of fossil fuel use include the recurring occurrence of oil crises, greenhouse gas emissions, and the increasingly major environmental concerns produced by industrial emissions. Nevertheless, the future biorefineries provide the sustainable solutions to manufacturing more eco-friendly products and processes (Sheldon, 2014). In a sustainable biorefinery scheme, renewable sources are turned into high-value products such as biodegradable polymers and bioplastics (Sudarsanam et al., 2020). This means that the whole process is eco-friendly and highly environmentally sustainable. This is done by combining different processes that allow different biological wastes to be used while reducing the energy use, raw material consumption, waste production, and greenhouse gases (Karthikeyan et al., 2017).

Similarly, in the petrochemical sector, the success of the bio-based chemical industry depends on its ability to produce a wide range of raw materials and final products. Globally, many scientists, and chemical companies are looking for alternatives, such as biomass-based materials for producing wide variety of basic chemicals and intermediates. Recently, many funding is now being provided and supported by regional, national, and international agencies in R&D on bio-based and sustainable chemicals. It is anticipated that the platform molecules will play a significant role in the ongoing bio-based revolution.

Globally, different of wastes and biowastes are generated from human and industrial activities. For example, food waste including leftovers, expired, or precooked is discharged from various sources such as households, industries, and hospitals often ends up in landfills (especially in developing countries) in municipal sewage sludge. The United Nations Environment Programme's 2021 Food Waste Index has found that an estimated 931 million tons of food waste end up in the trash every year. According to the food and agriculture organization (FAO), one-third of the world's food is wasted. Circular bioeconomy aims at integrating this waste processing into green technologies to make economically viable products. Food waste which contains biomolecules such as carbohydrates, proteins, and lipids can further be broken down into simpler compounds such as amino acids, glucose, alcohols, and fatty acids.

Non-renewable fossil fuels are becoming more geological issues, price fluctuations, scarcity in most of the regions, climate change, and air pollution problems; thus, lignocellulosic biomass in particular is being exploited as a potential raw material in producing energy, fuels, and chemicals. There are many thermal and non-thermal conversion process of biomass such as pyrolysis, liquefaction, gasification, high-pressure and high-temperature hydrothermal treatment, and other methods are some of the most important thermo-chemical and/or biological processes (Thallada et al., 2021). Moreover, gaseous, liquid, and solid forms of bio-products are subsequently separated after the first step of processing.

The structural complexity of the biomass raw material is somewhat difficult to evaluate and each entity is valuable building blocks. Thus, transforming biomass into platform molecules and building blocks are critical to keep intact and improve the atom efficiency. The US DOE summed up a list of 12 possible bio-based platform chemicals that can be made in high quantities from biomass through biological or chemical processes (Kover et al., 2022). These building block chemicals could be turned into a wide range of useful bio-based chemicals, intermediates, and final materials in the future. The most challenging part of the whole process is, of course, finding the promising catalytic systems, process conditions and cost-effective methods.

Over the last two decades, researchers have paid considerable attention to perfecting biorefinery methods for producing sustainable energy and fuels from biomass. However, it has needed a great deal of work to get to commercialization owing to operational difficulties, poor margins and the need for large manufacturing volumes. Nevertheless, the development of the cutting-edge tools necessary to convert biomass into usable chemical precursors has received comparatively little scholarly interest. The production of biofuels has a greater impact on the environment than the manufacturing of chemicals from biomass, which has several advantages (Calcio Gaudino et al., 2019). As an example, chemicals have a greater value, making small or moderate-scale manufacturing profitable. After a standard treatment of lignocellulosic-type biomass thru biological (e.g., fermentation) or even both thermochemical and catalytic processes, distinct organic compounds, such as sugars, polyols, furanics, and acids like levulinic acid, succinic acid, itaconic acid, and 3-hydroxypropionic acid, could be easily created (Kover et al., 2022). To keep producing a wide variety of chemical products, including additives and constituents for petroleum products for cars, solvents and paints, and new monomer structures for the polymer industry, the aforementioned biomolecules could be used as novel precursor material and intermediate products (Platform Chemicals) (Dulie et al., 2021; Cho et al., 2020). According to market research report, significant CAGR of 7% from 2021 to 2029 was foreseen in valorization of waste and biowastes based on the impact score. The bio-renewable chemicals and intermediates in bio-plastics, bio-solvents, bio-cleaners, and detergents will replace the conventional chemicals which are directly or indirectly produced by using fossil-based resources (Source: Maximize Market Research Pvt Ltd, USA).

Further, biological treatment methods, such as composting and anaerobic digestion, have gained a lot of popularity in recent years (Nguyen et al., 2021). The byproduct of anaerobic digestion is biogas, where food waste could generate biogas and solid manure. However, researchers noted that waste composition and the use of detergents might affect the yield of biogas generation. Numerous studies suggested that enzymes' capacity to digest the waste will be diminished due to the insufficient nutrients. Moreover, reducing the landfill wastes can reduce the GHG emissions in the future. By doing so, greenhouse gas emissions may be reduced or eliminated altogether if trash is kept out of landfills.

In this chapter, the primary goal is to provide a brief introduction to biomass conversion technologies to the respective building blocks and platform molecules, various biomass conversion technologies, techniques that efficiently transform biomass-derived platform molecules into valuable products chemicals and fuels, along with the associated studies that looked at environmental impacts throughout the entire life cycle of these processes. The complete scheme on biowastes conversion to different end products such as liquid and gaseous fuels, chemicals and intermediates are displayed in Fig. 18.1.

18.2 Biomass conversion technologies

Several processing methods are used to transform the biomass into platform molecules, the most frequent and practical being thermo-chemical, chemo-catalytic, and biological methods (Fig. 18.1).

Several pathways exist in extracting valuable fuels and chemicals from biomass-based waste materials through thermochemical conversion (Burguete et al., 2016; Zheng et al., 2021). High temperatures ($>200°C$) are required

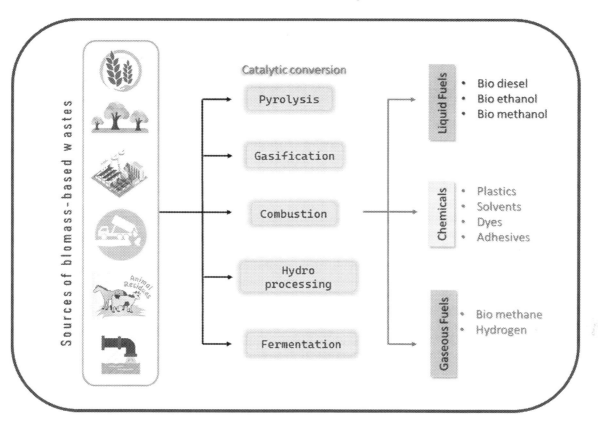

FIGURE 18.1 An overview of catalytic conversion of biomass-based waste into useful products.

to crack the biomolecule, either in the presence of oxygen (e.g., gasification) or without oxygen (e.g., pyrolysis). Thermochemical conversion process such as pyrolysis process, which includes all the chemical reactions that take place by heating the material in the absence of oxygen. The typical pyrolyzed products are steam, charcoal, organic oils/tars, and high amounts of gases including methane, hydrogen, carbon monoxide, and carbon dioxide. The pyrolysis is done two different ways, one approach is slow and other being fast pyrolysis and solely depends on many factors such as end products, which one choose (Hammerschmidt et al., 2011).

Traditionally, the production of charcoal for the use as fuel has been the fundamental motivation for the slow pyrolysis of biomass (Perveen et al., 2022). Whereas in fast pyrolysis, involves the rapid and controlled heating of biomass to target gaseous products. The fast pyrolysis gained popularity in the 20th century due to the heating rate significantly increase the liquid yield. Moreover, fast pyrolysis may result in high liquid yields above 70 wt.%, and the resultant bioliquid can be further processed to produce a wide range of platform molecules.

Gasification is a thermal process that converts solid biomass into gaseous combustible producer gas with varied concentrations, and it is an extension of pyrolysis and optimizes carbon and energy yields in the gas phase. Typically, syngas is produced by the controlled partial burning of biomass at high pressures and temperatures (more than 700°C). To increase the syngas' energy density, the Fischer–Tropsch technique is used to upgrade it further to hydrocarbon fuels.

Chemical catalytic processes are those that include one or more chemical agents. These reactions occur at moderate temperatures, which are often far lower than the temperatures necessary for thermal treatments. Reduction or oxidation of saccharides, acid treatment of saccharides to create 5-(chloromethyl) furfural, transesterification of triglycerides, deacetylation and depolymerization of chitin, and hydrolytic hydrogenation of lignin are examples of chemical-catalytic processes (Priya et al., 2016).

Using isolated enzymes and microorganisms constitutes a biological process (Mohanty et al., 2022). Biological treatments have shown promising pretreatment before additional processing stages. Advances in the biological breakdown of cellulose to glucose, for example, might pave the way for cheaper production in reducing the sugars such as sorbitol and xylitol from biomass. The need for diverse pretreatment methods for lignocellulose feed stock is an important factor in biological processing of biomass. Products recovered after fermentation might also had cost increase due to demand of high purity.

18.3 Utilization of biological-waste to value-added products

Anaerobic digestion has the potential to save energy costs and produce power locally. It may be implemented and integrated the biogas in wastewater treatment plants for saving fuel and energy cost to run the plant. Incineration in energy recovery is another method of waste valorization. Therefore, a number of facilities are installed worldwide, especially devoted to treat the biological wastes (Kapoor et al., 2020). However, the process of burning biological waste is energy-demanding process and has the potential to destroy useful functional groups in the feedstocks that are refined from it. Meanwhile, through bio-refinery approach, a number of value-added industrial chemicals, fuels, bio-products could be produced (Roy et al., 2023). Waste valorization has the ability to generate materials, chemicals, and products, as will many more will be discussed in this section.

To combat the daily waste, for example, food waste, scientists have been focusing on bio-processing to turn stale bread into consumer goods (e.g., waste bread/potatoes into bioethanol). Solid-state fermentation by fungi converts complex carbohydrates from unprocessed food waste into simpler sugars, which may then be used to make bioplastics like poly(3-hydroxybutyrate) and platform chemicals (e.g., SA). Another wastes from bakeries/confectionaries will be employed in the proposed biotechnological process since they may include starch, fructose, free amino nitrogen, and small quantities of supplementary nutrients which can be processed further to turn them into valuable molecules.

Several methods are mentioned that may be used to convert trash into valuable compounds (Fig. 18.2). Another fascinating option to increase the valorization procedures is the chemical exploitation of waste raw materials to create high-value chemicals. In a recent study, a solid acid catalyst was produced in transforming the high free fatty acids (FFA) waste cooking oil into biofuels with properties similar to that of biodiesel. The developed solid acid catalyst may simultaneously esterify FFA in waste oil and transesterify residual triglycerides. With this method, two types of food wastes can be used to make two useful things: a cheap solid acid catalyst and biofuels similar to biodiesel (e.g., corncobs and waste cooking oils can be potential feedstocks). Several approaches were used to characterize the solid acid catalysts. FTIR indicated various chemical bonding present, that is, $C=O$, C-O, C-S, and aromatic $C=C$ bonds in the waste materials. Solids' catalyzed activity converted the waste cooking oils into

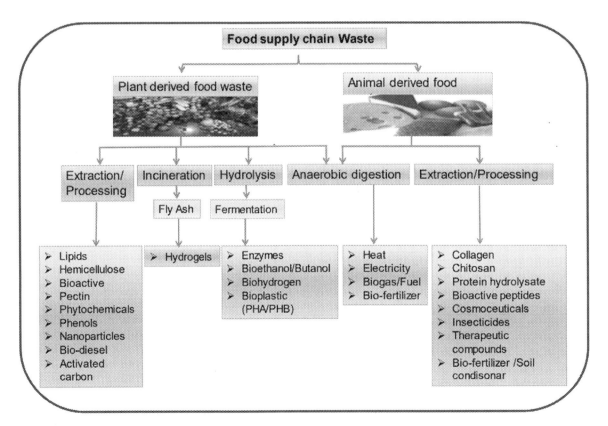

FIGURE 18.2 Bioprocessing methods of whole food value chain wastes to value added chemical (Karthikeyan et al., 2017). *Copyright permission from Elsevier (Exploitation of Food Industry Waste for High-Value Products Rajeev Ravindran, Amit K. Jaiswal Publication: Trends in Biotechnology, January 2016).*

biodiesel-like biofuels. Without filtration, oils might generate 98% methyl esters which is significantly a high yield. Transesterification kinetics were slower than oil FFA esterification. Despite minimal $-SO_3H$ loading, catalytic activity was higher, thus suggesting the distinct surface functionality is required. However, there is a challenge in improving the Solid acid catalysts' recyclability. After two reaction cycles, catalysts deactivate fast owing to water breakdown and hydrolysis of sulfonated groups. Moreover, the temperature, carbonizing atmosphere, and pressure should be examined to increase the catalyst's stability and robustness. Electrochemically generating a relatively pure biofuel from a residue reveals encouraging concept and evidence of the potency of an embedded valorization of numerous waste raw materials into the beneficial final products. Thus, biofuels without the need for complex and expensive pre-treatment of oils and customarily necessary to eliminate the high FFA content to enable the conventional based-transesterification process to be feasible (Mainali, 2012).

Further, nanocellulose, a biomaterial made from cellulose-rich sources, has gained popularity lately. Nano celluloses have great mechanical strength, high specific surface area, extensive surface modification capability, extremely low gas permeability, high biodegradability, and low toxicity. Cellulose nanocrystals (CNCs) and cellulose nanofibers (CNFs) may be differentiated by shape, dimension, and size of enzyme action. Several different types of cellulosic waste, including mandarin peel waste, rice straw, grape skin, corncobs, and sawdust, may be used to create nanocellulose. In order to mass-produce nanocellulose from inexpensive feedstocks, the American researchers developed a novel manufacturing process with low-energy CNFs or CNCs were obtained by fractionating the biomass with sulfur and aqueous ethanol to produce cellulose-rich solids, and then mechanically treating the cellulose-rich stream.

In addition, considering that only 7% of the oil used every year (3×10^8 t) is needed to make chemicals. Moreover, 1.3×10^9 tons of food waste generated every year around the world, which is a renewable resource can be used in making chemicals that should not be ignored, at least in terms of capacity. Raw material production and storage (mostly in developing countries), human consumption, industrial and municipal waste disposal (including residential wastes, expired food, wasted items from restaurants, supermarkets, and caterers) are the primary sources of food waste. The process of recovering value from food scraps is gaining momentum, and it may end up being an important factor in the production of chemicals and biopolymers for the bio-economy. In terms of water, carbohydrates, proteins, and oils, below is the breakdown of the most common byproducts and wastes from the food industry (Fig. 18.2). Carbohydrates and lipids are potential carbon sources, whereas proteins may be employed to provide nitrogen in fermentation processes. However, using food waste alone in fermentation-based chemical and biopolymer production may not be economically feasible. Making ethanol from olive pulp processing waste streams, for instance, will only be cost-effective if it is integrated into the production of many other value-added products and energy. Because of this, the optimization of feedstock use needs to be accomplished via the development of various approaches and methods in future biorefineries. In this way, it is necessary to first extract valuable phytochemicals before creating a fermentation media that makes efficient use of these compounds. For example, researchers were able to extract phytochemicals like hesperidin and limonene from orange peel residues by using microwaves. Several flavonoids and anthocyanidins are also found in many fruit wastes. These flavonoids can be extracted from the waste and by-product streams that come from processing these fruit wastes. The remaining parts, such as the extra protein, need to be processed further and sold in different markets. There has been a lot of work done recently to figure out how to best employ streams of food waste and various fractionation and valorization procedures in the construction of biorefineries. For example, Oreopoulou and Tzia developed a fractionation process to separate carotenoids, antioxidants, fibers, and pectins from apple pomace. Research is being conducted to see whether biotechnological processes can be used to extract ethanol, organic acids, and enzymes from olive mill waste waters. The separation process might provide useful isolates and a source of fermentable lactose. Hence, due to the rising global production of such waste residues, waste valorization has emerged as an appealing notion. That is why scientists are not only working on better valuation methods; they are also putting in extra effort to create greener materials by incorporating various green technologies. Therefore, a number of facilities are installed worldwide, especially devoted to treat the food waste. Meanwhile, through bio-refinery approach, a number of value-added industrial chemicals/feed stocks/fuels/bio-products could be produced.

18.4 Efficient conversion process of biomass derived platform molecules into fine chemicals and fuels

Several reviews on reactions of biomass platform chemicals have been published (Peixoto et al., 2022). Due to the need for renewable energy resources, this subject has expansive potential. Biomass valorization uses homogeneous and heterogeneous catalysts. In homogeneous catalysis for biomass transformation, gaining selectivity towards desired products is challenging because substrate-catalyst interactions are difficult to manage. Separation and purification make large-scale homogeneous catalysis more challenging and expensive. Heterogeneous catalysts having active sites and reactants in various phases may offer consistent catalyst durability and simple product separation in biomass conversion (Serrano-Ruiz et al., 2010).

Further, when it comes to heterogeneous catalysts, metal oxide catalysts have been found to be the best. This is because their macroporous structures can speed up organic reactions without a solvent, which leads to excellent catalytic activity. Even though metal oxide catalysts have come a long way and are used a lot in making biofuels, the reaction is still not selective. Low reaction activity is likely due to unstable bulk characteristics of active metal species. This bulk-character catalyst often has few exposed active sites. This is because the metal-support interface, defect active sites, and angular and angular-to-edge active sites are all depleted in these regions. Spreading the active metal atom by atom over the catalyst support might solve this problem.

Chemical or biotechnological processes may be employed to convert renewable carbon into chemicals. Complete analysis of sustainability, resource availability, and resource transportation will inform the final choice. Research into chemical or biotechnological methods (using microorganisms or enzymes) for the production of chemicals and products from biomass is the focus of "green chemistry" and "white biotechnology," which are two distinct areas of study. This chapter discusses recent achievements and future goals for the fermentation-based manufacturing of bio-based chemicals and polymers.

18.4.1 Aqueous-phase catalytic conversions of biomass platform chemicals

Water may operate as a reactant, source of hydrogen, or catalyst in addition to being a solvent, unlike comparatively inert organic solvents such as toluene and dioxane. Therefore, the aqueous-phase conversions need a more in-depth consideration than the organic-phase conversions or the gas-phase conversions since catalysts display totally diverse activity and stabilities in each of these phases. Heterogeneous catalysts for the aqueous phase conversion of biomass to chemicals are particularly well-suited to materials with a high surface modification and specific surface areas, such as alumina, silica, carbon compounds, and metal oxides. Gas-phase hydrocarbon conversion relies heavily on heterogeneous catalysts, which are widely used because they can withstand high temperatures, work with non-polar molecules, and are therefore useful to the petrochemical industry. There is a discussion of new findings about heterogeneous catalytic processes in the aqueous phase of biomass platform chemicals, such as dehydration and hydrogenation.

High-value compounds of furan compounds, such as furfural and HMF are produced by the dehydrating the abundant renewable base chemicals, that is, ketose and pentose (Morales et al., 2020). There is a significant need for dehydration reactions in aqueous-phase catalytic processing in the generation of oxygenated hydrocarbons from biomass as a feedstock in the manufacturing of high-value chemicals and liquid fuels. Aqueous-phase dehydration is improved by solid acid catalysts because of their low cost, low environmental impact, and high recycling potential. Therefore, many solid acid catalysts have been studied including phosphates and nanosized mixed oxides.

In order to mitigate the drawbacks of limited large-scale use and arduous manufacturing procedure, researchers have concentrated on developing effective, reduced cost, simplified structure, and heterogeneous solid acid phosphate catalysts. In addition to efficiently converting xylose to furfural, the phosphate catalyst also benefits from the synergetic effect of Lewis and Brønsted acid sites. Pentose dehydrating to furfural was examined at the molecular level using HCl as a Brønsted acid catalyst in aqueous circumstances, as described by Vinit Choudhary et al. We present a series of steps that, when followed in order, will efficiently transform xylose into furfural in a batch reactor. To convert xylose to xylulose, a Lewis acid, CrCl3, is used, and to convert xylulose to furfural, a Brønsted acid, HCl, is used. Hydrogenation is a necessary process for the majority of biomass platform compounds. Hydrogenation, hydrogenolysis, and hydrodeoxygenation involve hydrogenating $C=O$, $C=C$, or $C-O$ bonds; $C=O$ links are easier to activate. In current hydrogenation research, VIII group metal catalysts, primarily Ni, Pt, Pd, or Ru, are investigated; noble metal catalysts typically exhibit enhanced catalytic activity; nevertheless, more effort is required to end the conversion at desired products, rather than excessive hydrogenation. Noble metal assisted catalysts attracted the most investigation due to their strong aqueous hydrogenation activity. Furfural and LA hydrogenated with approximately 100% conversion and selectivity using noble metal catalysts. Li et al. investigated the reaction processes of Cu and group VIII metals in converting furfural to furfural alcohol.

18.4.2 Transformation of platform molecules utilizing MOF based catalyst

Metal-organic frameworks (MOFs) and MOF-based materials have seen widespread usage for gas separation and adsorption hydrogen storage, and drug delivery as research into heterogeneous catalysis has advanced.

The first attempt at cellulose hydrolysis using a MOF-based acidic catalyst was reported, marking a significant milestone in the development of MOFs toward their eventual use in biomass valorization (Herbst & Janiak, 2017). Following that, a significant number of MOF-based catalysts were developed, manufactured, and used in biomass transformation due to the fast development of synthetic methodologies and characterization tools.

18.4.3 Conversion process utilizing single atom catalyst

The building blocks of single-atom catalysts are individual atoms that are dispersed over and/or coordinated with the surface atoms of suitable hosts (Lu et al., 2021). This provides an additional mechanism for regulating the activity and selectivity of catalytic processes (Fig. 18.3), in addition to increasing the efficiency of metal atoms (Xu et al., 2022). Activation of catalytic active sites in single-atom alloys is a natural consequence of the strong metallic contacts between isolated single atoms and the metal nanostructure/support (host). Maximum active sites with the highest surface free energy may be created by atomically scattering isolated single atoms or alloys on the catalyst support. As a matter of fact, various reaction studies have used either isolated atoms or alloys as catalysts.

A thorough understanding of how single atoms or alloys affect biofuel responses is, however, still missing. Although there is a scarcity of single atoms or alloys catalyzed processes for biofuel generation, this review will summarize the most recent developments in this field. This is very important for evaluating the viability of single atoms or alloys in biofuel production. A recent study shows that most single atom alloys catalysts are catalytic in a range of chemical processes (e.g., hydrogenation, hydrogenolysis, and H_2 evolution).

18.4.4 Fermentative process of important platform chemicals

Fermentative synthesis of chemicals and biopolymers in the period of the bio-economy should depend on the use of renewable carbon sources. The following sections detail the many types of industrial waste and byproducts that are generated by manufacturing sectors. Such renewable carbon sources may be used in the fermentation of chemicals and biopolymers. Concepts for a biorefinery that may produce chemicals through fermentation and value-added products via resource fractionation are also discussed (Table 18.1).

Extraction, hydrolysis, and thermal treatment of wood may provide chemicals and fuels. Both exudates and solvent extraction procedures may be used to get wood's non-structural components. Numerous phenolic chemicals,

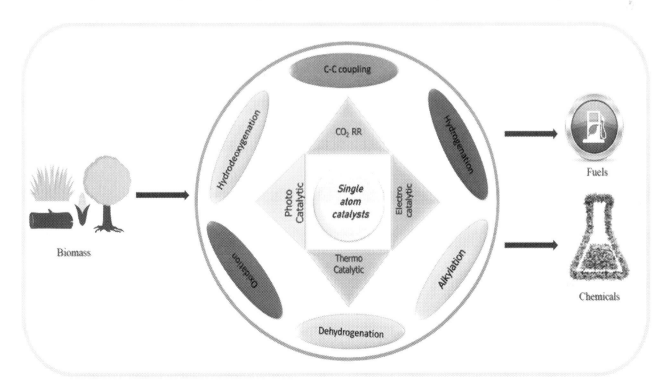

FIGURE 18.3 Schematic representation of various single atom catalytic biomass valorization.

TABLE 18.1 Food waste source and techniques used in bioenergy production (Mohanty et al., 2022).

Source	Microorganism	Parameters	Technique	Product
Fruit and vegetable fruit waste (FVW), co-digestion (CD), food waste (FW)	Acidogenic, acetogenic, and methanogenic microorganism	One stage batch reactor, Volatilesolids-8.74%, pH-7.24, OLR-2.12 g VS/L·d, HRT-30 days, temperature-37°C, reactor volume-0.00117 m	Anaerobic digestion	Biogas, highest production for FVW, FW, and CD is 2.2 L/g VS, 1.42.2 L/g VS and 1.22.2 L/g VS
Kitchen waste	Acidogenic, acetogenic, and methanogenic microorganism	Four digesters, kitchen manure/poultry manure-loading rate-300 mg/L with cow manure 50 g, mesophilic temperature-37°C	Anaerobic digestion	Biogas compressed biogas (920 11 mL) and methane content-48% in D3
Fruit and vegetable fruit waste	Acidogenic, acetogenic, and methanogenic microorganism	Biomass-465.12 kg/day, co digestion with cassava, temperature-37°C	Anaerobic digestion	Biogas 10 m^3/day
Palm oil mill effluent (POME) and fruit branches (FB)	S-AD Synergistaceae, Lachnospiraceae, and Caldicoprobacteraceae L-AD Porphyromonadaceae, Ruminococcaceae, Methanosaetaceae, Methanobacteriaceae	POME-liquid state AD, FB-solid state AD, HRT-30 days, Organic loading rate-1.66 g VS/L·d for L-AD and 6.03 g VS/L·d for S-AD	Anaerobic digestion	Biogas L-AD and S-AD increase biogas production by 29% and 71% compared to S-AD and L-AD alone
Organic waste substrate (rice, meat, beans, banana, lemon, pomegranate, apple and orange) and wheat straw	Ruminococcaceae, Synergistaceae, Spirochaetaceae, Cloacimonadaceae, Methanosaeta, Methanospirillu, Kosmotogaceae	pH for organic waste and inoculum 4.3 and 7.8, Carbon-44.8 ± 0.2%, Nitrogen-3.3 ± 0.03%, C/N-13.6 ± 0.06, OLR-4 g VS/L·d (80–100 days)	Anaerobic digestion	Biogas
Apple pomace	Kluyveromyces lactis, Kluyveromyces marxianus, Saccharomyces cerevisiae, and Lanchacea thermotolerans	Carbohydrate-16.64%, Hemicellulose-15.79%, lignin 19.8%, Total solid-865.6 g/kg, Volatile solid-720 g/kg, pH-7 to 8 (at the end)	Anaerobic digestion	Biogas and bioethanol CH4–596 L/g VS Methane yield makes it suitable for biogas production
Citrus waste	—	Optimized at 37°C for 5 days, highest VFA achieved-0.793 g VFA/VS at pH-6	Anaerobic digestion	Effect of VFA on biogas
Industrial food waste	Thermophilic condition Firmicutes, Synergistetes, Halanerobiateota, and Thermotogae Mesophilic condition Bacteriodetes, Firmicutes, and Cloacimonetes	Four industrial scale plug flow high solid treatment (HST), Volume ranging 340–650 m3, Initially operated at thermophilic condition-52°C followed by mesophilic temperature 38°C at later stages	Anaerobic digestion	Biogas 0.4–0.6 Nm^3 CH4/kg volatile acids Ammonia causes disturbance at thermophilic temperature
Grape must from 2 grape varieties (pinot gris and chardonnay)	Clostridium butyricum, Clostridium beijerinkii, Clostridium roseum and Clostridium diolis	Heat treatment-70°C for 1 h, mesophilic temperature-37°C, stirrer speed-220 rpm, initial pH5.7 and working pH-7	Dark Fermentation	Biohydrogen 205 ± 12 mL/L reactor h
Rice bran waste	Rhizopus oligosporus MTCC 556, Enterobacter aerogens MTCC 2822	Lignocellulose as substrate,180 rpm and 27°C ± 1°C, Media composition 0.05 g/L MgSO4·7H2O, yeast extract, 0.1 g/L KH2PO4, 0.05 g/L (NH4)2SO4, yeast extract	Dark Fermentation	Biohydrogen 5.4 mmol H2/g
Citrus peel	Escherichia and Clostridium species	pH-5.5–8.5, Temperature-30–40°C, Authochthonous inoculum-2.25 gTVS/L, Substrate-5–15 g/L, Nacl, peptone and CaCO3 (0–5 g/L)	Dark Fermentation, Plackett and Burman design	Highest biohydrogen production 13.29 mmol/L at pH-8.5 and substrate-15 g

Feedstock	Microorganism	Conditions	Method	Product/Yield
Sugarcane molasses	*Thermoanaerobacterium* and *Caproiciproducens* species	Thermophilic temperature-55°C, HRT-10 h, pH-5.38, Organic loading rate-86 kg-COD/m³ d, Substrate-6.21 ± 2.1 gCOD/gVSSd	Plackett and Burman design	Biohydrogen 8.5 NLH2/Ld
Banana waste	*Clostridium* and *Lactobacillus*	pH-5.5–7.7, temperature-30–44°C, headspace volume-40 to 60%, Inoculum-5%–15%, Carbohydrate 3–15 g/	Plackett and Burman design	Biohydrogen
Food waste	–	Ultrasound 1200 W, 20 kHz, time-5 min, total soild-2%, COD solubilization-1.98%	–	Biohydrogen at total solid 2% (5 min) of ultrasound treatment 38 mL/H2/VS
Food waste	*C. beijerinckii* NCIMB 8052	Carbohydrate-15%, Protein-5%, Fat-5%, Ash-1%, Moisture-74%	Acetone-butanolethanol fermentation (ABE)	Biobutanol 16.6 g/L butanol and 18.2 g/L hydrogen
Sweet sorghum	*C. acetobutylicum*	Lignocellulosic compounds as substrate, Drying-105 ± 3°C, 118 mL of serum bottles with 50 mL working volume, bottles filled with starch, sorghum grain powder, xylose, glucose	Acetone-butanolethanol fermentation	Biobutanol 8.3 g/L
Corn stover	*Clostridium saccharobutylicum* DSM 13864	Low acid treatment with nitric and sulfuric acid-0.89% at 125°C, detoxification of hydrolysate using adsorption resins, two stage gas stripping	Acetone-butanol ethanol fermentation (ABE), Response surface methodology (RSM)	Biobutanol and bioethanol 4.75 0.25 g/L acetone and 9.02 0.11 g/L butanol in 72 h
Food waste	*C. acetobutylicum* NRRL B-527	Batch and continuous operation using 20% (v/v) decanol in oleyl alcohol, Volumetric mass transfer coefficient (kLa)-0.025 1/min, NaOH (5%),	Acetone-butanol ethanol fermentation	Biobutanol
Sugarcane molasses	*C. beijerinckii* TISTR 1461	Rice bran, dried spent yeast and soya bean meal are used as substrate, Batch fermentation, Anaerobic condition	Response surface methodology (RSM)	Biobutanol 7 g/L, CaCO3 is substituted by eggshell as buffering agent
Banana peels	*Saccharomyces cerevisiae* and *Pichia* sp	Pre-treatment with sodium hydroxide and sulfuric acid	Acetone-butanolethanol fermentation	Biobutanol 90 g/L
Industrial food waste	*C. acetobutylicum* DP217 *C. acetobutylicum* ATCC 824 *C. beijerinckii* CC101 *C. acetobutylicum* B3	Recovery conditions *C. acetobutylicum* DP217 Pervaporation Silicalite 1 filled with PDMS/PAN composite membranes, Membrane area-0.0072 m2, Permeate side pressure-280 Pa	Acetone-butanolethanol fermentation	Biobutanol for *C. acetobutylicum* DP217 104.6 g/L Biobutanol for *C. acetobutylicum* ATCC 824 64 g/L

several polysaccharides, are among the most well-known components that may be extracted. Although the practice of removing some of these extractants dates back to ancient times, modern petrochemical technologies have largely supplanted the use of chemicals. Cellulose and hemicelluloses in wood wastes may be hydrolyzed under acidic circumstances or by enzyme-based treatment. Fermenting the sugars produces chemicals and biopolymers. Biochemical synthesis from wood biomass is a potential but challenging approach of valorizing this waste stream. Most important is cost-effectively extracting sugars from polysaccharides. The effectiveness of wood residue pretreatment, assuming enzymatic hydrolysis is used, influences all subsequent phases and the economics of the whole process. Simultaneous saccharification and fermentation may be used to intensify the process. The sugars that are synthesized may either be fermented immediately after enzymatic hydrolysis or saved for later use. Hydrolysis and fermentation can be performed in parallel for various reasons, including cost savings and improved efficiency (since the fermenting microorganisms are able to directly consume the sugars produced, end-product inhibition of the enzymes is reduced). Few researchers have focused on the bioproduction of other chemicals, such as lactic acid, 2,3-butanediol, butanol, fumaric, and succinic acid, from wood wastes. When triacylglycerols are reacted with methanol, 1 kg of glycerol is made for every 10 kg of biodiesel. Glycerol that has been transesterified is made neutral and then distilled to get rid of the methanol and water. The purity of the crude glycerol stream that comes out of making biodiesel depends on how it is made and what kind of oil is used. Depending on the catalyst employed in transesterification operations, crude glycerol may make up as much as 79% of a crude glycerol stream, with the remaining water, methanol, residual fatty acids and esters, sodium chloride (NaCl) or potassium sulfate (K2SO4), and so on varying from 5.3% to 14.2%. Glycerol is also a by-product of bioethanol synthesis, where the "thin stillage" left after ethanol distillation includes around 2% (w/v) glycerol. Sunflower meal is rich in antioxidants and protein, but its full potential is not realized when the meal is used in the manufacturing of fermentation vitamin supplements. Utilization of sunflower meal components was optimized, resulting in a state-of-the-art sunflower-based biorefinery that produced a high-purity protein isolate and an antioxidant-rich fraction. Sunflower meal was first fractionated using a sedimentation-flotation technique to separate it from a water-based solution. After completing this technique, three distinct fractions were obtained: one high in protein, one rich in lignocellulose, and one liquid fraction. Antioxidants were extracted from the protein- and lignocellulosic-rich fractions (mainly chlorogenic acid). The residual protein-rich fraction was subjected to acid and alkaline treatment to create a protein isolate (97% pure protein). This by-product has potential use in the manufacturing of biopolymers, amino acids, and food packaging. In addition, it may be hydrolyzed by enzymes, expanding its potential uses. Hydrolysis of the macromolecules in the residual sunflower streams relied on enzyme-rich fermented solids. The liquid byproduct of

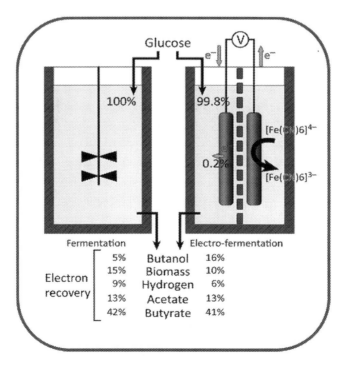

FIGURE 18.4 Electrochemically controlling microbial fermentative metabolism with electrodes (Moscoviz et al., 2016).

sunflower meal fractionation was used in enzymatic hydrolysis. When utilized as the sole fermentation medium, this nutrient-rich medium led to the formation of more than 40 g L1 PHB when combined with glycerol. Incorporating commercial levulinic acid into the fermentation medium also resulted in the production of P(3HB-co-3HV). Levulinic acid, which is used in the pharmaceutical industry, might be extracted from cellulose in the future using biorefineries. Alcoholic fermentation uses yeast to convert carbohydrates into ethanol and carbon dioxide. It's an anaerobic integration process since it occurs without oxygen. Heterotrophic algae or yeast may convert carbohydrates into lipids within their cells utilizing solvents. Using a combination of oligomerization and hydro-treatment, genetically modified bacteria can convert carbohydrates into short-chain gaseous alkenes, which may then be refined into jet fuel or gasoline. Electro-fermentation is a new fermentation pathway wherein microorganisms are stimulated using an electric field (Fig. 18.4), and it has the potential to do the following things well: (1) stabilize and maximize metabolisms of fermentation transition process by controlling redox and pH disparities; (2) induce whether breakdown or deformation of carbon chain via distinct oxidizing conditions; (3) increase the fabrication of adenosine triphosphate and enhance microbial biomass production; and (4) expel harmful byproducts.

18.5 Conclusions

Abundant biowaste are generated in millions of tons around the globe every day. There are environmental climate goals, and the objective of reducing waste is the highest priority to attain the sustainable development goals (SDGs). Waste can be efficiently processed and reused in many forms. There is huge scope of R&D that are needed to develop processing units in decentralized framework to treat and minimize the waste and its effects on the environment. The major energy supply of the world is composed of several different sources, but 9% ($\sim$ 51 EJ) comes from biomass, which is economical, dependable, and sustainable. Efficient conversion process of biomass derived biowastes to platform molecules into fine chemicals and fuels were explained in four routes, such as aqueous phase processing, conversion process utilizing MOF based catalyst, conversion process utilizing single atom catalyst, and fermentation process. The ultimate goal is to develop chemical processes that are both environmentally benign and scalable, and there is a lot of opportunity for development in the domain of continuous flow chemical processes for the conversion of biomass-derived biowastes to fuels and chemicals. The highlighted instances illustrate the diversity of chemical transformations available for biomass-derived platform molecules synthesized in a continuous flow setting using heterogeneous catalysts. Utilizing a wide range of chemical-biological techniques, it is possible to produce a wide range of high value-added chemicals and biofuel precursors that could be used to produce a number of sustainable alternatives to fossil-derived commodities in well-established markets and developed applications. We hope that the momentum presented in this chapter will encourage researchers and practitioners from both academia and industry to work together to develop effective flow processes for biomass processing scale suitable for manufacturing industrially.

References

Burguete, P., Corma, A., Hitzl, M., Modrego, R., Ponce, E., & Renz, M. (2016). Fuel and chemicals from wet lignocellulosic biomass waste streams by hydrothermal carbonization. *Green Chemistry: an International Journal and Green Chemistry Resource: GC, 18*, 1051–1060. Available from https://doi.org/10.1039/c5gc02296g.

Calcio Gaudino, E., Cravotto, G., Manzoli, M., & Tabasso, S. (2019). From waste biomass to chemicals and energy: Via microwave-assisted processes. *Green Chemistry: an International Journal and Green Chemistry Resource: GC, 21*, 1202–1235. Available from https://doi.org/10.1039/c8gc03908a.

Cho, E. J., Trinh, L. T. P., Song, Y., Lee, Y. G., & Bae, H. J. (2020). Bioconversion of biomass waste into high value chemicals. *Bioresource Technology, 298*, 122386. Available from https://doi.org/10.1016/j.biortech.2019.122386.

Dulie, N. W., Woldeyes, B., Demsash, H. D., & Jabasingh, A. S. (2021). An insight into the valorization of hemicellulose fraction of biomass into furfural: Catalytic conversion and product separation. *Waste and Biomass Valorization, 12*, 531–552. Available from https://doi.org/10.1007/s12649-020-00946-1.

Hammerschmidt, A., Boukis, N., Hauer, E., Galla, U., Dinjus, E., Hitzmann, B., Larsen, T., & Nygaard, S. D. (2011). Catalytic conversion of waste biomass by hydrothermal treatment. *Fuel, 90*, 555–562. Available from https://doi.org/10.1016/j.fuel.2010.10.007.

Herbst, A., & Janiak, C. (2017). MOF catalysts in biomass upgrading towards value-added fine chemicals. *CrystEngComm, 19*, 4092–4117. Available from https://doi.org/10.1039/c6ce01782g.

Kapoor, R., Ghosh, P., Kumar, M., Sengupta, S., Gupta, A., Kumar, S. S., Vijay, V., Kumar, V., Kumar Vijay, V., & Pant, D. (2020). Valorization of agricultural waste for biogas based circular economy in India: A research outlook. *Bioresource Technology, 304*, 123036. Available from https://doi.org/10.1016/j.biortech.2020.123036.

Karthikeyan, O. P., Mehariya, S., & Wong, J. W. Chung (2017). Bio-refining of food waste for fuel and value products. *Energy Procedia, 136*, 14–21. Available from https://doi.org/10.1016/j.egypro.2017.10.253.

Kover, A., Kraljić, D., Marinaro, R., & Rene, E. R. (2022). Processes for the valorization of food and agricultural wastes to value-added products: recent practices and perspectives. *Systems Microbiology and Biomanufacturing*, 2, 50–66. Available from https://doi.org/10.1007/s43393-021-00042-y.

Lu, Y., Zhang, Z., Wang, H., & Wang, Y. (2021). Toward efficient single-atom catalysts for renewable fuels and chemicals production from biomass and CO_2. *Applied Catalysis B: Environmental*, 292, 120162. Available from https://doi.org/10.1016/j.apcatb.2021.120162.

Mainali, K. (2012). Base catalytic transesterification of vegetable oil. *Science Progress*, 95, 50–72. Available from https://doi.org/10.3184/003685011X13205103999987.

Mohanty, A., Mankoti, M., Rout, P. R., Meena, S. S., Dewan, S., Kalia, B., Varjani, S., Wong, J. W. C., & Banu, J. R. (2022). Sustainable utilization of food waste for bioenergy production: A step towards circular bioeconomy. *International Journal of Food Microbiology*, 365, 109538. Available from https://doi.org/10.1016/j.ijfoodmicro.2022.109538.

Morales, G., Iglesias, J., & Melero, J. A. (2020). Sustainable catalytic conversion of biomass for the production of biofuels and bioproducts. *Catalysts*, 10, 4–7. Available from https://doi.org/10.3390/catal10050581.

Moscoviz, R., Toledo-Alarcón, J., Trably, E., & Bernet, N. (2016). Electro-fermentation: How to drive fermentation using electrochemical systems. *Trends in Biotechnology*, 34, 856–865. Available from https://doi.org/10.1016/j.tibtech.2016.04.009.

Nguyen, L. T., Phan, D. P., Sarwar, A., Tran, M. H., Lee, O. K., & Lee, E. Y. (2021). Valorization of industrial lignin to value-added chemicals by chemical depolymerization and biological conversion. *Industrial Crops and Products*, 161, 113219. Available from https://doi.org/10.1016/j.indcrop.2020.113219.

Peixoto, A. F., De Química, D., De Ciências, F., & Porto, U. (2022). *Biomass waste valorisation through sustainable catalytic processes: A biorefinery approach*, 9, 12020207.

Perveen, F., Farooq, M., Naeem, A., Humayun, M., Saeed, T., Khan, I. W., & Abid, G. (2022). Catalytic conversion of agricultural waste biomass into valued chemical using bifunctional heterogeneous catalyst: A sustainable approach. *Catalysis Communications*, 171, 106516. Available from https://doi.org/10.1016/j.catcom.2022.106516.

Priya, S. S., Bhanuchander, P., Kumar, V. P., Bhargava, S. K., & Chary, K. V. R. (2016). Activity and selectivity of platinum-copper bimetallic catalysts supported on mordenite for glycerol hydrogenolysis to 1,3-propanediol. *Industrial & Engineering Chemistry Research*, 55, 4461–4472. Available from https://doi.org/10.1021/acs.iecr.6b00161.

Roy, P., Mohanty, A. K., Dick, P., & Misra, M. (2023). A review on the challenges and choices for food waste valorization: environmental and economic impacts. *ACS Environmental Au*. Available from https://doi.org/10.1021/acsenvironau.2c00050.

Serrano-Ruiz, J. C., West, R. M., & Dumesic, J. A. (2010). Catalytic conversion of renewable biomass resources to fuels and chemicals. *Annual Review of Chemical and Biomolecular Engineering*, 1, 79–100. Available from https://doi.org/10.1146/annurev-chembioeng-073009-100935.

Sheldon, R. A. (2014). Green and sustainable manufacture of chemicals from biomass: State of the art. *Green Chemistry: an International Journal and Green Chemistry Resource: GC*, 16, 950–963. Available from https://doi.org/10.1039/c3gc41935e.

Sudarsanam, P., Duolikun, T., Babu, P. S., Rokhum, L., & Johan, M. R. (2020). Recent developments in selective catalytic conversion of lignin into aromatics and their derivatives. *Biomass Convers. Biorefinery*, 10, 873–883. Available from https://doi.org/10.1007/s13399-019-00530-1.

Thallada, B., Kumar, A., Jindal, M., & Maharana, S. (2021). Lignin biorefinery: New horizons in catalytic hydrodeoxygenation for the production of chemicals. *Energy and Fuels*, 35, 16965–16994. Available from https://doi.org/10.1021/acs.energyfuels.1c01651.

Tong, X., Charles Xu, C., Mu, X., & Weckhuysen, B. M. (2019). Preface: Catalysis for valorization of biomass and biomass-derived platform molecules (18th NCC. *Catalysis Today*, 319, 1. Available from https://doi.org/10.1016/j.cattod.2018.09.016.

Ventura, M., Marinas, A., & Domine, M. E. (2020). Catalytic processes for biomass-derived platform molecules valorisation. *Topics in Catalysis*, 63, 846–865. Available from https://doi.org/10.1007/s11244-020-01309-9.

Xu, H., Zhao, Y., Wang, Q., He, G., & Chen, H. (2022). Supports promote single-atom catalysts toward advanced electrocatalysis. *Coordination Chemistry Reviews*, 451, 214261. Available from https://doi.org/10.1016/j.ccr.2021.214261.

Zheng, A., Xia, S., Cao, F., Liu, S., Yang, X., Zhao, Z., Tian, Y., & Li, H. (2021). Directional valorization of eucalyptus waste into value-added chemicals by a novel two-staged controllable pyrolysis process. *Chemical Engineering Journal*, 404, 127045. Available from https://doi.org/10.1016/j.cej.2020.127045.

C H A P T E R

19

Thermal digestion process—a novel technique for converting solid organic waste into nutrient-rich organic fertilizer

Nitin Kumar and Sunil Kumar Gupta

Department of Environmental Science & Engineering, Indian Institute of Technology (Indian School of Mines), Dhanbad, Jharkhand, India

19.1 Introduction

Rapid urbanization, economic development, and exponential population growth are considered as the key contributors to the huge upsurge in the generation of municipal solid waste (MSW). The proper management of MSW is a challenging issue as it could contribute to severe environmental problems involving soil, water, and air pollution. Currently, around 2.1 billion tons per year of MSW are generated across the globe, and it is estimated to rise up to 3.4 billion tons per year by 2050 (Rafew & Rafizul, 2021). Solid organic waste (SOW) is the putrescible fraction of the municipal waste, which includes mainly food waste, yard waste, animal wastes, and agricultural wastes (Chen et al., 2020). A major fraction (up to 60%) of the MSW generated from the societies is composed up of SOW. This is found rich in plant essential nutrients, which makes it a suitable feedstock for the production of soil conditioners. In this outlook, researchers are putting efforts to develop sustainable techniques for the management of SOW and utilization of its inherent nutrient value (Singh & Kumari, 2019). Recycling SOW as soil conditioners could provide essential nutrients to support agriculture and reduce the dependency on chemical fertilizer (Wainaina et al., 2020). Composting is the widely used processing technology for converting the SOW into organic fertilizer. However, this process is time-intensive and also led to nutrient losses in the form of leachate and gaseous emissions. Therefore an innovative technique is needed to obviate these drawbacks and facilitate rapid conversion of SOW into soil conditioners.

Composters equipped with a heating system could accelerate the degradation of organic matter and reduce the duration of composting (Zaman et al., 2021). This method of on-site treatment of SOW also relieves the burdens on the treatment and disposal facilities. Ajmal et al. (2020) revealed that aeration-assisted high temperature (65°C) process accelerated the degradation and mineralization rate of agricultural waste. Pandey et al. (2016) found that the digestion at elevated temperature (60°C) converted food waste into pathogen-free soil amendment. Zhou et al. (2020) discovered that microbial-assisted temperature-controlled (60°C–70°C) composting device felicitated rapid composting of food waste. All of these studies were based on temperature-controlled microbial-assisted aerobic digestion. Whereas the thermal digestion process effectively degrades the organics into simpler forms without any microbial assistance (Kumar & Gupta, 2021). However, the detailed aspects of thermal digestion for the rapid treatment of SOW are still unexplored. Therefore an innovative digester based on the thermal digestion process was designed and tested. This chapter explores the detailed design aspects and operation of a compact thermal digester that could efficiently be utilized by municipalities and urban societies for the on-site and rapid treatment of SOW.

19.2 Qualitative and quantitative analysis of SOW

The composition of MSW varies from location to location, which depends upon socioeconomic status, lifestyle habits, cultural traditions, waste management strategies, and other factors. SOW is the most dominant fraction of the total MSW generated around the globe (Sharma & Jain, 2020). The composition of the waste stream changes with the income level of the countries. The share of organic waste in lower-income groups (LIG) is 64%, and this value reduces to 28% for upper-income group (UIG) countries. In comparison to LIG and UIG, the middle-income group (MIG) countries typically generate higher organic waste per annum (Fig. 19.1) (Kaza et al., 2018). Solid organic waste mainly includes food waste, yard waste, agricultural residues, and livestock waste. One of the highest fractions in municipal SOW is composed of food waste. The common constituents of SOW are shown in Table 19.1. Recycling and reuse strategies as well as innovative solutions are required for efficient handling and management of these wastes. Circular bioeconomy is based on transforming the SOW into useful products. In this context, wastes that had been a burden to societies are now being treated as a source for recovering valuable resources. These wastes are rich in organic matter and essential nutrients and could efficiently be used for the recovery of value-added products (Mahjoub & Domscheit, 2020).

The detailed characterization of SOW is needed to identify their biochemical properties, which would help in identifying suitable valorization technologies. Despite the high variability in the composition of SOW, few main physicochemical characteristics such as proximate, elemental, nutrients, and bromatological composition help in identifying suitable treatment options. The general range of their composition is summarized in Table 19.2.

19.3 Existing technologies on the nutrient recovery from solid organic waste

Improper disposal of SOW led to loss in considerable amounts of macronutrients, that is, N, P, and K, which are locked in it. These losses of valuable nutrients could be recuperated through suitable valorization technologies, and these nutrient-rich organic fertilizers could be employed in agricultural practices to improve soil fertility (Soobhany, 2019). The most commonly applied techniques used for the recovery of nutrients from SOW involve biological (composting and anaerobic digestion) and thermal conversion techniques (pyrolysis) (Fig. 19.2).

19.3.1 Composting

The composts derived from SOW could supply organic matter and essential nutrients to improve soil fertility (Soobhany et al., 2015). The macronutrients such as nitrogen (N), phosphorus (P), and potassium (K) are recognized as the essential nutrients for plants to secure proper growth and development (Roy et al., 2006). During composting, a large amount of total N is lost due to the transformation of nitrogen in the form of gaseous emissions

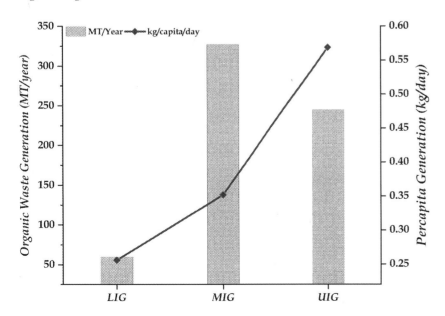

FIGURE 19.1 Current generation rate of solid organic waste by countries of different income groups.

TABLE 19.1 Types of solid organic waste and their major constituents.

Type of organic waste	Major constituents	References
Food waste	Cooked food, uncooked food, spoiled meat, fish, bones, eggshells, shells, fruit and vegetable peelings, rotten foods, tea leaves and coffee residues, etc.	Nanda and Berruti (2021)
Yard waste	Fallen leaves, trimmings from lawns, carvings of grass and tree, branches, bark and woods etc.	Nanda and Berruti (2021)
Agricultural residues	Straw from rice, sorghum, wheat, and other crops, sugarcane bagasse, corn stover, shells of nuts, stalks of cotton, soybean, sunflower, husks of rice, maize, groundnut, soybean etc.	FAO (2015)
Livestock waste	Livestock includes goats, pigs, dairy cattle, buffaloes, horses, sheep, chickens, ducks, quails, pigeons, and rabbits etc.	Dhanya et al. (2020)

TABLE 19.2 Physicochemical characterization of the solid organic waste.

Physicochemical properties	General range (%)	References
Proximate analysis		
Moisture content	48–90	Fu et al. (2019); Yadav et al. (2016); Saqib et al. (2018)
Volatile matter	62–78	
Ash content	4–14	
Fixed carbon	15–32	
Elemental analysis		
C	36–48	Yadav et al. (2016); Parra-Orobio et al. (2018); Abdul Samad et al. (2017)
H	4–7	
N	1.2–3.5	
S	0.1–0.6	
O	35–41	
Nutrients analysis		
N (%)	1.2–3.5	Kumar and Gupta (2022); Reyes-Torres et al. (2018)
P (%) as P_2O_5	0.2–1.4	
K (%) as K_2O	0.8–2.6	
Bromatological analysis		
Proteins	7–32	Dong et al. (2010); Alibardi & Cossu (2015); Poggio et al. (2016)
Carbohydrates	34–65	
Lipids	10–18	

(primarily NH_3 and N_2O). It has been reported that NH_3 emissions accounted for 79%–94% of the total TN losses (Shan et al., 2021). In addition, N_2O emissions account for 0.2%–9.9% of TN losses (Cáceres et al., 2018). Therefore the nitrogen loss due to NH_3 emission is the major reason for reducing compost quality and odor pollution in the composting process (Yang et al., 2019). Besides N, other nutrients such as P and K are also lost during composting, owing to leaching and runoff. P and K losses were 23%–39% and 20%–52% of their respective initial value (Soobhany, 2019). The nutrient recovery efficiency is given by Eq. (19.1) (Swarnam et al., 2016):

$$NRE = \frac{RR_t}{RR_c} \tag{19.1}$$

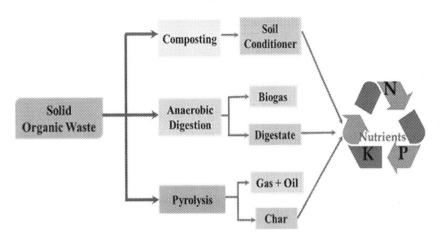

FIGURE 19.2 Existing technologies on the nutrient recovery from solid organic waste.

where RR_t is the recovery rate of "ith" nutrient in the treatment, and RR_c is the recovery rate of nutrient in the control. The recovery rate (RR_t) could be computed using Eq. (19.2)

$$RR = \left[\left(\frac{N_{ei}}{N_{fi}}\right) \times \frac{1}{T}\right] \times 100 \qquad (19.2)$$

where N_{fi} and N_{ei} are the "ith" nutrient content in the initial feedstock and the end product, respectively, T is the time duration of composting.

19.3.2 Anaerobic digestion

Anaerobic digestion (AD) converts organic matter into biogas and a nutrient-rich digestate. This digestate has the potential to be used as a soil conditioner to benefit crop production (Lee et al., 2021). The fertilizing value of digestate depends on the nutrients present in the feedstock. During AD, hydrolysis of the organic matter takes place that enhances the bioavailability of the nutrients. For example, a part of the organic nitrogen supplied with the feedstock is converted to ammonium. The liquid phase of digestate is usually rich in plant-available nutrients such as ammonium (NH_4^+), phosphate, and potassium (K). The NH_4^+ in the liquid phase is plant-available and may be directly comparable with synthetic fertilizer (Gutser et al., 2005). Whereas P and K in the feedstock are generally retained during the AD process. The liquid phase of the digestate consists of inorganic phosphate, whereas the solid phase contains both inorganic and organic P (Kataki et al., 2019). In organic waste, K is retained in cationic form; hence, K in digestate is directly available in ionic form in both solid and liquid phases (Sogn et al., 2018).

19.3.3 Pyrolysis

Pyrolysis is the thermal degradation of biomass by heating it into an anoxic environment (Akdeniz, 2019). The physical and chemical properties of the biochar depend on the initial characteristics of the feedstock and pyrolysis conditions, that is, temperature and duration (Novak et al., 2009). The biochar derived from plant-based material generally has low nitrogen, phosphorous, and potassium contents and is considered disadvantageous for supplying essential crop nutrients (Cantrell et al., 2012). The N content in biochar decreases as the pyrolysis temperature increases from 300°C to 700°C. The loss of N initiates at 200°C and, at 50% of N, is volatilized at ~700°C. The total N loss in biochar is mainly due to the volatilization of N into gaseous pyrolysis products, such as NH_3, HCN, and HCNO (Tian et al., 2017). Whereas the total P generally shows an increasing trend with the increases in pyrolysis temperature (Iqbal et al., 2015). The P content in the biochar was higher than in the raw material, suggesting that pyrolysis is efficient in retaining the P content (Wang et al., 2012). About 90% of K in the organic material is present in the form of ion-exchangeable structures. The total K increases with an increase in pyrolytic temperature. However, it begins to volatilize at a temperature above 600°C.

19.4 Thermal digestion process

An efficient, rapid, and eco-friendly method of waste treatment is the need of the hour. Thermal digestion emerged as a novel technology for the management of biowaste, resulting into the rapid production of organic fertilizer. In this method, SOW is shredded, heated, and rapidly decomposed in a thermal digester to obtain a nutrient-rich organic fertilizer as an end product of the operation. The method is quick, user-friendly, hygienic, and requires a small area for its operation, making it suitable for big cities and metros. It provides a decentralized treatment option, i.e., on-site processing of the SOW at the source itself, thus saving the transportation and dumping costs.

19.4.1 Process description and design of the digester

In the thermal digestion technique, the heat is transferred to the SOW through the convection mode of heating, that is, the heated air is uniformly circulated in the digestion chamber through a blower (Fig. 19.3). The digester works on a batch mode and was designed to handle 150 kg of SOW in a single batch. A dual-shaft shredder reduces the particle size of the inlet SOW into less than 10 mm, thus providing a larger surface area to accelerate the digestion process. Even though shredding of organic matter facilitates in accelerating the digestion process, the formation of fine paste is not advisable because reducing the size to fine paste led to sagging and compaction of the material (Zhang et al., 2019). An agitator shaft with agitating blades is also provided inside the reaction chamber to provide uniform and continuous mixing during the process. The continuous agitation of the shredded SOW inside the apparatus assists in shortening the duration of the drying process.

19.4.2 Optimization of process parameters

The performance of a thermal digester strongly depends upon its (1) drying temperature, (2) airflow velocity, and (3) agitational speed of the material. Studies reported that the drying time primarily depends upon the drying temperature and airflow velocity (Gupta et al., 2013). Whereas continuous agitation assists in homogeneous mixing of the material and also helps in improving the drying rate (Puspasari et al., 2012). In addition, the frictional force and the collision of the particles during the agitation reduce the particle size, which assists in the rapid drying of the material (Hou et al., 2017). The thermal digester is recommended to operate in the temperature range of 100°C–160°C. The drying temperature above 100°C led to the degradation of complex organic molecules. Moreover, the operating temperature must be kept below 160°C to avoid burning of the material and restrict the emissions of CO, CO_2, and volatile matters. While the airflow velocity and rotational speed of the agitator should be kept between 1 and 7 m/s and 10–40 rpm, respectively. The higher airflow rate and agitational speed beyond the recommended limits don't significantly affect the drying rate and unnecessarily consume

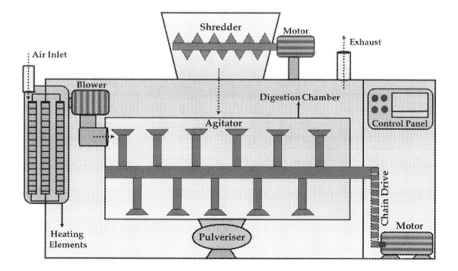

FIGURE 19.3 Design of the thermal digester.

higher specific energy. The optimum condition for digestion was selected on the basis of the minimum requirement of digestion time and energy consumption. The optimum condition was obtained at 140°C with an airflow velocity of 5 m/s and a uniform agitation of material at 20 rpm (Fig. 19.4).

19.4.3 Transformation of the nutrients in thermal digestion

The end product was recorded with a near-neutral pH, while the EC and bulk density value were also in compliance with the delineated norms of Fertilizer Association of India (FAI, 2019). The nutrient characterization of the end product revealed that the total N decreased by 10.3% (Table 19.3). This reduction in the value of N might be due to the volatilization of ammonical nitrogen during the drying. However, in thermally digested solids, the organic N is converted into a relatively stable form due to the creation of a dry bond structure (Septien et al., 2020). Surprisingly, the concentrations of P & K got increased by 13.8% and 15.17%. The increased concentration of P might be because of the conversion of organic P into inorganic P (Ajiboye et al., 2004). Whereas K ions get accumulated owing to the disintegration of organic matter, leading to the solubilization of insoluble K. The transformation of nutrients indicated that the end product is expected to act as a slow-release fertilizer on soil application. The detailed characteristics of the end product complied with the norms for organic fertilizer as recommended by the Fertilizer Association of India, 2019.

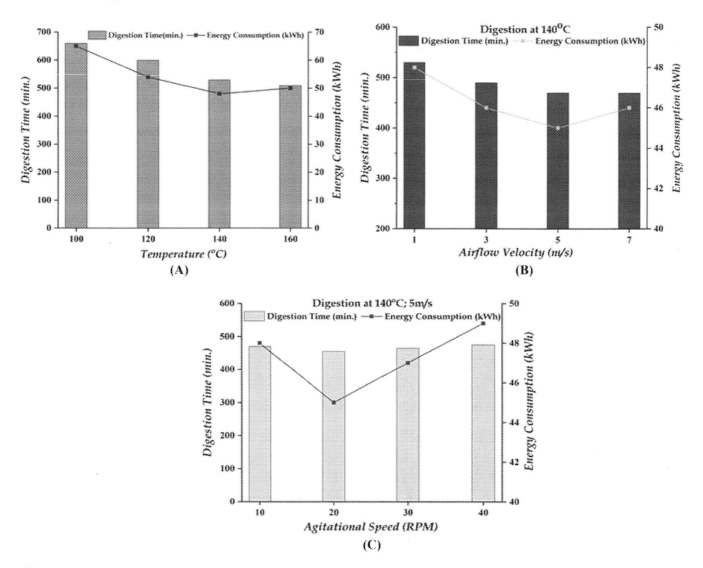

FIGURE 19.4 Effect of (A) temperature, (B) airflow velocity, and (C) agitational speed on digestion time and energy consumption.

TABLE 19.3 Manuring characteristics of the feedstock and end product.

Characteristics	Values[a]		Recommended values (FAI, 2019)
	Feedstock	End product	
pH (*1:10*)	7.23 ± 0.06	6.56 ± 0.06	6.5–7.5
EC (mS/cm)	3.89 ± 0.05	3.70 ± 0.05	<4
Bulk density (g/cm^3)	0.91 ± 0.03	0.81 ± 0.02	<1.0
C (%)	41.48 ± 0.34	36.42 ± 0.18	≥12%
N (%)	2.04 ± 0.04	1.83 ± 0.03	–
P (%)	0.36 ± 0.01	0.41 ± 0.01	–
K (%)	1.45 ± 0.02	1.67 ± 0.03	–
Total NPK (%)	3.85 ± 0.04	3.91 ± 0.04	≥1.2%
C:N	20.33 ± 0.56	18.92 ± 0.42	<20:1
Ca (mg/kg)	174.27 ± 0.41	134.68 ± 0.34	–
Na (mg/kg)	158.75 ± 0.28	127.35 ± 0.21	–

[a]*Values reported as mean ± S.D; -: not mentioned.*

19.4.4 Process feasibility analysis

Cost–benefit analysis was performed to check the economic feasibility of the process. The economic viability and the feasibility of the process were determined on the basis of the revenue generated by this system. The following expenditures associated with this system were identified:

1. *Fabrication cost:* Total cost incurred for the fabrication of the digester was approx. US$4500.
2. *Operational cost*: The operational cost includes the power consumed by the digester for processing the waste. The digester consumed approx. 45 kWh of energy for completing one batch of 150 kg. Hence, about US$ 3.1 was incurred per day for the treatment of 150 kg of SOW (US$0.02/kg).

The digester produces around 35 kg of organic fertilizer at the end of a batch. Hence, total operational cost to produce 1 tonne of organic fertilizer would be around US$90. The nutrient-rich soil conditioner can be sold at the rate of US$0.91/kg (Sudharmaidevi et al., 2017). Even if the output is sold at a much lower cost of US$ 0.2/kg, still a net profit of US$110 per month could be generated. This proves that the thermal digestion process is economically viable and could prove to be a revenue-generating option for the bulk waste generators.

19.5 Comparative assessment with various technologies

Comparative assessment of the existing nutrient recovery technologies with the thermal digestion process in terms of waste reduction and nutrient recovery potential, treatment time, economic and environmental aspects is given in Table 19.4. Biological treatment methods, that is, composting and anaerobic digestion, have lower treatment cost. However, it has the inherent disadvantages of longer treatment time, larger environmental footprints, nutrient losses, little volume reduction, and the possibility of bacteria inhibition. Moreover, they require stringent process monitoring, control, and optimization (Pham et al., 2015). Whereas pyrolysis is considered energy-intensive process. Pyrolysis is favorable in terms of reducing the treatment time. However, the high moisture contents and highly heterogeneous nature of SOW result in technoeconomic issues while using pyrolysis (Elkhalifa et al., 2019). Thermal digestion process operates at a higher temperature (>100°C) compared with other biological conversion techniques. However, the heat requirement for this process can be offset by using energy-efficient, that is, solar/renewable energy sources, and through the revenue generated by marketing of the end product. The higher temperature assists in rapidly reducing the huge volume, digesting the organics, and eliminating the pathogens. Moreover, the treatment time of thermal digestion is around 4–6 h,

TABLE 190.4 Assessment of technological options for the treatment of solid organic waste.

Criteria	Different technological options			
	Composting	Anaerobic digestion	Pyrolysis	Thermal digestion
Volume reduction potential	**	**	****	****
Treatment time	****	****	**	**
Nutrient recovery	**	***	***	****
Treatment cost	**	**	****	**
Technology readiness level	****	****	***	****
Potential environmental impact	****	****	**	*

*Low ** Low to moderate ***Moderate to high ****High.

which is much shorter than the conventional methods and thus considerably improves the throughput of end products (Kumar & Gupta, 2021). Among all the nutrient recovery techniques, thermal digestion seems to be an attractive option for converting SOW into the nutrient-rich end product.

19.6 Summary and conclusions

This chapter explored the feasibility of the thermal digestion process for the rapid treatment of SOW. The thermal digester was found effective in degrading the organics and converting the SOW into an end product rich in nutrients. The optimization of the operating parameters of a thermal digester revealed that the optimal condition for digestion was obtained at 140°C with an airflow velocity of 5 m/s and a uniform agitation of material at 20 rpm. The characteristics of the end product obtained from thermal digestion process resembled the properties of organic fertilizer and comply with all required norms as recommended by the Fertilizer Association of India. Comparative assessment of various technologies used for nutrient recovery from SOW depicted that thermal digestion proved to be a more efficient technology for rapidly converting the SOW into nutrient-rich organic fertilizer. The technoeconomic feasibility of the process revealed that it could prove to be a revenue-generating alternative for the bulk waste generators.

References

Abdul Samad, N. A. F., Jamin, N. A., & Saleh, S. (2017). Torrefaction of Municipal Solid Waste in Malaysia. *Energy Procedia* (Vol. *138*, pp. 313–318). Malaysia: Elsevier Ltd. Available from https://doi.org/10.1016/j.egypro.2017.10.106, http://www.sciencedirect.com/science/journal/18766102. 18766102.

Ajiboye B., Akinremi, O. O., & Racz, G. J. (2004). Canada Laboratory characterization of phosphorus in fresh and oven-dried organic amendments. *Journal of Environmental Quality* (Vol. *33*(3), pp. 1062–1069). Available from https://doi.org/10.2134/jeq2004.1062, https://acsess.onlinelibrary.wiley.com/loi/15372537 33. 00472425 ASA/CSSA/SSSA.

Ajmal, M., Aiping, S., Awais, M., Ullah, M. S., Saeed, R., Uddin, S., Ahmad, I., Zhou, B., & Zihao, X. (2020). Optimization of pilot-scale in-vessel composting process for various agricultural wastes on elevated temperature by using Taguchi technique and compost quality assessment. *Process Safety and Environmental Protection*, *140*, 34–45. Available from https://doi.org/10.1016/j.psep.2020.05.001, http://www.elsevier.com/wps/find/journaldescription.cws_home/713889/description#description.

Akdeniz, N. (2019). A systematic review of biochar use in animal waste composting. *Waste Management*, *88*, 291–300. Available from https://doi.org/10.1016/j.wasman.2019.03.054, http://www.elsevier.com/locate/wasman.

Alibardi, L., & Cossu, R. (2015). Composition variability of the organic fraction of municipal solid waste and effects on hydrogen and methane production potentials. *Waste management*, *36*, 147–155.

Cantrell, K. B., Hunt, P. G., Uchimiya, M., Novak, J. M., & Ro, K. S. (2012). Impact of pyrolysis temperature and manure source on physicochemical characteristics of biochar. *Bioresource Technology*, *107*, 419–428. Available from https://doi.org/10.1016/j.biortech.2011.11.084.

Chen, T., Zhang, S., & Yuan, Z. (2020). Adoption of solid organic waste composting products: A critical review. *Journal of Cleaner Production*, 272122712. Available from https://doi.org/10.1016/j.jclepro.2020.122712.

Cáceres, R., Malińska, K., & Marfà, O. (2018). Nitrification within composting: A review. *Waste Management*, *72*, 119–137. Available from https://doi.org/10.1016/j.wasman.2017.10.049, http://www.elsevier.com/locate/wasman.

Dhanya, B. S., Mishra, A., Chandel, A. K., & Verma, M. L. (2020). Development of sustainable approaches for converting the organic waste to bioenergy. *Science of the Total Environment* (723). Available from https://doi.org/10.1016/j.scitotenv.2020.138109, http://www.elsevier.com/locate/scitotenv.

Dong, L., Zhenhong, Y., & Yongming, S. (2010). Semi-dry mesophilic anaerobic digestion of water sorted organic fraction of municipal solid waste (WS-OFMSW). *Bioresource Technology*, *101*(8), 2722–2728.

References

Elkhalifa, S., Al-Ansari, T., Mackey, H. R., & McKay, G. (2019). Food waste to biochars through pyrolysis: A review. *Resources, Conservation and Recycling, 144*, 310−320. Available from https://doi.org/10.1016/j.resconrec.2019.01.024, http://www.elsevier.com/locate/resconrec.

FAI (2019). 2019 Shaheed Jit Singh Marg The Fertiliser (Control) Order 1985. The Fertiliser Association of India 10.

FAO (2015). FAO BEFS Technical Consultation on the Availability of Agricultural Residues in Turkey.

Fu, M. M., Mo, C. H., Li, H., Zhang, Y. N., Huang, W. X., & Wong, M. H. (2019). Comparison of physicochemical properties of biochars and hydrochars produced from food wastes. *Journal of Cleaner Production* (236). Available from https://doi.org/10.1016/j.jclepro.2019.117637, https://www.journals.elsevier.com/journal-of-cleaner-production.

Gupta, M. K., Sehgal, V. K., & Arora, S. (2013). Optimization of drying process parameters for cauliflower drying. *Journal of Food Science and Technology, 50*(1), 62−69. Available from https://doi.org/10.1007/s13197-011-0231-5.

Gutser, R., Ebertseder, T., Weber, A., Schraml, M., & Schmidhalter, U. (2005). Short-term and residual availability of nitrogen after long-term application of organic fertilizers on arable land. *Journal of Plant Nutrition and Soil Science, 168*(4), 439−446. Available from https://doi.org/10.1002/jpln.200520510.

Hou, J., Li, M., Xi, B., Tan, W., Ding, J., Hao, Y., Liu, D., & Liu, H. (2017). Short-duration hydrothermal fermentation of food waste: Preparation of soil conditioner for amending organic-matter-impoverished arable soils. *Environmental Science and Pollution Research, 24*(26), 21283−21297. Available from https://doi.org/10.1007/s11356-017-9514-3, http://www.springerlink.com/content/0944-1344.

Iqbal, H., Garcia-Perez, M., & Flury, M. (2015). Effect of biochar on leaching of organic carbon, nitrogen, and phosphorus from compost in bioretention systems. *Science of the Total Environment, 521−522*: 37−45. Available from https://doi.org/10.1016/j.scitotenv.2015.03.060, http://www.elsevier.com/locate/scitotenv.

Kataki, S., Hazarika, S., & Baruah, D. C. (2019). By-products of bioenergy systems (anaerobic digestion and gasification) as sources of plant nutrients: Scope of processed application and effect on soil and crop. *Journal of Material Cycles and Waste Management, 21*(3), 556−572. Available from https://doi.org/10.1007/s10163-018-00816-y, http://link.springer.de/link/service/journals/10163/index.htm.

Kaza, S., Yao, L., & Bhada-Tata, P., Woerden, F. V. (2018). What a waste 2.0: A global snapshot of solid waste management to 2050. World Bank Publications.

Kumar, N., & Gupta, S. K. (2022). Exploring drying kinetics and fate of nutrients in thermal digestion of solid organic waste. *Science of the Total Environment* (837). Available from https://doi.org/10.1016/j.scitotenv.2022.155804, http://www.elsevier.com/locate/scitotenv.

Kumar, N., & Gupta, S. K. (2021). Exploring the feasibility of thermal digestion process: A novel technique, for the rapid treatment and reuse of solid organic waste as organic fertilizer. *Journal of Cleaner Production* (318). Available from https://doi.org/10.1016/j.jclepro.2021.128600, https://www.journals.elsevier.com/journal-of-cleaner-production.

Lee, M. E., Steiman, M. W., & Angelo, S. K., St. (2021). Biogas digestate as a renewable fertilizer: Effects of digestate application on crop growth and nutrient composition. *Renewable Agriculture and Food Systems, 36*(2), 173−181. Available from https://doi.org/10.1017/S1742170520000186, http://journals.cambridge.org/RAF.

Mahjoub, B., & Domscheit, E. (2020). Chances and challenges of an organic waste−based bioeconomy. *Current Opinion in Green and Sustainable Chemistry, 25*100388. Available from https://doi.org/10.1016/j.cogsc.2020.100388.

Nanda, S., & Berruti, F. (2021). Municipal solid waste management and landfilling technologies: A review. *Environmental Chemistry Letters, 19*(2), 1433−1456. Available from https://doi.org/10.1007/s10311-020-01100-y, http://springerlink.metapress.com/app/home/journal.asp?wasp = d86tgdwvtg0yvw9gvkwp&referrer = parent&backto = browsepublicationsresults,140,541.

Novak, J. M., Lima, I., Xing, B., Gaskin, J. W., Steiner, C., Das, K. C., & Schomberg, H. (2009). , Characterization of designer biochar produced at different temperatures and their effects on a loamy sand. *Ann. Environmental Sciences: An International Journal of Environmental Physiology and Toxicology, 3*(2), 195−206.

Pandey, P. K., Vaddella, V., Cao, W., Biswas, S., Chiu, C., & Hunter, S. (2016). In-vessel composting system for converting food and green wastes into pathogen free soil amendment for sustainable agriculture. *Journal of Cleaner Production, 139*, 407−415. Available from https://doi.org/10.1016/j.jclepro.2016.08.034.

Parra-Orobio, B. A., Donoso-Bravo, A., Ruiz-Sánchez, J. C., Valencia-Molina, K. J., & Torres-Lozada, P. (2018). Effect of inoculum on the anaerobic digestion of food waste accounting for the concentration of trace elements. *Waste Management, 71*, 342−349. Available from https://doi.org/10.1016/j.wasman.2017.09.040, http://www.elsevier.com/locate/wasman.

Pham, T. P. T., Kaushik, R., Parshetti, G. K., Mahmood, R., & Balasubramanian, R. (2015). Food waste-to-energy conversion technologies: Current status and future directions. *Waste Management, 38*(1), 399−408. Available from https://doi.org/10.1016/j.wasman.2014.12.004, http://www.elsevier.com/locate/wasman.

Poggio, D., Walker, M., Nimmo, W., Ma, L., & Pourkashanian, M. (2016). Modelling the anaerobic digestion of solid organic waste−substrate characterisation method for ADM1 using a combined biochemical and kinetic parameter estimation approach. *Waste management, 53*, 40−54.

Puspasari, I., Talib, M. Z. M., Daud, W. R. W., & Tasirin, S. M. (2012). Drying kinetics of oil palm frond particles in an agitated fluidized bed dryer. *Drying Technology, 30*(6), 619−630. Available from https://doi.org/10.1080/07373937.2012.654873.

Rafew, S. M., & Rafizul, I. M. (2021). Application of system dynamics model for municipal solid waste management in Khulna city of Bangladesh. *Waste Management, 129*, 1−19. Available from https://doi.org/10.1016/j.wasman.2021.04.059, http://www.elsevier.com/locate/wasman.

Reyes-Torres, M., Oviedo-Ocaña, E. R., Dominguez, I., Komilis, D., & Sánchez, A. (2018). A systematic review on the composting of green waste: Feedstock quality and optimization strategies. *Waste Management, 77*, 486−499. Available from https://doi.org/10.1016/j.wasman.2018.04.037, http://www.elsevier.com/locate/wasman.

Roy, R. N., Finck, A., Blair, G. J., & Tandon, H. L. S. (2006). Plant nutrition for food security. A guide for integrated nutrient management. *FAO Fertilizer and Plant Nutrition Bulletin, 16*.

Saqib, N. U., Baroutian, S., & Sarmah, A. K. (2018). Physicochemical, structural and combustion characterization of food waste hydrochar obtained by hydrothermal carbonization. *Bioresource Technology, 266*, 357−363. Available from https://doi.org/10.1016/j.biortech.2018.06.112, http://www.elsevier.com/locate/biortech.

Septien, S., Mirara, S. W., Makununika, B. S. N., Singh, A., Pocock, J., Velkushanova, K., & Buckley, C. A. (2020). Effect of drying on the physical and chemical properties of faecal sludge for its reuse. *Journal of Environmental Chemical Engineering*, 8(1)103652. Available from https://doi.org/10.1016/j.jece.2019.103652.

Shan, G., Li, W., Gao, Y., Tan, W., & Xi, B. (2021). Additives for reducing nitrogen loss during composting: A review. *Journal of Cleaner Production*, 307127308. Available from https://doi.org/10.1016/j.jclepro.2021.127308.

Sharma, K. D., & Jain, S. (2020). Municipal solid waste generation, composition, and management: The global scenario. *Social Responsibility Journal*, 16(6), 917–948. Available from https://doi.org/10.1108/SRJ-06-2019-0210, http://www.emeraldinsight.com/info/journals/srj/srj.jsp.

Singh, A., & Kumari, K. (2019). An inclusive approach for organic waste treatment and valorisation using Black Soldier Fly larvae: A review. *Journal of Environmental Management*, 251109569. Available from https://doi.org/10.1016/j.jenvman.2019.109569.

Sogn, T. A., Dragicevic, I., Linjordet, R., Krogstad, T., Eijsink, V. G. H., & Eich-Greatorex, S. (2018). Recycling of biogas digestates in plant production: NPK fertilizer value and risk of leaching. *International Journal of Recycling of Organic Waste in Agriculture*, 7(1), 49–58. Available from https://doi.org/10.1007/s40093-017-0188-0, http://www.springer.com/environment/pollution + and + remediation/journal/40093.

Soobhany, N. (2019). Insight into the recovery of nutrients from organic solid waste through biochemical conversion processes for fertilizer production: A review. *Journal of Cleaner Production*, 241118413. Available from https://doi.org/10.1016/j.jclepro.2019.118413.

Soobhany, N., Mohee, R., & Garg, V. K. (2015). Recovery of nutrient from Municipal Solid Waste by composting and vermicomposting using earthworm Eudrilus eugeniae. *Journal of Environmental Chemical Engineering*, 3(4), 2931–2942. Available from https://doi.org/10.1016/j.jece.2015.10.025, http://www.journals.elsevier.com/journal-of-environmental-chemical-engineering/.

Sudharmaidevi, C. R., Thampatti, K. C. M., & Saifudeen, N. (2017). Rapid production of organic fertilizer from degradable waste by thermochemical processing. *International Journal of Recycling of Organic Waste in Agriculture*, 6(1), 1–11. Available from https://doi.org/10.1007/s40093-016-0147-1, http://ijrowa.khuisf.ac.ir/.

Swarnam, T. P., Velmurugan, A., Pandey, S. K., & Dam Roy, S. (2016). Enhancing nutrient recovery and compost maturity of coconut husk by vermicomposting technology. *Bioresource Technology*, 207, 76–84. Available from https://doi.org/10.1016/j.biortech.2016.01.046, http://www.elsevier.com/locate/biortech.

Tian, S., Tan, Z., Kasiulienė, A., & Ai, P. (2017). Transformation mechanism of nutrient elements in the process of biochar preparation for returning biochar to soil. *Chinese Journal of Chemical Engineering*, 25(4), 477–486. Available from https://doi.org/10.1016/j.cjche.2016.09.009.

Wainaina, S., Awasthi, M. K., Sarsaiya, S., Chen, H., Singh, E., Kumar, A., Ravindran, B., Awasthi, S. K., Liu, T., Duan, Y., Kumar, S., Zhang, Z., & Taherzadeh, M. J. (2020). Resource recovery and circular economy from organic solid waste using aerobic and anaerobic digestion technologies. *Bioresource Technology* (301). Available from https://doi.org/10.1016/j.biortech.2020.122778, http://www.elsevier.com/locate/biortech.

Wang, T., Camps-Arbestain, M., Hedley, M., & Bishop, P. (2012). Predicting phosphorus bioavailability from high-ash biochars. *Plant and Soil*, 357(1), 173–187. Available from https://doi.org/10.1007/s11104-012-1131-9.

Yadav, D., Barbora, L., Rangan, L., & Mahanta, P. (2016). Tea waste and food waste as a potential feedstock for biogas production. *Environmental Progress and Sustainable Energy*, 35(5), 1247–1253. Available from https://doi.org/10.1002/ep.12337, http://www3.interscience.wiley.com/journal/121640218/issueyeargroup?year = 2009&CRETRY = 1&SRETRY = 0.

Yang, F., Li, Y., Han, Y., Qian, W., Li, G., & Luo, W. (2019). Performance of mature compost to control gaseous emissions in kitchen waste composting. *Science of the Total Environment*, 657, 262–269. Available from https://doi.org/10.1016/j.scitotenv.2018.12.030, http://www.elsevier.com/locate/scitotenv.

Zaman, B., Hardyanti, N., & Purwono, P. (2021). Fast composting of food waste using thermal composter. *IOP Conference Series: Earth and Environmental Science*, 896(1)012013. Available from https://doi.org/10.1088/1755-1315/896/1/012013.

Zhang, Y., Kusch-Brandt, S., Gu, S., & Heaven, S. (2019). Particle size distribution in municipal solid waste pre-treated for bioprocessing. *Resources*, 8(4), 166. Available from https://doi.org/10.3390/resources8040166.

Zhou, X., Yang, J., Xu, S., Wang, J., Zhou, Q., Li, Y., & Tong, X. (2020). Rapid in-situ composting of household food waste. *Process Safety and Environmental Protection*, 141, 259–266. Available from https://doi.org/10.1016/j.psep.2020.05.039, http://www.elsevier.com/wps/find/journaldescription.cws_home/713889/description#description.

CHAPTER 20

Waste biomass conversion to energy storage material

Glaydson Simões Dos Reis[1], Sari Tuomikoski[2], Davide Bergna[2,3], Sylvia Larsson[1], Mikael Thyrel[1], Helinando Pequeno de Oliveira[4], Palanivel Molaiyan[2] and Ulla Lassi[2,3]

[1]Department of Forest Biomaterials and Technology, Swedish University of Agricultural Sciences, Biomass Technology Centre, Umeå, Sweden [2]Research Unit of Sustainable Chemistry, Faculty of Technology, University of Oulu, Oulu, Finland [3]Unit of Applied Chemistry, University of Jyvaskyla, Kokkola University Consortium Chydenius, Kokkola, Finland [4]Institute of Materials Science, Federal University of São Francisco Valley, Juazeiro, BA, Brazil

20.1 Introduction

The progressive consumption of the resources and the following step of waste generation are a consequence of several factors, such as each country's industrial, technological, and populational activity (Okolie et al., 2022; Yrjälä et al., 2022). However, the potential environmental load, the scarcity of crude oil–based resources, and the ever-growing consumption of energy are driving forces that motivate the use of waste as the source of resources, including the recovery of elements such phosphorus and nitrogen from energy production, for example, incineration and in the production of biogas and solid fuels and to synthesize materials for a wide range of applications (Kwon et al., 2020; Seow et al., 2022). The utilization of biomass residues as the main raw material for the preparation of nanostructured carbon-based porous materials is a winning strategy because it would have at least three simultaneous impacts (Kwon et al., 2020); (1) environmental pollution control, (2) wealth creation (through the synthesis of hi-tech materials), and (3) if the biomass is composed of agricultural residues, there is no competition with food crops. The biomass residues can be converted into biomass-derived materials and could be a source of electrode materials for batteries and SCs (dos Reis et al., 2020, 2021; Lima et al., 2022; Reis et al., 2021).

The typical strategy is based on thermochemical conversion (pyrolysis), using several chemical reagents (acid, basic, CO_2, and metal salts) as an activation process (Engamba Esso et al., 2022; Lua et al., 2004). This process is essential since it develops the material's specific surface area (SSA) and pore structures, and the available surface area for carbon structures is a prerequisite for the development of more effective electrodes for energy storage systems (Lin et al., 2009; Zhang et al., 2022).

Pyrolysis is the most efficient and standard method to synthesize biomass-derived carbon materials and is performed at elevated temperatures without oxygen (dos Reis et al., 2020; Engamba Esso et al., 2022). The quality and properties of the carbon materials are severely dependent on the pyrolysis conditions, such as temperature, ramping rate, holding time, and used catalyst. The conditions can influence the pore structure of the biomass-derived carbon materials, structural properties, for example, degree of crystallinity, and the presence of functional groups on carbons' surfaces (Collard & Blin, 2014), which are effectively beneficial to electrochemical applications.

The optimization in the conversion of the biomass residues into functional derived carbon materials for electrochemistry application is essential due to the inherent biomass precursor characteristics of different kinds of starting materials. As a standard procedure, the biomass can be first pretreated and characterized for main components (cellulose, hemicellulose, and lignin). Based on the composition, the most suitable thermochemical treatment route can be selected to produce high-value material with different characteristics in terms of physicochemical properties (Scheme 20.1).

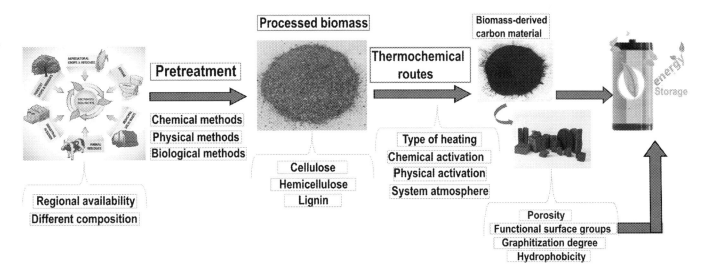

SCHEME 20.1 Biomass generation, processing, and utilization into carbon materials for energy storage applications.

20.2 Suitable sources of biomass for electrode preparation

Despite a variety of carbon forms, including graphite, carbon nanotubes, graphene, carbon fibers and carbon composites (Ahmed Abd-Elsalam, 2020; Mubarik et al., 2021) are being considered for several applications; unfortunately, often, these carbon materials have the drawbacks of high production costs, elaborate large-scale fabrication setups, and/or non-eco-friendly routes/processes. Thus, it is crucial to develop novel, eco-friendly, cheap carbon-based functional materials with scalable synthesis/fabrication processes. In this sense, carbon materials derived from biomass resources have gathered huge interest due to their worldwide availability and abundance, easy processing and handling, nontoxicity, and ease of being turned into the hierarchical porous structures to be employed in a wide range of applications such as adsorbents and catalysts for waters and gases treatments and energy applications (batteries and supercapacitor technologies) (dos Reis et al., 2020; Reis et al., 2021; Tang et al., 2021).

Carbon materials fabricated from biomass can have well-developed porosity with high specific surface areas (SSA), different and hierarchical pore structures, and an abundant number of functionalities on their surfaces, all of these properties allow the biomass-derived carbon materials to be used as the negative electrode (anode) with enhanced electrochemical properties. These enhanced properties might depend on electrolyte penetration, minimization of ion diffusion distances, and generating additional active sites, which are essential for obtaining energy storage systems with high electrochemical performances (dos Reis et al., 2020; Reis et al., 2021; Tang et al., 2021).

Biomass-derived carbons are obtained from any living or recently living organisms such as animals and plants, which include trees and residues, croops, domestic and industrial wastes, etc., which all can be successfully used as sustainable, eco-friendly, and low-cost resources to produce electrode materials for efficient electrochemical energy devices (dos Reis et al., 2020, 2021).

In the past few years, the international literature has brought several publications that summarize biomass as suitable and efficient precursor for sustainable materials for energy storage systems (EES) (dos Reis et al., 2020, 2021; Dubey et al., 2020; Senthil & Lee, 2021), including supercapacitors and lithium-ion batteries. For the preparation of biomass-derived carbon materials with suitable properties to be employed as anode for EES, some prerequisites such as large SSA and abundance of surface functionalities must be met. The characteristics of the carbon anode materials are severely influenced by the type and composition of the biomass

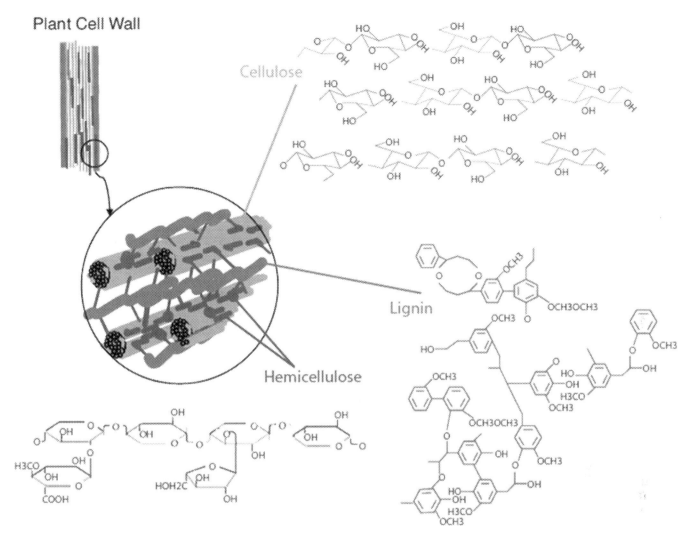

FIGURE 20.1 Simple structures of cellulose, hemicellulose, and lignin: major components of biomass (Wang et al., 2019). Source: *Reprinted with permission from Wang B., Sun Y.-C., Sun R.-C., Fractionational and structural characterisation of lignin and its modification as biosorbents for efficient removal of chromium from wastewater: a review, J. Leather Sci. Eng. 1 (2019) 5. https://doi.org/10.1186/s42825-019-0003-y Copyright 2019, SpringerOpen.*

precursor, which in turn reflect on its electrochemical metrics (dos Reis et al., 2020, 2021; Dubey et al., 2020; Senthil & Lee, 2021). Thus biomass composition is an important factor to be deeply studied before applying it as a carbon anode precursor. Biomass is mainly composed of hemicellulose, cellulose, and lignin, as shown in Fig. 20.1.

Biomass rich in cellulose can be used with success to synthesize carbon electrodes (Sermyagina et al., 2020). Pristine cellulose exhibits significantly lower electrochemical properties than modified cellulose, and its performance can be enhanced by modification, usually employing thermochemical routes. Cellulose content significantly impacts the microporosity development of the biomass-derived carbons (Sermyagina et al., 2020). da Silva Lacerda et al. (2015) reported that artificially increasing the cellulose content in the precursors increased the SSA values and led to improved porosity of the carbon materials, properties that play important roles in the electrochemical performances.

Xue et al. (2018) reported that high cellulose content plays a crucial role in mesopore structure formation in carbon material properties while lignin, for example, can boost the creation of a layered structure rich in micropores; these different properties can promote an important influence on electrochemical performances for biomass-derived carbon materials. Compared with cellulose and hemicellulose, lignin presents a much more complex structure with denser carbon chains and aromatic structures (Titirici et al., 2012), which can be easily converted into 3D hierarchical carbon structures during the pyrolysis process, as highlighted in Fig. 20.2.

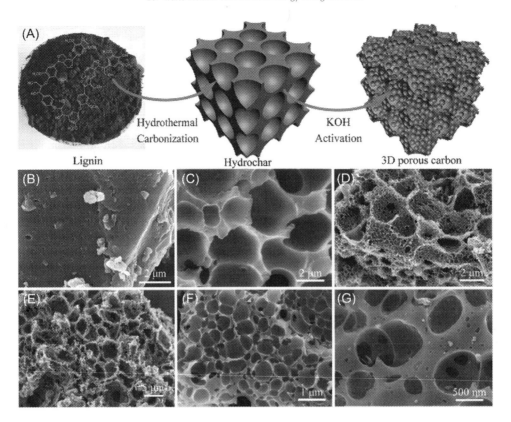

FIGURE 20.2 (A) Illustration of turning lignin into highly porous carbon material; (B) SEM micrograph of lignin pristine form; (C) SEM micrograph of lignin hydrothermally treated at 180°C; (D) and (E) SEM micrographs of pyrolyzed lignin; (F) and (G) the magnified micrographs of pyrolyzed lignin (Guo et al., 2017). Source: *Reprinted with permission from Guo N., Li M., Sun X., Wang F., Yang R., Enzymatic hydrolysis lignin derived hierarchical porous carbon for supercapacitors in ionic liquids with high power and energy densities, Green Chem. 19 (2017) 2595–2602. https://doi.org/10.1039/C7GC00506G. Copyright 2017, RSC Publishing.*

20.3 Production and properties of biomass-derived anode materials

The need to reduce CO_2 emissions, considered the main factor of the anthropic-induced global warming, has pushed the governments to move the world through progressive electrification of mobility and productive activities. How, in principle, electricity iproduced is a key factor in making the process of removing fossil fuels viable. There is a large debate about the opportunity of introducing different types of low carbon footprint technologies such as photovoltaic or Eolic power plants. These technologies rely on an intermittent source of energy that influences their relatively low-capacity factor (Alves, 2022). To increase the capacity factor of these plants, it is necessary to store the energy produced during times of surplus compared with the grid request to deliver the energy at a condition where a deficit in the production takes place. Some countries are considering an opportune mix of nuclear and renewable to keep the stability of the grid, but, undoubtedly, a need of a higher energy storage capacity will be required in the future, particularly considering that the EU has banned the production of ICE (internal combustion engine) starting from 2035. The trend is also to electrify the building's heating system. The most common method to store energy is to convert it to potential energy as in the case of pumped-storage hydroelectricity, but that requires an opportune orography and water availability. Another technology to store the energy is using H_2 produced from electrolysis as a carbon-free vector, but the overall process is relatively inefficient and deals with pressurized flammable gas. Battery-based energy storage seems to be the best candidate for the so-called electrification revolution because it is a rather mature and reliable technology (Bakeer et al., 2022).

The exponential need for batteries has caused a surge in the battery-containing materials, posing severe stress on the supply chains worldwide. Lately, graphite has risen as a problematic strategic material both for the commercial production control (Pandey, 2022) (China being by far the main producer (Garside, 2022) and for the quantity being the most abundant material present by mass in a Li-ion battery).

For all these reasons, researchers recently considered using biomass-derived carbon as an alternative way of producing carbon suitable for battery production. Typically, biomass-derived carbon has an amorphous structure that, in some studies, showed a potential for Na-ion batteries due to the better intercalation compared with graphitic carbons (Legrain et al., 2015a,b).

In any case, the degree of graphitisation has also been investigated in K-ion batteries (Wrogemann et al., 2022), showing that the interaction between the ions and the graphitic carbon structure and morphology is an important factor for the anode performance. Electrical conductivity plays a key role; in this sense, graphitic structure enhances the carbon capability of "moving" the electrons (Shi et al., 2017). Thus, tuning and understanding the interaction of the carbon structure with the ions are an important part of developing new battery technology. Subsequently, the main methods of graphitic carbon production will be reported.

The use of natural graphite is most common in battery production mainly due to the historically relatively low cost of production. When natural graphite is used for anode coating, high levels of carbon impurities require intensive preprocessing. Moreover, the anisotropic structure limits the random transportation and diffusion of the Li-ions in the structure, causing the battery's relatively poor life cyclability (Huang et al., 2013; Lin et al., 2014).

The second most important method to produce graphite is thermal graphitization, commonly called synthetic graphite. Typical feedstock materials are petroleum coke, bitumen, and also biomass (Kamal et al., 2020). Through this process, the basic structural unit of carbon is rearranged from a nonordered distribution. Following the Franklin model for nongraphitizing carbons, the one produced from biomass to a graphitic-like ordered structure (Ion et al., 2006; Liu, Zhang et al., 2019). This process happens through different steps, as reported in Fig. 20.3.

The industry prefers synthetic graphite in producing anode batteries mainly because of the achievable high purity and the controlled production method that allows better life cycling performance in the final battery (Xing et al., 2018). The high energy costs of production and the necessity to have specific graphitisation furnaces made of pure graphite capable of withstanding the high temperatures involved in this process are the biggest issue with this technology. The use of microwave heating might improve the heat transfer efficiency, but it presents a problem on industrial scales (Kim et al., 2016).

Catalytic graphitization is a process to produce graphitic carbons starting from nongraphitizing carbon precursors (such as biomass-derived carbon) at lower temperatures compared with the classic thermal graphitization (Fig. 20.4). The technology has been known for over 40 years (Oya & Otani, 1979) and consists of adding a catalyst that lowers the crystallization energy. Typically, transition metals have shown this capability, particularly Fe, Ni, and Co (Maldonado-Hódar et al., 2000). Several mechanisms have been proposed to describe the catalytic reactions, for example, the "solution-precipitation" mechanism, which occurs when the temperatures are higher than the melting point of the specific catalyst metal (Liu & Loper, 1991). This mechanism, however, doesn't explain why the catalytic conversion of nongraphitic carbon takes place at lower temperatures (<1000°C) compared, for example, with the melting point of Fe; thus, another theory considers the formation of supersaturated FeC droplets (Feng et al., 2011; Krivoruchko, 1998). The droplets dissolve amorphous carbon and precipitate it in a more ordered structure (Gomez-Martin et al., 2021).

Another relatively new method of producing pure graphite is the use of chemical vapor deposition. In this process, the source of carbon material is typically an organic gas (e.g., biomethane or purified syngas)

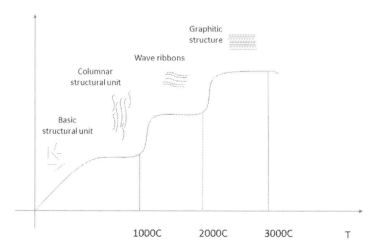

FIGURE 20.3 Thermal graphitization for production of synthetic graphite scheme.

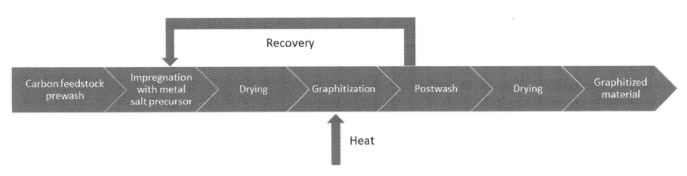

FIGURE 20.4 Example of catalytic graphitisation process from carbon feedstock material (e.g., biomass).

(Kairi et al., 2018; Saputri et al., 2020), which is diffused under predetermined levels of temperature and pressure over a metal catalyst (e.g., Cu, Ni, and Co). The catalytic reaction determines the split of the molecules depositing carbon atoms in layers over the metal catalyst surface, thus creating the graphitic structure in a "bottom-up" process. This process is also used to produce pristine graphene and carbon nanotubes. The limit of this process is the high cost and limited scalability, which can find uses in specific applications but seldomly in the battery mass production.

The flash Joule heating method is a recently invented process used to graphitize the carbon and utilize the Joule effect with an intense current (over 400 A), bringing the temperature for a fraction of seconds over 3000°C. The technology is still at the beginning of development and promises to bypass the high cost of the classical thermal synthetic graphitization due to the small-time exposure (Dong et al., 2022; Liu, Li et al., 2019).

20.4 Applications of biomass-derived carbon materials for energy storage systems

20.4.1 Biomass-based materials for supercapacitors

The progressive phasing out of fossil fuel–based energy sources introduces a large requirement for research in new and more effective energy storage devices. The rapid development of small portable electronics, wearables, and hybrid vehicles reinforces this aspect since more effective and flexible energy storage devices represent a big challenge for developing the integrated solution, as observed in the Internet of Things.

Supercapacitors appear as promising candidates for integration with the abovementioned devices due to the advantages relative to the resulting materials' power and energy density. As shown in Fig. 20.5, conventional batteries present outstanding performance in energy density associated with lower power density, while conventional capacitors offer superior power density and low energy density. Consequently, it is essential to fill up the gap in the Ragone plot by producing high-power and energy-density devices that cover the drawback of conventional batteries and capacitors.

Supercapacitors have been considered a newborn generation of conventional capacitors with a superior energy density that is a consequence of more effective charging mechanisms, classified into three different types, as follows: (1) Electrical double-layer capacitance (EDLC), (2) Pseudocapacitance (PC), and (3) Hybrid supercapacitors.

The EDLC mechanism is established at interfaces of a charged surface and liquid with the violation of the electroneutrality established by an external electric excitation. Consequently, a rearrangement of charges in the form of layers was modeled by Khademi and Barz (2020) in an equivalent circuit. At the near interface of a charged surface and the electrolyte, an immobile layer of adsorbed solvated ions of opposite charge (Stern layer) is in contact with a mobile diffuse layer (DL) and a bulk region in which free species are minimally affected by electrostatic forces with the electrodes. Capacitors represent bounded species, while parallel RC circuits represent the diffusive species into the bulk of the region between electrodes, as the Gouy–Chapman–Stern model explains the non-Faradic interactions (the electrostatically based process of energy storage).

On the other hand, pseudocapacitive processes are Faradaically conducted through oxidation–reduction reactions, displaying ions extrusion/insertion and underpotential deposition as important steps for charge separation. Hybrid supercapacitors use asymmetric configuration with battery-like and capacitor-like electrodes disposed of

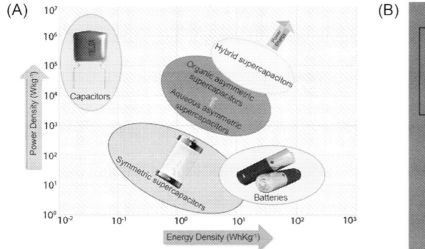 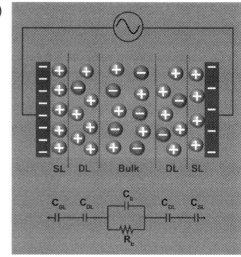

FIGURE 20.5 (A) Ragone plot and schematic position of devices in terms of their power density and energy density. (B) Different zones immobile to mobile species adsorbed species into EDLC structures. Source: *Reprinted with permission from (A) Khademi M., & Barz D. P. J., Structure of the electrical double layer revisited: Electrode capacitance in aqueous solutions, Langmuir. 36 (2020) 4250–4260. https://doi.org/10.1021/acs.langmuir.0c00024. (B) Forouzandeh P., Kumaravel V., Pillai S. C., Electrode materials for supercapacitors: A review of recent advances, Catal. 10 (2020). https://doi.org/10.3390/catal10090969. Copyright 2020, American Chemical Society.*

as anode and cathode, respectively, exploiting the advantages of Faradaic and non-Faradaic processes, resulting in the extension in the voltage window for both aqueous and ionic liquid–based electrolytes (Forouzandeh et al., 2020). The three types are schematically drawn in Fig. 20.6.

In addition to the configuration of devices, the composition of electrodes and separators plays a key role in the overall performance of devices, with the separator having a primary function in avoiding electron short circuits while in contact with the electrolyte and at the same time allowing ionic transport. EDLC electrodes are highly dependent on the surface area due to the required sites for charge accumulation in non-Faradaic processes. Consequently, carbonaceous materials such as activated carbon, graphene oxide, and carbon nanotubes are preferable due to the intrinsic high surface area, pore structure, and porosity.

However, the poor specific capacitance of EDLCs represents a drawback that can be circumvented by the incorporation of metal oxide and conducting polymers (PCs) that increase the energy density but introduce disadvantages concerning cyclability with a reduction in the capacitive retention due to the successive electrochemical reactions in these materials (Ghosh et al., 2019; Khademi & Barz, 2020).

In addition to these aspects, an important factor to be considered is the scale-up and the cost of production. Based on these requirements, activated carbon has been considered a promising candidate acquired from the carbonization of several precursors such as coconut shell, charcoal, and cellulose introducing advantages concerning the chemical modification of more costly components such as graphene and carbon nanotubes.

For this, different processes for carbonization and physical and chemical activation lead to the formation of open-pore structures, strategies based on the hydrothermal and microwave treatment have been explored. In particular, the electrochemical performance of SC electrodes is affected by the degree of porosity and heteroatom doping state. For the production of a highly porous structure, thermal treatment followed by chemical activation can be a viable strategy to reach desirable pore size distribution in an adequate condition of large channels that facilitate ion transportation. Despite these promising properties, the limitation of biomass-based electrodes for supercapacitors is the low energy density (Azwar et al., 2018).

Therefore additional strategies are required to improve the specific capacitance and potential operation window for these materials. Several methods for developing hierarchically porous structures doped with nitrogen, boron, sulfur, and phosphorous to facilitate the electrolyte permeation (distribution, connectivity) have been shown in the literature (Rawat et al., 2022; Yang et al., 2019). These heteroatom doping processes introduce pseudocapacitive components with direct consequences on the wettability of carbon and enhancement in the electrical conductivity of the modified material. The external doping procedure is conducted by mixing biomass with additives (such as urea, melamine, and phosphoric acid) and involves several steps. These processes have been

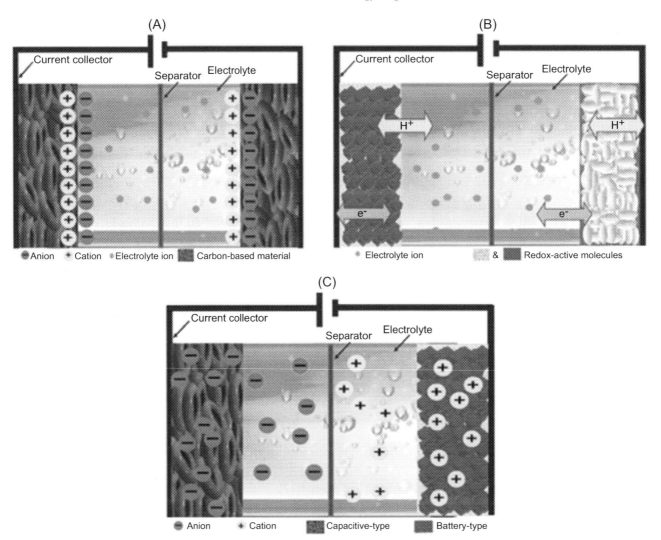

FIGURE 20.6 Scheme of typical supercapacitors composed of the current collector, electrodes, electrolyte, and separator exploring the mechanisms of (A) electrical double-layer capacitors (EDLC), (B) pseudocapacitors, and (C) hybrid supercapacitors. Source: *Reprinted with permission from Forouzandeh P., Kumaravel V., Pillai S.C., Electrode Materials for supercapacitors: A review of recent advances, Catal. 10 (2020). https://doi.org/10.3390/catal10090969. Copyright 2020, MDPI.*

progressively substituted by self-doping steps in which the reagents are intrinsically doped with heteroatoms, as in the case of materials rich in proteins as a source of N-doped porous carbon derivatives (Gopalakrishnan & Badhulika, 2020).

Considered a green method of doping, the nitrogen heteroatom insertion into the carbon matrix has been considered the most effective doping procedure due to the proximity of C and N in the periodic table that enables the electron donor activity of nitrogen to carbon, reducing the lattice mismatching, and allowing C-N bond structures to act as an n-type semiconductor (Gopalakrishnan & Badhulika, 2020). As alternatives to the N-based self-doping procedure, multiple heteroatoms doping levels result from the use of biomass from plants and microorganisms.

20.4.2 Biomass-based materials for batteries

Graphite is the standard anode material for commercial lithium-ion batteries (LIBs) providing a theoretical capacity of 372 mAh/g (Jache et al., 2016), which results in severe limitations in meeting the high-energy-devices market demands, such as those dictated by electrified vehicles. When using graphite in sodium-ion batteries (NIBs), its capacity for storing Na is even more limited, mainly due to its larger ionic size compared with Li^+ and the low thermodynamic stability of the sodium–graphite intercalation, which is reflected in poor electrochemical performance (Jache et al., 2016; Li, Hu, Li et al., 2016). To overcome these limitations, significant efforts have been put on amorphous and nongraphitic bio-based carbons to be used as suitable anodes for many battery configuration systems, which often outperform the graphite electrochemically (Guan et al., 2019; Yu et al., 2019).

Biomass-derived carbon materials can be made from any biomass, having excellent and adaptable structural and functional properties, which make them suitable to be used as anodes for LIBs, SIBs, potassium-ion (KIBs), and other battery configurations (Guan et al., 2019; Jache et al., 2016; Li et al., 2014; Li, Hu, Li et al., 2016; Yu et al., 2019). To be a good candidate for efficient anode materials, there are three basic requirements: (1) the potential of ion (Li, Na, or K) insertion and extraction in the anode versus ion must be as low as possible; (2) the amount of ion that the bio-based carbon anode can accommodate should be as large as possible to reach higher specific capacities; (3) the anode hosts should be able to withstand repeated ion insertion and extraction without any physicochemical and structural damages to achieve long cyclability life. To date, plenty of research deals with biomass-derived carbon anodes for LIBs, NIBs, and KIBs. Their inherent and diverse porous structures are suitable for various battery system applications. The biomass-derived carbon anodes' suitability and advantages in terms of their structures and properties could be simplified into four issues: (1) large interlayer spaces that enable good mechanical stability for the anodes during the ion insertion/extraction process. (2) Active surfaces, due to a large number of functional groups, provide improved chemical reactions with other active materials to assist ion transfer. (3) High SSA, well-developed pore strctures with different nano-sizes, and good thermal stability facilitate; (4) wide availability/accessibility facilitates the development of multiple carbon anodes using different types of biomass carbons and composites by simple synthesis methods.

The microstructure of carbon dictates its electrochemical performance, and the biomass precursors, synthesis strategies, and surface modifications play a significant role in the carbon's microstructure. To better understand the effects of bio-based carbon anode on battery performance, Li et al. (2014) investigated several biochars in this regard, indicating that the biocarbon anode with higher carbon content (in %) provided the best performance.

20.4.2.1 Lithium-ion battery (LIB)

Lithium-ion batteries (LIBs) are the established and fastest-growing battery technology and are already the most prominent technology in our society. It has been an enabling technology for portable electronic devices and is revolutionizing the automotive industry. High-energy-dense batteries are intensively researched worldwide to achieve a better energy transition, and LIBs have emerged as the most credible batteries among all due to their large energy density. Due to the very high demand and unequal geographic Li distribution worldwide, it is becoming an ever-increasing strategic resource, as well as other elements used in LIBs such as graphite, copper, and silicon, which increases the pressure to develop next-generation LIBs with more sustainable materials. Biomass-derived anodes appear promising materials, thus avoiding using critical silicon and graphite. Also, using graphite as an anode for LIBs carries a large CO_2 footprint and high costs because the graphite is derived from mining (Cavers et al., 2022). Therefore extensive research is conducted to develop innovative concepts where various biomass side streams are employed as precursors for carbon-based materials as anodes for LIBs (Campbell et al., 2015; Dou et al., 2019; Dou et al., 2021; Drews et al., 2021; Hernández-Rentero et al., 2020; Kietisirirojana et al., 2022; Kim et al., 2018; Lotfabad et al., 2014; Luna-Lama et al., 2019; Panda et al., 2021; Pramanik et al., 2021; Yokokura et al., 2020; Yu et al., 2021; Zhang et al., 2009; Zheng et al., 2021). Biomass carbon-derived materials have good and adaptable structural and functional properties, making them suitable for use as anodes for LIBs (Campbell et al., 2015; Dou et al., 2019; Dou et al., 2021; Drews et al., 2021; Hernández-Rentero et al., 2020; Kietisirirojana et al., 2022; Kim et al., 2018; Lotfabad et al., 2014; Luna-Lama et al., 2019; Panda et al., 2021; Pramanik et al., 2021; Yokokura et al., 2020; Yu et al., 2021; Zhang et al., 2009; Zheng et al., 2021). In addition, using biomass residues (and not graphite) to produce efficient anodes would be both environmentally friendly and economically advantageous (Guan et al., 2019; Li et al., 2014; Yu et al., 2019).

Dou et al. (2019) synthesized porous carbons derived from jute fiber using $CuCl_2$ activation and tested them as anodes for LIBs. The carbon materials were named according to the amount of $CuCl_2$ JFC-0 (no $CuCl_2$), JFC-6 (ratio of 1:6 of biomass: $CuCl_2$), JFC-8 (ratio 1:8), and JFC-10 (1:10). Porous carbon with BET up to 2043 m^2/g was obtained (JFC-8). From the Raman analysis, the JFC-8 presented the highest I_D/I_G value among all carbons,

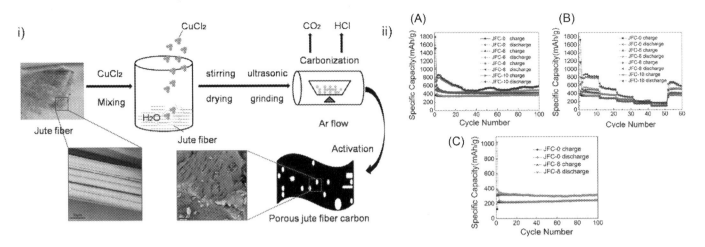

FIGURE 20.7 (i) Preparation method of jute fibers—derived porous carbon. (ii) (A) Charge—discharge profiles of JFC-0, JFC-6, JFC-8, and JFC-10 samples; (B) rate performance of four samples at 0.2 C, 0.5 C, 1 C, 2 C, 5 C and back to 0.2 C; (C) charge—discharge profiles of JFC-0 and JFC-8 samples at 1 C after 100 cycles. Source: *Reprinted with permission from Dou Y., Liu X., Yu K., Wang X., Liu W., Liang J., Liang C., Biomass porous carbon derived from jute fiber as anode materials for lithium-ion batteries, Diam. Relat. Mater. 98 (2019) 107514. https://doi.org/10.1016/j.diamond.2019.107514. Copyright 2019, Elsevier.*

highlighting that its structure has many more defects. Defects are desired because they increase the contact area of the electrode/electrolyte interface, which could increase the electrochemical performances of the electrodes. Interestingly, JFC-8 delivered a specific capacity equal to 580.4 mA h/g at the current density of 0.2 C after 100 cycles, the highest among all samples (see Fig. 20.7). The authors reported that this high capacity was due to the (1) high BET area that increases the active electrode/electrolyte contact area, accelerating mass diffusion; (2) the presence of macropores that improves the transport of electrolyte and enhances the Li^+ diffusion; (3) the high presence of micropores that acted as deep trap sites for Li^+ storage, thus improving the capacity of Li^+ storage; and (4) the presence of mesopores provided ion highways for ion transfer. These factors contributed to the high Li^+ storage on the jute fiber carbon anode.

Table 20.1 shows that researchers have made great efforts to develop anode materials for LIBs using biomass as the main precursor. Although the profile of each biomass anode differs from one to another, all of them have advantages and disadvantages. Table 20.1 shows the performance of various biomass anodes, and all of them showed good electrochemical metrics regardless of their different physicochemical characteristics such as surface area, porous structure, amorphous or crystalline structure, and degree of graphitization.

20.4.2.2 Sodium-ion battery (NIB)

NIB is one of the promising and viable alternatives for replacing LIB, because of not only the abundance of Na^+ resources, which significantly lower cost compared with LIBs, but also because NIBs can deliver very high storage capacities. Compared with Li, Na is considerably more abundant and much more evenly distributed on earth; for instance, 70% of Li is in South America while Na is everywhere, which results in costs far less for the Na compared with Li (Balogun et al., 2016). One of the biggest issues encountered in the development of sustainable and cost-effective NIBs is the low effectiveness of appropriate anode material because of the large size of Na ion and its slow kinetics. Further issues related to high operational potential lead to low energy density and high volume expansion that results in the destruction of anode materials during cycling, which in turn reflects in very poor coloumbic efficiencies (Palomares et al., 2012). In the past decade, intensive research on various hard carbon-based anodes suitable for sodium-ion intercalation has led to a broad spectrum of carbon yields, charge capacities, cyclabilities, and coulombic efficiencies.

Unlike LIB, graphite exhibits low efficiency in NIB application because Na hardly forms staged graphite intercalation compounds. This has trigged the research on other types of carbon-based material anodes such as hard carbons from biomass precursors that due to their larger interlayer carbon spacing can be extremely effective in NIB application (Dou et al., 2021; Luna-Lama et al., 2019). Stevens and Dahn (2000) published the first work that experimentally demonstrated the similarity between lithium and sodium insertion mechanisms in carbon materials. They showed that a hard carbon delivered an extraordinarily

TABLE 20.1 Comparative electrochemical performance of biomass-based carbon material anodes for LIB.

Biomass source	Synthesis method and morphology	Specific surface area (S_{BET}, m²/g)	Potential (V vs Li^+/Li)	Current rate (mA/g)	Initial capacity (Discharge/Charge) (mA h/g)	Capacity retention (mA h/g)/ (cycles)	Rate test (mA/g), (cycle)/ capacity (mA h/g)	Ref.
Coffee oil	Dry autoclaving method, Micrometer diameter spheroidal carbon particles	5	0.01–2.0	100	~440/281.8	290 (200)	500 (5) ~150	Kim et al. (2018)
Banana peel	High-temperature KOH activation method, High dense banana peel pseudo graphite	217	0.001–2.8	100	~2150/1075	800 (300)	10000 (10) ~100	Lotfabad et al. (2014)
Bagasse	Hydrothermal activation method, N, P codoped bagasse-based sheet-like mesoporous carbon	1307.21–2118.59	0.01–3.0	100	2347.56/1186.59	816.36 (50)	2000 (200) 592.38	Zheng et al. (2021)
Portobello mushroom	Carbon nanoribbon as free-standing, binder-free, and current collector-free Li-ion battery anodes	19.6	0.01–3.0	50	771.3/280	~260 (700)	–	Campbell et al. (2015)
Mustard seed waste	Hydrothermal method, high porous spherical carbon nanostructures in-situ doped of heteroatoms (N, S)	618	0.005–3.0	100	~822/617	~714 (550)	500 (10) 280	Pramanik et al. (2021)
Tamarind plant Seeds	High temperature KOH activation method, porous carbon	103.51	0.01–2.5	200	~1037/414	~370 (100)	–	Panda et al. (2021)
Rice Straws	High porous carbon, high temperature KOH activation method	3315	0.005–3.0	37.2	2041/986	–	744 (5) 257	Zhang et al. (2009)
Cherry Pit	KOH and H_3PO_4 activation, Disordered carbons,	1662	0.01–2.8	124	~1300/300	200 (200)	1860 (5) 70	Hernández-Rentero et al. (2020)
Jute Fiber	Micro-mesoporous carbon material using zinc chloride as an activator under high temperature carbonization in open atmosphere	1028.614	0.02–3.0	74.4	1173.3/534.1	427.2 (100)	1860 (10) 171.6	Dou et al. (2021)
Coffee waste grounds	Mechanochemical dry milling of spent coffee grounds followed by further carbonisation at 800 °C, Non-porous carbonaceous materials	<10	0.0–3.0	100	764/~380	285 ± 5 (100)	1000 (10) 150	Luna-Lama et al. (2019)
Spruce wood	Pyrolysis and ball milling method, Spruce hard carbon	61	0.01–3.0	37.2	~400/250	300 (400)	1488 (10) ~110	Drews et al. (2021)
Gold beard grass pollen	Pyrolysis and KOH activation method, Mesoporous carbon powder from bee-collected pollens	1107.447	0.01–3.0	37.2	788.99	297.283 (200) @ 2000 mA/g	5000 (-) 334.10	Kietisirirojana et al. (2022)
Avocado seeds	Pyrolysis, sulfuric acid treatment, Non-graphitic carbonaceous anodes	–	0.0–2.0	100	~420/380	~320 (100)	400 (5) ~200	Yokokura et al. (2020)
Wheat straw cellulose	KOH activation agent followed pyrolysis, Porous amorphous carbon	628	0.1–3.0	74.4	2750/~580	1420.5 (100)	1860 (10) 810	Yu et al. (2021)

TABLE 20.2 Comparative electrochemical performance of biomass-based carbon material anodes for NIB.

Biomass source	Synthesis method and morphology	Specific surface area (S_{BET}, m²/g)	Potential (V vs Na⁺/Na)	Current rate (mA/g)	Initial capacity (Discharge/Charge) (mA h/g)	Capacity retention (mA h/g)/ (cycles)	Rate test (mA/g), (cycle)/ capacity (mA h/g)	Ref.
Camphor wood residues	Carbonization followed by pyrolysis method, porous morphology	3.74	0.0–2.0	20	~324.6/391.8	268.1 (200)	20 (50) ~319	Guo et al. (2022)
Drug residue	Heat treatment method, porous structure	989.7	0.01–3.0	100	~801/ -	402 (500)	–	Wan et al. (2021)
Seaweed-derived carbon	High porous carbon by high-temperature KOH activation method, sheet-like carbon structure	1641	0.005–3.0	100	1342/287	192 (500)	800 (10) 114	Senthil et al. (2022)
Borassus flabellifer male inflorescences	High temperature treatment method	18.87	0.01–2.5	20	413/358	~322 (500)	2000 (10) 117	Kumaresan et al. (2021)
Chickpea husk	Sonochemical activation method, honeycomb-like morphology	1599	0.0–3.0	20	~800/330	~125 (500)	1000 (500) 125	Ghani et al. (2022)
Tea tomenta	High temperature treatment method, rod like morphology	13.92	0.0–3.0	28	~326.1	~262.4 (100)	–	Wang et al. (2022)
Cotton roll	Carbonization followed by pyrolysis method, braided fibrous morphology and hollow structure	38	0.0–2.0	30	~315	~262.4 (100)	–	Li, Hu, Titirici et al. (2016)
Cherry petals	High temperature treatment method, honeycomb-like morphology	1.86	0.01–3.0	20	461.1/310.2	~298.1 (100)	500 (500) 131.5	Zhu et al. (2018)
Cucumber stem	Carbonization followed by KOH activation method	1988.90	0.01–3.0	50	~1200/458.6	198.6 (500)	–	Li et al. (2019)

high capacity of 300 mAh/g, suggesting that a similar Na^+ insertion mechanism of Li^+ in graphite took place. Since then, numerous carbon-based anodes have been investigated as anodes of NIB, but still, there is a lot to learn in terms of its properties and its electrochemical performances (Table 20.2). Thus the development of carbonaceous materials with expanded interlayers is required for optimal use as anodes for SIBs.

There are many strategies to boost the Na^+ storage capacity of biomass-based anodes. The most promising ones are related to the adjustment of porosity of the biomass anodes such as pore size distribution and simultaneously increasing the SSA values as the number of defects in its structures. It is known that hierarchical pore structures (micro-, meso-, and macropores) facilitate the accessibility of active sites for Na ions, shortening the ion diffusion distance and improving charge transfer kinetics (Guo et al., 2022; Senthil et al., 2022; Wan et al., 2021). Table 20.2 exhibits the performance of various biomass-derived anodes, and all of them showed good electrochemical metrics regardless of their different physicochemical characteristics such as high surface area, porous structure, amorphous or crystalline structure, and degree of graphitization.

20.4.2.3 Potassium-ion battery (KIB)

The search of advanced battery systems with high energy density and low cost is the main objective of energy storage research groups, industry, and customers. Potassium-ion batteries (KIBs) have gathered massive attention because they meet the need for scalable power sources in academic and industrial energy-harvesting areas and can deliver high-energy capacities. KIBs represent promising candidates mainly due to some important advantages compared with LIBs, such as (1) K is an earth-abundant element compared with lithium, it occupies 1.5 mass% while Li has 0.0017 mass%. Moreover, K costs only one-tenth of the cost of Li (Pramudita et al., 2017). (2) The electrode potential of K is lower than that of Na and close to that of Li, which allows a broader range of choices for possible cathode materials and higher voltage of batteries, assuring relatively more promising energy density of KIBs. (3) Aluminum as the anode current collector for KIBs is much lighter than copper, which may enable higher gravimetric energy density of KIBs than that of LIBs. (4) K^+ has weaker Lewis's acidity and smaller solvation ion size compared with Li^+, thereby displaying an improved ionic conductivity and mobility (Yuan et al., 2021). (5) The redox potential of K^+/K is nearly equal in aqueous electrolytes or even lower in nonaqueous electrolytes (ethylene carbonate and propylene carbonate) compared with Li^+/Li, which allows KIBs to maintain higher output voltages and specific energy (Wu et al., 2019).

Fig. 20.8 displays the KIBs' different components and charge/discharge process for charge storage. KIBs have operating principles that are similar to that of LIBs. The electrons flow from the cathode to the biomass-derived anode when the cell is being charged. During discharging, K^+ ions commute between the cathode and anode in the opposite path, and electrons flow out of the anode to the cathode through the external circuit to maintain electrical neutrality. The KIBs technology should meet some basic requirements to store energy by the abovementioned processes. For instance, the bio-based anode must have suitable lattice sites or spaces to accommodate and release the K^+ freely in a reversible way to obtain high specific capacities. Also, the carbon anode should have a lower operating potential to achieve a high voltage and better energy in the full cell format. To show desired cycling performance, the biomass-derived anode

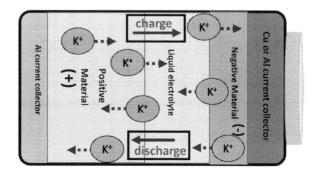

FIGURE 20.8 Schematic representation of working principle of KIBs.

must be physicochemically stable to support the volume changes during the reversible K ion intercalation/deintercalation.

Many factors affect the KIBs' performance. For instance, K^+ insertion/extraction affects the specific capacity of bio-based anodes, which can be improved by a structure that has sufficient space to accommodate K^+ ions. The rate performance strongly depends on the anode reaction dynamics, which are determined by the dynamic of K^+ diffusion and the electrical conductivity of the bio-based anodes that are influenced by their physicochemical features. Yang et al. (2020) prepared biomass-derived carbon anodes for KIBs and reported high specific capacity (407 mAh/g (50 mA/g) and 163.8 mAh/g (200 mA/g) after 50 and 100 cycles with different current densities, respectively). They concluded that the high battery efficiency was related to the anode physicochemical properties. The crystal structures of biocarbons were no-graphitic carbons with enlarged interlayer distance and versatile pores/defectives/edges, which helped alleviate volume change and improved the cycling performance. The carbon anodes with hierarchical porous structures had a high amorphous degree, which enabled the rapid K^+ diffusion and improved electrolyte wettability of the anodes for easier K^+ insertion/extraction.

Wu et al. (2022) evaluated the effect of lignin structure (molecular weight's role) and pyrolysis conditions of lignin-activated carbons on K-ion storage mechanisms. It was shown that K^+ storage in lignin carbons depends on bulk-insertion and surface-adsorption sites, which are influenced by pyrolysis conditions and lignin precursors. The authors reported that the insertion of K ions occurred in the interlayer spaces of graphite-like nanocrystals. However, surface adsorption of K ions took place mainly in disordered carbon areas in the lignin-activated carbons. The authors concluded that during the pyrolysis process, the lignin with low molecular weight tends to generate graphite-like structures because of easier depolymerization, higher amounts of fragments, and therefore, more graphitized. However, low-molecular-weight lignin carbons have less disordered carbon structures. Thus the role of surface-adsorption K ions in the carbon anode (Fig. 20.9A) is lower, negatively impacting the total K-ion storage capacity (Fig. 20.9A).

In contrast, a precursor with high-molecular-weight lignin generated carbons with a more disordered structure because long lignin chains are more difficult to be depolymerized, fragmented, and aromatized. The carbon with high disordered structures enhances the K-ion storage via the surface-adsorption mechanism (Fig. 20.9C). However, carbon anodes produced from lignin precursor with medium molecular weight possessed a balanced combination of disordered and graphite-like structures, which was reflected in better electrochemical performance by high K-ion storage capacity through the combined surface adsorption and bulk insertion (Fig. 20.9B).

The authors highlighted that the above findings would be inspirable by biomass-derived carbon anodes for battery application (KIBs, LIBs, and SIBs). The importance of the lignin molecular weight (and possibly the biomass molecular weight as well) in enhancing K-ion storage performance. Moreover, proving that lignin structures play an important role in the electrochemical performances can also expand the different biomass type precursors and their structures. For instance, a representative selection of biomass precursors and their anodes' morphological, structural, and electrochemical metrics on KIBs are shown in Table 20.3. The table shows that the diverse biomass precursors generated different surface morphologies and porosity structures, which impacted their electrochemical metrics. The physicochemical features of the biomass-derived anodes play a crucial role in its electrochemical metrics, due to the accumulation of ions at the micropores, transport, diffusion, and storage of ions at the mesopores and macropores. The different hierarchical pores and surface properties enable the anodes to maximize their specific capacity and show positive routes for future developments.

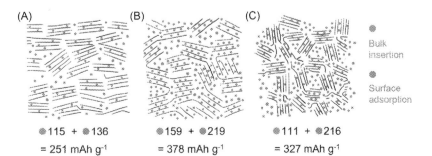

FIGURE 20.9 Proposed K-ion storage mechanisms in L700 (A), M700 (B), and H700 (C). *Source: Reprinted with permission from Wu Z., Zou J., Zhang Y., Lin X., Fry D., Wang L., Liu J., Lignin-derived hard carbon anode for potassium-ion batteries: Interplay among lignin molecular weight, material structures, and storage mechanisms, Chem. Eng. J. 427 (2022) 131547. https://doi.org/10.1016/j.cej.2021.131547. Copyright 2022, Elsevier.*

TABLE 20.3 Comparative electrochemical performance of bio-based carbon material anodes for KIB.

Biomass source	Synthesis method and morphology	Specific surface area (S_{BET}, m^2/g)	Potential (V vs K/K$^+$)	Current rate (mA/g)	Initial capacity (Discharge/charge) (mA h/g)	Capacity retention (mA h/g)/ (cycles)	Ref.
Bamboo	Carbonization and KOH-treated activation. medium porosity carbon anodes with crystalline structures with a high presence of defects and low graphitization degree.	339	0.01–2.8	50	~450	204 (300)	Tian et al. (2020)
Sugarcane bagasse	Biomass activated with NiCl$_2$. Highly amorphous and microporous materials doped with nitrogen.	~467	0.1–3.0	100	505	100.4 (400)	Deng et al. (2021)
Chitin	NaOH/urea activated carbón anodes. Meso-macro porous chitin microspheres-like materials	563	0.01–3.0	0.12 to 36 Coulombic	~320	180 (4000)	Chen et al. (2017)
Corn silk	Hydrothermal treatment and one-step carbonization. The carbon materials possess larger lattice spacing, amorphous structures, and a very low degree of graphitization with many defects.	—	0.01–3.0	100	~886	~121 (2600)	Zhang et al. (2021)
Hemp core	One-step carbonization with fluorine doping process. Material wrapped by fluorine-containing nanotubes rich in defects with widen pore sizes and low degree of graphitization.	780	0.001–3.0	200	~822/617	369.6 (500)	Wang et al. (2021)
Maple leaves	Carbonization and HNO$_3$-treated activation. Low porosry carbon anodes containing O/N functional groups with higher graphitization degree.	62.6	0.01–3.0	50	~934	~142 (1000)	Liu, Jing et al. (2019)
Cyanobacteria powder	Two-step carbonization with NaCl and KCl activation. Nitrogen/oxygen codoped hierarchically mesoporous carbon. Amorphous structures with high degree of graphitization.	473	0.01–3.0	50	912	~104 (1000)	Sun et al. (2019)
Rice husk	Carbonization temperatures from 900 to 1500°C. Highly amorphous microstructure rich in mesopores.	365	0.01–3.0	30	~250	104 (500)	Li et al. (2020)
Ganoderma lucidum spore	Anode prepared by one-step carbonization with no activation. Hollow-cage structure with no high porosity and high degree of graphitization with typical carbon amorphous structure.	104.4	0.02–3.0	1000	~450	~125 (700)	Yang et al. (2020)

20.5 Conclusions, challenges, and prospects

Biomass-derived materials ranging from multidimensional carbons can contribute to sustainable battery systems and sustainable battery components. This chapter also offers a comprehensive overview of the fabrication and application of biomass-derived materials in rechargeable batteries. In more efficient and economical ways, battery materials may be available from biomass-derived materials. The development of alternate electrode materials is a hot research topic, and many research groups investigate the possibilities for new active materials suitable for lithium-, sodium-, or potassium-ion batteries. More attention should be paid to upgrading the processing technologies to maximize biomass-based materials utilization with high efficiency and low cost to accelerate industrial applications. However, inorganic materials are often preferable for high-energy-density applications, with severe drawbacks regarding their sustainability. Switching to fully bio-derived energy storage devices in the future is highly prioritized.

References

Ahmed Abd-Elsalam Kamel (2020). *Carbon nanomaterials for agri-food and environmental applications.* https://www.elsevier.com/books/carbon-nanomaterials-for-agri-food-and-environmental-applications/abd-elsalam/978-0-12-819786-8.

Alves, B. (2022). Capacity factors for selected energy sources in the United States in 2021,

Azwar, E., Wan Mahari, W. A., Chuah, J. H., Vo, D.-V. N., Ma, N. L., Lam, W. H., & Lam, S. S. (2018). Transformation of biomass into carbon nanofiber for supercapacitor application — A review. *International Journal of Hydrogen Energy, 43*, 20811–20821. Available from https://doi.org/10.1016/j.ijhydene.2018.09.111.

Bakeer, A., Chub, A., Shen, Y., & Sangwongwanich, A. (2022). Reliability analysis of battery energy storage system for various stationary applications. *Journal of Energy Storage, 50*. Available from https://doi.org/10.1016/j.est.2022.104217.

Balogun, M.-S., Luo, Y., Qiu, W., Liu, P., & Tong, Y. (2016). A review of carbon materials and their composites with alloy metals for sodium ion battery anodes. *Carbon N. Y, 98*, 162–178. Available from https://doi.org/10.1016/j.carbon.2015.09.091.

Campbell, B., Ionescu, R., Favors, Z., Ozkan, C. S., & Ozkan, M. (2015). Bio-derived, binderless, hierarchically porous carbon anodes for Li-ion batteries. *Scientific Reports, 5*, 14575. Available from https://doi.org/10.1038/srep14575.

Cavers, H., Molaiyan, P., Abdollahifar, M., Lassi, U., & Kwade, A. (2022). Perspectives on improving the safety and sustainability of high voltage lithium-ion batteries through the electrolyte and separator region. *Advanced Energy Materials*, 2200147. Available from https://doi.org/10.1002/aenm.202200147.

Chen, C., Wang, Z., Zhang, B., Miao, L., Cai, J., Peng, L., Huang, Y., Jiang, J., Huang, Y., Zhang, L., & Xie, J. (2017). Nitrogen-rich hard carbon as a highly durable anode for high-power potassium-ion batteries. *Energy Storage Materials, 8*, 161–168. Available from https://doi.org/10.1016/j.ensm.2017.05.010.

Collard, F.-X., & Blin, J. (2014). A review on pyrolysis of biomass constituents: Mechanisms and composition of the products obtained from the conversion of cellulose, hemicelluloses and lignin. *Renewable and Sustainable Energy Reviews, 38*, 594–608. Available from https://doi.org/10.1016/j.rser.2014.06.013.

da Silva Lacerda, V., López-Sotelo, J. B., Correa-Guimarães, A., Hernández-Navarro, S., Sánchez-Báscones, M., Navas-Gracia, L. M., Martín-Ramos, P., & Martín-Gil, J. (2015). Rhodamine B removal with activated carbons obtained from lignocellulosic waste. *Journal of Environmental Management., 155*, 67–76. Available from https://doi.org/10.1016/j.jenvman.2015.03.007.

Deng, Q., Liu, H., Zhou, Y., Luo, Z., Wang, Y., Zhao, Z., & Yang, R. (2021). N-doped three-dimensional porous carbon materials derived from bagasse biomass as an anode material for K-ion batteries. *Journal of Electroanalytical Chemistry, 899*, 115668. Available from https://doi.org/10.1016/j.jelechem.2021.115668.

Dong, S., Song, Y., Ye, K., Yan, J., Wang, G., Zhu, K., & Cao, D. (2022). Ultra-fast, low-cost, and green regeneration of graphite anode using flash joule heating method. *EcoMat. n/a* (n.d.), e12212. Available from https://doi.org/10.1002/eom2.12212.

dos Reis, G. S., Larsson, S. H., de Oliveira, H. P., Thyrel, M., & Claudio Lima, E. (2020). Sustainable biomass activated carbons as electrodes for battery and supercapacitors—A mini-review. *Nanomater, 10*. Available from https://doi.org/10.3390/nano10071398.

dos Reis, G. S., Pinheiro Lima, R. M. A., Larsson, S. H., Subramaniyam, C. M., Dinh, V. M., Thyrel, M., & de Oliveira, H. P. (2021). Flexible supercapacitors of biomass-based activated carbon-polypyrrole on eggshell membranes. *Journal of Environmental Chemical Engineering, 9*, 106155. Available from https://doi.org/10.1016/j.jece.2021.106155.

Dou, Y., Liu, X., Wang, X., Yu, K., & Liang, C. (2021). Jute fiber based micro-mesoporous carbon: A biomass derived anode material with high-performance for lithium-ion batteries. *Materials Science and Engineering B, 265*, 115015. Available from https://doi.org/10.1016/j.mseb.2020.115015.

Dou, Y., Liu, X., Yu, K., Wang, X., Liu, W., Liang, J., & Liang, C. (2019). Biomass porous carbon derived from jute fiber as anode materials for lithium-ion batteries. *Diamond and Related Materials, 98*, 107514. Available from https://doi.org/10.1016/j.diamond.2019.107514.

Drews, M., Büttner, J., Bauer, M., Ahmed, J., Sahu, R., Scheu, C., Vierrath, S., Fischer, A., & Biro, D. (2021). Spruce hard carbon anodes for lithium-ion batteries. *ChemElectroChem, 8*, 4750–4761. Available from https://doi.org/10.1002/celc.202101174.

Dubey, P., Shrivastav, V., Maheshwari, P. H., & Sundriyal, S. (2020). Recent advances in biomass derived activated carbon electrodes for hybrid electrochemical capacitor applications: Challenges and opportunities. *Carbon N. Y, 170*, 1–29. Available from https://doi.org/10.1016/j.carbon.2020.07.056.

Engamba Esso, S. B., Xiong, Z., Chaiwat, W., Kamara, M. F., Longfei, X., Xu, J., Ebako, J., Jiang, L., Su, S., Hu, S., Wang, Y., & Xiang, J. (2022). Review on synergistic effects during co-pyrolysis of biomass and plastic waste: Significance of operating conditions and interaction mechanism. *Biomass and Bioenergy, 159*, 106415. Available from https://doi.org/10.1016/j.biombioe.2022.106415.

Feng, X., Chee, S. W., Sharma, R., Liu, K., Xie, X., Li, Q., Fan, S., & Jiang, K. (2011). In Situ TEM observation of the gasification and growth of carbon nanotubes using iron catalysts. *Nano Research*, 4. Available from https://doi.org/10.1007/s12274-011-0133-x.

Forouzandeh, P., Kumaravel, V., & Pillai, S. C. (2020). Electrode materials for supercapacitors: A review of recent advances. *Catalysts*, 10. Available from https://doi.org/10.3390/catal10090969.

Garside, M. (2022). *Major countries in worldwide graphite mine production in 2021*, Statista.Com. https://www.statista.com/statistics/267366/world-graphite-production/.

Ghani, U., Iqbal, N., Aboalhassan, A. A., Liu, B., Aftab, T., Zada, I., Ullah, F., Gu, J., Li, Y., Zhu, S., & Liu, Q. (2022). One-step sonochemical fabrication of biomass-derived porous hard carbons; towards tuned-surface anodes of sodium-ion batteries. *Journal of Colloid and Interface Science*, 611, 578–587. Available from https://doi.org/10.1016/j.jcis.2021.12.104.

Ghosh, S., Santhosh, R., Jeniffer, S., Raghavan, V., Jacob, G., Nanaji, K., Kollu, P., Jeong, S. K., & Grace, A. N. (2019). Natural biomass derived hard carbon and activated carbons as electrochemical supercapacitor electrodes. *Scientific Reports*, 9, 16315. Available from https://doi.org/10.1038/s41598-019-52006-x.

Gomez-Martin, A., Schnepp, Z., & Ramirez-Rico, J. (2021). Structural evolution in iron-catalyzed graphitization of hard carbons. *Chemistry of Materials*, 33. Available from https://doi.org/10.1021/acs.chemmater.0c04385.

Gopalakrishnan, A., & Badhulika, S. (2020). Effect of self-doped heteroatoms on the performance of biomass-derived carbon for supercapacitor applications. *Journal of Power Sources*, 480, 228830. Available from https://doi.org/10.1016/j.jpowsour.2020.228830.

Guan, Z., Guan, Z., Li, Z., Liu, J., & Yu, K. (2019). Characterization and preparation of nano-porous carbon derived from hemp stems as anode for lithium-ion batteries. *Nanoscale Research Letters*, 14, 338. Available from https://doi.org/10.1186/s11671-019-3161-1.

Guo, N., Li, M., Sun, X., Wang, F., & Yang, R. (2017). Enzymatic hydrolysis lignin derived hierarchical porous carbon for supercapacitors in ionic liquids with high power and energy densities. *Green Chemistry*, 19, 2595–2602. Available from https://doi.org/10.1039/C7GC00506G.

Guo, S., Chen, Y., Tong, L., Cao, Y., Jiao, H., long, Z., & Qiu, X. (2022). Biomass hard carbon of high initial coulombic efficiency for sodium-ion batteries: Preparation and application. *Electrochimica Acta*, 410, 140017. Available from https://doi.org/10.1016/j.electacta.2022.140017.

Hernández-Rentero, C., Marangon, V., Olivares-Marín, M., Gómez-Serrano, V., Caballero, Á., Morales, J., & Hassoun, J. (2020). Alternative lithium-ion battery using biomass-derived carbons as environmentally sustainable anode. *Journal of Colloid and Interface Science*, 573, 396–408. Available from https://doi.org/10.1016/j.jcis.2020.03.092.

Huang, S., Guo, H., Li, X., Wang, Z., Gan, L., Wang, J., & Xiao, W. (2013). Carbonization and graphitisation of pitch applied for anode materials of high power lithium ion batteries. *Journal of Solid State Electrochemistry*, 17. Available from https://doi.org/10.1007/s10008-013-2003-9.

Ion, I., Bondar, A. M., Kovalev, Y., Banciu, C., & Pasuk, I. (2006). Modification of the structural parameters of coal tar pitch induced by addition of nanocarbon-coated iron at primary carbonisation. *Journal of Optoelectronics and Advanced Materials*.

Jache, B., Binder, J. O., Abe, T., & Adelhelm, P. (2016). A comparative study on the impact of different glymes and their derivatives as electrolyte solvents for graphite co-intercalation electrodes in lithium-ion and sodium-ion batteries. *Physical Chemistry Chemical Physics*, 18, 14299–14316. Available from https://doi.org/10.1039/C6CP00651E.

Kairi, M. I., Zuhan, M. K. N. M., Khavarian, M., Vigolo, B., Bakar, S. A., & Mohamed, A. R. (2018). Co-synthesis of large-area graphene and syngas via CVD method from greenhouse gases. *Materials Letters*, 227. Available from https://doi.org/10.1016/j.matlet.2018.05.031.

Kamal, A. S., Othman, R., & Jabarullah, N. H. (2020). Preparation and synthesis of synthetic graphite from biomass waste: A review. *Systematic Reviews in Pharmacy*, 11.

Khademi, M., & Barz, D. P. J. (2020). Structure of the electrical double layer revisited: Electrode capacitance in aqueous solutions. *Langmuir*, 36, 4250–4260. Available from https://doi.org/10.1021/acs.langmuir.0c00024.

Kietisirirojana, N., Tunkasiri, T., Pengpat, K., Khamman, O., Intatha, U., & Eitssayeam, S. (2022). Synthesis of mesoporous carbon powder from gold beard grass pollen for use as an anode for lithium-ion batteries. *Microporous and Mesoporous Materials*, 331, 111565. Available from https://doi.org/10.1016/j.micromeso.2021.111565.

Kim, K., Adams, R. A., Kim, P. J., Arora, A., Martinez, E., Youngblood, J. P., & Pol, V. G. (2018). Li-ion storage in an amorphous, solid, spheroidal carbon anode produced by dry-autoclaving of coffee oil. *Carbon N. Y*, 133, 62–68. Available from https://doi.org/10.1016/j.carbon.2018.03.013.

Kim, T., Lee, J., & Lee, K. H. (2016). Full graphitisation of amorphous carbon by microwave heating. *RSC Advances*, 6. Available from https://doi.org/10.1039/c6ra01989g.

Krivoruchko, O. P. (1998). A new phenomenon involving the formation of liquid mobile metal-carbon particles in the low-temperature catalytic graphitisation of amorphous carbon by metallic Fe, Co and Ni. *Mendeleev Communications*, 8. Available from https://doi.org/10.1070/MC1998v008n03ABEH000944.

Kumaresan, T. K., Masilamani, S. A., Raman, K., Karazhanov, S. Z., & Subashchandrabose, R. (2021). High performance sodium-ion battery anode using biomass derived hard carbon with engineered defective sites. *Electrochimica Acta*, 368, 137574. Available from https://doi.org/10.1016/j.electacta.2020.137574.

Kwon, G., Bhatnagar, A., Wang, H., Kwon, E. E., & Song, H. (2020). A review of recent advancements in utilisation of biomass and industrial wastes into engineered biochar. *Journal of Hazardous Materials*, 400, 123242. Available from https://doi.org/10.1016/j.jhazmat.2020.123242.

Legrain, F., Kotsis, K., & Manzhos, S. (2015a). Amorphous carbon a promising material for sodium ion battery anodes: a first principles study. *The Journal of Physical Chemistry A*.

Legrain, F., Sottmann, J., Kotsis, K., Gorantla, S., Sartori, S., & Manzhos, S. (2015b). Amorphous (Glassy) carbon, a promising material for sodium ion battery anodes: A combined first-principles and experimental study. *The Journal of Physical Chemistry C*, 119. Available from https://doi.org/10.1021/acs.jpcc.5b03407.

Li, B., Zai, J., Xiao, Y., Han, Q., & Qian, X. (2014). SnO2/C composites fabricated by a biotemplating method from cotton and their electrochemical performances. *CrystEngComm*, 16, 3318–3322. Available from https://doi.org/10.1039/C3CE42659A.

Li, C., Li, J., Zhang, Y., Cui, X., Lei, H., & Li, G. (2019). Heteroatom-doped hierarchically porous carbons derived from cucumber stem as high-performance anodes for sodium-ion batteries. *Journal of Materials Science*, 54, 5641–5657. Available from https://doi.org/10.1007/s10853-018-03229-2.

Li, W., Li, Z., Zhang, C., Liu, W., Han, C., Yan, B., An, S., & Qiu, X. (2020). Hard carbon derived from rice husk as anode material for high performance potassium-ion batteries. *Solid State Ionics*, 351, 115319. Available from https://doi.org/10.1016/j.ssi.2020.115319.

Li, Y., Hu, Y.-S., Li, H., Chen, L., & Huang, X. (2016). A superior low-cost amorphous carbon anode made from pitch and lignin for sodium-ion batteries. *Journal of Materials Chemistry A*, 4, 96–104. Available from https://doi.org/10.1039/C5TA08601A.

Li, Y., Hu, Y.-S., Titirici, M.-M., Chen, L., & Huang, X. (2016). Hard carbon microtubes made from renewable cotton as high-performance anode material for sodium-ion batteries. *Advanced Energy Materials*, 6, 1600659. Available from https://doi.org/10.1002/aenm.201600659.

Lima, R. M., dos Reis, G. S., Thyrel, M., Alcaraz-Espinoza, J. J., Larsson, S. H., & de Oliveira, H. P. (2022). Facile synthesis of sustainable biomass-derived porous biochars as promising electrode materials for high-performance supercapacitor applications. *Nanomater*, 12. Available from https://doi.org/10.3390/nano12050866.

Lin, H., Ji, X., Chen, Q., Zhou, Y., Banks, C. E., & Wu, K. (2009). Mesoporous-TiO2 nanoparticles based carbon paste electrodes exhibit enhanced electrochemical sensitivity for phenols. *Electrochemistry Communications*, 11, 1990–1995. Available from https://doi.org/10.1016/j.elecom.2009.08.034.

Lin, Y., Huang, Z. H., Yu, X., Shen, W., Zheng, Y., & Kang, F. (2014). Mildly expanded graphite for anode materials of lithium ion battery synthesised with perchloric acid. *Electrochimica Acta*, 116. Available from https://doi.org/10.1016/j.electacta.2013.11.057.

Liu, C., Zhang, L., Yuan, X., Li, X., Wu, Y., & Wang, X. (2019). Effect of ZrC formation on graphitization of carbon phase in polymer derived ZrC−C ceramics. *Materials (Basel)*, 12. Available from https://doi.org/10.3390/ma12244153.

Liu, M., Jing, D., Shi, Y., & Zhuang, Q. (2019). Superior potassium storage in natural O/N−doped hard carbon derived from maple leaves. *Journal of Materials Science: Materials in Electronics*, 30, 8911–8919. Available from https://doi.org/10.1007/s10854-019-01219-x.

Liu, S., & Loper, C. R. (1991). The formation of kish graphite. *Carbon N. Y*, 29. Available from https://doi.org/10.1016/0008-6223(91)90119-4.

Liu, Y., Li, P., Wang, F., Fang, W., Xu, Z., Gao, W., & Gao, C. (2019). Rapid roll-to-roll production of graphene film using intensive Joule heating. *Carbon N. Y*, 155. Available from https://doi.org/10.1016/j.carbon.2019.09.021.

Lotfabad, E. M., Ding, J., Cui, K., Kohandehghan, A., Kalisvaart, W. P., Hazelton, M., & Mitlin, D. (2014). High-density sodium and lithium ion battery anodes from banana peels. *ACS Nano*, 8, 7115–7129. Available from https://doi.org/10.1021/nn502045y.

Lua, A. C., Yang, T., & Guo, J. (2004). Effects of pyrolysis conditions on the properties of activated carbons prepared from pistachio-nut shells. *Journal of Analytical and Applied Pyrolysis.*, 72, 279–287. Available from https://doi.org/10.1016/j.jaap.2004.08.001.

Luna-Lama, F., Rodríguez-Padrón, D., Puente-Santiago, A. R., Muñoz-Batista, M. J., Caballero, A., Balu, A. M., Romero, A. A., & Luque, R. (2019). Non-porous carbonaceous materials derived from coffee waste grounds as highly sustainable anodes for lithium-ion batteries. *Journal of Cleaner Production*, 207, 411–417. Available from https://doi.org/10.1016/j.jclepro.2018.10.024.

Maldonado-Hódar, F. J., Moreno-Castilla, C., Rivera-Utrilla, J., Hanzawa, Y., & Yamada, Y. (2000). Catalytic graphitisation of carbon aerogels by transition metals. *Langmuir*, 16. Available from https://doi.org/10.1021/la991080r.

Mubarik, S., Qureshi, N., Sattar, Z., Shaheen, A., Kalsoom, A., Imran, M., & Hanif, F. (2021). Synthetic approach to rice waste-derived carbon-based nanomaterials and their applications. *Nanomanufacturing*, 1. Available from https://doi.org/10.3390/nanomanufacturing1030010.

Okolie, J. A., Epelle, E. I., Tabat, M. E., Orivri, U., Amenaghawon, A. N., Okoye, P. U., & Gunes, B. (2022). Waste biomass valorisation for the production of biofuels and value-added products: A comprehensive review of thermochemical, biological and integrated processes. *Process Safety and Environmental Protection*, 159, 323–344. Available from https://doi.org/10.1016/j.psep.2021.12.049.

Oya, A., & Otani, S. (1979). Catalytic graphitisation of carbons by various metals. *Carbon N. Y*. Available from https://doi.org/10.1016/0008-6223(79)90020-4.

Palomares, V., Serras, P., Villaluenga, I., Hueso, K. B., Carretero-González, J., & Rojo, T. (2012). Na-ion batteries, recent advances and present challenges to become low cost energy storage systems. *Energy & Environmental Science*, 5, 5884–5901. Available from https://doi.org/10.1039/C2EE02781J.

Panda, M. R., Kathribail, A. R., Modak, B., Sau, S., Dutta, D. P., & Mitra, S. (2021). Electrochemical properties of biomass-derived carbon and its composite along with $Na_2Ti_3O_7$ as potential high-performance anodes for Na-ion and Li-ion batteries. *Electrochimica Acta*, 392, 139026. Available from https://doi.org/10.1016/j.electacta.2021.139026.

Pandey Ashutosh (2022). *Chinese graphite dominance threatens electric car ambitions*, Dw.Com. https://www.dw.com/en/chinese-graphite-dominance-threatens-electric-car-ambitions/a-60888876.

Pramanik, A., Chattopadhyay, S., De, G., & Mahanty, S. (2021). Efficient energy storage in mustard husk derived porous spherical carbon nanostructures. *Materials Advances*, 2, 7463–7472. Available from https://doi.org/10.1039/D1MA00679G.

Pramudita, J. C., Sehrawat, D., Goonetilleke, D., & Sharma, N. (2017). An initial review of the status of electrode materials for potassium-ion batteries. *Advanced Energy Materials*, 7, 1602911. Available from https://doi.org/10.1002/aenm.201602911.

Rawat, S., Mishra, R. K., & Bhaskar, T. (2022). Biomass derived functional carbon materials for supercapacitor applications. *Chemosphere*, 286, 131961. Available from https://doi.org/10.1016/j.chemosphere.2021.131961.

Reis, G. S., Oliveira, H. P., Larsson, S. H., Thyrel, M., & Claudio Lima, E. (2021). A short review on the electrochemical performance of hierarchical and nitrogen-doped activated biocarbon-based electrodes for supercapacitors. *Nanomater*, 11. Available from https://doi.org/10.3390/nano11020424.

Saputri, D. D., Jan'ah, A. M., & Saraswati, T. E. (2020). Synthesis of carbon nanotubes (CNT) by chemical vapor deposition (CVD) using a biogas-based carbon precursor: A review. *IOP Conference Series: Materials Science and Engineering*. Available from https://doi.org/10.1088/1757-899X/959/1/012019.

Senthil, C., & Lee, C. W. (2021). Biomass-derived biochar materials as sustainable energy sources for electrochemical energy storage devices. *Renewable and Sustainable Energy Reviews*, 137, 110464. Available from https://doi.org/10.1016/j.rser.2020.110464.

Senthil, C., Park, J. W., Shaji, N., Sim, G. S., & Lee, C. W. (2022). Biomass seaweed-derived nitrogen self-doped porous carbon anodes for sodium-ion batteries: Insights into the structure and electrochemical activity. *Journal of Energy Chemistry*, 64, 286–295. Available from https://doi.org/10.1016/j.jechem.2021.04.060.

Seow, Y. X., Tan, Y. H., Mubarak, N. M., Kansedo, J., Khalid, M., Ibrahim, M. L., & Ghasemi, M. (2022). A review on biochar production from different biomass wastes by recent carbonisation technologies and its sustainable applications. *Journal of Environmental Chemical Engineering*, 10, 107017. Available from https://doi.org/10.1016/j.jece.2021.107017.

Sermyagina, E., Murashko, K., Nevstrueva, D., Pihlajamäki, A., & Vakkilainen, E. (2020). Conversion of cellulose to activated carbons for high-performance supercapacitors. *Agronomy Research*. Available from https://doi.org/10.15159/ar.20.163.

Shi, S., Zhou, X., Chen, W., Chen, M., Nguyen, T., Wang, X., & Zhang, W. (2017). Improvement of structure and electrical conductivity of activated carbon by catalytic graphitisation using N2 plasma pretreatment and iron(III) loading. *RSC Advances*. Available from https://doi.org/10.1039/c7ra07328c.

Stevens, D. A., & Dahn, J. R. (2000). High capacity anode materials for rechargeable sodium-ion batteries. *Journal of the Electrochemical Society*, 147, 1271. Available from https://doi.org/10.1149/1.1393348.

Sun, Y., Xiao, H., Li, H., He, Y., Zhang, Y., Hu, Y., Ju, Z., Zhuang, Q., & Cui, Y. (2019). Nitrogen/oxygen co-doped hierarchically porous carbon for high-performance potassium storage. *Chemistry – A European Journal*, 25, 7359–7365. Available from https://doi.org/10.1002/chem.201900448.

Tang, X., Liu, D., Wang, Y.-J., Cui, L., Ignaszak, A., Yu, Y., & Zhang, J. (2021). Research advances in biomass-derived nanostructured carbons and their composite materials for electrochemical energy technologies. *Progress in Materials Science*, 118, 100770. Available from https://doi.org/10.1016/j.pmatsci.2020.100770.

Tian, S., Guan, D., Lu, J., Zhang, Y., Liu, T., Zhao, X., Yang, C., & Nan, J. (2020). Synthesis of the electrochemically stable sulfur-doped bamboo charcoal as the anode material of potassium-ion batteries. *Journal of Power Sources*, 448, 227572. Available from https://doi.org/10.1016/j.jpowsour.2019.227572.

Titirici, M.-M., White, R. J., Falco, C., & Sevilla, M. (2012). Black perspectives for a green future: Hydrothermal carbons for environment protection and energy storage. *Journal of Colloid and Interface Science*, 5, 6796–6822. Available from https://doi.org/10.1039/C2EE21166A.

Wan, H., Shen, X., Jiang, H., Zhang, C., Jiang, K., Chen, T., Shi, L., Dong, L., He, C., Xu, Y., Li, J., & Chen, Y. (2021). Biomass-derived N/S dual-doped porous hard-carbon as high-capacity anodes for lithium/sodium ions batteries. *Energy*, 231, 121102. Available from https://doi.org/10.1016/j.energy.2021.121102.

Wang, B., Sun, Y.-C., & Sun, R.-C. (2019). Fractionational and structural characterisation of lignin and its modification as biosorbents for efficient removal of chromium from wastewater: A review. *J Journal of Leather Science and Engineering*, 1, 5. Available from https://doi.org/10.1186/s42825-019-0003-y.

Wang, H., Chen, H., Chen, C., Li, M., Xie, Y., Zhang, X., Wu, X., Zhang, Q., & Lu, C. (2022). Tea-derived carbon materials as anode for high-performance sodium ion batteries. *Chinese Chemical Letters*. Available from https://doi.org/10.1016/j.cclet.2022.04.063.

Wang, P., Gong, Z., Wang, D., Hu, R., Ye, K., Gao, Y., Zhu, K., Yan, J., Wang, G., & Cao, D. (2021). Facile fabrication of F-doped biomass carbon as high-performance anode material for potassium-ion batteries. *Electrochimica Acta*, 389, 138799. Available from https://doi.org/10.1016/j.electacta.2021.138799.

Wrogemann, J. M., Fromm, O., Deckwirth, F., Beltrop, K., Heckmann, A., Winter, M., & Placke, T. (2022). Impact of degree of graphitization, surface properties and particle size distribution on electrochemical performance of carbon anodes for potassium-ion. *Batteries, Batter. Supercapsitors*, 5. Available from https://doi.org/10.1002/batt.202200045.

Wu, X., Chen, Y., Xing, Z., Lam, C. W. K., Pang, S.-S., Zhang, W., & Ju, Z. (2019). Advanced carbon-based anodes for potassium-ion batteries. *Advanced Energy Materials*, 9, 1900343. Available from https://doi.org/10.1002/aenm.201900343.

Wu, Z., Zou, J., Zhang, Y., Lin, X., Fry, D., Wang, L., & Liu, J. (2022). Lignin-derived hard carbon anode for potassium-ion batteries: Interplay among lignin molecular weight, material structures, and storage mechanisms. *Chemical Engineering Journal*, 427, 131547. Available from https://doi.org/10.1016/j.cej.2021.131547.

Xing, B., Zhang, C., Cao, Y., Huang, G., Liu, Q., Zhang, C., Chen, Z., Yi, G., Chen, L., & Yu, J. (2018). Preparation of synthetic graphite from bituminous coal as anode materials for high performance lithium-ion batteries. *Fuel Processing Technology*, 172. Available from https://doi.org/10.1016/j.fuproc.2017.12.018.

Xue, Y., Du, C., Wu, Z., & Zhang, L. (2018). Relationship of cellulose and lignin contents in biomass to the structure and RB-19 adsorption behavior of activated carbon. *New Journal of Chemistry*, 42, 16493–16502. Available from https://doi.org/10.1039/C8NJ03007C.

Yang, H., Ye, S., Zhou, J., & Liang, T. (2019). Biomass-derived porous carbon materials for supercapacitor. *Frontiers in Chemistry*, 7. Available from https://doi.org/10.3389/fchem.2019.00274.

Yang, M., Dai, J., He, M., Duan, T., & Yao, W. (2020). Biomass-derived carbon from Ganoderma lucidum spore as a promising anode material for rapid potassium-ion storage. *Journal of Colloid and Interface Science*, 567, 256–263. Available from https://doi.org/10.1016/j.jcis.2020.02.023.

Yokokura, T. J., Rodriguez, J. R., & Pol, V. G. (2020). Waste biomass-derived carbon anode for enhanced lithium storage. *ACS Omega*, 5, 19715–19720. Available from https://doi.org/10.1021/acsomega.0c02389.

Yrjälä, K., Ramakrishnan, M., & Salo, E. (2022). Agricultural waste streams as resource in circular economy for biochar production towards carbon neutrality. *Current Opinion in Environmental Science & Health*, 26, 100339. Available from https://doi.org/10.1016/j.coesh.2022.100339.

Yu, K., Wang, B., Bai, P., Liang, C., & Jin, W. (2021). Wheat straw cellulose amorphous porous carbon used as anode material for a lithium-ion battery. *Journal of Electronic Materials*, 50, 6438–6447. Available from https://doi.org/10.1007/s11664-021-09173-3.

Yu, K., Wang, J., Song, K., Wang, X., Liang, C., & Dou, Y. (2019). Hydrothermal synthesis of cellulose-derived carbon nanospheres from corn straw as anode materials for lithium ion batteries. *Nanomater*, 9. Available from https://doi.org/10.3390/nano9010093.

Yuan, X., Zhu, B., Feng, J., Wang, C., Cai, X., & Qin, R. (2021). Recent advance of biomass-derived carbon as anode for sustainable potassium ion battery. *Chemical Engineering Journal*, 405, 126897. Available from https://doi.org/10.1016/j.cej.2020.126897.

Zhang, F., Wang, K.-X., Li, G.-D., & Chen, J.-S. (2009). Hierarchical porous carbon derived from rice straw for lithium ion batteries with high-rate performance. *Electrochemistry Communications*, 11, 130–133. Available from https://doi.org/10.1016/j.elecom.2008.10.041.

Zhang, M., Zhang, J., Ran, S., Sun, W., & Zhu, Z. (2022). Biomass-derived sustainable carbon materials in energy conversion and storage applications: Status and opportunities. A mini review. *Electrochemistry Communications*, 138, 107283. Available from https://doi.org/10.1016/j.elecom.2022.107283.

Zhang, Y., Zhao, R., Li, Y., Zhu, X., Zhang, B., Lang, X., Zhao, L., Jin, B., Zhu, Y., & Jiang, Q. (2021). Potassium-ion batteries with novel N, O enriched corn silk-derived carbon as anode exhibiting excellent rate performance. *Journal of Power Sources*, *481*, 228644. Available from https://doi.org/10.1016/j.jpowsour.2020.228644.

Zheng, S., Luo, Y., Zhang, K., Liu, H., Hu, G., & Qin, A. (2021). Nitrogen and phosphorus co-doped mesoporous carbon nanosheets derived from bagasse for lithium-ion batteries. *Materials Letters*, *290*, 129459. Available from https://doi.org/10.1016/j.matlet.2021.129459.

Zhu, Z., Liang, F., Zhou, Z., Zeng, X., Wang, D., Dong, P., Zhao, J., Sun, S., Zhang, Y., & Li, X. (2018). Expanded biomass-derived hard carbon with ultra-stable performance in sodium-ion batteries. *Journal of Materials Chemistry A*, *6*, 1513–1522. Available from https://doi.org/10.1039/C7TA07951F.

C H A P T E R

21

Sewage waste as substrate for value

Rahul Ranjan, Rohit Rai, Vikash Kumar and Prodyut Dhar

School of Biochemical Engineering, Indian Institute of Technology (BHU), Varanasi, Uttar Pradesh, India

21.1 Introduction

Sewage sludge (SS) is a semisolid by-product obtained from wastewater produced globally. Domestic and industrial discharges are the two major sources of SS production, and with increasing population, the annual sewage production has increased in every nation (Grobelak et al., 2019). Due to rising SS pollution, there is a need to manage organic waste present in wastewater sludge using innovative and sustainable technologies. About 1.5 million tons of SS is produced every year in the United States (Svanström et al., 2004). SS due to its abundance requires treatment technologies that are scalable as well as scattered to environmental protection following the sustainable development goals. Municipal sewage generally consists of a variety of street runoff and industrial wastes in addition to residues from households and human excreta (Zuloaga et al., 2012). Historically, the disposal of wastewater and sludge had not been considered significant until the industrial revolution. During the era of first human communities, the waste produced was decomposed by natural cycles, because of which, waste disposal was not a significant problem (Vuorinen et al., 2007). Mesopotamian Empire was the first civilization to come with a solution for sanitation problems. In the ruins of Babylonia and Ur, the remains of homes have drainage systems for wastes (Cooper, 2001). In Indus valley civilization, houses were connected to drainage systems. The sewage was treated using sumps in which the solids used to settle, before flowing into the drainage channels. This was the first attempt to treat the sewage on record (Webster, 1962). For Egyptian civilization, the mention of limestone toilets, toilet stools with ceramic bowls and clay pot beneath (Breasted, 1906). However, Greeks were the forerunners of modern sanitation technology. The presence of public latrines conveying wastewater and stormwater outside the city is found in ancient Greece (Tölle-Kastenbein, 1993). Significant technical developments in sanitation happened during the roman era. The water pipes and sewer were not the Roman inventions; however, Romans perfected the use of these equipment. The rich had their own baths and toilets, whereas the commoners used to share public baths, toilets, and water fountains. The designing of sewage system in ancient Rome was comparatively complex than previous civilizations. It included many smaller sewers that used to merge into a large common sewer (Cooper, 2001). After the collapse of Roman civilization, no significant development occurred in the sewage system from middle age to industrial revolution. The water systems adapted and developed by Romans were not followed during these times, and the water was drawn from river and wells and disposed without treatment. This led to multiple hygiene and sanitation problems. The cesspits have been mentioned in the historical texts as reserves to store wastes, which were sold to farmers as fertilizers (Aiello et al., 2008). However, the "era of sanitary enlightenment" began with high industrialization rate during 18th century. In London, a collection of sewers and pumping station wastewater was made and discharged into the Thames River (Sori, 2001). In Germany, a sewer system was constructed in 1842; however, Garman cities began construction in 1867 (Seeger, 1999). Development of sewage systems occurred throughout the Europe, and during 20th century, revolution in wastewater management occurred due to societal and environmental concerns of wastewater pollution (Shifrin, 2005). The processes used for treatment of water, that is, primary, secondary, and tertiary treatment, also evolved during the history. For primary treatment, Fosses Mouras were used in year 1860, which evolved to septic tanks in 1895 and Imhoff tank in 1902 (Chatzakis et al., 2006). For secondary treatment, trickling filters in 1895 and later rotating biological contractor (RBC) (1960) and membrane bioreactors (MBRs) (1980) came into existence (Tchobanoglus et al., 2003). Tertiary treatment for wastewater came into existence by 1950s and later upgraded with phosphorus removal in 1970 (Gayman, 2008). The sewage sludge contains N, P, K that can be used as fertilizers, biochar, and hydrochar, etc. However, application of sewage sludge to agriculture can be harmful if SS contains heavy metals, toxic organic chemicals, and pathogens that can deteriorate the soil quality (Grobelak et al., 2019).

SS sludge has a semisolid form, which consists of four major components: (1) heavy metals [lead (Pb) (0.01–38 mg/kg), cadmium (Cd) (0.01–1.99 mg/kg), Cr (0.075–21.4 mg/kg), Zn (30–425 mg/kg)] (Azizi et al., 2013) (2) toxic organic chemicals [Acrylonitrile (0.036–82.3 mg/kg), Toluene (1.13–324 mg/kg) (Harrison et al., 2006)]; (3) nutrients [phosphorus (P) (1.06%–1.53% of SS), nitrogen (N) (2.5%–6.3% of SS) (Singh & Agrawal, 2008; Bridle & Pritchard, 2004)]; and (iv) pathogens [*Escherichia coli* 1.3–8 CFU/m^3 (Straub et al., 1993), *Staphylococcus* spp. 10^2–10^5 bacteria/g dry matter (DM) (Dumontet et al., 2001)] as shown in Fig. 21.1. A brief description of SS components has been provided below.

1. *Heavy metals (HMs)*: HMs are generally present in high concentrations in sewage sludge, which are known to be toxic to plants and animals. However, optimal concentration of the HM is found to be essential for the growth of flora and fauna. Therefore, concentration of HMs in sewage is a key factor for its use on land because higher quantity may damage the crops and enter the food chain instead of improving soil fertility (McGrath et al., 1994). The highest contribution of heavy metals in sewage comes from industrial wastes. According to the US standards for sewage sludge, a clean sludge should contain 100 mg arsenic (As), 2000 mg chromium (Cr), 1200 mg copper (Cu), 500 mg Nickel (Ni) per kg dry weight of sewage sludge (Rechcigl, 1995). Table 21.1 presents the regulatory limits for heavy metals in different countries across the globe. Across the globe, Germany has comparatively strict guidelines for heavy metals, whereas Poland has less strict guidelines for regulatory limits.

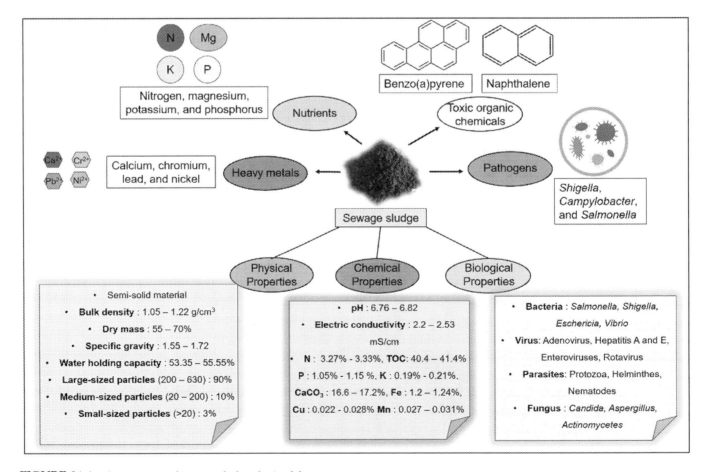

FIGURE 21.1 Composition of sewage sludge obtained from wastewater.

TABLE 21.1 Regulatory limits of heavy metals around different nations in the world.

Country	Cd	Cr	Cu	Hg	Ni	Pb	Zn	Ref.
Austria	2–10	50–500	300–500	2–10	25–100	100–500	1500–200	ENV (2000)
Belgium	10	500	600	10	100	500	2000	ENV (2000)
China	5	600	800	5	100	300	2000	LeBlanc et al. (2009)
Denmark	0.8	100	1000	0.8	30	120	4000	ENV (2000)
Finland	3	300	600	2	100	150	1500	ENV (2000)
Germany	2	80	600	1,4	60	100	1500	ENV (2000)
Greece	20–40	500	1000–1750	16–25	300–400	750–1200	2500–4000	ENV (2000)
Hungary	10	1000	1000	10	200	750	2500	ENV (2000)
India	5	50	300	0.15	Not limited	100	1000	Parida et al. (2019)
Italy	20	Not limited	1000	10	300	750	2500	ENV (2000)
Japan	5	500	Not limited	2	300	100	Not limited	LeBlanc et al. (2009)
The Netherlands	1.25	75	75	0.75	30	100	300	ENV (2000)
Poland	20–50	500–2500	1000–2000	16–25	300–500	750–1500	2500–5000	ENV (2000)
Portugal	20	1000	1000	16	300	750	2500	ENV (2000)
Romania	10	500	500	5	100	300	2000	ENV (2000)
Russia	15	500	750	7.5	200	250	1750	LeBlanc et al. (2009)
Slovenia	0.5	40	30	0.2	30	40	100	ENV (2000)
Spain	20–40	1000–1750	1000–1750	16–25	300–400	750–1200	2500–4000	ENV (2000)
Sweden	2	100	600	2.5	50	100	800	ENV (2000)
United States	39	No Ceiling	1500	17	420	300	2800	Mininni (2013)

2. *Toxic organic chemicals*: Pesticides, household chemicals, and industrial wastes are key contributors of toxic organic chemicals to the sewage wastes. Toxic chemicals from these sources are a matter of concern because their variability and unknown hazard affect the environment and public health. Evaluation of organic toxicants is difficult because of varying frequency of occurrence, chemical nature, and toxicity potential. Some of the well-known organic chemicals are benzo(a)pyrene, chlorobenzene, naphthalene, and Aroclor 1260 (Hue, 1995). For proper utilization of sewage wastes, estimation and disposal of toxic chemical wastes are important. Physical, chemical, and biological methods can be used for treatment of toxic organic chemicals in sewage sludge. Physical process such as sorption onto the solid surface, and volatilization, chemical degradation, such as hydrolysis and biodegradation can be used to remove the toxic organic chemicals present in the sewage sludge. The methods such as land disposal, pyrolysis, and marine disposal are used for degradation of toxic chemicals adsorbed on the adsorbent in physical treatment. For chemicals degradation, foam separation, chemical precipitation, ultrafiltration, and ion exchange can be used. During biodegradation, anaerobic and aerobic metabolism of microbes can be used for the treatment of SS (Rogers, 1996).

3. *Nutrients*: Presence of macro- and micronutrients varies greatly in the sewage systems. The typical sewage contains 2.5%–3.43% nitrogen, 1.06%–1.4% phosphorous, 0.08%–0.23% potassium, 0.4%–3% magnesium, and 1.6%–2.7% calcium (Hue, 1995). The nitrogen and phosphorus present in the sewage sludge can act as fertilizers and can be utilized for agriculture. The advantage of using organic nitrogen in sludge that it causes less ground water pollution compared with nitrogen in chemical fertilizers. However, the sewage sludge with high nitrogen content can be harmful to the surrounding water systems by causing eutrophication (Mtshali et al., 2014). Phosphorus, similar to nitrogen, can be utilized as a fertilizer, and its

availability in sludge is independent of the treatment process. Another macronutrient, that is, potassium is also the part of fertilizer systems. The nitrogen, phosphorus, and potassium come together and form NPK fertilizer system (Haby et al., 1990). The proportion of these nutrients is similar to animal manure except potassium. The SS is composed of significant amounts of phosphorus (0.8%–0.23%) and nitrogen (2.5%–3.4%), but very low concentration of potassium (0.4%–3%) (Mtshali et al., 2014). Magnesium as a macronutrient has a role in participating in nucleophilic reactions, aggregation of ribosomes, and is found in the center of chlorophyll. Magnesium also helps in enzyme function for RNA polymerases, ATPases, phosphatases, and carboxylases (Shaul, 2002). Calcium present in the sewage sludge has structural role in cell membranes and cell wall. Calcium also participates in organic and inorganic reactions and acts as intracellular messenger for plants. However, access calcium reduces the growth of plant communities on the calcareous soils and thus must be present in low concentration in the sewage for optimum plant growth (White & Broadley, 2003).

4. *Pathogens*: Viruses, bacteria, protozoa, parasitic worms are present as pathogens abundantly in the SS. The factors that decide the number and types of microbes in SS are sludge treatment, sanitation, and population density of the surrounding area (Hue, 1995). The methods available for treatment of pathogens in sludge are air drying, mesophilic aerobic digestion, and anaerobic digestion (Hue, 1995). However, these methods reduce the pathogenic microbial load but do not eliminate them completely and are called as "process to significantly reduce pathogens (PSRP)" (Yeager & O'brien, 1983). Thermophilic treatment applied after PSRP destroys the remaining pathogens and thus called "process to further reduce pathogens (PFRP)" (Yeager & O'brien, 1983). The common pathogens present in the sewage systems are *Shigella*, *Campylobacter*, and *Salmonella*. These microorganisms such as *Shigella* causes dysentery, *Campylobacter* results into gastroenteritis, and *Salmonella* causes salmonellosis. About 110 varieties of viruses are present in raw sewage including Poliovirus, Echovirus, Hepatitis virus, Adenovirus, and Reovirus, which can cause diseases such as hepatitis, meningitis, and respiratory infections (Lewis & Gattie, 2002). Due to the strong adsorption of virus onto the sludge, their numbers remain undercounted. The sorbed viruses survive longer and remain more active than free virus in the sewage stream; however, infection can be induced only when virus is not bound to the sludge (Hue, 1995). Among all the different protozoan species present in the wastewater, only three pathogenic species are of major concern, i.e., *Giardia lambia*, *Balantidium coli*, and *Entamoeba histolytica* due to their ability of inducing mild-to-severe diarrhea (Romdhana et al., 2009). Eggs laid by helminth parasites such as *Taenia saginata* (tapeworm) and *Ascaris lumbricoides* tend to settle out with the solids during primary treatment of wastewater (Jimínez et al., 2002). Protozoa survive the sludge treatment process and thus are a major concern because of their ability to infect animals and humans even when present in small numbers. The chemical and biological complexity of SS is a global concern, and its conversion to value-added products may prove sustainable and profitable.

21.1.1 Physicochemical properties of sewage sludge

Physical, chemical, and biological properties of SS are determined by evaluation of the characteristics of wastewater (Świerczek et al., 2018). Supplementing SS to the soil improves the physiochemical and biological properties of soil such as bulk density, porosity, aggregate stability, soil fertility, water movement, and retention. The physical, chemical, and biological properties of SS are thus an important parameter in its utilization for applications such as agriculture, value-added products, and biomaterials.

21.1.1.1 Physical properties of sewage sludge

The SS is a residual, semisolid material in wastewater, released from industries or wastewater treatment plants. For estimation of physical properties of sewage sludge, sludge dry mass, water content, water holding capacity, bulk density, particle size distribution are evaluated (El-Nahhal et al., 2014). Unprocessed SS has a water content of 99%, whereas dehydrated SS has water content of 80%–85% with organic content. For evaluation of sludge biomass and water content, the water-containing sludge samples are dried at 105°C for 24 hours. The dried samples are then measured for weight difference and will be used for calculation of DM and water content for the sludge (Wilke, 2005). For water holding capacity, about 0.5 L of air-dried sludge is put in small columns (10 cm diameter). The columns are soaked with 0.25 L distilled water, and the weight difference is measured for determining water holding capacity (El-Nahhal et al., 1998). Particle size distribution is obtained using a hydrometer. The dried and sieved samples (2 mm) are homogenized with 250 mL distilled water and 50 mL Kalgon solution. Hydrometer is then gently immersed into the sedimentation cylinder and allowed to stabilize. The hydrometer readings are used to calculate the particle size distribution (Wen et al., 2002). For determining hydrophobicity of the sludge, water drip penetration time method is used. In this procedure, a drop of water is placed onto the sludge surface, and the time required for its penetration is calculated. About 50 μL of water drop is placed on the surface of sludge on Jencons Sealpette instrument (Letey, 1969). The resulting bulk density of sewage sludge is 1.05–1.22 g/cm^3, dry mass of 55%–70%, specific gravity of 1.55–1.72. The moisture content of sludge is 1.75%–1.87%, water holding capacity of 53.35%–55.55%. The particles of solid sludge are majorly (90%) composed of large particles (200–630 μm), whereas medium-sized (20–200 μm) and fine-sized particles (<20 μm) are present in approximately 10% and approximately 3% abundance. Water drip penetration time shows a value of 98.1–131. 4 seconds (El-Nahhal et al., 2014). The physical properties of sewage sludge are important parameter for its utilization in different areas complemented by chemical and biological properties.

21.1.1.2 Chemical properties of SS

Chemical properties of SS include characteristics such as pH, electric conductivity, presence of organic materials, heavy metals (Zn, Cu, Ni, Cd), and nutrients (NPK). Total organic carbon and total nitrogen present in the sewage sludge can be determined by Total Organic Carbon (TOC) analyzer instrument and Kjeldahl method (Khanmohammadi et al., 2015). TOC analyzer measures the organic carbon present in the sample by completely oxidizing the organic molecules to CO_2 and measuring the CO_2 concentration. To determine the Na, K, and P, the sludge is first combusted at 550°C, and the resulting ash was extracted and determined using 2 M HCl. Total P in the sludge is determined using ascorbic-NH_4-molybdate blue colorimetry at 880 nm. Total K and Na can be measured using flame photometry. Calcium carbonate is measured by back-titration method with NaOH (Helmke & Sparks, 1996). The concentration of trace elements such as Cd, Co, Pb, Cr, Ni, Mn, Cu, Zn, and Fe is determined using atomic absorption according to the US-EPA Method3050B (US Environmental Protection Agency, 1996) (Lindsay & Norvell, 1978). The pH of the sewage sludge ranges from 6.76 to 6.82, electric conductivity ranges from 2.2 to 2.53 mS/cm. Total nitrogen of sewage sludge ranges from 3.27% to 3.33%, TOC ranges from 40.2% to 41.4%, phosphate ranges from 1.05% to 1.15%, potassium ranges from 0.19% to 0.21%, sodium is present in 0.2% of quantity, and calcium carbonate is present in range of 16.6%–17.2%. Heavy metal concentration in the sewage sludge is: Fe in the range of 1.2%–1.24%, Zn 0.09%, Cu in the range of 0.022%–0.028%, Mn in the range of 0.027%–0.031%, Ni in the range of 0.006, Cr 0.009, Pb in the range of 0.009%–0.011% (Khanmohammadi et al., 2015). Chemical composition of sewage sludge is important for its treatment, utilization, extraction of elements, and to access the toxicity associated with it.

21.1.1.3 Biological properties of SS

The pathogens present in the SS can be categorized into four major groups: bacteria, viruses, fungi, and parasites (Black, 1974). The bacteria present in the sludge majorly consist of pathogenic bacteria. One of the most common bacterial group present in SS is *Salmonella*, and among this group, *Salmonella typhi* causes most fatal infections such as typhoid. *Shigella* is often found in wastewater samples and comes from poultry, milk products, raw vegetable, and human sewage. Four *Shigella* species are present abundantly in the SS, that is, *Shigella sonnei*, *Shigella boydii*, *Shigella flexneri*, and *Shigella dysenteriae* causing shigellosis in humans (Tchobanoglus et al., 2003). Other bacterial groups present in the SS are *E. coli*, *Yersinia*, *Leptospira*, *Clostridium*, *Mycobacterium*, and *Vibrio*. Among the 28 species, *Vibrio cholera* is known to cause human cholera. Yersinia enterocolitica present in the sewage cause diarrhea, fever, abdominal pain, headache, and vomiting (Romdhana et al., 2009).

The second major group of microorganisms present in the sewage sludge are viruses. The most common groups of viruses found in the SS are Rotavirus, Adenovirus, Hepatitis A and E virus, and Enterovirus (Coxsackie and Poliovirus). Enteric viruses cause common-to-severe forms of infections in the fecal-oral route and respiratory routes. Adenovirus and Reovirus cause gastroenteritis, eye infection, and respiratory illness (Tchobanoglus et al., 2003). Rotavirus are waterborne pathogens, which cause diarrhea. Hepatitis A and E cause faintness and anorexia, whereas Enterovirus cause disease such as myocarditis, meningitis, respiratory infections, skin and eye infections (Pepper et al., 2006).

Parasites such as Helminthes, nematodes, and protozoa are present in SS. Most significant protozoan in the SS are *G. lambia*, *Cyclospora*, and *Cryptosporidium parvum* that can infect humans with compromised immune system (Tchobanoglus et al., 2003). Helminthic parasites such as Tapeworms and Flatworms present in SS can affect the intestine and localize the eye, ear, and heart (Pepper et al., 2006). Fungal pathogens in sewage sludge belong to *Candida*, *Aspergillus*, and *Actinomyces* that can generate mycotoxins in the SS (Black, 1974).

21.1.2 Global production of sewage sludge

Sewage sludge coming out of wastewater system is a concern for both the developing and underdeveloped countries. From the total wastewater produced in 2010, 70% of the wastewater came from domestic sources. Global annual domestic water withdrawals in 2010 accounted for about 390 km^3, which was 477 km^3 in year 2000 (Mateo-Sagasta et al., 2015). Every year, European countries produce about 10 million tons DM of SS. With modifications in systems for collection and treatment of wastewater, the total amount of SS has also increased (Bianchini et al., 2016), which can be disposed through multiple routes. These routes must be designed, and the sludge must be distributed according to the Waste Framework Directive (EU Directive, 2008). According to the directive, the preferential order of strategies used for SS is: prevention, minimization, reuse, recycling, energy recovery, and landfilling (Directive, 2008). Landfilling of SS is considered as final and undesired treatment. According to the Landfill Directive (EC Directive, 1999), the amount of SS subjected to landfills is expected to reduce in the upcoming years. The directive also obliges the member states to decrease the quantity of biodegradable wastes in SS to decrease disposal through landfilling routes (Commission, 1999). The data for SS production in the year 2014 suggest Germany as the highest producer of SS (1958 k tons DM/year), whereas the minimum sludge was produced by Malta (6 k tons DM/year) in Europe (Bianchini et al., 2016). About 56% of the sludge produced in Germany is disposed through incineration, whereas composting of sludge is the least preferred route (2010). For Malta, the statistics vary as 100% of the total sewage produced is disposed through landfilling, which is not considered for sludge disposal in countries such as Germany (Bianchini et al., 2016).

In China, daily wastewater treatment capacity improved rapidly during the year 2007 and 2013 due to which the amount of SS generated and treated also increased. In 2013, 6.25 million tons DM sludge was produced, and per capita sludge production was 4.6 kg dry solids (DSs), which is significantly lower than that in the first-world countries (Yang et al., 2015). The annual sludge production of countries such as USA, Iran, Turkey, Canada, and Brazil is 6510, 650, 580, 550, and 370 thousand metric tonnes, respectively (Singh et al., 2020). In countries such as United States (US), Japan, and European Union (EU), the SS is generally disposed in landfills, incinerated, or reused for agricultural activities. However, countries such as Spain, Denmark, and France utilize maximum part of the generated sludge (>50%) for the agricultural usage (Samolada & Zabaniotou, 2014). Luxemburg and Greece still dispose 80%−90% of the SS in landfills (Samolada & Zabaniotou, 2014). However, countries such as Japan incinerate more than 70% of their waste as there is scarcity of land for dumping sites (Samolada & Zabaniotou, 2014). SS is a worldwide concern, and proper sludge management strategies are crucial for minimizing the hazards and maximizing value-added products obtained from SS. Most of the developing and underdeveloped nations are still struggling to adopt strategies for proper management of SS and end up disposing its maximum part to landfills. The introduction of innovative ways of SS disposal is thus a novel research area with added benefit of obtaining value-added products as discussed later in the chapter.

21.1.3 Laws and regulations associated with sewage sludge utilization

Due to rapid increase in the global population, different nations are adopting strategies to counter the rising issues of SS depending on their geographical conditions, technologies, agricultural norms, etc. Some of the examples of laws and regulations for disposal of sewage in developed and developing nations are mentioned in Fig. 21.2.

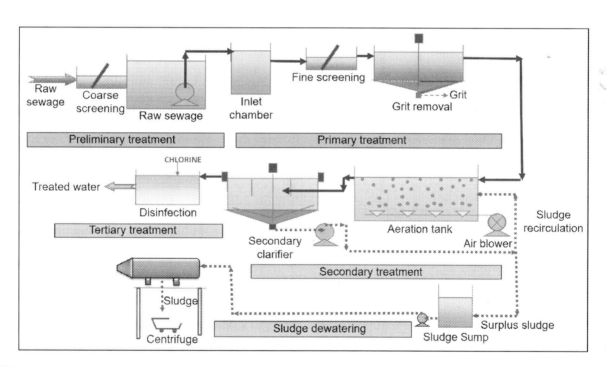

FIGURE 21.2 Processing of raw sludge for the treatment of wastewater to obtain sewage sludge.

21.1.3.1 Japan

Japan is one of the world's most densely inhabited countries. Due to low land availability, landfilling as a sewage disposal method was abandoned. Instead of this gasification, incineration, melting, carbonization, and drying of SS have been introduced in the last three decades (Spinosa, 2007). The laws regulating thermal processing of SS are Waste Management and Public Cleaning Law (1970) for handling the incineration ash, the Air Pollution Control Law (1968), and Environmental Pollution Prevention Act (1965). The use of SS for cement production and aggregate products is determined by the Soil Contamination Countermeasures Law (1968), which states the limits for contaminants determined by toxic elements leachate tests (LeBlanc et al., 2009). The fertilizer regulation act involves the criteria for HMs in SS to be used as fertilizer and recommends record-keeping and reporting of production rate as well as producer's data (LeBlanc et al., 2009). Since the start of 21st century, Japan has initiated innovative projects such as LOTUS and SPIRIT21 (Ref?) for the utilization and recycling of sewage to achieve: (1) the "zero sludge" disposal scenario and (2) the maximum exploitation and most efficient use of renewable energy from sludge known as "green energy." Use of anaerobically digested food waste and SS as fertilizers in several Japanese municipalities has to meets the Fertilizer Law's quality standards (Christodoulou & Stamatelatou, 2016). SS is a waste product, but with regulatory simplifications, it can be utilized as fertilizer and energy source.

21.1.3.2 USA

Since the 1970s, US waste policy has been changed to encourage the use of SS biosolids on land. The "Standard for the Use or Disposal of Sewage Sludge," 1993, specifies a number of factors for SS pollutants, as well as their treatment, utilization, and pathogen reduction. It also includes a set of standards and laws designed to promote better management and disposal practices and techniques (Venkatesan et al., 2015). According to the rule, biosolids are not waste but a nutrient source due to which application of SS in agriculture became very flexible. However, the public pressure and doubts about utilization of SS have restricted its valorization; SS incinerators are renamed as waste incinerators because of public skepticism. The stricter regulations have posed multiple barriers against energy recovery from SS (Christodoulou & Stamatelatou, 2016).

21.1.3.3 Australia and New Zealand

Australia and New Zealand's legislation has mostly centered the application of SS in land replenishment. The EPA's 503 rule limits the usage of SS because of the presence of organic contaminants in addition to pathogens and heavy metals for utilization in land application. The simplification of guidelines was suggested by AWA mentioning overregulation as the reason behind costly compliance (Mukheibir et al., 2012). In Australia, biosolids were considered as waste; however, New South Wales revised waste regulation in the year 2014 by introducing "Resource Recovery Exemption." This reform made utilization of SS easy for the stakeholders (Christodoulou & Stamatelatou, 2016).

21.1.3.4 European Union

The EU legislative framework includes Directives that state the standard rules and conditions for the participating states to comply and implement in their national legislation. EU policy in the current century is centered toward reviewing biowaste, and waste management guidelines, and their application in the energy recovery (LeBlanc et al., 2009). The first EU legislation toward SS management came in 1986 with SS Directive, which promoted agriculture complying with all the provisions regarding harmful effects on living organisms. The directive sets maximum level of toxic limits for potentially hazardous compounds in SS and soil (Sewage sludge Internet, 2022). EU legislation of categorizing SS as waste is the most rigid, stringent, and conservative when compared with other nations. The limits for harmful components in the sewage are generally stricter than EPA's rules and regulations. The additional restriction of Directives is that it does not allow the use of SS as fertilizers in organic farming. Many German states have banned land application of SS, and all the SS is processed through incineration. SS use in agricultural land is regulated with Fertilizer Law, which states provisions for organic raw materials to synthesize fertilizers (Lehmphul, 2022). The national legislative framework on utilization of SS, which complies with the 86/278/EEC Directive, established in 1992, defines the heavy metal concentration limits in soil. The Federal Emission Control Act sets out the obligations of thermal disposal of SS. Germany is going to become the first country in EU to make amendments in legislative in order to recover phosphates from SS; however, further revision in legislation is in progress to restrict the use of SS in landfill. The application of the SS Directive in the Greek national framework defines "Methods, Conditions, and Restrictions on the Use in Agriculture of Sludge Derived from Treatment of Domestic and Municipal Wastewater" (MD 80568/4225/1991). The SS Directive's specified restrictions, pollutant concentration limits, and provisions have been implemented. Greece has really implemented the SS Directive's upper limit value for all HMs (Christodoulou & Stamatelatou, 2016). National SS legislation is being amended with the purpose of decreasing SS landfilling while simultaneously investigating the possibility of recycling and utilizing it as an alternative energy source. The SS is being researched for its potential value-added products as discussed in the chapter, which will reduce the cost associated with its disposal and will improve its commercial value.

21.1.3.4.1 Strategies for sewage treatment

Sewage is a combination of wastes from residential, institutional, commercial, and industrial enterprises that comprises over 99% water. The method of removing impurities and wastes from wastewater, specifically domestic sewage, is called sewage treatment. To eliminate these toxins and generate ecologically acceptable treated wastewater, various physical, chemical, and biological methods are applied (Demirbas et al., 2017). In other words, sewage treatment is the planned removal of dissolved particles, pollutants, and toxic substances from wastewater. Several methods are used for treatment, which have evolved over the time. Traditionally methods including septic tank, trickling filters, radial flow tank were used (Lofrano & Brown, 2010). Currently, several types of reactors are used along with the updated traditional methods. The reactors and methods are discussed later in this chapter. Factors such as source, physical and chemical characteristics of the wastewater are the ones that play important role in determining the final step for the sewage treatment. According to Central Pollution Control Board (CPCB), India, around 6190 million liters of sludge is treated every day. The sewage treatment process includes four steps, namely Preliminary, Primary, Secondary, and Tertiary treatment.

21.1.4 Preliminary sewage treatment

In the preliminary phase, the large coarse particles, grits (heavy solids), etc., are screened and removed as shown in Fig. 21.2. The ultimate aim is to remove components that may cause operational or maintenance issues in the treatment process (Demirbas et al., 2017). It involves the removal of floating debris such as broken tree branches, animal corpses, rags, cans, plastic bottles, broken glass, pieces of paper, wood, etc., and heavily settled inorganic solids after passing through a bar screen (Topare et al., 2011). Bar screens are used for coarse screening, and they are of various sizes. For bar spacing, the 1-inch aperture size is sufficient; however, 0.5–1.5-inch openings can also be employed. For fine screening, 0.25-inch aperture size is also used. The typical practice is to utilize 0.32-inch × 2-inch bars up to 6 feet in length and 0.375-inch × 2-inch or 0.375-inch × 2.5-inch bars up to 12 feet in length. The bar needs to be at least 9 inches above the maximum sewage level. Flow velocity is around 2 feet per second and maximum it can go upto 3 feet per second (Oakley, 2018; Partners et al., 2011).

Inorganic materials such as sand, cinders, and gravel, together known as grit, are commonly found in wastewater. The volume of a given effluent relies mostly on whether the collecting sewage system is sanitary or mixed (Partners et al., 2011). Grit will abrade pumps and cause major operating problems in sedimentation tanks and sludge digesters by accumulating around and clogging outlets and pump suctions. As a result, it is usual practice to remove this debris using grit chambers. Grit chambers are often situated ahead of pumps and, if mechanically cleaned, should be preceded by coarse bar rack screens (Oakley, 2018; Partners et al., 2011). The velocity in these channels is low enough to deposit heavy inorganic materials while retaining organic wastes in suspension. The length of channel grit chambers should be constructed to be 50% longer than theoretically necessary to allow for agitation and outlet dispersion. The chamber floor is kept far enough below (not less than 2.5 inches) the weir crest. This allows the buildup of about 2.5 cubic feet of grit in every 10 days per million gallons of wastewater. The detention time is often between 20 seconds and 1 minute. For sewer system carrying wastewater or sewage from multiple sources, there should be enough storage for 30 cubic feet of grit for every million gallons of flow. In case of separate sewer system, the minimum storage should be 10 cubic feet of grit for every million gallons of flow (Oakley, 2018).

Preaeration grit tanks are also used in which compressed air is flushed in the sewage. This helps in several ways such as removing suspended solids in sedimentation tanks, extraction of grease and oil from effluent, and reducing BOD. Compressed air is flushed or forced in the sewage water for around 30 minutes at the rate of 0.1 cubic feet per gallon of water. Typically, air is delivered at a rate of 1.0–4.0 cubic feet per minute, to ensure proper agitation (Oakley, 2018). The collected solid can be then disposed of in a landfill or recycled as per the need. The preliminary treatment also involves the removal of oil and greases from the sewage waste via floatation or skimming process. Skimming is done in special chambers, which are designed in such a way that floating particles, such as oil, fat, and grease, rise to the surface of waste water until evacuated, whereas the sewage passes out constantly through dividers or barriers (Partners et al., 2011). The preliminary treatment also helps in reducing the BOD of sewage waste by up to 15%–30% (Topare et al., 2011). The reduction in BOD is achieved by chlorination of waste water (prechlorination). A decrease of 15%–35% in BOD may result by chlorinating raw wastewater to create a residual of 0.2–0.5 mg/L following 15 minutes of contact. Typically, for every mg/L of chlorine, a decrease of at least 2 mg/L of 5-day BOD is achieved. After the preliminary phase, the sewage waste is processed for the next step, that is, primary treatment.

21.1.5 Primary sewage treatment

During primary treatment, the sewage water is stored in large settling tanks known as primary settlers or sedimentation tanks (Bajpai, 2018). These settling tanks or primary clarifiers are of rectangular shape mostly with length: width ratio of 3:1–5:1 and the length can range from 50 to 200 feet with minimum depth of 8–13 feet. Circular-shaped tanks are also used for the same purpose having diameter of 50–200 feet and minimum depth of 9–14 feet (Partners et al., 2011). This is done to get the heavier solids (organic solids, grits) settled down (sedimentation) and lighter material to float up (Poblete et al., 2022). The minimum detention time for which sewage water is kept in settling tanks is around 2 hours (Partners et al., 2011).

Primary treatment is used as a pretreatment rather than as a stand-alone treatment. Sludge is produced during primary treatment, which involves energy usage for pumping and other processes. Using screens and settling tanks, the first treatment mechanically removes suspended solids such as debris, fibers, bark particles, filler, and coating materials. Screens of different sizes ranging from 0.2 1 inches are used in the sieving/filtration process to remove bigger particles (Partners et al., 2011). For primary treatment, sedimentation is the most popular mechanical purification process, in which materials are separated out by settling in a basin. During sedimentation, the particles settle at the bottom of the settling tank only, under the influence of gravity, generating a sludge (primary sludge) that must be taken out from basin. Scraping, pumping, and sucking sludge out of the basin are the common methods used on a regular basis (Demirbas et al., 2017; Bajpai, 2018; Poblete et al., 2022). Sedimentation can remove up to 70% of total suspended solids (TSS) from the sewage.

Methods such as primary clarifiers, sedimentation tanks, and septic tanks are traditional methods used in primary treatment.

Some of the advanced methods include micro-solid sludge dissolved air flotation (DAF), which was introduced in 1970 (Bajpai, 2018; Poblete et al., 2022). During DAF, the air is pumped in sewage water with a pressure of around 3–9 bar in order to improve primary treatment. This air expansion causes the formation of microbubbles, which get attached to particles and give swirling motion to water. Microbubbles attached to particles make them float to the surface. The air circulation also increases the oxygen content in water, which can later be used by microbes for further treatment (Poblete et al., 2022). Although DAF units can effectively eliminate 80%–98% of suspended solids (TSS) or agglomerates and take away finely dispersed organic materials, the decrease of dissolved organic material (COD) is limited to roughly 20% (Bajpai, 2018). DAF devices have been proven to be cost-effective for treating big water drifts with sediment concentrations ranging from 1.1 gram per gallon to 18.9 gram per gallon (Bajpai, 2018).

Coagulants such as aluminum and iron salts are often used during the DAF process to shorten the flocculation time for floating particles. The typical DAF residence time is 10–20 minutes (Poblete et al., 2022). DAF's primary drawback in comparison with basic clarifiers is its superior mechanical design and higher power expenses owing to compressed air. The first large-scale DAF for primary treatment was described by Johnson et al. al, 2014 (Johnson et al., 2014). They did, however, observe a small improvement in performance in comparison to traditional primary treatment (primary clarifiers only), with removal of 50%–75% of suspended particles and 30%–50% COD. The DAF has been used to treat sewage from various sectors such as dairy, poultry, domestic and municipal, and industrial showing removal of 60%–88% TSS, 60%–98% BOD, and 48%–99% COD (Muñoz-Alegría et al., 2021).

In the 1990s, methods such as charged bubble flotation (CBF) and primary effluent filtration (PEF) were introduced. The CBF technique involves production of a 40%–50% volumetric air content by suspending micron-sized (about 7–50 μm) bubbles in water that are externally created (Eddy et al., 2014). A thin soap coating formed of an electrically charged surfactant covers (encapsulates) each bubble, depending on the use, either cationic or anionic. For the absorption of charged and hydrophobic compounds, the charged bubbles offer a significant contact area. The wastewater from primary sedimentation tanks is filtered as part of primary effluent filtration (PEF). Utilizing cutting-edge filter technology, such as the WesTech disc cloth filter and the fuzzy filter (compressible filter media), the filtration is done in PEF (Eddy et al., 2014). However, in comparison to CBF, the PEF is not so common nowadays.

Chemically enhanced primary treatment (CEPT) is also one of the advanced methods that involves use of certain chemicals such as ferric chloride, aluminum sulfate, and ferric sulfate. Addition of these chemicals during primary pretreatment causes precipitation of the TSS, as well as colloidal solids (Eddy et al., 2014). CEPT can reduce the BOD and TSS up to 50%—80% and 80%—90%, respectively.

The remaining suspended solids along with scum formed during the whole process are collected together and are referred to as primary sludge. This primary sludge is collected and frequently stabilized by anaerobic decomposition in a digestion tank (Demirbas et al., 2017; Topare et al., 2011). The waste is recycled or utilized as a soil conditioner. Sometimes, the preliminary and primary treatments are considered one and often termed as the primary treatment only.

21.1.6 Secondary sewage treatment

Secondary treatment is the further processing of the effluent obtained from primary treatment. The removal of suspended particles and biodegradable organic materials is referred to as secondary treatment. During secondary treatment, the microorganisms are utilized to digest the dissolved organic material or any biological content to reduce the BOD/COD of the sewage water; hence, it is also called as biological treatment. The secondary treatment can reduce the BOD up to 80%—90% (Bajpai, 2018). The microbes act on the effluent, digest the organic matter, and break it into simple organic or inorganic components depending on the aerobic or anaerobic mode of respiration (Demirbas et al., 2017).

21.1.6.1 Aerobic treatment

When microbes use oxygen to act on sewage waste, it is known as the aerobic treatment. The microbes involved in aerobic treatment include proteobacteria, actinobacteria, beta proteobacteria, delta proteobacteria, xanthomonadales, sphingomonadales, flavobacterium, *Nitrospira* (nitrite-oxidizing), *Nitrosomonas* (ammonium oxidizing), etc (Świątczak & Cydzik-Kwiatkowska, 2018). These microbes digest organic matter into carboxylic acids and finally into carbon dioxide (Demirbas et al., 2017). Organic matter such as carbohydrates, proteins, and lipids serves as a food source for biomass growth, and microbes convert them into their simple monomeric form and finally to carbon dioxide through respiration and metabolism. Organic contaminants such as aromatic compounds, chlorobenzenes, and polyaromatic hydrocarbons are broken down using specific pathways to be used as carbon source (Bouwer, 1993; Lu et al., 2019). Along with microbes, some annelids, rotifers, and protozoans such as the *Ameba* and *Paramecium* are also present, which remove the fine particulate matter from water. Aerobic treatment can occur in three ways:

1. Biofiltration

    Biofiltration is done to remove the suspended particles, which do not get eliminated in secondary treatment. Microbes such as *Nitrosomonas* and *Nitrosococcus* are used for nitrate removal and denitrification (Rajta et al., 2020). Different filters such as intermittent filters and trickling filters are used during biofiltration. Microorganisms form thin biofilm on the filter surface, known as the zoogleal film, and then capture the organic materials and aerobically break them down. Proteobacteria are the prominent bacteria group, which aerobically consumes the organic matter. Even certain bacteria that are resistant to antibiotics, such as *Pseudomonas, Acinetobacter, Acidovorax*, and *Sphingomonas*, are also active in biofiltration process (Lu et al., 2019; Guarin & Pagilla, 2021). The elimination of BOD using low-rate trickling filters is around 85%. A bench-scale bioelectrochemical trickling filter was demonstrated by Liang et al., 2020, that removes 97% COD (Liang et al., 2020). The COD reduction was studied using electrodes, which were semiimmersed in sewage water for a period of 2 months, with an OLR (organic load ratio) of 1 gCOD/L/day. Rehman et al. (2020), used a pilot-scale trickling filter for the treatment of household sewage. They investigated the effects of the hydraulic retention time (HRT) and observed 70.9% removal of COD and 82.5% removal of the bacterial count for an HRT of 2 days. Forbis-Stokes et al. (2020), used three-staged filters including a trickling filter comprising crushed charcoal, a scrap iron filter, and a bamboo woodchip filter and observed removal of 73%—82% turbidity, 31%—50% phosphorus, and 67%—75% COD.

    Trickling filters offer several advantages such as low-energy requirement, simple operation, less or no maintenance, and improved sludge thickening properties. However, there are certain disadvantages also such as less removal of TSS content in the effluent, increased sensitivity to colder temperatures. generation of odors, and incidents of uncontrolled solids shedding (Eddy et al., 2014).

2. Aerated activated sludge (AAS)

In this process, the effluent after initial treatment having high COD (450–600 mg/L) content is treated to generate slurry with a low COD content (40–60 mg/L) (Orhon & Sözen, 2020). The aerated bioreactor tank has a settler that separates the sludge solids from the sewage water (Orhon et al., 2009). A portion of the settled activated sludge is recycled into the aerated bioreactor to restart the AAS process. The bulk activated sludge goes through a succession of treatment operations, including microbial aerobic digestion, dewatering, incineration, and biofertilizer reuse (Demirbas et al., 2017). The AAS bioreactor is supplied with primary effluent, recycled activated sludge, and pumped with compressed air. The AAS effluent is sent to the secondary settler after hydraulic retention duration of 1–10 days, where 80%–90% COD is eliminated from the effluent.

There are several advancements that have been introduced in AAS process. Some of the advancements include membrane bioreactor (MBR), moving-bed bioreactor (MBBR), integrated fixed-film activated sludge (IFAS), etc (Orhon & Sözen, 2020; Waqas et al., 2020). The MBR involves the replacement of gravitational settling with membrane filtration. In general, MBR is a sort of combination of filtration and conventional activated sludge process. In the integrated fixed-film activated sludge (IFAS) process, the attached and suspended growth systems are combined. With a considerable level of nitrification–denitrification, IFAS is successful at removing dissolved organic carbon (Waqas et al., 2020).

High treatment efficiency, controllability of the process, especially oxygen usage, low space demand are some of the pros of the AAS methods. The considerable cons of AAS methods are high operational expenses, vulnerability to disruptions, and production of large quantities of sludge from biological waste. Without any safeguards in place, such as an equalization basin, there is a risk of operational instability (Bajpai, 2018).

3. Oxidation ponds

It involves the use of natural water bodies such as ponds and lagoons, through which wastewater is allowed to run for a specific time period before being held for 2–3 weeks. The organic components are broken down in this pond by the interaction of microbes in the presence of sunlight and oxygen. Photosynthetic microbes are primary communities involved in degradation of organic compounds in these types of open ponds. These microbes involve various photosynthetic bacteria (chorobiaceae, chloroflexaceae, rhodospirillaceae, etc.), algae, cyanobacteria (blue green algae). These microbes utilize different organic compounds as food source or carbon source and break them into smaller molecular forms. The construction of lagoons requires large land areas (≈ 100 hectares or more) (Ho & Goethals, 2020). Treatment of wastes in lagoons produces less quantity of sludge as compared with the activated sludge process (Bajpai, 2018). The BOD reduction for these types of ponds is around 95%, and detention time is of 2–6 days.

A treatment or holding pond is an aerated lagoon or aerated basin equipped with artificial aeration to aid biological wastewater oxidation. It has a huge size (≥ 100 ha), but shallow in depth (0.2–1 m), and uses active microorganisms (algae and cyanobacteria) and mechanical aeration to cleanse wastewater biologically. The aerated lagoons require lot of energy for aeration and designed to stay in place for 5–7 days. However, it is simple to use and maintain (Bajpai, 2018).

4. Anaerobic treatment

When the microbes digest organic material present in the sewage in the absence of oxygen, it is known as anaerobic treatment. The anaerobic treatment process takes place in four stages: hydrolysis, acidogenesis, acetogenesis, and methanogenesis. The microbes involved in the anaerobic treatment are *Clostridium*, *Desulphobolus*, *Treponema*, methanotrophs, methanogens, *Bifidobacterium*, and facultative bacteria such as *E. coli*, *Tetrasphaera*, fermicutes, and *Methylobacterium* (Cyprowski et al., 2018; Nascimento et al., 2018).

21.1.6.1.1 Hydroysis

During hydrolysis, the organic matter is broken down into simple and soluble forms of proteins (amino acids), lipids (fatty acids), and carbohydrates (monosaccharides). *Chloroflexi, Bacteroidetes*, and *fermicutes* are major group of microbes that are active during this phase and transform the organic matter into their monomeric forms (Cyprowski et al., 2018; Pasalari et al., 2021; Menzel et al., 2020).

21.1.6.1.2 Acidogenesis

This phase involves the conversion of the hydrolysis product into short-chain carboxylic acids and alcohols. Volatile fatty acids such as propionic acid (*Propionicimonas, Veilonella*), butyric acid (*Clostridium, Butyribacterium*), acetic acid (*Clostridium*), and ethanol (*Clostridium zymomonas, Acetobacterium*) are major products of this phase.

21.1.6.1.3 Acetogenesis

In this phase, the organic acids produced are converted into acetate by microbes such as *Clostridium, Pelotomaculum, Syntrophobacter*, and *Acetobacterium* (Menzel et al., 2020; Wang et al., 2019). Acetic acid is a major product of this phase; hence, it is named as acetogenesis.

21.1.6.1.4 Methanogenesis

This is the final phase and is marked by production of methane. Acetate serves as the substrate for aceticlastic methanogens such as *Methanosaeta* and *Methanosarcina*, which separate it into methyl group and CO_2 before reducing the latter to methane. Additionally, hydrogenotrophic methanogens (*Methanothermobacter, Methanomassiliicoccales*, and *Methanobacterium*) use CO_2 as a carbon source and energy source to reduce methane in the presence of H_2.

The end product of the anaerobic treatment can be methane, hydrogen gas, organic acids, etc., depending on the type of microbial species present. This technique has high initial expenses, but unlike aerobic treatment, it does not require oxygen, resulting in considerable energy savings over time (Bajpai, 2018).

For the same amount of BOD reduction, the anaerobic treatment produces sludge in less amount compared with aerobic treatment. Similarly, this method may decompose the solid organic matter and convert it to biogas, which could then be utilized as an alternative energy source, lowering the expenses of sludge disposal significantly (Bajpai, 2018, 2017). As anaerobic treatment is carried out in the absence of oxygen and is better suited for the treatment of wastewaters with high COD (100,000 mg COD/l) (Bajpai, 2017).

After desulfurization, the biogas generated can be utilized to generate electricity. In most cases, a flare may be used to burn biogas in an emergency. There are a variety of process designs to choose from including upflow anaerobic sludge blanket (UASB), fixed-bed, expanded granular sludge blanket (EGSB), internal circulation (IC) reactors, and sludge contact process have all been considered for anaerobic degradation of effluents having low substrate concentrations and high sulfate concentrations.

Upflow anaerobic sludge blanket (UASB): In this reactor, there is a tank that encloses the sludge bed. An equal distribution of flow, a high enough flow velocity, and the agitation brought on by gas production work together to achieve mixing between the feedstock and sludge. In the process of turning sludge into high-density granules, a blanket or granular matrix is created, which is maintained in suspension by regulated upflow velocity (Bajpai, 2017).

Fixed-bed reactor: This is also known as anaerobic filter (AF). It consists of a packing material filled column. This packing material encourages biomass retention in the reactor because it allows for microbe attachment or entrapment. Microorganisms live both as those adhering to the packing material and as granules or flocs in suspension (Dutta et al., 2018).

Expanded granular sludge blanket (EGSB): This is an upgraded form of UASB. The bed expands when it is run with a faster upflow velocity. This may enhance the contact between the sludge and the wastewater, which minimizes the issue brought on by a lack of substrate. High solid retention durations and reduced fat buildup are two benefits of the EGSB over UASB (Bajpai, 2017; Dutta et al., 2018).

Internal circulation reactor (IC): In IC reactor, the generated biogas is sent upward through a pipe to a degasifier unit after being separated from the liquid halfway through the reactor by a gas/liquid separator mechanism. The separated gas is taken out of the reactor, and a different line returns the liquid and sludge combination to its bottom. The interaction between the sludge and wastewater is enhanced as a result of the gas lift transfer.

The so-called tower reactor, which uses the upflow anaerobic sludge blanket (UASB) technique, is now the most popular form of reactor (Bajpai, 2018, 2017).

21.1.7 Tertiary treatment

After secondary treatment, the tertiary treatment comprises a few extra processes to minimize turbidity, organics, phosphorus, nitrogen, metals, and microbial pathogens (Demirbas et al., 2017; Topare et al., 2011). Most methods include physical and chemical treatments such as coagulation, filtration, flocculation, organics adsorption on activated carbon, and further disinfection (Bajpai, 2018; Poblete et al., 2022).

Physical adsorption by means of activated carbon is a well-known method for removing organic micropollutants. Powdered and granular are the two forms of activated carbon that are available (Marsh & Reinoso, 2006). Adsorption beds with downstream filters for turbidity reduction are widely used with granular activated carbon and should be reactivated on a regular basis. Powdered activated carbon (PAC) is directly dosed into gravity filters and clarifiers and is more effective than granular activated carbon (GAC) (Boehler et al., 2012). This is because of smaller size of particles and larger adsorption area. With the help of PAC at concentration of 0.01–0.02 g/L in the tertiary filter with 10 h filtering time of a SWTP, Hu et al. (2016), reported better organic micropollutant reduction for secondary effluent at 0.012 g COD/L, using PAC at 0.01–0.02 g/L with a contact duration of 15–30 minutes

Chemical treatments, such as flocculation and precipitation, are commonly used methods. They are distinct because of the use of different purifying methods. The destabilizing of colloidal particles known as coagulation is caused by the presence of a coagulant, a chemical reagent. Coagulants also adhere to the colloidal particles' surface and neutralize the overall negative surface charge, causing formation of bigger colloidal particles, which can be removed by sedimentation (Braz et al., 2010). Various aluminum and ferrous salts are used as coagulants. Sher et al. (2021), observed removal of 84.4%–91.3% of suspended micro solids, with a $FeCl_3$ coagulant increasing the removal effectiveness to 88.9–93.6%. $Al_2(SO_4)_3$, $FeSO_4$, $FeCl_3$, and $Ca(OH)_2$ were used together to treat wastewater from wine industries (Braz et al., 2010). With the help of those chemicals, more than 95% of reduction in TSS was observed. Ma et al. (2021), observed more than 85% and 89% of phosphate and nitrate removal after adding Fe(II). Due to the repelling effect of the negatively charged surface of water, very tiny colloidal particles are not removed during sedimentation. When these added metal ions react with water, hydroxides are generated. The colloidal particles are then adsorbed to hydroxides via sweep coagulation, forming bigger flocks, which are easier to settle.

Adding cations such as ferric and aluminum ions or long-chain polymers in the effluents causes flocculation. Destabilized particles clump together as microfloc and then into large floccules that may be settled and are known as floc. The creation of the floc may be aided by the addition of a second reagent known as a flocculant (Braz et al., 2010). $Na_2SiO_3 \cdot 9H_2O$ and $Al_2(SO_4)_3 \cdot 18H_2O$ were used for flocculation in combination with starch, and about 99% of turbidity was removed (Wang et al., 2020). Using ferrous and aluminum salts, Favero et al. (2020) observed % COD reduction. Adding long-chain polymers as flocculating agents works similarly by linking the colloids to the polymer. Polymers such as Poly(DAC-co-AM-co-BA), Poly(AM-co-poly(ethylene oxide methyl ether methacrylate)), Poly(MAPTAC-co-AM), Poly(DAC-co-AM-co-BA) are some of the examples (Vajihinejad et al., 2019). Some natural polymers such as chitosan, cellulose, and alginate are also used to develop long-chained polymer flocculants (Vajihinejad et al., 2019). To avoid chemical overdose, the water pH must be regulated, and mixing should be done quickly at first. Chemical precipitation is a well-known and widely used process in municipal wastewater treatment, particularly for phosphorus removal. It involves the addition of aluminum, calcium, or iron metal salts to change the physical state of dissolved particles, which make sedimentation easier.

Precipitation, subsequent filtration, or clarifying separates the dissolved organic compounds from wastewater. Aluminum and ferric salts or lime are commonly used for precipitation. Chemical precipitation is used in tertiary treatment to lower nutrient levels, particularly phosphorus. Phosphorus, nitrogen, and COD levels are decreased by 80%–90%, 30%–60%, and 80%–90%, respectively, when biological treatment is coupled with chemical precipitation. Barua et al. (2019), with the help of calcium and magnesium salts, observed about 85% phosphate removal following precipitation. The precipitation of organic and inorganic materials (chemical flocs) is followed by sedimentation or flotation. Polyelectrolytes are utilized during mixing phase to improve flocculation.

The aim of tertiary treatment is to improve the quality of sewage water to meet standards for industrial and household usage or any particular discharge criteria. In the context of municipally treated water, the tertiary treatment involves the removal of pathogens, to make sure that the water can be consumed safely. It's much more prevalent when the wastewater is going to be reused for irrigation (e.g., food crops, golf courses), recreation (e.g., lakes, estuaries), or drinking water (Demirbas et al., 2017).

The sequential treatment of sewage water helps in bringing wastewater into reusable form. Preliminary, primary, secondary, and tertiary treatments involve several processes, which all together help in reducing BOD, COD, TSS, and dissolved solids from sewage water. At the end of the process, the water can be used for activities such as irrigation, agriculture, or can be discharged in water body such as river or lake.

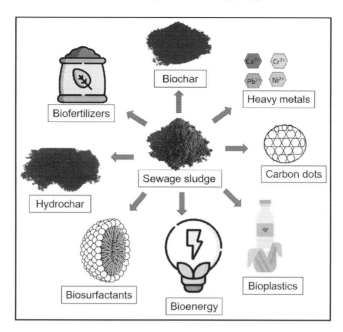

FIGURE 21.3 Examples of value-added products obtained using sewage sludge.

21.2 Value-added products from sewage sludge

In the recent years, the research toward valorization of sewage sludge has increased rapidly. New and improved methodologies toward successful conversion of sewage to value-added products have been devised. Some examples of products obtained from sewage sludge are biochar, heavy metals, biofertilizers, hydrochar, carbon dots, biosurfactants, and bioplastics as shown in Fig. 21.3.

21.2.1 Biochar

Sewage sludge (SS) is undigested waste, which consists of more than 50% of total solid contents, which may include plants and animal remains, microorganisms, and other inorganic and organic matters. This waste organic and inorganic matter from wastewater treatment plants is used as raw material for the development of biochar. Several definitions have been proposed for biochar in the literature. Biochar may be defined as the thermally pyrolyzed product of SS, or it is the product produced by the heating of biomass in the absence or limited supply of oxygen referred to as charring or pyrolysis (Racek et al., 2020; Lehmann & Joseph, 2015). International Biochar Initiative (IBI 2003) defines biochar as the product generated from a thermochemical process where biomass is heated in a limited supply of oxygen. Biochar can be generated from a variety of feedstocks, which leads to its different characteristics such as carbon fractions (Wei et al., 2019a), moisture content, and elemental fractions (Xu et al., 2019). Significant variations in feedstock characteristics that are pertinent to the qualities of biochar as a soil supplement include: (Grobelak et al., 2019) amount of cellulose, holocellulose, hemicellulose, lignin, and other minor organic components, (Svanström et al., 2004) amount of inorganic components such as salts, alkali earth metals and alkaline earth metals, ash contents, (Zuloaga et al., 2012) amount of material other than organic and inorganic components such as plastic in case of paper sewage sludge, (Vuorinen et al., 2007) bulk and true density, particle size, and moisture contents. Alipour et.al in their study utilized sewage sludge and sludge cake where HTC and pyrolysis were used to produce biochar and hydrochar. Morphological structure of the raw feedstock and pyrolyzed and HTC-mediated biochar were studied where porosity and surface area of biochar increased as compared with raw sludge. pH of the hydrochar decreased and biochar increased after HTC and pyrolysis process. Concentration of heavy metals such as Ni, Cu, Zn, and pb increased in biochar and hydrochar (Alipour et al., 2021). Pariyar et al. investigated the changes in properties caused by biochar synthesis utilizing feedstock such as rice husk, sawdust, food waste, and chicken litter. Biochar generated from poultry waste and paper waste has been found to have a pH rise of 6–10. Biochar generated from rice husk rose in surface area from 3.39 to 443.79 m^2/g and from 11.61 to 280.97 m^2/g, respectively. The highest ash percentage was reported in biochar generated from poultry manure (44.10%) and paper waste (57.20%). Biochar created from sawdust and rice husk has been found to be more stable and porous (Pariyar et al., 2020). Zang et al. in their research utilized SS (Lanzhou, China) to produce biochar. After the formation of biochar, pH increased from 5.8 to 10.2, surface area also improved from 5.26 to 15.23 m^2/g as compared with SS (Zhang et al., 2022). Wang et al. in their work synthesized biochar utilizing sewage sludge. pH of the biochar increased as compared with SS. Large amount of -OH groups present in the SS gets decomposed as the peak at 3400 cm^{-1} in the FTIR spectra got stretched. Amount of HMs such as Cd, Zn, Ni, Cu, and Pb increased in biochar as compared with SS (Wang et al., 2021). The feedstock may be processed (SS, manures), unprocessed (rice husk, hay), or a combination of both, or maybe a combination of biomasses (contaminants should not be more than 2% of dry weight) and dilutants. Dilutants can be added upto 10% of dry weight or more in the feedstock and must be considered as one of the feedstock ingredients (Grobelak et al., 2019) IBI has also established three basic biochar material testing types. Type A tests are related to the utility of biochar where three properties are being considered for testing such as pH, electrical conductivity, and liming potential. pH is most important parameter to understand how biochar and soil interact with each other. Biochar supplementation into soil helps in altering the acidic nature, and increased pH in soils is responsible for micronutrients deficits and agricultural yields (Haq et al., 2021). The testing period varies depending on the ratio of biochar and the solution made from distilled water, deionized water, calcium chloride, or potassium chloride. In the experiment, 50 mL of distilled water was added to a centrifuge tube containing 5 g of dry biochar, which was then mechanically shaken for 60 minutes. After letting the biochar suspension sit for 30 minutes, the pH of the suspension was measured. Salt concentration of the biochar is an important parameter to study. Supplementing large amount of biochar into soil may adversely affect the growth and development of salt-sensitive plants (Vasconcelos ACF de, 2020). Concentration and the type of salt in the solution are the two factors to which electrical conductivity depends on. Higher the slat concentration in the solution, higher will be the electrical conductivity, which is also used to determine the salinity of the soils (Pansu & Gautheyrou, 2006). Test procedures include weighing of 5 g of the dried biochar followed by transferring into 100 mL centrifuge tube. Then 50 mL of distilled water was added and mechanically shaken for 60 minutes. Then the sample was kept still for 30 minutes. After the completion of settling process of the biochar suspension, electrical conductivity was measured (Singh et al., 2017a). Amount of alkaline oxide and carbonates present in the biochar ashes contributes to alkalinity of the soils. Alkalinity of the biochar ashes depends on the nature of feedstock. The amount of alkaline oxide and carbonates defines the use of biochar ashes as liming agents and acidic soil supplements. To propose the proper amount of lime, prior information of the pH-buffering capacity of the soil and the liming potential of the biochar is required. Test procedure involves the weighing of 500 mg of dried biochar in a test tube followed by addition of 10 mL HCl (2 M) solution. Then the suspension was incubated for 16 hour under stirring condition. Then the suspension was titrated using 0.5 M sodium hydroxide solution till pH reached 7, and amount of sodium hydroxide consumed was recorded (a mL). Blank titration also needs to be calculated using 10 mL, 1 M HCl solution titrated with sodium hydroxide in the absence of biochar (b mL). The batch should contain a reference sample of $CaCO_3$ powder, which should be dried at 105°C for 1 hour (Singh et al., 2017b). Then percentage $CaCO_3$ equivalent was calculated using the following formula:

$$\text{Percentage } CaCO_3 \text{ equivalent} = \frac{M \times (b-a) \times 100.09 \times 100 \; 10^{-3}}{2 \times W}$$

where M is the molarity of sodium hydroxide, a is the volume of sodium hydroxide used during biochar titration, b is the volume of sodium hydroxide consumed during blank titration, and W is the weight of biochar.

The type B tests are for the assessment of toxic and harmful components of biochar, and type C tests are for soil quality improvement. SS is one of the most promising and abundantly available feedstocks for agriculture and allied fields. (SS) is regarded as a legitimate nutritional source. This fundamental quality prompted numerous research studies to examine the effects of applying this biosolid to the agricultural field on soil health and plant growth (Boudjabi & Chenchouni, 2021). It contains compounds such as organic carbon (OC), phosphorous (P), nitrogen (N), and potassium (K), which are the important macronutrients for agriculture. Some toxic and harmful compounds such as xenobiotics, monobutylin, dibutylin, and polyaromatic hydrocarbons are also present, which can be remediated using process such as bioremediation, photocatalysis, biostimulation, membrane-based filtration, and electrokinetic remediation (Fijalkowski et al., 2017; Patel et al., 2020). Adding SS to soil can improve its nutritional content, availability of nutrients, transform its physical and chemical properties, and promote agricultural production and yields. For examples, Grobelak et.al, utilized of SS from the food sector to soil to improve its properties by improving the protective buffering function. One-time applications of SS are beneficial for improving fertilizing capacities and ingested minerals that simulate plant growth and development. SS addition to the soil also minimizes the accumulation of heavy metals in the shots of plants, which ultimately improves the plant productivity. These effects can be seen after 5 years of using SS (Grobelak et al., 2017). Mantovi et al. conducted a comparison study of SS and mineral fertilizer on crops such as wheat, beet, maize, and sugarcane cultivated on silty-loamy soil in Italy. Continuous SS addition over 12 years improves soil fertility by increasing organic matter, nitrogen percentage, and decreasing soil alkalinity. However, heavy elements such as Cu and Zn were shown to accumulate (Mantovi et al., 2005). Biochar contains micronutrients (N, P, and K), which is sustainable source of for proving agricultural fertility. Biochar can minimize the uses of chemical fertilizers and also a source of nutrients from sludge. Porous morphology of the biochar can act as protective habitat for the microorganism (*Rhizobium, Mycorrhizae, and Lactobacterium*) responsible for improving soil fertility and also helps in nutrients transformation (Biochar, 2012; Atkinson et al., 2010). Productivity of the acidic and sandy soil can be improved by supplementing biochar. Biochar contains little amount of organic carbon and large amount of ash content, which has liming effects that improve the soil physical properties as well as ability to absorb nutrients (Dai et al., 2020). SS feedstock containing less prorportion of carbon and nitrogen found to be more efficient in enhancing the phosphorous yield (Gao et al., 2019).

21.2.1.1 Methods of biochar production

Based on development and modernization, the processes used to produce biochar can be categorized into traditional and modern processes. Traditional approaches include burning of biomass, slow pyrolysis and fast pyrolysis while modern approach involves process such as gasification, torrefaction, flash and vaccum pyrolysis, microwave pyrolysis, hydrothermal carbonization, electromodified and magnetic biochar.

21.2.1.2 Burning of biomass

According to archeological evidence, humankind first produced and used biochar more than 2500 years ago. The initial proof of the discovery came from the Amazon Basin in South America, which is popularly known as "Terra Preta soils" and contains three times the soil organic carbon content and nutrient concentration. Historically, people used to stack up the wood in the soil and burned it slowly in limited supply of air or no air. Other methods include the partial burning of wood in open atmosphere and then covering it up in the soil (Reilly, 2022). Biochar was not only employed as a soil amendment during the prehistoric period; it was also used to make liquid, which was then used for the preservation of dead bodies and meats, as coloring material, cork to seal barrels, and as arrowheads in shafts (Reilly, 2022). Furthermore, development and advancement in the technology improved the process for biomass production.

21.2.1.3 Pyrolysis process

Pyrolysis can be defined as thermal degradation of biomass into char or biochar under limited supply or in the absence of oxygen (Gabhane et al., 2020). Biomass or SS originated from agricultural sources mainly composed of organic matter (cellulose, hemicellulose, and lignin). Thermal degradation of cellulose occurs at 350°C, and lignin degrades at the temperature above 350°C so the pyrolysis temperature is generally kept in between 350°C and 700°C (Czernik & Bridgwater, 2004). Despite the fact that pyrolysis itself is an anaerobic heating process, it requires some other heating process such as oxidation and partial oxidation, hot gases, and solids, etc. Since the process of pyrolysis has been developed decades ago, although parameters such as heating rate (rate at which temperature is raised for pyrolysis generally expressed as °C per minute), temperature, and residence time further categorize pyrolysis into fast pyrolysis and slow pyrolysis.

21.2.1.4 Slow pyrolysis process

As the name suggests, slow pyrolysis is conventional process that requires large time for the thermal degradation of feedstock into biochar in the temperature ranges of 300°C–700°C with the heating rate of 5°C–7°C per minutes (Lai, 2022). The primary product of pyrolysis is biochar, which makes up about 45% of total production; however, other products such as bio-oil (25%–35%) and syngas (20%) are also produced (Méndez et al., 2015; Lai, 2022; Verma, 2012).

21.2.1.5 Fast pyrolysis process

Fast pyrolysis process is highly effective thermochemical process of conversion of biomass into biofuels (Dai et al., 2017). It is generally operated at the temperature grater than 500°C at the heating rate of 300°C/min in the absence of oxygen (Huang et al., 2016). Short retention time and high product recovery are the main benefits of this process. Bio-oil (~60% recovery) is the primary product of fast pyrolysis method; however, syngas (~20%) and biochar (~20%) are also produced (Huang et al., 2016; Liu et al., 2016).

21.2.1.6 Gasification process

Gasification is one of the most commonly used and effective thermochemical processes for the conversion of biomass or SS into syngas (Guan et al., 2016). Generally, the process of gasification is operated at higher temperature (> 700°C) under the limited supply of oxygen and steam. Hydrogen is the primary product; however, little amount of biochar is also produced as waste (Turner, 2022).

21.2.1.7 Torrefaction process

Torrefaction is a thermochemical process that occurs in the absence of oxygen at temperatures ranging from 230°C to 300°C. During this process, hemicellulosic biomass components are broken down, resulting in stable, carbon-rich torrefied wood and volatiles, which can act as biochar (Prins et al., 2006).

21.2.1.8 Flash pyrolysis

Flash pyrolysis is a modified rapid pyrolysis process that is carried out at temperatures ranging from 900°C to 1200°C. Thermal breakdown of the biomass was achieved in less than 1 minute with a heating rate of 1000°C/min. Because of the fast-heating rate and short residence duration, bio-oil production increases and biochar production decreases.

21.2.1.9 Vacuum pyrolysis

Vacuum pyrolysis is thermal degradation process where biomass is being heated at 450°C–600°C under the low vacuum pressure of 0.05–0.02 Mpa (Tripathi, 2022). This process produces good quality of biochar with high porosity, which can be utilized for the adsorption of minerals as well as supplements for soil.

21.2.1.10 Hydrothermal carbonization

Biomass containing large amount of moisture is directly being used for the production of biochar. In this process, additional drying of biomass is not required. Wet biomass is directly heated at temperature range of 220°C–240°C under high pressure of 2–10 Mpa, which results in production of biochar with phosphorous and nitrogen content and can serve as supplement for soil.

21.2.1.11 Microwave pyrolysis

One of the more sophisticated pyrolysis methods uses the dielectric heating process to produce thermal energy. Due to its superior thermal characteristics, such as quick, selective, and uniform heating with a lower sintering temperature that favours steam gasification, this technique currently has the attention of researchers (Gabhane et al., 2020). When compared with previous pyrolysis methods, this procedure reduces the required temperature by 200°C. The frequency and power employed are 400 W and 2450 MHz. This method produces biochar, which has a high capacity to absorb carbon dioxide.

21.2.1.12 Electromodified biochar

Altering the surface of the biochar by mixing chemical Fe-salt, Mg-salt, and Al-salt under applied electric field for 2–10 hours (Fang et al., 2014; Jung et al., 2015). These modifications impart functional groups on the surface of biochar and help in improved adsorption. It is simple, cost-effective, rapid method to produce biochar, which enhances the surface area of biochar. Hypochlorous acid, hypochlorite ions, and aluminum ions can be produced on the surface of biochar by chemical reaction or aluminum electrode (Fang et al., 2014; Hu et al., 2015).

21.2.1.13 Magnetic biochar

Magnetic biochar is produced by coating the surface of biochar with Fe-ions and heating it in a microwave at 450°C with a binding agent obtained from chestnut shell. It can also be produced by adding gelation and iron ions into the surface of biochar (Zhou, 2017). Incorporating Fe_2O_3, $FeSO_4.7H_2O$, and $FeCl_{3to}.6H_2O$ onto the

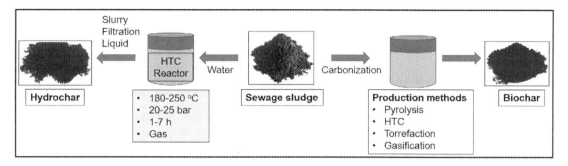

FIGURE 21.4 Methods for processing sewage sludge to obtain biochar and hydrochar.

surface of biochar under inert nitrogen condition in a muffle furnace at 1000°C (Lai et al., 2023; Zhou, 2017). This modification helps in easy recovery of toxic metal ions such as arsenic (+II), lead (+II), and copper (+2).

A brief outline of biochar and hydrochar preparation from sewage sludge is shown in Fig. 21.4.

The chemical properties of biochar are controlled by pyrolysis types (flash pyrolysis, torrefaction, microwave-assisted pyrolysis, and fast pyrolysis), feedstock composition, processing time, and temperature, as well as diluents and additives (Gabhane et al., 2020). Biochar is produced from SS heterogeneous material and contains different types of inorganic compounds, thus making it difficult to analyze its exact composition. Chemical properties include elemental composition such as C, O, N, P, H, and S, atomic ratio, energy content, fixed carbon and volatile matter, structural composition, and functionality (Pariyar et al., 2020). The concept of biochar has completely transformed the paradigm of SS management and is now employed as a multifunctional substance for enhancing soil fertility, reducing greenhouse gas emissions, acting as a pollutant adsorbent, bioenergy production, climate change mitigation, and retaining water (Lehmann & Joseph, 2015). Yue et al. produced biochar from SS and used it to improve soil quality, resulting in a 50% increase in grass growth and a 74% increase in biomass in pot experiments. The concentrations of N, P, OC, K, and black carbon were also increased by 1.5, 5.6, 1.9, 0.4, and 4.5 times, respectively (Yue et al., 2017). Huang et al. created biochar from SS, saw husk, and rice stalk using the pyrolysis technique (300°C–700°C). The addition of rice stem and husk reduced biochar production and heavy metal levels (Cu, Zn, and Ni) while increasing the amount of organic matter (Huang et al., 2017). In their study, Li et al. synthesized biochar from SS and treated it with lanthanum via pyrolysis (600°C). Synthesized biochar was used for phosphate adsorption, with maximum adsorption of 93.91 mg/g (Li et al., 2020). Fan et al. prepared biochar from SS, which could adsorb 89% of methylene blue dye onto its surface (Fan et al., 2017). Ifthikar et al. synthesized magnetic biochar from municipal SS by suspending biochar in ferric chloride solution for the adsorption of lead (Pb II). Maximum adsorption was 249 mg/g in 200 ppm, and equilibrium was achieved in just 1 hour (Ifthikar et al., 2017). Méndez et al. tried to study the effects of SS and SS biochar on peat (a component of horticulture growth media) as well as on the growth of *Lactuca sativa*. Biochar addition improved the percentage of N, P, and K in the media, and biochar supplementation to peat at a volumetric rate of 10% improved the biomass production of *L. sativa* by 184%–270% and shoot length by 137%–147% (Méndez et al., 2016). Biochar from SS has also been used for the adsorption of emerging contaminants such as 2,4-dichlorophenol (Kalderis et al., 2017), tetracycline (Ref?), and doxycycline (Tang et al., 2018a; Wei et al., 2019b). A summarized study of the different source of biomass, process involved, and their characteristics is mentioned in Table 21.2

some other applications of biochar derived from SS are mentioned in Table 21.3.

21.2.2 Hydrochar

Hydrochar (char-like material) is a hydrothermally carbonized (HTC) product obtained from SS (anaerobically digested effluents) at high temperature and pressure (Huezo et al., 2021). The HTC method can utilize a wide range of feedstock without requiring considerable pretreatment. In the HTC process, biomass temperature is raised between 180°C and 260°C to achieve fully carbonized materials. Temperature below 180°C could not carbonize the feedstock, and temperature above 260°C is the temperature, and pressure lies between 1.5 and 26 MPa for hydrothermal liquefaction. During the HTC process, water acts as a catalyst and polar solvent, which helps in the transfer of mass as well as accelerates the rate of reactions (Funke & Ziegler, 2010). A gaseous product formed during HTC is carbon

TABLE 21.2 Methodologies for the treatment of sewage sludge, physical parameters, with improved properties and applications. treating.

Sources	Process and temperature (°C)	Physical parameters	Characteristics, yield (%)	References
Deinking sludge	Slow pyrolysis, 300°C	Heating rate (HR) — 10°C/min, Residence time (RT) -2 h	Improved porosity, water holding capacity of peat	Méndez et al. (2015)
Wastewater sludge	Slow pyrolysis, 300°C and 600°C	HR-2°C/min and 30°C/min RT - 2 h	Increase surface aera and porosity with increase in temperature, Carbon content-0.40 and 0.73 kg/kg dry sludge, *Lepidium sativum* growth	Vilas-Boas et al. (2021)
SS and microalgae mixture	Slow pyrolysis, 500°C and 350°C	HR-30°C/ min	80% biochar yield w.r.t initial weight of the substrate	Bolognesi et al. (2021)
	Slow pyrolysis 600°C	HR-13°C/min, RT - 30 min	Improved surface area (16.90 to 307.10 m^2/g) and pore volume (0.057–0.18 cm^3/g), increased ash content	Sahoo et al. (2021)
SS and manures	Slow pyrolysis 300°C	HR-7°C/min, RT -30 min	Yield - 39.2%–80.9%, carbon content - 237.5–514.7 g/kg, specific surface area - 96.06–110.83 m^2/g	Hossain et al. (2021)
Corn stover and scum	Microwave-assisted catalytic copyrolysis, 500°C	HR-15°C/min, RT - 20 min	Improved bio-oil and biochar yield	Liu et al. (2016)
Pine wood chips	Fast pyrolysis, 424°C	RT — 20–30 s	Porous structure, surface areas - 20.9, salicyclic adsorption capacity 22.70 mg/g	Essandoh et al. (2015)
Rapeseed, SS, miscanthus, and wood	Gasification, 750°C	Gasification time 1–30 min	Porous structure in case of rapeseed and miscanthus, Hydrogen yield — 35–45 volume %, Highest ash content in case of SS ~355 g/kg	Sattar et al. (2014)
Pine sawdust	Gasification, 600°C–850°C	steam flow rate —0.357 g/min/g of biomass, reaction time — 15 min	Ash content 83.2%, carbon conversion efficiency of 95.78%	Yan et al. (2010)
Ananas comosus and *Annona squamosa*	Torrefaction 210°C–300°C	HR-18°C/min, Reaction time - 30 min	*Ananas comosus* biochar higher heating value 27.7 MJ/kg, energy return on investment — 22.9, fuel ratio — 0.36 and *Annona squamosa* maximum energy density of 1.58	Lin et al. (2021a)
Washingtonia filifera petiol and *Sterculia foetida* follicle	Torrefaction 210°C–300°C	HR-30°C/min, Reaction time - 30 min	Energy density ~ 1.33, energy return on investment — 23.2	Lin et al. (2021a)
Walnut shell and pearl millet	Torrefaction 210°C–300°C	HR-25°C/min, RT — 30–90 min	Yield — 41%–91%, gross calorific value 22–27 MJ/kg, rough and nonuniform surface, large holes on biochar surface,	Abdullah etal. (2022)
Pine wood and SS	Torrefaction 80°C–520°C	HR - -268.15°C/min, -263.15°C /min	-	Piersa et al. (2021)
Banana split	Flash light irradiation	Xenon lamp with 575 V-pulses, energy density of 13.1 J/cm^2 for 14.5 Ms, reaction time — 12 s	Rapid and economical process, energy balance output - 4.09 MJ/kg, 33 wt.% biochar yield	O. Silva et al. (2022)
Phoenix dactylifera L	Flash pyrolysis, 400°C–700°C	Retention time — 3 to 9 min	Specific area - 140.3 to 292.8 m^2/g, smooth surface and no visible pores, carbon content improve, 40% bio-oil production	Ateş and Yaşar (2021)
Albizia odoratissima	Flash pyrolysis 350°C–550°C	HR-20°C–40°C/min	Heating value for biochar - 23.47 MJ/kg	Sowmya Dhanalakshmi et al. (2022)

(Continued)

TABLE 21.2 (Continued)

Sources	Process and temperature (°C)	Physical parameters	Characteristics, yield (%)	References
Palm kernel shell	Microwave vacuum pyrolysis power 750 W	Heating time – 35 min	Porous, High specific surface - 270 cm^2/g, biochar yield ~ 23 wt.% improved moisture content,	Wan Mahari et al. (2020)
Prosopis juliflora	Vacuum pyrolysis, 500°C	Pressure - 10–12 kPa, RT – 40–60 min, HR-10°C–30°C/min	Rough surface, lack of pores, Specific surface improved, lower activation energy- 22–33 kJ/mol	Pawar and Panwar (2022)
Syagrus coronata	Vacuum pyrolysis, 400°C–700°C	HR - 10°C/min, TR – 2 h, vacuum 100 mm Hg	Presence of micropores, smooth surface, 50 wt.% biochar yield at 400°C	Santos et al. (2020)
Cassava stem	Microwave vacuum pyrolysis, 550 W and 750 W, 430°C	Vacuum pressure – 8–15 kPa for 40 min	70–75 wt.% biochar yield, calorific value ~ 20 MJ/kg, large number of pores on the surface of biochar and can be used as solid fuel	Foong et al. (2020)
Rice husk	Hydrothermal carbonization, 180°C	Pressure – 70 bar, reaction time – 20 min	~57% biochar yield, ~14% silica oxide reduction, nonporous and rigid morphology	Hossain et al. (2020)
Park biowaste and hydrochar	Hydrothermal carbonization, 200°C	HR - 15°C/min	More carbon content in biomass-Fe- biochar w.r.t. hydrochar-Fe-biochar, ~70% wt.% ash content in biomass-Fe-biochar and ~41% in hydrochar-Fe-biochar, microporous morphology in both biochar	Álvarez et al. (2020)
Municipal waste sludge	Hydrothermal carbonization 100°C, 150°C and 180°C (group 1), 180°C (group 2)	Retention time h hour (group 1), 5 h (group 2)	High heating - ~15 MJ/kg, appearance of carbon microsphere	Chen et al. (2020)
Oil palm fiber	Microwave (MW) pyrolysis	MW frequency - ~2450 MHz, power – 200–1400 W	Porous biochar structures, high heating value ~ 20–23 MJ/kg, higher carbon content > 60 wt.%	Arafat Hossain et al. (2017)
Empty fruit bunch	MW pyrolysis	MW frequency ~ 2.45 GHz, power – 3.38 kW, resident time – 30 min	Biochar yield ~50.0 wt.%, higher heating value -25.19 MJ/kg,	Md Said et al. (2022)
Oil palm waste	MW pyrolysis	Power 500–700 W	Highly porous and rough biochar structure, carbon content ~ 51%, 38 wt.% biochar yield, specific surface area - 210 m^2/g	Liew et al. (2018)
Marine microalgae	Modified biochar, 450°C	Current density ~ 94 mA/cm^2, tome – 5 min	Improved surface area, maximum phosphorous adsorption capacity - 31.28 mg/g	Jung et al. (2015)

dioxide and methane along with H_2O, hydrochar, and liquors (Huezo et al., 2021). Some novel uses of hydrochar are solid fuel, adsorbent, biomass for anaerobic digestion, and supplement for soil (Fang et al., 2014; Ferrentino et al., 2020a; Kim et al., 2014; Parshetti et al., 2013).

Huezo et al. generated hydrochar from anaerobically digested SS in their study (Ohio, US). The experiment lasted for 30–70 minutes at temperatures ranging from 180°C to 260°C and pH levels ranging from 6.8 to 8.2. The maximum yields of biochar at pH 8.2 and 6.8 were 70–80 wt.%% and 62–73 wt.%, respectively. When the temperature exceeds 260°C, poor yield of hydrochar was obtained (Huezo et al., 2021). Ferrentino et al. studied the coupling effect of HTC and AD for the treatment of SS. SS was used for the generation of hydrochar and liquor, which were recycled back to HTC helping in the improvement of the yield of biomethane up to 187 ± 18 mL $CH_4g^{-1}COD$ when 45% of hydrochar was supplemented with original feedstock (Ferrentino et al., 2020a). Zhang et.al used a hydrothermal autoclave at 200°C for 2 hours to make hydrochar from rice straw and SS. Hydrothermal treatment improved the fuel characteristics of hydrochar HHVs (14.57 MJ/kg). The elimination of alkali during the reaction improved the temperature of the fusion ash (Zhang et al., 2020). Hydrochar production from SS can be a sustainable process, which can minimize the dependency on chemical fertilizers since it is used as supplements for soil, which can improve the soil quality as well as productivity. Hydrochar can also be

TABLE 21.3 Examples of studies preparing biochar from waste sewage sludge.

Source	Process (temperature and time)	Application	Ref.
SS and tea waste	Co-pyrolysis (300°C, 2 h)	90% MB removal @ 10 g/mL	Wei et al. (2019b)
SS and rice husk	Co-pyrolysis (500°C, 2 h)	Removal of direct red, acid orange, react blue 19, MB	Chen et al. (2019)
SS	Microwave pyrolysis	Eosin and safranin adsorption	Zhang et al. (2018)
SS	Liquefaction process (260°C–380°C)	Malachite green and MB	Leng et al. (2015)
SS	Pyrolysis (600°C, 1 h)	X-GRL	Xiao et al. (2018)
SS	Pyrolysis (500°C, 4 h)	Cu^{+2} adsorption	Shen et al. (2018)
SS	Pyrolysis (300°C–600°C, 4 h) electromagnetic induction	Pb (II), Cd (II) removal	Xue et al. (2019)
SS	Pyrolysis (600°C, 2 h)	Pb (II), Cd (II) adsorption	Ni et al. (2019)
SS	Persulfate oxidation process	Removal of tetracycline (82.2%)	Yu et al. (2019)
SS and fly ash	Pyrolysis (1200°C, 2 h)	Microbial fuel cell (current density of 25 A/m^2)	Jia et al. (2018)
SS and dolomite	Pyrolysis (800°C, 2 h)	PO4-P adsorption (29.18 mg/g)	Li et al. (2019)

TABLE 21.4 Examples of studies preparing hydrochar from waste sewage sludge.

Source	Process	Temperature and time	Application	Ref.
Municipal sludge ash	HTC	250°C, 3 h	MB removal (95%), in 25 min	Ferrentino et al. (2020b)
SS	HTC	Power 1 kW, 1	Removal of nitrogen	Xu et al. (2020)
SS with fruits and vegetable waste	HTC	220°C, 12 h	2.sixfold improvement in calorific value of SS	He et al. (2019)
SS and rice straw	Co-HTC	220°C, 2 h	Higher heating volume (HHV) obtained 14.57 MJ/kg	Zhang et al. (2020)
SS and lignocellulose (saw & fir dust) and charcoal	Co-HTC	200°C, 2 h	HHV increased by 6%, CC increased to 34.9 to 37.9% and decreased volatile matter from 56.4 to 40.7%	Wilk et al. (2021)
SS	HTC	260°C, 1 h	Nitrogen retention and nitrogen content improved	Chu et al. (2020)
Centrifuged SS	HTC	200°C, 1 h, 3 h, 5 h and 8 h	3 h HTC, recovered ~70% humic acid and ~80% phosphate phosphorous	Malhotra and Garg (2020)
SS and sawdust	Co-HTC	200°C, 1 h	Gasification efficiency improved (77.73%), HVV- 8.15 MJ/Nm^3	Ma et al. (2019)
SS	Microwave - HTC	150°C–250°C, 0–2 h	Improved nitrogen, phosphorous, and potassium contents well as ignition temperature	Wang et al. (2022)
SS and PVC	Co-HTC	230°C, 1 h	Zinc and manganese were removed, nickel and arsenic toxicity were reduced, dichlorination efficiency improved	Lu et al. (2021)

utilized to produce biofuel such as biomethane, which can be part of circular economy and sustainable development goal. Some of the novel and recent updates in applications of hydrochar are tabulated in Table 21.4.

21.2.3 Carbon dots

Carbon dots (CDs), also known as carbon quantum dots, (diameter less than 10 nm) were discovered by Xiaoyou Xu in 2004 while purifying and segregating single-walled nanotubes produced using arc-discharge method (Sun & Lei, 2017; Xu et al., 2004). CDs are chemically composed of three basic elements including carbon, hydrogen, and oxygen (Bhartiya et al., 2016). Generally, carbon-based nanomaterials are classified into zero-dimensional and one-dimensional materials. Carbon dots are regarded as zero-dimensional material and have point-like structures and quasi spherical carbon dots (Lin et al., 2021b; Qu et al., 2013). CDs are typically amorphous spherical structures with sp^2 and sp^3 hybridization (Sciortino et al., 2018). In general, C-dots are synthesized via two approaches (Grobelak et al., 2019): top-down approach and (Svanström et al., 2004) bottom-up approach. In the top-down approach, materials are first fragmented and decomposed to produce CDs (Sharma & Das, 2019). In bottom-up approach, CDs are treated as bulk material, which is hydrothermally transformed into "quantum-sized dots" (Sendão et al., 2020; Kashani et al., 2019). CDs synthesized from sustainable raw materials, SS, and biomass are cheaper, and it is sustainable eco-friendly source for production of CDs (Khairol Anuar et al., 2021). It also helps in reutilization of waste into useful products. Several synthetic methods that are being used for the synthesis of CDs include chemical synthesis (Khairol Anuar et al., 2021), hydrothermal synthesis, ultrasonic synthesis, microwave-assisted synthesis (Rodríguez-Padrón et al., 2018), and pyrolysis and carbonization. Among all the mentioned methods, hydrothermal synthesis is widely practiced because of its simple, clean, and cheap synthesis approach (Funke & Ziegler, 2010). In this method, biomass is subjected to an optimum condition in a hydrothermal autoclave, the synthesized product is washed with distilled water. Mainly, three types of feedstocks can be used for the synthesis of CDs such as animal and plant biomass, waste biomass, and proteins. Some studies have also been carried out using SS for the synthesis of CDs. For example, Hu et al. utilized SS to synthesize CDs utilizing microwave techniques in their study. The sample (2 g) and water were microwaved for 30 minutes @ 700 W, and the photoluminescent properties of synthetic CDs are used to detect p-nitrophenols (Hu & Gao, 2020). The material was washed and dried for 24 hours at 60°C before being pyrolyzed for 2 hours at 900°C with nitrogen supplementation labeled as MSC-A. The second approach involved dissolving 5 g of municipal SS in 40 mL of distilled water and placing it in a hydrothermal autoclave for 24 hours at 180°C labeled as MSC-B. In synthesized N-doped CDs, CO_2 is transformed into methane, ethanol, and methanol. Faraday efficiency was 90% at 0.71 V, with remarkable stability throughout the reaction (Deng et al., 2022). Biowaste serves as the substrate for the production of CDs and CDs have wide application in the fields such as sensing, bioimaging, photocatalysis, drug delivery so waste biomass can be used for producing value-added products, which may shape the economy.

21.2.4 Biosurfactants

Biosurfactants are amphiphilic in nature with two separate parts: one hydrophobic and the other hydrophilic, with the ability to change the surface or interfacial tension (Yuan et al., 2014). The surfactants can be anionic (carboxyl and sulfonate), cationic (phosphoniumand quaternary ammonium), and nonionic (ethylene oxide). This classification is mainly based on the charge present on the hydrophilic part and has high molecular weight and complex structure. Biosurfactants have recently received a lot of attention since they can be used for pollutant adsorption (mainly hydrophobic compounds) (Gaubert et al., 2016). Microorganisms such as *Ochrobactrum anthropic* (Ibrahim, 2018; Zarinviarsagh et al., 2017; Tripathi et al., 2020), *S. maltophilia* (Tripathi et al., 2020; Nogueira et al., 2020), *Microbacterium esteraromaticum* (Tripathi et al., 2020; Kumari et al., 2018; Wongbunmak et al., 2017), *Pseudomonas mendocina* (Sivasankar & Suresh Kumar, 2017; Maia et al., 2019), and *Pseudomonas aeruginosa* (Deivakumari et al., 2020; Rehman et al., 2021), *Rhodococcus* sp (Pi et al., 2017; Kuyukina & Ivshina, 2019). and *Baccilus* sp. are responsible for degradation of hydrophobic contaminants, which release biosurfactants (Braz et al., 2010; Hu et al., 2016). Produced biosurfactant has wide application in the field of soil washing (Pei et al., 2018; Mao et al., 2015; Lai et al., 2009), degradation of crude oil (Sharma et al., 2019; Lee et al., 2018), degradation of hydrophobic compounds (Silva et al., 2014), removal of heavy metals, food emulsifier (Campos et al., 2015; Gaur et al., 2019; Satpute et al., 2018), and antimicrobial agents (Mesbaiah et al., 2016; Chen et al., 2017). SS is used as the source for the production of biosurfactant. Ghurye et al. in his study used sewage surfactant (Texas) as a mixed microbial source, acclimatized in molasses solution (20 g/L, 7 days, pH-6.6), and incubated in a master culture reactor for the production of biosurfactants. The results suggest that with an increase in the critical micelle dilution, biosurfactant production increases. However, in the whole experiment, emulsification ability increased resulting in increase in biosurfactant accumulation (Ghurye et al., 1994). Partovi et al. in their studies used waste from soyabean oil refinery for the production of biosurfactant using microbial culture of *P. aeruginosa* Mr01. The maximum yield was 1.6 times on soyabean supplement with 89.6 purity; however, surface tension and emulsification were continuously changing (Partovi et al., 2013). Tripathi et.al utilized oil sludge in their study for the production of biosurfactants. Different microorganisms were exploited such as *Ochrobactrum anthropic* IITR07, *S. maltophilia* IITR87, *M. esteraromaticum* IITR47, *P. mendocina* IITR46, and *P. aeruginosa* IITR48. Maximum yield was achieved by *M. esteraromaticum* IITR47 utilizing naphthalene as a carbon source. The biosurfactants produced were identified as glycolipid and rhamnolipid (Tripathi et al., 2020). Chen et al. in their study utilized kitchen waste for the production of biosurfactant assisted by microbial culture *P. aeruginosa*. The maximum of rhamnolipids was 50.86% with LC-Ms, and the critical micelle concentration was 55.87 mg/L. According to the study, rather than decomposing or landfilling SS, it can be used to produce biosurfactants as a renewable and inexpensive carbon source.

21.2.5 Bioplastics

Bioplastic is biologically synthesized material from biological sources such as polysaccharides (cellulose and starch) and proteins. Commonly used bioplastics are PLA (polylactic acid), PPT (poly (p-phenylene)), formic acid diol ester (PTT), and PHT (polyhydroxyalkanoates). Bioplastics are synthesized by two approaches: (1) biological synthesis and (2) chemical synthesis. For example, PHA is synthesized via a biological route using microorganisms as their energy storage component, whereas PLA and PPT are chemically synthesized (Liu et al., 2019). In general, bioplastics are categorized as (1) all bioplastic (completely biologically derived) and (Svanström et al., 2004) part bioplastic (modified with some petroleum product). Further, bioplastic can also be categorized as biodegradable and nonbiodegradable. The plastic we use every day is typically sourced from petroleum, which is one of the primary contributors to environmental imbalance. So, there is a need for some alternative materials that should not be harmful and toxic to our ecosystem. Bioplastic could be an option for replacing plastic; however, its production is very expensive, unavailability of a good recycling system, and incompetent waste management system. Number of studies have been carried out for the production of PHA from monoculture, but a significant portion of the production cost is spent on media and reactor maintenance (Kumar et al., 2018). As a result, a less expensive feedstock for bioplastic synthesis is currently being explored. SS with a diverse microbial community could be used as a feedstock for bioplastic synthesis, particularly PHA. Using SS as a feedstock will also help to reduce the environmental impact of waste disposal (Venkateswar Reddy & Venkata Mohan, 2012). Rapid population growth and economic development have increased the load of SS, which can be a more cost-effective source of bioplastics (Kumar et al., 2018). Microbial species such as yeast, bacteria, and fungi are capable of utilizing carbon from SS and can synthesize triglycerol and polylactic acid (Kumar et al., 2017).

Tu and colleagues investigated the impact of phosphorus on PHA production using sewage. Thermal hydrolysis (155°C–175°C, 6 bar, 30 min) and acidogenic fermentation (110°C, 1 h) were used to treat the sludge. The resultant substrate was centrifuged at 10,000 rpm for 10 minutes after acidogenic fermentation before being used to generate PAH. When the P-content was lowered to 1.35 mg/L, the formation of PAH rose from 23 wt.% to 51 wt.%. Zhang et al. examined raw sludge to thermally hydrolyzed sludge (THS) (155°C–175°C, 30 min.) for the production of volatile fatty acids (VFAs), which was subsequently used as a substrate for synthesis of PHA. THS generated 44.6% more VFA than raw sludge. At 55°C, raw sludge output increased by 15.7%, although THS production declined by 12.2%. PHA output was 34.6% PHAs/DCW overall (Zhang et al., 2019).

21.2.5.1 Biofertilizers

Nowadays, biofertilizers have become an indispensable part of sustainable economy and eco-friendly agriculture practices. It also helps in nutrient management in agriculture and also aids in minimizing the use of chemically synthesized fertilizers. The practice of using biofertilizers is very old. In the late 19th century, first biofertilizer was manufactured by Nobbe and Hiltner by culturing *Rhizobia*, later on *Azobacter* and blue-green algae were discovered and used for the commercial production of biofertilizers. In the context of India, the first commercial application of biofertilizer started in the mid-20th century by N V Joshi using *Rhizobium*. It is well known that for increasing the per capita yield of agriculture, chemical fertilizers are extensively used resulting in detrimental effects on the environment as well as humans and animals; thus biofertilizers could be a hope to minimize the damage and can be fertilizers for future. SS is converted into biofertilizers with the supplementation of different microorganisms (bacteria, yeast, algae). Biofertilizers also contain a wide range of nitrogen-fixing microbes (*Mycobacterium, Desulfovibrio, Rhizobium, Corynebacterium, Acetobacter*), phosphorous and potassium solubilizers (*Pseudomonas, Rhizobiumand, Bacilli*).

Recent developments have been initiated in the synthesis of biofertilizers from SS. Ma et al., for example, used activated sludge and food waste in their study to make biofertilizers and biogas. To digest sludge and food waste simultaneously, two steps were followed. In the first step, the substrate was mixed and digested anaerobically with processed fungal biomass (*Aspergillus oryzae*) to produce hydrolysate. The second step involved converting digested materials into biofertilizers while liquid components were extracted and codigested with sludge to generate biomethane. The biomethane produced in the second step is 1.8 times that produced in the first step (Ma et al., 2017). Paliya et.al, in their study, utilized sludge ash (waste from dewatered SS) for the production of *Rhizobium*-based biofertilizer by inoculating *Rhizobium* inoculums into sludge ash. This preparation was used to assess the lentil seeds, and its effect was evaluated by assessing the plant growth. Lentil seeds supplemented with *Rhizobium*-based biofertilizer improved the plant growth, seed yield, root length, and nodules by 10%, 53.4%, 19%, and 42%, respectively (Paliya et al., 2019). Asses et.al in their work utilized poultry slaughterhouse waste along with SS, wood dust, and activated compost for the removal of microbes as well as production of biofertilizers. The reaction was carried out for 90 days in 300 m^3 vessels. After the 20 days of operation, the temperature exceeded 65°C, and a significant reduction in the population of pathogenic microbes was seen such as *Coliforms* (guts), *Streptococci*, and *E. coli*. *Salmonella* were completely removed from the system. Composting poultry slaughterhouse waste generated hygienic compost with adequate organic matter (49.1%) having C/N ratio of 13.92% at a slightly basic pH (7.7). Germination index reached 91% at the end of the process. Supplementing compost with peat improved the stem and leaf length by 63.8% and 57.9%, respectively. Dry biomass and fresh biomass of *Zea mays* improved by 65.1% and 66.6%, respectively. Lastly, it showed no effect of phytotoxicity (Asses et al., 2019). All the above studies suggest that SS has great potential application in agriculture and can minimize the environmental load after incineration and help in boosting the economy when used as biofertilizers.

21.3 Applications of sewage sludge

21.3.1 *Heavy metal extraction from sewage wastes*

Heavy metals are primarily found in effluent discharged from industrial wastes, household garbage, food waste, pharmaceuticals waste, cosmetics, animal and human excreta, all contribute HMs to wastewater (Sörme & Lagerkvist, 2002; Chanaka Udayanga et al., 2018). These contaminated wastewaters are transported to treatment plants, where the majority of the HMs and other contaminants end up in the generated SS. Physiochemical and biological interactions are responsible for the accumulation of 50%–80% of HMs in the SS (Yang et al., 2020). The availability of organic matter (proteins, saccharides, and humid matter) in SS is very high, which easily interacts with HMs. According to the experts, HMs are transferred from sewage to SS because of biomass adsorption (Chojnacka, 2010). There are two steps in adsorption: (1) passive sorption and (2) bioaccumulation. Passive sorption does not depend on the metabolism of the cell but rather depends upon the functional group and chemical structure on the surface of biomass (Bădescu et al., 2018; Choińska-Pulit et al., 2018; Qin et al., 2020). The three basic processes that occur during the adsorption process are ion exchange, complexation (mixing of solution with charged atoms), and inorganic microprecipitation (precipitation of small particles) reactions. Ion exchange takes place in the presence of functional groups such as carboxylic, alcohol, amide, and amines. These functional groups are influenced by multivalent metal ions (Ca^{2+}, Cd^{2+}, Pb^{2+} Ni^{2+}, Cr^{2+}) (Escudero et al., 2016). Alidoust et al. discovered that Cd^{+2} can be ion-exchanged with Ca^{+2} in their study (Alidoust et al., 2015). Yang et al. (2017) observed that only a small proportion of adsorbed ions can be ion-exchanged, showing that ion exchange is not the major absorption process. The second stage is known for the active uptake of HMs by microorganisms known as bioaccumulation. Metabolic activities and protein carriers help in the transportation of HMs from the surface of the cell to inside the cell where bioaccumulation occurs due to biochemical reactions between HMs and chemicals (proteins) present inside the cell (Pagnanelli et al., 2009). For example, according to Huang et al., heavy metals accumulated in *Bacillus cereus* RC-1 are via active transport and metabolic activity.

Electrokinetic remediation and bioleaching technology are the two promising technologies for the extraction of heavy metals from SS; however, ultrasound-assisted remediation can also be used (Pathak et al., 2009).

Electrokinetic remediation is based on the application of low direct current ranges between 0 and 1 A on the electrodes inserted into contaminated SS, leads to the migration of electron species with production of electric fields, which helps in the migration of HMs to the electrode surface (cathode and anode). Migration of ions occurs through three processes: (1) electroosmosis, (2) electrophoresis, and (3) electromigration (Xu et al., 2017; Tang et al., 2018b). In the case of toxic HMs, generally electromigration of ions takes place on the oppositely charged electrode. Several weak (acetic acid, carboxylic acid, oxalic acid) and strong acids (HCl, HNO_3, H_2SO_4) and some complex compounds such as NTA and EDTA have also been utilized for the extraction of HMs from SS. Several studies have been concluded to remove heavy metals from SS. For example, Bezzina et al. in their research tried to remove Cu^{+2}, Zn^{+2}, Pb^{+2}, and Fe^{+2} from SS via ion exchange resins of weak acid media. Deng et al. studied the effect of ultrasound (0–20 min) assisted nitric acid (0–0.2 M) technique for the extraction/removal of HMs like Cu^{+2}, Zn^{+2}, Pb^{+2}. Percentage removal of Cu^{+2}, Zn^{+2}, and Pb^{+2} was 9.5%, 82.2%, and 87.3%, respectively. Pie et al. used a modified electrokinetic method (using a chelating agent ethylenediamine, 18 V and 4 days) for the extraction of HMs. Removal efficiency of Cu^{+2}, Zn^{+2}, and Pb^{+2} was 68%, 58%, and 87%, respectively. Peng et al. applied the electrokinetic (32 V) method along with bioleaching for the extraction of Cu^{2+} and Zn^{2+}. The maximum removal efficiency of Cu^{2+} and Zn^{2+} was 99% and 79%, respectively (Peng et al., 2011). Chao et.al in their work used supercritical fluid (TBP/HNO_3 and C_2H_5OH P = 20 MPa, $T = 30°C$) for the extraction of Pb, Cu, and Cd. Maximum removal efficiencies for Pb, Cu, and Cd were 63.2%, 78.3%, and 69.5%, respectively. Suanon et al. in their research used chemicals such as citric acid and N, N-Bis(carboxymethyl) glutamic acid for the removal of HMs such as Cr, Cd, Cu, Zn, Ni, and Co with maximum removal efficiencies of 90.2%, 83.9%, 81.2%, 85.6%, 89.3%, and 87.3%, respectively. Xia et al. in their work utilized inorganic salt NaCl to extract HMs from SS at a temperature of 700°C for 4 hours. Maximum removal efficiencies of HMs such as Pb, Cu, Ni, Zn, Cr, Cd, and Mn were 91%, 86%, 29%, 99%, 19%, 95%, and 93%.

21.3.2 Bioenergy

Fossil fuels are the major source of energy, but the increasing demand cannot be met with limited supply of nonrenewable sources of energy. At the same time, harnessing energy from fossils has very detrimental effects on humans as well as the ecosystem. For the same reason, focus has shifted toward renewable energy sources. Municipal SS contains a large quantity of organic matter (Djandja et al., 2020), and it has been reported that 1000 kg of SS contains nearly 200 kg of organic compounds including nitrogen (6 kg), phosphorous (8 kg), and soluble salts (10 kg) (Iticescu et al., 2018). When SS is pyrolyzed at temperatures above 600°C, a variety of compounds are produced, including carbon monoxide, methane, hydrogen, carbon dioxide, and bio-oils, all of which can be used as fuels (Tang, 2017). Different pyrolysis methods can be used to produce the desired product. For example, slow pyrolysis of SS leads to the formation of biochar, whereas fast pyrolysis forms liquid products (Arazo et al., 2017). Pyrolysis oils are produced via the liquefaction method while carbonization and gasification lead to the formation of solid and gaseous products. Bio-oil is a dark-colored highly oxygenated compound. Chemical nature of bio-oil is very complex and composed of organic compounds such as phenols, aldehydes, alcohols, and acids, water molecules, and ash particles (Isahak et al., 2012). Bio-oil has high oxygen contents and oxygenated compounds (acids, phenols, alcohols), which improves the viscosity and stability (Czernik & Bridgwater, 2004). Bio-oil is composed of two phases (Grobelak et al., 2019) aqueous phase and (Svanström et al., 2004) nonaqueous phase. Aqueous phase is composed of low-molecular-weight oxygenated compounds, whereas nonaqueous phase high-molecular-weight oxygenated aromatic (benzene, toluene, indene) and polyaromatic compounds (naphthalene, fluorine, and phenanthrene) (Virmond et al., 2013). Physiochemical properties and composition of bio-oil depend on the following factors (Grobelak et al., 2019) source and origin of raw material or precursor, (Svanström et al., 2004) moisture content in the biomass, (Zuloaga et al., 2012) reactor and process used, (Vuorinen et al., 2007) reaction parameters, and product collection methods (Dai et al., 2019). Calorific value of bio-oil ranges between 33 and 36 MJ/kg, density (1.11–1.13 kg/dm^3), HHV (14–19 kg/dm), and Water (20–30 wt.%) (Mujahid et al., 2020). Currently bio-oil is being produced on the commercial scale in the countries such as Canada, USA, Finland, and Europe. Industries such as AE Cote-Nord Bioenergy/Ensyn and Red Arrows-Ensyn (Canada), Genting (Malaysia), Pyrovac (Canada), UPM refinery and Fortum-Valmet (Finland), BTG-BTL/Empyro (Netherland), and Kior (US) (Hu & Gholizadeh, 2020). Bio-oil can be utilized for the application such as biofuel, heavy fuel, hydrogen, polyurethane, plastic, bio pitch, and wood vinegar production (Hu & Gholizadeh, 2020). Approximately 50–80 wt. %bio-oil can be recovered, which contains a large amount of water (15%–35%), organic compounds, and oxygen (~ 50%). About 50 wt.% char can be produced as a pyrolyzed product from SS. Pyrolysis of SS or any biomass at higher temperature produces syngas

TABLE 21.5 Examples of studies preparing bioenergy from waste sewage sludge.

Methods	Products	Temperature (°C)	Residence time	References
Carbonization	Char	400		
Fast pyrolysis	Bio-oil	450	0.55 s	Arazo et al. (2017)
Fast pyrolysis	Bio-oil	525	1.5 s	Arazo et al. (2017)
Fast pyrolysis	Bio-oil	540	-	Arazo et al. (2017)
Fast pyrolysis	Bio-oil	500	low	Alvarez et al. (2015)
Slow pyrolysis	Bio-oil	700	23 min	Gao et al. (2017)

(carbon monoxide, methane, hydrogen, carbon dioxide) and hydrocarbons (ethane, propane, ethene) as well as some inorganic compounds such as hydrogen sulfide, ammonia, cyanide, and water vapor. Syngas can also be used to produce heat and electricity. Pyrolyzed fuel can be used to run the boiler and can be used as a transportation fuel in the future (Trinh et al., 2013; Gao et al., 2017). Methanization of syngas such as CO or H_2 can produce natural gas (Djandja et al., 2020). Some of the products produced via pyrolysis based on different temperatures, residence times, and pyrolysis types are tabulated in Table 21.5.

Biodiesel is another very interesting biofuel produced from SS. Qi et.al in their study utilized anaerobic−anoxic−oxic (A2/O) and membrane bioreactor and in situ transesterification was applied to SS. Obtained products from reactors were used for the production of lipids, which ultimately transesterified into biodiesel. The conditions for biodiesel production were optimized using orthogonal (3-factors, 4 levels) and single-factor tests. When the methanol-to-sludge mass ratio was 10:1 and 8:1, the biodiesel output was 16.6% and 4.2%, respectively, with the purity of 96.7% and 92.7% at temperatures of 60°C and 50°C, respectively, while the H_2SO_4 concentration was held constant (5% v/v) (Qi et al., 2016). Wang and colleagues produced biodiesel from scum sludge in their research. Scum sludge was converted into lipids before being transesterified into biodiesel. The highest yield observed was 22.7% (Wang et al., 2016). Ghodke et.al in their research work pyrolyzed SS between the temperature range of 250°C and 700°C to produce a wide range of valuable products in fluidized-bed reactors. Products such as bio-oil, biochar, and pyrolysis gases were produced with a maximum yield of 22.4 wt.%, 58.7 wt.%, and 18.9 wt.%, respectively. Both bio-oil and pyrolyzed gas were reported to have potential applications in engine fuel (Ghodke et al., 2021).

21.4 Conclusion

Sewage sludge is a waste obtained from domestic and industrial wastewater, which contains nutrients, toxic organic chemicals, pathogens, and heavy metals. The wastewater is processed through preliminary, primary, secondary, and tertiary sewage treatment to obtain sewage sludge. Sewage sludge worldwide is usually discarded through strategies such as landfilling and incineration, which makes this an energy-intensive and space-consuming process. In the recent years, there has been a significant advancement toward the valorization of sewage sludge to obtain value-added products. Biochar, heavy metals, biofertilizers, bioenergy, hydrochar, carbon dots, biosurfactants, and bioplastics are some examples of the products obtained from waste sewage sludge. The chapter also discusses laws and policies for disposal of sewage across different nations depending on the geographical location and resources. The research toward conversion of sewage sludge to value-added products is still a blooming subject and is expected to yield more products in the near future.

Acknowledgment

Dhar is thankful to Department of Biotechnology (DBT), Govt. of India, for the Ramalingaswami fellowship and research funding (BT/HRD/35/02/2006) to carry out this work. Ranjan and Kumar are grateful to Ministry of Human Resource Development (MHRD), Govt. of India. Rai is thankful for the junior research fellowship from Department of Biotechnology, Govt. of India (DBTHRDPMU/JRF/BET-20/I/2020/AL/07). The authors are grateful to School of Biochemical engineering, Indian Institute of Technology (IIT-BHU), Varanasi for their kind support, which made this work possible.

References

Αρχική σελίδα - βάση Δεδομένων Παρακολούθησης Λειτουργίας Ε.Ε.Λ. [Internet]. [cited 2022 May 24]. Available from: http://astikalimata.ypeka.gr/.

Abdullah, I., Ahmad, N., Hussain, M., Ahmed, A., Ahmed, U., & Park, Y. K. (2022). Conversion of biomass blends (walnut shell and pearl millet) for the production of solid biofuel via torrefaction under different conditions. *Chemosphere, 295*, 133894, May 1.

Aiello, A. E., Larson, E. L., & Sedlak, R. (2008). Hidden heroes of the health revolution sanitation and personal hygiene. *American Journal of Infection Control, 36*(10), S128–S151.

Alidoust, D., Kawahigashi, M., Yoshizawa, S., Sumida, H., & Watanabe, M. (2015). Mechanism of cadmium biosorption from aqueous solutions using calcined oyster shells. *Journal of Environmental Management, 150*, 103–110, Mar 1.

Alipour, M., Asadi, H., Chen, C., & Rashti, M. R. (2021). Bioavailability and eco-toxicity of heavy metals in chars produced from municipal sewage sludge decreased during pyrolysis and hydrothermal carbonization. *Ecological Engineering, 162*, 106173, Apr 1.

Alvarez, J., Amutio, M., Lopez, G., Bilbao, J., & Olazar, M. (2015). Fast co-pyrolysis of sewage sludge and lignocellulosic biomass in a conical spouted bed reactor. *Fuel, 159*, 810–818, Jul 27.

Álvarez, M. L., Gascó, G., Palacios, T., Paz-Ferreiro, J., & Méndez, A. (2020). Fe oxides-biochar composites produced by hydrothermal carbonization and pyrolysis of biomass waste. *Journal of Analytical and Applied Pyrolysis, 151*, 104893, Oct 1.

Arafat Hossain, M., Ganesan, P., Jewaratnam, J., & Chinna, K. (2017). Optimization of process parameters for microwave pyrolysis of oil palm fiber (OPF) for hydrogen and biochar production. *Energy Conversion and Management, 133*, 349–362, Feb 1.

Arazo, R. O., Genuino, D. A. D., de Luna, M. D. G., & Capareda, S. C. (2017). Bio-oil production from dry sewage sludge by fast pyrolysis in an electrically-heated fluidized bed reactor. *Sustainable Environment Research, 27*(1), 7–14.

Asses, N., Farhat, W., Hamdi, M., & Bouallagui, H. (2019). Large scale composting of poultry slaughterhouse processing waste: Microbial removal and agricultural biofertilizer application. *Process Safety and Environmental Protection, 124*, 128–136, Apr 1.

Ateş, F., & Yaşar, B. (2021). Utilization of date palm stones for bio-oil and char production using flash and fast pyrolysis. *Biomass Conv Bioref*. Available from https://doi.org/10.1007/s13399-021-01350-y, [Internet].

Atkinson, C. J., Fitzgerald, J. D., & Hipps, N. A. (2010). Potential mechanisms for achieving agricultural benefits from biochar application to temperate soils: A review. *Plant Soil, 337*(1), 1–18.

Azizi, A. B., Lim, M. P. M., Noor, Z. M., & Abdullah, N. (2013). Vermiremoval of heavy metal in sewage sludge by utilising Lumbricus rubellus. *Ecotoxicology and Environmental Safety, 90*, 13–20, Apr.

Bădescu, I. S., Bulgariu, D., Ahmad, I., & Bulgariu, L. (2018). Valorisation possibilities of exhausted biosorbents loaded with metal ions – A review. *Journal of Environmental Management, 224*, 288–297, Oct 15.

Bajpai P. (2017). Anaerobic reactors used for waste water treatment. In *Anaerobic technology in pulp and paper industry* [Internet]. Singapore: Springer Singapore; [cited 2022 Jul 14]. p. 37–53. (SpringerBriefs in Applied Sciences and Technology). Available from: http://link.springer.com/10.1007/978-981-10-4130-3_5.

Bajpai, P. (2018). Biermann's handbook of pulp and paper. *Paper and board making* (Volume 2). Elsevier.

Bajpai, P. (2017). *Pulp and paper industry: Emerging waste water treatment technologies* (p. 230) . Elsevier.

Barua, S., Zakaria, B. S., Chung, T., Hai, F. I., Haile, T., Al-Mamun, A., et al. (2019). Microbial electrolysis followed by chemical precipitation for effective nutrients recovery from digested sludge centrate in WWTPs. *Chemical Engineering Journal, 361*, 256–265, Apr 1.

Bhartiya, P., Singh, A., Kumar, H., Jain, T., Singh, B. K., & Dutta, P. K. (2016). Carbon dots: Chemistry, properties and applications. *Journal of the Indian Chemical Society, 93*(7), 759–766.

Bianchini, A., Bonfiglioli, L., Pellegrini, M., & Saccani, C. (2016). Sewage sludge management in Europe: a critical analysis of data quality. *International Journal of Environment and Waste Management, 18*(3), 226–238.

Black, C. (1974). *Eidsness, transfer USEPAO of T. Process design manual for sludge treatment and disposal. Technology Transfer.* US Environmental Protection Agency.

Boehler, M., Zwickenpflug, B., Hollender, J., Ternes, T., Joss, A., & Siegrist, H. (2012). Removal of micropollutants in municipal wastewater treatment plants by powder-activated carbon. *Water Science and Technology, 66*(10), 2115–2121.

Bolognesi, S., Bernardi, G., Callegari, A., Dondi, D., & Capodaglio, A. G. (2021). Biochar production from sewage sludge and microalgae mixtures: Properties, sustainability and possible role in circular economy. *Biomass Conv Bioref, 11*(2), 289–299.

Boudjabi, S., & Chenchouni, H. (2021). On the sustainability of land applications of sewage sludge: How to apply the sewage biosolid in order to improve soil fertility and increase crop yield? *Chemosphere, 282*, 131122, Nov 1.

Bouwer, E. (1993). Bioremediation of organic compounds? Putting microbial metabolism to work. *Trends in Biotechnology, 11*(8), 360–367.

Braz, R., Pirra, A., Lucas, M. S., & Peres, J. A. (2010). Combination of long term aerated storage and chemical coagulation/flocculation to winery wastewater treatment. *Desalination, 263*(1–3), 226–232.

Breasted, J. H. (1906). Ancient records of Egypt; historical documents from the earliest times to the Persian conquest, collected, edited, and translated with commentary. Ancient records 2nd series.

Bridle, T., & Pritchard, D. (2004). Energy and nutrient recovery from sewage sludge via pyrolysis. *Water Science and Technology: A Journal of the International Association on Water Pollution Research, 50*, 169–175, Feb 1.

Campos, J. M., Stamford, T. L. M., Rufino, R. D., Luna, J. M., Stamford, T. C. M., & Sarubbo, L. A. (2015). Formulation of mayonnaise with the addition of a bioemulsifier isolated from Candida utilis. *Toxicology Reports, 2*, 1164–1170, Jan 1.

Chanaka Udayanga, W. D., Veksha, A., Giannis, A., Lisak, G., Chang, V. W. C., & Lim, T. T. (2018). Fate and distribution of heavy metals during thermal processing of sewage sludge. *Fuel, 226*, 721–744, Aug 15.

Chatzakis, M. K., Lyrintzis, A. G., Mara, D. D., & Angelakis, A. N. (2006). Sedimentation tanks through the ages. In *Proceedings of the 1st IWA international symposium on water and wastewater technologies in ancient civilizations*. p. 755–61.

Chen, C., Liu, G., An, Q., Lin, L., Shang, Y., & Wan, C. (2020). From wasted sludge to valuable biochar by low temperature hydrothermal carbonization treatment: Insight into the surface characteristics. *Journal of Cleaner Production, 263*, 121600, Aug 1.

Chen, J., Wu, Q., Hua, Y., Chen, J., Zhang, H., & Wang, H. (2017). Potential applications of biosurfactant rhamnolipids in agriculture and biomedicine. *Applied Microbiology and Biotechnology, 101*(23), 8309–8319.

Chen, S., Qin, C., Wang, T., Chen, F., Li, X., Hou, H., et al. (2019). Study on the adsorption of dyestuffs with different properties by sludge-rice husk biochar: Adsorption capacity, isotherm, kinetic, thermodynamics and mechanism. *Journal of Molecular Liquids, 285*, 62–74, Jul 1.

Choińska-Pulit, A., Sobolczyk-Bednarek, J., & Łaba, W. (2018). Optimization of copper, lead and cadmium biosorption onto newly isolated bacterium using a Box-Behnken design. *Ecotoxicology and Environmental Safety, 149*, 275–283. Available from https://www.sciencedirect.com/science/article/pii/S0147651317308369.

Chojnacka, K. (2010). Biosorption and bioaccumulation — the prospects for practical applications. *Environment International, 36*(3), 299–307.

Christodoulou, A., & Stamatelatou, K. (2016). Overview of legislation on sewage sludge management in developed countries worldwide. *Water Science and Technology, 73*(3), 453–462.

Chu, Q., Xue, L., Singh, B. P., Yu, S., Müller, K., Wang, H., et al. (2020). Sewage sludge-derived hydrochar that inhibits ammonia volatilization, improves soil nitrogen retention and rice nitrogen utilization. *Chemosphere, 245*, 125558, Apr 1.

Commission, E. (1999). Council Directive 1999/31/EC of 26 April 1999 on the landfill of waste. *Official Journal, L, 182*, 1–19.

Cooper, P. F. (2001). Historical aspects of wastewater treatment. *Decentralised Sanitation and Reuse*.

Cyprowski, M., Stobnicka-Kupiec, A., Ławniczek-Wałczyk, A., Bakal-Kijek, A., Gołofit-Szymczak, M., & Górny, R. L. (2018). Anaerobic bacteria in wastewater treatment plant. *International Archives of Occupational and Environmental Health, 91*(5), 571.

Czernik, S., & Bridgwater, A. V. (2004). Overview of applications of biomass fast pyrolysis oil. *Energy Fuels, 18*(2), 590–598.

Dai, L., Fan, L., Liu, Y., Ruan, R., Wang, Y., Zhou, Y., et al. (2017). Production of bio-oil and biochar from soapstock via microwave-assisted co-catalytic fast pyrolysis. *Bioresource Technology, 225*, 1–8. Available from https://www.sciencedirect.com/science/article/pii/S096085241631522X.

Dai, L., Wang, Y., Liu, Y., Ruan, R., He, C., Yu, Z., et al. (2019). Integrated process of lignocellulosic biomass torrefaction and pyrolysis for upgrading bio-oil production: A state-of-the-art review. *Renewable and Sustainable Energy Reviews, 107*, 20–36, Jun 1.

Dai, Y., Zheng, H., Jiang, Z., & Xing, B. (2020). Combined effects of biochar properties and soil conditions on plant growth: A meta-analysis. *Science of The Total Environment, 713*, 136635, Apr 15.

Deivakumari, M., Sanjivkumar, M., Suganya, A. M., Prabakaran, J. R., Palavesam, A., & Immanuel, G. (2020). Studies on reclamation of crude oil polluted soil by biosurfactant producing Pseudomonas aeruginosa (DKB1). *Biocatalysis and Agricultural Biotechnology, 29*, 101773, Oct 1.

Demirbas, A., Edris, G., & Alalayah, W. M. (2017). Sludge production from municipal wastewater treatment in sewage treatment plant. *Energy Sources, Part A: Recovery, Utilization, and Environmental Effects, 39*(10), 999–1006.

Deng, L., Yuan, H., Qian, X., Lu, Q., Wang, L., Hu, H., et al. (2022). Municipal sludge-derived carbon dots-decorated, N-doped hierarchical biocarbon for the electrochemical reduction of carbon dioxide. *Resources, Conservation and Recycling, 177*, 105980, Feb 1.

Directive, E. C. (2008). Directive 2008/98/EC of the European Parliament and of the Council of 19 November 2008 on waste and repealing certain Directives. *Official Journal of the European Union L, 312*(3).

Djandja, O. S., Wang, Z. C., Wang, F., Xu, Y. P., & Duan, P. G. (2020). Pyrolysis of municipal sewage sludge for biofuel production: A review. *Ind Eng Chem Res, 59*(39), 16939–16956, Sep 30.

Dumontet, S., Scopa, A., Kerje, S., & Krovacek, K. (2001). The importance of pathogenic organisms in sewage and sewage sludge. *Journal of the Air & Waste Management Association, 51*(6), 848–860.

Dutta, A., Davies, C., & Ikumi, D. S. (2018). Performance of upflow anaerobic sludge blanket (UASB) reactor and other anaerobic reactor configurations for wastewater treatment: a comparative review and critical updates. *Journal of Water Supply: Research and Technology-Aqua, 67*(8), 858–884.

Eddy, M. &, Abu-Orf, M., Bowden, G., Burton, F. L., Pfrang, W., Stensel, H. D., et al. (2014). *Wastewater engineering: Treatment and resource recovery*. McGraw Hill Education.

El-Nahhal, I. Y., Al-Najar, H. M., & El-Nahhal, Y. (2014). Physicochemical properties of sewage sludge from Gaza. *International Journal of Geosciences, 5*(6).

El-Nahhal, Y., Nir, S., Polubesova, T., Margulies, L., & Rubin, B. (1998). Leaching, phytotoxicity, and weed control of new formulations of alachlor. *Journal of Agricultural and Food Chemistry, 46*(8), 3305–3313.

ENV E. Working Document on Sludge, 3rd Draft. DG ENV E. 2000;3.

Escudero, L. B., Maniero, M. Á., Agostini, E., & Smichowski, P. N. (2016). Biological substrates: Green alternatives in trace elemental preconcentration and speciation analysis. *TrAC Trends in Analytical Chemistry, 80*, 531–546, Jun 1.

Essandoh, M., Kunwar, B., Pittman, C. U., Mohan, D., & Mlsna, T. (2015). Sorptive removal of salicylic acid and ibuprofen from aqueous solutions using pine wood fast pyrolysis biochar. *Chemical Engineering Journal, 265*, 219–227, Apr 1.

Fan, S., Wang, Y., Wang, Z., Tang, J., Tang, J., & Li, X. (2017). Removal of methylene blue from aqueous solution by sewage sludge-derived biochar: Adsorption kinetics, equilibrium, thermodynamics and mechanism. *Journal of Environmental Chemical Engineering, 5*(1), 601–611.

Fang, C., Zhang, T., Li, P., Jiang, R. F., & Wang, Y. c. (2014). Application of Magnesium Modified Corn Biochar for Phosphorus Removal and Recovery from Swine Wastewater. *International Journal of Environmental Research and Public Health, 11*(9), 9217–9237. Available from https://www.mdpi.com/1660-4601/11/9/9217.

Favero, B. M., Favero, A. C., Taffarel, S. R., & Souza, F. S. (2020). Evaluation of the efficiency of coagulation/flocculation and Fenton process in reduction of colour, turbidity and COD of a textile effluent. *Null, 41*(12), 1580–1589.

Ferrentino, R., Merzari, F., Fiori, L., & Andreottola, G. (2020a). Coupling hydrothermal carbonization with anaerobic digestion for sewage sludge treatment: Influence of HTC liquor and hydrochar on biomethane production. *Energies, 13*(23), 6262.

Ferrentino, R., Ceccato, R., Marchetti, V., Andreottola, G., & Fiori, L. (2020b). Sewage sludge hydrochar: An option for removal of methylene blue from wastewater. *Applied Sciences, 10*(10), 3445.

Fijalkowski, K., Rorat, A., Grobelak, A., & Kacprzak, M. J. (2017). The presence of contaminations in sewage sludge — The current situation. *Journal of Environmental Management, 203*, 1126–1136, Dec 1.

Foong, S. Y., Abdul Latiff, N. S., Liew, R. K., Yek, P. N. Y., & Lam, S. S. (2020). Production of biochar for potential catalytic and energy applications via microwave vacuum pyrolysis conversion of cassava stem. *Materials Science for Energy Technologies, 3*, 728–733, Jan 1.

Forbis-Stokes, A. A., Miller, G. H., Segretain, A., Rabarison, F., Andriambololona, T., & Deshusses, M. A. (2020). Nutrient removal from human fecal sludge digestate in full-scale biological filters. *Chemosphere, 257*, 127219.

Funke, A., & Ziegler, F. (2010). Hydrothermal carbonization of biomass: A summary and discussion of chemical mechanisms for process engineering. *Biofuels, Bioproducts and Biorefining, 4*(2), 160–177.

Gabhane, J. W., Bhange, V. P., Patil, P. D., Bankar, S. T., & Kumar, S. (2020). Recent trends in biochar production methods and its application as a soil health conditioner: A review. *SN Applied Sciences*, 2(7), 1307.

Gao, N., Quan, C., Liu, B., Li, Z., Wu, C., & Li, A. (2017). Continuous pyrolysis of sewage sludge in a screw-feeding reactor: Products characterization and ecological risk assessment of heavy metals. *Energy Fuels*, 31(5), 5063–5072.

Gao, S., DeLuca, T. H., & Cleveland, C. C. (2019). Biochar additions alter phosphorus and nitrogen availability in agricultural ecosystems: A meta-analysis. *Science of the Total Environment*, 654, 463–472, Mar 1.

Gaubert, A., Clement, Y., Bonhomme, A., Burger, B., Jouan-Rimbaud Bouveresse, D., Rutledge, D., et al. (2016). Characterization of surfactant complex mixtures using Raman spectroscopy and signal extraction methods: Application to laundry detergent deformulation. *Analytica Chimica Acta*, 915, 36–48, Apr 7.

Gaur, V. K., Regar, R. K., Dhiman, N., Gautam, K., Srivastava, J. K., Patnaik, S., et al. (2019). Biosynthesis and characterization of sophorolipid biosurfactant by Candida spp.: Application as food emulsifier and antibacterial agent. *Bioresource Technology*, 285, 121314, Aug 1.

Gayman, M. (2008). A glimpse into London's early sewers. *Cleaner Magazine*.

Ghodke, P. K., Sharma, A. K., Pandey, J. K., Chen, W. H., Patel, A., & Ashokkumar, V. (2021). Pyrolysis of sewage sludge for sustainable biofuels and value-added biochar production. *Journal of Environmental Management*, 298, 113450, Nov 15.

Ghurye, G. L., Vipulanandan, C., & Willson, R. C. (1994). A practical approach to biosurfactant production using nonaseptic fermentation of mixed cultures. *Biotechnology and Bioengineering*, 44(5), 661–666.

Grobelak, A., Czerwińska, K., & Murtaś, A. (2019). General considerations on sludge disposal, industrial and municipal sludge. *Industrial and municipal sludge* (pp. 135–153). Elsevier.

Grobelak, A., Placek, A., Grosser, A., Singh, B. R., Almås, Å. R., Napora, A., et al. (2017). Effects of single sewage sludge application on soil phytoremediation. *Journal of Cleaner Production*, 155, 189–197, Jul 1.

Guan, G., Kaewpanha, M., Hao, X., & Abudula, A. (2016). Catalytic steam reforming of biomass tar: Prospects and challenges. *Renewable and Sustainable Energy Reviews*, 58, 450–461. Available from https://www.sciencedirect.com/science/article/pii/S1364032115016998.

Guarin, T. C., & Pagilla, K. R. (2021). Microbial community in biofilters for water reuse applications: A critical review. *Science of the Total Environment*, 773, 145655.

Haby, V. A., Russelle, M. P., & Skogley, E. O. (1990). Testing soils for potassium, calcium, and magnesium. *Soil Testing and Plant Analysis*, 3, 181–227.

Haq, I., Singh, A., Devki, & Kalamdhad, A. S. (2021). application of biochar for sustainable development in agriculture and environmental remediation. In: Haq, I. Kalamdhad, A. S, editors. Emerging treatment technologies for waste management [Internet]. Singapore: Springer [cited 2022 Jul 13]. p. 133–53. Available from: https://doi.org/10.1007/978-981-16-2015-7_6.

Harrison, E. Z., Oakes, S. R., Hysell, M., & Hay, A. (2006). Organic chemicals in sewage sludges. *Science of the Total Environment*, 367(2–3), 481–497.

He, C., Zhang, Z., Ge, C., Liu, W., Tang, Y., Zhuang, X., et al. (2019). Synergistic effect of hydrothermal co-carbonization of sewage sludge with fruit and agricultural wastes on hydrochar fuel quality and combustion behavior. *Waste Management*, 100, 171–181, Dec 1.

Helmke, P. A., & Sparks, D. L. (1996). Lithium, sodium, potassium, rubidium, and cesium. *Methods of Soil Analysis: Part 3 Chemical Methods*, 5, 551–574.

Ho, L., & Goethals, P. L. (2020). Municipal wastewater treatment with pond technology: Historical review and future outlook. *Ecological Engineering*, 148, 105791.

Hossain, M. Z., Bahar, M. M., Sarkar, B., Donne, S. W., Wade, P., & Bolan, N. (2021). Assessment of the fertilizer potential of biochars produced from slow pyrolysis of biosolid and animal manures. *Journal of Analytical and Applied Pyrolysis*, 155, 105043, May 1.

Hossain, N., Nizamuddin, S., Griffin, G., Selvakannan, P., Mubarak, N. M., & Mahlia, T. M. I. (2020). Synthesis and characterization of rice husk biochar via hydrothermal carbonization for wastewater treatment and biofuel production. *Scientific Reports*, 10(1), 18851.

Hu, J., Aarts, A., Shang, R., Heijman, B., & Rietveld, L. (2016). Integrating powdered activated carbon into wastewater tertiary filter for micropollutant removal. *Journal of Environmental Management*, 177, 45–52, Jul 15.

Hu, X., & Gholizadeh, M. (2020). Progress of the applications of bio-oil. *Renewable and Sustainable Energy Reviews*, 134, 110124, Dec 1.

Hu, X., Ding, Z., Zimmerman, A. R., Wang, S., & Gao, B. (2015). Batch and column sorption of arsenic onto iron-impregnated biochar synthesized through hydrolysis. *Water Research*, 68, 206–216. Available from https://www.sciencedirect.com/science/article/pii/S0043135414007040.

Hu, J., & Gao, Z. (2020). Sewage sludge in microwave oven: A sustainable synthetic approach toward carbon dots for fluorescent sensing of para-Nitrophenol. *Journal of Hazardous Materials*, 382, 121048, Jan 15.

Huang, H. J., Yang, T., Lai, F. Y., & Wu, G. Q. (2017). Co-pyrolysis of sewage sludge and sawdust/rice straw for the production of biochar. *Journal of Analytical and Applied Pyrolysis*, 125, 61–68, May 1.

Hue, N. V. (1995). Sewage sludge. *Soil Amendments and Environmental Quality*, 199–247.

Huezo, L., Vasco-Correa, J., & Shah, A. (2021). Hydrothermal carbonization of anaerobically digested sewage sludge for hydrochar production. *Bioresource Technology Reports*, 15, 100795, Sep 1.

Huang, Y. F., Chiueh, P. T., & Lo, S. L. (2016). A review on microwave pyrolysis of lignocellulosic biomass. *Sustainable Environment Research*, 26(3), 103–109.

Ibrahim, H. M. M. (2018). Characterization of biosurfactants produced by novel strains of Ochrobactrum anthropi HM-1 and Citrobacter freundii HM-2 from used engine oil-contaminated soil. *Egyptian Journal of Petroleum*, 27(1), 21–29.

Ifthikar, J., Wang, J., Wang, Q., Wang, T., Wang, H., Khan, A., et al. (2017). Highly efficient lead distribution by magnetic sewage sludge biochar: Sorption mechanisms and bench applications. *Bioresource Technology*, 238, 399–406, Aug 1.

Isahak, W., Hisham, M., Yarmo, A., & Taufiq-Yap, Y. H. (2012). A review on bio-oil production from biomass by using pyrolysis method. *Renewable and Sustainable Energy Reviews*, 16, 5910–5923, Oct 1.

Iticescu, C., Georgescu, L. P., Murariu, G., Circiumaru, A., & Timofti, M. (2018). The characteristics of sewage sludge used on agricultural lands. *AIP Conference Proceedings*, 2022(1), 020001, Nov 6.

Jia, Y., Feng, H., Shen, D., Zhou, Y., Chen, T., Wang, M., et al. (2018). High-performance microbial fuel cell anodes obtained from sewage sludge mixed with fly ash. *Journal of Hazardous Materials*, 354, 27–32, Jul 15.

Jimínez, B., Maya, C., Sánchez, E., Romero, A., Lira, L., & Barrios, J. A. (2002). Comparison of the quantity and quality of the microbiological content of sludge in countries with low and high content of pathogens. *Water Science and Technology*, 46(10), 17–24.

Johnson, B. R., Phillips, J., Bauer, T., Smith, G., Smith, G., & Sherlock, J. (2014). Startup and performance of the world's first large scale primary dissolved air floatation clarifier. *Proceedings of the Water Environment Federation*, 2014(6), 712–721.

Jung, K. W., Hwang, M. J., Jeong, T. U., & Ahn, K. H. (2015). A novel approach for preparation of modified-biochar derived from marine macroalgae: Dual purpose electro-modification for improvement of surface area and metal impregnation. *Bioresource Technology*, 191, 342–345.

Jung, K. W., Hwang, M. J., Jeong, T. U., & Ahn, K. H. (2015). A novel approach for preparation of modified-biochar derived from marine macro-algae: Dual purpose electro-modification for improvement of surface area and metal impregnation. *Bioresource Technology*, 191, 342–345.

Kalderis, D., Kayan, B., Akay, S., Kulaksız, E., & Gözmen, B. (2017). Adsorption of 2,4-dichlorophenol on paper sludge/wheat husk biochar: Process optimization and comparison with biochars prepared from wood chips, sewage sludge and hog fuel/demolition waste. *Journal of Environmental Chemical Engineering*, 5(3), 2222–2231.

Kashani, H. M., Madrakian, T., Afkhami, A., Mahjoubi, F., & Moosavi, M. A. (2019). Bottom-up and green-synthesis route of amino functionalized graphene quantum dot as a novel biocompatible and label-free fluorescence probe for in vitro cellular imaging of human ACHN cell lines. *Materials Science and Engineering: B*, 251, 114452, Dec 1.

Khairol Anuar, N. K., Tan, H. L., Lim, Y. P., So'aib, M. S., & Abu Bakar, N. F. (2021). A review on multifunctional carbon-dots synthesized from biomass waste: Design/ fabrication, characterization and applications. *Frontiers in Energy Research*, 9, [Internet]. Available from https://www.frontiersin.org/article/10.3389/fenrg.2021.626549.

Khanmohammadi, Z., Afyuni, M., & Mosaddeghi, M. R. (2015). Effect of pyrolysis temperature on chemical and physical properties of sewage sludge biochar. *Waste Management & Research*, 33(3), 275–283.

Kim, D., Lee, K., & Park, K. Y. (2014). Hydrothermal carbonization of anaerobically digested sludge for solid fuel production and energy recovery. *Fuel*, 130, 120–125, Aug 15.

Kumar, M., Ghosh, P., Khosla, K., & Thakur, I. S. (2018). Recovery of polyhydroxyalkanoates from municipal secondary wastewater sludge. *Bioresource Technology*, 255, 111–115, May 1.

Kumar, M., Morya, R., Gnansounou, E., Larroche, C., & Thakur, I. S. (2017). Characterization of carbon dioxide concentrating chemolithotrophic bacterium Serratia sp. ISTD04 for production of biodiesel. *Bioresource Technology*, 243, 893–897, Nov 1.

Kumari, S., Regar, R. K., & Manickam, N. (2018). Improved polycyclic aromatic hydrocarbon degradation in a crude oil by individual and a consortium of bacteria. *Bioresource Technology*, 254, 174–179, Apr 1.

Kuyukina, M. S., & Ivshina, I. B. (2019). Production of trehalolipid biosurfactants by Rhodococcus, [Internet]In H. M. Alvarez (Ed.), *Biology of Rhodococcus* (pp. 271–298). Cham: Springer International Publishing. Available from https://doi.org/10.1007/978-3-030-11461-9_10.

Lai, C. C., Huang, Y. C., Wei, Y. H., & Chang, J. S. (2009). Biosurfactant-enhanced removal of total petroleum hydrocarbons from contaminated soil. *Journal of Hazardous Materials*, 167(1), 609–614.

Lai, W. Y., Lai, C. M., Ke, G. R., Chung, R. S., Chen, C. T., & Cheng C. H., et al. (2023). The effects of woodchip biochar application on crop yield, carbon sequestration and greenhouse gas emissions from soils planted with rice or leaf beet. *Journal of the Taiwan Institute of Chemical Engineers*, 44(6), 1039–1044.

Lai, W. Y., Lai, C. M., Ke, G. R., Chung, R. S., Chen, C. T., Cheng, C. H., et al. (2023). The effects of woodchip biochar application on crop yield, carbon sequestration and greenhouse gas emissions from soils planted with rice or leaf beet. *Journal of the Taiwan Institute of Chemical Engineers*, 44(6), 1039–1044.

LeBlanc, R. J., Matthews, P., & Richard, R. P. (2009). Global atlas of excreta, wastewater sludge, and biosolids management: moving forward the sustainable and welcome uses of a global resource. *Un-habitat*.

Lee, D. W., Lee, H., Kwon, B. O., Khim, J. S., Yim, U. H., Kim, B. S., et al. (2018). Biosurfactant-assisted bioremediation of crude oil by indigenous bacteria isolated from Taean beach sediment. *Environmental Pollution*, 241, 254–264, Oct 1.

Lehmann, J., & Joseph, S. (2015). Biochar for environmental management: Science, technology and implementation (p. 977) Routledge.

Lehmphul, K. (2015). Sewage sludge management in Germany [Internet]. Umweltbundesamt [cited 2022 May 24]. Available from: https://www.umweltbundesamt.de/en/publikationen/sewage-sludge-management-in-germany.

Leng, L., Yuan, X., Huang, H., Shao, J., Wang, H., Chen, X., et al. (2015). Bio-char derived from sewage sludge by liquefaction: Characterization and application for dye adsorption. *Applied Surface Science*, 346, 223–231, Aug 15.

Letey, J. (1969). Measurement of contact angle, water drop penetration time, and critical surface tension. 1969;

Lewis, D. L., & Gattie, D. K. (2002). Peer reviewed: pathogen risks from applying sewage sludge to land. *Environmental Science & Technology*, 36(13), 286A–293A.

Li, J., Li, B., Huang, H., Lv, X., Zhao, N., Guo, G., et al. (2019). Removal of phosphate from aqueous solution by dolomite-modified biochar derived from urban dewatered sewage sludge. *Science of the Total Environment*, 687, 460–469, Oct 15.

Li, J., Li, B., Huang, H., Zhao, N., Zhang, M., & Cao, L. (2020). Investigation into lanthanum-coated biochar obtained from urban dewatered sewage sludge for enhanced phosphate adsorption. *Science of the Total Environment*, 714, 136839, Apr 20.

Liang, Q., Yamashita, T., Koike, K., Matsuura, N., Honda, R., Hara-Yamamura, H., et al. (2020). A bioelectrochemical-system-based trickling filter reactor for wastewater treatment. *Bioresource Technology*, 315, 123798.

Liew, R. K., Nam, W. L., Chong, M. Y., Phang, X. Y., Su, M. H., Yek, P. N. Y., et al. (2018). Oil palm waste: An abundant and promising feedstock for microwave pyrolysis conversion into good quality biochar with potential multi-applications. *Process Safety and Environmental Protection*, 115, 57–69, Apr 1.

Lin, X., Xiong, M., Zhang, J., He, C., Ma, X., Zhang, H., et al. (2021b). Carbon dots based on natural resources: Synthesis and applications in sensors. *Microchemical Journal*, 160, 105604, Jan 1.

Lin, Y. L., Zheng, N. Y., & Hsu, C. H. (2021a). Torrefaction of fruit peel waste to produce environmentally friendly biofuel. *Journal of Cleaner Production*, 284, 124676, Feb 15.

Lin, Y. L., Zheng, N. Y., & Lin, C. S. (2021a). Repurposing Washingtonia filifera petiole and Sterculia foetida follicle waste biomass for renewable energy through torrefaction. *Energy, 223*, 120101, May 15.

Lindsay, W. L., & Norvell, Wa (1978). Development of a DTPA soil test for zinc, iron, manganese, and copper. *Soil Science Society of America Journal, 42*(3), 421–428.

Liu, F., Li, J., & Zhang, X. L. (2019). Bioplastic production from wastewater sludge and application. *IOP Conference Series: Earth and Environmental Science, 344*(1), 012071.

Liu, S., Xie, Q., Zhang, B., Cheng, Y., Liu, Y., Chen, P., et al. (2016). Fast microwave-assisted catalytic co-pyrolysis of corn stover and scum for bio-oil production with CaO and HZSM-5 as the catalyst. *Bioresource Technology, 204*, 164–170.

Lofrano, G., & Brown, J. (2010). Wastewater management through the ages: A history of mankind. *Science of the Total Environment, 408*(22), 5254–5264.

Lu, H., Zhang, G., Zheng, Z., Meng, F., Du, T., & He, S. (2019). Bio-conversion of photosynthetic bacteria from non-toxic wastewater to realize wastewater treatment and bioresource recovery: A review. *Bioresource Technology, 278*, 383–399.

Lu, X., Ma, X., Qin, Z., Chen, X., Chen, L., & Tian, Y. (2021). Co-hydrothermal carbonization of sewage sludge and polyvinyl chloride: Hydrochar properties and fate of chlorine and heavy metals. *Journal of Environmental Chemical Engineering, 9*(5), 106143.

Ma, H., Gao, X., Chen, Y., Zhu, J., & Liu, T. (2021). Fe (II) enhances simultaneous phosphorus removal and denitrification in heterotrophic denitrification by chemical precipitation and stimulating denitrifiers activity. *Environmental Pollution, 287*, 117668.

Ma, J., Chen, M., Yang, T., Liu, Z., Jiao, W., Li, D., et al. (2019). Gasification performance of the hydrochar derived from co-hydrothermal carbonization of sewage sludge and sawdust. *Energy, 173*, 732–739, Apr 15.

Ma, Y., Yin, Y., & Liu, Y. (2017). New insights into co-digestion of activated sludge and food waste: Biogas versus biofertilizer. *Bioresource Technology, 241*, 448–453, Oct 1.

Maia, M., Capão, A., & Procópio, L. (2019). Biosurfactant produced by oil-degrading Pseudomonas putida AM-b1 strain with potential for microbial enhanced oil recovery. Bioremediation. *Bioremediation Journal, 23*(4), 302–310.

Malhotra, M., & Garg, A. (2020). Hydrothermal carbonization of centrifuged sewage sludge: Determination of resource recovery from liquid fraction and thermal behaviour of hydrochar. *Waste Management, 117*, 114–123, Nov 1.

Mantovi, P., Baldoni, G., & Toderi, G. (2005). Reuse of liquid, dewatered, and composted sewage sludge on agricultural land: effects of long-term application on soil and crop. *Water Research, 39*(2), 289–296.

Mao, X., Jiang, R., Xiao, W., & Yu, J. (2015). Use of surfactants for the remediation of contaminated soils: A review. *Journal of Hazardous Materials, 285*, 419–435, Mar 21.

Marsh, H., & Reinoso, F. R. (2006). *Activated carbon* (p. 555). Elsevier.

Mateo-Sagasta, J., Raschid-Sally, L., & Thebo, A. (2015). *Global wastewater and sludge production, treatment and use. Wastewater* (pp. 15–38). Springer.

McGrath, S. P., Chang, A. C., Page, A. L., & Witter, E. (1994). Land application of sewage sludge: scientific perspectives of heavy metal loading limits in Europe and the United States. *Environmental Reviews, 2*(1), 108–118.

Md Said, M. S., Azni, A. A., Wan Ab Karim Ghani, W. A., Idris, A., Ja'afar, M. F. Z., & Mohd Salleh, M. A. (2022). Production of biochar from microwave pyrolysis of empty fruit bunch in an alumina susceptor. *Energy, 240*, 122710, Feb 1.

Méndez, A., Cárdenas-Aguiar, E., Paz-Ferreiro, J., Plaza, C., & Gascó, G. (2016). The effect of sewage sludge biochar on peat-based growing media. *Biological Agriculture & Horticulture* [Internet]. Available from https://www.tandfonline.com/doi/abs/10.1080/01448765.2016.1185645.

Méndez, A., Paz-Ferreiro, J., Gil, E., & Gascó, G. (2015). The effect of paper sludge and biochar addition on brown peat and coir based growing media properties. *Scientia Horticulturae, 193*, 225–230, Sep 22.

Méndez, A., Paz-Ferreiro, J., Gil, E., & Gascó, G. (2015). The effect of paper sludge and biochar addition on brown peat and coir based growing media properties. *Scientia Horticulturae, 193*, 225–230.

Menzel, T., Neubauer, P., & Junne, S. (2020). Role of microbial hydrolysis in anaerobic digestion. *Energies, 13*(21), 5555.

Mesbaiah, F. Z., Eddouaouda, K., Badis, A., Chebbi, A., Hentati, D., Sayadi, S., et al. (2016). Preliminary characterization of biosurfactant produced by a PAH-degrading Paenibacillus sp. under thermophilic conditions. *Environmental Science and Pollution Research, 23*(14), 14221–14230.

Mininni, G., & Dentel, S. (2013). Highlights of current legislation on sludge and bio-waste in EU member states and in the United States. In: Conférence Internationale '"Gestion Innovante des Boues d'Epuration al'Echelle Europé enne,"'Charleroi Espace Meeting Européen, Charleroi, Belgium.

Mtshali, J. S., Tiruneh, A. T., & Fadiran, A. O. (2014). Characterization of sewage sludge generated from wastewater treatment plants in Swaziland in relation to agricultural uses. *Resources and Environment, 4*(4), 190–199.

Mujahid, R., Riaz, A., Insyani, R., & Kim, J. (2020). A centrifugation-first approach for recovering high-yield bio-oil with high calorific values in biomass liquefaction: A case study of sewage sludge. *Fuel, 262*, 116628, Feb 15.

Mukheibir, P., Mitchell, C. A., McKibbin, J. L., Komatsu, R., Ryan, H., & Fitzgerald, C. (2012). Adaptive planning for resilient urban water systems under an uncertain future. In: *Australian Water Association Convention-Ozwater*. Australian Water Association (AWA).

Muñoz-Alegría, J. A., Muñoz-España, E., & Flórez-Marulanda, J. F. (2021). Dissolved air flotation: A review from the perspective of system parameters and uses in wastewater treatment. *Technology, 24*(52), e2111.

Nascimento, A. L., Souza, A. J., Andrade, P. A. M., Andreote, F. D., Coscione, A. R., Oliveira, F. C., et al. (2018). Sewage sludge microbial structures and relations to their sources, treatments, and chemical attributes. *Frontiers in Microbiology, 9*, [Internet]. Available from https://www.frontiersin.org/article/10.3389/fmicb.2018.01462.

Ni, B. J., Huang, Q. S., Wang, C., Ni, T. Y., Sun, J., & Wei, W. (2019). Competitive adsorption of heavy metals in aqueous solution onto biochar derived from anaerobically digested sludge. *Chemosphere, 219*, 351–357, Mar 1.

Nogueira, I. B., Rodríguez, D. M., da Silva Andradade, R. F., Lins, A. B., Bione, A. P., da Silva, I. G. S., et al. (2020). Bioconversion of agroindustrial waste in the production of bioemulsifier by Stenotrophomonas maltophilia UCP 1601 and application in bioremediation process. *International Journal of Chemical Engineering, 2020*, e9434059, Jan 31.

Silva, W. O., Nagar, B., Soutrenon, M., & Girault, H. H. (2022). Banana split: biomass splitting with flash light irradiation. *Chemical Science, 13*(6), 1774–1779.

Oakley, S. (2018). Preliminary treatment and primary sedimentation. Water and Sanitation for the 21st Century: Health and Microbiological Aspects of Excreta and Wastewater Management (Global Water Pathogen Project).

Orhon, D., Babuna, F. G., & Karahan, O. (2009). *Industrial wastewater treatment by activated sludge*. IWA Publishing.

Orhon, D., & Sözen, S. (2020). Reshaping the activated sludge process: has the time come or passed? *Journal of Chemical Technology & Biotechnology, 95*(6), 1632–1639.

Pagnanelli, F., Mainelli, S., Bornoroni, L., Dionisi, D., & Toro, L. (2009). Mechanisms of heavy-metal removal by activated sludge. *Chemosphere, 75*(8), 1028–1034.

Paliya, S., Mandpe, A., Kumar, S., & Kumar, M. S. (2019). Enhanced nodulation and higher germination using sludge ash as a carrier for bio-fertilizer production. *Journal of Environmental Management, 250*, 109523, Nov 15.

Pansu, M., Gautheyrou, J. (2006). Handbook of Soil Analysis. Mineralogical, Organic and Inorganic Methods. In: Handbook of Soil Analysis: Mineralogical, Organic and Inorganic Methods.

Parida, A., Capoor, M. R., & Bhowmik, K. T. (2019). Knowledge, attitude, and practices of Bio-medical Waste Management rules, 2016; Bio-medical Waste Management (amendment) rules, 2018; and Solid Waste Rules, 2016, among health-care workers in a tertiary care setup. *Journal of Laboratory Physicians, 11*(4), 292–296.

Pariyar, P., Kumari, K., Jain, M. K., & Jadhao, P. S. (2020). Evaluation of change in biochar properties derived from different feedstock and pyrolysis temperature for environmental and agricultural application. *Science of the Total Environment, 713*, 136433, Apr 15.

Parshetti, G. K., Liu, Z., Jain, A., Srinivasan, M. P., & Balasubramanian, R. (2013). Hydrothermal carbonization of sewage sludge for energy production with coal. *Fuel, 111*, 201–210, Sep 1.

Partners, G., Macero, E., & Guyer, J. P. (2011). An Introduction to Preliminary Wastewater Treatment, 23.

Partovi, M., Lotfabad, T. B., Roostaazad, R., Bahmaei, M., & Tayyebi, S. (2013). Management of soybean oil refinery wastes through recycling them for producing biosurfactant using Pseudomonas aeruginosa MR01. *World Journal of Microbiology & Biotechnology, 29*(6), 1039–1047.

Pasalari, H., Gholami, M., Rezaee, A., Esrafili, A., & Farzadkia, M. (2021). Perspectives on microbial community in anaerobic digestion with emphasis on environmental parameters: A systematic review. *Chemosphere, 270*, 128618.

Patel, A. B., Shaikh, S., Jain, K. R., Desai, C., & Madamwar, D. (2020). Polycyclic aromatic hydrocarbons: Sources, toxicity, and remediation approaches. *Frontiers in Microbiology, 11*, 562813, Nov 5.

Pathak, A., Dastidar, M. G., & Sreekrishnan, T. R. (2009). Bioleaching of heavy metals from sewage sludge: A review. *Journal of Environmental Management, 90*(8), 2343–2353.

Pawar, A., & Panwar, N. L. (2022). A comparative study on morphology, composition, kinetics, thermal behaviour and thermodynamic parameters of Prosopis Juliflora and its biochar derived from vacuum pyrolysis. *Bioresource Technology Reports, 18*, 101053, Jun 1.

Pei, G., Sun, C., Zhu, Y., Shi, W., & Li, H. (2018). Biosurfactant-enhanced removal of o,p-dichlorobenzene from contaminated soil. *Environmental Science and Pollution Research, 25*(1), 18–26.

Peng, G., Tian, G., Liu, J., Bao, Q., & Zang, L. (2011). Removal of heavy metals from sewage sludge with a combination of bioleaching and electrokinetic remediation technology. *Desalination, 271*(1), 100–104.

Pepper, I. L., Brooks, J. P., & Gerba, C. P. (2006). Pathogens in biosolids. *Advances in Agronomy, 90*, 1–41.

Pi, Y., Chen, B., Bao, M., Fan, F., Cai, Q., Ze, L., et al. (2017). Microbial degradation of four crude oil by biosurfactant producing strain Rhodococcus sp. *Bioresource Technology, 232*, 263–269, May 1.

Piersa, P., Szufa, S., Czerwińska, J., & Ünyay, H. (2021). Adrian Ł, Wielgosinski G, et al. Pine wood and sewage sludge torrefaction process for production renewable solid biofuels and biochar as carbon carrier for fertilizers. *Energies, 14*(23), 8176.

Poblete, I. B. S., Araujo, O. de Q. F., & de Medeiros, J. L. (2022). Sewage-water treatment and sewage-sludge management with power production as bioenergy with carbon capture system: A review. *Processes, 10*(4), 788.

Prins, M. J., Ptasinski, K. J., & Janssen, F. J. J. G. (2006). Torrefaction of wood: Part 2. Analysis of products. *Journal of Analytical and Applied Pyrolysis, 77*(1), 35–40.

Qi, J., Zhu, F., Wei, X., Zhao, L., Xiong, Y., Wu, X., et al. (2016). Comparison of biodiesel production from sewage sludge obtained from the A2/O and MBR processes by in situ transesterification. *Waste Management, 49*, 212–220, Mar 1.

Qin, H., Hu, T., Zhai, Y., Lu, N., & Aliyeva, J. (2020). The improved methods of heavy metals removal by biosorbents: A review. *Environmental Pollution, 258*, 113777, Mar 1.

Qu, J., Luo, C., Zhang, Q., Cong, Q., & Yuan, X. (2013). Easy synthesis of graphene sheets from alfalfa plants by treatment of nitric acid. *Materials Science and Engineering: B, 178*(6), 380–382, Apr 1.

Racek, J., Sevcik, J., Chorazy, T., Kucerik, J., & Hlavinek, P. (2020). Biochar – Recovery material from pyrolysis of sewage sludge: A review. *Waste Biomass Valor, 11*(7), 3677–3709.

Rajta, A., Bhatia, R., Setia, H., & Pathania, P. (2020). Role of heterotrophic aerobic denitrifying bacteria in nitrate removal from wastewater. *Journal of Applied Microbiology, 128*(5), 1261–1278.

Rechcigl, J. E. (1995). *Soil amendments and environmental quality* (2). CRC Press.

Rehman, A., Ayub, N., Naz, I., Perveen, I., & Ahmed, S. (2020). Effects of hydraulic retention time (HRT) on the performance of a pilot-scale trickling filter system tteating low-strength domestic wastewater. *Polish Journal of Environmental Studies, 29*(1).

Rehman, R., Ali, M. I., Ali, N., Badshah, M., Iqbal, M., Jamal, A., et al. (2021). Crude oil biodegradation potential of biosurfactant-producing Pseudomonas aeruginosa and Meyerozyma sp. *Journal of Hazardous Materials, 418*, 126276, Sep 15.

Reilly, J. (1925). The technology of wood distillation: With special reference to the methods of obtaining the intermediate and finished products from the primary distillate. *Nature, 116*, 779–780 Google Scholar [Internet]. [cited 2022 Jul 14]. Available from: https://scholar.google.com/scholar_lookup?title = The%20technology%20of%20wood%20distillation%3A%20with%20special%20reference%20to%20the%20methods%20of%20obtaining%20the%20intermediate%20and%20finished%20products%20from%20the%20primary%20distillate&journal = Nature&doi = 10.1038%2F116779a0&volume = 116&pages = 779-780&publication_year = 1925&author = Reilly%2CJ.

Rodríguez-Padrón, D., Algarra, M., Tarelho, L. A. C., Frade, J., Franco, A., de Miguel, G., et al. (2018). Catalyzed Microwave-Assisted Preparation of Carbon Quantum Dots from Lignocellulosic Residues. *ACS Sustainable Chemistry & Engineering, 6*(6), 7200–7205.

Rogers, H. R. (1996). Sources, behaviour and fate of organic contaminants during sewage treatment and in sewage sludges. *Science of the Total Environment*, 185(1–3), 3–26.

Romdhana, M. H., Lecomte, D., Ladevie, B., & Sablayrolles, C. (2009). Monitoring of pathogenic microorganisms contamination during heat drying process of sewage sludge. *Process Safety and Environmental Protection*, 87(6), 377–386.

Sahoo, S. S., Vijay, V. K., Chandra, R., & Kumar, H. (2021). Production and characterization of biochar produced from slow pyrolysis of pigeon pea stalk and bamboo. *Cleaner Engineering and Technology*, 3, 100101, Jul 1.

Samolada, M. C., & Zabaniotou, A. A. (2014). Comparative assessment of municipal sewage sludge incineration, gasification and pyrolysis for a sustainable sludge-to-energy management in Greece. *Waste Management*, 34(2), 411–420.

Santos, L. E. R., Meili, L., Soletti, J. I., de Carvalho, S. H. V., Ribeiro, L. M. O., Duarte, J. L. S., et al. (2020). Impact of temperature on vacuum pyrolysis of Syagrus coronata for biochar production. *Journal of Material Cycles and Waste Management*, 22(3), 878–886.

Satpute, S. K., Zinjarde, S. S., & Banat, I. M. (2018). Recent updates on biosurfactant/s in Food industry. *Microbial cell factories* (pp. 1–20). Taylor & Francis.

Sattar, A., Leeke, G. A., Hornung, A., & Wood, J. (2014). Steam gasification of rapeseed, wood, sewage sludge and miscanthus biochars for the production of a hydrogen-rich syngas. *Biomass and Bioenergy*, 69, 276–286, Oct 1.

Sciortino A., Cannizzo A., Messina F. (2018). Carbon nanodots: A review—From the current understanding of the fundamental photophysics to the full control of the optical response. C. Dec;4(4):67.

Seeger, H. (1999). The history of German waste water treatment. *European Water Management*, 2, 51–56.

Sendão, Ricardo, Yuso, M. del V. M. de, Algarra, M., Esteves da Silva, J. C. G., & Pinto da Silva, L. (2020). Comparative life cycle assessment of bottom-up synthesis routes for carbon dots derived from citric acid and urea. *Journal of Cleaner Production*, 254, 120080, May 1.

Sewage sludge [Internet]. [cited 2022 May 24]. Available from: https://ec.europa.eu/environment/topics/waste-and-recycling/sewage-sludge_en.

Sharma, A., & Das, J. (2019). Small molecules derived carbon dots: synthesis and applications in sensing, catalysis, imaging, and biomedicine. *Journal of Nanobiotechnology*, 17(1), 92.

Sharma, S., Verma, R., & Pandey, L. M. (2019). Crude oil degradation and biosurfactant production abilities of isolated Agrobacterium fabrum SLAJ731. Biocatalysis and Agricultural. *Biotechnology*, 21, 101322, Sep 1.

Shaul, O. (2002). Magnesium transport and function in plants: the tip of the iceberg. *Biometals*, 15(3), 307–321.

Shen, T., Tang, Y., Lu, X. Y., & Meng, Z. (2018). Mechanisms of copper stabilization by mineral constituents in sewage sludge biochar. *Journal of Cleaner Production*, 193, 185–193, Aug 20.

Sher, F., Hanif, K., Rafey, A., Khalid, U., Zafar, A., Ameen, M., et al. (2021). Removal of micropollutants from municipal wastewater using different types of activated carbons. *Journal of Environmental Management*, 278, 111302, Jan 15.

Shifrin, N. S. (2005). Pollution management in twentieth century. *Journal of Environmental Engineering ASCEnt*, 131, 676–691.

Silva, E. J., Rocha e Silva, N. M. P., Rufino, R. D., Luna, J. M., Silva, R. O., & Sarubbo, L. A. (2014). Characterization of a biosurfactant produced by Pseudomonas cepacia CCT6659 in the presence of industrial wastes and its application in the biodegradation of hydrophobic compounds in soil. *Colloids and Surfaces B: Biointerfaces*, 117, 36–41, May 1.

Singh, B., Camps-Arbestain, M., Lehmann, J., & CSIRO (Australia). (2017a). Biochar: a guide to analytical methods [Internet] [cited 2022 Jul 13]. Available from: https://search.ebscohost.com/login.aspx?direct = true&scope = site&db = nlebk&db = nlabk&AN = 1488495.

Singh, B., Shen, Q., & Camps Arbestain, M. (2017b). Chapter 3. Biochar pH, electrical conductivity and liming potential. p. 23–38.

Singh, R. P., & Agrawal, M. (2008). Potential benefits and risks of land application of sewage sludge. *Waste Management*, 28(2), 347–358.

Singh, V., Phuleria, H. C., & Chandel, M. K. (2020). Estimation of energy recovery potential of sewage sludge in India: Waste to watt approach. *Journal of Cleaner Production*, 276, 122538.

Sivasankar, P., & Suresh Kumar, G. (2017). Influence of pH on dynamics of microbial enhanced oil recovery processes using biosurfactant producing Pseudomonas putida: Mathematical modelling and numerical simulation. *Bioresource Technology*, 224, 498–508, Jan 1.

Sori, E. (2001). *La città ei rifiuti: ecologia urbana dal Medioevo al primo Novecento* (537). Il mulino.

Sörme, L., & Lagerkvist, R. (2002). Sources of heavy metals in urban wastewater in Stockholm. *Science of the Total Environment*, 298(1), 131–145.

Sowmya Dhanalakshmi, C., Kaliappan, S., Mohammed Ali, H., Sekar, S., Depoures, M. V., Patil, P. P., et al. (2022). Flash pyrolysis experiment on Albizia odoratissima biomass under different operating conditions: A comparative study on bio-oil, biochar, and noncondensable gas products. *Journal of Chemistry*, 2022, e9084029, Jul 9.

Spinosa, L. (2007). Wastewater sludge: A global overview of the current status and future 414. *Prospects*.

Spokas, K. A., Cantrell, K. B., Novak, J. M., Archer, D. W., Ippolito, J. A., Collins, H. P., et al. (2012). Biochar: A synthesis of its agronomic impact beyond carbon sequestration. *Journal of Environmental Quality* - Wiley Online Library [Internet]. [cited 2022 Jul 13]. Available from: https://acsess.onlinelibrary.wiley.com/doi/abs/10.2134/jeq2011.0069.

Straub, T. M., Pepper, I. L., & Gerba, C. P. (1993). Hazards from pathogenic microorganisms in land-disposed sewage sludge, [Internet] In G. W. Ware (Ed.), *Reviews of environmental contamination and toxicology* (pp. 55–91). New York, NY: Springer New York. Available from http://link.springer.com/10.1007/978-1-4684-7065-9_3.

Sun, X., & Lei, Y. (2017). Fluorescent carbon dots and their sensing applications. *TrAC Trends in Analytical Chemistry*, 89, 163–180, Apr 1.

Svanström, M., Fröling, M., Modell, M., Peters, W. A., & Tester, J. (2004). Environmental assessment of supercritical water oxidation of sewage sludge. *Resources, Conservation and Recycling*, 41(4), 321–338.

Świątczak, P., & Cydzik-Kwiatkowska, A. (2018). Performance and microbial characteristics of biomass in a full-scale aerobic granular sludge wastewater treatment plant. *Environmental Science and Pollution Research*, 25(2), 1655–1669.

Świerczek, L., Cieślik, B. M., & Konieczka, P. (2018). The potential of raw sewage sludge in construction industry—a review. *Journal of Cleaner Production*, 200, 342–356.

Tang, L., Yu, J., Pang, Y., Zeng, G., Deng, Y., Wang, J., et al. (2018a). Sustainable efficient adsorbent: Alkali-acid modified magnetic biochar derived from sewage sludge for aqueous organic contaminant removal. *Chemical Engineering Journal*, 336, 160–169, Mar 15.

Tang, J., He, J., Xin, X., Hu, H., & Liu, T. (2018b). Biosurfactants enhanced heavy metals removal from sludge in the electrokinetic treatment. *Chemical Engineering Journal, 334*, 2579–2592, Feb 15.

Tang, S., Tian, S., Zheng, C., & Zhang, Z. (2017). Effect of calcium hydroxide on the pyrolysis behavior of sewage sludge: reaction characteristics and kinetics. *Energy Fuels, 31*(5), 5079–5087 [Internet]. [cited 2022 May 24]. Available from https://pubs.acs.org/doi/full/10.1021/acs.energyfuels.6b03256.

Tchobanoglus, G., Burton, F., & Stensel, H. D. (2003). Wastewater engineering: treatment and reuse. *American Water Works Association Journal, 95*(5), 201.

The effects of woodchip biochar application on crop yield, carbon sequestration and greenhouse gas emissions from soils planted with rice or leaf beet - ScienceDirect [Internet]. [cited 2022 Jul 14]. Available from: https://www.sciencedirect.com/science/article/pii/S1876107013001740.

Tölle-Kastenbein, R. (1993). Archeologia dell'acqua. *La cultura idraulica nel mondo classico*, 51.

Topare N. S., Attar S .J., & Manfe M. M. (2011). Sewage/wastewater treatment technologies: A review.8.

Trinh, T., Jensen, P., Sarossy, Z., Dam-Johansen, K., Knudsen, N., Sørensen, H., et al. (2013). Fast pyrolysis of lignin using a pyrolysis centrifuge reactor. *Energy & Fuels, 27*, 3802–3810, Jun 21.

Tripathi, V., Gaur, V. K., Dhiman, N., Gautam, K., & Manickam, N. (2020). Characterization and properties of the biosurfactant produced by PAH-degrading bacteria isolated from contaminated oily sludge environment. *Environ Sci Pollut Res, 27*(22), 27268–27278.

Tripathi, M., Sahu, J. N., & Ganesan, P. (2016). Effect of process parameters on production of biochar from biomass waste through pyrolysis: A review. *Renewable and Sustainable Energy Reviews, 55*, 467–481 Google Scholar [Internet]. [cited 2022 Jul 14]. Available from: https://scholar.google.com/scholar_lookup?title = Effect%20of%20process%20parameters%20on%20production%20of%20biochar%20from%20biomass%20waste%20through%20pyrolysis%3A%20a%20review&journal = Renew%20Sustain%20Energy%20Rev&doi = 10.1016%2FJ.RSER.2015.10.122&volume = 55&pages = 467-481&publication_year = 2016&author = Tripathi%2CM&author = Sahu%2CJN&author = Ganesan%2CP.

Turner, J., Sverdrup, G., Mann, M. K., Maness, P. C., Kroposki, B., Ghirardi, M., Evans, R. J., & Blake, D. (2008). Renewable hydrogen production. International Journal of Energy Research, Google Scholar [Internet]. [cited 2022 Jul 14]. Available from: https://scholar.google.com/scholar_lookup?title = Renewable%20hydrogen%20production&journal = Int%20J%20Energy%20Res&doi = 10.1002%2Fer.1372&volume = 32&pages = 379-407&publication_year = 2008&author = Turner%2CJ&author = Sverdrup%2CG&author = Mann%2CMK&author = Maness%2CP-C&author = Kroposki%2CB&author = Ghirardi%2CM&author = Evans%2CRJ&author = Blake%2CD.

Vajihinejad, V., Gumfekar, S. P., Bazoubandi, B., Rostami Najafabadi, Z., & Soares, J. B. P. (2019). Water Soluble Polymer Flocculants: Synthesis, Characterization, and Performance Assessment. *Macromolecular Materials and Engineering, 304*(2), 1800526.

de Vasconcelos, A. C. F. (2020). Biochar effects on amelioration of adverse salinity effects in soils [Internet]. *Applications of Biochar for Environmental Safety*. IntechOpen [cited 2022 Jul 13]. Available from: https://www.intechopen.com/chapters/undefined/state.item.id.

Venkatesan, A. K., Done, H. Y., & Halden, R. U. (2015). United States National Sewage Sludge Repository at Arizona State University—a new resource and research tool for environmental scientists, engineers, and epidemiologists. *Environmental Science and Pollution Research, 22*(3), 1577–1586.

Venkateswar Reddy, M., & Venkata Mohan, S. (2012). Effect of substrate load and nutrients concentration on the polyhydroxyalkanoates (PHA) production using mixed consortia through wastewater treatment. *Bioresource Technology, 114*, 573–582, Jun 1.

Verma, M., Godbout, S., Brar, S. K., Solomatnikova, O., Lemay, S. P., & Larouche, J. P., (2012). Biofuels Production from Biomass by Thermochemical Conversion Technologies. *Int. J. Chem. Eng. 2012*, e542426. Available from: https://doi.org/10.1155/2012/542426.

Vilas-Boas, A. C. M., Tarelho, L. A. C., Kamali, M., Hauschild, T., Pio, D. T., Jahanianfard, D., et al. (2021). Biochar from slow pyrolysis of biological sludge from wastewater treatment: characteristics and effect as soil amendment. *Biofuels, Bioproducts and Biorefining, 15*(4), 1054–1072.

Virmond, E., Rocha, J. D., Moreira, R. F. P. M., & José, H. J. (2013). Valorization of agroindustrial solid residues and residues from biofuel production chains by thermochemical conversion: a review, citing Brazil as a case study. *Brazilian Journal of Chemical Engineering, 30*, 197–230, Jun.

Vuorinen, H. S., Juuti, P. S., & Katko, T. S. (2007). History of water and health from ancient civilizations to modern times. *Water Supply, 7*(1), 49–57, Mar 1.

Wan Mahari, W. A., Nam, W. L., Sonne, C., Peng, W., Phang, X. Y., Liew, R. K., et al. (2020). Applying microwave vacuum pyrolysis to design moisture retention and pH neutralizing palm kernel shell biochar for mushroom production. *Bioresource Technology, 312*, 123572, Sep 1.

Wang, N. X., Lu, X. Y., Tsang, Y. F., Mao, Y., Tsang, C. W., & Yueng, V. A. (2019). A comprehensive review of anaerobic digestion of organic solid wastes in relation to microbial community and enhancement process: A comprehensive review of anaerobic digestion of organic solid wastes. *Journal of the Science of Food and Agriculture, 99*(2), 507–516.

Wang, R., Zhang, H., Lian, L., Wang, X., Zhu, B., & Lou, D. (2020). Flocculant Containing Silicon, Aluminum, and Starch for Sewage Treatment. *Journal of Chemical Engineering of Japan, 53*(10), 592–598.

Wang, X., Chang, V. W. C., Li, Z., Chen, Z., & Wang, Y. (2021). Co-pyrolysis of sewage sludge and organic fractions of municipal solid waste: Synergistic effects on biochar properties and the environmental risk of heavy metals. *Journal of Hazardous Materials, 412*, 125200, Jun 15.

Wang, Y. jie, Yu, Y., Huang, H. jun, Yu, C. long, Fang, H. sun, Zhou, C. huo, et al. (2022). Efficient conversion of sewage sludge into hydrochar by microwave-assisted hydrothermal carbonization. *Science of the Total Environment, 803*, 149874, Jan 10.

Wang, Y., Feng, S., Bai, X., Zhao, J., & Xia, S. (2016). Scum sludge as a potential feedstock for biodiesel production from wastewater treatment plants. *Waste Management, 47*, 91–97, Jan 1.

Waqas, S., Bilad, M. R., Man, Z., Wibisono, Y., Jaafar, J., Mahlia, T. M. I., et al. (2020). Recent progress in integrated fixed-film activated sludge process for wastewater treatment: A review. *Journal of Environmental Management, 268*, 110718.

Webster, C. (1962). The sewers of Mohenjo-Daro. *Journal (Water Pollution Control Federation)*, 116–123.

Wei, S., Zhu, M., Fan, X., Song, J., Peng, P., Li, K., et al. (2019a). Influence of pyrolysis temperature and feedstock on carbon fractions of biochar produced from pyrolysis of rice straw, pine wood, pig manure and sewage sludge. *Chemosphere, 218*, 624–631, Mar 1.

Wei, J., Liu, Y., Li, J., Zhu, Y., Yu, H., & Peng, Y. (2019b). Adsorption and co-adsorption of tetracycline and doxycycline by one-step synthesized iron loaded sludge biochar. *Chemosphere, 236*, 124254, Dec 1.

Wen, B., Aydin, A., & Duzgoren-Aydin, N. S. (2002). A comparative study of particle size analyses by sieve-hydrometer and laser diffraction methods. *Geotechnical Testing Journal, 25*(4), 434–442.

White, P. J., & Broadley, M. R. (2003). Calcium in plants. *Annals of Botany, 92*(4), 487–511.

Wilk, M., Śliz, M., & Lubieniecki, B. (2021). Hydrothermal co-carbonization of sewage sludge and fuel additives: Combustion performance of hydrochar. *Renewable Energy, 178*, 1046–1056, Nov 1.

Wilke, B. M. (2005). Determination of chemical and physical soil properties. *Monitoring and assessing soil bioremediation. Springer*, 47–95.

Wongbunmak, A., Khiawjan, S., Suphantharika, M., & Pongtharangkul, T. (2017). BTEX- and naphthalene-degrading bacterium Microbacterium esteraromaticum strain SBS1–7 isolated from estuarine sediment. *Journal of Hazardous Materials, 339*, 82–90, Oct 5.

Xiao, B., Dai, Q., Yu, X., Yu, P., Zhai, S., Liu, R., et al. (2018). Effects of sludge thermal-alkaline pretreatment on cationic red X-GRL adsorption onto pyrolysis biochar of sewage sludge. *Journal of Hazardous Materials, 343*, 347–355, Feb 5.

Xu, D., Cao, J., Li, Y., Howard, A., & Yu, K. (2019). Effect of pyrolysis temperature on characteristics of biochars derived from different feedstocks: A case study on ammonium adsorption capacity. *Waste Management, 87*, 652–660, Mar 15.

Xu, X., Ray, R., Gu, Y., Ploehn, H. J., Gearheart, L., Raker, K., et al. (2004). Electrophoretic Analysis and Purification of Fluorescent Single-Walled Carbon Nanotube Fragments. *Journal of the American Chemical Society, 126*(40), 12736–12737.

Xu, Y., Zhang, C., Zhao, M., Rong, H., Zhang, K., & Chen, Q. (2017). Comparison of bioleaching and electrokinetic remediation processes for removal of heavy metals from wastewater treatment sludge. *Chemosphere, 168*, 1152–1157, Feb 1.

Xu, Z. X., Song, H., Li, P. J., Zhu, X., Zhang, S., Wang, Q., et al. (2020). A new method for removal of nitrogen in sewage sludge-derived hydrochar with hydrotalcite as the catalyst. *Journal of Hazardous Materials, 398*, 122833, Nov 5.

Xue, Y., Wang, C., Hu, Z., Zhou, Y., Xiao, Y., & Wang, T. (2019). Pyrolysis of sewage sludge by electromagnetic induction: Biochar properties and application in adsorption removal of Pb(II), Cd(II) from aqueous solution. *Waste Management, 89*, 48–56, Apr 15.

Yan, F., Luo, S. y., Hu, Z. quan, Xiao, B., & Cheng, G. (2010). Hydrogen-rich gas production by steam gasification of char from biomass fast pyrolysis in a fixed-bed reactor: Influence of temperature and steam on hydrogen yield and syngas composition. *Bioresource Technology, 101*(14), 5633–5637.

Yang, G., Zhang, G., & Wang, H. (2015). Current state of sludge production, management, treatment and disposal in China. *Water Research, 78*, 60–73.

Yang, K., Zhu, Y., Shan, R., Shao, Y., & Tian, C. (2017). Heavy metals in sludge during anaerobic sanitary landfill: Speciation transformation and phytotoxicity. *Journal of Environmental Management, 189*, 58–66, Mar 15.

Yang, W., Song, W., Li, J., & Zhang, X. (2020). Bioleaching of heavy metals from wastewater sludge with the aim of land application. *Chemosphere, 249*, 126134, Jun 1.

Yeager, J. G., & O'brien, R. T. (1983). Irradiation as a means to minimize public health risks from sludge-borne pathogens. *Journal (Water Pollution Control Federation)*, 977–983.

Yu, J., Tang, L., Pang, Y., Zeng, G., Wang, J., Deng, Y., et al. (2019). Magnetic nitrogen-doped sludge-derived biochar catalysts for persulfate activation: Internal electron transfer mechanism. *Chemical Engineering Journal, 364*, 146–159, May 15.

Yuan, C., Xu, Z. Z., Fan, M. X., Liu, H. Y., Xie, Y., & Zhu, T. (2014). Study on characteristics and harm of surfactants. *Journal of Chemical and Pharmaceutical Research, 6*, 2233–2237.

Yue, Y., Cui, L., Lin, Q., Li, G., & Zhao, X. (2017). Efficiency of sewage sludge biochar in improving urban soil properties and promoting grass growth. *Chemosphere, 173*, 551–556, Apr 1.

Zarinviarsagh, M., Ebrahimipour, G., & Sadeghi, H. (2017). Lipase and biosurfactant from Ochrobactrum intermedium strain MZV101 isolated by washing powder for detergent application. *Lipids Health Dis, 16*(1), 177.

Zhang, C., Liu, L., Zhao, M., Rong, H., & Xu, Y. (2018). The environmental characteristics and applications of biochar. *Environmental Science and Pollution Research, 25*(22), 21525–21534.

Zhang, D., Jiang, H., Chang, J., Sun, J., Tu, W., & Wang, H. (2019). Effect of thermal hydrolysis pretreatment on volatile fatty acids production in sludge acidification and subsequent polyhydroxyalkanoates production. *Bioresource Technology, 279*, 92–100, May 1.

Zhang, S., Pi, M., Su, Y., Xu, D., Xiong, Y., & Zhang, H. (2020). Physiochemical properties and pyrolysis behavior evaluations of hydrochar from co-hydrothermal treatment of rice straw and sewage sludge. *Biomass and Bioenergy, 140*, 105664, Sep 1.

Zhang, X., Zhao, B., Liu, H., Zhao, Y., & Li, L. (2022). Effects of pyrolysis temperature on biochar's characteristics and speciation and environmental risks of heavy metals in sewage sludge biochars. *Environmental Technology & Innovation, 26*, 102288, May 1.

Zhou, Z., Liu, Y., Liu, S., Liu, H., Zeng, G., Tan, X., Yang, C., Ding, Y., Yan, Z., Cai, X. (2017). Sorption performance and mechanisms of arsenic (V) removal by magnetic gelatin-modified biochar. *Chem. Eng. J. 314*, 223–231. Available from: https://doi.org/10.1016/j.cej.2016.12.113.

Zuloaga, O., Navarro, P., Bizkarguenaga, E., Iparraguirre, A., Vallejo, A., Olivares, M., et al. (2012). Overview of extraction, clean-up and detection techniques for the determination of organic pollutants in sewage sludge: A review. *Analytica Chimica Acta, 736*, 7–29, Jul 29.

C H A P T E R

22

Converting biomass waste to water treatment chemicals

Tatiana Samarina[1,2], Varsha Srivastava[2], Outi Laatikainen[1] and Sari Tuomikoski[2]

[1]School of Engineering, Kajaani University of Applied Sciences, Kajaani, Finland [2]Research Unit of Sustainable Chemistry, Faculty of Technology, University of Oulu, Oulu, Finland

22.1 Introduction

Over the past decades, rising human population and rapid urbanization led to increasing agricultural and industrial activities globally. The water pollution associated with anthropogenic activities has become more serious than ever. At a worldwide scale, the water pollution causes clean water deficiency, endangers aquatic living organisms, and jeopardizes the water security in highly populated areas.

There are many types of contaminants in water. Traditionally, heavy metals, excessive nutrients, biological agents, and organic substances, such as dyes and pharmaceuticals, are seen as major contributors deteriorating water quality. However, new categories of contaminants such as pharmaceuticals and personal care products, endocrine disruptors, and microplastics are emerging constantly and demanding attention and improvement of water treatment practices. Meanwhile, there are many treatment technologies available to supplement conventional methods of contaminant removal from water, for example, coagulation and flocculation, advanced oxidation processes (AOPs), (electro)chemical precipitation, adsorption, and membrane techniques. Although the above-mentioned methods understandably have their own advantages and limitations, sustainability becomes a crucial parameter for technology employing and development.

There are four main types of chemicals traditionally used to improve the water quality: pH neutralizers, antifoaming agents, coagulants, and flocculants. Coagulation with iron and other polyvalent metals are widely used in municipal water and wastewater treatment practices, while neutralisation with lime or acids is common to industrial wastewater treatment. Advanced or tertiary treatment steps become increasingly common for both due to stricter regulations imposed to effluent pollutants concentration and the new challenges associated with the need to deal with emerging pollutants. Thus, new filtering and adsorbent materials, sources of additional carbon, or chemicals enhancing the removal of phosphorous and nitrogen are needed. The major drawback of all these substances is that some of them are toxic, mostly non-degradable and have tendency to leave residues in the treated water. Recently, bio-based chemicals have received a lot of attention due to their advantages over conventional synthetic or inorganic agents. Biodegradable and sustainable substances inhibiting corrosion, preventing scale and removing pollutants could become environmentally friendly alternatives, when facing condition allowing compatible performance and cost-efficiency with synthetic chemicals.

Recently, bio-based chemicals have received a lot of attention due to their advantages over conventional synthetic or inorganic agents. Biodegradable and sustainable substances that inhibit corrosion, prevent scaling, and remove pollutants could become environmentally friendly alternatives, on the creterion that they work at least as well as conventional ones and cost-efficient enough from the point of view the whole chain of treatment.

Although bioalcohol is used as a chemical in a ternary treatment, its production aspects were excluded from this chapter due to more in-depth discussion in Chapter 10.

22.2 Biomass feedstock for recovery of chemicals

Biomass on land is typically defined as the mass of living organism and dry biomass includes about 50% of carbon (Houghton et al., 2008). Biomass can be classified by several ways. One general classification for different kinds of biomass groups are (1) wood and woody biomass, (2) herbaceous and agricultural biomass, (3) aquatic biomass, (4) animal and human biomass wastes, (5) contaminated biomass and industrial biomass wastes (semibiomass), and (6) biomass mixtures (Vassilev et al., 2010). The usage of biomass as a raw material for several applications will be increased because it is a one solution to control global warming and to get diversity of different raw materials (Briens et al., 2008). Biomass is typically used as an energy source or as a raw material for biofuels. Biomass has also been used for the production of several kinds of biochemicals and/or bioproducts. Based on the estimation, approximately 50 Mt of different biomass-derived and polymeric materials are produced annually (Antar et al., 2021). This chapter focuses on the recovery of chemicals from wood and woody biomass, agricultural biomass, and industrial side-streams. In Fig. 22.1 is presented different waste biomass raw materials and their potential usage applications for (waste)water treatment.

Agricultural residues have potential for sustainable use because they have only low impact on land use changes (Hochman et al., 2014). They have also other positive properties, for example sorghum straw has short harvest time and easy availability and red algae has high photosynthetic efficiency (Antar et al., 2021). In the agricultural production, biomass is obtained as residues in the several process stages. Crops from the horticultural and arable fields, for example, are formed in the production stage of crops and fruits in agriculture. Straw, leaves, husks, and cobs are the examples of biomass-based residues formed during the harvesting of the main crops (Avcıoğlu et al., 2019). Properties of the agricultural biomass are different compared to the woody biomass because agricultural biomass contains less carbon and hydrogen. In addition, agricultural residues include typically higher amounts of ash and inorganic elements (nitrogen, sulfur, and chlorine) than woody biomass (Fournel et al., 2015). Different agricultural biomass can be used in several applications. Table 22.1 presents some examples about variety of waste feedstocks and their potential utilization applications.

Table 22.2 presents the list of different biomass sources and their valorization methods to the different applications or products.

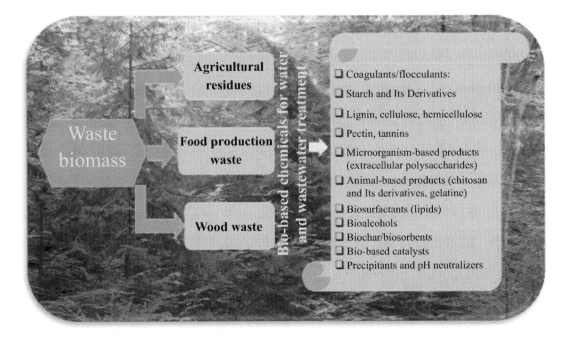

FIGURE 22.1 Valorization waste biomass as bio-based chemicals for water and wastewater treatment Valorization waste biomass as bio-based chemicals for water and wastewater treatment.

TABLE 22.1 Agricultural biomass feedstock and examples related to their usage applications.

Waste feedstock	Application	Reference
Coconut and peanut shell waste	Biocatalyst synthesis for enzyme immobilization	Rao and Rathod (2019)
Grass biomass	Sugars obtained are used as a substrate for mcl-PHA (polyhydroxyalkanoates) production	Davis et al. (2013)
Leaves, peels	Extraction of phytochemical, enzymes	Rao and Rathod (2019)
Orange peels	Isolate peroxidases	Nadar et al. (2018)
Orange peel powder	Raw material was modified to magnetic adsorbent towards cadmium removal by using co-precipitation with Fe_3O_4	Bhatnagar et al. (2015)
Peanut husk powder	Copper adsorption from wastewater	Rao and Rathod (2019)
Rice straw	The amine groups were added to the surface and produced material was used as an adsorbent to remove Cr(VI) and Ni(II)	Wu et al. (2016)

TABLE 22.2 Conversion of waste biomass into biochemicals via various valorization techniques.

Source of biomass	Valorization method	Application	Product	Yield	Ref
Biochemicals					
Tomato peels	Acid hydrolysis	—	Cellulose	10–13	Jiang and Hsieh (2015)
Cassava peel	Hydrolysis	Dual coagulant to remove bacteria and suspended solids from dam water	Starch	—	Asharuddin et al. (2019)
Mango kernel	Solvent extraction	As secondary raw material	Starch	39.3–45	Ferraz et al. (2019)
Yellow skin potato	Water extraction	As secondary raw material	Starch	16.5	Altemimi (2018)
Shrimp shell waste	Ultrasound-assisted deacetylation	As an antioxidant or antimicrobial agent	Chitosan	17	Hafsa et al. (2016)
Shrimp shell waste	Fermentation by using protease-producing Bacillus species	Reduction of Fe^{3+} to Fe^{2+}	Chitin	Even 95% (deproteinization rate)	Ghorbel-Bellaaj et al. (2012)
Sweet potato	Ultrasonic degradation	Antioxidant	Pectin	—	Ogutu and Mu (2017)
Potato pulp	Citric acid extraction	Natural emulsifier	Pectin	14.2	Yang et al. (2018)
Orange peel	Hydrolysis	Flocculant	Pectin	—	Mohan (2014)
Sugar-beet pulp	Salt solution extraction	—	Pectin	21.8	Arslan (1995)
Pomelo peel	Water extraction	Flocculant	Pectin	6.5	Piriyaprasarth and Sriamornsak (2011)
Pomelo peel	Acidic or alkaline extraction	—	Pectin	12–24	Wandee et al. (2019)
Citrus pectin	Alkylation	Gel formation agent	Pectin	—	Liu et al. (2017)
Calcined paper mill sludge and fly ash from biomass combustion	Paper mill sludge was calcined and fly ash used without pre-treatments	Chemical precipitation (ammonium and phosphate removal from anaerobic digestion plant reject water)	ash as precipitation agent	—	Myllymäki et al. (2020)
Wood ash	Unburned carbon particles were removed before use	Chemical precipitation of struvite (magnesium ammonium phosphate) precipitation from urine	ash as precipitation agent	—	Sakthivel et al. (2012)
Filtering materials/sorbents					
Carbon residue from biomass gasification	Chemical activation	Nitrate and phosphate removal	Adsorbent	—	Kilpimaa et al. (2014)
Rice straw biochars and fly ash	Alkali-fusion of fly ash, mixing with biochars	Methylene blue removal	Adsorbent	—	Wang et al. (2020)
Biomass fly ash	Geopolymerization	Methylene blue removal	Adsorbent	—	Novais et al. (2018)
Rice hush ash and bagasse fly ash	Sieving to specific particle size	Diuron removal	Adsorbent	—	Deokar et al. (2016)

22.3 Recovery of biochemicals from biomass

Inorganic coagulants and organic polymer flocculants are essential chemicals of potable water and wastewater treatment practices. Inorganic coagulants are mostly represented by aluminum and iron salts, with less frequent use of hydrated lime and magnesium carbonate Despite the high efficiency of treatment with aluminum and iron salts, pre-hydrolyzed inorganic coagulants are increasingly used due to wider range of possible application conditions and more controlled manner of metal hydrolysis species formation. While inorganic coagulants are commonly used, they are not devoid of shortcomings. For instance, chemical residuals, which could pose health risks or speed up the aging of equipment, are produced within the treatment process. Moreover, during the sludge management not all routes could be used due to residual chemicals, limiting a circularity potential of this material flow.

The extraction procedures of biochemicals from biomass waste feedstock should be optimized in a case-specific basis in order to get maximum yield and desirable composition of extracts. Although no universal recommendations could be provided, the methods and stages commonly described in literature are shown on the Fig. 22.2. The yield and cost of the resulting product depends on three main stages: (1) the prepossessing procedure of the production source, that is fresh, frozen or dried state the raw material, milling, cutting and homogenization; (2) particular extraction procedure (solvent used, extraction time and temperature, solid-liquid ratio, and physical assistance); purification step.

22.3.1 Bio-based coagulants, coagulant aids/flocculants, and other materials for advanced treatment practices

Starch and its derivativeshave been adopted for wastewater treatment among the first from the class of bio-based coagulants and coagulant aids (Jiang et al., 2021). Chemically, starches are composed of a mixture of two polymers from the polysaccharide family: amylopectin and amylose. Thus, the physical and chemical characteristics of the resulting coagulants and coagulant aids strongly depend on amylopectin and amylose fraction and their structures in the recovered product. The polymers' fractions in the resulting product, in its turn, greatly rely on a method of extraction.

Being one of the most abundant plant-based polymers in plants, starch can be extracted from the agro- or food waste like parts of a plant (leaves, stalk, fruit skin, seeds, or roots) or from whole fruits/grain/tuber (unripe, culled or wasted). Traditional sources of starch such as corn, potato, rice, and cassava could be supplemented by an unconventional ones (Ahmad et al., 2022; Ashogbon & Akintayo, 2014; Makroo et al., 2021). Moreover, waste biomass serving as an additional source of starch may provide cost reduction both the primer products for food industry and supplementing products for non-food applications such as water and wastewater treatment. As a basic recovery process, the acid hydrolysis of plant feedstock is widely used, however the tendency towards bioconversion (through enzymatic processes) is observed (Cho et al., 2020). Starches extracted from potato peels (Altemimi, 2018), mango kernel (Ferraz et al., 2019), and cassava peels (Asharuddin et al., 2019) could be applied as sole coagulants or as raw materials for modification starches and their grafting with conventional polymers.

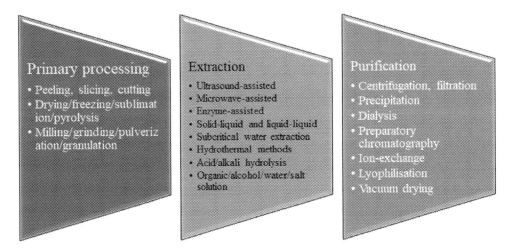

FIGURE 22.2 Preparation bio-based chemicals for water and wastewater treatment Preparation bio-based chemicals for water and wastewater treatment.

22.3.1.1 Lignin and cellulose

Biomass (softwood and brewer's spent grains) pretreatment for lignin extraction using deep eutectic solvent was investigated and the impact of temperature and hydrogen bond donor (lactic acid and Glycerol) on the extraction process was evaluated (Provost et al., 2022). Higher extraction was obtained at 80°C temperature. Additionally, the processing of biomass for lignin recovery was done using the ionic liquid 1-butyl-3-methylimidazolium chloride (Guiao et al., 2022). In another study, bamboo biomass feedstock was used for lignin extraction using nitrile-based ionic liquid (Muhammad et al., 2013). Formic acid-based pretreatment was evaluated for lignin extraction from diverse nonwood cellulosic biomass (Watkins et al., 2015). Interestingly, the extraction of lignin from steam-exploded biomass was tested using microwave-assisted extraction (Angelini et al., 2017). Further, in another investigation, an artificial neural network modeling was used to investigate lignin extraction from oil palm biomass (Rashid et al., 2021).

Lignin and lignin-derived products have been extensively investigated for water remediation. Lignin carbon nanofiber with 589 m^2/g specific surface area was developed for methylene blue adsorption (Beck et al., 2017). Further, in another report, the applicability of lignin hydrogel was discussed for metal and dye removal (Thakur et al., 2017). Gao and Courtney Moore (2021) investigated the effectiveness of cationic lignin polymers as flocculants for treating municipal wastewater. In this study, lignin-based flocculants were developed by the radical polymerization of kraft lignin. Approximately 17%−23% reduction in COD and 51%−60% in TOC were noted for lignin-based flocculants. In another investigation, lignin-grafted cationic polyacrylamide flocculant was developed for water flocculation (Chen et al., 2020). Further, paper mill sludge was utilized for the development of four lignin-based flocculants and tested for dye-containing wastewater (Guo et al., 2019). It was observed that flocculants with higher charge density performed better than those with lower charge densities. Fang et al. (2010) synthesized lignin-based cationic flocculant for the removal of anionic azo-dyes from wastewater. The developed flocculant was able to reduce COD > 89%. Various parameters such as pH, dye concentration, dose of flocculants, and settling time were optimized.

He et al. (2016) developed water-soluble anionic kraft lignin via oxidation and sulfomethylation and applied it as a flocculant for dye removal. It was mentioned that the charge density significantly affects the flocculant's effectiveness. In addition, the pH of dye solutions had an impact on COD elimination. In another investigation, Kraft Lignin-based sorbent materials were developed for cetirizine dihydrochloride removal (Stanisz et al., 2021). Higher removal (90%) was observed at pH 6.2 when the concentration of cetirizine dihydrochloride was taken as 10 mg/L. Due to the availability of both polar and nonpolar functional groups, lignin-based sorbents are very efficient in pollutant adsorption. Further, lignin-derived carbon materials for Azure B dye removal (Zhu et al., 2022). Carboxymethylated lignin-based flocculants were tested for treatment of wastewater contaminated with copper (Zeng et al., 2022).

Cellulose is one of the most prevalent biopolymers on the earth, and its efficient exploitation aims to contribute to long-term, sustainable growth (Akhlamadi et al., 2021; Chopra & Manikanika, 2022; Galiwango et al., 2019; Janaswamy et al., 2022). Cellulose is the major component of a plant cell wall, which exists together with hemicellulose and lignin (Doh et al., 2020; Galiwango et al., 2019). Cellulose consists of a linear chain of glucopyranose units connected by β-(1,4) glucosidic bonds that form a regular intramolecular and intermolecular hydrogen bonding network (Akhlamadi et al., 2021; Akinjokun et al., 2021; Almashhadani et al., 2022). The typical formula for cellulose is $(C_6H_{10}O_5)_n$, where n is the number of monomeric β-(1−4)-d-glucopyranose units, which varies depending on the cellulose source (Galiwango et al., 2019). Cellulose is categorized into four categories based on its crystalline structure: type I, II, III, and IV (He et al., 2021). Natural cellulose derived from biomasses can be converted into micro-and nanoscale materials, culminating in microcrystalline cellulose, microfibrillar cellulose, and nanocellulose (Collazo-Bigliardi et al., 2018). Nanocellulose (NC) materials can be categorized into cellulose nanofibers (CNFs) and cellulose nanocrystals (CNCs) based on their size, aspect ratio, and crystallization (Akhlamadi et al., 2021). CNFs have a low degree of crystallinity (60%−80%), with widths of nanometers and lengths of microns, while CNCs have a higher degree of crystallinity (85%) with widths of 3−20 nm and lengths of 100−350 nm (Akhlamadi et al., 2021; Akinjokun et al., 2021). The nature of the cellulose resource has a significant impact on the characteristics and morphologies of extracted cellulose (Beck-Candanedo et al., 2005).

Corn stalks, wheat straw, rice straw, and soy stalks are all effective sources of cellulose in agriculture (Janaswamy et al., 2022). Several seaweed species were found to have considerable content of cellulose (Baghel et al., 2021). Biomasses such as oil palm (Kumneadklang et al., 2019), mengkuang leaves (Sheltami et al., 2012), agar waste biomass (Martínez-Sanz et al., 2020), arecanut husk fiber (C.S. et al., 2016), rice husk (Collazo-Bigliardi et al., 2018), date palm (Galiwango et al., 2019), cocoa pod husk (Akinjokun et al., 2021), black spruce (Beck-Candanedo et al., 2005), banana plants (Mueller et al., 2014), coffee husk (Collazo-Bigliardi et al., 2018), seaweed biomass (Baghel et al., 2021; Doh et al., 2020), *Thespesia populnea* barks (Kathirselvam et al., 2019), sugar cane bagasse (Sriwong & Sukyai, 2022), agro-industrial waste (Henrique et al., 2013), and pineapple leaf (Ravindran et al., 2019) have been explored for cellulose extraction. Characteristics and yields of the obtained products are listed in Table 22.3.

TABLE 22.3 Properties of cellulose-based materials extracted from various biomass.

Cellulose source	Properties of extracted cellulose	Reference
Waste pencil chips	Nanocrystalline cellulose: needle/rod like structure. Crystallinity of CNCs (91%) Specific surface area 486.430 m^2/g Pore volume- 46.383 (cm^3/g)	Akhlamadi et al. (2021)
Cotton	Cellulose nanocrystals (Spherical/Elliptical) Crystallinity index 81.23% Crystallite size 10–50 nm Average particle size 221 nm	Pandi et al. (2021)
Cocoa husk waste	Cellulose nanocrystals Diameter 10–60 nm Length 41–155 nm Crystallinity index 67.60% CNC yield–25%	Akinjokun et al. (2021)
Oil palm mesocarp fiber	Crystallinity index 46.81% Cellulose yield 64.0%	Azlan et al. (2021)
Arecanut husk fiber	Nanocellulose fibrils Diameter- 3–5 nm Crystallinity index-73% Yield 22%–26%	C.S. et al. (2016)
Elaeis guineensis (empty fruit bunch)	Nanocellulose Crystallinity index 80.3% (by One-pot process) Crystallinity index 75.4% (by Multistep process) Yield 42.0% (by One-pot process) Yield 27.5% (by Multistep process) Diameter 51.6 ± 15.4 nm (by One-pot process) Diameter 47.9 ± 23.7 nm (by Multistep process)	Chen et al. (2017)
Cofee husk	Cellulose nanocrystals Crystallinity index—50% Diameter-39nm	Collazo-Bigliardi et al. (2018)
Date palm biomass	Cellulose and α-cellulose Crystallinity index 52.27% Cellulose yield—74.70%, 71.50%, and 73.82% for Rachis, leaflet, and fiber parts, respectively α-Cellulose—78.63%, 75.64%, and 70.40% for Rachis, leaflet, and fiber parts, respectively	Galiwango et al. (2019)
Rice straw	Cellulose nanofibers Yield 27.19 ± 0.99% (rice straw, particle size < 75 μm) Yield 38.31 ± 0.85% (rice straw, particle size 150 to 250 μm)	Ratnakumar et al. (2022)
Cactus fruit waste seeds	Cellulose microfibers (CMFs) and CNCs CMFs diameter–11 μm CMF-Crystallinity index –72%, CMF-Yield–25% CNCs diameter-13 ± 3 nm CNCs-needle like structure CNCs-Length 419 ± 48 nm CNCs-Crystallinity index—72%, CNCs-Yield–25% Crystallinity index of CNCs—86% Crystallinity index CMF–72%	Ait Benhamou et al. (2022)
Apple pomace	Cellulose nanocrystals: needle-like structure Yield 27.96 ± 0.78% Averages diameter of 7.9 ± 1.25 nm Length 28 ± 2.03 nm Crystallinity index 78%	Melikoğlu et al. (2019)

Cellulose-based materials derived from natural sources are becoming increasingly significant for water treatment (Collazo-Bigliardi et al., 2018). Due to their biodegradable nature, non-toxic, low cost, and abundant availability, cellulose and cellulose-derived materials have garnered considerable attention in the development of bio-nanocomposite materials, adsorbents, catalysts, and a variety of chemicals for water treatment (Akhlamadi et al., 2021; Akinjokun et al., 2021; Almashhadani et al., 2022; Chen et al., 2017; Doh et al., 2020; Galiwango et al., 2019; Oksman et al., 2011; Collazo-Bigliardi et al., 2018).

Cellulose and cellulose-derived materialshave been extensively applied for water remediation. Wood-based cellulose nanocrystals (sulfated and carboxylated) were synthesized for the removal of cationic dye Auramine (Pinto et al., 2020). It was reported that synthesized CNCs were able to remove 82% of dye within 30 minutes of contact time and showed 20 mg/g adsorption capacity. Furthermore, cellulose-based adsorbent was synthesized via radiation grafting and tested for textile wastewater treatment containing Acid Blue 25 and Acid Blue 74 (Goel et al., 2015). The grafted cellulose adsorbent showed 540.0 mg/g adsorption capacity for Acid Blue 25 while 340.0 mg/g was recorded for Acid Blue 74. In another investigation, cellulose grafted calcium hydroxide was developed for dye removal (Zhu et al., 2021). Additionally, cellulose-derived polyols was examined for the removal of boron and organic pollutants (Hong et al., 2021). Cellulose-derived polyols showed approximately 70%–98% removal of selected organic pollutants. Pei et al. (2021) investigated Cr (VI) removal using cellulose-based composites. Further in another work, cellulose nanofibrils-based foam was developed for Cr (VI) removal (da Silva & Rosa, 2022). In this study, the foam was prepared using freeze-drying and the effect of citric acid was assessed on cellulose foam properties. Jaihan et al. developed cellulose-based bio-sorbents from papaya peel waste and examined them for Pb (II) removal (Jaihan et al., 2022). Liu et al. (2022) developed bifunctional cellulose-based adsorbent using corn stalk and evaluated the adsorption efficiency for Cu (II) and Pb (II). The applicability of cellulose-derived photocatalysts in water treatment has been explored by many researchers (Mohamed et al., 2017; Mohamed Noor et al., 2020).

Cellulose/γ-Fe_2O_3-ZrO_2 photocatalyst was synthesized for Congo red degradation (Helmiyati et al., 2022). Rice straw was utilized for the cellulose source. The developed photocatalyst was very efficient and gave 98.5% degradation in 30 minutes. Further in another study, g-C_3N_4/cellulose (CN/CE) hybrid photocatalyst was synthesized for the photocatalytic degradation of methylene blue and showed 99.8% degradation (Bai et al., 2020). The efficacy of modified cellulose was investigated for the photocatalytic degradation of carbamazepine from water (Ledezma-Espinoza et al., 2022). Further, in another investigation, palm oil wastewater was treated using cellulose-based flocculant (Mohamed Noor et al., 2020). In this study, polyacrylamide was grafted on magnetic cellulose. It was demonstrated that 1.5 g/L of flocculant dose and pH 8.0 were best for flocculation and approximately 88.62% COD removal was achieved under optimum conditions.

Another plant-based polymer from polysaccharide family that could be originated from waste biomass valorization is pectin. This linear polysaccharide could be extracted from the plant cell walls, for instance, citrus peels (Mohan, 2014) or sweet potato pulp (Ogutu & Mu, 2017; Yang et al., 2018). Being biocompatible and nontoxic, pectins are anionic in nature unlike starches. Pectin mainly consists of joined d-galacturonic acid molecules with a certain amount of neutral sugars presented as branch chains (Chen et al., 2015). Similar to others natural coagulants, final product properties depend on primary source and extraction procedure applied for its recovery. Pectin, its derivatives, and hybrids grafted with polyacrylamide have been used recently for preparation of cationic and non-ionic polyelectrolytes (Bolto & Gregory, 2007), green catalysts (Budarin et al., 2008), adsorbents for removal persistent contaminants (Nasrollahzadeh et al., 2021), and applied as flocculants (Piriyaprasarth & Sriamornsak, 2011; Ho et al., 2009).

Tannins, natural and ubiquitous polyphenolic compounds, are highly reactive due to their phenolic nature. Tannins could be extracted from waste biomass such as bark, fruits, and leaves. For instance, species from South America such as *Acacia* and *Schinopsis* are well-known tannin feedstocks. Commercial flocculant products such as TANFLOC originated from *Acacia* (Beltrán-Heredia & Sánchez-Martín, 2009) and some others as Ecotan or Polyacqua (Tomasi et al., 2022) already presented on the market. However, species with tannin-rich barks such as *Castanea, Quercus ilex, Suber or Robur, Pinus pinaster* could be found across the EU and used to recover tannins (Bacelo et al., 2016; Sánchez-Martín et al., 2010). Özacar and Şengil used tannins obtained from *valonia* (an autoctonous tree) as coagulation aid (Özacar & Şengil, 2003; Özacar & Şengil, 2000). Tannin bioflocculant has been always combined with $Al_2(SO_4)_3$ in order to enhance further filtration of the formed sludge (Özacar & Şengil, 2003; Özacar & Şengıl, 2000). A modification of tannins via polymerization with formaldehyde as cross-linking agent could be used to preparation of an insoluble non-linear polymers and biosorbents (Bacelo et al., 2018; Bacelo et al., 2016).

Not only plants could be a source of waste biomass for production of biochemicals. Microorganism-based coagulants obtained from bacteria, fungi, or algae are the great examples of value-added products gained via waste biomass valorization.

Alginates probably are produced in the largest quantities amount all of microorganism-based chemicals. Alginates are composed of blocks of mannuronic acid residues (M-blocks), blocks of guluronic acid residues (G-blocks), and blocks with alternating M and G residues (MG-blocks). The order of the blocks determines the properties of the resulting materials. On the own way, the ratio of the M/G-blocks defined by feature of particular species and by method of the extraction (Nechita, 2017). Marine brown algae (seaweed) and certain strains of bacteria (*Azotobacter and Pseudomonas*) contain these polysaccharides as structural components of the cells. Commercially alginates are extracted exclusively from seaweeds. The extraction of commercial alginates can be obtained via calcium alginate process or alginic acid process. The drawback of the latter is a more complicated separation of alginic acid, which is precipitated in form of gel. The calcium alginate process is preferable since separation of calcium salts precipitated in fibrous form is easy. Recently, three-step extraction approach for extraction of sodium alginate from waste *Sargassum natans* have been proposed (Mohammed et al., 2018) with maximum purity and yield up to 74% and 16%, respectively. The microbial fermentation is also technically and economically feasible for extraction of the alginates (Abu Bakar et al., 2021; Al-Wasify et al., 2015; Shahadat et al., 2017).

The unique physicochemical properties promote the use of this biopolymer in water treatment applications. Ionic or covalent cross-linking and formation of alginic acid gels (Nasrollahzadeh et al., 2021) of initial alginates allow to apply the coagulant aid in wide pH region, while further chemical modification could bring new properties to resulting materials. Oxidation, sulfation, and esterification are the main ways of derivatisation for introducing the variety of amphiphilic alginates.

Animal-based coagulants up until recently have mostly been presented by chitin and its derivative chitosan. However, gelatin, gums, and meat and bone meal are promising biochemicals reported recently, which could be used in wastewater treatment sector.

Chitosan, along with cellulose and starch, is one of the most abundant biopolymers. Belonging to the class of polysaccharide flocculants, chitosan is a deacetylated derivative of chitin mainly originated from crustacean shells (crabs, lobsters, shrimps, etc.). This polysaccharide consists of a linear copolymer of D-glucosamine and N-acetyl-D-glucosamine (Nechita, 2017). Its properties conditioned by the presence of amino and hydroxyl functional groups in its structure attracted attention of water and wastewater professionals.

The degree of deacetylation, molecular weight, and origin of the raw material define flocculation performance of the product. Even more important role plays pH value of application due to chitosan insolubility in neutral and alkaline media. Although chitosan in its unmodified form rarely used, some authors have applied it directly (Al-Manhel et al., 2018; Lichtfouse et al., 2019; Pontius, 2016). Moreover, chitin without deacetylation was used by Saritha et al. (2017) as a coagulant, and showed results comparable to those of aluminum sulfate. The use of chitin is an attractive option that allows not only to reduce the cost of flocculant production, but also to avoid chemical processing (deacetylation, modification, grafting) and, as a result, environmental stress. Valorization methods and possible applications of biopolymer are listed in Table 22.2.

Similarly to starch (and some other recovered biopolymers), chitin and chitosan are insoluble in water and must be modified to increase their applicability and flocculating activity. Synthetic modification of chitosan with 3-chloro-2-hydroxypropyl trimethylammonium chloride and carboxymethyl groups improve water solubility over wider pH conditions (Cainglet et al., 2020). Obtained amphoteric materials possess favorable charge neutralization of negatively charged species and colloidal suspensions. Those properties allow to use the modified flocculant for phosphorous and turbidity removal from natural waters (Agbovi & Wilson, 2018).

Other animal-based biopolymers having potential as bioflocculants are **gelatine and collagen**. These materials composed of protein (up to 85%–92%) have abundant functional groups in their polymer chains, which opens up the ability to modify and tune their properties. Gelatine and meat and bone meal (MBM) extracts are other promising sources for valorization of waste biomass as bio-based flocculants (Piazza & Garcia, 2010).

Extraction procedure is simple and eco-friendly (e.g., acid or alkaline hydroxylation), and MBM as a raw material is inexpensive since MBM subjects to incineration in the EU. Thus, the resulting bioflocculants could have a low cost, and be used as potential replacement of anionic polyacrylamide (Wang et al., 2011). However, application of these biochemicals for use in water treatment is at an early laboratory stage. The derivatives and composites of gelatin for heavy metal and dyes removal have been reviewed by (El-Gaayda et al., 2021).

22.3.2 Sustainable chemicals for advance treatment

With progress in environmental risk assessment and research on the health risks associated with known and emerging pollutants, the need for advanced treatment methods is becoming more and more evident. Tightening water quality regulations for both drinking water and wastewater treatment are prompting researchers and industry to seek for advanced practices via integration an additional treatment steps into conventional schemes. Adsorption, biofiltration, precipitation, and AOPs could be used as preliminary or ternary treatment stages and supplied chemicals for these methods should be preferably produced in a sustainable manner.

Biomass (lignocellulosic and agricultural, for example) can be used as a raw material for the production of biochar, activated carbon and variety of functional materials. They have applications in soil improving, adsorption and catalytical applications as catalysts or catalysts support materials. However, biomass-derived catalysts materials and preparation of carbon materials are excluded from this chapter because it has been discussed more detail in Chapter 23.

Biomass-derived carbon used in soil improvement are called biochars. Biochars can be produced via biomass gasification or slow pyrolysis and they have potential to be used for soil improvement (Fryda & Visser, 2015). Different agricultural or biomass-derived wastes have been used as a precursor for activated carbon production and used as an adsorbent for the removal of pharmaceuticals, metal(loid)s and dyes. Some examples about biomass-derived carbons in pharmaceuticals removal are presented in Table 22.4.

In addition to pharmaceuticals, biomass-derived carbons can be used in the removal of metal(loid)s and dyes. Giri et al. (2012) used Eichhornia crassipes root in the removal of Cr(VI). This material was pyrolyzed at 600°C by using H_2SO_4 as chemical activator and the surface area of produced activated carbon was 109.23 m^2/g (Giri et al., 2012). Pine wood sawdust was used as a precursor for the biomass-based activated carbon that has been modified by sodium hydroxide and that material was used as an adsorbent for the methylene blue removal (Zhu et al., 2017). Chemical modification without pyrolysis step can also be used. By using this production method for biomass-based adsorbent, product is called biosorbent. Arsenic can be removed via chemically modified water melon rind, for example (Shakoor et al., 2018).

Alkaline fly ashes from combustion process have several potential utilization applications. It has potential to be used as a neutralizer in acid soils, forest soils and tropical acid soils, and act as an alternative choice as soil liming agent to substitute commercial lime (Kilpimaa et al., 2014; Ohno & Susan Erich, 1990; Demeyer et al., 2001; Kahl et al., 1996). Wood-derived ashes has also potential to act as a fertiliser for example, as a source of potassium (Naylor & Schmidt, 1986). Alkaline ashes have also ability to affect the electrical conductivity in pore water, and with increased soil pH, these properties enhance the number of soil bacteria

TABLE 22.4 Agricultural or biomass-derived wastes as a precursor for activated carbon to be used as an adsorbent.

Biomass waste	Pharmaceutical	Pyrolysis temperature [°C]	Activation method	Biochar properties	Reference
Pine sawdust	Sulfamethoxazole	650	$FeCl_2$ + KOH + KNO_3	Surface area: 125.8 m^2/g Total pore volume 0.14 cm^3/g	Reguyal and Sarmah (2018)
Tea waste	Sulfamethazine	700	Steam	Surface area: 576.1 m^2/g Total pore volume 0.109 cm^3/g	Rajapaksha et al. (2016)
Cauliflowers roots	Chlortetracycline	500	None	Surface area: 232.15 m^2/g Total pore volume 0.15 cm^3/g	Qin et al. (2017)
Olive stones	Paracetamol	500	H_3PO_4	Surface area: 990 m^2/g Total pore volume 0.91 cm^3/g	García-Mateos et al. (2015)
Wheat straw	Ketoprofen	700	HCl	Surface area: 605 m^2/g Total pore volume 0.421 cm^3/g	Wu et al. (2018)
Fruit shell	Ibuprofen	650	steam	Surface area: 308 m^2/g Total pore volume 0.384 cm^3/g	Chakraborty et al. (2018)
Pomelo peel	Carbamazepine	600–900	KOH	Surface area: 904.1 m^2/g Total pore volume 0.506 cm^3/g	Chen et al. (2017)

(Bang-Andreasen et al., 2017). Ashes have also possibilities to be used as a neutralisation of the solutions (Cheremisinoff, 1995).

Wood-derived ashes have also potential to be applied in the removal of nutrients via chemical precipitation and/or adsorption. Lu et al. (2009) studied the removal mechanisms of phosphate by using fly ash. They found that adsorption and precipitation are the main phosphate removal mechanisms by using fly ash (Lu et al., 2009). At high pH values, the precipitation of calcium phosphate is the main mechanism. Calcium ions from fly ash can also react with phosphate and forms hydroxylapatite (Schwartz and Lopes, 2014). Wood ashes include typically calcium, potassium, magnesium, manganese, phosphate, sulfate, and metals (aluminum, iron, chromium, lead, for example). Struvite, magnesium ammonium phosphate (MAP), precipitation is suitable way to recover ammonium and phosphate from waters by using magnesium as a precipitant. Magnesium salts are typically expensive and it is also listed as critical raw material in EU level (Sakthivel et al., 2012). Therefore, ashes have been showed to be suitable magnesium source for struvite precipitation (Myllymäki et al., 2020). Sakthivel et al. (2012) studied wood ash produced in India as a magnesium source to precipitate MAP from source-separated urine. Experiments were done in batch mode. Results showed that the precipitate was not pure struvite due to the low phosphorus and high calcite content. Myllymäki et al. (2020) used biomass-derived fly ash from power plant to precipitate MAP. They found quite good removal capacities for ammonium nitrogen (74%) and phosphate (59%). Formed MAP have potential to be used as a recycled fertilizer. Pesonen et al. (2020) was optimized the struvite precipitation by preparing $MgSO_4$ solution from biomass-derived fly ash by treating ash with sulfuric acid. At this stage, calcium and magnesium oxides or carbonates of ash form insoluble calcium carbonate and soluble magnesium sulfate (Pesonen et al., 2020). Even 97% removal for phosphate was obtained and pure struvite was formed based on XRD diffractograms.

In addition to bio-based adsorbents and bio-derived precipitants, other conventional chemicals in water and wastewater sector can be replaced with bio-based ones. Various surface-active chemical substances are one of them. Lipids were investigated as de- and antifoaming agents to suppress foaming (Kougias et al., 2015). Biosurfactants could be co-produced during the treatment of polluted wastewater in agrosector (Fegade et al., 2021). Agro-industrial waste and its by-products is widely used for production of biosurfactants via fermentation (Mohanty et al., 2021), especially starch-rich waste as cassava, potato, rice straw, or soybean flour (Solanki et al., 2021). All in all, production of biosurfactants and their application as detergents in households and industry might decrease environmental burden caused by petroleum-based chemicals (Padma Ishwarya & Nisha, 2021; Sarubbo et al., 2022).

22.4 Technology readiness level and application of bio-based chemicals in water and wastewater treatment practices

While research on bio-based chemicals for water and wastewater treatment has made immense progress over the past decade, testing the new biochemicals beyond lab-scale are still at early stages. Mostly, TRL of the technologies referred in the literature both for biochemicals and their applications is close to 5–6 (technology validated/demonstrated in relevant environment). As for manufacturing of biocoagulants/flocculants, tannin and cationic starches are the only examples of commercialized products. The production starch-based biocomposites from *Posidonia oceanica* waste biomass at demo scale has been reported recently (Benito-González et al., 2021). Nevertheless, bright examples of pilot- and semi-, and even full-scale applications of biochemical for water treatment are represented in literature.

Saleem and Bachmann (2019) have collected examples of pilotings as well as more developed schemes of plant-based chemicals for water and wastewater applications. Primarily, *M. oleifera* seed extracts have been used to treat surface waters (river and stream) across the world. As for semi- and full-scale applications, same coagulant has been used in Malawi and Nigeria (Saleem & Bachmann, 2019). Application of plant-based chemicals for wastewater management and sludge dewatering have been reviewed by Das et al. (2021), and pilot plant for sewage treatment with *Mangifera indica* seed extracts mentioned. Examples of application of animal-based coagulant, chitosan, for direct wastewater treatment provided by Lichtfouse et al. (2019). Effluents from the sugar industry, palm oil mill effluents, and post-treatment of sanitary landfill leachate have been purified at pilot-scale. Commercialized tannin coagulant was tested at pilot-scale plant for municipal wastewater treatment by Hameed et al. (2016); Roselet et al. (2016); Sánchez-Martín et al. (2010). Results showed than biocoagulants have a competing performance compared to polyacrylamide flocculant in terms of turbidity, BOD/COD removal, size, and structure of flocs in resulting sludge characteristics. Iron mineral processing wastewater has been treated with microbial bioflocculant at pilot-scale application (Liu et al., 2019). Kitchen waste was a raw material for bioflocculant production. The flocculation rate of 92% was achieved, while the pH of wastewater was only slightly changed to 7.7.

22.5 Challenges and future perspective

To meet current sustainability requirements, bio-based water and wastewater treatment chemicals need to have adequate ability to improve the cumulative characteristics of water, such as removing turbidity, suspended solids, organic and inorganic pollutants, color, and BOD and COD. Also manufacturing process should be carried out in a sustainable and environment-friendly manner. Being biodegradable and renewable with adjustable properties, and sometimes contributing to decreased sludge formation, these chemicals could take a prominent place on family of water and wastewater treatment tools. However, application of biochemicals faces own challenges and limitations.

For coagulants and flocculants, operational parameters play the critical impact on the treatment performance. As with conventional types of these chemicals, the optimization of mixing speed and time, temperature, and a dosage on real waters and in future operational conditions prior the on-site application is highly recommended. Optimization of the operational parameters is necessary to conduct on real water matrixes also when precipitation or adsorptive materials are applied. Using simulated waters, research generally limits the application of new products due to inadequate and over-prediction of performance in real case scenarios.

Besides enhancing technological performance, other challenges should be addressed to bring biochemicals to wider use. Taking as an example the plant-based coagulants produced from *M. oleifera*, Saleem and Bachmann (2019) listed several issues for the commercialization of biocoagulants.

The major constraints are abundance of raw material supply for biocoagulant production and manufacturing of sufficient amounts of biocoagulants for replacement of traditional ones. Even in case the adequate amount of raw material would be found with combining dispersed material sources, collection of dispersed raw material feedstock correlates with increased need for transportation, which further contributes both to high OPEX costs and large CO_2 footprint for manufacturing. Dispersed raw material sources also correlate with inhomogenous material quality, which sets challenges to production processes. Another challenge is associated with the cost-effective production of biocoagulants with steady quality. The most effective extraction and purification techniques such as ion exchange, lyophilisation, supercritical solvent extraction, or dialysis are both cost- and energy-intensive. Extraction and purification processes might also involve strong chemicals, which cause challenges for process wastewater purification processes. Efficient yet low-cost and eco-friendly procedures to liberate a pure coagulant fraction from an initial biomass are required.

Approximate prices for main classes of bio-based coagulants produced from raw materials have been estimated by El-Gaayda et al. (2021) and Kurniawan et al. (2020). The authors highlighted that economic benefits could be derived if low-cost biomass used as a source of raw material for biocoagulant production. However, it should be noted that collection of biomass waste, its transportation, storage, pretreatment/preservation, and expenses on supplemented chemicals are the hidden costs, which weaken an overall economic performance of biochemicals.

Saleem and Bachmann (2019) have indicated gaps in legislation concerning bio-based coagulants as another constraint to commercialization. To date, several levels or regulatory constraints related to waste-based material productions need to be taken into account. Regulation-based end-of-waste certification processes are to date criticized as demanding and complicated. The aim to increase agility in end-of-waste processes in order to face, for example, the EU aim for circular economy has been recognized and work for reshaping the processes is ongoing. In addition, various new regulation to support circular economy processes is under-development. Thus, the EU suggestions for eco-design directive alterations would support in ensuring the raw material recycling in future. Lack of toxicological studies on new biochemicals is one of the topics that should be addressed in the future work to help overcome this challenge. The residues of conventional coagulants/flocculants, i.e., acrylamide or aluminum ion emissions are known drawbacks of the technology; however, biochemical are also not free from this disadvantage. The increase of BOD and COD of water treated biochemically indicated the possibility of secondary pollution of receiving waters by residual concentrations.

Along with toxicological studies, reduction in sludge formation and biogas potential remain uncovered, and discussion needs to be launched to discover the benefits and drawbacks of new biochemicals in potential applications. The handling of residual sludge is one of the key challenges in the water and wastewater sector both in rural and densely populated areas. Unsustainable sludge management practices cause additional expenses for treatment facilities, lost profit, and substantial pressure on environment. The presence of aluminum or iron in large quantities in the sludge significantly impedes anaerobic digestion, which is the preferred method of sludge treatment in terms of energy cogeneration, sanitation, and the production of soil improvers. The substitution of conventional chemicals with bio-based solutions with enhanced biodegradability may result in improvement in biogas production via better digestion performance.

Addressing the abovementioned topics and rethinking of treatment practises might strengthen the economic performance of biochemicals for the whole sector. In return, acceptance of biochemicals for the cleantech sector by end users, authorities, and potential investors will make a breakthrough in the commercialization of new products. As stated by El-Gaayda et al. (2021), raising public awareness of the environmental issues associated with water treatment and supporting local governments can be important factors in a paradigm shift toward application the biochemicals.

22.6 Conclusions

The biomass waste generated globally has been intensified for the past century: 1,6 billion metric tons/year of food waste (De Clercq et al., 2017), 18 million tons of wood industry waste (Wood US EPA OLEM, 2017), about 1 billion metric tons/year biomass ash in EU (IEA Bioenergy Annual Report, 2019). The waste biomass is considered as a sustainable source for various biorefinery platforms to recover biofuels (gas and diesel). But recovery and production other bio-based value-added products such as bioalcohols, antioxidants, fertilizers, and other biochemicals should be supported more extensively in order to promote transition to greener and safer future.

Consisting of proteins, lipids, polysaccharides along with macronutrients (nitrogen, phosphorus, calcium, and potassium), the biomass waste should be seen as a valuable resource to support transition from linear to circular economy. Valorization of biomass waste via eco-friendly techniques should be prioritized to minimize environmental burden and GHG footprint for the approach. Multi-platform biorefineries incorporating energy cogeneration and recovery of biochemicals could create a massive flow of recycled materials for the pharmaceutical, cosmetics, chemical, and clean-tech industries.

Bio-based chemicals listed are important for both upstream (production of biochemicals) and downstream processes (biogas and biofuel generation, fertiliser production). Such biochemicals are a critical driver for greener water and wastewater operations with reduced GHG emission, minimizing sludge generation, or obtaining the sludges with enhanced biogas and valorization potential. However, the competitiveness of new bio-based chemicals in water treatment sector mostly depends on their primary performance compared to traditional ones (fossil- and petroleum-based). Also, legislation and other levels of regulation have essential part in either supporting or limiting the possibilities to create, sell, and use new biowaste-based chemicals.

Acknowledgments

Authors would like to thank the Academy of Finland project (decision number 346537). The study was conducted as part of the REMAC project "Renewing Sludge Management Concepts" (Karelia CBC, KA11000). Authors gratefully acknowledge the Administration of Kajaani city for supporting the work via Measurepolis project.

References

Abu Bakar, Siti Nur Hatika, Abu Hasan, Hassimi, Sheikh Abdullah, Siti Rozaimah, Kasan, Nor Azman, Muhamad, Mohd Hafizuddin, & Kurniawan, Setyo Budi (2021). A review of the production process of bacteria-based polymeric flocculants. *Journal of Water Process Engineering*, 40, 101915. Available from https://doi.org/10.1016/j.jwpe.2021.101915, https://www.sciencedirect.com/science/article/pii/S2214714421000027.

Agbovi, Henry K., & Wilson, Lee D. (2018). Design of amphoteric chitosan flocculants for phosphate and turbidity removal in wastewater. *Carbohydrate Polymers*, 189, 360–370. Available from https://doi.org/10.1016/j.carbpol.2018.02.024, https://www.sciencedirect.com/science/article/pii/S0144861718301620.

Ahmad, Azmi, Kurniawan, Setyo Budi, Sheikh Abdullah, Siti Rozaimah, Othman, Ahmad Razi, & Hasan, Hassimi Abu (2022). Exploring the extraction methods for plant-based coagulants and their future approaches. *The Science of the Total Environment*, 818, 151668. Available from https://doi.org/10.1016/j.scitotenv.2021.151668.

Ait Benhamou, Anass, Kassab, Zineb, Boussetta, Abdelghani, Salim, Mohamed Hamid, Ablouh, El-Houssaine, Nadifiyine, Mehdi, QAISS, Abou El. Kacem, Moubarik, Amine, & Achaby, Mounir El (2022). Beneficiation of cactus fruit waste seeds for the production of cellulose nanostructures: Extraction and properties. *International Journal of Biological Macromolecules*, 203, 302–311. Available from https://doi.org/10.1016/j.ijbiomac.2022.01.163, https://www.sciencedirect.com/science/article/pii/S0141813022001842.

References

Akhlamadi, Golnoosh, Goharshadi, Elaheh K., & Saghir, Siavosh Vojdani (2021). Extraction of cellulose nanocrystals and fabrication of high alumina refractory bricks using pencil chips as a waste biomass source. *Ceramics International*, 47(19), 27042−27049. Available from https://doi.org/10.1016/j.ceramint.2021.06.117, https://www.sciencedirect.com/science/article/pii/S0272884221018629.

Akinjokun, Adebola Iyabode, Petrik, Leslie Felicia, Ogunfowokan, Aderemi Okunola, Ajao, John, & Ojumu, Tunde Victor (2021). Isolation and characterization of nanocrystalline cellulose from cocoa pod husk (CPH) biomass wastes. *Heliyon*, 7(4), e06680. Available from https://doi.org/10.1016/j.heliyon.2021.e06680, https://www.sciencedirect.com/science/article/pii/S2405844021007830.

Al-Manhel, Alaa Jabbar, Al-Hilphy, Asaad Rehman Saeed, & Niamah, Alaa Kareem (2018). Extraction of chitosan, characterisation and its use for water purification. *Journal of the Saudi Society of Agricultural Sciences*, 17(2), 186−190. Available from https://doi.org/10.1016/j.jssas.2016.04.001, https://www.sciencedirect.com/science/article/pii/S1658077X16300224.

Al-Wasify, Raed, Al-Sayed, A.-S. A., Saleh, S. M., & Abouelwafa, Ahmed (2015). *Bacterial exopolysaccharides as new natural coagulants for surface water treatment*, 8, 198−207.

Almashhadani, Abdulsalam Q., Leh, Cheu Peng, Chan, Siok-Yee, Lee, Chong Yew, & Goh, Choon Fu (2022). Nanocrystalline cellulose isolation via acid hydrolysis from non-woody biomass: Importance of hydrolysis parameters. *Carbohydrate Polymers*, 286, 119285. Available from https://doi.org/10.1016/j.carbpol.2022.119285, https://www.sciencedirect.com/science/article/pii/S0144861722001898.

Altemimi, Ammar B. (2018). Extraction and Optimization of Potato Starch and Its Application as a Stabilizer in Yogurt Manufacturing. *Foods*, 7(2), 14. Available from https://doi.org/10.3390/foods7020014, https://www.mdpi.com/2304-8158/7/2/14.

Angelini, Stefania, Ingles, David, Gelosia, Mattia, Cerruti, Pierfrancesco, Pompili, Enrico, Scarinzi, Gennaro, Cavaglio, Gianluca, Cotana, Franco, & Malinconico, Mario (2017). One-pot lignin extraction and modification in γ-valerolactone from steam explosion pre-treated lignocellulosic biomass. *Journal of Cleaner Production*, 151, 152−162. Available from https://doi.org/10.1016/j.jclepro.2017.03.062, https://www.sciencedirect.com/science/article/pii/S0959652617304973.

Antar, Mohammed, Lyu, Dongmei, Nazari, Mahtab, Shah, Ateeq, Zhou, Xiaomin, & Smith, Donald L. (2021). Biomass for a sustainable bioeconomy: An overview of world biomass production and utilization. *Renewable and Sustainable Energy Reviews*, 139, 110691. Available from https://doi.org/10.1016/j.rser.2020.110691, https://www.sciencedirect.com/science/article/pii/S1364032120309758.

Arslan, N. (1995). Extraction of pectin from sugar-beet pulp and intrinsic viscosity molecular weight relationship of pectin solutions. *Journal of Food Science and Technology-Mysore*, 32(5), 381−385. Available from http://www.webofscience.com/wos/woscc/full-record/WOS:A1995TP99800005.

Asharuddin, Syazwani Mohd, Othman, Norzila, Zin, Nur Shaylinda Mohd, Tajarudin, Husnul Azan, & Md Din, Mohd Fadhil (2019). Flocculation and antibacterial performance of dual coagulant system of modified cassava peel starch and alum. *Journal of Water Process Engineering*, 31, 100888. Available from https://doi.org/10.1016/j.jwpe.2019.100888, https://www.sciencedirect.com/science/article/pii/S2214714419301369.

Ashogbon, Adeleke Omodunbi, & Akintayo, Emmanuel Temitope (2014). Recent trend in the physical and chemical modification of starches from different botanical sources: A review. *Starch - Stärke*, 66(1−2), 41−57. Available from https://doi.org/10.1002/star.201300106, https://onlinelibrary.wiley.com/doi/abs/10.1002/star.201300106.

Avcıoğlu, A. O., Dayıoğlu, M. A., & Türker, U. (2019). Assessment of the energy potential of agricultural biomass residues in Turkey. *Renewable Energy*, 138, 610−619. Available from https://doi.org/10.1016/j.renene.2019.01.053, https://www.sciencedirect.com/science/article/pii/S0960148119300539.

Azlan, Nadiah Syafiqah Mohd, Yap, Chiew Lin, Gan, Suyin, & Abdul Rahman, Mohd Basyaruddin (2021). Effectiveness of various solvents in the microwave-assisted extraction of cellulose from oil palm mesocarp fiber. *Materials Today: Proceedings*. Available from https://doi.org/10.1016/j.matpr.2021.12.086, https://www.sciencedirect.com/science/article/pii/S2214785321077610.

Bacelo, Hugo, Vieira, B. árbara R. C., Santos, S. ílvia C. R., Boaventura, Rui A. R., & Botelho, Cidália M. S. (2018). Recovery and valorization of tannins from a forest waste as an adsorbent for antimony uptake. *Journal of Cleaner Production*, 198, 1324−1335. Available from https://doi.org/10.1016/j.jclepro.2018.07.086, https://www.sciencedirect.com/science/article/pii/S0959652618320596.

Bacelo, Hugo A. M., Santos, S. ílvia C. R., & Botelho, Cidália M. S. (2016). Tannin-based biosorbents for environmental applications − A review. *Chemical Engineering Journal*, 303, 575−587. Available from https://doi.org/10.1016/j.cej.2016.06.044, https://www.sciencedirect.com/science/article/pii/S1385894716308518.

Baghel, Ravi S., Reddy, C. R. K., & Singh, Ravindra Pal (2021). Seaweed-based cellulose: Applications, and future perspectives. *Carbohydrate Polymers*, 267, 118241. Available from https://doi.org/10.1016/j.carbpol.2021.118241, https://www.sciencedirect.com/science/article/pii/S0144861721006287.

Bai, Wending, Yang, Xiaogang, Du, Xiaolin, Qian, Zhouqi, Zhang, Yong, Liu, Lin, & Yao, Juming (2020). Robust and recyclable macroscopic g-C3N4/cellulose hybrid photocatalysts with enhanced visible light photocatalytic activity. *Applied Surface Science*, 504, 144179. Available from https://doi.org/10.1016/j.apsusc.2019.144179, https://www.sciencedirect.com/science/article/pii/S0169433219329952.

Bang-Andreasen, Toke, Nielsen, Jeppe T., Voriskova, Jana, Heise, Janine, Rønn, Regin, Kjøller, Rasmus, Hansen, Hans C. B., & Jacobsen, Carsten S. (2017). Wood Ash Induced pH Changes Strongly Affect Soil Bacterial Numbers and Community Composition. *Frontiers in Microbiology*, 8. Available from https://www.frontiersin.org/article/10.3389/fmicb.2017.01400.

Beck, Rika J., Zhao, Yong, Fong, Hao, & Menkhaus, Todd J. (2017). Electrospun lignin carbon nanofiber membranes with large pores for highly efficient adsorptive water treatment applications. *Journal of Water Process Engineering*, 16, 240−248. Available from https://doi.org/10.1016/j.jwpe.2017.02.002, https://www.sciencedirect.com/science/article/pii/S2214714416306584.

Beck-Candanedo, Stephanie, Roman, Maren, & Gray, Derek G. (2005). Effect of Reaction Conditions on the Properties and Behavior of Wood Cellulose Nanocrystal Suspensions. *Biomacromolecules*, 6(2), 1048−1054. Available from https://doi.org/10.1021/bm049300p, https://doi.org/10.1021/bm049300p.

Beltrán-Heredia, J., & Sánchez-Martín, J. (2009). Municipal wastewater treatment by modified tannin flocculant agent. *Desalination*, 249(1), 353−358. Available from https://doi.org/10.1016/j.desal.2009.01.039, https://www.sciencedirect.com/science/article/pii/S0011916409009138.

Benito-González, Isaac, Göksen, G. ülden, Pérez-Bassart, Zaida, López-Rubio, Amparo, Sánchez, Rafael, Alonso, José María, Gavara, Rafael, Gallur, Miriam, & Martínez-Sanz, Marta (2021). Pilot plant scale-up of the production of optimized starch-based biocomposites loaded with cellulosic nanocrystals from Posidonia oceanica waste biomass. *Food Packaging and Shelf Life*, 30, 100730. Available from https://doi.org/10.1016/j.fpsl.2021.100730, https://www.sciencedirect.com/science/article/pii/S2214289421000983.

Bhatnagar, Amit, Sillanpää, Mika, & Witek-Krowiak, Anna (2015). Agricultural waste peels as versatile biomass for water purification – A review. *Chemical Engineering Journal*, 270, 244–271. Available from https://doi.org/10.1016/j.cej.2015.01.135, https://www.sciencedirect.com/science/article/pii/S1385894715001746.

Bolto, Brian, & Gregory, John (2007). Organic polyelectrolytes in water treatment. *Water Research*, 41(11), 2301–2324. Available from https://doi.org/10.1016/j.watres.2007.03.012, https://www.sciencedirect.com/science/article/pii/S0043135407001881.

Briens, Cedric, Piskorz, Jan, & Berruti, Franco (2008). Biomass Valorization for Fuel and Chemicals Production – A Review. *International Journal of Chemical Reactor Engineering*, 6(1). Available from https://doi.org/10.2202/1542-6580.1674, https://www.degruyter.com/document/doi/10.2202/1542-6580.1674/html.

Budarin, Vitaly L., Clark, James H., Luque, Rafael, Macquarrie, Duncan J., & White, Robin J. (2008). Palladium nanoparticles on polysaccharide-derived mesoporous materials and their catalytic performance in C–C coupling reactions. *Green Chemistry*, 10(4), 382–387. Available from https://doi.org/10.1039/B715508E, https://pubs.rsc.org/en/content/articlelanding/2008/gc/b715508e.

C.S., Julie Chandra, George, Neena, & Narayanankutty, Sunil K. (2016). Isolation and characterization of cellulose nanofibrils from arecanut husk fibre. *Carbohydrate Polymers*, 142, 158–166. Available from https://doi.org/10.1016/j.carbpol.2016.01.015, https://www.sciencedirect.com/science/article/pii/S0144861716000308.

Cainglet, Annaliza, Tesfamariam, Axumawit, & Heiderscheidt, Elisangela (2020). Organic polyelectrolytes as the sole precipitation agent in municipal wastewater treatment. *Journal of Environmental Management*, 271, 111002. Available from https://doi.org/10.1016/j.jenvman.2020.111002.

Chakraborty, Prasenjit, Banerjee, Soumya, Kumar, Sumit, Sadhukhan, Sutonu, & Halder, Gopinath (2018). Elucidation of ibuprofen uptake capability of raw and steam activated biochar of Aegle marmelos shell: Isotherm, kinetics, thermodynamics and cost estimation. *Process Safety and Environmental Protection*, 118, 10–23. Available from https://doi.org/10.1016/j.psep.2018.06.015, https://www.sciencedirect.com/science/article/pii/S0957582018303331.

Chen, Dezhi, Xie, Shasha, Chen, Caiqin, Quan, Hongying, Hua, Li, Luo, Xubiao, & Guo, Lin (2017). Activated biochar derived from pomelo peel as a high-capacity sorbent for removal of carbamazepine from aqueous solution. *RSC Advances*, 7(87), 54969–54979. Available from https://doi.org/10.1039/C7RA10805B, https://pubs.rsc.org/en/content/articlelanding/2017/ra/c7ra10805b.

Chen, Jun, Liu, Wei, Liu, Cheng-Mei, Li, Ti, Liang, Rui-Hong, & Luo, Shun-Jing (2015). Pectin Modifications: A Review. *Critical Reviews in Food Science and Nutrition*, 55(12), 1684–1698. Available from https://doi.org/10.1080/10408398.2012.718722, https://doi.org/10.1080/10408398.2012.718722.

Chen, Nian, Liu, Weifeng, Huang, Jinhao, & Qiu, Xueqing (2020). Preparation of octopus-like lignin-grafted cationic polyacrylamide flocculant and its application for water flocculation. *International Journal of Biological Macromolecules*, 146, 9–17. Available from https://doi.org/10.1016/j.ijbiomac.2019.12.245, https://www.sciencedirect.com/science/article/pii/S0141813019393912.

Cheremisinoff, Paul N. (1995). *Waste Minimization and Cost Reduction for the Process Industries* (p. 351) Elsevier.

Cho, Eun Jin, Trinh, Ly. Thi Phi, Song, Younho, Lee, Yoon Gyo, & Bae, Hyeun-Jong (2020). Bioconversion of biomass waste into high value chemicals. *Bioresource Technology*, 298, 122386. Available from https://doi.org/10.1016/j.biortech.2019.122386, https://www.sciencedirect.com/science/article/pii/S0960852419316165.

Chopra, Lalita, & Manikanika. (2022). Extraction of cellulosic fibers from the natural resources: A short review. *Materials Today: Proceedings*, 48, 1265–1270. Available from https://doi.org/10.1016/j.matpr.2021.08.267, https://www.sciencedirect.com/science/article/pii/S2214785321057114.

Collazo-Bigliardi, Sofía, Ortega-Toro, Rodrigo, & Chiralt Boix, Amparo (2018). Isolation and characterisation of microcrystalline cellulose and cellulose nanocrystals from coffee husk and comparative study with rice husk. *Carbohydrate Polymers*, 191, 205–215. Available from https://doi.org/10.1016/j.carbpol.2018.03.022, https://www.sciencedirect.com/science/article/pii/S0144861718302789.

Das, Nilanjana, Ojha, Nupur, & Mandal, Sanjeeb Kumar (2021). Wastewater treatment using plant-derived bioflocculants: green chemistry approach for safe environment. *Water Science and Technology: A Journal of the International Association on Water Pollution Research*, 83(8), 1797–1812. Available from https://doi.org/10.2166/wst.2021.100.

Davis, Reeta, Kataria, Rashmi, Cerrone, Federico, Woods, Trevor, Kenny, Shane, O'Donovan, Anthonia, Guzik, Maciej, Shaikh, Hamid, Duane, Gearoid, Gupta, Vijai Kumar, Tuohy, Maria G., Padamatti, Ramesh Babu, Casey, Eoin, & O'Connor, Kevin E. (2013). Conversion of grass biomass into fermentable sugars and its utilization for medium chain length polyhydroxyalkanoate (mcl-PHA) production by Pseudomonas strains. *Bioresource Technology*, 150, 202–209. Available from https://doi.org/10.1016/j.biortech.2013.10.001, https://www.sciencedirect.com/science/article/pii/S0960852413015812.

De Clercq, Djavan, Wen, Zongguo, Gottfried, Oliver, Schmidt, Franziska, & Fei, Fan (2017). A review of global strategies promoting the conversion of food waste to bioenergy via anaerobic digestion. *Renewable and Sustainable Energy Reviews*, 79, 204–221. Available from https://doi.org/10.1016/j.rser.2017.05.047, https://www.sciencedirect.com/science/article/pii/S1364032117306792.

Deokar, Sunil K., Singh, Diksha, Modak, Sweta, Mandavgane, Sachin A., & Kulkarni, Bhaskar D. (2016). Adsorptive removal of diuron on biomass ashes: a comparative study using rice husk ash and bagasse fly ash as adsorbents. *Desalination and Water Treatment*, 57(47), 22378–22391. Available from https://doi.org/10.1080/19443994.2015.1132394, https://doi.org/10.1080/19443994.2015.1132394.

Doh, Hansol, Lee, Min Hyeock, & Whiteside, William Scott (2020). Physicochemical characteristics of cellulose nanocrystals isolated from seaweed biomass. *Food Hydrocolloids*, 102, 105542. Available from https://doi.org/10.1016/j.foodhyd.2019.105542, https://www.sciencedirect.com/science/article/pii/S0268005X19315516.

El-Gaayda, Jamila, Titchou, Fatima Ezzahra, Oukhrib, Rachid, Yap, Pow-Seng, Liu, Tianqi, Hamdani, Mohamed, & Akbour, Rachid Ait (2021). Natural flocculants for the treatment of wastewaters containing dyes or heavy metals: A state-of-the-art review. *Journal of Environmental Chemical Engineering*, 9(5), 106060. Available from https://doi.org/10.1016/j.jece.2021.106060, https://www.sciencedirect.com/science/article/pii/S221334372101037X.

Fegade, Umesh, Inamuddin., & Adetunji, Charles Oluwaseun (2021). Chapter 1 - Application of biosurfactant for treatment of effluent waste, polluted wastewater treatment, and sewage sludge Green Sustainable Process for Chemical and Environmental Engineering and Science (pp. 1–19). Elsevier. Available from https://www.sciencedirect.com/science/article/pii/B9780128226964000206.

Ferraz, Clara A., Fontes, Roseli L. S., Fontes-Sant'Ana, Gizele C., Calado, Verônica, López, Elvis O., & Rocha-Leão, Maria H. M. (2019). Extraction, Modification, and Chemical, Thermal and Morphological Characterization of Starch From the Agro-Industrial Residue of

Mango (Mangifera indica L) var. *Ubá*. *Starch - Stärke*, 71(1-2), 1800023. Available from https://doi.org/10.1002/star.201800023, https://onlinelibrary-wiley-com.pc124152.oulu.fi:9443/doi/10.1002/star.201800023.

Fournel, S., Marcos, B., Godbout, S., & Heitz, M. (2015). Predicting gaseous emissions from small-scale combustion of agricultural biomass fuels. *Bioresource Technology*, 179, 165−172. Available from https://doi.org/10.1016/j.biortech.2014.11.100, https://www.sciencedirect.com/science/article/pii/S0960852414017143.

Fryda, Lydia, & Visser, Rianne (2015). Biochar for Soil Improvement: Evaluation of Biochar from Gasification and Slow Pyrolysis. *Agriculture*, 5(4), 1076−1115. Available from https://doi.org/10.3390/agriculture5041076, https://www.mdpi.com/2077-0472/5/4/1076.

Galiwango, Emmanuel, Abdel Rahman, Nour S., Al-Marzouqi, Ali H., Abu-Omar, Mahdi M., & Khaleel, Abbas A. (2019). Isolation and characterization of cellulose and α-cellulose from date palm biomass waste. *Heliyon*, 5(12), e02937. Available from https://doi.org/10.1016/j.heliyon.2019.e02937, https://www.sciencedirect.com/science/article/pii/S240584401936596X.

Gao, Weijue, & Courtney Moore, Pedram Fatehi (2021). Cationic Lignin Polymers as Flocculant for Municipal Wastewater. *Polymers*, 13, 3871.

García-Mateos, F. J., Ruiz-Rosas, R., Marqués, M. D., Cotoruelo, L. M., Rodríguez-Mirasol, J., & Cordero, T. (2015). Removal of paracetamol on biomass-derived activated carbon: Modeling the fixed bed breakthrough curves using batch adsorption experiments. *Chemical Engineering Journal*, 279, 18−30. Available from https://doi.org/10.1016/j.cej.2015.04.144, https://www.sciencedirect.com/science/article/pii/S1385894715006348.

Ghorbel-Bellaaj, Olfa, Younes, Islem, Maâlej, Hana, Hajji, Sawssen, & Nasri, Moncef (2012). Chitin extraction from shrimp shell waste using Bacillus bacteria. *International Journal of Biological Macromolecules*, 51(5), 1196−1201. Available from https://doi.org/10.1016/j.ijbiomac.2012.08.034, https://www.sciencedirect.com/science/article/pii/S0141813012003509.

Giri, Anil Kumar, Patel, Rajkishore, & Mandal, Sandip (2012). Removal of Cr (VI) from aqueous solution by Eichhornia crassipes root biomass-derived activated carbon. *Chemical Engineering Journal*, 185-186, 71−81. Available from https://doi.org/10.1016/j.cej.2012.01.025, https://www.sciencedirect.com/science/article/pii/S1385894712000289.

Goel, Narender Kumar, Kumar, Virendra, Misra, Nilanjal, & Varshney, Lalit (2015). Cellulose based cationic adsorbent fabricated via radiation grafting process for treatment of dyes waste water. *Carbohydrate Polymers*, 132, 444−451. Available from https://doi.org/10.1016/j.carbpol.2015.06.054, http://www.sciencedirect.com/science/article/pii/S0144861715005639.

Guiao, Karelle S., Tzoganakis, Costas, & Mekonnen, Tizazu H. (2022). Green mechano-chemical processing of lignocellulosic biomass for lignin recovery. *Chemosphere*, 293, 133647. Available from https://doi.org/10.1016/j.chemosphere.2022.133647, https://www.sciencedirect.com/science/article/pii/S0045653522001400.

Guo, Kangying, Gao, Baoyu, Wang, Wenyu, Yue, Qinyan, & Xu, Xing (2019). Evaluation of molecular weight, chain architectures and charge densities of various lignin-based flocculants for dye wastewater treatment. *Chemosphere*, 215, 214−226. Available from https://doi.org/10.1016/j.chemosphere.2018.10.048, https://www.sciencedirect.com/science/article/pii/S0045653518319027.

Hafsa, J., Smach, M. A., Charfeddine, B., Limem, K., Majdoub, H., & Rouatbi, S. (2016). Antioxidant and antimicrobial proprieties of chitin and chitosan extracted from Parapenaeus Longirostris shrimp shell waste. *Annales Pharmaceutiques Françaises*, 74(1), 27−33. Available from https://doi.org/10.1016/j.pharma.2015.07.005, https://www.sciencedirect.com/science/article/pii/S0003450915000875.

Hameed, Yasir Talib, Idris, Azni, Hussain, Siti Aslina, & Abdullah, Norhafizah (2016). A tannin-based agent for coagulation and flocculation of municipal wastewater: Chemical composition, performance assessment compared to Polyaluminum chloride, and application in a pilot plant. *Journal of Environmental Management*, 184, 494−503. Available from https://doi.org/10.1016/j.jenvman.2016.10.033, https://www.sciencedirect.com/science/article/pii/S0301479716308179.

He, Hong, An, Fengping, Wang, Yiwei, Wu, Wanying, Huang, Zhiwei, & Song, Hongbo (2021). Effects of pretreatment, NaOH concentration, and extraction temperature on the cellulose from Lophatherum gracile Brongn. *International Journal of Biological Macromolecules*, 190, 810−818. Available from https://doi.org/10.1016/j.ijbiomac.2021.09.041, https://www.sciencedirect.com/science/article/pii/S0141813021019607.

He, Wenming, Zhang, Yiqian, & Fatehi, Pedram (2016). Sulfomethylated kraft lignin as a flocculant for cationic dye. *Colloids and Surfaces A: Physicochemical and Engineering Aspects*, 503, 19−27. Available from https://doi.org/10.1016/j.colsurfa.2016.05.009, https://www.sciencedirect.com/science/article/pii/S092777571630317X.

Helmiyati, Helmiyati, Fitriana, Nurani, Chaerani, Metha Listia, & Dini, Fitriyah Wulan (2022). Green hybrid photocatalyst containing cellulose and γ−Fe_2O_3−ZrO_2 heterojunction for improved visible-light driven degradation of Congo red. *Optical Materials*, 124, 111982. Available from https://doi.org/10.1016/j.optmat.2022.111982, https://www.sciencedirect.com/science/article/pii/S0925346722000167.

Henrique, Mariana Alves, Silvério, Hudson Alves, Neto, Wilson Pires Flauzino, & Pasquini, Daniel (2013). Valorization of an agro-industrial waste, mango seed, by the extraction and characterization of its cellulose nanocrystals. *Journal of Environmental Management*, 121, 202−209. Available from https://doi.org/10.1016/j.jenvman.2013.02.054, https://www.sciencedirect.com/science/article/pii/S0301479713001527.

Ho, Y. C., Norli, I., Alkarkhi, Abbas F. M., & Morad, N. (2009). Analysis and optimization of flocculation activity and turbidity reduction in kaolin suspension using pectin as a biopolymer flocculant. *Water Science and Technology*, 60(3), 771−781. Available from https://doi.org/10.2166/wst.2009.303, https://doi.org/10.2166/wst.2009.303.

Hochman, Gal, Rajagopal, Deepak, Timilsina, Govinda R., Zilberman, David, Timilsina, Govinda R., & Zilberman, David (2014). Impacts of Biofuels on Food Prices The Impacts of Biofuels on the Economy, Environment, and Poverty (pp. 47−64). New York, New York, NY: Springer. Available from http://link.springer.com/10.1007/978-1-4939-0518-8_4.

Hong, Mei, Li, Die, Wang, Bingyu, Zhang, Jingyu, Peng, Bin, Xu, Xiaoling, Wang, Yan, Bao, Chunyang, Chen, Jing, & Zhang, Qiang (2021). Cellulose-derived polyols as high-capacity adsorbents for rapid boron and organic pollutants removal from water. *Journal of Hazardous Materials*, 419, 126503. Available from https://doi.org/10.1016/j.jhazmat.2021.126503, https://www.sciencedirect.com/science/article/pii/S0304389421014680.

Houghton, R. A., Jørgensen, Sven Erik, & Fath, Brian D. (2008). Biomass Encyclopedia of Ecology. Academic Press. *Oxford*, 448−453. Available from https://www.sciencedirect.com/science/article/pii/B9780080454054004626.

IEA Bioenergy. IEA Bioenergy Annual Report. (2019). This publication was produced by IEA Bioenergy. The IEA Bioenergy Technology Collaboration Programme (IEA Bioenergy TCP) is organised under the auspices of the International Energy Agency (IEA) but is functionally and legally autonomous. Available from: https://www.ieabioenergy.com/wp-content/uploads/2020/05/IEA-Bioenergy-Annual-Report-2019.pdf. Accessed 01.11.23.

Jaihan, Wilavan, Mohdee, Vanee, Sanongraj, Sompop, Pancharoen, Ura, & Nootong, Kasidit (2022). Biosorption of lead (II) from aqueous solution using Cellulose-based Bio-adsorbents prepared from unripe papaya (Carica papaya) peel waste: Removal Efficiency, Thermodynamics, kinetics and isotherm analysis. *Arabian Journal of Chemistry*, 15(7), 103883. Available from https://doi.org/10.1016/j.arabjc.2022.103883, https://www.sciencedirect.com/science/article/pii/S187853522200199X.

Janaswamy, Srinivas, Yadav, Madhav P., Hoque, Mominul, Bhattarai, Sajal, & Ahmed, Shafaet (2022). Cellulosic fraction from agricultural biomass as a viable alternative for plastics and plastic products. *Industrial Crops and Products*, 179, 114692. Available from https://doi.org/10.1016/j.indcrop.2022.114692, https://www.sciencedirect.com/science/article/pii/S0926669022001753.

Jiang, Feng, & Hsieh, You-Lo (2015). Cellulose nanocrystal isolation from tomato peels and assembled nanofibers. *Carbohydrate Polymers*, 122, 60–68. Available from https://doi.org/10.1016/j.carbpol.2014.12.064, https://www.sciencedirect.com/science/article/pii/S0144861714012673.

Jiang, Xincheng, Li, Yisen, Tang, Xiaohui, Jiang, Junyi, He, Qiang, Xiong, Zikang, & Zheng, Huaili (2021). Biopolymer-based flocculants: a review of recent technologies. *Environmental Science and Pollution Research*, 28(34), 46934–46963. Available from https://doi.org/10.1007/s11356-021-15299-y, https://doi.org/10.1007/s11356-021-15299-y.

Kathirselvam, M., Kumaravel, A., Arthanarieswaran, V. P., & Saravanakumar, S. S. (2019). Isolation and characterization of cellulose fibers from Thespesia populnea barks: A study on physicochemical and structural properties. *International Journal of Biological Macromolecules*, 129, 396–406. Available from https://doi.org/10.1016/j.ijbiomac.2019.02.044, https://www.sciencedirect.com/science/article/pii/S014181301836656X.

Kilpimaa, Sari, Runtti, Hanna, Kangas, Teija, Lassi, Ulla, & Kuokkanen, Toivo (2014). Removal of phosphate and nitrate over a modified carbon residue from biomass gasification. *Chemical Engineering Research and Design*, 92(10), 1923–1933. Available from https://doi.org/10.1016/j.cherd.2014.03.019, https://www.sciencedirect.com/science/article/pii/S0263876214001506.

Kougias, P. G., Boe, K., Einarsdottir, E. S., & Angelidaki, I. (2015). Counteracting foaming caused by lipids or proteins in biogas reactors using rapeseed oil or oleic acid as antifoaming agents. *Water Research*, 79, 119–127. Available from https://doi.org/10.1016/j.watres.2015.04.034, https://www.sciencedirect.com/science/article/pii/S0043135415002717.

Kumneadklang, Sureeporn, O-Thong, Sompong, & Larpkiattaworn, Siriporn (2019). Characterization of cellulose fiber isolated from oil palm frond biomass. *Materials Today: Proceedings*, 17, 1995–2001. Available from https://doi.org/10.1016/j.matpr.2019.06.247, https://www.sciencedirect.com/science/article/pii/S2214785319316141.

Kurniawan, Setyo Budi, Abdullah, Siti Rozaimah Sheikh, Imron, Muhammad Fauzul, Mohd Said, Nor Sakinah, 'Izzati Ismail, Nur, Hasan, Hassimi Abu, Othman, Ahmad Razi, & Purwanti, Ipung Fitri (2020). Challenges and Opportunities of Biocoagulant/Bioflocculant Application for Drinking Water and Wastewater Treatment and Its Potential for Sludge Recovery. *International Journal of Environmental Research and Public Health*, 17(24), 9312. Available from https://doi.org/10.3390/ijerph17249312, https://www.ncbi.nlm.nih.gov/pmc/articles/PMC7764310/.

Ledezma-Espinoza, Aura, Rodríguez-Quesada, Laria, Araya-Leitón, María, Avendaño-Soto, Esteban D., & Starbird-Perez, Ricardo (2022). Modified cellulose/poly(3,4-ethylenedioxythiophene) composite as photocatalyst for the removal of sulindac and carbamazepine from water. *Environmental Technology & Innovation*, 27, 102483. Available from https://doi.org/10.1016/j.eti.2022.102483, https://www.sciencedirect.com/science/article/pii/S2352186422001092.

Lichtfouse, Eric, Morin-Crini, Nadia, Fourmentin, Marc, Zemmouri, Hassiba, Nascimento, Inara Oliveira do Carmo, Queiroz, Luciano Matos, Tadza, Mohd Yuhyi, Picos-Corrales, Lorenzo A., Pei, Haiyan, Wilson, Lee D., & Crini, Grégorio (2019). Chitosan for direct bioflocculation of wastewater. *Environmental Chemistry Letters*, 17(4), 1603–1621. Available from https://doi.org/10.1007/s10311-019-00900-1, https://hal.archives-ouvertes.fr/hal-02381712.

Liu, Cheng-mei, Guo, Xiao-juan, Liang, Rui-hong, Liu, Wei, & Chen, Jun (2017). Alkylated pectin: Molecular characterization, conformational change and gel property. *Food Hydrocolloids*, 69, 341–349. Available from https://doi.org/10.1016/j.foodhyd.2017.03.008, https://www.sciencedirect.com/science/article/pii/S0268005X16308025.

Liu, Weijie, Dong, Zhen, Sun, Di, Chen, Ying, Wang, Shiwei, Zhu, Jingrong, & Liu, Cong (2019). Bioconversion of kitchen wastes into bioflocculant and its pilot-scale application in treating iron mineral processing wastewater. *Bioresource Technology*, 288, 121505. Available from https://doi.org/10.1016/j.biortech.2019.121505, https://www.sciencedirect.com/science/article/pii/S0960852419307357.

Liu, Yi, Fan, Hongying, Wang, Xuan, Zhang, Jian, Li, Wenting, & Wang, Rong (2022). Controllable synthesis of bifunctional corn stalk cellulose as a novel adsorbent for efficient removal of Cu^{2+} and Pb^{2+} from wastewater. *Carbohydrate Polymers*, 276, 118763. Available from https://doi.org/10.1016/j.carbpol.2021.118763, https://www.sciencedirect.com/science/article/pii/S0144861721011504.

Lu, S. G., Bai, S. Q., Zhu, L., & Shan, H. D. (2009). Removal mechanism of phosphate from aqueous solution by fly ash. *Journal of Hazardous Materials*, 161(1), 95–101. Available from https://doi.org/10.1016/j.jhazmat.2008.02.123, https://www.sciencedirect.com/science/article/pii/S0304389408004391.

Makroo, H. A., Naqash, S., Saxena, J., Sharma, Savita, Majid, D., & Dar, B. N. (2021). Recovery and characteristics of starches from unconventional sources and their potential applications: A review. *Applied Food Research*, 1(1), 100001. Available from https://doi.org/10.1016/j.afres.2021.100001, https://www.sciencedirect.com/science/article/pii/S2772502221000019.

Martínez-Sanz, Marta, Cebrián-Lloret, Vera, Mazarro-Ruiz, Jesús, & López-Rubio, Amparo (2020). Improved performance of less purified cellulosic films obtained from agar waste biomass. *Carbohydrate Polymers*, 233, 115887. Available from https://doi.org/10.1016/j.carbpol.2020.115887, https://www.sciencedirect.com/science/article/pii/S0144861720300618.

Melikoğlu, Arzu Yalçın, Bilek, Seda Ersus, & Cesur, Serap (2019). Optimum alkaline treatment parameters for the extraction of cellulose and production of cellulose nanocrystals from apple pomace. *Carbohydrate Polymers*, 215, 330–337. Available from https://doi.org/10.1016/j.carbpol.2019.03.103, https://www.sciencedirect.com/science/article/pii/S0144861719303820.

Mohamed, Mohamad Azuwa, Mutalib, Muhazri Abd, Mohd Hir, Zul Adlan, Zain, M. F. M., Mohamad, Abu Bakar, Minggu, Lorna Jeffery, Asikin Awang, Nor, & Salleh, W. N. W. (2017). An overview on cellulose-based material in tailoring bio-hybrid nanostructured photocatalysts for water treatment and renewable energy applications. *International Journal of Biological Macromolecules*, 103, 1232–1256. Available from https://doi.org/10.1016/j.ijbiomac.2017.05.181, http://www.sciencedirect.com/science/article/pii/S0141813017312722.

Mohamed Noor, Mohamed Hizam, Ngadi, Norzita, Mohammed Inuwa, Ibrahim, Opotu, Lawal Anako, & Mohd Nawawi, Mohd Ghazali (2020). Synthesis and application of polyacrylamide grafted magnetic cellulose flocculant for palm oil wastewater treatment. *Journal of Environmental Chemical Engineering*, 8(4), 104014. Available from https://doi.org/10.1016/j.jece.2020.104014, https://www.sciencedirect.com/science/article/pii/S2213343720303626.

Mohammed, Akeem, Bissoon, Rakesh, Bajnath, Elisheba, Mohammed, Kristy, Lee, Thérèse, Bissram, Meera, John, Nigel, Jalsa, NigelK., Lee, Koon-Yang, & Ward, Keeran (2018). Multistage extraction and purification of waste Sargassum natans to produce sodium alginate: An optimization approach. *Carbohydrate Polymers*, *198*, 109−118. Available from https://doi.org/10.1016/j.carbpol.2018.06.067, https://www.sciencedirect.com/science/article/pii/S0144861718307215.

Mohan, S. Mariraj (2014). Use of naturalized coagulants in removing laundry waste surfactant using various unit processes in lab-scale. *Journal of Environmental Management*, *136*, 103−111. Available from https://doi.org/10.1016/j.jenvman.2014.02.004, https://www.sciencedirect.com/science/article/pii/S0301479714000632.

Mohanty, Swayansu Sabyasachi, Koul, Yamini, Varjani, Sunita, Pandey, Ashok, Ngo, Huu Hao, Chang, Jo-Shu, Wong, Jonathan W. C., & Bui, Xuan-Thanh (2021). A critical review on various feedstocks as sustainable substrates for biosurfactants production: a way towards cleaner production. *Microbial Cell Factories*, *20*(1), 120. Available from https://doi.org/10.1186/s12934-021-01613-3, https://doi.org/10.1186/s12934-021-01613-3.

Mueller, Silvana, Weder, Christoph, & Foster, E. Johan (2014). Isolation of cellulose nanocrystals from pseudostems of banana plants. *RSC Advances*, *4*(2), 907−915. Available from https://doi.org/10.1039/C3RA46390G, http://doi.org/10.1039/C3RA46390G.

Muhammad, Nawshad, Man, Zakaria, Azmi Bustam, M., Mutalib, M. I. Abdul, & Rafiq, Sikander (2013). Investigations of novel nitrile-based ionic liquids as pre-treatment solvent for extraction of lignin from bamboo biomass. *Journal of Industrial and Engineering Chemistry*, *19*(1), 207−214. Available from https://doi.org/10.1016/j.jiec.2012.08.003, https://www.sciencedirect.com/science/article/pii/S1226086X12002626.

Myllymäki, Pekka, Pesonen, Janne, Nurmesniemi, Emma-Tuulia, Romar, Henrik, Tynjälä, Pekka, Hu, Tao, & Lassi, Ulla (2020). The Use of Industrial Waste Materials for the Simultaneous Removal of Ammonium Nitrogen and Phosphate from the Anaerobic Digestion Reject Water. *Waste and Biomass Valorization*, *11*(8), 4013−4024. Available from https://doi.org/10.1007/s12649-019-00724-8, https://doi.org/10.1007/s12649-019-00724-8.

Nadar, Shamraja S., Rao, Priyanka, & Rathod, Virendra K. (2018). Enzyme assisted extraction of biomolecules as an approach to novel extraction technology: A review. *Food Research International*, *108*, 309−330. Available from https://doi.org/10.1016/j.foodres.2018.03.006, https://www.sciencedirect.com/science/article/pii/S0963996918301741.

Nasrollahzadeh, Mahmoud, Sajjadi, Mohaddeseh, Iravani, Siavash, & Varma, Rajender S. (2021). Starch, cellulose, pectin, gum, alginate, chitin and chitosan derived (nano)materials for sustainable water treatment: A review. *Carbohydrate Polymers*, *251*, 116986. Available from https://doi.org/10.1016/j.carbpol.2020.116986, https://www.sciencedirect.com/science/article/pii/S0144861720311590.

Naylor, L. M., & Schmidt, E. J. (1986). Agricultural use of wood ash as a fertilizer and liming material. *Tappi; (United States)*, *69*, 10. Available from https://www.osti.gov/biblio/6379258.

Nechita, Petronela (2017). *Applications of Chitosan in Wastewater Treatment*. IntechOpen. Available from https://www.intechopen.com/chapters/undefined/state.item.id.

Novais, Rui M., Ascensão, Guilherme, Tobaldi, David M., Seabra, Maria P., & Labrincha, João A. (2018). Biomass fly ash geopolymer monoliths for effective methylene blue removal from wastewaters. *Journal of Cleaner Production*, *171*, 783−794. Available from https://doi.org/10.1016/j.jclepro.2017.10.078, https://www.sciencedirect.com/science/article/pii/S0959652617323715.

Ogutu, Fredrick Onyango, & Mu, Tai-Hua (2017). Ultrasonic degradation of sweet potato pectin and its antioxidant activity. *Ultrasonics Sonochemistry*, *38*, 726−734. Available from https://doi.org/10.1016/j.ultsonch.2016.08.014, https://www.sciencedirect.com/science/article/pii/S1350417716302851.

Ohno, Tsutomu, & Susan Erich, M. (1990). Effect of wood ash application on soil pH and soil test nutrient levels. *Agriculture, Ecosystems & Environment*, *32*(3), 223−239. Available from https://doi.org/10.1016/0167-8809(90)90162-7, https://www.sciencedirect.com/science/article/pii/0167880990901627.

Oksman, Kristiina, Etang, Jackson A., Mathew, Aji P., & Jonoobi, Mehdi (2011). Cellulose nanowhiskers separated from a bio-residue from wood bioethanol production. *Biomass and Bioenergy*, *35*(1), 146−152. Available from https://doi.org/10.1016/j.biombioe.2010.08.021, https://www.sciencedirect.com/science/article/pii/S0961953410002783.

Özacar, Mahmut, & Şengil, İ. Ayhan (2003). Evaluation of tannin biopolymer as a coagulant aid for coagulation of colloidal particles. *Colloids and Surfaces A: Physicochemical and Engineering Aspects*, *229*(1), 85−96. Available from https://doi.org/10.1016/j.colsurfa.2003.07.006, https://www.sciencedirect.com/science/article/pii/S092777570300462X.

Özacar, Mahmut, & Şengıl, Ayhan (2000). Effectiveness of tannins obtained from valonia as a coagulant aid for dewatering of sludge. *Water Research*, *34*(4), 1407−1412. Available from https://doi.org/10.1016/S0043-1354(99)00276-6, https://www.sciencedirect.com/science/article/pii/S0043135499002766.

Padma Ishwarya, S., & Nisha, P. (2021). Foaming agents from spent coffee grounds: A mechanistic understanding of the modes of foaming and the role of coffee oil as antifoam. *Food Hydrocolloids*, *112*, 106354. Available from https://doi.org/10.1016/j.foodhyd.2020.106354, https://www.sciencedirect.com/science/article/pii/S0268005X20310626.

Pandi, Narsimha, Sonawane, Shirish H., & Kishore, K. Anand (2021). Synthesis of cellulose nanocrystals (CNCs) from cotton using ultrasound-assisted acid hydrolysis. *Ultrasonics Sonochemistry*, *70*, 105353. Available from https://doi.org/10.1016/j.ultsonch.2020.105353, https://www.sciencedirect.com/science/article/pii/S1350417720308178.

Pei, Yanbo, Li, Menglin, Li, Wei, Su, Kai, Chen, Junmin, Yang, Hongwei, Hu, Daiyan, & Zhang, Shengli (2021). Cr(VI) removal by cellulose-based composite adsorbent with a double-network structure. *Colloids and Surfaces A: Physicochemical and Engineering Aspects*, *625*, 126963. Available from https://doi.org/10.1016/j.colsurfa.2021.126963, https://www.sciencedirect.com/science/article/pii/S0927775721008323.

Pesonen, Janne, Sauvola, Emilia, Hu, Tao, & Tuomikoski, Sari (2020). Desalination and Water Treatment Use of side stream-based $MgSO_4$ as chemical precipitant in the simultaneous removal of nitrogen and phosphorus from wastewaters. *Desalination and water treatment*, *194*, 389−395. Available from https://doi.org/10.5004/dwt.2020.26037.

Piazza, G. J., & Garcia, R. A. (2010). Meat & bone meal extract and gelatin as renewable flocculants. *Bioresource Technology*, *101*(2), 781−787. Available from https://doi.org/10.1016/j.biortech.2009.03.078, https://www.sciencedirect.com/science/article/pii/S0960852409003629.

Pinto, Alexandre H., Taylor, Jeffrey K., Chandradat, Richard, Lam, Edmond, Liu, Yali, Leung, Alfred C. W., Keating, Michael, & Sunasee, Rajesh (2020). Wood-based cellulose nanocrystals as adsorbent of cationic toxic dye, Auramine O, for water treatment. *Journal of Environmental Chemical Engineering*, *8*(5), 104187. Available from https://doi.org/10.1016/j.jece.2020.104187, https://www.sciencedirect.com/science/article/pii/S2213343720305364.

Piriyaprasarth, Suchada, & Sriamornsak, Pornsak (2011). Flocculating and suspending properties of commercial citrus pectin and pectin extracted from pomelo (Citrus maxima) peel. *Carbohydrate Polymers, 83*(2), 561–568. Available from https://doi.org/10.1016/j.carbpol.2010.08.018, https://www.sciencedirect.com/science/article/pii/S0144861710006387.

Pontius, Frederick W. (2016). Chitosan as a Drinking Water Treatment Coagulant. *American Journal of Civil Engineering, 4*(5), 205. Available from https://doi.org/10.11648/j.ajce.20160405.11, https://www.sciencepublishinggroup.com/journal/paperinfo?journalid = 229&doi = 10.11648/j.ajce.20160405.11.

Provost, V., Dumarcay, S., Ziegler-Devin, I., Boltoeva, M., Trébouet, D., & Villain-Gambier, M. (2022). Deep eutectic solvent pretreatment of biomass: Influence of hydrogen bond donor and temperature on lignin extraction with high 3-O-4 content. *Bioresource Technology, 349*, 126837. Available from https://doi.org/10.1016/j.biortech.2022.126837, https://www.sciencedirect.com/science/article/pii/S0960852422001663.

Qin, Tingting, Wang, Zhaowei, Xie, Xiaoyun, Xie, Chaoran, Zhu, Junmin, & Li, Yan (2017). A novel biochar derived from cauliflower (Brassica oleracea L.) roots could remove norfloxacin and chlortetracycline efficiently. *Water Science and Technology, 76*(12), 3307–3318. Available from https://doi.org/10.2166/wst.2017.494, https://iwaponline.com/wst/article/76/12/3307/38386/A-novel-biochar-derived-from-cauliflower-Brassica.

Rajapaksha, Anushka Upamali, Vithanage, Meththika, Lee, Sang Soo, Seo, Dong-Cheol, Tsang, Daniel C. W., & Ok, Yong Sik (2016). Steam activation of biochars facilitates kinetics and pH-resilience of sulfamethazine sorption. *Journal of Soils and Sediments, 16*(3), 889–895. Available from https://doi.org/10.1007/s11368-015-1325-x, https://doi.org/10.1007/s11368-015-1325-x.

Rao, Priyanka, & Rathod, Virendra (2019). Valorization of Food and Agricultural Waste: A Step towards Greener Future. *The Chemical Record, 19*(9), 1858–1871. Available from https://doi.org/10.1002/tcr.201800094, https://onlinelibrary.wiley.com/doi/abs/10.1002/tcr.201800094.

Rashid, Tazien, Taqvi, Syed Ali Ammar, Sher, Farooq, Rubab, Saddaf, Thanabalan, Murugesan, Bilal, Muhammad, & Islam, Badar ul (2021). Enhanced lignin extraction and optimisation from oil palm biomass using neural network modelling. *Fuel, 293*, 120485. Available from https://doi.org/10.1016/j.fuel.2021.120485, https://www.sciencedirect.com/science/article/pii/S0016236121003616.

Ratnakumar, A., Samarasekara, A. M. P. B., Amarasinghe, D. A. S., & Karunanayake, L. (2022). The influence of particle size on the extraction of cellulose nanofibers using chemical-ultrasonic process. *Materials Today: Proceedings*. Available from https://doi.org/10.1016/j.matpr.2022.04.518, https://www.sciencedirect.com/science/article/pii/S2214785322026499.

Ravindran, Lakshmipriya, Sreekala, M. S., & Thomas, Sabu (2019). Novel processing parameters for the extraction of cellulose nanofibres (CNF) from environmentally benign pineapple leaf fibres (PALF): Structure-property relationships. *International Journal of Biological Macromolecules, 131*, 858–870. Available from https://doi.org/10.1016/j.ijbiomac.2019.03.134, https://www.sciencedirect.com/science/article/pii/S0141813018365097.

Reguyal, Febelyn, & Sarmah, Ajit K. (2018). Adsorption of sulfamethoxazole by magnetic biochar: Effects of pH, ionic strength, natural organic matter and 17α-ethinylestradiol. *Science of The Total Environment, 628-629*, 722–730. Available from https://doi.org/10.1016/j.scitotenv.2018.01.323, https://www.sciencedirect.com/science/article/pii/S0048969718303656.

Roselet, Fabio, Burkert, Janaína, & Abreu, Paulo Cesar (2016). Flocculation of Nannochloropsis oculata using a tannin-based polymer: Bench scale optimization and pilot scale reproducibility. *Biomass and Bioenergy, 87*, 55–60. Available from https://doi.org/10.1016/j.biombioe.2016.02.015, https://www.sciencedirect.com/science/article/pii/S0961953416300368.

Sakthivel, S. Ramesh, Tilley, Elizabeth, & Udert, Kai M. (2012). Wood ash as a magnesium source for phosphorus recovery from source-separated urine. *Science of The Total Environment, 419*, 68–75. Available from https://doi.org/10.1016/j.scitotenv.2011.12.065, https://linkinghub.elsevier.com/retrieve/pii/S0048969711015312.

Saleem, Mussarat, & Bachmann, Robert Thomas (2019). A contemporary review on plant-based coagulants for applications in water treatment. *Journal of Industrial and Engineering Chemistry, 72*, 281–297. Available from https://doi.org/10.1016/j.jiec.2018.12.029, https://www.sciencedirect.com/science/article/pii/S1226086X18314965.

Saritha, V., Srinivas, N., & Vuppala, N. V. Srikanth (2017). Analysis and optimization of coagulation and flocculation process. *Applied Water Science, 7*(1), 451–460. Available from https://doi.org/10.1007/s13201-014-0262-y, https://doi.org/10.1007/s13201-014-0262-y.

Sarubbo, Leonie A., Silva, Maria da Gloria C., Durval, Italo José B., Bezerra, K. áren Gercyane O., Ribeiro, Beatriz G., Silva, Ivison A., Twigg, Matthew S., & Banat, Ibrahim M. (2022). Biosurfactants: Production, properties, applications, trends, and general perspectives. *Biochemical Engineering Journal, 181*, 108377. Available from https://doi.org/10.1016/j.bej.2022.108377, https://www.sciencedirect.com/science/article/pii/S1369703X22000468.

Shahadat, Mohammad, Teng, Tjoon Tow, Rafatullah, Mohd, Shaikh, Z. A., Sreekrishnan, T. R., & Ali, S. Wazed (2017). Bacterial bioflocculants: A review of recent advances and perspectives. *Chemical Engineering Journal, 328*, 1139–1152. Available from https://doi.org/10.1016/j.cej.2017.07.105, https://www.sciencedirect.com/science/article/pii/S1385894717312548.

Shakoor, Muhammad Bilal, Niazi, Nabeel Khan, Bibi, Irshad, Shahid, Muhammad, Sharif, Fakhra, Bashir, Safdar, Shaheen, Sabry M., Wang, Hailong, Tsang, Daniel C. W., Ok, Yong Sik, & Rinklebe, J. örg (2018). Arsenic removal by natural and chemically modified water melon rind in aqueous solutions and groundwater. *Science of The Total Environment, 645*, 1444–1455. Available from https://doi.org/10.1016/j.scitotenv.2018.07.218, https://www.sciencedirect.com/science/article/pii/S0048969718327104.

Sheltami, Rasha M., Abdullah, Ibrahim, Ahmad, Ishak, Dufresne, Alain, & Kargarzadeh, Hanieh (2012). Extraction of cellulose nanocrystals from mengkuang leaves (Pandanus tectorius). *Carbohydrate Polymers, 88*(2), 772–779. Available from https://doi.org/10.1016/j.carbpol.2012.01.062, https://www.sciencedirect.com/science/article/pii/S0144861712000823.

da Silva, Daniel J., & Rosa, Derval S. (2022). Chromium removal capability, water resistance and mechanical behavior of foams based on cellulose nanofibrils with citric acid. *Polymer, 253*, 125023. Available from https://doi.org/10.1016/j.polymer.2022.125023, https://www.sciencedirect.com/science/article/pii/S0032386122005110.

Solanki, Jyoti D., Patel, Dhaval T., Patel, Kamlesh C., Nataraj, M., Inamuddin., & Adetunji, Charles Oluwaseun (2021). *Chapter 10 - Production of biosurfactants using agroindustrial wastes as substrates Green Sustainable Process for Chemical and Environmental Engineering and Science* (pp. 185–210). Elsevier. Available from https://www.sciencedirect.com/science/article/pii/B9780128226964000127.

Sriwong, Chotiwit, & Sukyai, Prakit (2022). Simulated elephant colon for cellulose extraction from sugarcane bagasse: An effective pretreatment to reduce chemical use. *Science of The Total Environment, 835*, 155281. Available from https://doi.org/10.1016/j.scitotenv.2022.155281, https://www.sciencedirect.com/science/article/pii/S0048969722023749.

Stanisz, Ma. Łgorzata, Smułek, Wojciech, Popielski, Krzysztof, Klapiszewski, Łukasz, Kaczorek, Ewa, & Jesionowski, Teofil (2021). Sustainable design of lignin-based spherical particles with the use of green surfactants and its application as sorbents in wastewater treatment. *Chemical Engineering Research and Design, 172*, 34–42. Available from https://doi.org/10.1016/j.cherd.2021.05.028, https://www.sciencedirect.com/science/article/pii/S0263876221002318.

Sánchez-Martín, J., Beltrán-Heredia, J., & Solera-Hernández, C. (2010). Surface water and wastewater treatment using a new tannin-based coagulant. *Pilot plant trials. Journal of Environmental Management, 91*(10), 2051–2058. Available from https://doi.org/10.1016/j.jenvman.2010.05.013, https://www.sciencedirect.com/science/article/pii/S0301479710001428.

Thakur, Sourbh, Govender, Penny P., Mamo, Messai A., Tamulevicius, Sigitas, Mishra, Yogendra Kumar, & Thakur, Vijay Kumar (2017). Progress in lignin hydrogels and nanocomposites for water purification: Future perspectives. *Vacuum, 146*, 342–355. Available from https://doi.org/10.1016/j.vacuum.2017.08.011, https://www.sciencedirect.com/science/article/pii/S0042207X17306048.

Tomasi, Isabella T., Machado, Cláudia A., Boaventura, Rui A. R., Botelho, Cidália M. S., & Santos, S. ílvia C. R. (2022). Tannin-based coagulants: Current development and prospects on synthesis and uses. *The Science of the Total Environment, 822*, 153454. Available from https://doi.org/10.1016/j.scitotenv.2022.153454.

Vassilev, Stanislav V., Baxter, David, Andersen, Lars K., & Vassileva, Christina G. (2010). An overview of the chemical composition of biomass. *Fuel, 89*(5), 913–933. Available from https://doi.org/10.1016/j.fuel.2009.10.022, https://www.sciencedirect.com/science/article/pii/S0016236109004967.

Wandee, Yuree, Uttapap, Dudsadee, & Mischnick, Petra (2019). Yield and structural composition of pomelo peel pectins extracted under acidic and alkaline conditions. *Food Hydrocolloids, 87*, 237–244. Available from https://doi.org/10.1016/j.foodhyd.2018.08.017, https://www.sciencedirect.com/science/article/pii/S0268005X18304946.

Kaifeng Wang, Na Peng, Jianteng Sun, Guining Lu, Meiqin Chen, Fucai Deng, Rongni Dou, Lijun Nie, Yongming Zhong, Synthesis of silica-composited biochars from alkali-fused fly ash and agricultural wastes for enhanced adsorption of methylene blue. Science of The Total Environment. 729 (2020), 139055. Available from: https://doi.org/10.1016/j.scitotenv.2020.139055, https://www.sciencedirect.com/science/article/pii/S0048969720325729.

Wang, Xue Chuan, Zhang, Sha, Zhou, Liang, & Ren, Long Fang (2011). The Hydrolysis of Gelatin and its Treatment for Waste Drilling Fluid. *Advanced Materials Research, 281*, 141–146. Available from https://doi.org/10.4028/http://www.scientific.net/AMR.281.141, https://www.scientific.net/AMR.281.141.

Watkins, Dereca, Nuruddin, Md, Hosur, Mahesh, Tcherbi-Narteh, Alfred, & Jeelani, Shaik (2015). Extraction and characterization of lignin from different biomass resources. *Journal of Materials Research and Technology, 4*(1), 26–32. Available from https://doi.org/10.1016/j.jmrt.2014.10.009, https://www.sciencedirect.com/science/article/pii/S2238785414000982.

Wood US EPA OLEM 2017 2017 2022 6 30 2022/06/30/08:46:51 2017 9 12 2017/09/12/T14:45:09-04:00 This page describes the generation, recycling, combustion with energy recovery, and landfilling of wood materials, and explains how EPA classifies such material. Wood: Material-Specific Data Wood https://www.epa.gov/facts-and-figures-about-materials-waste-and-recycling/wood-material-specific-data.

Wu, Lin, Yang, Ningwei, Li, Binghua, & Bi, Erping (2018). Roles of hydrophobic and hydrophilic fractions of dissolved organic matter in sorption of ketoprofen to biochars. *Environmental Science and Pollution Research, 25*(31), 31486–31496. Available from https://doi.org/10.1007/s11356-018-3071-2, https://doi.org/10.1007/s11356-018-3071-2.

Wu, Yunhai, Fan, Yiang, Zhang, Meili, Ming, Zhu, Yang, Shengxin, Arkin, Aynigar, & Fang, Peng (2016). Functionalized agricultural biomass as a low-cost adsorbent: Utilization of rice straw incorporated with amine groups for the adsorption of Cr(VI) and Ni(II) from single and binary systems. *Biochemical Engineering Journal, 105*, 27–35. Available from https://doi.org/10.1016/j.bej.2015.08.017, https://www.sciencedirect.com/science/article/pii/S1369703X15300449.

Yang, Jin-Shu, Mu, Tai-Hua, & Ma, Meng-Mei (2018). Extraction, structure, and emulsifying properties of pectin from potato pulp. *Food Chemistry, 244*, 197–205. Available from https://doi.org/10.1016/j.foodchem.2017.10.059, https://www.sciencedirect.com/science/article/pii/S0308814617316898.

Zeng, Jia, Zhang, Dongqiao, Liu, Weifeng, Huang, Jinhao, Yang, Dongjie, Qiu, Xueqing, & Li, Shenhui (2022). Preparation of carboxymethylated lignin-based multifunctional flocculant and its application for copper-containing wastewater. *European Polymer Journal, 164*, 110967. Available from https://doi.org/10.1016/j.eurpolymj.2021.110967, https://www.sciencedirect.com/science/article/pii/S0014305721007011.

Zhu, Guoting, Xing, Xianjun, Wang, Jiaquan, & Zhang, Xianwen (2017). Effect of acid and hydrothermal treatments on the dye adsorption properties of biomass-derived activated carbon. *Journal of Materials Science, 52*(13), 7664–7676. Available from https://doi.org/10.1007/s10853-017-1055-0, https://doi.org/10.1007/s10853-017-1055-0.

Zhu, Shiyun, Xu, Jun, Wang, Bin, Xie, Junxian, Ying, Guangdong, Li, Jinpeng, Cheng, Zheng, Li, Jun, & Chen, Kefu (2022). Highly efficient and rapid purification of organic dye wastewater using lignin-derived hierarchical porous carbon. *Journal of Colloid and Interface Science, 625*, 158–168. Available from https://doi.org/10.1016/j.jcis.2022.06.019, https://www.sciencedirect.com/science/article/pii/S0021979722009833.

Zhu, Tian, Li, Yijing, Yang, Hao, Liu, Jingguang, Tao, Yanzhi, Gan, Weixing, Wang, Shuangfei, & Nong, Guangzai (2021). Preparati on of an amphoteric adsorbent from cellulose for wastewater treatment. *Reactive and Functional Polymers, 169*, 105086. Available from https://doi.org/10.1016/j.reactfunctpolym.2021.105086, https://www.sciencedirect.com/science/article/pii/S1381514821002789.

CHAPTER 23

Waste-biomass-derived potential catalyst materials for water reclamation

Varsha Srivastava, Anne Heponiemi, Sari Tuomikoski, Riikka Kupila, Davide Bergna and Ulla Lassi

Research Unit of Sustainable Chemistry, Faculty of Technology, University of Oulu, Oulu, Finland

Highlights

1. Lignocellulosic biomass-derived carbon materials are economic and active catalytic materials for advanced oxidation processes (AOPs)
2. The composition of biomass and its conversion technique have profound impacts on the properties of resulting carbonaceous materials.
3. The inherent functional groups in biomass-derived carbonaceous materials stimulate catalytic activity.
4. The photodegradation of recalcitrant pollutants is improved in the heterojunction of carbonaceous materials with semiconductors.

Abbreviations

AC	Activated carbon
AOPs	Advanced oxidation processes
BC	Biochar
BET	Brunauer, Emmett, and Teller
BM	Bagasse of malt
BOD	Biological oxygen demand
BPA	Bisphenol A
BPMC	Buthylphenylmethyl carbamate
BS	Bagasse of sugarcane
BTU	British thermal unit
CB	Conduction band
COD	Chemical oxygen demand
CR	Carbon residue
CBRN	Chemical–biological–radiological and nuclear
CW	Willow carbon
CWAO	Catalytic wet air oxidation
CWPO	Catalytic wet peroxide oxidation
DFT	Density functional theory
EAC	Extruded activated carbon
EDS	Energy-dispersive X-ray spectroscopy
EC	Emerging Contaminants
Eg	Band gap energy
FTIR	Fourier-transform infrared spectroscopy
GAC	Granulated activated carbon

5-HMF	5 − (hydroxymethyl)furfural
HPLC	High-performance liquid chromatography
HTC	Hydrothermal carbonization
ICP-MS	Inductively coupled plasma−mass spectrometry
ICP-OES	Inductively coupled plasma−optical emission spectrometry
IUPAC	International Union of Pure and Applied Chemistry
MB	Methylene blue
MO	Methyl orange
O_2^-	Superoxide radical
OH	Hydroxyl radicals
PAC	Powdered activated carbon
PVA	Polyvinyl alcohol
ROS	Reactive oxygen species
SC	Seed of chia
SSA	Specific Surface Area
TBO	Toluidine blue O
TC	Tetracycline
TNP	2,4,6-trinitrophenol
TOC	Total organic carbon
TPD	Temperature-programmed desorption
TEM	Transmission electron microscopy
SEM	Scanning electron microscopy
XPS	X-ray photoelectron spectroscopy

23.1 Introduction

Biomass can be classified as renewable organic material. The usage of renewable materials is the basis for sustainable development. The use of biomass is important because it avoids the carbon dioxide emission formation from fossil fuels. Biomass use gives also diversity to the markets, enabling the use of different raw materials from alternative sources and providing employment opportunities, especially in local and rural areas (Goldemberg & Teixeira Coelho, 2004). Lignocellulosic biomass, also known as lignocellulose, is the most abundant, highly renewable, and most economical natural resource in the world (Ge et al., 2018; Qian, 2014; Yousuf et al., 2020). Lignocellulosic biomass can be defined as material that is plant or plant-based, and it has no potential to be used as food or feed. Lignocellulosic biomass is typically divided into biomass, virgin biomass, and energy crops. Lignocellulosic biomass includes mainly different agricultural and forest residues, energy crops, and yard trimmings (Table 23.1). Cellulose (9%−80%), hemicellulose (10%−50%), and lignin (5%−35%) are the main polymer components of lignocellulosic biomass and remaining fractions include, e.g., ash, oils, and proteins (Wei et al., 2017; Xu & Li, 2017).

About 340 million mt of lignocellulosic biomass was available for bioenergy production in the United States in 2012 (Xu & Li, 2017; Yousuf et al., 2020) while 4532 trillion BTU (British thermal unit) biomass was used for energy consumption in the United States in 2020 (Jaganmohan, 2022). According to the European legislation following the Waste Framework Directive in the waste management, "every member should find possible reuse of unused materials coming from a primary feedstock creating the concept of secondary raw material." The resource recovery is a fundamental part and is correlated with waste valorization through different processes.

A definition of material recovery should be introduced to cover forms of recovery other than energy recovery and other than the reprocessing of waste into materials used as fuels or other means to generate energy. It includes preparing for reuse, recycling, backfilling, and other forms of material recovery such as the reprocessing of waste into secondary raw materials for engineering purposes in the construction of roads or other infrastructure. Depending on the specific factual circumstances, such reprocessing can fulfill the definition of recycling if the use of materials is based on proper quality control and meets all relevant standards, norms, specifications, and environmental and health protection requirements for the specific use (European Parliament, European Council, 2018).

In the European Union level, annual biomass produced in the land-based sectors (agriculture and forestry) is on average 1466 mt in dry matter (956 mt in agriculture and 510 mt in forestries). Biomass harvested and used in 2013 from EU agricultural and forestry sectors was ca 805 mt dry matter (578 mt agriculture and 227 mt forestry). The difference in numbers is caused by the fact that part of biomass remains in the field, for example, as a carbon sink (Camia et al., 2018).

23.1 Introduction

TABLE 23.1 Lignocellulosic biomass waste composition (wt. %).

Lignocellulosic biomass	Cellulose %	Hemicellulose %	Lignin %	References
Olive stone	30.80	17.10	32.6	González et al. (2009)
Rice husk	37.10	29.40	24.14	Kalita et al. (2015)
Pomegranate seeds	26.98	25.52	39.67	Uçar et al. (2009)
Rice straw	28.00	32.15	19.64	Shawky et al. (2011)
Citrus peel waste	25.40	9.40	23.60	John et al. (2017)
Corn stalks	33.30	29.80	16.65	Shawky et al. (2011)
Walnut shell	40.10	20.70	18.20	González et al. (2009)
Soybean straw	34.09	16.05	21.60	Wan et al. (2011)
Sunflower residue	58.30	29.80	11.90	Lee and Park (2020)
Coconut fiber	31.60	26.33	25.02	Gonçalves et al. (2019)
Oak sawdust	41.90	6.30	26.20	Atila (2019)
Sweet sorghum bagasse	41.60	25.10	20.30	Kim and Day (2011)
Cherry Stones	29.40	14.70	30.70	González et al. (2003)
Orange peel	14.40	10.90	1.33	Bicu and Mustata (2011)
Palm shell	29.70	47.70	53.40	Adinata et al. (2007)
Miscanthus	41.40	19.70	22.60	Ivanovski et al. (2022)
Oak waste wood	38.30	25.50	22.00	Ivanovski et al. (2022)
Groundnut shells	38.31	27.62	21.10	Bano and Negi (2017)
Almond tree pruning	33.70	20.10	25.00	González et al. (2009)
Eucalyptus wood	34.70	27.30	35.80	Elyounssi et al. (2012)
Aspen	47.14	19.64	22.11	Pedersen et al. (2016)
Sunflower stalks	34.00	20.80	29.70	Monlau et al. (2012)
Napier grass	47.10	31.20	21.60	Reddy et al. (2018)
Pine wood	42.00	23.00	24.00	Shemfe et al. (2015)
Coconut shell	19.80	68.70	30.10	Adinata et al. (2007)
Mixed waste wood	37.20	23.80	27.00	Ivanovski et al. (2022)
Almond shell	32.50	25.50	24.80	González et al. (2009)

Every year, a massive chunk of lignocellulosic biomass waste is generated, which demands further treatment due to environmental constraints. Lignocellulosic biomass waste might originate from a single source or multiple sources. Different by-products such as gas and tar residue generated during the conversion of lignocellulosic biomass waste to carbonaceous materials require additional treatment. Various ways for the transformation of biomass waste into valuable products have been investigated for the exploitation of biomass waste (Adams et al., 2018; Xu & Li, 2017).

23.1.1 Biomass conversion methods

There are several applications for biomass use, and it can be converted into different products by using different conversion technologies, which include mainly biochemical, thermochemical, and physicochemical processes (Adams et al., 2018). Biochemical conversion includes both anaerobic digestion and fermentation. Four main processes for thermochemical biomass conversion are pyrolysis, gasification, combustion, and hydrothermal processing called also liquefaction. Physicochemical conversion includes mainly extraction (Adams et al., 2018).

Several kinds of technologies are available to produce different products from biomass such as chemicals, carbonaceous materials, and functional materials. Lignocellulosic biomass has been used for the production of several platform chemicals such as furfural and 5-HMF or in the production of ethanol (Ge et al., 2018; Wei et al., 2017). It can also be used for bioenergy production, biogas production, and as a bulking agent for composting (Xu & Li, 2017). Bioenergy production includes all solid, liquid, or gaseous form fuels that can be produced by using different technologies (Adams et al., 2018).

The aim of this book chapter is to provide an overview of the possible uses of side stream lignocellulosic biomass waste to produce catalyst materials useful for water reclamation. Among all, thermochemical conversion technologies are the most relevant approaches for the conversion of biomass hence discussed in more detail in this book chapter.

The first stage in all thermochemical processes is pyrolysis, which is carried out without oxygen. In pyrolysis, several chemical reactions are happening forming products, which are solid, liquid, and gaseous form (Adams et al., 2018).

Pyrolysis can be divided into fast, slow, and flash pyrolysis (Adams et al., 2018; Patel et al., 2016). Typically slow pyrolysis is performed at the heating rate of 0.1 °C–1 °C /s and temperatures of 277 °C–897 °C (Wang et al., 2017) while fast pyrolysis is made at the heating rate of 10 °C–200 °C/s and temperatures of 577 °C–977 °C

Flash pyrolysis performed at a heating rate of >727 °C/s at a temperature interval of 650 °C–130 °C and residence time less than 0.5 seconds favors a yield-efficient production of liquid condensates (>75% wt.). The reactor designs are similar to the fast pyrolysis ones, but for achieving a complete thermal decomposition in such a fast period of time, the raw material particle size must be smaller (105–250 μm) than in fast pyrolysis (Li et al., 2013).

23.2 Biomass-derived catalyst materials

Novel biomass-derived porous carbons are attractive candidates for the preparation of carbon-based materials with a wide range of catalytic applications. Carbonaceous catalysts are environmentally benign and inexpensive materials, especially when prepared from waste or side stream biomass materials, and this could provide a cost-competitive advantage as compared with existing heterogeneous catalysts (De et al., 2015). Many types of carbon materials have been used to prepare carbon-based catalysts such as graphite, carbon black, activated carbon (AC), carbon fibers, pyrolytic carbon, polymer-derived carbon, fullerenes, and nanotubes. However, AC with its high surface area is the choice for most carbon-supported catalysts (Serp & Figueiredo, 2008; Serp & Machado, 2015).

Biomass-based raw materials such as wood, straws, polymers, and palm and nut shells have been used in the preparation of carbon materials for various catalytic applications (Dias et al., 2007; González-García, 2018; Mohamad Nor et al., 2013). A large portion of carbon materials had been synthesized from fossil fuel sources before the wider exploration of renewable resources (De et al., 2015). Lignocellulosic biomass, as a naturally abundant, carbon-containing, renewable resource, is considered a suitable carbon raw material for the synthesis of functional carbon materials (Liu et al., 2015).

Biomass-derived carbonaceous catalysts have great potential to act as supports and active materials in heterogeneously catalyzed processes including water treatment and electrocatalysis (Davies et al., 2021) and carbonaceous materials have a long history of application in heterogeneous catalysis. Their primary application has been a support to active metal or metal oxide, or they can also act directly as catalysts, known as "carbocatalysts," in a wide range of industrial processes. Carbonaceous materials such as AC, pyrolytic carbon, soot, biochar, and cokes can be used as catalyst material or as a support.

According to the *International Union of Pure and Applied Chemistry*, IUPAC Goldbook AC can be defined as "A porous carbon material, a char which has been subjected to reaction with gases, sometimes with the addition of chemicals, for example, $ZnCl_2$, before, during or after carbonization in order to increase its adsorptive properties" (McNaught & Wilkinson, 2008). ACs can also be defined as "non-hazardous, processed, carbonaceous products, having a porous structure and a large internal surface area. These materials can adsorb a wide variety of substances, that is, they can attract molecules to their internal surface and are therefore called adsorbents. The volume of pores of the ACs is greater than 0.2 cm^3/g. The internal surface area (BET) is greater than 400 m^2/g. The width of the pores ranges from 0.3 to several thousand nanometers" (European council of Chemical Manufacturers federations, 1986).

Further distinction can be made according to the AC particle size. In this case, it is possible to divide

1. Powdered activated carbon (PAC) according to ASTM 2652-94 (reapproved 1999) occurs when the size of the particle is less than 0.18 mm (80USmesh). This distinction can be further made in powder AC >0.045 mm and fine mesh AC $0.045 < x < 0.18$ mm. According to EC COUNCIL REGULATION, No 649/2008 of July 8, 2008, the PAC is defined as 90% mass (w/w) with a particle size less than 0.5 mm.
2. Extruded activated carbon (EAC) is produced by mixing the AC with an opportune binder and subsequently mechanically pressed to obtain a specific shape typically a cylinder. This treatment consent to have some advantages in specific applications where, for instance, an excessive flow load drop can be detrimental. It also enhances the mechanical resistance of carbon.
3. Granulated activated carbon (GAC) is activated carbon with particle size bigger than 0.18 mm.

23.2.1 Activated carbon

The AC is produced at a relatively elevated temperature from nonrenewable raw materials such as coal, lignite, petroleum residues, and peat. Moreover, wood and the lignocellulosic wastes from forestry and agriculture are well-suited raw materials for AC preparation (Khezami et al., 2005). The preparation of the AC is performed through slow carbonization of the raw material and activation with a physical or chemical activation agent. The choice of activating agent and the preparation temperature are major factors in controlling the physical and chemical structures and can affect the performance and applicability of the AC produced. The purpose of activation is to improve the surface area and pore volume of AC by opening new pores and developing existing ones. The specific surface areas (SSAs) of the porous structure of AC can involve approximately 500–3000 m^2/g of buildup from channels with varying diameters (Radovic, 2008). In addition to its high SSA, AC contains a certain amount of heteroatoms on the surface (Rodríguez-reinoso, 1998).

23.2.1.1 Physical activation

The synthesis of AC by physical activation can be performed through two different steps: carbonization and activation. Carbonization is the thermal decomposition (i.e., pyrolysis) and the removal of noncarbon species from the raw material, in which volatile matter is removed by producing high-tenor solid carbon. In the activation process, the removal of the carbon atoms from the nanostructure of the precursor accessing and interconnecting the inherent structural porosity occurs. During this phase, the carbon changes physically; additionally, the carbon changes chemically by opening the carbon matrix and changing its functional groups. The quantity of activating agent used is a function of the type and the mass of carbon source material. In physical activation, the carbonized raw material is activated by the physical activating agent. Typically, the activating agents used are water vapor (steam) or carbon dioxide and those mixtures. The activation temperature varies between 600 °C and 900 °C. The choice of the activation agent and the activation time depends on the type of raw material and the pore size distribution needed. Generally, CO_2 determines a more microporous structure, while steam favors the production of mesopores (Rodríguez-Reinoso et al., 1995). In Fig. 23.1, a typical temperature activation profile is presented. From a chemical point of view, during steam activation, the following compounds exist: O_2, CO_2, H_2, and H_2O. Simplified reactions for steam activation are listed in Table 23.2.

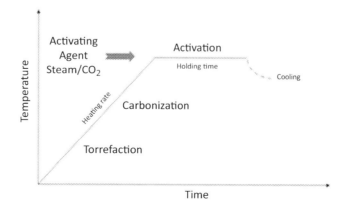

FIGURE 23.1 Scheme of physical activation of carbon.

TABLE 23.2 Simplified reactions occurring during steam activation and relative enthalpy (Bergna et al., 2020).

Reactants	Products	Enthalpy of reaction
$C_x(H_2O)_y$	$xC(s) + yH_2O$	
$C + H_2O$	$CO + H_2$	$\Delta H = 31.14$ kcal/mol
$CO + H_2O$	$CO_2 + H_2$	$\Delta H = -9.65$ kcal/mol
$C + CO_2$	$2CO$	$\Delta H = 40.79$ kcal/mol
$C + 2H_2$	CH_4	$\Delta H = -17.87$ kcal/mol

23.2.1.2 Chemical activation

Chemical activation can be considered a single-step process in which the carbonization of raw material and activation are done simultaneously. In chemical activation, an activating agent, for example acid, alkali, or neutral (salt) (e.g., H_3PO_4, KOH, NaOH or $ZnCl_2$), is impregnated in the raw material before the heat treatment process (Ahmadpour & Do, 1996). After impregnation, chemical activation proceeds as a thermal treatment in an inert atmosphere at different temperatures, depending on the activating agent chosen (Ateş & Özcan, 2018). In chemical activation, impregnated samples are dried in an oven and then heated in a furnace, at temperatures that might be lower like in the case of $ZnCl_2$ (400 °C–600 °C) compared with physical activation (from 600 °C–900 °C), under an inert atmosphere (e.g., nitrogen) (Varila et al., 2017).

Chemicals during the activation step can act as a dehydrating agent (e.g., $ZnCl_2$ and H_3PO_4), and these types of chemicals are commonly used for the activation of especially lignocellulosic materials. Alkali hydroxides are used as chemical activating agents for the activation of coal or chars. Typically chemical activation can determine a different pore size distribution. When $ZnCl_2$ is employed as an activation agent, mesoporous structure is created, whereas KOH activation produces microporous carbon material (Chun & Whitacre, 2017). In the case of chemical activation, important factors include the correct mass-to-carbon ratio needed to avoid excessive consumption of reagents and the proper activation temperature, which can vary according to the type of chemical agent used (Bergna et al., 2022). Differently from physical activation, a final washing phase to remove the precursor material from the chemical activating agent and to uncover the porosity is necessary.

23.2.1.3 Activated carbon characteristics

The most important properties of AC are porosity, SSA, and functional groups on the surface. These properties affect, for example, the mechanism of how the active metal can be impregnated to the AC-supported catalysts. Table 23.3 presents case studies in which lignocellulosic biomass has been activated chemically and/or physically to prepare porous materials and how the raw material and preparation method affect the characteristics of the AC.

23.2.2 Other carbonaceous materials derived from lignocellulosic biomass

In addition to AC, biomass can be converted into a variety of carbonaceous compounds, including biochar, hydrochar, and carbon residue. The definition of biochar has been related mainly to the final use of carbonized biomass. Recently the definition of biochar is bounded as carbon used only as an amendment in the soil or carbon sequestration (Singh et al., 2010). Different standard test methods have been defined to characterize different types of biochar to define the quality and the suitable applications in soil improvement ("A biochar classification system & associated test methods," 2019). The standard quality of the biochar has recently surged high importance also because it can be traded in the EU as C-sink credit for CO_2 uptake. Through the pyrolysis, the biomass is modified, only a partial part of the material is converted to CO_2 while a part becomes stable solid carbon that doesn't undergo further decomposition and change of matter phase (EBC, 2020).

Other biomass-derived carbonaceous material used for catalysis is, for example, hydrochar. Hydrochar is a carbonaceous material with coal-like properties produced in the pressurized low-temperature thermal conversion of a feedstock for example biomass waste sources, and the process is known as hydrothermal carbonization (HTC). In the HTC process, the feedstock is submerged in water in subcritical conditions and heated in the range of 180 °C–260 °C in a closed system at 20–100 bar (Davies et al., 2021). As-synthesized hydrochars often have lower than desired SSA for catalytic applications (20–50m^2/g), alongside a low pore volume; however, they can be modified further by varying process conditions or posttreatment (Davies et al., 2021).

TABLE 23.3 Reaction conditions during chemical or physical activation of carbon.

Lignocellulosic biomass	Activation agent	Temperature [°C]	Retention time [h]	Characteristics	Functional groups (According to FTIR analysis)	References
Biomass fiber waste	Steam	800	2.0	S_{BET} 840 m^2/g, mainly mesoporous structure	Not analyzed	Williams and Reed (2006)
Biomass fiber waste	ZnCl$_2$	450	6.9	S_{BET} 2400 m^2/g, mainly microporous structure	Not analyzed	Williams and Reed (2006)
Barley straw	Steam	700	1.0	S_{BET} 552 m^2/g, V_{mic} 0.2304 cm^3/g, V_{total} 0.2576 cm^3/g	C=O stretching in aldehydes, ketones and esters, carboxylic groups and lactones, phenolic groups, hydroxyl groups	Pallarés et al. (2018)
Barley straw	Steam	800	1.0	S_{BET} 534 m^2/g, V_{mic} 0.2186 cm^3/g, V_{total} 0.2994 cm^3/g	Not analyzed	Pallarés et al. (2018)
Barley straw	CO$_2$	700	1.0	S_{BET} 211 m^2/g, V_{mic} 0.0830 cm^3/g, V_{total} 0.0938 cm^3/g	Not analyzed	Pallarés et al. (2018)
Orange peels	K$_2$CO$_3$	700	1.0	S_{BET} 477 m^2/g, V_{mic} 0.21 cm^3/g, V_{total} 0.23 cm^3/g	C-O stretching vibration of carboxylic acids and alcohols, methyl, methylene and methoxy groups. O-H stretching vibrations related to the lignin, hemicellulose, cellulose, pectin and absorbed water	Köseplu and Akmil-Bb_ar (2015)
Orange peels	K$_2$CO$_3$	950	1.0	S_{BET} 1352 m^2/g, V_{mic} 0.22 cm^3/g, V_{total} 0.79 cm^3/g	O-H stretching vibrations related to the lignin, hemicellulose, cellulose, pectin and absorbed water C-O stretching vibration of carboxylic acids and alcohols, methyl, methylene and methoxy groups.	Köseplu and Akmil-Bb_ar (2015)
Orange peels	ZnCl$_2$	700	1.0	S_{BET} 822 m^2/g, V_{mic} 0.09 cm^3/g, V_{total} 0.50 cm^3/g	O-H stretching vibrations are related to the lignin, hemicellulose, cellulose, pectin, and absorbed water. C-O stretching related to carboxylic acids and alcohols	Köseplu and Akmil-Bb_ar (2015)
Soybean oil cake	K$_2$CO$_3$	600	1.0	S_{BET} 643.54 m^2/g, V_{mic} 0.272 cm^3/g, V_{total} 0.336 cm^3/g	Not analyzed	Tay et al. (2009)
Soybean oil cake	KOH	800	1.0	S_{BET} 618.54 m^2/g, V_{mic} 0.143 cm^3/g, V_{total} 0.291 cm^3/g	Not analyzed	Tay et al. (2009)
Chestnut shell	H$_3$PO$_4$	750	0.33	S_{BET} 1137.88 m^2/g, V_{mic} 0.424 cm^3/g, V_{total} 0.601 cm^3/g	O-H stretching vibrations or adsorbed water. C=C aromatic ring stretching and -C-H stretching was also found as well as -P-O and -P=O functional groups.	Duan et al. (2021)
Chestnut shell	H$_3$PO$_4$	850	0.33	S_{BET} 1413.02 m^2/g, V_{mic} 0.562 cm^3/g, V_{total} 0.738 cm^3/g	O-H stretching vibrations or adsorbed water. C=C aromatic ring stretching and -C-H stretching was also found as well as -P-O and -P=O functional groups	Duan et al. (2021)

Another source of material for AC preparation is the carbon residue that is formed as a side stream during biomass gasification. Gasification is a process designed to produce syngas, a mix of CO, CO_2, CH_4, and H_2 in a gasifier that can be used to generate heat and power or as a platform for other chemicals (e.g., Fischer–Tropsch). The composition of the syngas may vary depending on the used starting material, the temperature, and the typology of gasifier (Heidenreich & Foscolo, 2015), but in general, the process can be summarized as follows:

Biomass + air + heat = Syngas + ash + carbon residue

During the gasification, the O_2 content is kept low to avoid complete combustion of the starting material. The carbon residue produced can be used as source material to produce ACs (Maneerung et al., 2016). These ACs have shown good performances in the removal of phosphate, nitrate, heavy metals, and organic pollutants from wastewater (Kilpimaa et al., 2014).

23.3 Carbon-based catalyst preparation and properties

Wide ranges of carbon-based materials are used for catalytic applications. However, the most frequent materials utilized as adsorbents and in catalytic applications are ACs with high SSA and different functional groups (Bandosz, 2008). Carbon material's bulk and surface properties depend on their physical and chemical structure, which in turn depends on the used raw material and preparation methods (Radovic, 2008). The unique properties of carbon materials can be tailored to specific purposes, and knowledge of their existence and chemistry is essential to many processes in technology (Boehm, 1966). Properties affecting the preparation and usage of carbon-based materials as catalyst materials are factors such as SSA, porosity, functionality, purity, that is, mineral matter, and stability of the material (Arunajatesan et al., 2008). Carbon materials can contain inorganic mineral matter, which is usually given as ash content and depends on the raw material and preparation process.

High mineral matter content arising from the raw material or introduced during the preparation process by chemical modification can be a drawback for adsorptive and catalytic behavior as it can act as an active metal in catalytic application affecting the selectivity of the process (Bandosz, 2008; Rodríguez-reinoso, 1998).

The heterogeneous surface of carbons consists of the faces of graphene sheets and the edges of such layers. The edge sites are much more reactive than the atoms in the interior of the graphene sheets, and functional groups are predominantly located on the edges (Boehm, 2002). The high SSA of carbon materials, especially in ACs, is built up by pores between the graphene sheets. Pores are classified by the IUPAC based on their diameter as micropores (up to 2 nm), mesopores (from 2 to 50 nm), and macropores (50 nm or greater). Larger pores may be more suitable in some catalytic applications than smaller pores with diffusion limitations especially for liquid applications if large organic molecules are involved in the application (Boehm, 2002; Marsh & Rodríguez-Reinoso, 2006; Rodríguez-reinoso, 1998). For example, Gurrath et al. (2000) observed diffusion limitations on AC catalysts due to high microporosity and noticed that even molecules such as cyclohexene couldn't reach the metal particles in the catalyst's narrow micropores. The functional groups on the carbon surface play an important role in the heterogeneous reaction mechanism (Rodríguez-reinoso, 1998).

The most common heteroatoms on carbon are oxygen, nitrogen, hydrogen, and sulfur. Also, phosphorus and chlorine have been detected on carbon surfaces in lower amounts. Depending on the raw material and preparation method used, approximately 70–98 wt.% carbon, 2–25 wt.% oxygen, 1–5 wt.% hydrogen, 0–5 wt.% of nitrogen, and 0–5 wt.% of sulfur content can be obtained from elemental analysis of ACs (Bandosz, 2008; Boehm et al., 1964). These elements form organic functional groups bound onto the edges of the graphite-like layers such as carboxylic acids, lactones, phenols, carbonyls, aldehydes, ethers, amines, nitro compounds, and phosphates (Fig. 23.2) (Salame & Bandosz, 2001).

Oxygen is the most common heteroatom on the carbon surface, with at least half of it presents in the form of surface oxide groups (Boehm et al., 1964). Surface oxides can be acidic, basic, or neutral in character. The acidic character of carbon surfaces is closely related to the oxygen-containing surface groups, which are formed when the carbon surface is exposed to oxygen by reactions with oxidizing agents from the gas phase or solutions, whether at room temperature or at high temperatures (Barton et al., 1997; Boehm, 2002; Serp & Figueiredo, 2008). Generally, carboxyl groups, anhydrides, hydroxyls lactones, and lactols groups are considered as acidic surface groups. Carbonyl and ether groups are considered as neutral groups on the surface or may form basic structures such as chromenes,

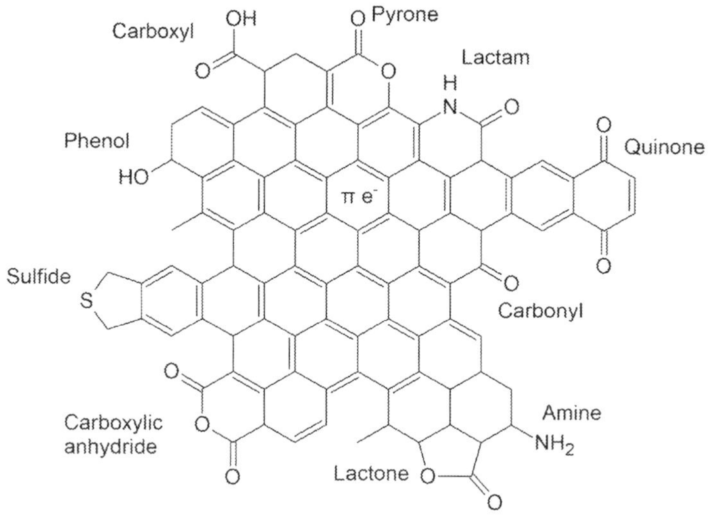

FIGURE 23.2 Typical functional groups present on the carbon surface.

pyrones, and quinones. In addition, the π-electron density of carbon basal planes is considered as a chemically active site that accepts or donates electrons (Figueiredo & Pereira, 2008; Serp & Figueiredo, 2008).

Nitrogen content on carbon materials is usually very low, and these functionalities are not formed spontaneously on the carbon surfaces. The introduction of nitrogen functionalities is usually done by treatments with nitrogen-containing reagents such as ammonia, melamine, or urea, for example (Serp & Figueiredo, 2008). Treatments can be done by carbonizing nitrogen-containing organic materials or mixing raw materials with nitrogen-containing precursors or by treatments of carbon material at high temperatures with nitrogen-containing gases. The type of nitrogen functionalities on the carbon surface depends on the treatment applied and can be for example lactams, imines, amines, pyridinic, or pyrrole-type structures (Serp & Figueiredo, 2008). Sulfur can be present in carbon as elemental sulfur, inorganic or organosulfur compounds. Sulfur complexes have been used sometimes for the adsorption of metal ions. Carbon surfaces can be modified to sulfur-containing functionalities by treatments with compounds such as H_2S, CS_2, or SO_2 at various temperatures (Bandosz, 2008).

The precise properties of the resulting carbon materials depend on the type of raw material, the activation method, and the chemical treatment used during or after the preparation step. The variety of active sites on carbon materials and the ease of modification make carbonaceous materials versatile catalysts. Among the many reactions in which carbon catalysts are active are oxidations, reductions, hydrogenations, dehydrogenations, dehydration, and other bond-forming and bond-cleaving reactions (Davies et al., 2021; Figueiredo & Pereira, 2008).

Functionalization of carbon materials can be achieved by chemical modification during the preparation of carbon material by carbonization or modification afterward by chemical treatment as discussed previously in this chapter. Most often used techniques are gaseous or liquid-phase treatments such as oxidation by common oxidizing agents, for example, hydrogen peroxide, ozone, and permanganate. Nitric acid or HNO_3/H_2SO_4 oxidizing treatments are the most widely used methods for metal nanoparticle anchoring onto the carbon support as they can affect carbon wetting properties and hydrophilicity (Rodríguez-reinoso, 1998; Serp & Figueiredo, 2008).

The impregnation is the simplest method to prepare supported catalysts. Carbon-supported catalysts are in many cases prepared by the impregnation method known as dry impregnation (also incipient wetness impregnation) or wet impregnation. In the dry impregnation method, the volume of the metal precursor solution is equal to the pore volume of the support material, in contrast to the wet impregnation where excess volumes are used. Other methods used to prepare carbon-supported metal catalysts are, for example, ion exchange and precipitation methods (Bitter & de Jong, 2008; Iwanow et al., 2020).

Physical and chemical properties of the carbon materials and catalysts can be characterized by using a variety of techniques. Characterization methods frequently used are such as physisorption and chemisorption methods, scanning electron microscopy (SEM), transmission electron microscopy (TEM), energy-dispersive X-ray spectroscopy (EDS), Fourier-transform infrared spectroscopy (FTIR), Boehm titrations, temperature-programmed desorption (TPD), CHNS/O elemental analysis, X-ray photoelectron spectroscopy (XPS), and inductively coupled plasma methods (ICP-Ms or OES) to determine their physical or chemical properties.

23.4 Application of biomass-derived catalysts for water treatment

Lignocellulosic biomasses are readily available and often utilized to produce carbonaceous materials such as carbon, AC, and biochar. Biomass-derived carbonaceous materials have gained tremendous attention in the fabrication of catalysts due to their excellent attributes including high thermal stability as well as high resistance toward acid/base (Park et al., 2021). Waste lignocellulosic biomasses are low-cost and readily available feedstock to produce carbonaceous support material for the fabrication of heterogeneous catalysts. Lignocellulosic biomass composition and their conversion techniques for obtaining carbonaceous materials have a direct impact on the electrical conductivity of carbon materials (Adinaveen et al., 2016). Higher pyrolysis temperatures could culminate in the conversion of sp^3-hybrid carbon to sp^2-hybrid, enabling the carbon skeleton to collapse and enhancing the defect structure of biochar (Do Minh et al., 2020; Zhao et al., 2021). Excellent conductivity of carbon materials enhances the electron flow and hole/electron separation efficiency. The heterojunction of carbonaceous materials with semiconductors has the benefit of preventing electron–hole pair recombination. Additionally, the presence of sp^2-hybridized carbon and electron-rich surface functional groups (i.e., carboxyl, carbonyl, hydroxyl, phenolic hydroxyl, amines groups) can improve the catalytic activity (Başer et al., 2021). Further, heteroatoms doping in pristine carbon can generate new sites for catalytic activity (Zhu et al., 2020). The creation of N-functionalities in biochar has been shown to accelerate the degradation of organic pollutants in photocatalytic treatment (Başer et al., 2021).

In this section, we have focused on the application of lignocellulosic biomass-derived catalytic materials for wastewater treatment by advanced oxidation processes (AOPs). Main emphasis has been given to photocatalytic treatment, catalytic wet peroxide oxidation (CWPO), and catalytic wet air oxidation (CWAO) due to the wide application of lignocellulosic biomass-derived catalytic materials in these approaches. Heterogeneous catalyst-based AOPs have received immense attention in recent years as they can convert toxic contaminants into low or nontoxic compounds. Furthermore, organic contaminants can be mineralized into CO_2 and H_2O.

23.4.1 Photocatalytic treatment of wastewater by biomass-derived catalysts

In this section, we have discussed the photocatalytic efficiency of pristine, modified, and heterojunctions based on biomass-derived carbonaceous materials in heterogenous photocatalytic degradation of organic pollutants (Table 23.4). In heterogeneous photocatalysis, under UV/visible light irradiation, catalysts are activated to produce reactive oxygen species (ROS) such as hydroxyl radicals ($^\bullet OH$), superoxide anion radicals ($O_2^{\bullet -}$), singlet oxygen ($^1O_2$), which can attack on recalcitrant organic compounds and destruct the organic compounds into carbon dioxide, water, and small inorganic compounds (Babu et al., 2019). Various studies, for example, bisphenol A (Mohamed et al., 2018), tetracycline (Wang et al., 2018); imazapic and imazapyr herbicides (Yavari et al., 2019), sulfadimidine (Chen et al., 2017), rhodamine B (Li et al., 2019), methyl orange (Shan et al., 2020) demonstrated that heterogeneous catalysts developed with the integration of lignocellulosic biomass-derived carbonaceous materials were very efficient in the removal of a variety of organic pollutants.

AC derived from different lignocellulosic biomass waste has been extensively studied. AC developed from rice husk powder was utilized for catalyst synthesis (Singh et al., 2022). The carbonization of biowaste and polymerization were accomplished in acidic conditions. A ternary nanocomposite (ZCP) was fabricated using zinc oxide (ZnO), polypyrrole(PPy), and AC. The photocatalytic activity of ZCP was examined for methylene blue (MB) degradation. The ZCP composite was able to degrade 98% of MB dye within 20 minutes under visible light. Synthesized catalyst was effective for MB degradation up to 10 successive cycles.

Further, in another study, pistachio shells–based AC was employed for the development of solar light–driven photocatalyst (Sane et al., 2022). Pistachio shells sample was activated with $ZnCl_2$ activation agent and then carbonized to produce AC. Developed AC and Bi_5O_7I were used for the synthesis of heterojunction. Photocatalysts with 1%, 5%, 10%, and 20% w/w AC-Bi_5O_7I were synthesized, and the efficiency of heterojunction was examined for the methyl orange (MO) degradation under solar light. Bandgap energy of heterojunction (AC-Bi_5O_7I) was reported to be 2.10 eV. Due to AC loading, the bandgap of the photocatalyst was decreased while the surface area was increased due to AC loading. Extremely low dye removal was obtained in the absence of catalyst, which increased with the increase of catalyst dose, and dye removal was also increased. Higher degradation was achieved for lower concentrations. 5AC-Bi_5O_7I (with 5% AC) displayed the optimum photocatalytic activity (99.9% reduction within 80 min) in comparison to pristine Bi_5O_7I.

TABLE 23.4 Photocatalytic degradation of organic pollutants by biomass-derived catalysts.

Biomass	Biomass based catalyst	Characterization	Organic pollutant	Analysis of treated sample	Experimental conditions	Rate constant	References
Walnut	Nanocomposites of bismuth molybdate and waste walnut shells derived biochar (BMOC15)	S_{BET} 18.7 m^2/g	MB	UV-Vis spectrophotometer	- pH neutral -Photocatalyst dose 20 mg -Vol of MB solution-20 mL, -Conc. of MB solution 2 × 10^{-5} M -Light source: Photoreactor- 40 W and two 15 W white light emitting LED lamps	$k_1 = 0.01267$ min^{-1}	Kumari et al. (2021)
Wheat straw pellets (WSP)& oil seed rape straw(OSR)	Iodine doped (I-WSP and I-OSR) biochar	S_{BET} of I-WSP (36.71 m^2/g) and I-OSR (29.14 m^2/g) E_g values for I-WSP 0.93 eV and I-OSR-0.90 eV	Phenol & TC	HPLC for phenol analysis; UV–Vis spectrophotometer for TC analysis	-Conc. of phenol or TC (50 mg/L) -Vol of solution 50 mL -Light source: Visible light 300-W Xenon lamp	-phenol degradation kinetics on I-WSP ($k_1 = 1.30 × 10^{-2}$ min^{-1}) and I-OSR ($k_1 = 8.05 × 10^{-3}$ min^{-1}) -photocatalytic TC degradation on I-WSP ($k_1 = 1.78 × 10^{-3}$ min^{-1}) and I-OSR ($k_1 = 9.85 × 10^{-4}$ min^{-1})	Wang et al. (2021)
Rice straw	g-MoS$_2$_ PGBC (porous graphite biochar) nanocomposites	S_{BET} 266.8 m^2/g E_g values 1.91 eV	TC	HPLC and UV-Vis spectrophotometer	-Catalyst dose-20 -TC Conc 20 mg/L -Volume of TC 50 mL -Light source- 300 W Xe lamp	$k_1 = 0.0049$ min^{-1}	Ye et al. (2019)
Walnut shell	Ag/TiO$_2$/biochar composites	S_{BET} 35.212 m^2/g E_g values 3.28 eV	MO	UPLC system coupled quadrupole and time-of-flight technology (QTOF)	-Catalyst dose-10 mg -MO volume- 40 mL -MO conc. 20 mg/L -Irradiation time- 60 min. -Irradiation source- 500 W -long arc mercury-vapor lamp -wavelength of 360 nm.	$k_1 = 6.29 × 10^{-2}$ min^{-1}	Shan et al. (2020)
Sunflower straw powder-based biochar	Biochar-based supramolecular self-assembled g-C$_3$N$_4$	E_g values 2.28 eV	Phenanthrene	UV–Vis spectrophotometer	Catalyst dose- 20 mg -Volume of phenanthrene 100 mL -Concentration of phenanthrene 1.0 mg/L -pH range of 7 ± 0.5 Light source: 250-W high-pressure sodium lamp with a wavelength range of 400–800 nm	$k_1 = 0.355$ h^{-1}	Lin et al. (2022)
Peanut shells derived biochar	TiO$_2$/biochar composite	S_{BET} 118.540 m^2/g	MB	UV–Vis spectrophotometer	-Catalyst dose- 40 mg -Volume of MB solution- 40 mL, Conc of MB solution-30 mg/L MB Light source: 300 W Hg and 350 W Xe lamps to evaluate the UV- and Vis-photocatalytic irradiation respectively	$k_1 = 0.040$ min^{-1}	Luo et al. (2022)

Biochar or modified biochar, in addition to AC, has been developed to be particularly successful in the remediation of numerous organic contaminants (Do Minh et al., 2020) (Table 23.4). Wang et al. (2021) reported pyrolysis of wheat straw pellets (WSP) and oilseed rape straw (OSR) biomass at 700 °C for biochar production. Both WSP and OSR were further functionalized by the facile iodine doping method and named as I-WSP and I-OSR using hydrothermal treatment. Photodegradation of phenol and tetracycline (TC) under visible light was examined using both I-WSP and I-OSR. The pristine biochar was acidic while the I-doped biochar showed alkaline nature. Both I-WSP and I-OSR were able to degrade phenol up to nearly zero within 240 minutes under visible light illumination. The availability of micro and mesopores in I-doped biochar eventually boosted the degradation of phenol and TC. Additionally, reduced charge carrier transfer and enhanced photo-induced excitation affected the photocatalytic performance of I-WSP and I-OSR. The photocatalytic performance of iodine-doped biochar was amplified because of the high-density photo-induced charge carriers as well as the low electron transfer resistance attributed to iodine.

A visible light photocatalyst was synthesized by modified biochar-based supramolecular self-assembled g-C_3N_4 for phenanthrene degradation (Lin et al., 2022). Sunflower straw powder was used for biochar synthesis (@ 500 °C). It was demonstrated that the synthesized catalyst (A-BC/g–C_3N_4–D) was able to remove 77% phenanthrene and followed a first-order reaction rate (0.355 h^{-1}). The radical quenching experiment verified the presence of reactive species $^{\bullet}OH$, $^{\bullet}O_2^-$, and h^+; however, $^{\bullet}O_2^-$ played the dominant role in phenanthrene degradation. It was also stated that $^{\bullet}OH$ and h^+ might also participate in the degradation process. Electron−hole pairs were produced from g-C_3N_4 activation under visible light, and biochar acted as a terminal electron acceptor due to its strong electron storage capacity.

Luo et al. synthesized graphene-like biochar from peanut shells, which was further used for the synthesis of biochar/TiO_2 composite (Luo et al., 2022). It was examined that biochar/TiO_2 composite was more efficient in the photodegradation of MB in comparison to TiO_2. The improved photocatalytic activity of the biochar/TiO_2 composite can be credited to the synergistic impact among the graphene-like biochar, the self-introduced N and O heteroatoms, as well as the coupling between TiO_2 and biochar. In biochar/TiO_2 composite, graphene-like biochar and TiO_2 can generate photoelectrons. The holes and photoelectrons can easily transfer to the conduction band (CB) of TiO_2. Because of charge transfer, recombination is efficiently diminished resulting in increased stability. The combination of TiO_2 and graphene-like biochar improves the adsorptive and photocatalytic performances of composites.

Walnut shell is a readily available biomass and consists of high lignin content by mass (Nishide et al., 2021). Walnut-derived biochar support was utilized for the development of composite catalyst material for the catalytic degradation of MB (Kumari et al., 2021). For this study, biochar was synthesized using a hydrothermal process and then coupled with bismuth molybdate (Bi_2MoO_6) (BMO) in three different wt.%, namely BC 5%, 10%, and 15%. Synthesized nanocomposite photocatalyst was named as BMOC. Different % of biochar was coupled with Bi_2MoO_6 to examine the effect of carbonaceous support materials. XPS investigation confirmed the existence of functional groups C−C, C−O and O=C−O, Bi−O, Mo−O, and O−H in the BMOC. The SSA of the BMOC sample (15 wt.% BC) was higher (18.7 m^2/g) in contrast to BMO (5.7 m^2/g). Under visible light, 90% degradation of MB was achieved in optimum conditions. By coupling different ratios of BC with Bi_2MoO_6, morphological variation was observed, which additionally affected the degradation of MB.

Additionally, walnut biochar was utilized as the support material for the development of Ag/TiO_2-based catalyst (Shan et al., 2020). In this study, walnut shell biochar was prepared at 700 °C (WB700). Ag/TiO_2/biochar composite was synthesized for methyl orange (MO) elimination. The SSA of WB700 was reported to be 66.06 m^2/g, while a reduced surface area (35.21 m^2/g) was noticed in Ag/TiO_2/biochar composite. The synergetic impact of Ag, TiO_2, and biochar was assessed on MO degradation. The highest decolorization and mineralization efficiencies for Ag/TiO_2/biochar were 97.48% and 85.38%, respectively under ultraviolet light irradiation. The MO degradation followed the pseudo-first-order kinetic model (rate constant of 6.29×10^{-2} min^{-1}). The bandgap energy E_g was reported 3.28 eV for Ag/TiO_2/biochar composite. It was revealed that Ag, along with TiO_2, served as an electron donor, whereas biochar served as an electron acceptor, resulting in improved separation of photogenerated e^- and h^+ during MO degradation. Further, rice straw−derived biochar was utilized for the synthesis of porous graphite biochar (PGBC) followed by the incorporation of g-MoS_2 on the surface of PGBC (Ye et al., 2019). The bandgap energy of g-MoS_2/PGBC was calculated as 1.91 eV. It was illustrated that in the hydrothermal process, g-MoS_2 nanosheets expand disorderly on the tubular carbon wall. This behavior may be linked to the interaction of biochar's functional groups with precursors' Mo^{4+}. The BET SSA of g-MoS_2/PGBC was established to be 266.8 m^2/g. The photocatalytic efficacy of the g-MoS_2/PGBC composite was examined by the photodegradation of TC (20 mg/L) under visible light irradiation.

It is very clear from the preceding discussions of Section 23.4.1 that the lignocellulosic-based carbonaceous materials are potential materials for photocatalyst development and can provide an efficient solution for water reclamation by eliminating recalcitrant organic contaminants.

23.4.2 Biomass-based catalysts in catalytic wet peroxide oxidation

Catalytic wet peroxide oxidation (CWPO), also called heterogeneous Fenton oxidation, is a water treatment technology in which liquid hydrogen peroxide is used to decompose organic compounds of the treated effluent to CO_2 and H_2O. The process operates at a temperature range of 20 °C–165 °C while metallic ion (typically Fe^{2+} or Cu^+) is decomposed by H_2O_2 generation reactive hydroxyl radicals (Márquez et al., 2018). The decomposition of H_2O_2 can also happen on the surface of carbon materials through an electron transfer (Gomes et al., 2010), and hence, activated carbon and related materials have been widely studied in CWPO (Kurian, 2021). Nowadays, biomass-based carbons have gained more attention due to their environmental friendliness compared with coal-based carbons (Table 23.5).

Liu et al. (2012) applied pine bark as a raw material for AC catalyst in CWPO of phenol. They studied several preparation methods for Fe/AC catalysts. Demineralization of pine bark before active metal impregnation and carbonization of the material with a microwave oven has a positive effect on the catalyst activity. The activation of the catalyst was performed at four different temperatures (350 °C, 400 °C, 450 °C, and 500 °C), and the catalyst prepared at 500 °C showed the highest activity in the CWPO of phenol.

Esteves et al. (2020) studied fitted biochars from residues of the olive oil industry as support materials for Fe-catalysts in the CWPO of synthetic olive oil industry wastewater. Biochars were prepared from the olive tree sawdust and olive stones by carbonization under N_2 150cm^3/min at 800 °C for 2 hours. Biochars were activated physically by CO_2 and chemically by KOH, and Fe was impregnated on the prepared supports by incipient wetness impregnation. After 4 hours of oxidation reaction, around 55%, phenolic compounds and 35% TOC removals were achieved with the most active biochar-based catalyst. Fe/biochar catalyst prepared from olive stone and activated with KOH performed slightly better than the same material with CO_2 activation. However, with commercial AC-supported Fe catalysts, over 90% of phenolic compounds abatement was achieved, but the adsorption of phenolics was responsible for over 30% of the removal and remarkable leaching of active metal occurred during the CWPO.

Date stems were used to produce lignocellulosic-based AC for CWPO of toluidine blue O dye (TBO) (Samir et al., 2021). Raw material was first activated with $ZnCl_2$ and further pyrolyzed at 600 °C for 1 hours and finally, incipient wetness impregnation was used to add active metal to the support. Prepared catalyst showed almost 100% TBO removal already after few minutes of CWPO, and moreover, after four consecutive tests, still 80% of TBO and around 55% TOC conversions were achieved with date stem pyrolyzed AC with Fe as an active metal.

Diaz De Tuesta et al. (2021) treated bagasse of sugarcane (BS), bagasse of malt (BM), and seed of chia (SC) by pyrolysis, HTC and sequential HTC, and pyrolysis to achieve pyrochars, hydropyrochars, and ACs. Pyrochars were prepared from three raw materials by thermal treatment under N_2 flow (100 cm^3/min) in the tubular furnace first 1 hour at three different temperatures (120 °C, 400 °C, 600 °C) and finally at 800 °C for 4 hours. Samples were named BS-C, BM-C, and SC-C. In HTC treatment, raw materials were first suspended in three different solutions, that is, distilled water, 2.5 M $FeCl_3$, and 2.5 M H_2SO_4. The experiments were performed at 200 °C for 3 hours in a stainless-steel reactor. Prepared hydrochars were collected with filtration, washed with distilled water, and dried at 60 °C. Produced samples were named according to their raw material (BS, BM, SC) and used suspend solution (H_2O, H_2SO_4, $FeCl_3$). Moreover, ACs were prepared from produced hydrochars by pyrolysis of samples at 800 °C for 4 hours. Totally six samples were prepared, namely BS-H_2O-C, BS-H_2SO_4-C, BS-Fe-C, BM-H_2O-C, BM-H_2SO_4-C, and BM-Fe-C. Seeds of chia were not used for the preparation of ACs due to their low porosity potential. Prepared materials were studied in the CWPO of caffeine aqueous solution. All produced materials were active in the removal of caffeine after 24 hours reaction, and almost total conversion of the pollutant was achieved with the majority of tested catalysts. The highest conversion of caffeine (80%–99%) was observed with hydrochars prepared in the presence of iron and ACs produced by sequential HTC assisted with $FeCl_3$ or H_2SO_4 and pyrolysis (BS-Fe-C and BM-Fe-C, BS-H_2SO_4-C and BM-H_2SO_4-C). Materials that were prepared with iron solution removed the caffeine faster, but the leaching of iron in the treated water was significant, and therefore, ACs prepared with H_2SO_4 were the most active and durable catalyst in the research.

In the research of Karthikeyan et al. (2013), nanoporous AC was prepared from rice husk. The material was first precarbonized at 400 °C and activated by phosphoric acid at 700 °C, 800 °C, and 900 °C. Before use, materials

TABLE 23.5 Catalytic wet peroxide oxidation studies with lignocellulosic-based biomass catalysts.

Biomass	Biomass based catalysts	Characterization	Organic pollutant	Reaction conditions	Analysis of treated sample	Results	References
Pine bark	Fe/AC prepared from demineralized pine bark in microwave reactor	S_{BET}: 131–313 m^2/g	Phenol	c [phenol]: 100 mg/L, T: 50 °C, c [H$_2$O$_2$]: stoichiometric amount, c [catalyst]: 4 g/L	UV-Vis at 254 nm, TOC	Almost 100% phenol removal, nearly 80% TOC conversion	Liu et al. (2012)
Olive tree sawdust/olive stones	Fe/OSC biochar prepared from olive stone Fe/SDC biochar prepared from olive sawdust	S_{BET}: 176–777 m^2/g, pH$_{pzc}$: 2.0–11.5	Synthetic phenolic wastewater	COD 770 mg/L, TOC 210 mg/L, T: 25 °C, c [H$_2$O$_2$]: 1 g/L, c [catalyst]: 0.5 g/L	HPLC (phenolic content), TOC	50%–56% total phenolic removal, 35% TOC conversion	Esteves et al. (2020)
Date stems	Fe/AC pyrolyzed from date stem, activated with ZnCl$_2$	S_{BET}: 1069 m^2/g, pore size 3.33 nm	Toluidine blue O dye (TBO)	c [TBO]: 50–400 mg/L T: 20 °C c [H$_2$O$_2$]: 1–6 mmol/L c [catalyst]: 0.1–1.8 g/L	UV-Vis (TBO), TOC	99% TBO removal, 64% TOC conversion	Samir et al. (2021)
Bagasse of sugarcane (BC), bagasse of malt (BM), seed of chia (SC)	Activates carbons (ACs) Pyrochars (PCs) Hydrochars (HCs)	S_{BET} (m^2/g): ACs: 322–447 PCs: 6–214 HCs: 7–75	Caffeine	c [caffeine]: 1–100 mg/L T: 80 °C, pH 3 c [H$_2$O$_2$]: 1–6 mmol/L c [catalyst]: 2.5 g/L	UV-Vis (H$_2$O$_2$), HPLC (caffeine), TOC	100% caffeine removal with Fe-HC catalyst and BM-based ACs	Diaz De Tuesta et al. (2021)
Rice husk	ACs used as fluidized bed	S_{BET}: 180–298 m^2/g	o, p- and m-cresols	T: 30 °C, pH 2.5–4.5 c [H$_2$O$_2$]: 2–15 mmol/L c [catalyst]: 2–25 g/L	BOD$_5$, COD, TOC	TOC removals: 41% (homogeneous) 85% (heterogeneous)	Karthikeyan et al. (2013)
Patchouli	ZnCr$_2$O$_4$/CNS composites calcined at 400 and 800 W	XRD, FTIR, SEM	Buthylphenylmethyl carbamate pesticide (BPMC)	c [BPMC]: 500 mg/L T: RT, c [H$_2$O$_2$]: 0.15% c [catalyst]: 0.2 g/L	TOC	ZnCr$_2$O$_4$/CNS composite calcined at 800 W showed three times higher activity in the BPMC removal than sample calcined at 400 W	Setianingsih et al. (2021)
Carbon residue (CR) formed in biomass gasification	CR, 2.5Fe/CR, 5.0Fe/CR, 33Fe/CR	S_{BET}: 17–91 m^2/g	Bisphenol A (BPA)	c [BPA]: 60 mg/L, T: 50 °C, c [H$_2$O$_2$]: 1.5 g/L, c [catalyst]: 1–2 g/L	HPLC, TOC	63% BPA removal, 31% TOC conversion with the most stable catalyst (5.0Fe/CR)	Juhola et al. (2017)
Carbon residue (CR) formed in biomass gasification	Carbon composite catalysts: CR + CaO/, metakaolin + cement + NaOH/KOH	S_{BET}: 152–205 m^2/g	Bisphenol A (BPA)	c [BPA]: 60 mg/L, T: 50 °C, c [H$_2$O$_2$]: 1.5 g/L, c [catalyst]: 1 g/L	HPLC, TOC	50% BPA removal, 48% TOC conversion with the most durable and stable composite	Juhola et al. (2019)

were washed with distilled water and dried at 100 °C for 1 hour. According to pore size distribution analysis, the phosphoric acid activation at 800 °C produced the largest pore volume for the pore diameter of 38 Å. Therefore it was chosen as the fluidized bed for the Fenton oxidation of the *o*-, *p*-, and *m*-cresols. Both homogeneous and fluidized-bed Fenton oxidation experiments were performed while $FeSO_4 \times 7H_2O$ was used as a homogeneous catalyst in the reaction. Nanoporous AC performed higher efficiency in heterogenous Fenton reactions than in the homogeneous experiments. Authors assumed that the available active sites of the AC performed as adsorbent for both hydroxyl radicals and cresols and therefore the reaction rate was increased compared with the homogeneous oxidation reaction. With nanoporous AC, around 80% COD, TOC, and BOD removal was achieved after 6 hours oxidation of cresols while without AC organic concentrations, abatements were only 40%.

In the research of Setianingsih et al. (2021), patchouli biomass was investigated as a potential source for the production of carbon-based catalysts. The biomass was dried, sieved, and mixed with $ZnCl_2$, and the mixture was pyrolyzed in a microwave oven with the power of 400, 600, and 800 W for 50 minutes. Formed products were mixed with water to obtain colloids and further evaporated to produce ZnO/CNS nanocomposites. ZnO/CNS were further calcined at 400 and 800 W and after that mixed with KOH, $ZnCl_2$, $CrCl_3 \times 6H_2O$, and distilled water and finally calcined at 600 W for 5 minutes to produce $ZnCr_2O_4$/CNS composite. The activity of the prepared material was studied in the catalytic H_2O_2 oxidation of buthylphenylmethyl carbamate pesticide (BPMC). The $ZnCr_2O_4$/CNS composite calcined at 800 W showed three times higher catalytic activity in the degradation of BPMC than the composite, which was calcined at 400 W.

Juhola et al. (2017) examined the activity of biomass-based carbon residue (CR) as a support material for iron catalysts in the CWPO of bisphenol A (BPA). The used CR was formed in the gasification process of the Finnish wood biomass. Before use, CR was dried at 105 °C, ground and sieved to the particle size $< 150 \mu m$. After sieving, CR was washed with deionized water to the neutral pH and finally, dried at 105 °C. Iron was added as $FeCl_3$ to support incipient wet impregnation or wet impregnation with the target concentration of Fe 2.5, 5.0, and 33 wt.%. After impregnation, the catalysts were dried, grounded, and sieved to particle size $< 150 \mu m$, and finally calcined at 280 °C for 5 hours under N_2 15 dm^3/min. All prepared materials were active in the CWPO of BPA, and 100% BPA removal was achieved with 2.5Fe/CR and 33Fe/CR. However, leaching of the iron occurred during oxidation experiments, and therefore 5.0Fe/CR was the most active and durable catalyst in the studied reaction conditions.

In another study, Juhola et al. (2019) studied the suitability of CRs as raw materials for composite catalysts. CRs were produced in the gasification process of willow at 400 °C–500 °C in the pilot-scale willow-carbon production process (willow carbon, CW) and in the biomass gasification process of a mixture of pine and birch at 1000 °C (CR). Both materials were pretreated by washing with 1 N solution of HCl: H_2SO_4 for 24 hours, filtered and washed with distilled water for 1 hour. In addition, CW and CR without acid treatment were used and washed with distilled water until a neutral pH was achieved. All samples were filtered and dried overnight at 105 °C. For the preparation of carbon composite catalysts, a wet granulation process was applied by using a rotary drum granulator.

Carbon material and binging agents (CaO, metakaolin, and/or cement) were mixed in the granulation mixer, and solvent (5 M KOH or NaOH) was added gradually to the mixer with an agitation rate of 1200 rpm. The total mixing time was varied between 500 and 1000 seconds. Only CR-based materials were kept in their granulated structure in the stability tests, and those were further studied in the CWPO of BPA. With powdered CR, almost 100% BPA removal was achieved after 3 hours oxidation; however, CR was not catalytically active while BPA abatement was pure adsorption. Granulated CRs performed good catalytic activity at around 65% BPA degradation and 50% TOC conversion with the most active catalyst, CR2 prepared with binging agents CaO and metakaolin by using NaOH as a solvent.

23.4.3 Biomass-based catalysts in catalytic wet air oxidation

Catalytic wet air oxidation (CWAO) is a water treatment technique, which operates at elevated temperatures (125 °C–200 °C) and pressures (0.5–5 MPa). In the process, highly concentrated wastewater with toxic and hazardous compounds are oxidized with air or molecular oxygen to CO_2 and H_2O. Both homogeneous and heterogeneous catalysts have been widely used in the process, and among them, carbon materials have been also examined (Jing et al., 2016). Table 23.6 provides examples of the use of biomass-derived carbons in the CWAO.

Morales-Torres et al. (2010) were probably the first ones who used biomass-based carbon catalysts in the CWAO. Carbons were prepared from olive stones by chemical activation with KOH. First, the raw material was

TABLE 23.6 Catalytic wet air oxidation studies with lignocellulosic-based biomass catalysts.

Biomass	Biomass based catalyst	Characterization	Organic pollutant	Reaction conditions	Analysis of treated sample	Results	References
Olive stones	AC	S_{BET}: 121–1530 m^2/g, pH_{pzc}: 6.6–7.5	2,4,6-trinitrophenol (TNP)	Batch reactor: c [TNP]: 125 mg/L, T: 200 °C, p [air]: 5.5 MPa, c [catalyst]: 0.2 g/L	HPLC, DOC	Total conversion of TNP	Morales-Torres et al. (2010)
Walnut shells	Cu-loaded carbon (AC, PI)	S_{BET}: 422–474 m^2/g	Phenol	Fixed bed reactor: c [phenol]: 1000 mg/L, T: 140 °C–220 °C, p [total]: 3 MPa (O_2 flow 9 l/h, liquid flow 0.03 L/h) c [catalyst]: 2 g	UV-Vis (phenol), COD	95 and 85% conversions of phenol and COD after 8.5 h with PI catalyst	Wang et al. (2020)
Cellulose fibers	N-doped char	S_{BET}: 1305 m^2/g	Phenol	Batch reactor: c [phenol]: 1000 mg/L, T: 190 °C–260 °C, p [O_2]: 1 MPa, c [catalyst]: 0.2 wt.%,	HPLC	Total removal of phenol at 220 °C after 45 min	Tews et al. (2021)
Sawdust	PSD/MK_Fe: granulated (metakaolin, $CaCO_3$) carbon catalyst with Fe as active metal	S_{BET}: 299 m^2/g, pHzpc: 2.4	BPA	Batch reactor: c [BPA]: 60 mg/L, T: 160 °C, p [air]: 2 MPa, c [catalyst]: 1 g/L, pH 5–6	HPLC, TOC	> 98% BPA removal, 70% TOC conversion	Juhola et al. (2021)

sieved and then carbonized at three different temperatures (500 °C, 600 °C, and 800 °C) for 1 hour under N_2 flow. Carbonized samples were demineralized with acidic solution and after that process sieved and chemically activated with different portions of KOH. After KOH treatment, carbons were first heated to 60 °C, then to 110 °C, pyrolyzed at 300 °C for 3 hours, and finally at 800 °C for 2 hours in N_2 flow (Ubago-Pérez et al., 2006). The characterization results showed that SSA and the microporosity of the samples increased while the carbonization temperature increased. The catalytic activity of prepared carbons was studied in the CWAO of 2,4,6-trinitrophenol (TNP), and the total conversion of TNP was achieved after 2 hours reaction. Microporosity enhanced the adsorption of the TNP, and little oxidation occurred in the meso- and macropores while with the sample carbonized at a lower temperature (i.e., sample with lower SSA), the removal of TNP was faster due to the larger amount of meso and macropores.

Wang et al. (2020) used walnut shells as a raw material for catalysts for the reaction. Walnut shells were first ground, sieved, washed, dried, and then mixed with prepared copper-ammonia solution for 24 hours. After filtration and drying, the walnut shell-copper-ammonia sample was carbonized under N_2 at 800 °C for 2 hours. The activation of carbonized sample was performed with CO_2 at 800 °C for 2 hours and cooled to room temperature under N_2 (sample named PI). For comparison, walnut shells were carbonized as described but without the addition of copper-ammonia. For this sample copper-ammonia solution was added after carbonization (labeled as AC) and after impregnation, the dried AC was calcined under N_2 at 350 °C for 4 hours (Cu/AC catalyst). Both catalysts with different Cu loading showed good catalytic activity in the CWAO of phenol. According to XPS analysis in PI catalysts, Cu was present as Cu^{2+} but also as Cu^+ and Cu^0. The $Cu^+ + Cu^0$ atomic percentage of the PI catalysts was high, which promoted the Cu^+/Cu^{2+} redox conversion and further the formation of free radicals in the reaction. Therefore, PI catalysts showed higher catalytic activity in the CWAO of phenol than ACs.

Tews et al. (2021) studied also the CWAO of phenol. They produced nitrogen-doped char by carbonizing cellulose fibers first under N_2 at 850 °C for 1 hours and then with NH_3 for 1 hours. Finally, the sample was cooled to room temperature under N_2. XPS was used to analyze the nitrogen-containing functional groups of the N-doped char, and totally six nitrogen groups were identified. It was found that over 50 atom% of the analyzed groups consisted of pyridinic compounds, which have a positive effect on the wet oxidation reactions (Soares et al., 2016). The catalytic activity of N-doped char was studied in the CWAO of phenol in various reaction conditions, and the produced material showed comparable efficiency with commercial copper catalyst. Authors performed density functional theory (DFT) to observe which N functional group has the most significant impact on the free radical formation in the oxidation reaction. According to DFT, dipyridinic functional groups were the most important ones in facilitating the formation of free hydroxyl radicals. Juhola et al. (2021) carbonized pine sawdust under N_2 (200 mL/min) at 800 °C followed by the activation at 800 °C for 2 hours in steam (120 g/h). The carbonized material was oxidized with 6 M HNO_3 for 3 hours at 85 °C, washed with deionized water to neutral pH, and dried overnight at 105 °C. Produced carbon was used as a raw material to prepare granulates.

At the beginning of the granulation process, carbon and binders (metakaolin and $CaCO_3$) were wet ground and mixed in a ball mill, filtered, and dried at 105 °C. Dried mixture was mixed further with water, phosphoric acid, and polymeric organic resin PVA while glycerol was added to prevent agglomeration and embrittlement. The mixture was put in the ultrasound bath for 4 hours at 40 °C, and finally formed pasta was shaped into small granules and immersed in the solution of $B(OH)_3$ and $Fe(NO_3)_3 \times 9H_2O$ for 24 hours at 70 °C with smooth mixing. Finally, materials were filtered, washed, dried at 60 °C, and calcinated at 280 °C for 3 hours under N_2 flow. The catalytic activity of granulated biomass-based carbon catalyst (namely PSD/MK_Fe) was studied in the CWAO of BPA. After 3 hours of reaction at 160 °C almost complete BPA abatement and 70%, TOC conversion was achieved with this catalyst. The catalytic activity of PSD/MK_Fe was related to the high SSA and surface acidity. Moreover, the consecutive CWAO of BPA with this catalyst showed that the granular composite catalyst could be regenerated and reused without the loss of catalytic activity.

23.5 Challenges and future perspectives

According to a recent analysis, the AC market's global value in 2021 was 5689.7 million USD, and it was projected to be 10,947.3 million USD in 2028 (Vantage Market Research, 2022). The use of AC is widespread in many production processes varying from the classic gaseous stream and water treatment to more specific applications such as pharmacology and chemical-biological-radiological and nuclear (CBRN) protection. The properties of the ACs are hugely different depending on the functionalization of the SSA. The potential growth in the market for carbon-based catalysts depends on a better understanding of carbon surfaces and microstructures and improvements in quality control and production methods. Although relatively good knowledge of conventional oxide support (silica, alumina, zeolites) surface chemistry has already allowed the design of catalysts, such examples are still rare for carbon materials (Serp & Machado, 2015). The development of characterization techniques, which are identified as a limiting factor, should be improved. If the scalable and standardized production of catalytic materials from biomass can be realized, it would provide a significant advancement in the development of sustainable chemical industry (Davies et al., 2021). Consequently, more effort is needed to develop greener synthetic protocols to design biomass-derived carbon materials for catalytic applications (De et al., 2015).

One of the challenges in producing carbon-based catalyst materials from biomass is reproducibility. It is a challenging task to maintain and regulate the physicochemical properties, as well as the quality of carbon materials derived from biomass. Biomass feedstock, for example, may comprise metallic impurities and other pollutants such as pesticides, which might alter the characteristics of carbon materials produced. Moreover, if the carbon supports are made from pyrolysis of different batches of feedstock, it is difficult to control the catalytic properties due to variations in the physicochemical properties of the pristine carbon support. Thermal and chemical modifications are usually carried out for the enhancement of the catalytic activity of biomass-derived carbonaceous materials. Chemical treatments, coupling of semiconductors, and doping of heteroatoms are commonly used to modify AC and biochar produced from various feedstocks, which incur additional costs in catalyst manufacturing. Metals with biomass-carbon-based support are commonly found in manufactured catalysts. If the pH is in the acidic range, metal leaching may occur during the catalytic degradation of organic contaminants. Metal leaching issues can be alleviated by synthesizing a non-metal-based catalyst on carbon support (Wang et al., 2021).

Furthermore, when it comes to the utilization of lignocellulosic biomass waste-derived carbon materials for catalyst preparation and application in water reclamation, the important factors that need to be considered are the cost of its production, tailoring of the catalysts with carbon materials incorporation, the properties and characteristics of catalyst, the composition and impurities of biomass feedstock. Moreover, when compared with metal dopants, nonmetal dopants exhibit less photo-corrosion. Catalyst recovery concerns can be addressed by designing catalytic membrane and electrode materials that can reduce catalyst loss during treatment along with addressing recovery issues.

Typically, carbons produced from biomass have been applied in catalytic processes in powder form. However, the use of catalysts in granules produces several advantages. For example separation of granules from the treated effluent is easy, and the flow properties of the granules are better. Granules can be produced from carbonized materials; however, it is rather challenging due to the high cellulose content, which causes the fibrous and brittle structure of the granules. Therefore, binders such as metakaolin, starch, calcium carbonate, and resins can be used to produce carbon granules, but it should be noted that these additives bring new challenges to characterize carbon materials. Optimization of the process parameters is needed, and a life cycle assessment should be performed. Therefore the economic and technical points of view should be carefully considered before biomass waste-based catalyst materials can be applied to industrial use (Ijagwe et al., 2021).

23.6 Conclusions

The widely available lignocellulosic biomasses have drawn considerable attention to the development of carbonaceous materials, and they've been explored as potential replacements for fossil carbon-based materials. For the efficient conversion of biomass into carbonaceous material, various techniques have been studied. In recent years, biomass-derived AC and biochar have been extensively researched and evaluated as catalyst supports for the removal of a wide range of contaminants from wastewater. Biomass-derived carbonaceous material is excellent for catalyst support owing to its inherent functional groups, high stability, and electrical conductivity. However, because pristine carbonaceous material is ineffective, numerous studies have claimed that by modifying it adequately, enhanced catalytic activity can be obtained. The photocatalytic activity of the resultant catalyst materials can also be improved by heterojunction with semiconductors and doping of heteroatoms N, P, and S. Moreover, for CWPO and CWAO, carbonized materials have been activated chemically or physically with acids/bases or gases such as N_2. Importantly, the catalytic reaction requires a catalyst with a greater SSA, which enables the catalysts to adsorb the pollutants more easily, increasing their catalytic effectiveness. The catalyst's efficiency can be improved by employing carbon-based materials. Different functional groups, high porosity, and large SSA of carbon-based supports provide additional active sites for pollutant adsorption and degradation. The more frequently encountered difficulties in catalytic degradation of contaminants are environmental sustainability and efficiency. The primary requirement for advancement in the treatment technology for wastewater containing recalcitrant contaminants is the minimization of the total cost and time of the treatment processes. To improve photocatalytic performance, a potentially affordable, efficient, sustainable, and easy-to-prepare nanocomposite is urgently needed. Moreover, the economic and environmental point of view should be always considered when side stream-based materials are employed as more high-value products.

Acknowledgments

The authors acknowledge the funding from the Academy of Finland for project EXPO-STV-decision number 346537 and project CO2MetMES-decision number 329228.

References

A biochar classification system and associated test methods (2019). In *Biochar for environmental management*. https://doi.org/10.4324/9780203762264-15.

Adams, P., Bridgwater, T., Lea-Langton, A., Ross, A., & Watson, I. (2018). In P. Thornley, & P. B. T.-G. G. B. of B. S. Adams (Eds.), *Chapter 8 - Biomass conversion technologies* (pp. 107–139). Academic Press. Available from https://doi.org/10.1016/B978-0-08-101036-5.00008-2.

Adinata, D., Wan Daud, W. M. A., & Aroua, M. K. (2007). Preparation and characterization of activated carbon from palm shell by chemical activation with K2CO3. *Bioresour. Technol., 98*, 145–149. Available from https://doi.org/10.1016/j.biortech.2005.11.006.

Adinaveen, T., Vijaya, J. J., & Kennedy, L. J. (2016). Comparative study of electrical conductivity on activated carbons prepared from various cellulose materials. *Arabian Journal for Science and Engineering, 41*, 55–65. Available from https://doi.org/10.1007/s13369-014-1516-6.

Ahmadpour, A., & Do, D. D. (1996). The preparation of active carbons from coal by chemical and physical activation. *Carbon*. Available from https://doi.org/10.1016/0008-6223(95)00204-9.

Arunajatesan, V., Chen, B., Möbus, K., Ostgard, D. J., Tacke, T., & Wolf, D. (2008). Carbon-supported catalysts for the chemical industry. *Carbon Materials as Catalyst*. Available from https://doi.org/10.1002/9780470403709.ch15, Wiley Online Books.

Ateş, F., & Özcan, Ö. (2018). Preparation and characterization of activated carbon from poplar sawdust by chemical activation: Comparison of different activating agents and carbonization temperature. *European Journal of Engineering and Technology Research*, 3. Available from https://doi.org/10.24018/ejers.2018.3.11.939.

Atila, F. (2019). Compositional changes in lignocellulosic content of some agro-wastes during the production cycle of shiitake mushroom. *Sci. Hortic. (Amsterdam)., 245*, 263–268. Available from https://doi.org/10.1016/j.scienta.2018.10.029.

Babu, D. S., Srivastava, V., Nidheesh, P. V., & Kumar, M. S. (2019). Detoxification of water and wastewater by advanced oxidation processes. *The Science of the Total Environment, 696*, 133961. Available from https://doi.org/10.1016/j.scitotenv.2019.133961.

Bandosz, T. J. (2008). Surface chemistry of carbon materials. *Carbon Materials as Catalyst*. Available from https://doi.org/10.1002/9780470403709.ch2, Wiley Online Books.

Bano, S., & Negi, Y. S. (2017). Studies on cellulose nanocrystals isolated from groundnut shells. *Carbohydr. Polym., 157*, 1041–1049. Available from https://doi.org/10.1016/j.carbpol.2016.10.069.

Barton, S. S., Evans, M. J. B., Halliop, E., & MacDonald, J. A. F. (1997). Acidic and basic sites on the surface of porous carbon. *Carbon, 35*, 1361–1366. Available from https://doi.org/10.1016/S0008-6223(97)00080-8.

Başer, B., Yousaf, B., Yetis, U., Abbas, Q., Kwon, E. E., Wang, S., Bolan, N. S., & Rinklebe, J. (2021). Formation of nitrogen functionalities in biochar materials and their role in the mitigation of hazardous emerging organic pollutants from wastewater. *Journal of Hazardous Materials, 416*, 126131. Available from https://doi.org/10.1016/j.jhazmat.2021.126131.

Bergna, D., Hu, T., Prokkola, H., Romar, H., & Lassi, U. (2020). Effect of some process parameters on the main properties of activated carbon produced from peat in a lab-scale process. *Waste and Biomass Valorization*, 11. Available from https://doi.org/10.1007/s12649-019-00584-2.

Bergna, D., Varila, T., Romar, H., & Lassi, U. (2022). Activated carbon from hydrolysis lignin: Effect of activation method on carbon properties. *Biomass and Bioenergy, 159*, 106387. Available from https://doi.org/10.1016/j.biombioe.2022.106387.

Bicu, I., & Mustata, F. (2011). Cellulose extraction from orange peel using sulfite digestion reagents. *Bioresour. Technol., 102*, 10013–10019. Available from https://doi.org/10.1016/j.biortech.2011.08.041.

Bitter, J. H., & de Jong, K. P. (2008). Preparation of carbon-supported metal catalysts. *Carbon Materials as Catalyst*. Available from https://doi.org/10.1002/9780470403709.ch5, Wiley Online Books.

Boehm, H. P. (1966). Chemical identification of surface groups. *Advances in Catalysis, 16*, 179–274. Available from https://doi.org/10.1016/S0360-0564(08)60354-5.

Boehm, H. P. (2002). Surface oxides on carbon and their analysis: A critical assessment. *Carbon, 40*, 145–149. Available from https://doi.org/10.1016/S0008-6223(01)00165-8.

Boehm, H.-P., Diehl, E., Heck, W., & Sappok, R. (1964). Surface oxides of carbon. *Angewandte Chemie International Edition, 3*, 669–677. Available from https://doi.org/10.1002/anie.196406691.

Camia, A., Robert, N., Jonsson, K., Pilli, R., Garcia Condado, S., et al. (2018). *Biomass production, supply, uses and flows in the European Union: First results from an integrated assessment*, EUR 28993, 2018. EN, Publ. Off. Eur. Union, Luxemb.

Chen, N., Huang, Y., Hou, X., Ai, Z., & Zhang, L. (2017). Photochemistry of hydrochar: Reactive oxygen species generation and sulfadimidine degradation. *Environmental Science & Technology, 51*, 11278–11287. Available from https://doi.org/10.1021/acs.est.7b02740.

Chun, S. E., & Whitacre, J. F. (2017). Formation of micro/mesopores during chemical activation in tailor-made nongraphitic carbons. *Microporous Mesoporous Mater*, 251. Available from https://doi.org/10.1016/j.micromeso.2017.05.038.

Davies, G., El Sheikh, A., Collett, C., Yakub, I., & McGregor, J. (2021). In S. B. T.-E. C. M. for C. Sadjadi (Ed.), *Chapter 5 - Catalytic carbon materials from biomass* (pp. 161–195). Elsevier. Available from https://doi.org/10.1016/B978-0-12-817561-3.00005-6.

De, S., Balu, A. M., van der Waal, J. C., & Luque, R. (2015). Biomass-derived porous carbon materials: synthesis and catalytic applications. *ChemCatChem, 7*, 1608–1629. Available from https://doi.org/10.1002/cctc.201500081.

Dias, J. M., Alvim-Ferraz, M. C. M., Almeida, M. F., Rivera-Utrilla, J., & Sánchez-Polo, M. (2007). Waste materials for activated carbon preparation and its use in aqueous-phase treatment: A review. *Journal of Environmental Management, 85*, 833–846. Available from https://doi.org/10.1016/j.jenvman.2007.07.031.

Diaz De Tuesta, J. L., Saviotti, M. C., Roman, F. F., Pantuzza, G. F., Sartori, H. J. F., Shinibekova, A., Kalmakhanova, M. S., Massalimova, B. K., Pietrobelli, J. M. T. A., Lenzi, G. G., & Gomes, H. T. (2021). Assisted hydrothermal carbonization of agroindustrial byproducts as effective step in the production of activated carbon catalysts for wet peroxide oxidation of micro-pollutants. *Journal of Environmental Chemical Engineering, 9*, 105004. Available from https://doi.org/10.1016/J.JECE.2020.105004.

Do Minh, T., Song, J., Deb, A., Cha, L., Srivastava, V., & Sillanpää, M. (2020). Biochar based catalysts for the abatement of emerging pollutants: A review. *Chemical Engineering Journal*, 124856. Available from https://doi.org/10.1016/j.cej.2020.124856.

Duan, D., Feng, Z., Dong, X., Chen, X., Zhang, Y., Wan, K., et al. (2021). Improving bio-oil quality from low-density polyethylene pyrolysis: Effects of varying activation and pyrolysis parameters. *Energy, 232*, 121090. Available from https://doi.org/10.1016/j.energy.2021.121090.

EBC (2020). *Certification of the carbon sink potential of biochar*, Ithaka Institute.

Elyounssi, K., Collard, F.-X., Mateke, J. N., & Blin, J. (2012). Improvement of charcoal yield by two-step pyrolysis on eucalyptus wood: A thermogravimetric study. *Fuel, 96*, 161–167. Available from https://doi.org/10.1016/j.fuel.2012.01.030.

Esteves, Bruno M., Morales-Torres, S., Maldonado-Hódar, F. J., & Madeira, L. M. (2020). Fitting biochars and activated carbons from residues of the olive oil industry as supports of Fe-catalysts for the heterogeneous Fenton-like treatment of simulated olive mill wastewater. *Nanomaterials, 10*, 1–26. Available from https://doi.org/10.3390/nano10050876.

European council of Chemical Manufacturers Federations (1986). *Test methods for activated carbon* [WWW Document]. Cefic. URL https://activatedcarbon.org/images/Test_method_for_Activated_Carbon_86.pdf.

European Parliament, European Council (2018). Directive (EU) 2018/851 of the European Parliament and of the Council of 30 May 2018 amending Directive 2008/98/EC on waste. *Off. J. Eur. Union*.

Figueiredo, J. L., & Pereira, M. F. R. (2008). Carbon as catalyst. *Carbon Materials as Catalyst*. Available from https://doi.org/10.1002/9780470403709.ch6, Wiley Online Books.

Ge, X., Chang, C., Zhang, L., Cui, S., Luo, X., Hu, S., Qin, Y., & Li, Y. (2018). In Y. Li, & X. B. T.-A. in B. Ge (Eds.), *Chapter five - Conversion of lignocellulosic biomass into platform chemicals for biobased polyurethane application* (pp. 161–213). Elsevier. Available from https://doi.org/10.1016/bs.aibe.2018.03.002.

Goldemberg, J., & Teixeira Coelho, S. (2004). Renewable energy—traditional biomass vs. modern biomass. *Energy Policy, 32*, 711–714. Available from https://doi.org/10.1016/S0301-4215(02)00340-3.

Gomes, H. T., Miranda, S. M., Sampaio, M. J., Silva, A. M. T., & Faria, J. L. (2010). Activated carbons treated with sulphuric acid: Catalysts for catalytic wet peroxide oxidation. *Catalysis Today, 151*, 153–158. Available from https://doi.org/10.1016/j.cattod.2010.01.017.

González-García, P. (2018). Activated carbon from lignocellulosics precursors: A review of the synthesis methods, characterization techniques and applications. *Renewable and Sustainable Energy Reviews, 82*, 1393–1414. Available from https://doi.org/10.1016/j.rser.2017.04.117.

Gonçalves, F. A., Ruiz, H. A., dos Santos, E. S., Teixeira, J. A., & de Macedo, G. R. (2019). Valorization, Comparison and Characterization of Coconuts Waste and Cactus in a Biorefinery Context Using NaClO2–C2H4O2 and Sequential NaClO2–C2H4O2/Autohydrolysis Pretreatment. *Waste and Biomass Valorization, 10*, 2249–2262. Available from https://doi.org/10.1007/s12649-018-0229-6.

González, J. F., Encinar, J. M., Canito, J. L., Sabio, E., & Chacón, M. (2003). Pyrolysis of cherry stones: energy uses of the different fractions and kinetic study. *J. Anal. Appl. Pyrolysis, 67*, 165–190. Available from https://doi.org/10.1016/S0165-2370(02)00060-8.

González, J. F., Román, S., Encinar, J. M., & Martínez, G. (2009). Pyrolysis of various biomass residues and char utilization for the production of activated carbons. *J. Anal. Appl. Pyrolysis, 85*, 134–141. Available from https://doi.org/10.1016/j.jaap.2008.11.035.

Gurrath, M., Kuretzky, T., Boehm, H. P., Okhlopkova, L. B., Lisitsyn, A. S., & Likholobov, V. A. (2000). Palladium catalysts on activated carbon supports: Influence of reduction temperature, origin of the support and pretreatments of the carbon surface. *Carbon, 38*, 1241–1255. Available from https://doi.org/10.1016/S0008-6223(00)00026-9.

Heidenreich, S., & Foscolo, P. U. (2015). New concepts in biomass gasification. *Progress in Energy and Combustion Science*. Available from https://doi.org/10.1016/j.pecs.2014.06.002.

Ivanovski, M., Goricanec, D., Krope, J., & Urbancl, D. (2022). Torrefaction pretreatment of lignocellulosic biomass for sustainable solid biofuel production. *Energy, 240*, 122483. Available from https://doi.org/10.1016/j.energy.2021.122483.

Iwanow, M., Gärtner, T., Sieber, V., & König, B. (2020). Activated carbon as catalyst support: Precursors, preparation, modification and characterization. *Beilstein Journal of Organic Chemistry, 16*, 1188–1202. Available from https://doi.org/10.3762/bjoc.16.104.

Jaganmohan, M. (2022). *Biomass energy consumption in the United States from 2006 to 2021*. Stastica.

Jing, G., Luan, M., & Chen, T. (2016). Progress of catalytic wet air oxidation technology. *Arabian Journal of Chemistry, 9*, S1208–S1213. Available from https://doi.org/10.1016/J.ARABJC.2012.01.001.

Jjagwe, J., Olupot, P. W., Menya, E., & Kalibbala, H. M. (2021). Synthesis and application of granular activated carbon from biomass waste materials for water treatment: A review. *Journal of Bioresources and Bioproducts, 6*, 292–322. Available from https://doi.org/10.1016/J.JOBAB.2021.03.003.

John, I., Yaragarla, P., Muthaiah, P., Ponnusamy, K., & Appusamy, A. (2017). Statistical optimization of acid catalyzed steam pretreatment of citrus peel waste for bioethanol production. *Resour. Technol., 3*, 429–433. Available from https://doi.org/10.1016/j.reffit.2017.04.001.

Juhola, R., Heponiemi, A., Tuomikoski, S., Hu, T., Huuhtanen, M., Bergna, D., & Lassi, U. (2021). Preparation of granulated biomass carbon catalysts—Structure tailoring, characterization, and use in catalytic wet air oxidation of bisphenol A. *Catalysts, 11*, 1–20. Available from https://doi.org/10.3390/catal11020251.

Juhola, R., Heponiemi, A., Tuomikoski, S., Hu, T., Prokkola, H., Romar, H., & Lassi, U. (2019). Biomass-based composite catalysts for catalytic wet peroxide oxidation of bisphenol A: Preparation and characterization studies. *Journal of Environmental Chemical Engineering, 7*, 103127. Available from https://doi.org/10.1016/j.jece.2019.103127.

Juhola, R., Heponiemi, A., Tuomikoski, S., Hu, T., Vielma, T., & Lassi, U. (2017). Preparation of novel fe catalysts from industrial by-products: Catalytic wet peroxide oxidation of bisphenol A. *Topics in Catalysis, 60*, 1387–1400. Available from https://doi.org/10.1007/s11244-017-0829-6.

Kalita, E., Nath, B. K., Deb, P., Agan, F., Islam, M. R., & Saikia, K. (2015). High quality fluorescent cellulose nanofibers from endemic rice husk: Isolation and characterization. *Carbohydr. Polym., 122*, 308–313. Available from https://doi.org/10.1016/j.carbpol.2014.12.075.

Karthikeyan, S., Sekaran, G., & Gupta, V. K. (2013). Nanoporous activated carbon fluidized bed catalytic oxidations of aqueous o, p and m-cresols: Kinetic and thermodynamic studies. *Environmental Science and Pollution Research, 20*, 4790–4806. Available from https://doi.org/10.1007/s11356-012-1380-4.

Khezami, L., Chetouani, A., Taouk, B., & Capart, R. (2005). Production and characterisation of activated carbon from wood components in powder: Cellulose, lignin, xylan. *Powder Technology, 157*, 48–56. Available from https://doi.org/10.1016/j.powtec.2005.05.009.

Kilpimaa, S., Runtti, H., Kangas, T., Lassi, U., & Kuokkanen, T. (2014). Removal of phosphate and nitrate over a modified carbon residue from biomass gasification. *Chemical Engineering Research and Design, 92*. Available from https://doi.org/10.1016/j.cherd.2014.03.019.

Kösep lu, E., & Akmil-Bb_ar, C. (2015). Preparation, structural evaluation and adsorptive properties of activated carbon from agricultural waste biomass. *Adv. Powder Technol., 26*, 811–818.

Kim, M., & Day, D. F. (2011). Composition of sugar cane, energy cane, and sweet sorghum suitable for ethanol production at Louisiana sugar mills. *J. Ind. Microbiol. Biotechnol., 38*, 803–807. Available from https://doi.org/10.1007/s10295-010-0812-8.

Kumari, N., Chhabra, T., Kumar, A., & Krishnan, V. (2021). Bioderived carbon supported bismuth molybdate nanocomposites as bifunctional catalysts for removal of organic pollutants: Adsorption and photocatalytic studies. *Materials Letters, 302*130455. Available from https://doi.org/10.1016/j.matlet.2021.130455.

Kurian, M. (2021). Advanced oxidation processes and nanomaterials -A review. *Cleaner Engineering and Technology, 2*, 100090. Available from https://doi.org/10.1016/j.clet.2021.100090.

Lee, J., & Park, K. Y. (2020). Impact of hydrothermal pretreatment on anaerobic digestion efficiency for lignocellulosic biomass: Influence of pretreatment temperature on the formation of biomass-degrading byproducts. *Chemosphere, 256*, 127116. Available from https://doi.org/10.1016/j.chemosphere.2020.127116.

Li, H., Chen, Y., Zhou, W., Jiang, H., Liu, H., Chen, X., & Guohui, T. (2019). $WO_3/BiVO_4/BiOCl$ porous nanosheet composites from a biomass template for photocatalytic organic pollutant degradation. *Journal of Alloys and Compounds, 802*, 76–85. Available from https://doi.org/10.1016/j.jallcom.2019.06.187.

Li, L., Rowbotham, J. S., Christopher Greenwell, H., & Dyer, P. W. (2013). An introduction to pyrolysis and catalytic pyrolysis: Versatile techniques for biomass conversion, in:*New and Future Developments in Catalysis: Catalytic Biomass Conversion*. Available from https://doi.org/10.1016/B978-0-444-53878-9.00009-6.

Lin, M., Li, F., Cheng, W., Rong, X., & Wang, W. (2022). Facile preparation of a novel modified biochar-based supramolecular self-assembled g-C_3N_4 for enhanced visible light photocatalytic degradation of phenanthrene. *Chemosphere, 288*, 132620. Available from https://doi.org/10.1016/j.chemosphere.2021.132620.

Liu, W.-J., Jiang, H., & Yu, H.-Q. (2015). Thermochemical conversion of lignin to functional materials: A review and future directions. *Green Chemistry: an International Journal and Green Chemistry Resource: GC, 17*, 4888–4907. Available from https://doi.org/10.1039/C5GC01054C.

Liu, X., Bi, X. T., Liu, C., & Liu, Y. (2012). Performance of Fe/AC catalyst prepared from demineralized pine bark particles in a microwave reactor. *Chemical Engineering Journal, 193–194*, 187–195. Available from https://doi.org/10.1016/J.CEJ.2012.04.039.

Luo, H., Yu, S., Zhong, M., Han, Y., Su, B., & Lei, Z. (2022). Waste biomass-assisted synthesis of TiO_2 and N/O-contained graphene-like biochar composites for enhanced adsorptive and photocatalytic performances. *Journal of Alloys and Compounds, 899*, 163287. Available from https://doi.org/10.1016/j.jallcom.2021.163287.

Maneerung, T., Liew, J., Dai, Y., Kawi, S., Chong, C., & Wang, C. (2016). Activated carbon derived from carbon residue from biomass gasification and its application for dye adsorption: Kinetics, isotherms and thermodynamic studies. *Bioresource Technology, 200*, 350–359.

Márquez, J. J. R., Levchuk, I., & Sillanpää, M. (2018). Application of catalytic wet peroxide oxidation for industrial and urban wastewater treatment: A review. *Catalysts, 8*. Available from https://doi.org/10.3390/CATAL8120673.

Marsh, H., & Rodríguez-Reinoso, F. (2006). *Activated carbon*. Elsevier. Available from https://doi.org/10.1016/B978-0-08-044463-5.X5013-4.

McNaught, D.A. A., & Wilkinson, A. (2008). IUPAC Goldbook [WWW Document]. IUPAC Compend. Chem. Terminol.

Mohamad Nor, N., Lau, L. C., Lee, K. T., & Mohamed, A. R. (2013). Synthesis of activated carbon from lignocellulosic biomass and its applications in air pollution control—A review. *Journal of Environmental Chemical Engineering, 1*, 658–666. Available from https://doi.org/10.1016/j.jece.2013.09.017.

Mohamed, M. A., Zain, M. M. F., Jeffery Minggu, L., Kassim, M. B., Saidina Amin, N. A., Salleh, W. N. W., Salehmin, M. N. I., Md Nasir, M. F., & Mohd Hir, Z. A. (2018). Constructing bio-templated 3D porous microtubular C-doped g-C_3N_4 with tunable band structure and enhanced charge carrier separation. *Applied Catalysis B: Environmental, 236*, 265–279. Available from https://doi.org/10.1016/j.apcatb.2018.05.037.

Monlau, F., Barakat, A., Steyer, J. P., & Carrere, H. (2012). Comparison of seven types of thermo-chemical pretreatments on the structural features and anaerobic digestion of sunflower stalks. *Bioresour. Technol., 120*, 241–247. Available from https://doi.org/10.1016/j.biortech.2012.06.040.

Morales-Torres, S., Silva, A. M. T., Pérez-Cadenas, A. F., Faria, J. L., Maldonado-Hódar, F. J., Figueiredo, J. L., & Carrasco-Marín, F. (2010). Wet air oxidation of trinitrophenol with activated carbon catalysts: Effect of textural properties on the mechanism of degradation. *Applied Catalysis B: Environmental, 100*, 310–317. Available from https://doi.org/10.1016/J.APCATB.2010.08.007.

Nishide, R. N., Truong, J. H., & Abu-Omar, M. M. (2021). Organosolv fractionation of walnut shell biomass to isolate lignocellulosic components for chemical upgrading of lignin to aromatics. *ACS Omega, 6*, 8142–8150. Available from https://doi.org/10.1021/acsomega.0c05936.

Pallarés, J., González-Cencerrado, A., & Arauzo, I. (2018). Production and characterization of activated carbon from barley straw by physical activation with carbon dioxide and steam. *Biomass and Bioenergy, 115*, 64–73. Available from https://doi.org/10.1016/j.biombioe.2018.04.015.

Park, J.-H., Wang, J. J., & Seo, D.-C. (2021). Comparison of catalytic activity for treating recalcitrant organic pollutant in heterogeneous Fenton oxidation with iron-impregnated biochar and activated carbon. *Journal of Water Process Engineering, 42*, 102141. Available from https://doi.org/10.1016/j.jwpe.2021.102141.

Patel, M., Zhang, X., & Kumar, A. (2016). Techno-economic and life cycle assessment on lignocellulosic biomass thermochemical conversion technologies: A review. *Renewable and Sustainable Energy Reviews, 53*, 1486–1499. Available from https://doi.org/10.1016/j.rser.2015.09.070.

Pedersen, T. H., Grigoras, I. F., Hoffmann, J., Toor, S. S., Daraban, I. M., Jensen, C. U., et al. (2016). Continuous hydrothermal co-liquefaction of aspen wood and glycerol with water phase recirculation. *Appl. Energy*, 162, 1034–1041. Available from https://doi.org/10.1016/j.apenergy.2015.10.165.

Qian, E. W. (2014). In S. Tojo, & T. B. T.-R. A. to S. B. S. Hirasawa (Eds.), *Chapter 7 - Pretreatment and saccharification of lignocellulosic biomass* (pp. 181–204). Boston: Academic Press. Available from https://doi.org/10.1016/B978-0-12-404609-2.00007-6.

Radovic, L. R. (2008). Physicochemical properties of carbon materials: A brief overview. *Carbon Materials as Catalyst*. Available from https://doi.org/10.1002/9780470403709.ch1, Wiley Online Books.

Reddy, K. O., Maheswari, C. U., Dhlamini, M. S., Mothudi, B. M., Kommula, V. P., Zhang, J., et al. (2018). Extraction and characterization of cellulose single fibers from native african napier grass. *Carbohydr. Polym.*, 188, 85–91. Available from https://doi.org/10.1016/j.carbpol.2018.01.110.

Rodríguez-reinoso, F. (1998). The role of carbon materials in heterogeneous catalysis. *Carbon*, 36, 159–175. Available from https://doi.org/10.1016/S0008-6223(97)00173-5.

Rodríguez-Reinoso, F., Molina-Sabio, M., & González, M. T. (1995). The use of steam and CO_2 as activating agents in the preparation of activated carbons. *Carbon*. Available from https://doi.org/10.1016/0008-6223(94)00100-E.

Salame, I. I., & Bandosz, T. J. (2001). Surface chemistry of activated carbons: Combining the results of temperature-programmed desorption, boehm, and potentiometric titrations. *Journal of Colloid and Interface Science*, 240, 252–258. Available from https://doi.org/10.1006/jcis.2001.7596.

Samir, B., Bakhta, S., Bouazizi, N., Sadaoui, Z., Allalou, O., Le Derf, F., & Vieillard, J. (2021). TBO degradation by heterogeneous fenton-like reaction using Fe supported over activated carbon. *Catalysts*, 11, 1456. Available from https://doi.org/10.3390/catal11121456.

Sane, P. K., Rakte, D., Tambat, S., Bhalinge, R., Sontakke, S. M., & Nemade, P. (2022). Enhancing solar photocatalytic activity of Bi_5O_7I photocatalyst with activated carbon heterojunction. *Advanced Powder Technology*, 33, 103357. Available from https://doi.org/10.1016/j.apt.2021.11.009.

Serp, P., & Figueiredo, J. L. (2008). *Carbon materials for catalysis, carbon materials for catalysis*. John Wiley and Sons. Available from https://doi.org/10.1002/9780470403709.

Serp, P., & Machado, B. (2015). *Carbon (Nano)materials for catalysis. Nanostructured carbon materials for catalysis* (pp. 1–45). The Royal Society of Chemistry. Available from https://doi.org/10.1039/9781782622567-00001.

Setianingsih, T., Purwonugroho, D., & Prananto, Y.P. (2021). *Synthesis of CNS, ZnO/CNS and $ZnCr_2O_4$/CNS composites from patchouli biomass by using microwave for remediation of pesticide contaminated surface water in paddy field*. IOP Conf. Ser. Earth Environ. Sci. 930. https://doi.org/10.1088/1755-1315/930/1/012020.

Shan, R., Lu, L., Gu, J., Zhang, Y., Yuan, H., Chen, Y., & Luo, B. (2020). Photocatalytic degradation of methyl orange by Ag/TiO_2/biochar composite catalysts in aqueous solutions. *Materials Science in Semiconductor Processing*, 114, 105088. Available from https://doi.org/10.1016/j.mssp.2020.105088.

Shawky, B. T., Mahmoud, M. G., Ghazy, E. A., Asker, M. M. S., & Ibrahim, G. S. (2011). Enzymatic hydrolysis of rice straw and corn stalks for monosugars production. *J. Genet. Eng. Biotechnol.*, 9, 59–63. Available from https://doi.org/10.1016/j.jgeb.2011.05.001.

Shemfe, M. B., Gu, S., & Ranganathan, P. (2015). Techno-economic performance analysis of biofuel production and miniature electric power generation from biomass fast pyrolysis and bio-oil upgrading. *Fuel*, 143, 361–372. Available from https://doi.org/10.1016/j.fuel.2014.11.078.

Singh, A. R., Dhumal, P. S., Bhakare, M. A., Lokhande, K. D., Bondarde, M. P., & Some, S. (2022). In-situ synthesis of metal oxide and polymer decorated activated carbon-based photocatalyst for organic pollutants degradation. *Separation and Purification Technology*, 286, 120380. Available from https://doi.org/10.1016/j.seppur.2021.120380.

Singh, B., Singh, B. P., & Cowie, A. L. (2010). Characterisation and evaluation of biochars for their application as a soil amendment, in: *Australian Journal of Soil Research*. Available from https://doi.org/10.1071/SR10058.

Soares, O. S. G. P., Rocha, R. P., Gonçalves, A. G., Figueiredo, J. L., Órfão, J. J. M., & Pereira, M. F. R. (2016). Highly active N-doped carbon nanotubes prepared by an easy ball milling method for advanced oxidation processes. *Applied Catalysis B: Environmental*, 192, 296–303. Available from https://doi.org/10.1016/j.APCATB.2016.03.069.

Tay, T., Ucar, S., & Karagöz, S. (2009). Preparation and characterization of activated carbon from waste biomass. *J. Hazard. Mater.*, 165, 481–485. Available from https://doi.org/10.1016/j.jhazmat.2008.10.011.

Tews, I., Garcia, A., Ayiania, M., Mood, S. H., Mainali, K., McEwen, J.-S., & Garcia-Perez, M. (2021). Nitrogen-doped char as a catalyst for wet oxidation of phenol-contaminated water. *Biomass Conversion and Biorefinery*. Available from https://doi.org/10.1007/s13399-020-01184-0.

Ubago-Pérez, R., Carrasco-Marín, F., Fairén-Jiménez, D., & Moreno-Castilla, C. (2006). Granular and monolithic activated carbons from KOH-activation of olive stones. *Microporous Mesoporous Mater*, 92, 64–70. Available from https://doi.org/10.1016/j.micromeso.2006.01.002.

Uçar, S., Erdem, M., Tay, T., & Karagöz, S. (2009). Preparation and characterization of activated carbon produced from pomegranate seeds by $ZnCl_2$ activation. *Appl. Surf. Sci.*, 255, 8890–8896. Available from https://doi.org/10.1016/j.apsusc.2009.06.080.

Vantage Market Research (2022). Global activated carbon market report 2022: Market is expected to record a value of USD 10947.3 Million in 2028, Rising at a CAGR of 9.8%; Says Vantage Market Research [WWW Document]. URL https://www.globenewswire.com/news-release/2022/01/21/2370723/0/en/Global-Activated-Carbon-Market-Report-2022-Market-is-Expected-to-Record-a-Value-of-USD-10947-3-Million-in-2028-Rising-at-a-CAGR-of-9-8-Says-Vantage-Market-Research.html.

Varila, T., Bergna, D., Lahti, R., Romar, H., Hu, T., & Lassi, U. (2017). Activated carbon production from peat using $ZnCl_2$: Characterization and applications. *Bioresources*, 12, 8078–8092. Available from https://doi.org/10.15376/BIORES.12.4.8078-8092.

Wan, C., Zhou, Y., & Li, Y. (2011). Liquid hot water and alkaline pretreatment of soybean straw for improving cellulose digestibility. *Bioresour. Technol.*, 102, 6254–6259. Available from https://doi.org/10.1016/j.biortech.2011.02.075.

Wang, H., Li, G., Zhang, S., Li, Y., Zhao, Y., Duan, L., & Zhang, Y. (2020). Preparation of Cu-loaded biomass-derived activated carbon catalysts for catalytic wet air oxidation of phenol. *Industrial & Engineering Chemistry Research*. Available from https://doi.org/10.1021/acs.iecr.9b05750.

Wang, S., Dai, G., Yang, H., & Luo, Z. (2017). Lignocellulosic biomass pyrolysis mechanism: A state-of-the-art review. *Progress in Energy and Combustion Science*. Available from https://doi.org/10.1016/j.pecs.2017.05.004.

Wang, T., Dissanayake, P. D., Sun, M., Tao, Z., Han, W., An, N., Gu, Q., Xia, D., Tian, B., Ok, Y. S., & Shang, J. (2021). Adsorption and visible-light photocatalytic degradation of organic pollutants by functionalized biochar: Role of iodine doping and reactive species. *Environmental Research, 197*, 111026. Available from https://doi.org/10.1016/j.envres.2021.111026.

Wang, T., Liu, X., Ma, C., Liu, Y., Dong, H., Ma, W., Liu, Z., Wei, M., Li, C., & Yan, Y. (2018). A two step hydrothermal process to prepare carbon spheres from bamboo for construction of core−shell non-metallic photocatalysts. *New Journal of Chemistry, 42*, 6515−6524. Available from https://doi.org/10.1039/C8NJ00953H.

Wei, H., Yingting, Y., Jingjing, G., Wenshi, Y., & Junhong, T. (2017). In M. A. B. T.-E. of S. T. Abraham (Ed.), *Lignocellulosic biomass valorization: production of ethanol* (pp. 601−604). Oxford: Elsevier. Available from https://doi.org/10.1016/B978-0-12-409548-9.10239-8.

Williams, P. T., & Reed, A. R. (2006). Development of activated carbon pore structure via physical and chemical activation of biomass fibre waste. *Biomass and Bioenergy, 30*, 144−152. Available from https://doi.org/10.1016/j.biombioe.2005.11.006.

Xu, F., & Li, Y. (2017). In M. A. B. T.-E. of S. T. Abraham (Ed.), *Biomass digestion* (pp. 197−204). Oxford: Elsevier. Available from https://doi.org/10.1016/B978-0-12-409548-9.10108-3.

Yavari, Saba, Sapari, N. B., Malakahmad, A., & Yavari, Sara (2019). Degradation of imazapic and imazapyr herbicides in the presence of optimized oil palm empty fruit bunch and rice husk biochars in soil. *Journal of Hazardous Materials, 366*, 636−642. Available from https://doi.org/10.1016/j.jhazmat.2018.12.022.

Ye, S., Yan, M., Tan, X., Liang, J., Zeng, G., Wu, H., Song, B., Zhou, C., Yang, Y., & Wang, H. (2019). Facile assembled biochar-based nanocomposite with improved graphitization for efficient photocatalytic activity driven by visible light. *Applied Catalysis B: Environmental, 250*, 78−88. Available from https://doi.org/10.1016/j.apcatb.2019.03.004.

Yousuf, A., Pirozzi, D., & Sannino, F. (2020). In A. Yousuf, D. Pirozzi, & F. B. T.-L. B. to L. B. Sannino (Eds.), *Chapter 1 - Fundamentals of lignocellulosic biomass* (pp. 1−15). Academic Press. Available from https://doi.org/10.1016/B978-0-12-815936-1.00001-0.

Zhao, C., Shao, B., Yan, M., Liu, Z., Liang, Q., He, Q., Wu, T., Liu, Y., Pan, Y., Huang, J., Wang, J., Liang, J., & Tang, L. (2021). Activation of peroxymonosulfate by biochar-based catalysts and applications in the degradation of organic contaminants: A review. *Chemical Engineering Journal, 416*, 128829. Available from https://doi.org/10.1016/j.cej.2021.128829.

Zhu, K., Bin, Q., Shen, Y., Huang, J., He, D., & Chen, W. (2020). In-situ formed N-doped bamboo-like carbon nanotubes encapsulated with Fe nanoparticles supported by biochar as highly efficient catalyst for activation of persulfate (PS) toward degradation of organic pollutants. *Chemical Engineering Journal, 402*, 126090. Available from https://doi.org/10.1016/j.cej.2020.126090.

C H A P T E R 24

Technoeconomic feasibility analysis of waste bioprocessing

V.M. Jaganathan[1], Joseph Sekhar Santhappan[2], Rajalingam Arumuganainar[2], M. Edwin[3] and Godwin Glivin[4]

[1]Department of Energy and Environment, National Institute of Technology, Tiruchirappalli, Tamil Nadu, India
[2]Engineering Department, College of Engineering and Technology, University of Technology and Applied Sciences, Shinas, Oman [3]Department of Mechanical Engineering, University College of Engineering Nagercoil, Anna University Constituent College, Nagercoil, Tamil Nadu, India [4]Department of Energy and Environment, National Institute of Technology, Tiruchirappalli, Tamil Nadu, India

24.1 Introduction

24.1.1 Demand of renewable energy

Throughout the last few decades, population and economic growth increased global energy demand, fossil fuel depletion, and temperature rise. This necessitates large-scale energy source modifications. To fulfill future primary energy demand, the "energy shift" must be sustainable. Government agencies and research institutes are developing new energy policies and eco-friendly technology as energy and its impact on the global economy are inevitable. The IPCC special report summarizes three energy scenarios for a sustainable future based on steps to prevent the 1.5°C global temperature rise (Masson et al., 2020). Energy transition priority determines classifications. Fig. 24.1 shows the principal energy supply of the next few decades. The 2020–50 energy scenario integration pathways predict a higher renewable energy contribution than fossil fuels. In 2050, fossil fuel contributions varied from coal, oil, and gas. The severity of greenhouse gas mitigation reduced coal's share

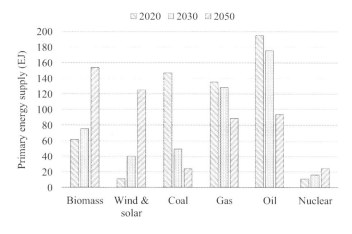

FIGURE 24.1 Global primary energy supply scenario (Masson et al., 2020).

to primary energy supply, but oil's contribution remained strong. Renewables—biomass, wind, and solar—are rising. Bioenergy will also supply primary energy for electricity, liquid transportation fuels, and industrial heating, according to the figure. The low and high action paths reflect this renewable transition to limit global warming. "Renewable energy transition" is inevitable.

Due to fossil fuel depletion and climate change, all energy strategies and reports in this decade call for renewable primary energy deployment. Hence, renewable energy technologies for energy generation have grown significantly in recent decades. Renewable usage hiked 5% in 2017 (Source: https://www.iea.org). Biomass dominated 2017 renewable energy supply from various locations. Africa, Asia, America, and Europe contributed 96%, 65%, 59%, and 59% biomass. This trend suggests that biomass will be a key renewable energy source. The International Energy Agency (IEA) stated that the bioenergy contributed 50% of renewable energy in 2017, mostly in heat, power, and transport. In 2023, 30% of bioenergy growth is projected from heating, which uses biomass as solid, liquid, and gaseous fuels. India, China, EU, and USA predicted this growth. Half of the growth is predicted from Asian countries, notably India and China. While renewable energy is predicted to contribute two-thirds of global primary energy supply in 2050, the energy shift is still far off. Primary energy supply may rise from 15% to 63% in three decades (IEA, 2022; International Energy Agency, 2020). Consequently, to achieve energy transition, renewables and energy efficiency must be increased to eliminate 94% of predicted emissions as per the Paris Climate Agreement (Kalaiselvan et al., 2022).

24.1.2 Potential of biogas as future energy

Biogas could address modern society's growing organic waste and the need to minimize global greenhouse gas (GHG) emissions. This new International Energy Agency (IEA) paper examines the potential for biogas in the global energy system, the opportunities and risks, and what policymakers and industry can do to support sustainable growth in this sector. Biogas production shows a world where resources are continuously used and repurposed to meet rising energy demand while also benefiting the environment. Biogas decreases solid biomass use for cooking in underdeveloped nations, increasing health and economy. Biogas delivers clean cooking to 200 million people by 2040. In the sustainable development scenario (SDS), half of whom are Africans. Furthermore, biogas is more valuable in SDS like the IEA that achieves global targets to combat climate change, improves air quality, and provides modern energy. Biogas can assist the SDS's goals of universal energy access, better air, and keeping global temperatures far below 2°C and pursuing measures to limit the same to 1.5°C in accordance with the Paris Agreement. In 2018, biogas production was 35 Mtoe, a quarter of the anticipated potential. Sustainable potential might provide 20% of global petrol demand. Biogas may be produced worldwide, and sustainable feedstock will expand 40% by 2040. North and South America, Europe, Africa, and Asia Pacific, where natural gas consumption and imports have grown quickly, offer the greatest opportunity. Based on growing feedstock availability in a larger global economy and improved waste management and collection programs in many emerging nations, the overall potential will grow quickly over the next two decades (IEA, 2022; International Energy Agency, 2020).

24.1.3 Importance of biogas plants

Intensive development in garbage disposal and management is a priority research area today to overcome the environmental issues. Anaerobic digestion (AD) is a promising way to generate energy from organic waste. It's cheaper than traditional fuels. However, successful implementation of this needs analysis of many aspects, including waste type, organic matter content, process parameter during anaerobic digestion, OLR, temperature, waste collection, and sludge use. An appropriate biodigester must be used for this conversion due to the complex nature of anaerobic digestion process. There are two sizes of biogas facilities: small and large. Domestic, household, decentralized, farm, and community biogas plants are all examples of small-scale biogas facilities. These plants are commonly used in rural settings because of their minimal initial investment requirements. Dome plants, floating drum plants, and balloon/bag digesters are the three main types of residential biogas systems. There is a plentiful supply of organic feedstock available to the family, including animal waste (such cow, pig, and chicken dung), human waste, and food scraps. Although digestate is a high-quality liquid fertilizer, biogas is used as a hob cooking fuel. Many social, environmental, health, and economic benefits support households, especially in emerging and developing economies. Biogas saves time and money by replacing the need for collecting and preparing firewood, traditionally a female task (Makara et al., 2021). Nevertheless, commercial biogas plants need a lot of feedstock, which might come from things such as agricultural waste, municipal organic waste, industrial waste, or energy crops on a massive scale. Depending on their size, commercial biogas plants require considerable financial outlays. Building a biogas plant is not something to get into without careful consideration of the financial, economic, legal, environmental, and social implications (IEA, 2022). Feasibility studies from a biogas consultant can predict biogas and energy generation, payback period, and total investment cost based on the available feedstocks. In response to fluctuating energy needs, biogas is frequently converted into both power and heat. Biogas used as a transportation fuel must have at least 95% of its volume converted to methane before it may be used. Distribution of biogas or biomethane for culinary purposes requires loading and transporting the gas in biogas backpacks, high-pressure gas cylinders, or biogas pipelines, and then delivering it to individual residences. Biogas's by-product, organic fertilizer, or soil improver puts vital nutrients back into the ground. The commonly used biogas plant models are Khadi Village Industrial Commission (KVIC), Deenabandhu, JANATHA, mesophilic tube digester, and so on. For the successful operation of biogas plants, many factors are considered during their selection, construction, and maintenance.

24.1.4 Benefits of biogas plants

Biogas has many benefits. Farms, dairies, and factories can turn the expense of waste treatment into a source of income by installing a biogas system. Reducing reliance on foreign oil imports, lowering greenhouse gas emissions, improving environmental quality, and creating local jobs are all possible outcomes of recycling garbage into electricity, heat, or vehicle fuel. The use of petrochemical and mined fertilizers can be decreased thanks to biogas systems because of the chance to recycle nutrients back into the food supply system. Biogas provides an opportunity for the agricultural and technology supply industries by increasing the value of organic wastes and by-products and adding to the country's energy supply. Users, farmers, investors, and the general public can all reap rewards from the utilization of biogas facilities.

The amount of carbon dioxide released during combustion of biogas is equivalent to the amount of carbon dioxide utilized to produce the organic material transformed during anaerobic digestion. Using gas from trash as an energy source is fantastic for fighting global warming because it releases no greenhouse gases into the atmosphere. Because of people's growing awareness of environmental issues, biogas has grown increasingly popular. Production of biogas lessens demands on nonrenewable energy sources. Biogas production is a strong backer of the climate protection goal (lower GHG emissions and lessening of global warming). Work is needed in the areas of AD feedstock production, collection, and transportation, technical equipment production, biogas plant building, operation, and maintenance. As a result, a corporation will have the chance to launch other biogas-related ventures. On the other hand, a project's return on investment and energy expenses can both profit from the generation of electricity and organic fertilizer from waste materials.

24.1.5 Biogas plants for rural economy

Farmers and residents of rural areas without access to the power grid can generate their own electricity and heat using biogas produced from organic wastes such as biomass and manure. In locations with biogas plants, garbage collection and management are much enhanced. As AD kills off parasites, worm eggs, and flies, it helps keep water cleaner and safer for human consumption. Overflowing landfills that release unpleasant odors and allow toxic liquids to seep into subterranean water sources will be reduced, resulting in smaller landfill regions. Organic (digestate) is a by-product of biogas production that can replace chemical fertilizers. The digester's waste fertilizer can hasten plant development and strengthen immunity to disease. Contrarily, commercial fertilizers are full of chemicals that might have detrimental effects and even lead to food illness (Baredar et al., 2020). Digestate from animal manure has increased uniformity and nutrient availability, making it a more effective fertilizer. Rich in nitrogen, phosphorous, potassium, and micronutrients, this digestate can be applied to soils using the same methods as liquid manure. There would be no need for mineral fertilizer if plants were used as cosubstrates in biogas production and the wastes were reused in farming. By recycling their nutrients, nitrate leaching can be minimized. Because of the increased compatibility and health of the plants, biogas production is subsidized in many countries, allowing farmers to earn more money.

24.1.6 Challenges in the development of biogas plants

Due to low technology and high investment prices, biogas technology competes with solar systems, wind power, and other renewable energy. Today's biogas technology isn't universal. Biogas technology is hard to spread. Lagoon biogas systems require lots of acreage. Biogas generation is not profitable for all organic wastes. Few can become inexpensive biogas systems. Advanced biogas systems are expensive and complicated. The tiny biogas plant's investment cost is higher than fossil fuel diesel. Smaller biogas systems have a long payback period, which deters farmers and investors. Some equipment is short-lived and expensive to use and maintain. Biogas power may not compete with grid electricity depending on local conditions and feedstocks. Market demand for upgrading and bottling biogas limits its use and marketing (Makara et al., 2021).

24.1.7 Costs involved in biogas generation

As mentioned above, anaerobic digestion turns organic wastes into energy that can be used in numerous ways. To assess a region's economic viability, the same must be implemented. Economic analysis helps choose a biogas plant size and model that uses the most biowaste. The rural families' lack of biogas digester awareness is one of AD technology's biggest barriers. A cost–benefit analysis is used to determine whether rural households might use biodigesters for cooking, lighting, and fertilizers. The analysis of two biogas power plants, 4 m^3 and 6 m^3, shows the breakeven of 10 years and 8 years for wood and dung replacement, respectively. Biogas consumption increases investment risk without subsidies, despite positive net present value. The scientists also recommended using slurry fertilizers to make biogas technology succeed. Modifying the substrate, upgrading procedures, processing capabilities, and adding biogas through wastewater sludge treatment revealed that the biogas generation can be lucrative. Several issues with garbage collection, transport, land filling fees, and low-capacity operation may hurt its profitability. The economic viability study is conducted for small, combined heat and power and combined cycle systems using a 5 MW biogas-fueled gas turbine. After 20 years, combined heat and power could have a higher NPV than combination cycle. The technoeconomic study of a solid-state fermentation-based bioprocess for fermented hydrogen generation from food waste shows that the system is practical with a 5-year PBP, 26.75% ROI, and 24.07% IRR (IEA, 2022; Tan et al., 2022). The simulation of an up-low anaerobic sludge blanket (UASB) reactors is performed in Aspen Icarus Process Evaluator (Aspen IPE) with all AD technology equipment costs. The base case used 200,000 tonnes of spruce wood per year for ethanol production. Aspen IPE shows that the process is economically and environmentally viable. For 8–16 m^3 biogas facilities, biogas power plant is viable with a PBP of 1.17–1.01 years and NPV of 4500–9500 dollars in Uganda (Walekhwa et al., 2014). This technology's discount rates, capital, operating, and maintenance costs are identified for hindering its economic feasibility.

The organic municipal solid waste (OMSW) biogas generation with a new biogas digestion has been studied for its technoeconomic feasibility with sensitivity analysis. This 2 m^3 biodigester could fuel 4–6 people. In the cost analysis of biochemical conversion after combined thermo-chemo-sonic disintegration of Waste Activated Sludge (WAS), the profit is shown as 42.6 USD per tonne of sludge (Makara et al., 2021). The biogas plant in a hot food business is found to be environmentally and economically beneficial, and it could support EU's climate targets (Broberg Viklund & Lindkvist, 2015). The evaluation of biogas power plants demonstrates that organic wastes have a great biogas production potential, and many unused biowastes should be investigated for biogas digester installation in many places. Temperature, OLR, HRT, pH, and other operating parameters should be regulated to improve digester performance, according to studies. Many articles confirm the theoretical and experimental study procedure. Pretreatments and codigestion can increase biogas quality. One of the finest biodigester simulators is ADM1. Most research studies, whether steady or transient, have maintained the organic loading rate constant, although this may not be practicable in real life because biowaste output is not constant in any residential building organization. To apply this technology widely, the effects of organic loading rate modification on digester performance and economics must be explored. The literature also shows that this technique requires public awareness and cooperation.

24.1.8 Motivation and objectives of the study

Many investigations on the production of biogas as well as its application in a wide variety of contexts have been carried out. The majority of the earlier research that has been done for a steady organic loading in biodigesters has looked into the feasibility, both technically and economically, of biogas plants for the use of biogas in a wide variety of applications. There haven't been many studies done to determine whether or not, it will be feasible to use biogas technology in the educational sector, despite the fact that there is a substantial possibility that educational institutions may make use of biogas in the future. The academic activities are not uniform throughout a year in any educational institution. This is mostly the result of the fluctuating number of students and staff members throughout the course of the academic year. In addition, there could be an inconsistency in the amount of organic waste available for biodigesters. The following objectives need to be accomplished in order to conduct a scientific investigation into the possibility of making use of the biowaste that is currently available in educational institutions in the chosen region, as well as to determine the impact of nonuniform loading on the efficiency of the biogas plant and the commercial feasibility of the endeavor: to identify and characterize the organic wastes that are currently available in the selected academic institutions in the area under investigation that have inconsistent student populations; to investigate the influences of nonuniform charging of organic wastes. In addition, a suitable simulation model is utilized in order to make a prediction regarding the fluctuation in biogas yield and quality that is brought about by the inconsistent loading of organic waste. The prototype experimental facility is the source of the data that are used for the modeling as well as the economic studies.

24.2 Materials and methods

24.2.1 Selection of organic wastes

The generation of biofuel through the anaerobic digestion of organic wastes in educational institutions would play a significant role in the promotion of both rural and urban prosperity. Because of this, the institutions in the southern part of India and the areas around it were chosen for this study. The overall methodology is represented in Fig. 24.2. The schools in this area have been put into groups based on how many students attend each

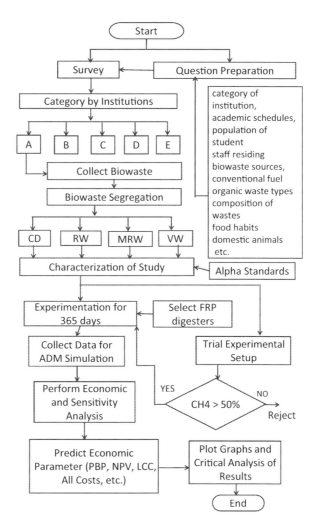

FIGURE 24.2 The procedure followed to group biowastes and the selection of biogas plant models.

one, as well as how likely they are to have biowaste and how consistent it is throughout the year. A survey was conducted using the necessary questionnaire in order to select the biowaste for the study. This survey was able to reveal the specifics of the biowaste, such as the quantity, content, and varieties of biowaste. The format of the questionnaire was determined in part by the type of educational establishment being polled, the academic schedules being followed, the population of students and staff living on and off campus, the sources of biowaste generation, the conventional fuel being used for cooking, and other factors (Kalaiselvan et al., 2022). Based on the student strength during an academic year, they are grouped as A, B, C, D, and E. In this, the lowest average population is taken as A, and the highest is taken as E. Other categories have values between A and E. In each institution, parameters such as the total population, the density of the livestock population, the waste disposal technology, the biowaste suitable for anaerobic digestion (kg/day), the quantity of dung production (kg/day), and the quantity of conventional fuel (LPG) used for cooking (kg/day) were collected.

The following types of organic waste are typically available at each institution: mixed rice trash, cooked rice waste, vegetable waste (both cooked and uncooked), tea leaves, spent coffee waste, waste oil, fruit waste, and so on. The study showed that there was no discernible difference between the two groups' variations. Because of this, in addition to cow manure, the three categories chosen for this study are rice waste (RW), vegetable waste (VW), and mixed rice waste (MRW). The selection of these three categories was based on the quantity of waste and how easily it could be sorted. Because of the highly restricted supply, all of the other types of biowaste, with the exception of the waste from uncooked vegetables and rice, are combined with the MRW. This decision was made because of the exceedingly limited quantity.

24.2.2 Physical and chemical characteristics of organic wastes

The process of anaerobic digestion makes use of a wide variety of biowastes derived from a wide variety of sources, such as municipal solid waste, domestic waste, and industrial garbage. Temperature, pH, TS, and VS affect biogas production. Finding the right characteristics could make biowaste selection easier. Temperature is a crucial factor in food waste biogas production. At 15°C, 25°C, 35°C, 45°C, 55°C, and 65°C, food waste solubilized at 47.5%, 62.2%, 70.0%, 72.7%, 56.1%, and 45.9%. Biogas generation was higher in mesophilic (35°C–45°C) than thermophilic (55°C–65°C) settings. Thermophilic conditions can be improved by lowering hydraulic retention time (HRT) and digester volume. The digestion of fruit vegetable waste (FVW) under thermophilic (55°C), psychrophilic (20°C), and mesophilic (35°C) temperatures has been examined in a laboratory-scale tubular anaerobic digester. The study demonstrates that thermophilic settings can produce 144% and 41% more biogas than the other two. For dairy cow slurry, anaerobic digestion is affected by the altering storage time and temperature. At 9°C, the 26-week-old sludge did not affect biogas production. After 8 weeks, the 20°C slurry reduced biogas output (Browne et al., 2015).

In the vegetable waste (VW) hydrolysis and methanogenesis in two-stage anaerobic digesters with changing OLR, at OLR of 2.6 g volatile solids per day, the biogas generation is from 1.2 to 4.4 L per day, in which the methane content is 27.4%–60.5%. Due to pH change and dilution, effluent cycling to two-stage reactor can increase biogas production. A study on the anaerobic digestion stability and performance with different HRT and OLR shows that the methane output is highest at 25 days of HRT and decreased with OLR and HRT (Aramrueang et al., 2016). Based on the analysis, a continuous digestion process kinetic model predicted methane yield. In the pig manure digestion for hydrogen production, the pH is 5.0, 5.5, and 6.0 in the first stage, whereas for the second stage, loading rate is 96.4, 48.2, and 32.1 kg of VS per cubic meter. For this text, the HRT is 12, 24, and 36 hours, respectively. Hydrogen content peaked at 12 hours HRT and 96.2 kg OLR. Pig dung substrate increases hydrogen production. Loading rates, HRTs, and pH levels exhibited good response; however, hydrogen generation is low compared with other substrates. In a mesophilic up-flow anaerobic staged reactor, acid-pretreated rice straw wastes increase biohydrogen production by 52%, which is due to the pH, contact time, and substrate concentration in biohydrogen synthesis. Under mesophilic (35°C) conditions, methane yield is inhibited without pH change in the codigestion of sewage sludge (SS) and sugar beet pulp lixiviation (SBPL). Optimal pH increases biodegradability and methane output. Therefore, in this presented work, the physical and chemical properties of the selected organic wastes are also studied as per the standard procedure.

pH, VS, and TS are three important factors that affect biogas generation from organic waste. Therefore, these properties have been measured experimentally. According to APHA standards. biomass kept in preweighed porcelain vessels is heated at 60°C for 24 hours and then at 103°C for 3 hours. A hot air oven was used for this purpose. The samples before and after heating in hot air oven are give in Fig. 24.3. The dried samples in the jar were

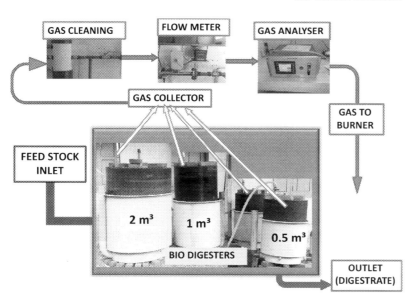

FIGURE 24.3 The experimental set of the organic waste digestion plant.

TABLE 24.1 Characteristic of selected organic waste (experimental and values reported in literature).

Property		Feedstock			
		CD	MRW	RW	VW
pH	Experimental (present study)	6.5	4.91	6.61	6.35
	From previous studies (Mane et al., 2015; Miah et al., 2016; Mirmohamadsadeghi et al., 2019)	6.3	7	7.1	7.1
TS (%)	Experimental (present study)	15.98	20.25	30.28	10.55
	From previous studies (Mane et al., 2015; Miah et al., 2016; Somashekar et al., 2014)	17	14.4	9.3	9.3
VS (%)	Experimental (present study)	64.99	90.15	90.11	90.45
	From previous studies (Miah et al., 2016; Zhang et al., 2007)	89	89.5	78–93	78–93

put on a scale. After the samples and the container dried, their weights were measured on a scale with an accuracy of 0.001 g. When figuring out the VS of the feed materials, the standard method was used. At least once a day, with an accuracy of 0.05%, the pH was measured. The VS of the cooled samples from the desiccator was calculated by weighing the samples. Standard equations from the literature are used to figure out the amounts of TS, VS, and pH. The details of the measuring process are already reported in the literature (Glivin & Sekhar, 2019, 2020). The chemical properties of the four types of biowaste used in this study are listed in Table 24.1 The experimental values are also compared with the results reported in previous studies in the table to confirm the validity of present study.

24.2.3 Types of biogas plants and its selection for the study

The selection of an appropriate digester is critical to the accomplishment of the mission of a power plant that is fueled by biogas. There are primarily two types of biodigesters, namely the fixed type and the mobile type (Alavi-Borazjani et al., 2020). There are no moving parts, making the design of fixed dome plants straightforward. It doesn't have any steel parts; thus, it won't rust. As a result, this plant should last for at least 20 years. These plants are erected underground, making them compact and less vulnerable to environmental hazards. However, the biodigester takes a long time to heat up. In comparison to the other types of domes on the market, the fixed dome type offers the most value for your money, with low maintenance costs and few problems. The Gobar gas plant, or floating drum-based biodigester, consists of an underground digester and a portable gas storage unit (Liang et al., 2021). The gas is collected in a drum that rotates in response to the pressure of the gas inside.

A steel drum is placed on top of the gas drum to separate the biogas production cycle from the gas accumulation. This helps keep the gas pressure steady. Still, this type of digester is expensive, and it needs to be cleaned and maintained every year to keep working. This makes it hard for a stand-alone plant to be a cost-effective choice for long-term use (Aramrueang et al., 2016). Polyethylene is used to make the balloon biodigester. It is made by putting a round polyethylene film on both ends of a polyvinyl chloride pipe and then wrapping it with used tire tube elastic bands (Hernández & Rodríguez, 2013). These two pipes are the entry and exit points. At the top of the cylinder, there is also a PVC pipe that lets the gas escape. As the cylinder of polyethylene is flexible, it is necessary to make a cradle to hold the reactor.

The anaerobic sequencing batch reactor (ASBR) has a drawing and filling unit that works with a single reservoir where all the steps and processes happen. Compared with other systems, the ASBR makes it easier to handle and increases the rate of production. The Continuous Stirred-Tank Reactor (CSTR) is a first-generation reactor that gives a good yield and is highly flexible. This widely used reactor is good at dealing with liquid manure from animals and organic waste from factories. The Advanced Candu Reactor (ACR) works well with wastes that have a lot of solids in suspension (Hernández & Rodríguez, 2013). This reactor is mostly held together by a bed of sludge at the bottom. The amount of granules in the sludge is a key factor in figuring out how well the UASB works. ABR has shown that it can perform well even when there is a high rate of loading and the input feeds are eco-friendly and don't cause complexes. Compared with other reactors, the reactor can handle longer holding times, less sludge formation, and organic loads (shock loads) (Mariano et al., 2020). Without any pretreatment, ARFs have a 90% chance of giving a yield of $0-0.405$ m^3 CH_4 per kg COD, mesophilic tube digester that produces biogas was examined, in which the feedstocks included are leftover fruits and vegetables. The performance of the digester was demonstrated by varying HRT and feed concentration.

During testing, the pH remained constant. With 6% TS feed and 20 days HRT, the digester operated at its optimum. In the anaerobic digestion of fruit vegetable wastes (FVWs), methanogenic bacteria activity and pH were reduced by rapid acidification and increased by synthesis of volatile fatty acids (VFAs). It has been established that the most effective FVW technology is continuous two-phase systems. A 5-L digester was examined for coating color, fabrication type, and environmental factors (indoor and outdoor). As comparison to tin, black-coated tin, translucent plastic, brown-colored plastic, transparent plastic, and glass biodigesters, black-painted plastic digesters produced 15%–22% more gas. Using agricultural resources, anaerobic digestion facilities that are thermophilic create more gas than those that are mesophilic, according to the optimization. Also, the current monitoring system needs to be changed so that it can evaluate process parameters using new sensors to find anomalies. Using a Mesophilic Two-stage ASBR, the Temperature-Phased Anaerobic Sequencing Batch Reactor (TPASBR) system with temperature control was analyzed (Kim et al., 2011). Fruit and vegetable wastes were codigested in single-phase and two-phase digesters. Methane output from ADSL jumped by 13.6%. The biodegradability of Metal Biogas Digesters (MBDs) and Fiber-Reinforced Plastic (FRP) digesters was examined and found suitable of small households (Nnamdi & Victor, 2015). For the Khadi and Village Industries Commission (KVIC)-type floating drum biogas plants in hilly regions, the digester and gas holder construction is harder than JANATA biogas facilities. KVIC-type floating drum biogas plants are profitable. In cold places, the performance of KVIC type of digester was improved by solar heaters with spiral coil heat exchangers. The important parameters to be considered to maintain the temperature and performance of KVIC digester with solar collector are heat capacity of slurry, the area of the solar heat collector, and the organic loading rate (OLR). To prevent upward heat loss, a greenhouse, water heater, and solar stills over the dome could be used (Tiwari et al., 1996).

JANATA biogas plant is a KVIC variant with fixed domes instead of floating drums. Bricks, stones, and cements make the dome cheaper than the KVIC type. This concept has a major drawback: creating a gas-tight dome. Poor construction causes dome cracks to leak. Consequently, these biogas facilities required qualified managers and laborers for construction. The rural and semiurban Indian households use FRP model biogas systems. Digester tanks, floating drums, and water jackets use reinforced polyesters. Inlet, exit, and GS central guide pipes are PVC. These smaller biogas plants are above ground. These biogas plants can only be $1-12$ m^3. Portable and easy-to-maintain FRP biogas plants. For small-scale applications, such models are cheaper. This model's size is a drawback. This model has an average 10-year lifespan.

24.2.4 Experimental setup and procedure

Fig. 24.3 illustrates the schematic diagram of the experimental setup with FRP biogas plants. The sizes of the digester are 2 m^3, 1 m^3, 0.25 m^3, and 0.25 m^3. The various components of the digester are depicted in Fig. 24.4, and the specifications are listed in Table 24.2. The digester tank's input hose loads biowaste. The floating drum

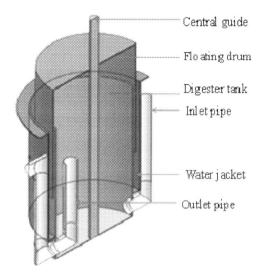

FIGURE 24.4 Orthographic view of the half-sectioned biogas plant.

TABLE 24.2 Specification of the digesters.

Specifications	D1	D2	D3 and D4
Type	FRP	FRP	FRP
OLR (kg)	20–50	10–25	1–6
Inner diameter (m)	1.1	0.77	0.45
Length (m)	1.1	1.17	0.65
Central guide (m)	0.0381	0.0381	0.0317
Inlet pipe diameter (m)	0.0762	0.0762	0.0762
Outlet pipe diameter (m)	0.1016	0.1016	0.0762

stores anaerobic digestion biogas. The water jacket prevents digester and floating drum biogas odors and leakage. A galvanized steel (GS) central guide pipe keeps the floating drum level. After digestion, exit pipes drain digestate. PVC inlets and outlets. The drain plug removes substrate from the digester tank during cleaning. This study used D1, D2, D3, and D4 digesters. Loading and draining the chosen biogas plants are simple. Biogas plants are easy to relocate. FRP digesters are better than fixed-type biogas systems. A pressure gage above the floating drum and RTD sensors in the substrate measure and maintain digester pressure and temperature. A biogas-calibrated multigas analyzer and thermal gas flow meter measure biogas quality and amount. Before each instrument, bypass lines prevent chocking. A cooking cooker burns biogas safely. To monitor pH and temperature during digestion, thermometers, and pH electrodes are submerged in the digester tank. All floating drums have control valves to link them to a manifold, and the outflow from the manifold is connected to a multigas analyzer (NUCON) with $\pm 0.5\%$ accuracy via a thermal gas flow meter. This study uses a manufacturer-calibrated gas flow meter, and the reading errors are checked every 3 months (Glivin & Sekhar, 2020; Glivin et al., 2018).

All four digesters were first loaded with a 1:1 mixture of cow dung and water for the growth of methanogenic bacteria with an HRT of 55 days. Once AD was confirmed to be complete, four different types of waste from an "A" category educational institution were added gradually over the course of 30 days with a consistent OLR. A multigas analyzer and a thermal gas flow meter were used to assess the purity and daily output of methane. At regular intervals, the feedstock's quality, pH, and temperature were measured and averaged over the course of the digesting process. The range of temperatures measured during the trial study was from 29°C to 34°C.

According to the experimental plan, the same digesters were employed for the pilot study and its subsequent 365-day duration. The pace of loading, however, changed daily in response to the accessibility of biowaste. Only 10% of each waste was taken daily, and the same had been utilized to load the digesters, because the total amount of biowastes exceeded the allowable OLR of the chosen digesters. The loading pattern represented the influence of nonuniform biogas generation. During an entire year, pH, temperature, TS, VS, biogas composition, and yield were measured in addition to the accessibility of biowaste. Almost 90% of the temperature readings were above 28°C with a range of 25°C–36°C throughout the year in the research region. Thus, it was hypothesized that mesophilic conditions prevailed during the digestive process. The table below demonstrates the quality and quantity of biogas produced every day for a full year from a sample reading of RW at different phases in a 0.25 m^3 digester. Based on academic scheduling, the entire academic year was divided into four parts.

24.2.5 Mathematical modeling

A simulation model examined biodigester performance for different biowaste types. Anaerobic Digestion Model 1 (ADM1) uses chemical and mathematical methodologies to model biological events during anaerobic digestion (Batstone et al., 2002). The usual AD model is ADM1, since it integrates all prior AD models. This toolbox identifies anaerobic digestion process parameters and shows system behavior. MATLAB was used for all analyses. This model can help design large or small biogas plants. Academic schedules generate the most biowaste in educational institutions. This biowaste availability variation influences methanogen bacteria loading and activity. Therefore by studying digester performance with accessible biowaste throughout the year, one may anticipate the minimum and maximum biogas output in various academic programs. Based on daily yield changes, this estimate can be used to determine biogas plant capacity. The total simulation of the anaerobic digestion is given in Fig. 24.8, which shows the simulation procedure used in ADM1 software tool. Biowastes input parameters, study quantity, and duration were provided. The ADM1 toolbox analyzes biological reactions and calculates digestion biogas characteristics and yield. The toolbox predicts biogas quality and quantity.

The inputs for the simulation are TS, VS, temperature, pH, and HRT. The values used in this analysis are given in Table 24.3. Four phases are selected based on the student population. In each phase, the average biowaste generation and weather conditions are almost the same. The toolbox has many predefined values to conduct the simulation of all stages inside the biodigester. The OLR may fluctuate, affecting the above qualities. ADM1 toolbox allows model runs for any number of days. The ADM1 toolbox dialog boxes specify biogas plant capacity and simulation duration. The simulation output was recorded afterward. The output parameters are CH_4, CO_2, H_2, TS, VS, pH, NH_4, and NH_3. To assess biogas quality, methane (CH_4) content was monitored daily (Fig. 24.5).

24.2.6 Technoeconomic analysis of biogas generation

Economic viability is important when choosing a nonuniform loading biogas plant. Thus the economic analysis was based on Capital Cost (CC), Annual Operation Cost, Payback Period (PBP), Net Present Value (NPV), and Life Cycle Cost (LCC). This analysis uses standard equations from previous studies. This economic analysis examined KVIC, JANATA, and FRP biogas plants. Due to their merits of design and installation, KVIC types are best for high-capacity facilities. JANATA model plants, made of bricks, resist rusting and are suggested. Due to mobility, FRP variants are appropriate for smaller capacities. Their original investment also matters. Owing to such issues, economic factors are examined. For this purpose, the institutions are categorized into A, B, C, D, and E based on the student strength, and the capacity of biodigester selected for those institutions is given in Table 24.4. Based on the student population, four phases such as 1–150, 151–225, 226–315, and 316 to 365 are taken in this analysis (Glivin, Vairavan, et al., 2021; Kalaiselvan et al., 2022). From the sample study in an institution, the biogas generation potential of other institutions was taken. The calculated per capita biogas production

TABLE 24.3 Input parameters of biowaste for the simulation.

Input parameters	FEED			
	CD	MRW	RW	VW
pH value	6.3–6.5	4.91–6.5	6.1–6.7	6–7
TS	15–17	13–20	9–31	4–22
VS	63–89	88–91	90–95	78–93
Temperature (°C)	29°C–33°C	29°C–33°C	29°C–33°C	29°C–33°C

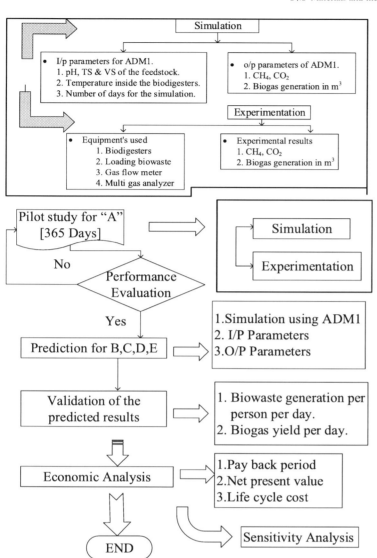

FIGURE 24.5 The overall simulation of the system in ADM1 toolbox.

TABLE 24.4 Institution categories and the total volume of digesters in each category.

Type	Student strength	Total volume of digester (m³)
Category-A	1000–2499	25
Category-B	2500–4999	50
Category-C	5000–8999	100
Category-D	9000–19999	170
Category-E	20000–40000	450

per day was between 14 and 19 L, and the mean was 15 L, besides 53% methane composition. Calculating biogas plant capacity for each category using the mean value. Biowaste availability and biogas yield determine an institution's biogas plant capacity. The biowaste availability in various categories of institutions over a year was calculated based on data from the pilot research. Table 24.4 shows the biogas plant's capacity based on the population's academic year average. The biogas plant capacity is 25–450 m³. Therefore all institutions cannot use the same biogas plant. Hence, Indian biogas plant specifics were examined.

The digester, installation, and government subsidies are all part of the Biogas Plant's initial investment. The operations and maintenance cost, as well as the annual depreciation value, make up a plant's running cost. It is estimated that 2% of a system's initial investment goes into maintenance. For the KVIC, JANATA, and FRP models, the expected life span of the energy infrastructure is 15, 20, and 10 years, respectively. The cost of transporting and disposing of the biowaste as well as the cost of labor is factored into the calculations. The net present value of an investment is the difference between the present value of the benefits and the present value of the costs resulting from the investment. For this calculation, the conventional procedures utilized in the literature (Glivin et al., 2018; Glivin, Kalaiselvan, et al., 2021) are applied. The acceptance criteria of an investment project, as determined by the NPV method, are accepted, and rejected if the NPV is greater than or less than zero. Another key economic indicator is the system's Life Cycle Cost (LCC), which accounts for all costs associated with the system during its lifetime while taking the value of money into consideration. The additional parameters for the economic analysis are also calculated. The annual operation cost is calculated by adding the costs of the energy source, operation, maintenance, and depreciation. The total income is calculated from the income from gas and slurry. Based on the above, the profit is calculated. The cost of LPG is considered to calculate the income from biogas, and the sale value of the slurry is taken as the income from slurry. The annual O&M cost is taken as 2% of the capital investment, and the interest rate is taken as 12%. For the NPV and LCC calculations, the lives of KVIC, JANATHA, and FRP are taken as 15, 20, and 10 years, respectively (Rajendran et al., 2014).

24.3 Results and discussion

Results from both theoretical and experimental examinations of the operation of anaerobic digesters in educational institutions with a nonuniform organic loading rate are provided. The gathered biowaste is analyzed theoretically for its qualities. The experimental values are used to verify the simulation results. Capital cost, operating cost, payback time, net present value, and life cycle cost were among the economic criteria examined, and results were reported in this chapter as well.

24.3.1 Verification of theoretical predictions

The theoretical study was a 30-day trial study compared with experimental values. Fig. 24.6 shows the average methane content in CD, MRW, RW, and VW from experimental and simulation work, on the 30th day, CD has 61.45% and 59.69% methane respectively as shown in Fig. 24.6A. According to multiple researchers, experimental and theoretical values vary by 5%. Therefore this strategy is valid. This simulation was used to forecast methane generation from MRW, RW, and VW, as illustrated in Figs. 24.6B–D, respectively. The simulation for all wastes in this study is valid because the variation is less than 5%. The amount of biogas produced is validated based on the anaerobic digestion of MRW during the 365-day pilot research. The calculated and plotted actual and expected biogas yields are presented in Fig. 24.7A. The graph demonstrates that the difference between anticipated and experimental values is low, with a deviation of less than 5%. Consequently, the validity of the existing methodology is demonstrated (Glivin & Sekhar, 2020).

Several assumptions employed to lower the simulation's complexity may account for this variance. The projected biogas yield for all the categories is given in Fig. 24.7B.

Because the theoretical and experimental results of biogas yield from the pilot study are closer to one another, as shown in Fig. 24.7A, this model can be used to predict the biogas yield with various loading rates, as shown in Fig. 24.7B, due to the fact that the theoretical and experimental results of biogas yield from the pilot study are closer to one another. The yield for each category is calculated using each institution's unique class schedule in conjunction with the amount of biowaste that is currently available (Glivin & Sekhar, 2016).

24.3.2 Pilot study: nonuniform loading rate effect

To test digester methane composition and biogas yield, 365 days of nonuniform biowaste generation in an educational institution is studied. Fig. 24.8A displays the loading rate of RW created in category "A" universities for 365 days to help comprehend academic calendars. RW availability is 7–46 kg in phase I, the first 150 days. As mentioned before, the digester may only hold 10% of biowaste or 0.07–4.6 kg. Due to digester size, comparable methods were used to load the additional digesters. Fig. 24.8B shows biogas yield. The average yield is 0.16 m^3,

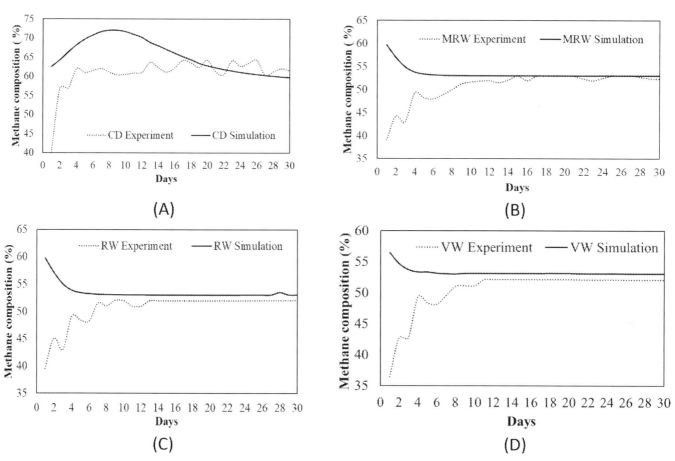

FIGURE 24.6 Methane composition in the biogas from (A) CD, (B) MRW, (C) RW, and (D) VW.

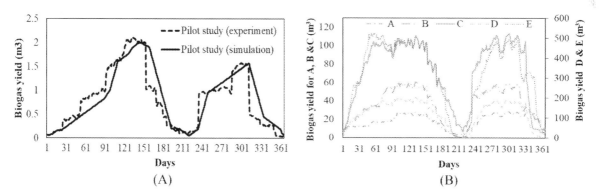

FIGURE 24.7 (A) Biogas yield of pilot study for 365 days. (B) Biogas yield for categories A, B, C, D, & E for 365 days.

and the maximum yield is 0.18 m^3. Exams and vacations make Phase II nonacademic. As indicated in Fig. 24.8A, the student population and RW availability were low, and the average loading was 1–3.2 kg. As shown in Fig. 24.8B, the biogas yield decreased slowly with the loading rate, reaching 0.05 m^3 on the 180th day and 0.01 m^3 on the 225th day. Biowaste increased due to student growth in phase III. Similar to phase I, biowaste addition increases biogas yield. The average loading rate is 1–4.9 kg, and the biogas production is 0.03–0.15 m^3. In phase IV, the nonacademic schedule reduced student enrollment, and biogas yield was between 0.01 and 0.08 m^3. A constant loading was assumed in Fig. 24.7A to anticipate this biogas yield's divergence from uniform loading. Fig. 24.7B shows the yield prediction at 0.09 m^3 for a uniform RW loading of 2.5 kg.

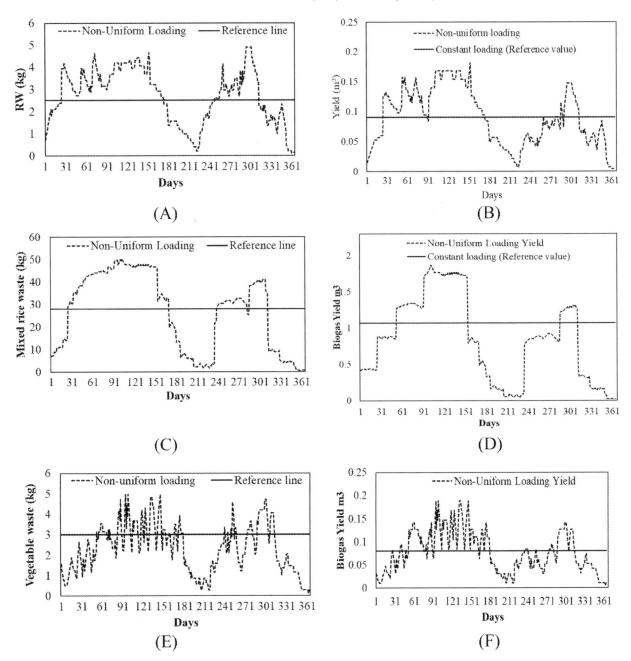

FIGURE 24.8 (A) Loading pattern of rice waste for 365. (B) Biogas yield from rice waste for 365 days. (C) Loading pattern of mixed rice waste for 365 days. (D) Biogas yield from mixed rice waste for 365 days. (E) Loading pattern of vegetable waste for 365 days. (F) Biogas yield from vegetable waste for 365 days.

The biogas yield for MRW started at 0.4 m³ on day 0 and peaked at 1.8 m³ on day 150 (Fig. 24.8C). While it was true that all biowastes experienced a delay in the beginning, it was also noticed that the biogas yield rose and eventually stabilized. The availability of MRW was greater than that of RW during phase II (the nonacademic schedule), but it was lower than that of RW in phase I. Fig. 24.8D shows that when the loading rate was changed from 0.7 to 32 kg, the biogas yield peaked at 0.3 m³ on day 180 and dropped to 0.1 m³ on day 225. During phase III, which is quite comparable with phase I, the average loading rate is 7–38 kg, and the average biogas yield is 0.8–1.3 m³. In phase IV, as in phase II, fewer students are enrolled because of their commitments to activities outside of the classroom. The loading rate of 0.7–10 kg resulted in a biogas production of 0.1–0.3 m³ during phase IV. Biogas yield for uniform loading is 1.07 m³ as shown in Fig. 24.8C. Fig. 24.8D displays the results of theoretical and experimental research showing that for uniform and nonuniform loading of MRW, respectively, the average methane content is 55.69% and 54.85%.

The VW's uniform and nonuniform loading rates are depicted in Fig. 24.8E. In this analysis, we assume a uniform loading of 2.5 kg, whereas the real highest and minimum loadings were 1 and 5 kg, respectively. During phase I, when the loading rate was between 1 and 5 kg. The biogas yield from VW ranged from 0.01 m^3 to 0.18 m^3. Fig. 24.8E shows that at a loading rate between 0.5 and 2 kg, biogas yields of 0.03 m^3 on day 180 and 0.01 m^3 on day 225 were achieved, reflecting the lower availability of VW during academic periods (a). The average biogas yield for VW throughout phase III was 0.04–0.14 m^3, and the average loading rate was 1–5 kg. Phase IV sees a drop in enrollment as students drop out to pursue other interests, with typical load sizes ranging from 1 to 5 kg. In the same scenario, the yield of biogas was calculated to be between 0.01 m^3 and 0.09 m^3. As shown in Fig. 24.8F, the range of biogas yield is 0.007–0.08 m^3. VW was not readily available in the dorms and cafeterias, in contrast to other biowastes. Proper digestion and high biogas generation are the results of methanogen bacteria that were produced during phase I with a uniform loading rate. Phase II's limited yield potential was caused by a lack of loading, which prevented enough methanogen bacteria from being produced for anaerobic digestion to take place. Phase III biogas yields suffer. In phase I, cow manure was loaded to generate methanogen bacteria; therefore even if the loading rate was similar, the biogas yield was not. Lower loading rates reduce anaerobic digesting activity and methanogen microorganisms. Similalr to phase II, phase IV had a lower loading rate and lower biogas yield (Feng et al., 2016; Glivin & Sekhar, 2020; Zhu et al., 2014). Hence, loading rate nonuniformity may affect biogas yield. Nonetheless, biogas methane content varies little.

24.3.3 Economic analysis of biogas production

Table 24.5 shows the total of building, installation, annual operating, and other costs of commercial type KVIC, JANATA, and FRP models (category A to category E) based on southern Indian market prices. Regardless of use, India subsidizes household digesters. Commercial digesters are only subsidized for power generation. This study excludes the subsidy.

The analysis focuses on nonuniform biogas plant model selection and LPG reduction. FRP has the highest cubic meter cost, followed by KVIC and JANATA. The tendency is due to plant size (12 m^3) and the necessity for several high-capacity units. Gas holder fabrication costs more for KVIC model than JANATA model. Due to the steel body's frequent maintenance and corrosion, KVIC petrol holders are expensive. Even replacing the gas holder with fiber reinforced polyester raises investment costs. As shown in Table 24.5, KVIC model installation costs decrease gradually from category A to category E, but JANATA model costs are almost the same for all categories (Fig. 24.9).

TABLE 24.5 Estimation of capital and installation costs of selected biogas plants.

Type	component	Cost (INR)				
		25 m^3	50 m^3	100 m^3	170 m^3	450 m^3
KVIC	Fabrication	315,235	470,308	793,285	1,231,235	2,874,339
	Labor	33,300	49,650	69,300	90,150	109,200
	Pipe, fittings, biogas stove	8000	8000	10,000	10,800	16,000
	Total Cost - KVIC	356,535	527,958	872,585	1,332,185	2,999,539
JANATHA	Fabrication	218,205.5	419,800	792,221	1,342,576	3,353,440
	Labor	33,300	49,650	69,300	90,150	109,200
	Pipe, fittings, biogas stove	8000	8000	10,000	10,800	16,000
	Total Cost - JANATHA	259,505	477,450	871,521	1,443,526	3,478,640
FRP	Fabrication	400,000	800,000	1,600,000	2,720,000	7,200,000
	Labor	33,300	49,650	69,300	90,150	109,200
	Pipe, fittings, biogas stove	8000	8000	10,000	10,800	16,000
	Total Cost (FRP)	441,300	857,650	1,679,300	2,820,950	7,325,200

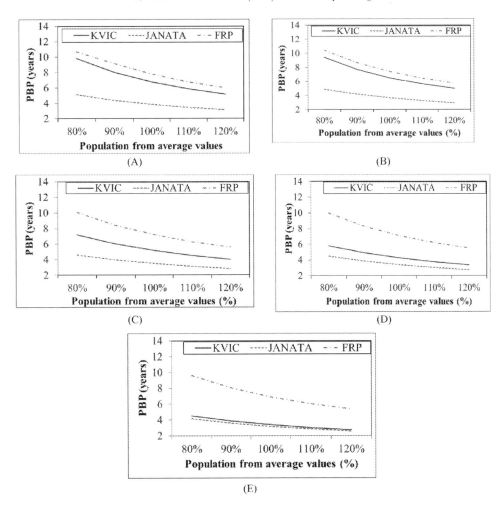

FIGURE 24.9 PBP of categories (A) A, (B) B, (C) C, (D) D, and (E) E.

24.3.4 Net present value

Table 10 shows the net present value for implementing biogas digesters in different types of institutions. The Net Present Value of an investment is the amount by which its benefits exceed its expenditures at the current time (Glivin et al., 2018). The results reveal that NPV rises when biogas plant sizes expand. It is also noted that all the categories in this analysis produce positive numbers. When the average population of institutions in categories A, B, and C using FRP-type digesters is lowered by 20%; however, a small negative value is detected. The biogas plant project may be the most desirable for implementation in educational institutions based on the NPV. The NPV for both uniform and nonuniform loading rates are displayed in Table 24.6. These data show that uniform loading is preferable to nonuniform loading. Similar results were found with nonuniform loading rates, which suggests that such digesters could be used effectively in schools with a wide range of class times.

24.3.5 Life cycle cost

Among competing choices that are technically suitable for the implementation, the most cost-effective one is identified using a Life Cycle Cost analysis (Tan et al., 2022). Hence, we calculated the LCC and plotted it in Fig. 24.10A, which demonstrates that the JANATA LCC is superior to the other two models. Fig. 24.10B displays the outcome obtained with constant loading rates. However, KVIC is recommended because of the complexity of the design and construction of bigger-size JANATA model biogas plants (Kalia & Singh, 1999).

TABLE 24.6 Net present value for the selected models and institution categories.

Models	Description	NPV for different categories (× 1000 INR)				
		A	B	C	D	E
KVIC	Nonuniform loading	243	504	1070	1880	5250
	Uniform loading	402	822	1700	2970	8105
JANATA	Nonuniform loading	289	596	1230	2100	5760
	Uniform loading	463	944	1920	3300	8891
FRP	Nonuniform loading	50	116	264	471	1400
	Uniform loading	182	379	790	1370	3769

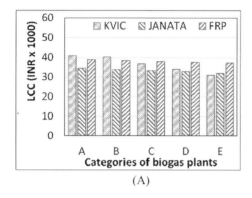
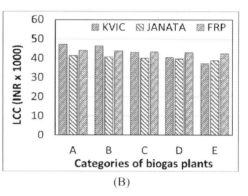

FIGURES 24.10 (A) LCC per m³ of different biogas plants (nonuniform loading). (B) LCC per m³ of different biogas plants (uniform loading).

TABLE 24.7 Cost of unit electricity from conventional and proposed energy systems.

Category of institution	Electricity Cost (INR/kWh)				
	JANATA	KVIC	FRP	LPG	Grid
A	3.55	7.55	10.42	7.09	6.60
B	3.08	7.07	10.07	7.08	6.60
C	2.57	5.07	9.63	7.06	6.60
D	2.28	3.54	9.36	7.04	6.60
E	1.75	1.87	8.99	7.03	6.60

24.3.6 Prices paid for individual units of electricity

Table 24.7 presents the results of annual calculations made to determine the various costs associated with the generation of electricity from biowaste that is readily available in an educational institution and its comparable quantity of LPG. The cost of grid electricity has been taken from the previous studies (Glivin & Sekhar, 2020; Glivin et al., 2018). The cost of producing one unit of electricity was determined by factoring in the investment cost, the cost of maintaining the system, and the total number of units produced each year.

24.3.7 Sensitivity analysis

The sensitivity analysis is carried out by following the approaches that are considered to be conventional in order to determine the effect that the modification in the relevant parameters have on the PBP and NPV.

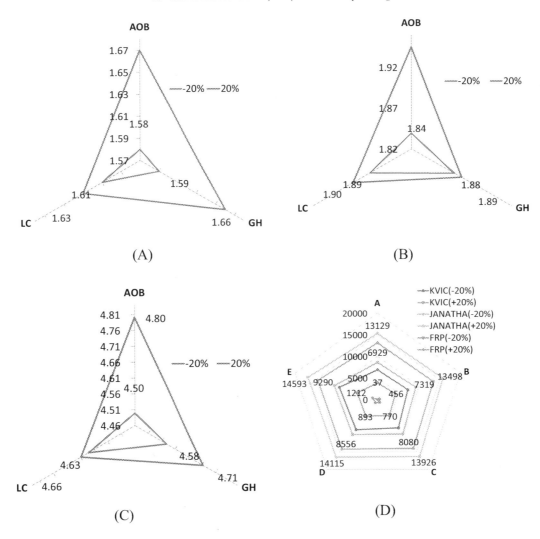

FIGURE 24.11 (A) Sensitivity of AOB, GH, and LC on PBP (KVIC). (B) Sensitivity of AOB, GH, and LC on PBP (JANATHA). (C) Sensitivity of AOB, GH, and LC on PBP (FRP). (D) Sensitivity of interest on NPV.

The impact of 20% increase or decrease in the major costs and interest rate are considered in this study (Glivin & Sekhar, 2016; Glivin et al., 2018). The PBP for the KVIC model biogas plant that falls under category E is depicted in Fig. 24.11A–C. In this model, the PBP is affected by both the GH and the AOC. It would appear that the other characteristics are not nearly as important. The sensitivity of interest on NPV of biogas plants with respect to interest rate is shown in Fig. 24.11D.

The figures display the PBP for category E of the JANATA model biogas plant for each of the three possible situations. Under the JANATA model, the AOC is the significant parameter that affects the PBP of BGP. Based on the three different possibilities, the AOC and GH parameters appear to be the most relevant ones that have an effect on the PBP of BGP. In both the PBP and the FRP models, the effects of the other factors are negligible at best.

24.4 Conclusion

The technoeconomic aspects of organic waste utilization have been studied through a combination of simulation and experimental investigation. The organic wastes commonly available in academic institutions in and around the study region, such as rice waste, mixed rice waste, and vegetable waste, were individually analyzed for five different categories of institutions. To predict the economic parameter, initial data were collected from experimental data, and the simulation results were validated. The predicted CH_4 in biogas deviates less than 5%

from the experimental values. From all biowaste biogas 52%–58% methane was generated. Therefore this study's biogas can be used for heating, electricity generating, and cooling. Population influenced biogas production in all four seasons of a year; however, CH4 quality was practically identical in all phases.

All models have PBPs 44%–57% greater than uniform loading. In institutes A, B, C, and D, JANATA biogas plants are recommended. E-category institutions can use JANATA and KVIC. All biogas plant models modify the expected PBP by 25%–50% for a mean population variation of ±20%. Furthermore, a models with NPVs ±20% from the mean have positive NPVs. All the tested categories have positive NPVs, indicating economic viability. Sensitivity study shows that input elements such as civil construction cost and petrol holder/dome cost affect PBP. This study did not address drainage or night soil as it is difficult to separate from soap and other biodigestive inhibitors. Campuses can plan a waste management system to collect waste from different areas and construct digesters near the main collection areas to implement this system. To reduce transportation and loading expenses, digesters should be developed so waste can be loaded at ground level. This profitable, eco-friendly technology requires campus-wide involvement. Therefore students, employees, and other stakeholders must be informed. A multistage digestion/codigestion system will be researched to increase biogas yield. Waste availability varies from place to place. This element can be included to the pilot study forecast. Moreover, the local trash and animal dung may also be used to lower PBP.

References

Alavi-Borazjani, S. A., Capela, I., & Tarelho, L. A. C. (2020). Over-acidification control strategies for enhanced biogas production from anaerobic digestion: A review. *Biomass and Bioenergy, 143*, 105833. Available from https://doi.org/10.1016/j.biombioe.2020.105833.

Aramrueang, N., Rapport, J., & Zhang, R. (2016). Effects of hydraulic retention time and organic loading rate on performance and stability of anaerobic digestion of Spirulina platensis. *Biosystems Engineering, 147*, 174–182. Available from https://doi.org/10.1016/j.biosystemseng.2016.04.006.

Baredar, P., Khare, V., & Nema, S. (2020). *Design and optimization of biogas energy systems*.

Batstone, D. J., Keller, J., Angelidaki, I., et al. (2002). The IWA anaerobic digestion model no 1 (ADM1). *Water Science and Technology, 45*, 65–73. Available from https://doi.org/10.2166/wst.2002.0292.

Broberg Viklund, S., & Lindkvist, E. (2015). Biogas production supported by excess heat – A systems analysis within the food industry. *Energy Conversion and Management, 91*, 249–258. Available from https://doi.org/10.1016/j.enconman.2014.12.017.

Browne, J. D., Gilkinson, S. R., & Frost, J. P. (2015). The effects of storage time and temperature on biogas production from dairy cow slurry. *Biosystems Engineering, 129*, 48–56. Available from https://doi.org/10.1016/j.biosystemseng.2014.09.008.

Feng, R., Li, J., Dong, T., & Li, X. (2016). Performance of a novel household solar heating thermostatic biogas system. *Applied Thermal Engineering, 96*, 519–526. Available from https://doi.org/10.1016/j.applthermaleng.2015.12.003.

Glivin, G., Edwin, M., & Sekhar, S. J. (2018). Techno-economic studies on the influences of nonuniform feeding in the biogas plants of educational institutions. *Environmental Progress & Sustainable Energy, 37*. Available from https://doi.org/10.1002/ep.12892.

Glivin, G., Kalaiselvan, N., Mariappan, V., et al. (2021). Conversion of biowaste to biogas: A review of current status on techno-economic challenges, policies, technologies and mitigation to environmental impacts. *Fuel, 302*, 121153. Available from https://doi.org/10.1016/j.fuel.2021.121153.

Glivin, G., & Sekhar, S. J. (2016). Experimental and analytical studies on the utilization of biowastes available in an educational institution in india. *Sustain, 8*. Available from https://doi.org/10.3390/su8111128.

Glivin, G., & Sekhar, S. J. (2019). Studies on the feasibility of producing biogas from rice waste. *Romanian Biotechnological Letters*, Doi 10.

Glivin, G., & Sekhar, S. J. (2020). Waste potential, barriers and economic benefits of implementing different models of biogas plants in a few Indian educational institutions. *BioEnergy Research, 13*, 668–682.

Glivin, G., Vairavan, M., Manickam, P., & Santhappan, J. S. (2021). Techno economic studies on the effective utilization of non-uniform biowaste generation for biogas production. *Anaerobic Digestion in Built Environments, 81*.

Hernández, M., & Rodríguez, M. (2013). Hydrogen production by anaerobic digestion of pig manure: Effect of operating conditions. *Renewable Energy, 53*, 187–192. Available from https://doi.org/10.1016/j.renene.2012.11.024.

IEA (2022) International Energy Agency (IEA) World Energy Outlook 2022. https://www.IeaOrg/Reports/World-Energy-Outlook-2022/Executive-Summary.

International Energy Agency (2020). *Outlook for biogas and biomethane*. Prospects for organic growth. IEA Publ 1–93.

Kalaiselvan, N., Glivin, G., Bakthavatsalam, A. K., et al. (2022). A waste to energy technology for Enrichment of biomethane generation: A review on operating parameters, types of biodigesters, solar assisted heating systems, socio economic benefits and challenges. *Chemosphere, 293*, 133486. Available from https://doi.org/10.1016/j.chemosphere.2021.133486.

Kalia, A. K., & Singh, S. P. (1999). Case study of 85 m3 floating drum biogas plant under hilly conditions. *Energy Conversion and Management, 40*, 693–702. Available from https://doi.org/10.1016/S0196-8904(98)00137-X.

Kim, H.-W., Nam, J.-Y., & Shin, H.-S. (2011). A comparison study on the high-rate co-digestion of sewage sludge and food waste using a temperature-phased anaerobic sequencing batch reactor system. *Bioresource Technology, 102*, 7272–7279. Available from https://doi.org/10.1016/j.biortech.2011.04.088.

Liang, T., Elmaadawy, K., Liu, B., et al. (2021). Anaerobic fermentation of waste activated sludge for volatile fatty acid production: Recent updates of pretreatment methods and the potential effect of humic and nutrients substances. *Process Safety and Environmental Protection, 145*, 321–339. Available from https://doi.org/10.1016/j.psep.2020.08.010.

Makara, L., Lytour, L., & Chanmakara, M. (2021). *Practical biogas plant development handbook: Potential biogas resources*, Legal Review, and Good Practice of Biogas Construction in Cambodia. 126.

Mane, A. B., Rao, B., & Rao, A. B. (2015). Characterisation of fruit and vegetable waste for maximizing the biogas yield. *International Journal of Advanced Engineering Science and Technological Research, 3*, 489–500.

Mariano, A. P. B., Unpaprom, Y., & Ramaraj, R. (2020). Hydrothermal pretreatment and acid hydrolysis of coconut pulp residue for fermentable sugar production. *Food and Bioproducts Processing, 122*, 31–40. Available from https://doi.org/10.1016/j.fbp.2020.04.003.

Masson, V., Lemonsu, A., Hidalgo, J., & Voogt, J. (2020). Urban climates and climate change. *Annual Review of Environment and Resources, 45*, 411–444. Available from https://doi.org/10.1146/annurev-environ-012320-083623.

Miah, M. R., Rahman, A. K. M. L., Akanda, M. R., et al. (2016). Production of biogas from poultry litter mixed with the co-substrate cow dung. *J Taibah Univ Sci, 10*, 497–504. Available from https://doi.org/10.1016/j.jtusci.2015.07.007.

Mirmohamadsadeghi, S., Karimi, K., Tabatabaei, M., & Aghbashlo, M. (2019). Biogas production from food wastes: A review on recent developments and future perspectives. *Bioresource Technology Reports, 7*, 100202. Available from https://doi.org/10.1016/j.biteb.2019.100202.

Nnamdi, M., & Victor, N. (2015). *Comparative evaluation of fiber-glass reinforced plastic and metal biogas digesters.* 38–44.

Rajendran, K., Kankanala, H. R., Martinsson, R., & Taherzadeh, M. J. (2014). Uncertainty over techno-economic potentials of biogas from municipal solid waste (MSW): A case study on an industrial process. *Applied Energy, 125*, 84–92. Available from https://doi.org/10.1016/j.apenergy.2014.03.041.

Somashekar, R. K., Verma, R., & Naik, M. A. (2014). Potential of biogas production from food waste in a uniquely designed reactor under lab conditions. *International Journal of Geology, 2*, 2348.

Tan, W. E., Liew, P. Y., Tan, L. S., et al. (2022). Life cycle assessment and techno-economic analysis for anaerobic digestion as cow manure management system. *Energies, 15*, 1–16. Available from https://doi.org/10.3390/en15249586.

Tiwari, G. N., Usmani, J. A., & Chandra, A. (1996). Determination of period for biogas production. *Energy Conversion and Management, 37*, 199–203. Available from https://doi.org/10.1016/0196-8904(95)00167-C.

Walekhwa, P. N., Lars, D., & Mugisha, J. (2014). Economic viability of biogas energy production from family-sized digesters in Uganda. *Biomass and Bioenergy, 70*, 26–39. Available from https://doi.org/10.1016/j.biombioe.2014.03.008.

Zhang, R., El-Mashad, H. M., Hartman, K., et al. (2007). Characterization of food waste as feedstock for anaerobic digestion. *Bioresource Technology, 98*, 929–935. Available from https://doi.org/10.1016/j.biortech.2006.02.039.

Zhu, G., Li, J., & Jha, A. K. (2014). Anaerobic treatment of organic waste for methane production under psychrophilic conditions. *International Journal of Agriculture And Biology, 16*, 1025–1030.

CHAPTER

25

Sustainability assessment method for waste biomass processing

Anusha Airi[1], Sari Piippo[2] and Eva Pongracz[1]

[1]University of Oulu, Water, Energy and Environmental Engineering Research Unit, Oulu, Finland [2]Finnish Environment Institute, Syke, Oulu, Finland

25.1 Introduction

The management of biowaste in the European Union (EU) has been improving over the years, but still there is loss of potential secondary raw materials. Biowaste is defined as biodegradable garden and park waste, food and kitchen waste from households, restaurants, caterers and retail premises, and comparable waste from food processing plants. The main environmental threat from biowaste is the production of methane from biowaste decomposing in landfills. The EU Landfill Directive of 1999 banned the landfill of biowaste, which significantly reduced the methane emission problem. However, the Landfill Directive did not prescribe specific treatment options for the diverted waste. Across the European Union, only about 40% of biowaste is recycled to compost and digestate. As up to 50% of municipal solid waste (MSW) is organic, and the MSW recycling target in the EU is 65% by 2035, there is a strong need to improve the recovery of waste biomass (ECN European Compost Network, 2022).

There are several environmentally favorable options for the management of biowaste diverted from landfills, and the choice can be made based on a number of considerations, such waste composition, quality and quantity, climatic and geographic conditions of the country, existing collection systems and supportive infrastructure, as well as the potential of use and markets for the waste-derived products. Utilization of biowaste can also contribute to closing nutrient cycles. Nutrients are an exchange of ions and molecules as food/inputs within living organisms in the ecosystems from one biosphere to another forming a cycle called nutrient cycle. Nutrient cycle is also known as biogeochemical cycle, which basically includes two phases: the organismic and the environmental. In the organismic part, the nutrients move from producers to consumers to microbes while in the environmental part, the nutrients are available in soil, air, and water and sometimes in two or more physical environments all at once (Chiras, 2016). The recovery of nutrients from biomass through bioconversion processes will allow restoring nutrients to soil, thereby closing the nutrient loop and maintaining the fertility of the soil (Rakshit, et al., 2015). Such advanced bioconversion matched with complementary biomass production will also contribute to a circular bioeconomy (EC, 2018b).

However, sustainability in the context of biowaste management refers to finding solutions that manage biowaste without adverse impacts to human health or the environment, while seeking synergy benefits with other environmental services, which are also economically viable and provide community benefits. It is also expected that strategies for the management of biowaste should be determined in a transparent manner.

This chapter presents a sustainability assessment method, which can be used to compare biowaste treatment methods across their economic, environmental, and social sustainability. The method aims to offer local stakeholders and authorities to understand the sustainability impacts of biowaste management projects. The objective is to help in decision-making and uncovering issues that might be hidden or otherwise remain unnoticed. The method is illustrated using a case study of a small Finnish municipality.

25.2 Sustainability

Facing grand challenges such as climate change, resources depletion, and global inequality, the importance of sustainability is amplified. Sustainable development as a concept was introduced in 1987 in the Brundtland report, meaning a development that meets the needs of the present without compromising the ability of future generations to meet their needs (Brundtland, 1987). This is indicating that natural resources should only be exploited within the limits of their regenerative capability. Sustainability then could be argued to be an equilibrium state, when both human needs and ecosystem needs are achieved in balance, as illustrated in Fig. 25.1.

Sustainability is a holistic concept that requires balancing environmental, social, and economic considerations in concert. Environmental sustainability concerns the well-being of the planet, including preserving ecosystem services, and exploiting natural resources within sustainable limits. As illustrated in Fig. 25.1, ecosystems needs range from "food" and habitat to diversity. Human activities should be conducted in such a manner that we do not encroach on the space of natural systems and do not distort species diversity.

Social sustainability is concerns from individual well-being to societal well-being. People need to satisfy their basic needs for food and shelter to higher needs of self-realization. On a societal level, the aim is the create peaceful, just, and equitable societies.

Economic sustainability considers the impacts of economic decisions, both in terms of natural resources use and creation of jobs and distribution of wealth. The aim is long-term economic growth that supports the achievement of human needs, without harming the ecological system and jeopardizing ecosystem services.

While many would consider this equilibrium view as an unattainable ideal, the purpose here is to indicate an outcome of sustainable development we should strive at.

25.2.1 Sustainability assessment

Although there are various international efforts on measuring sustainability, only a few of them have an integral approach taking into account environmental, economic, and social aspects. In most cases, the focus is on one of the three aspects. For example, Life Cycle Assessment (LCA) is used to evaluate the environmental performance of products, but it concentrates on environmental impacts only.

The most extensive work in terms of sustainability assessment has been done by the Global Reporting Initiative (GRI). GRI is a nongovernmental organization that aims at driving sustainability and has developed an Environmental, Social, and Governance (ESG) reporting framework to be used worldwide. GRI's Sustainability Reporting Guidelines define the principles and indicators that organizations can use to measure and report their economic, environmental, and social performance. While this is valuable, many companies use these indicators while publishing their annual or environmental reports. (GRI Global Reporting Initiative, 2022) GRI standards are used on a company level, and compiling an extensive report on a total of 34 indicators is time-consuming.

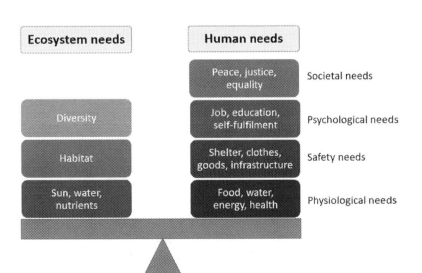

FIGURE 25.1 Sustainability as equilibrium of human needs and ecosystem needs. Balancing human needs and ecosystem needs.

25.2.2 Sustainability assessment method

In order to evaluate the most critical aspects sustainability, an easy-to-use sustainability assessment method was developed at the University of Oulu, Finland, for the Renewable Energy Empowerment in Northern territories (RECENT) project (Niemelä, 2016). In the project we evaluated renewable energy investments, which also included waste-to-energy (W2E) solutions. A particular focus of attention was on solutions that can derive synergy benefits, meaning that provide solutions to multiple environmental problems at the same time. For example, W2E solutions help solve waste problems but can also contribute to energy security. Similarly biowaste management can have multiple benefits, from reducing environmental impacts, through nutrient recycling, but can improve sustainability values, and some methods can also lead to energy recovery.

The method is based on assessing indicators on environment, economic, and social sustainability (Niemelä, 2016): CO_2 reduction, synergy advantages, land-use impacts, impacts on the environment, payback time, impact on citizen's health, teaching sustainable values, community impact, and energy security. The indicators were selected based on GRI indicators and were also inspired by the Energy Indicators for Sustainable Development proposed by IAEA (2005). The methodology deploys a semiquantitative assessment on the Likert scale (Likert, 1932), by the use a simple set of questions. The answers are then marked positive or negative from +2 to −2, depending upon the responses the questions listed in each criteria. The nine sustainability indicators and the related assessment tables are listed below:

Environmental sustainability

1. CO_2 reduction (Table 25.1)
2. Synergy advantages (Table 25.2)
3. Land use impacts (Table 25.3)
4. Impacts on the environment (Table 25.4)

TABLE 25.1 CO_2 reduction.

Indicator 1—CO_2 reduction				Total points	
1.1 How does the pilot contribute to CO_2 reduction?	Increases CO_2 emissions		Neutral effect	Decreases CO_2 emissions	
	Notably (−1p)	Little (−0,5p)	0p	Little (0,5p)	Notably (1p)
1.2. Does the chosen energy technology(ies) replace fossil fuel- based energy production?	No (-0,5p)			Yes (0,5p)	
1.3 Does the solution(s) utilize unused biomass, such as forest or agriculture biomass?	No (-0,5p)			Yes (0,5p)	

TABLE 25.2 Synergy advantages.

Indicator 2—synergy advantages					Total points
2.1 Synergy advantages – How many of the following challenges does the pilot contributes to? The solution considers all of the following: Waste, Energy, Climate Change and Transportation	None (−2p)	One (−1p)	Two (0p)	Three (1p)	Four or more (2p)

TABLE 25.3 Land use impacts.

				Total points	
3.1 Does the land area occupied by the pilot solution have significance, cultural value or other importance?	No (1p)			Yes (−1p)	
3.2 Estimate the impact of the pilot solution on the land area occupied	High negative (−1p)	Negative (−0,5p)	Neutral (0p)	Positive (0,5p)	High positive (1p)

TABLE 25.4 Impacts on the environment.

	Total points				
4.1 Effect on air quality?	High negative effect (−1p)	Negative effect (−0,5p)	Neutral effect (0p)	Positive effect (0,5p)	High positive effect (1p)
4.2 Does the solution decrease the quality of water and soil or does it have negative impact on biodiversity?	High negative effect (−1p)	Negative effect (−0,5p)	Neutral effect (0p)	Positive effect (0,5p)	High positive effect (1p)

TABLE 25.5 Payback time.

Indicator 5—payback time	Total points				
How long is the payback time of the investment?	+25 years (-2p)	17−25 years (-1p)	12−17 years (0p)	5−12 years (1 p)	<5 Years (2p)

TABLE 25.6 Impact on citizens health.

Indicator 6—impacts on citizens health	Total points	
6.1 Positive impacts on citizens health	Yes = 0,5p	No = 0p
	6.1.1 Solution is safe (does not pose danger or risks) to inhabitants nearby? 6.1.2 Solution ensures clean and healthy habitat for living? 6.1.3 Solution offers sustainable water treatment or waste management possibilities? 6.1.4 Does the solution enable citizen with safe, clean, renewable and reliable energy?	
6.2 Negative impacts on citizen health	Yes = −0,5p	No = 0p
	6.2.1 Does the solution emit noxious gasses in harmful quantities? 6.2.2 Does the solution release toxic compounds in harmful quantities? 6.2.3 Does the solution cause a significant risk of injury? 6.2.4 Does the solution cause significant noise or esthetic harm?	

TABLE 25.7 Sustainable value teaching.

Indicator 7—Does the solution support teaching sustainable values?	Total points	
7.1 Does pilot include implementation of clean or renewable energy technologies?	No: −0,5 p	Yes: 0,5p
7.2 Does the pilot promote the energy efficiency?	No: −0,5 p	Yes: 0,5p
7.3 Does the pilot promote participation of stakeholders?	No: −0,5 p	Yes: 0,5p
7.4 Is the solution visible?	No: −0,5 p	Yes: 0,5p

Economic sustainability

5. Payback time (Table 25.5)

Social sustainability

6. Impact on citizens health (Table 25.6)
7. Sustainable value teaching (Table 25.7)
8. Community impact (Table 25.8)
9. Energy security (Table 25.9)

The total sum of the values will provide information of the overall scale of impacts when considering all the nine indicators of the environmental, economic, and social sustainability (Table 25.10).

TABLE 25.8 Community impact.

Indicator 8—community impact	Total points	
8.1 Does the solution support social cohesion and interaction?	No: −0,5 p	Yes: 0,5p
8.2 Does the solution(s) improve the community's adaptation to climate change?	No: −0,5 p	Yes: 0,5p
8.3 Does the Pilot improve local job creation and local business?	No: −1 p	Yes: 1p

TABLE 25.9 Energy security.

Indicator 9—energy security	Total points	
9.1 To what degree does the solution contribute to energy needs of the community?	No: −0,5 p	Yes: 0,5p
9.2 How many months per year the solution functions due to seasonal variance?	No: −0,5 p	Yes: 0,5p
9.3 Is the solution prone to intermittency issues?	No: −0,5 p	Yes: 0,5p
9.4 Does the pilot offer energy storing capacity?	No: −0,5 p	Yes: 0,5p

TABLE 25.10 Scaling of the impact categories.

Impact/performance	Points
High positive	2
Positive	1
Neutral	0
Negative	−1
High negative	−2

This weighing method is based on the Likert scale. It should be noted that the main purpose of this method is to a preliminary assessment and make a comprehensive comparison of different treatment alternatives.

25.3 Case study

25.3.1 Assessment of biowaste management options in a small Finnish municipality

Finland is a Northern European country with the total population of the country of 5.5 million with an average density of 18.1 inhabitants per km^2. The population is sparsely distributed with the capital Helsinki being the most populous city and the Northernmost parts the least populated. The climate is from warm to extreme cold; with winter months (December, January and February) the coldest with limited sunlight hours, while the summer months (June, July and August) are pleasantly warm and sunny (experiencing midnight sun and nightless nights).

The sustainability assessment method is illustrated with a case study of Puolanka Municipality in the northwest of the Kainuu province (Fig. 25.2). The total area of the municipality is about 2599 km^2, and the population is 2597 (Puolangan Kunta, 2022). Under the Waste Act (646/2011), municipal waste management regulations are issued with an objective to promote healthy atmosphere for living and alleviate dangers of waste or waste management on environment and human health. Local municipal waste management organization in Kainuu is called "Ekokymppi," and it handles in collection, recycling, treatment, and final disposal of waste in the Kainuu region as well as provides consultation/information center regarding the waste matters (Ekokymppi, 2022a) (Fig. 25.2).

The following hypothetical scenarios for biowaste treatment were evaluated:

1. Scenario 1: Codigestion: Separately collected biowaste is transported to the nearest biogas plant (in Oulu, see Fig. 25.3)

FIGURE 25.2 Map of Finland and Kainuu region. Location of the research sites (Puolanka, Oulu and Kajaani) (Business Kainuu, 2023).

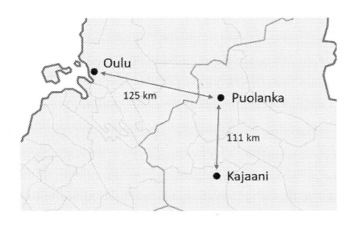

FIGURE 25.3 Distances of treatment facilities from Puolanka. Oulu and Kajaani marked in relation to Puolanka on the map of Finland.

2. Scenario 2: Composting: Separately collected biowaste is transported to the nearest composting plant (in Kajaani)
3. Scenario 3: Investing in small-scale digester: Separately collected biowaste to be treated in Puolanka locally, with the municipality building their own biogas facility

25.3.2 Scenarios

The sustainability assessment method is tested to assess these three scenarios to illustrate how to ascertain the most suitable option in terms of overall sustainability.

25.3.2.1 Scenario 1: codigestion

The nearest biowaste treating biogas plan to Puolanka operated by Gasum Oy is located in Oulu at the Rusko waste center. The transportation distance to Oulu is 125 km from Puolanka city center. The codigester is accepting separately collected biowaste, which is treated in an anaerobic digester, which produces biogas. Biogas can be used both as biofuel generating heat and electricity and purified to methane, a transport biofuel (Lukehurst et al., 2010).

In addition to biogas, the plant generates digestate as by-product. Digestate is nutrient-rich and serves the same purpose of increasing the quality of soil. The digestate from the Oulu plant is used as natural fertilizer replacing chemical fertilizers (Kiertokaari, 2023).

Anaerobic digestion is a very flexible technology, accepts wide range of inputs, produces energy and other valuable end products. As anaerobic digestion allows closing nutrient and energy cycles, it is considered more beneficial to composting and more in sync with circular economy (Chang & Pires, 2015; Wellinger, et al., 2013).

25.3.2.2 Scenario 2: composting

The Majasaari waste center is in Kajaani about 111 km by car from Puolanka city center. The waste center serves in sorting all the different wastes from biowaste to paper, plastics and sends them to the concerned waste management companies for further treatment. Other municipal waste (recyclables) and construction waste are received, sorted, and pretreated. Also, the special waste and hazardous waste are received by the center (Ekokymppi, 2022a). Biowaste is handled and treated there through windrow composting. The waste is crushed by a wheel loader and covered with the wooden chips. The compost is about 2 meters high and 50 meters long. The temperature is monitored, and the compost rows are turned regularly for maintaining the efficiency of the process. The mature end product is used in covering the soils nearby the waste center (Ekokymppi, 2021).

Composting processes include the conversion of organic solid waste with chemicals (bulking agents and amendments) in the presence of various bacteria and fungi, surplus supply of air to produce numerous chemical changes and water, which further breaks down into resistant substances called compost and water with the release of energy (David Border Composting Consultancy DBCC, 2002). The humus end product generated after the composting process is the compost. The quality of the compost is likely to depend upon feed substrates, pre- and postprocessing time and operating conditions and high-rate and curing design parameters maintained in the system.

25.3.2.3 Scenario 3: small-scale digestion

This scenario is based on the assumption that a small-scale biogas plant would be built within 10 kms in Puolanka, Biocenter of Puolanka (Puolangan Biokeskus). This investment project would promote recycling of biowaste in a sustainable way, generation of carbon-neutral fuel, as well as create jobs and employment locally (Kainuun Etu, 2022).

25.3.3 Data and calculations

The calculations are based on the data available on the official websites. A more detailed version of these calculations can also be found in Airi (2019).

The calculations are based on the assumption that the average annual separately collected biowaste per person in the Kainuu area is 58 kg (Table 25.11) (Ekokymppi, 2022a,b). Hence, it is calculated that the total amount of biowaste generated in Puolanka is 141 926 kg/a.

Using Google maps reference, the car distance from Puolanka city to Oulu (Kiertokaari Oy) and Kajaani (Majasaari Waste center) was 125 and 111 kms, respectively (Table 25.12). The proposed Puolangan biokeskus (anaerobic digestion plant) is about 10 kms. The gate fees used for Kiertokaari Oy, Oulu was 70 €/t (Kiertokaari Oy, 2022), Majasaari waste center was 140.12 €/t (Ekokymppi, 2022b) and for Puolangan biokeskus was assumed to be 50 €/t for calculative purpose.

The transportation costs and associated emissions are collected in Table 25.13. The diesel price of 2.507 €/l (Polttoaine, 2022) is taken to calculate the round trips for all three options (Table 25.4.). For local collection of biowaste (Scenario 3), delivery-type heavy weight lorry with maximum cargo volume of $8m^3$ was considered.

TABLE 25.11 Amount of separately collected biowaste in Puolanka.

Population	Amount of separately collected biowaste in Kainuu area per person per annum (kg/a/p)		Total amount of separately collected biowaste in Puolanka			
	in a year (kg/a)	per week (kg/wk)	in a year (t/a)		in a week (t/wk)	
2447	58		141,926	2729.35	141.93	2.73

TABLE 25.12 Gate fees and transportation distances of bio-waste from Puolanka to different treatment facilities.

Options	Destination	Distance from Puolanka (km)	Gate fees with taxes (€/t)
Scenario 1	Kiertokaari Oy, Oulu (Gasum Ltd.)	125	70
Scenario 2	Majasaari Waste Center, Kajaani	111	140.12
Scenario 3	Puolangan Biokeskus	10	50

TABLE 25.13 Transportation costs and CO_2 emissions (Lipasto, 2017).

Options	Distance (km) (round trip)	Distance (km/a)	Avg. fuel price (€/l)	Avg. fuel consumption (l/100 km)	Round-trip cost (€)	Yearly round-trip cost (€/a)	CO_2e (g/km)	CO_2e (t)
Local collection	20	3120	2.51	21.9	10.98	1712.98	521	1.63
Scenario 1	250	4870		33.7	222.2	3268.35	796	3.02
Scenario 2	222	4674			198.54	3102.76		2.86
Scenario 3	Calculations for the local collection for the bio-waste cover the transportation and emissions for Scenario 3							

The fuel consumption for the lorry is 21.9 l/100 km (an average of 18,3 l/100 km empty lorry and 25.5 l/100 km fully loaded). In total, 15.16 m³/wk (assuming volume weight of 180 kg/m³) of biowaste is generated in Puolanka. It is collected three times/week and the weekly cost would be 10.98 €, and annual 1712.98 €. For Scenarios 1 and 2, waste load is transported using semitrailer truck for highway driving. The fuel consumption is 33.7 l/100 km (an average value of 26.6 l/100 km for empty truck and fully loaded is 40.7 l/100 km) (Lipasto, 2017).

As the semitrailer truck has a maximum capacity of 25 tons, the truck would carry biowaste to the destinations only once in 8 weeks so, the yearly cost would be 6–7 times the round-trip cost (Table 25.3). But, in practice to avoid odor and other harmful impacts around the surrounding, the frequency of transporting the waste would be at least once a week. The truck collecting biowaste from neighboring municipalities could also collect the biowaste from Puolanka waste collection center in Puolanka. Since the wastes are collected from the Puolanka city first, local collection is added for calculating transportation costs in the yearly cost of Scenarios 1 and 2. Thus, the annual round-trip cost for Scenario 1 is calculated as 3268.35 € while Scenario 2 it was 3102.76 €.

Emissions from composting plant were calculated using the Eq. (25.1) (Research Triangle Institute RTI International, 2010)

$$E_{CO2} = EF_{compost} * \sum_{N=1}^{N} (M_{compost,n} * TS_n) \tag{25.1}$$

For Scenario 3 (similar as local collection), since the heavy lorry is used for transportation, an average of CO_2 emissions (empty 436 g/km and full load 606 g/km), i.e., 521 g/km is used (Lipasto, 2017). The annual amount of CO_2 emissions is calculated to be 1 625.52 kg of CO_2, that is, 1.63 t of CO_2 emissions.

For Scenarios 1 and 2, semitrailer truck is used for transporting the waste through the highways, an average of CO_2 emissions for empty 630 g/km and full load 962 g/km, that is, 796 g/km is used (Lipasto, 2017). Also, the emissions while collecting and transporting the wastes locally within the city are added. Thus the total annual transportation emissions for Scenarios 1 and 2 are 3.02 t of CO_2 and 2.86t of CO_2, respectively.

The following equations are used for calculating the results in Scenario 3:
The volume of the reactor is calculated using Eq. (25.2) (Biosantech, et al., 2008)

$$V_R = \frac{AF_v}{365/RT} * f \tag{25.2}$$

The total methane yield for biowaste is calculated using the Eq. (25.3) (Kiviluoma-Leskelä, 2010)

$$MP = AF_t * MP_t \tag{25.3}$$

The thermal output is calculated as Eq. (25.4) (Rutz et al., 2015)

Thermal output(kW) = methane production per hour(m³/h) ∗ calorific value of methane(kWh/m³) ∗ thermal efficiency of CHP (25.4)

The calculation of NPV is based on Vedenjuoksu (2009)

$$\text{NPV} = C_t \times \frac{(1+r)^t - 1}{r(1+r)^t} - C_o \times \frac{(1+r)^t - 1}{r(1+r)^t} - C_i \quad (25.5)$$

Based on Vierros (2009), PP is calculated as:

$$PP = \frac{C_i}{C_t - C_o} \quad (25.6)$$

25.3.4 Results

The results for physical parameters, technical data, and economy were assessed and listed in Table 25.14.

In this scenario assessment it was found that investing in an AD plant at the Puolanka municipality would be attainable. The payback period was calculated to be less than 2 years, which means high turnover for the investors. The biogas energy output for a year was calculated to be 1 42.4 MWh, which would be enough to replace fuel demand of 10.5 petrol-based cars and 15.1 diesel-based cars. The annual digestate sales were estimated to be 1 343.36 € and per year, and revenue from heat sales to district heating at the unit price of 83,5 €/MWh was estimated to be 11 890.4 €. The project would also create jobs and business opportunities in Puolanka. It was also found that the carbon emissions for AD plants in either Oulu (Scenario 1) or Puolanka (Scenario 3) would reduce carbon emissions, while composting (Scenario 2) would add to greenhouse gases emissions.

TABLE 25.14 Outputs of the research.

Physical parameters	
Temperature (°C)	35
Total solids (%)	25
Hydraulic retention time (HRT) (d)	21
Technical results	
Amount of biowaste in Puolanka (t/a)	141.926
Volume of the reactor (m³)	55
Biomethane production (m³/a)	14,192.6
Biogas production (m³/a)	21,974.62
Energy output from biogas (MWh/a)	142.4
Economic assessment	
Digestate sale (€) per year	1343.36
Heat production revenue (€) per year	11890.4
NPV (€)	−49151.165
PP (years)	−1.67
CO_2 emissions	(t CO_{2e}/year)
Scenario 1 (AD plant and transportation emissions)	−34.98
Scenario 2 (composting and transportation emissions)	+2603.02
Scenario 3 (AD plant and transportation emissions)	−36.37

25.3.5 Sustainability assessment of the case study

The three options were then assessed answering the questions of Tables 25.1–25.9, using the data and calculations from Table 25.14. The values for these indicators for the three scenarios are listed in Table 25.15.

The results can also be illustrated as a radar diagram, see Fig. 25.4. Higher points indicate environmental benefits, while negative points indicate environmental impacts.

TABLE 25.15 Summary of points for the three scenarios.

Dimension	Indicator	Scenario 1	Scenario 2	Scenario 3
Environmental	1. CO_2 reduction	1,5	0,5	1,5
	2. Synergy advantages	2	−1	2
	3. Land-use implication	1	1	1
	4. Impact on the environment	1,5	−1	2
Economic	5. Payback time	2	2	2
Social	6. Impact on citizens' health	2	1	2
	7. Does the solution teach sustainable values	2	0	2
	8. Community impact	0	−1	2
	9. Energy security	1	0	1

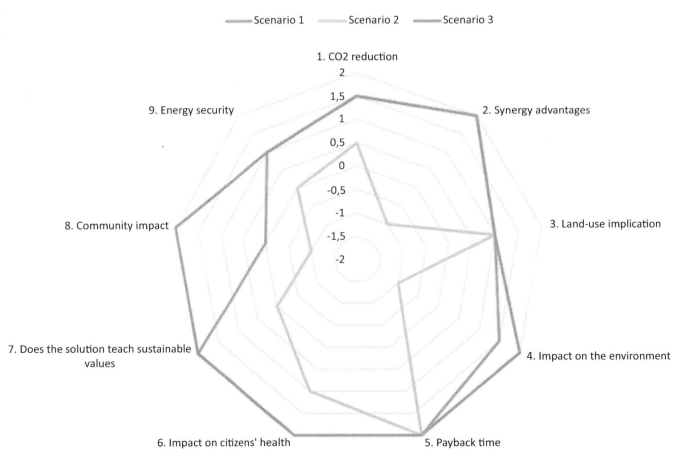

FIGURE 25.4 Radar diagram. Comparison of the three scenarios.

According to the results, Scenario 2; the transportation of source separated biowaste from the municipality of Puolanka to Kajaani Majasaari composting center, is the least favorable when considering sustainability assessments that have been done. The high amount of carbon dioxide emissions and no socioeconomic returns of past case made it worse than Scenario 1 despite being a rather economical option. The payback time for Scenario 1 (Puolanka to Kiertokaari Oy, Oulu) and Scenario 2 (composting in Kajaani Majasaari) are categorized equally high (2 point) because these waste centers are already existing and have been operating for a long time, and hence, their investment returns are not applicable for the case.

Scenario 3 (biowaste treatment in AD plant built locally in Puolanka) was found to be more advantageous than other options when considering an impact on environment and community as it would provides the community with clean renewable energy (biogas) for district heating and replace dependency on fossil fuel consumption. The advantage of a local AD plant is also lower gate fees and transportation costs and increase in job or business opportunities locally. Based on this sustainability assessment, when considering economic, environmental and social impacts in tandem, Scenario 3 emerges as the best solution for the source separated biowaste management in Puolanka municipality. The project would also help in closing nutrient cycle by providing organic fertilizer to be used on soils. Overall, the project would promote circular economy in Puolanka through utilizing the wastes as raw material to provide energy and products as long-term economic benefits.

25.4 Discussion and conclusions

When selecting the best possible waste management solutions, one should not consider only the economic and technical factors but also to have wider perspective of sustainability issues with both global and local points of view and by learning from the past and thinking about future. Sustainability in the context of biowaste management refers to finding solutions that manage biowaste without adverse impacts to the environment, while seeking synergy benefits with other environmental considerations, such as nutrient recovery, or revenues from the sale of energy recovered. Sustainability assessment requires considering environmental, economic, and social impacts in concert. In this chapter, we presented an easy-to-use semiquantitative assessment method, developed to compare practical solutions and to provide an overview of multiple sustainability indicators.

Sustainable biowaste management also requires the engagement of stakeholders, including business, communities and individuals and government agencies to find solutions in a collaborative effort. By working together, it is possible to find solutions that provide no hazard to human health and the environment, protect our resource base and the well-being of ecosystems, are economically viable, and contribute to sharing sustainable values.

Acknowledgments

The authors acknowledge the RECENT project (funded by the Northern Periphery and Arctic Programme) within which the sustainability template was originally developed, and Antton Niemelä for his contribution to the development of the first version of template as part of his diploma work in the RECENT project. The case study is based on the master's thesis of Anusha Airi. The work was financed by the Maa-ja vesitekniikan tuki (MVTT) Foundation and The Waste Management Association (JHY).

References

Airi, A. (2019). *Circular economy and closing nutrient cycles: planning sustainable bio-waste management system for Puolanka municipality*. University of Oulu, Faculty of Technology, Environmental Engineering. Master's thesis (tech). Available online at: http://jultika.oulu.fi/Record/nbnfioulu-201902061151.

Biosantech T. A. S., Rutz D., Prassl H., Köttner M., Finsterwalder T., Volk S., Janssen R. (2008). *Biogas Handbook*. Esbjerg. University of Southern Denmark. Available online at: http://www.lemvigbiogas.com/BiogasHandbook.pdf.

Brundtland, G.H. (1987) Our common future: report of the World Commission on Environment and Development. Geneva, UN-Dokument A/42/427.

Business Kainuu (2023). Kainuu map. Available at: https://businesskainuu.fi/wp-content/uploads/2022/02/kainuu-kartta.jpg.

Chang, I., & Pires, A. (2015). *Technology matrix for solid waste management. sustainable solid waste management: a systems engineering approach*. Hoboken, New Jersey: The Institute of Electrical and Electronics Engineers, Inc. John Wiley & Sons, Inc.

Chiras, D. D. (2016). *Environmental science (tenth ed.)*. Jones and Bartlett Learning. United States of America.

David Border Composting Consultancy (DBCC). (2002). *The composting process. Process and plant for waste composting and other aerobic treatment*. R&D Technical Report P1−311/TR.

EC. (2018b). *Closing nutrient cycles. Opportunities. Participant Portal. Research & Innovation*. European Commission. Available at: http://ec.europa.eu/research/participants/portal/desktop/en/opportunities/h2020/topics/ce-rur-08-2018-2019-2020.html.

ECN (European Compost Network). (2022). ECN Data Report 2022. https://www.compostnetwork.info/wordpress/wp-content/uploads/ECN-rapport-2022.pdf.

Ekokymppi. (2021). *Annual reports 2020*. Available online at: https://www.ekokymppi.fi/ekokymppi/vuosikertomukset.html.

Ekokymppi. (2022a). *Municipal Waste Authority of Kainuu. Kainuun jätehuollon kuntayhtymä*. Available online at: https://www.ekokymppi.fi/tietopankki/vieraat-kielet/in-english.html.

Ekokymppi. (2022b). *Majasaaren Jätekeskus*. Kainuun jätehuollon kuntayhtumä. Available online at: https://www.ekokymppi.fi/hinnastot/hinnasto-2022-majasaari.htmljultika.

GRI (Global Reporting Initiative) (2022) Consolidated Set of the GRI Standards. Available from https://www.globalreporting.org/.

IAEA (2005) Energy indicators for sustainable development: guidelines and methodologies, International Atomic Energy Agency, ISBN 92-0-116204-9.

Kainuun Etu. (2022). *Biocentre of Puolanka*. Available online at: http://kainuunetu.fi/puolangan-biokeskus-en.

Kiertokaari Oy. (2022). Gate price (*Household biowaste*). Available at: https://kiertokaari.fi/hinnat-ja-maksaminen/ruskon-jatekeskuksen-hinnasto/.

Kiertokaari Oy. (2023). *Kiertokaari Kestävän kiertotalouden edistäjä*. Yrityksemme. Available at: https://kiertokaari.fi/kiertokaari/yrityksemme/.

Kiviluoma-Leskelä L. (2010). *Biokaasun tuottaminen ja hyödyntäminen Lappeenrannassa*. (English: Generation and utilization of biogas in Lappeenranta.) Lappeenranta, Finland: Lappeenranta University of Technology, Faculty of Technology. Master's thesis. 122

Lipasto. (2017). *Lipasto unit emissions database*. Available online at: http://lipasto.vtt.fi/yksikkopaastot/tavaraliikennee/tieliikennee/kajaksuurijakelue.htm.

Niemelä, A. (2016). *Sustainability of small-scale renewable energy solutions in northern rural communities: case eco-district of Päivänpaisteenmaa*. Master's thesis. Energy and Environment Unit. University of Oulu.

Likert, R. (1932). A technique for the measurement of attitudes. *Arch. Psychological*, 140, 1–55.

Lukehurst, C. T., Frost, P., & Saedi, T. A. L. (2010). *Utilisation of digestate from biogas plants as biofertilizer. International Energy Agency (IEA) bioenergy*.

Polttoaine. (2022). *Average diesel prices*. Available at: https://www.polttoaine.net/Oulu.

Puolangan Kunta. (2022). *Municipality Info: Basic information about the municipality*. The municipality of Puolanka. Available online at: https://www.puolanka.fi/kuntainfo/perustietoa-kunnasta.html.

Rakshit, A., Singh, H. B., & Sen, A. (2015). *Nutrient use efficiency: from basics to advances* (pp. 21–26). Springer (India) Pvt. Ltd.

Research Triangle Institute (RTI) International. (2010). *Greenhouse gas emissions estimation methodologies for biogenic emissions from selected source categories: solid waste disposal wastewater treatment ethanol fermentation*. Draft submitted on December 14, 2010. EPA Contract No. EP-D-06-118.

Rutz D., Mergner R. and Janssen R. (2015). *Sustainable heat use of biogas plant – a handbook 2nd edition*. WIP Renewable Energies, Munich, Germany. Available online at: http://www.biogasheat.org/wp-content/uploads/2015/03/Handbook-2ed_2015-02-20cleanversion.pdf.

Vedenjuoksu T. 2009. *Talousmatematiikka*. (English: Commercial mathematics). Available online at: http://cc.oulu.fi/~tvedenju/talousmatematiikka/files/investointi_esim.pdf.

Vierros T. 2009. *Investointilaskelmat*. (English: Investment calculations). Available online at: https://wiki.aalto.fi/display/TU22/8.+Investointilaskelmat.

Wellinger, A., Murphy, J., & Baxter, D. (2013). *The biogas handbook: Science, production and applications. IEA Bioenergy* (52). Woodhead Publishing Series in Energy.

Index

Note: Page numbers followed by "*f*" and "*t*" refer to figures and tables, respectively

A

AAS. *See* Aerated activated sludge (AAS)
ABE. *See* Acetone-Butanol-Ethanol (ABE)
AC. *See* Activated carbon (AC)
Acacia, 347
Acetic acid, 317
Acetobacterium woodii, 258
Acetogenesis, 167, 317
Acetogens, 258
Acetone, 155
    as organic solvent, 34
Acetone-Butanol-Ethanol (ABE), 227
Acid Blue 25, 347
Acid Blue 74, 347
Acid hydrolysis, 122
Acid pretreatment, 43–44
Acid-catalyzed steam explosion process, 37
Acid-catalyzed transesterification, 157
Acidogenesis, 3, 12, 169, 317
Acidophiles, 206
Acidophilic enzymes, 206
Acidovorax, 315
Acinetobacter, 315
ACR. *See* Advanced Candu Reactor (ACR)
Activated carbon (AC), 6, 291, 364–366, 373
    characteristics, 366
    chemical activation, 366
    physical activation, 365
Activated sludge, 12
Activation, 365
AD. *See* Anaerobic digestion (AD)
Additives, 249
Adenovirus, 309
ADM1. *See* Anaerobic Digestion Model 1 (ADM1)
Adsorbents, 6, 364
Advanced Candu Reactor (ACR), 392
Advanced oxidation processes (AOPs), 370
Aerated activated sludge (AAS), 316
Aeration, 108–109
Aerobic digestion, 11–12
Aerobic treatment, 315–317
AFEX. *See* Ammonia fiber explosion (AFEX)
Affinity binding, 208
AFPs. *See* Antifreeze proteins (AFPs)
AFT. *See* Ash fusion temperature (AFT)
Agitational speed of material, 279–280
Agricultural/agriculture, 85, 117
    biomass, 84
        feedstock and examples related to usage applications, 343*t*
    waste, 10
    by-products, 178
    and food wastes, 166
    harvesting, 10
    and livestock waste, 73
    residues, 10, 30–31, 58, 85, 342
    wastes, 24, 30, 58, 73, 86, 227, 233–234
        biomass, 3
        sources, 17
Agro-industrial waste, 227
Agroindustrial residues, 233–234
Agrowastes, 215, 233–234
    agro/vegetable source, 154
        edible vegetable oil sources, 154
        nonedible vegetable oil sources, 154
    forms of, 234–235
    process of pigment production from, 240, 240*f*
    production of pigments
        from agrowastes without using microorganisms, 237
        from microorganisms using agrowastes as substrate, 238–239
    sources, 234, 234*f*
    types of pigments produced from, 235–236, 236*t*
Air pollutants, 188–189
Air Pollution Control Law (1968), 311–312
Airflow velocity, 279–280
Alcohols, 12, 33, 152
    pretreatment using alcohol as organic solvent, 33–34
        acetone as organic solvent, 34–35
        ethanol as organic solvent, 33–34
        methanol as organic solvent, 33
        peracetic acid as organic solvent, 34
        pretreatment by organic acids, 34
Aldehydes, 368
Algae, 86, 120
Algal biomass, 11
Algal cell lysis, 110
Algal strains, production of biofuel from waste water using, 203
Algal systems, 105
Alginates, 348
Alkali hydroxides, 366
Alkali pretreatment process, 23, 32, 44
Alkali technologies, 30
Alkaline extraction method, 118–119
Alkaline lipases, 206
Alkaline peroxide pretreatment process, 44
Alkaline pretreatment (AP), 30–32, 138
    technologies, 30–32, 44
    ammonia pretreatment, 31–32
    lime pretreatment, 32
    pretreatment with sodium hydroxide, calcium hydroxides, and sodium carbonate, 31
Alkaline proteases, 206
Alkaline-catalyzed production process, 160
Alkaliphiles, 206
Alkaliphilic enzymes, 206
Alkyl phenols, 249
Aluminum, 314
Amber acid. *See* Succinic acid (SA)
Ameba, 315–316
American Standard Test Method (ASTM), 161
Amines, 368
Ammonia fiber explosion (AFEX), 31, 37, 43
Ammonia pretreatment, 31–32
    types of, 31*t*
Amylases, 202, 206
Amylopectin, 253
Amylose, 253
Anaerobic digestion (AD), 2–3, 11–12, 58, 61–62, 123, 166, 264, 266, 278, 386–388, 390, 411
    comparison of different methods of waste management, 12*t*
    pretreatment, 91–92
    process, 94, 121–122, 166–168, 167*f*
Anaerobic Digestion Model 1 (ADM1), 394
Anaerobic filter (AF). *See* Fixed-bed reactors
Anaerobic organic waste digestion, 166
Anaerobic sequencing batch reactor (ASBR), 392
Anaerobic sludge digesters, 76–77
Anaerobic treatment, 316
Animal husbandry, biomass waste from, 11
Animal manure, 57–58
Animal residue, 155
Animal waste, 178, 386–387
Animal-based coagulants, 348
Announced Pledges Scenario (APS), 165
Annual Operation Cost, 394–395
Anthocyanidins, 267
Anthocyanins, 233–236
Anthropogenic processes, 71–72
Anthropogenic wastes, 73–74
Antifoaming agents, 341
Antifreeze proteins (AFPs), 205–206
AOPs. *See* Advanced oxidation processes (AOPs)
AP. *See* Alkaline pretreatment (AP)

APILs. *See* Aprotic ionic liquids (APILs)
Aprotic ionic liquids (APILs), 44
APS. *See* Announced Pledges Scenario (APS)
Aquatic organism, 110
Aqueous-phase catalytic conversions of biomass platform chemicals, 268
Areca catechu, 233–234
Aroclor 1260, 307
Aromatic compounds, 315–316
Aromatic ring, 27
Arsenic (As), 306
  As-synthesized hydrochars, 366
Artificial neural network modeling, 345
ASBR. *See* Anaerobic sequencing batch reactor (ASBR)
Ascaris lumbricoides, 308
Ascomycetes, 139
Ash fusion temperature (AFT), 125
Aspen Icarus Process Evaluator (Aspen IPE), 388
Aspen IPE. *See* Aspen Icarus Process Evaluator (Aspen IPE)
Aspergillus
  A. awamori, 216–217
  A. carbonarius, 216–217
  A. niger, 46
  A. oryzae, 233–234
  A. wentii, 216–217
ASTM. *See* American Standard Test Method (ASTM)
Autohydrolysis process, 36
Azaphilones, 233–234

B

Bacillus
  B. cereus, 46
  B. licheniformis, 216–217
  B. stearothermophilus, 205
  B. subtilis, 46, 202–203
Bacteria, 308–309
Bacterial pretreatment, 46
Bacteroides cellulosilyticus, 46
Bacteroidetes, 316
Bagasse of malt (BM), 373
Bagasse of sugarcane (BS), 373
Bakery waste, 215
Balantidium coli, 308
Balloon/bag digesters, 386–387
Bamboo biomass feedstock, 345
Bar screens, 313
Barrett–Joyner–Halenda method (BJH method), 119
Base Case Simulation software (BCS software), 159
Base-catalyzed transesterification, 156
Batch reactor, 27
Battery-based energy storage, 288
BCS software. *See* Base Case Simulation software (BCS software)
Benzo(a)pyrene, 307
BET method. *See* Brunauer–Emmett–Teller method (BET method)
BET-BJH method. *See* Brunauer–Emmett–Teller–Barrett–Joyner–Halenda method (BET-BJH method)

BFRs. *See* Brominated flame retardants (BFRs)
BG. *See* Biogas (BG)
2,2′ bicinchoninate method (BCA method), 119
Bifunctional cellulose-based adsorbent, 347
Bio-based chemicals, 341
  application in water and wastewater treatment practices, 350–351
Bio-based coagulants, 344–348
Bio-based production, 215
Bio-deterioration, 251
Bio-fragmentation, 251
Bio-oil, 330–331
Bioaccumulation, 329
Bioactive compounds, 15
Biobutanol, 116
Biocatalysts, 201
Biochar, 4, 6, 120, 126, 319–323, 366, 372
  burning of biomass, 321
  electromodified biochar, 322
  examples of studies, 326t
  fast pyrolysis process, 322
  flash pyrolysis, 322
  gasification process, 322
  hydrothermal carbonization, 322
  magnetic biochar, 322–323
  methods of biochar production, 321
  microwave pyrolysis, 322
  pyrolysis process, 321
  slow pyrolysis process, 321
  torrefaction process, 322
  vacuum pyrolysis, 322
Biochemical conversion, 3, 363
  anaerobic digestion, 123
  enzymatic conversion, 123–124
  fermentation, 124
  of waste biomass to bioenergy, 123–124
Biochemical methods, 3, 11–12, 88–89, 91–92
  aerobic digestion, 12
  anaerobic digestion, 11–12
  of bioenergy production, 89f
  bioprocessing of biomass waste, 4f
Biochemicals recovery from biomass, 344–350
  agricultural or biomass-derived wastes as precursor for activated carbon, 349t
  bio-based coagulants, coagulant aids/flocculants, and materials for advanced treatment practices, 344–348
  challenges and future perspective, 351–352
  preparation bio-based chemicals for water and wastewater treatment, 344f
  sustainable chemicals for advance treatment, 349–350
Biochemistry of lignocelluloses, 136–137
Biocoagulants, 350–351
Biocontrol agents, 17
Biodegradable garbage, 86
Biodegradable plastics, 245–246, 249–250
Biodiesel, 9, 11, 16, 110, 121, 127–128, 156, 158, 331
  challenges and bottlenecks of biodiesel commercialization, 160–161

  combustion, 151–152
  commercialization of, 159–160, 159f
  different techniques for large-scale production of, 156–157
    microemulsion, 157
    pyrolysis, 157
    transesterification, 156–157
  market and Indian scenario, 152–153
  potential of lipid-rich feedstocks for large-scale biodiesel production, 153–156, 154t
  production, 151, 202
  scale-up of biodiesel production and technoeconomic analysis, 158–159
Biodigesters, 391, 394
Bioeconomy, 129–130
Bioelectrochemical system, 191
Bioenergy, 41, 83, 115, 330–331, 385–386
  biochemical conversion of waste biomass to, 123–124
  biomass materials, 84–88
  catalytic conversion of waste biomass to, 122–123
  in combating climate change and greenhouse effects, 115
  conventional bioenergy production methods, 88–90
  lab-scale to large-scale opportunities, 100–102
  microwave application in bioenergy production, 90–100
  microwave-assisted bioenergy production, 96–100
  microwave-assisted pretreatment of biomass, 90–96
  need for, 83–84
  derivatives from biomass, 84f
  pretreatment
    in generating, 41–42
    of waste biomass, 42–47
  sources and classification of bioenergy from waste biomass, 119–122
    gaseous fuels, 121–122
    liquid fuels, 120–121
    solid fuels, 120
  systems, 110
  thermochemical conversion of waste biomass to, 124–126
  waste biomass as potential source of, 115–116
Bioethanol, 16, 120, 127
Biofertilizers, 328–329
Biofilm, 251
Biofiltration, 315
Biofuels, 15–16, 29–30, 110, 116, 152, 264
  biodiesel, 16
  bioethanol, 16
  biogas, 16
  production of biofuel from waste water using algal strains, 203
Biogas (BG), 16, 61–62, 121–122, 166, 171, 386, 410–411
  benefits of biogas plants, 387
  challenges in development of biogas plants, 388

costs involved in biogas generation, 388
importance of biogas plants, 386–387
plants for rural economy, 387
potential of biogas as future energy, 386
quantity measurement, 176–177
technologies, 11
types of biogas plants and selection for study, 391–392
worldwide development, 165–166
yield, 180–181
Biogeochemical cycle. See Nutrient cycle
Biohydrogen, 62, 116, 122, 129
production, 187, 190
Bioleaching technology, 329–330
Biological agents, 341
Biological conversion techniques, 60–62
anaerobic digestion, 61–62
composting, 60
fermentation, 62
vermicomposting, 60–61
Biological decay process, 138
Biological hydrolysis, 141
Biological oxygen demand (BOD), 106
Biological pretreatment, 46–50, 139
bacterial/microbial pretreatment, 46
enzymatic pretreatment, 46–47
methods, 23, 47–50, 139, 264
Biological treatments, 265
Biological waste
bioprocessing techniques, 71, 76–77
utilization to value-added products, 266–267
Biomass, 21–23, 30, 185–187, 362, 386
biomass-based catalysts in
catalytic wet air oxidation, 375–377
catalytic wet peroxide oxidation, 373–375
biomass-based fuels, 110
biomass-based materials for
batteries, 292–299
supercapacitors, 290–292
biomass-derived carbon materials, 289
applications of biomass-derived carbon materials for energy storage systems, 290–299
biomass-derived catalyst materials, 364–368
activated carbon, 365–366
carbonaceous materials, 366–368
biomass-derived wastes, 185–187
composition, 47
conversion methods, 1, 264–265, 363–364
feedstock for recovery of chemicals, 342–343
gasification process, 188
H_2 from biomass-derived biowastes, 185–187
potential biowaste sources, 185–187
hydrolysis, 23
materials, 84–88
microwave-assisted pretreatment of, 90–96
pretreatment methods of biomass for 2G bioethanol, 137
production and properties of biomass-derived anode materials, 288–290

properties, 86–88
composition, 87–88
heating values, 86
moisture content, 86
particle size and density, 88
residues, 117, 285
sources, 84–86, 85f
agriculture, 85
algae, 86
forest source, 84–85
waste, 85–86
supply chain, 41
torrefaction, 25–26
dry, 26
wet, 25–26
types of, 21–23
Biomass wastes, 1, 6, 15, 91
classification of, 9–11
agricultural biomass waste, 10
algal biomass, 11
biomass waste from animal husbandry, 11
municipal solid waste composition and recyclable waste, 10f
municipal waste, 11
types of biomass wastes and composition, 10t
functional materials from, 6
high-value bio-products from, 15–17
biocontrol agents, 17
biofuels, 15–16
biopolymers, 17
biosurfactants, 17
conversion of biowaste into value-added products, 15f
industrial biocatalyst, 16
microbial pigments, 17
Biomaterials, 115
Biomethane, 116, 121–122, 129
Biophotolysis, 190–191
Bioplastics, 328–329
biofertilizers, 328–329
Biopolymers, 17
Bioprocessing methods, 2–3
of whole food value chain wastes, 266f
Biorefinery, 41, 129–130
advance pretreatment in, 50
approach, 109, 266
of microalgal cultivation in municipal wastewater, 109–110
of microalgal municipality wastewater treatment, 109f
biorefinery-based BDO production, 223–224
bread waste, 224
farmland waste, 223–224
vegetables and fruits, 223
biorefinery-based ethanol production, 226
cotton-based waste, 226
MSW, 226
PPW, 226
process, 34
waste biomass as potential substrate for, 41
Biorefining, 3
Bioremediation, 105
Biosurfactants, 17, 327–328

Biotechnological/biotechnology, 204
approaches, 246
arbitration, 254
Biowastes, 11, 170, 187, 263–264, 405
assessment of biowaste management options in small Finnish municipality, 409–410
distances of treatment facilities from Puolanka, 410f
Finland and Kainuu region, 410f
data and calculations, 411–413
evaluation of biowaste characteristics, 172–174
hydrogen production processes from, 187–192
techniques for extraction of value-added products from, 11–15
Bismuth molybdate (BMO), 372
Bisphenol A (BPA), 370, 375
BJH method. See Barrett–Joyner–Halenda method (BJH method)
Blakeslea trispora MTCC 88, 238
BM. See Bagasse of malt (BM)
BMO. See Bismuth molybdate (BMO)
BOD. See Biological oxygen demand (BOD)
Boehm titrations, 369
Bole chips, 84–85
Bottleneck, 161
Bougainvillea spectabilis, 235–236
BPA. See Bisphenol A (BPA)
BPMC. See Buthylphenylmethyl carbamate pesticide (BPMC)
Bread waste, 224
British thermal unit (BTU), 362
Brominated flame retardants (BFRs), 249
Brunauer–Emmett–Teller method (BET method), 119
Brunauer–Emmett–Teller–Barrett–Joyner–Halenda method (BET-BJH method), 119
BS. See Bagasse of sugarcane (BS)
BTU. See British thermal unit (BTU)
Burning of biomass, 321
Butane-2,3-diol, 222–224, 223f
Butanedioic acid ($C_4H_6O_4$), 217
2,3-Butanediol (BDO), 222
applications, 224
biorefinery-based BDO production, 223–224
metabolic pathway, 224
Butanol, 227–228
applications, 228
isomers, 227f
metabolic pathway, 228
production using, 227–228
Buthylphenylmethyl carbamate pesticide (BPMC), 375
1-butyl-3-methylimidazolium chloride ([Bmim]Cl), 138
By–products, 187

C

CA. See Citric acid (CA)
CABBS. See Cascading Algal Biomethane-Biorefinery System (CABBS)

CAD. *See* Cis-aconitate dehydrogenase (CAD)
CAGR. *See* Compound annual growth rate (CAGR)
Calcium, 307–308
Calcium hydroxides, 31
  pretreatment with, 31
Campylobacter, 308
Candida
  C. shehatae, 142
  C. tropicalis, 203, 216–217
Candida bombicola. *See* Sophorolipid (*Candida bombicola*)
Capital Cost (CC), 394–395
Carbocatalysts, 364
Carbohydrate-Binding Domain (CBD), 123–124
Carbohydrates, 110, 187, 263–264, 267
Carbon black, 364
Carbon composites, 286
Carbon dioxide (CO_2), 3, 14, 62–63, 90, 129, 236, 256
  depressurization, 236
  reduction, 407t
Carbon dioxide Adsorption Metal Hydride Intermediate buffer (COA-MIB), 194–195
Carbon dioxide capture and storage technology (CCS), 185
Carbon dots (CDs), 327
Carbon fibers, 286, 364
Carbon materials, 285, 364
Carbon monoxide (CO), 90, 256
Carbon nanotubes, 286, 291
Carbon quantum dots. *See* Carbon dots (CDs)
Carbon residue (CR), 368, 375
Carbon-based adsorbents, 6
Carbon-based catalysts, 364
  preparation and properties, 368–369
Carbon-rich solid products, 63
Carbon-to-nitrogen ratio (C/N ratio), 170
Carbonaceous catalysts, 364
Carbonaceous materials, 291, 364
  derived from lignocellulosic biomass, 366–368
Carbonization, 365
Carbonyls, 368–369
Carboxylic acids, 368
Carboxymethylated lignin-based flocculants, 345
Carotenoids, 110, 233–236
Carthamidin, 235–236
Carthamin, 235–236
Cascading Algal Biomethane-Biorefinery System (CABBS), 109
Castanea, 347
Catalytic conversion
  of biomass-based wastes, 264, 265f
    aqueous-phase catalytic conversions of biomass platform chemicals, 268
    biomass conversion technologies, 264–265
    conversion process utilizing single atom catalyst, 269
    efficient conversion process of biomass derived platform molecules, 267–273
    fermentative process of important platform chemicals, 269–273
    transformation of platform molecules utilizing MOF based catalyst, 268–269
    utilization of biological-waste to value-added products, 266–267
  of waste biomass to bioenergy, 122–123
    microwave-irradiated catalytic conversion, 122
    oil-bath-mediated catalytic conversion, 122–123
Catalytic domain (CD), 123–124
Catalytic graphitization, 289, 290f
Catalytic wet air oxidation (CWAO), 370
Catalytic wet peroxide oxidation (CWPO), 370
  biomass-based catalysts in, 373–375
  biomass-based catalysts in, 375–377
  studies with lignocellulosic-based biomass catalysts, 374t, 376t
Catalyzed transesterification, 156–157
Cauliflower waste, 228
CB. *See* Conduction band (CB)
CBD. *See* Carbohydrate-Binding Domain (CBD)
CBF. *See* Charged bubble flotation (CBF)
CBH. *See* Cellobiohydrolases (CBH)
CBP. *See* Consolidated bioprocessing (CBP)
CBRN. *See* Chemical-biological-radiological and nuclear (CBRN)
CC. *See* Capital Cost (CC)
CCS. *See* Carbon dioxide capture and storage technology (CCS)
CD. *See* Catalytic domain (CD); Circular dichroism (CD)
CDs. *See* Carbon dots (CDs)
Cell recycle batch fermenter (CRBF), 124
Cellobiohydrolases (CBH), 47
Cellulases, 140, 201–202, 206
Cellulose, 24, 29–30, 42, 47, 87–88, 137, 201–202, 287f, 345–348, 362
  cellulose-based materials extracted from biomass, 346t
  cellulose-derived polyols, 347
  cellulose-rich lignocellulosic biomass, 43
  degrading enzymes, 201–202
  determination of, 118–119
  particles, 119
Cellulose nanocrystals (CNCs), 267, 345
Cellulose nanofibers (CNFs), 267, 345
Central Pollution Control Board (CPCB), 313
CEPT. *See* Chemically enhanced primary treatment (CEPT)
Cereals, 234
  waste, 235
Cesspits, 305–306
Charged bubble flotation (CBF), 314–315
Chemical oxygen demand (COD), 105–106
Chemical-biological-radiological and nuclear (CBRN), 377–378
Chemical(s), 341, 364
  activation, 366
    reaction conditions, 367t
  affinities, 237
  catalytic processes, 265
  hydrolysis, 140–141
  precipitation, 318
  pretreatment process, 23, 43
  residuals, 344
  treatments, 318
  vapor deposition, 289–290
Chemically enhanced primary treatment (CEPT), 315
Chemisorption methods, 369
Chemolithoautotrophic bacteria, 258
Chemolithotrophic bacteria, 258
Chipping process, 23
Chitin, 348
Chitosan, 348
Chlorella sorokiniana, 203
Chlorobenzenes, 307, 315–316
Chloroflexi, 316
5-Chloromethylfurfural (CMF), 128
CHNS/O elemental analysis, 369
CHP system. *See* Combined heat and power system (CHP system)
Chromium (Cr), 306
Chromobacterium vaccinii DSM 25150, 238
Circular bioeconomy, 263–264, 276
Circular dichroism (CD), 208
Circular economy, 246, 351
Cis-aconitate dehydrogenase (CAD), 221
Citizens health, 408t
Citobacter sp., 252
Citric acid (CA), 216–217
  biosynthetic production, 216–217
  global market with applications, 217
  metabolic pathway, 216, 219f
  physicochemical properties and brief history, 216
  production from microbial cell factories using renewable waste biomass, 217t
  structure, 216f
Climate change, 21, 26, 115
  bioenergy in combating, 115
Clostridium cellulolyticum, 146
Clostridium thermocellum, 46
CNCs. *See* Cellulose nanocrystals (CNCs)
CNFs. *See* Cellulose nanofibers (CNFs)
Co-pyrolysis, 256
COA-MIB. *See* Carbon dioxide Adsorption Metal Hydride Intermediate buffer (COA-MIB)
Coagulants, 314, 318, 341
  aids/flocculants, 344–348
Coagulation, 318, 341
Cobs, 342
COD. *See* Chemical oxygen demand (COD)
Codigestion process, 169–170, 410–411
Cold-Acclimation Proteins (Caps), 205–206
Collagen, 348
Combined heat and power system (CHP system), 1, 123, 197
Combustion, 1, 83, 125, 363
Commercial applications for bioenergy production, 100–101
Commercial biogas plants, 386–387
Commercially available pigments from agrowaste and drawbacks of using agrowastes, 240–241

Community impact, 409t
Composters, 275
Composting, 60, 264, 275–278, 411
Compound annual growth rate (CAGR), 152, 217
Concentrated acids, 44
Conducting polymers (PCs), 291
Conduction band (CB), 372
Congo red degradation, 347
Consolidated bioprocessing (CBP), 124, 142
Contaminants, 341
Contamination, 5
Continuous Stirred-Tank Reactor (CSTR), 392
Conventional bioenergy production methods, 88–90
   biochemical processes, 88–89
   thermochemical processes, 89–90
Conventional drying process, 92–94
Conventional methods, 55
Conventional sludge active system, 105
Conversion techniques, 116
Copper (Cu), 306
Corn, 85
   stalks, 345–346
Corynebacterium spp., 216–217
Cost–benefit analysis, 281, 388
Costs involved in biogas generation, 388
Cotton-based waste, 226
Covalent binding, 209
CPCB. *See* Central Pollution Control Board (CPCB)
CR. *See* Carbon residue (CR)
CRBF. *See* Cell recycle batch fermenter (CRBF)
CRISPR-Cas9-based gene manipulation technique, 76–77
CRISPR/Cas9 gene editing, 76–77
Crops, 166
   residues, 10, 85
Cross linking, 209
Cryogenic distillation method, 194
Cryptococcus citroformans, 216–217
CSTR. *See* Continuous Stirred-Tank Reactor (CSTR)
Cultivation of microalgae in municipal wastewater, 110–111
Cultured algal species, 108–109
Cupreavidue necator, 258
CWAO. *See* Catalytic wet air oxidation (CWAO)
CWPO. *See* Catalytic wet peroxide oxidation (CWPO)
Cyanobacteria, 120
Cyclic chlorinated hydrocarbons, 249

D

DAF. *See* Dissolved air flotation (DAF)
Dairy industries, 106
Decomposition process, 60
Deenabandhu, 386–387
Deep eutectic solvent pretreatment, 45
Deep Eutectic Solvents (DESs), 45
Degree of Polymerization (DP), 42, 117–118
   determination of, 119
Deinococcus radiodurans, 206

Dense metal membranes, 195
DESs. *See* Deep Eutectic Solvents (DESs)
Dibutylin, 320–321
Dichomitus squalens, 23
Diffuse layer (DL), 290
2,5-diformylfuran (DEF), 128
Digestate, 3, 386–387, 411
Digester volume, 390
Digesting process, 172
Dilute acid, 43
Dimethylformamide (DMF), 128
Dioxins, 249
Direct microbial fermentation (DMC), 124
Dissolved air flotation (DAF), 314
Dissolved oxygen (DO), 106
DL. *See* Diffuse layer (DL)
DLS approach. *See* Dynamic Light Scattering approach (DLS approach)
DM. *See* Dry matter (DM)
DMC. *See* Direct microbial fermentation (DMC)
DMF. *See* Dimethylformamide (DMF)
DO. *See* Dissolved oxygen (DO)
Dome plants, 386–387
Domestic discharges, 305–306
Domestic energy sources, 51
DP. *See* Degree of Polymerization (DP)
Drop-in fuels, 120, 126–127
Dry impregnation method, 369
Dry matter (DM), 306–308
Dry solids (DSs), 310
Dry torrefaction, 26
Drying, 92–94
   process, 42–43
   temperature, 279–280
DSs. *See* Dry solids (DSs)
Dual-shaft shredder, 279
Dunaliella salina, 204–205
Dyes, 341
Dynamic Light Scattering approach (DLS approach), 119

E

EAC. *See* Extruded activated carbon (EAC)
Earthworms, 60–61
EC. *See* Electrical conductivity (EC)
Economic analysis, 388
   of biogas production, 399
Economic sustainability, 406
Economic viability, 394–395
Edible oil, 156
Edible vegetable oil sources, 154
EDLC. *See* Electrical double-layer capacitance (EDLC)
EDS. *See* Energy-dispersive X-ray spectroscopy (EDS)
EES. *See* Energy storage systems (EES)
EGSB. *See* Expanded granular sludge blanket (EGSB)
Eichhornia crassipes, 135–136
EJ. *See* Exajoules (EJ)
Ekokymppi, 409
Electric power, 1
Electrical conductivity (EC), 106, 120, 289

Electrical double-layer capacitance (EDLC), 290
Electricity, incineration for, 13
Electro-fermentation, 269–273
Electroactive microbes, 76
Electrochemical hydrogen production, 192
Electrochemical process, 44, 192
Electrohydrogenesis method, 191
Electrokinetic remediation, 329–330
Electromicrobiomes, 76
Electromigration, 329–330
Electromodified biochar, 322
Electronic wastes (e-waste), 73–74
Electroosmosis, 329–330
Electrophoresis, 329–330
Electrospray ionization mass spectrometry, 109–110
ELMs. *See* Engineered Living Materials (ELMs)
EMF. *See* Ethoxymethylfurfural (EMF)
Emulsion, 157
Encapsulation, 209
Energy, 151
   crops, 178
   demand, 9
   energy-intensive drying process, 65
   resources, 116
   security, 409t
   shift, 385–386
   sources, 170
   storage material
      applications of biomass-derived carbon materials for energy storage systems, 290–299
      biomass generation, processing, and utilization into carbon materials, 286f
      production and properties of biomass-derived anode materials, 288–290
      suitable sources of biomass for electrode preparation, 286–287
Energy storage systems (EES), 286–287
Energy-dispersive X-ray spectroscopy (EDS), 369
Engineered Living Materials (ELMs), 76
Entamoeba histolytica, 308
Enterobacter ludwigii FMCC 204, 223
Enterovirus, 309
Entrapment, 209
Environmental, Social, and Governance (ESG), 406
Environmental pollution, 71
Environmental Pollution Prevention Act (1965), 311–312
Environmental Protection Agency (EPA), 16, 246
Environmental sustainability, 406–408
Enzymatic biodiesel synthesis, 155
Enzymatic conversion, 123–124
Enzymatic hydrolysis, 42, 50, 123, 141
Enzymatic pretreatment, 46–47, 139–140
Enzymatic transesterification, 158
Enzymes, 46, 139, 201, 207
   enzyme-catalyzed transesterification, 157
   modification using immobilization, 207–209
   pectinase, 202
   in plastic reduction, 252

Enzymes (*Continued*)
  for pretreatment and hydrolysis of lignocellulosic biomass, 140
EPA. *See* Environmental Protection Agency (EPA)
EPSs. *See* Extracellular polymeric substances (EPSs)
Equivalence ratio (ER), 90
ER. *See* Equivalence ratio (ER)
Escherichia coli, 146, 218
ESG. *See* Environmental, Social, and Governance (ESG)
Esterification, 158, 348
ETFE. *See* Poly-ethylene-co-tetrafluoroethylene (ETFE)
Ethanol, 224–227, 225f
  applications, 227
  biorefinery-based ethanol production, 226
  metabolic pathway, 226
  as organic solvent, 33–34
  pretreatment process, 34
Ethers, 368
  groups, 368–369
Ethoxymethylfurfural (EMF), 128
Ethyl alcohol, 224
EU. *See* European Union (EU)
European Union (EU), 246, 310
  Landfill Directive of 1999, 405
Exajoules (EJ), 165
Excessive nutrients, 341
Exothermic reactions, 90
Expanded granular sludge blanket (EGSB), 317
Experiment errors, 177
Extracellular polymeric substances (EPSs), 91–92, 251
Extraction, 233–234
  biochemical methods, 11–12
  mechanical methods, 13–15
    high voltage electric discharge, 14
    microwave-assisted extraction, 14–15
    pulse electric field, 13
    supercritical fluid extraction, 14
    ultrasound-assisted extraction, 14
  thermochemical methods, 12–13
  of value-added products from biowaste, 11–15
Extreme thermophiles, 205
Extremophiles, 205
Extremophilic enzymes, 204–207
  acidophilic enzymes, 206
  alkaliphilic enzymes, 206
  halophilic enzymes, 204–205
  psychrophilic enzymes, 205–206
  radiophilic enzymes, 206–207
    future prospects, 207
  thermophilic enzymes, 205
Extremophilic microbes, 76, 201
Extremozymes, 204–205
Extruded activated carbon (EAC), 365

F
Fabrication cost, 281
Face-centered cubic structure (FCC structure), 195
FAMEs. *See* Fatty acid methyl esters (FAMEs)
FAO. *See* Food and Agricultural Organization (FAO)
Farmland waste, 223–224
Fast pyrolysis process, 265, 322, 364
Fatty acid methyl esters (FAMEs), 127–128, 157
Fatty acids, 110
FCC structure. *See* Face-centered cubic structure (FCC structure)
FDCA. *See* Furandicarboxylic acid (FDCA)
Feasible raw materials, 123
Federation of Indian Chambers of Commerce and Industry (FICCI), 227
Fermentation, 3, 41–42, 124, 142
  evaluation of fermentation processes and end products, 62t
  of important platform chemicals, 269–273
  process, 62
Fermicutes, 316
FFAs. *See* Free fatty acids (FFAs)
Fiberglass-reinforced polyester (FRP), 174–175, 392
FICCI. *See* Federation of Indian Chambers of Commerce and Industry (FICCI)
First-generation biodiesel production, 160
First-generation bioethanol, 120
Fischer–Tropsch technique, 265
Fixed type biodigesters, 391
Fixed-bed reactors, 27, 317
Flash Joule heating method, 290
Flash pyrolysis, 322, 364
Flatworms, 310
Flavonoids, 267
Floating drum. *See* Gas holder
Floating drum-based biodigester, 391
Flocculants, 341
Flocculation, 318
Flowers, 234
  garbage, 235
  waste, 235
Fluids, 236
Food and Agricultural Organization (FAO), 233–234, 263–264
Food scraps, 386–387
Food waste (FW), 56–57, 60, 71–72, 178, 185–187, 215, 263–264, 270t
Food-grade biomass, 120
Forest management, 84–85
Forest source, 84–85
Forestry, 117
Formic acid (CH_2O_2), 140–141
  formic acid-based pretreatment, 345
Fossil fuels, 21, 115
Fourier-transform infrared spectroscopy (FTIR), 369
Franklin model, 289
Free fatty acids (FFAs), 155, 266–267
FRP. *See* Fiberglass-reinforced polyester (FRP)
Fruit vegetable wastes (FVWs), 390, 392
Fruits, 223, 234
  waste, 56, 235
FTIR. *See* Fourier-transform infrared spectroscopy (FTIR)
Fullerenes, 364
Functional materials, 364
  from biomass waste, 6
Functionalization of carbon materials, 369
Fungi, 23, 309
Furandicarboxylic acid (FDCA), 128
Furanic fuels, 121, 128
Furans, 42, 249
FVWs. *See* Fruit vegetable wastes (FVWs)
FW. *See* Food waste (FW)

G
GAC. *See* Granular activated carbon (GAC)
Galdiera sulphuraria, 76
Galvanized steel (GS), 174–175, 392–393
Garden waste (GW), 57
Gas holder, 175
Gaseous fuels, 121–122, 128–129
  biohydrogen, 122, 129
  biomethane, 122, 129
Gaseous sulfur dioxide, 37
Gasification, 1, 13, 63, 86, 90, 120, 188, 250, 265, 322, 363, 368
GCV. *See* Gross Calorific Value (GCV)
Gelatine, 348
Gene editing technologies, 76–77
Genetic manipulations, 76–77
Genome engineering techniques, 109–110
Genomic DNA, 201
Geobacter sulfurreducens, 76
GHG emissions. *See* Greenhouse gas emissions (GHG emissions)
GHs. *See* Glycosyl hydrolases (GHs)
Giardia lambia, 308
Global Reporting Initiative (GRI), 406
Global warming, 21, 165–166
Gloeophyllum trabeum, 46
Glucoamylase, 206
Glucose, 202
Glycerol, 269–273
Glycerol-choline chloride, 45
Glycol, 222–224
Glycosyl hydrolases (GHs), 210
Gobar gas plant, 391
Gouy–Chapman–Stern model, 290
Government agencies, 385–386
Granular activated carbon (GAC), 318, 365
Graphene, 286
Graphene oxide, 291
Graphite, 286, 288, 292–293, 364
Graphitic carbon, 6
Grass lignin, 137
Grease trap waste, 158
Green chemistry, 122
Green energy production, 10
Green hydrogen production, 185
Green technology, 190
Greenhouse effects, bioenergy in combating, 115
Greenhouse gas emissions (GHG emissions), 11, 26, 32–33, 60, 74–75, 83, 110, 120–121, 126–127, 129, 143, 249, 386
GRI. *See* Global Reporting Initiative (GRI)
Grinding processes, 23
Grit, 313

Gross Calorific Value (GCV), 86
GS. *See* Galvanized steel (GS)
5′-guanylic acid (5′-GMP), 204–205
GW. *See* Garden waste (GW)

H
Halophiles, 204–205
Halophilic enzymes, 204–205
HDPE. *See* High-density polyethylene (HDPE)
Heat
  incineration for, 13
  transfer, 36
Heating values, 86
  HHV of biomass fuels, 87t
Heavy metals (HMs), 306, 341
  extraction from sewage wastes, 329–330
Hemicellulases, 140, 201–202
Hemicelluloses, 24, 44, 118–119, 137, 287f, 362
  determination of, 118–119
  distillation, 33
Hepatitis A and E virus, 309
Heterogeneous catalysts, 267–268, 364
Heterogeneous Fenton oxidation. *See* Catalytic wet peroxide oxidation (CWPO)
Heterogeneous material, 1
Heterogeneous reactions, 63
Heterogeneous waste glass catalyst, 155
Heterotrophic bacterial species, 106
Hexachlorocyclohexane isomers (HCH isomers), 17
Hexose, 203
HHV. *See* High Heating Value (HHV)
Hibiscus sabdariffa, 235–236
High Heating Value (HHV), 86
High voltage electric discharge (HVED), 14
High-density polyethylene (HDPE), 246
High-value bio-products from biomass waste, 15–17
Histidine (His), 208
HMF. *See* Hydroxymethylfurfural (HMF)
HMs. *See* Heavy metals (HMs)
Hot water extraction, 25–26
Household chemicals, 307
Household organic waste, 56
  waste generation and compositional variation in, 56
HRT. *See* Hydraulic retention time (HRT)
HTC. *See* Hydrothermal carbonization (HTC)
HTC product. *See* Hydrothermally carbonized product (HTC product)
HTL. *See* Hydrothermal liquefaction (HTL)
Human waste, 386–387
Husks, 342
HVED. *See* High voltage electric discharge (HVED)
Hybrid supercapacitors, 290
Hydraulic retention time (HRT), 168–169, 315, 390
Hydrochar, 323–327, 326t, 366. *See also* Biochar
Hydrochloric acid (HCl), 140–141

Hydrogen (H_2), 185–187, 256, 368
  from biomass-derived biowastes, 185–187
  distribution, 196–197
  economy, 185, 187
  electrochemical production, 192
  end use, 196–197
  from industrial emissions and side-stream gases, 188–189
  production processes from biowastes and industrial waste gases, 187–192
  production technologies, 112, 188
  purification method, 193–194
  separation and purification, 193–196
    cryogenic distillation, 194
    dense metal membranes, 195
    metal hydrides, 194–195
    polymeric membranes, 195–196
    pressure swing adsorption, 193–194
Hydrogen peroxide, 118
Hydrogen sulfide (H_2S), 169
Hydrogen-to-carbon ratio (H/C ratio), 123
Hydrogenation, 250
Hydrolysis, 3, 41–42, 316
  enzymes for hydrolysis of lignocellulosic biomass, 140
  process, 24–26
Hydrolytic enzyme, 202
Hydropyrochars, 373
Hydrothermal carbonization (HTC), 4, 63, 322, 366
Hydrothermal conversion process, 25–26
Hydrothermal liquefaction (HTL), 13, 63, 90, 120, 125–126
Hydrothermal pretreatment, 94
Hydrothermal processing, 363
Hydrothermal treatment, 63, 92, 256–257
Hydrothermally carbonized product (HTC product), 323–325
Hydrous pyrolysis. *See* Hydrothermal liquefaction (HTL)
2-hydroxy-propane-1,2,3-tricarboxylic acid, 216
Hydroxyl radicals (•OH), 370
Hydroxymethylfurfural (HMF), 128, 146

I
IA. *See* Itaconic acid (IA)
IAEA. *See* International Atomic Energy Agency (IAEA)
IC. *See* Internal circulation reactor (IC)
ICP-Ms. *See* Inductively coupled plasma methods (ICP-Ms)
Ideonella sakaiensis, 252
IEA. *See* International Energy Agency (IEA)
IFAS. *See* Integrated fixed-film activated sludge (IFAS)
ILs. *See* Ionic liquids (ILs)
Imazapic herbicides, 370
Imazapyr herbicides, 370
Immobilization, 207–208
  enzyme modification using, 207–209
  modes, 208–209
    affinity binding, 208
    covalent binding, 209
    cross linking, 209
    encapsulation, 209
    entrapment, 209
    ionic bonding, 208
Impregnation, 369
Incenerated MSW (MSWI), 187
Incineration, 13, 62–63, 245–246, 250, 257
  for heat and electricity, 13
  process, 62–63
Inductively coupled plasma methods (ICP-Ms), 369
Industrial biocatalyst, 16
Industrial biodiesel, 158–159
Industrial discharges, 305–306
Industrial emissions and side-stream gases, hydrogen from, 188–189
Industrial source, 152
Industrial wastes, 73, 155, 307
  biological processes, 190–191
  chemical and thermal, 188–189
    hydrogen from industrial emissions and side-stream gases, 188–189
  electrochemical processes, 192
  hydrogen production processes from, 187–192
  photochemical processes, 191–192
Industrialization, 105
Industrially relevant enzymes, 203
  bioprospecting of, 201–203
  production of biofuel from waste water using algal strains, 203
  types, 201–203
    amylase, 202
    cellulase/hemicellulase, 201–202
    laccases, 202
    lipases, 202
    pectinase, 202
    proteases, 202–203
    yeast enzymes, 203
Inorganic carbon, 106
Inorganic coagulants, 344
Integrated fixed-film activated sludge (IFAS), 316
Internal circulation reactor (IC), 317
International Atomic Energy Agency (IAEA), 73
International Energy Agency (IEA), 135, 152, 386
Ionic bonding, 208
Ionic liquids (ILs), 29, 44, 138
  application, 29–30
  bioenergy precursor production from waste biomass using PIL, 45f
  pretreatment process, 29–30, 44–45
Ionic solution, 26
Iron salts, 314
Isoelectric points (pI), 206
Itaconic acid (IA), 220
  biosynthetic production, 221
  global market with applications, 221
  metabolic pathway, 221
  physicochemical properties and brief history, 220
  production from microbial cell factories using renewable waste biomass, 221t
  structure, 220f

J
JANATHA, 386–387, 392
Joule effect, 290

K
Khadi Village Industrial Commission (KVIC), 386–387, 392
Klebsiella oxytoca, 146
Kluyveromyces
  K. fragilis, 203
  K. lactis, 203
Kosakonia sp., 252
Kraft Lignin-based sorbent materials, 345
Kraft process, 31
KVIC. *See* Khadi Village Industrial Commission (KVIC)

L
5-L digester, 392
Laccases, 139–140, 202
Lacticaseibacillus casei, 146
Lactones, 368
Land disposal, 307
Land use impacts, 407t
Landfilling of SS, 310
Landfills, 64–65, 245–246
Lantana camara, 135–136
LCB. *See* Lignocellulosic biomass (LCB)
LCC. *See* Life Cycle Cost (LCC)
LDPE. *See* Low-density polyethylene (LDPE)
Leaves, 342
Levulinic acid, 269–273
LHV. *See* Low Heating Value (LHV)
LHW. *See* Liquid hot water (LHW)
LIBs. *See* Lithium-ion batteries (LIBs)
Life Cycle analyses/assessment (LCA), 120, 406
Life Cycle Cost (LCC), 394–396, 400
LIG. *See* Lower-income groups (LIG)
Lignin, 21, 24, 87–88, 136–137, 202, 287f, 345–348, 362
  determination of lignin content, 118–119
  removal, 42–43
Lignin peroxidase (LiP), 139–140
Ligninolytic enzymes, 139
Lignocellulose. *See* Lignocellulosic biomass (LCB)
Lignocellulosic biomass (LCB), 1, 21, 26, 30, 42, 87–88, 120, 137, 362, 364, 370
  enzymes for pretreatment and hydrolysis of, 140
  pretreatment process of, 22f
  waste composition, 363t
Lignocellulosic materials, 135
Lignocellulosic wastes, 73, 116. *See also* Industrial wastes
  biochemistry of lignocelluloses, 136–137
  biological pretreatment method, 139
  combined pretreatment, 138
  different enzymes for pretreatment and hydrolysis of lignocellulosic biomass, 140
  diversity and potentiality of lignocellulosic residue, 135–136, 136t
  enzymatic pretreatment, 139–140
  fermentation, 142
  future scope, 145–146
  lignocellulosic bioethanol, 143–144
  pilot plant, 142
  pretreatment methods of biomass for 2G bioethanol, 137
  saccharification, 140–141
  2G bioethanol production, 145–146
  supply chain management of lignocellulosic bioethanol, 144–145
Lignocellulosics, 135
"Like dissolves like" principle, 222
Likert scale, 407, 409
Lime pretreatment, 32
LiP. *See* Lignin peroxidase (LiP)
Lipases, 157, 202
Lipids, 156, 263–264, 267
  lipid-rich edible oil sources, 154
  lipid-rich feedstocks for large-scale biodiesel production
    agro/vegetable source, 154
    animal source, 155
    industrial waste, 155
    lipid-rich sources, 155–156
    microalgal source, 155
    potential of, 153–156, 154t
    waste oil, 156
    waste water, 155–156
  lipid-rich nonedible plants, 154
Liquefaction, 363
Liquid biofuels, 127
Liquid chromatography, 109–110
Liquid fuels, 120–121, 127–128
  biodiesel, 121, 127–128
  bioethanol, 120, 127
  furanic fuels, 121, 128
  synthesis of liquid biofuels from waste lignocellulosic biomass, 121f
Liquid hot water (LHW), 43, 138
Lithium-ion batteries (LIBs), 292–294
  comparative electrochemical performance of biomass-based carbon material anodes for, 295t
Livestock, 57–58
  waste, 57–58
Low Heating Value (LHV), 86
Low-density polyethylene (LDPE), 246
Lower-income groups (LIG), 276

M
MAAs. *See* Micosporine-like amino acids (MAAs)
Macroalgae, 117
MACT. *See* Microwave-assisted catalytic torrefaction (MACT)
MAE. *See* Microwave-assisted extraction (MAE)
MAG. *See* Microwave-assisted gasification (MAG)
Magnesium, 307–308
  salts, 350
Magnesium ammonium phosphate (MAP), 350
Magnetic biochar, 322–323
MAHT. *See* Microwave-assisted hydrothermal pretreatment (MAHT)
MAHTL. *See* Microwave-assisted hydrothermal liquefaction (MAHTL)
Majasaari waste center, 411
Management of waste biomass, 1
Manganese peroxidase (MnP), 139–140
Mangifera indica, 350–351
MAP. *See* Magnesium ammonium phosphate (MAP); Microwave-assisted pyrolysis (MAP)
Marine disposal, 307
MAT. *See* Microwave-assisted torrefaction (MAT)
Material recovery, 362
MB. *See* Methylene blue (MB)
MBBR. *See* Moving-bed bioreactor (MBBR)
MBDs. *See* Metal Biogas Digesters (MBDs)
MBM. *See* Meat and bone meal (MBM)
MBRs. *See* Membrane bioreactors (MBRs)
MCOs. *See* Multi-copper oxidases (MCOs)
Mealworm in plastic reduction, 252
Meat and bone meal (MBM), 348
MEC. *See* Microbial electrolytic cell (MEC)
Mechanical pretreatment, 42–43, 47–49
  drying, 42–43
  milling, 43
  sonication, 43
Mechanical recycling, 250
MEF. *See* Moderate electric field (MEF)
Melanin, 233–234
Membrane bioreactors (MBRs), 305–306, 316
Mesophilic digestion process, 169
Mesophilic tube digester, 386–387
Metabolic engineering approaches, 17–18
Metabolomics, 74
Metagenomics, 74, 210–212
  oxidoreductases, 211–212
  phosphatises, 210
  protease, 210
Metal Biogas Digesters (MBDs), 392
Metal hydrides (MHs), 194–195
Metal oxide catalysts, 268
Metal-organic frameworks (MOFs), 193–194, 268–269
  transformation of platform molecules utilizing MOF based catalyst, 268–269
Metaomics-based approaches, 76–77
Metaproteomics, 74
Metatranscriptomics, 74
Methane (CH_4), 3, 394
  production, 179
Methanization of syngas, 330–331
Methano pyruskandleri, 205
Methanobacterium, 46
Methanococcus, 46
Methanogenesis, 317
Methanogenic bacteria, 176
Methanol as organic solvent, 33
Methanosaeta, 317
Methanosarcina, 317
Methyl ester, 155
Methyl orange (MO), 372
Methylene blue (MB), 370
Methylfurfural (MF), 128

2-Methylidenebutanedioc acid ($C_5H_6O_4$), 220
MF. See Methylfurfural (MF)
MFCs. See Microbial fuel cells (MFCs)
MHET. See Mono (2-hydroxyethyl) terephthalic acid (MHET)
MHs. See Metal hydrides (MHs)
Micosporine-like amino acids (MAAs), 109–110
Microalgae, 11, 105, 117, 155, 203
   cultivation, 105
      different biorefinery approaches for, 111t
      in municipal wastewater, 110–111
Microalgal cultivation in municipal wastewater
   biorefinery approaches of, 109–110
   challenges in, 112
Microalgal source, 155
Microbe-based bionanomining, 77
Microbes, 60–61, 74
Microbial batteries, 76
Microbial biotechnology, 254
Microbial electrolytic cell (MEC), 129, 191
Microbial enzymes, 201, 210
Microbial fuel cells (MFCs), 76
Microbial metabolic pathways, 71
Microbial pigments, 17
Microbial pretreatment, 46
Microbial sources, 152
Microemulsion, 157
Microorganisms, 3, 23, 42, 50, 88
   microorganism-based coagulants, 348
   in plastic reduction, 251
   production of pigments
      from agrowastes without using, 237
      from microorganisms using agrowastes as substrate, 238–239, 239t
Microplastics, 245–246
Microwave plasma gasification system, 99
Microwave pyrolysis, 90, 322
Microwave technology, 122
Microwave-assisted bioenergy production, 96–100
   microwave-assisted gasification, 99–100
   microwave-assisted hydrothermal liquefaction, 100–102
   microwave-assisted pyrolysis, 96–97
   microwave-assisted torrefaction, 98–99
Microwave-assisted catalytic torrefaction (MACT), 94
Microwave-assisted drying, 92–94
Microwave-assisted extraction (MAE), 14–15
Microwave-assisted gasification (MAG), 99–100
Microwave-assisted heating, 83, 100–102
Microwave-assisted hydrothermal liquefaction (MAHTL), 100
Microwave-assisted hydrothermal pretreatment (MAHT), 94–96
Microwave-assisted pretreatment of biomass, 90–96
   biochemical process, 91–92
   thermochemical process, 92–96

Microwave-assisted pyrolysis (MAP), 96–97, 100–101
Microwave-assisted torrefaction (MAT), 94, 98–99
Microwave-irradiated catalytic conversion, 122
Microwave-irradiated technology, 122
Middle-income group (MIG), 276
MIG. See Middle-income group (MIG)
Milling process, 23, 43
Mineral oil, 122–123
Mitochondrial tricarboxylic transporter (MTTA), 221
Mixed matrix membrane (MMM), 196
Mixed rice waste (MRW), 171–172, 390
MMM. See Mixed matrix membrane (MMM)
MnP. See Manganese peroxidase (MnP)
MO. See Methyl orange (MO)
Mobile type biodigesters, 391
Moderate electric field (MEF), 236
MOFs. See Metal-organic frameworks (MOFs)
Moisture content, 27, 86
   effect of, 27–28
Molecular biotechnology tools for waste valorization, advances in, 75–77
Molecular hydrogen, 191
Monascus purpureus EBY3, 238
Monascus purpureus FTC5356, 238
Mono (2-hydroxyethyl) terephthalic acid (MHET), 252
Monoalkyl fatty acid esters, 151–152
Monobutylin, 320–321
Moorella thermoacetica, 258
Moving-bed bioreactor (MBBR), 316
MRW. See Mixed rice waste (MRW)
MSW. See Municipal solid waste (MSW)
MTTA. See Mitochondrial tricarboxylic transporter (MTTA)
Multi-copper oxidases (MCOs), 211
Multi-use plastics, 246
Municipal sewage, 305–306
Municipal solid waste (MSW), 1, 9, 55, 85–86, 185–187, 226, 275, 405
Municipal wastes, 11, 73–74
   biorefinery approaches of microalgal cultivation in municipal wastewater, 109–110
   challenges in microalgal cultivation in municipal wastewater, 112
   composition of municipal wastewater, 106–109
   cultivation of microalgae in municipal wastewater, 110–111
Municipal wastewater, 105–106
   physicochemical parameters of, 107t
2-mythylidenesuccinic acid. See Itaconic acid (IA)

N

Nanocellulose (NC), 267, 345
Nanomaterials, 110
Nanoparticles (nps), 50, 77, 110
Nanotechnology, 50
Nanotubes, 364

Naphthalene, 307
National grids, 166
Natural gas, 122
Natural graphite, 289
Natural resources, 71
NC. See Nanocellulose (NC)
NCV. See Net Calorific Value (NCV)
Net Calorific Value (NCV), 86
Net present value (NPV), 159, 394–395, 400
Next-generation biological conversion techniques, 55
NIBs. See Sodium-ion batteries (NIBs)
Nickel (Ni), 306
Nitric acid (HNO_3), 140–141, 369
Nitro compounds, 368
Nitrogen (N), 276–278, 307–308, 320–321, 368–369
Nitrogen oxides (NO_2), 63
Nitrosococcus, 315
Nitrosomonas, 315
NMVOCs. See Nonmethane VOCs (NMVOCs)
Non-polar polymers, 254
Non-renewable fossil fuels, 264
Noncatalyzed transesterification, 157
Noncorrosive catalysts, 122
Nonedible vegetable oil sources, 154
Nonmethane VOCs (NMVOCs), 188–189
Nonrenewable resources, 83
Nontoxic wastes, 215
Nonuniform loading rate effect, 396–399
Novel enzymes, bioprospecting of, 201–203
NPV. See Net present value (NPV)
Nutrient cycle, 405
Nutrients, 3, 307–308
   nutrient-rich environment, 203
   nutrient-rich fertilizers, 60
   nutrient-rich wastewater, 105
   recovery efficiency, 276–278

O

OC. See Organic carbon (OC)
Oil-bath-mediated catalytic conversion, 122–123
Oilseed rape straw (OSR), 372
OLR. See Organic loading rate (OLR); Rate of organic loading (OLR)
Omics, 17–18
Omics-based technologies, 76–77
OMSW. See Organic municipal solid waste (OMSW)
Operational cost, 281
OPH. See Organophosphorus hydrolase (OPH)
Orange peels, 234
Organic acids, 42, 215–221
   citric acid, 216–217
   pretreatment by, 34
   succinic acid, 217–220
Organic biomass, 3
Organic carbon (OC), 106, 320–321
Organic contaminants, 315–316
Organic debris, 3
Organic fertilizer, 275

Organic loading rate (OLR), 175–176, 315, 392
Organic materials, 4, 10
  residues, 85–86
Organic matter, 106
Organic molecules, 84
Organic municipal solid waste (OMSW), 388
Organic polymer flocculants, 344
Organic solid waste management, 11
Organic solvents, 221–228
  acetone as, 34
  butane-2,3-diol, 222–224
  ethanol as, 33–34
  methanol as, 33
  peracetic acid as, 34
  physical properties, 222t
  pretreatment process, 26, 32–35
    application of organic solvent, 33
    pretreatment using alcohol as organic solvent, 33–34
Organic substances, 341
Organic waste digestion (OWD), 166
Organic wastes, 56, 62–63, 71–72
  anaerobic digestion process, 166–168
  biogas quantity measurement, 176–177
  biogas worldwide development, 165–166
  characteristic, 391t
  experimental procedure, 175–176
    pilot study, 176
    trail study, 176
  experimental set, 391f
  materials and methods, 170–177
    evaluation of biowaste characteristics, 172–174
    experimental setup, 174–175
    feedstock survey, 170–172
  physical and chemical characteristics, 390–391
  results, 177–181
    chemical characterization of feedstocks, 178–179
    experimental findings of trial and pilot studies, 179
    effect of nonuniform loading rate in pilot study, 180–181
    physical characterization of feedstocks, 178
    sensitivity analysis, 181
  selection of, 389–390
  uncertainty analysis, 177
Organophosphorus hydrolase (OPH), 208
OSR. See Oilseed rape straw (OSR)
OWD. See Organic waste digestion (OWD)
Oxidation, 348
  ponds, 316
  process, 108
Oxidative radicals, 43
Oxidizing agent, 34
Oxidoreductases, 211–212
Oxo-degradation technology, 249–250
Oxygen, 368
  sensitivity, 190–191
  supply, 63
Oxygenated organic compounds, 192
Oxygenated VOCs, 189

Ozone, 26
  pretreatment, 27–28
Ozonolysis, 27
  application, 27
  influence of different process parameters, 27–28
    effect of moisture content, 27–28
    effect of particle size, 28
    effect of pH, 28
    effect of reactor design, 27
  pretreatment process, 26–28

P

PAC. See Powdered activated carbon (PAC)
Pachysolen tannophilus, 142
PAHs. See Polycyclic aromatic hydrocarbons (PAHs)
Paints, 233–234
Palletization process, 25
Palm and nut shells, 364
Palm oil, 153
Paper mill sludge, 345
Paramecium, 315–316
Parasites, 309
Parasitic worms, 308
Passive sorption, 329
Patchouli biomass, 375
Pathogens, 308
  contamination, 11
Payback Period (PBP), 394–395
Payback time (PBT), 159, 408t
PBAT. See Poly (butylene adipate-co-terephthalate) (PBAT)
PBB. See Polybrominated biphenyls (PBB)
PBP. See Payback Period (PBP)
PBT. See Payback time (PBT)
PC. See Production capacity (PC); Pseudocapacitance (PC)
PCR. See Polymerase chain reaction (PCR)
PCs. See Conducting polymers (PCs)
PDBE. See Polybrominated diphenyl ethers (PDBE)
PE. See Polyethylene (PE)
Pectin, 202, 347
Pectinases, 140, 202
Pectinatus sp., 218
PEF. See Primary effluent filtration (PEF); Pulse electric field (PEF)
Penicillium funiculosum, 46
PEP. See Phosphoenolpyruvate (PEP)
Peracetic acid as organic solvent, 34
Peroxidases, 140
Pesticides, 307
PET. See Polyethylene terephthalate (PET)
Petrochemical plastics, 249
Petroleum, 129
  industry wastewater, 73
  oils, 116
  refinery, 50
  wastes, 74–75
PFRP. See Process to further reduce pathogens (PFRP)
PGBC. See Porous graphite biochar (PGBC)
pH neutralizers, 341
PHA. See Polyhydroxyalkanoate (PHA)

Phaeodactylum tricornutum, 17–18
Phanerochaete chrysosporium, 23
Pharmaceuticals, 341
Pharmacology, 377–378
PHB. See Polyhydroxy butyrate (PHB)
PHC. See Polyhydroxy octanoate (PHC)
Phenolics, 42, 100
Phenols, 368
PHH. See Polyhydroxy hexanoate (PHH)
Phosphatases, 210
Phosphates, 368
Phosphoenolpyruvate (PEP), 218
Phosphoric acid (H_3PO_4), 140–141
Phosphorus (P), 276–278, 307–308, 320–321
Photo reforming, 191
Photoautotrophic growth, 106
Photocatalytic degrdation of organic pollutants by biomass-derived catalysts, 371t
Photocatalytic reaction, 192
Photocatalytic treatment of wastewater by biomass-derived catalysts, 370–373
Photochemical processes, 191–192
Photofermentation process, 190
Photogenerated electrons, 191
Photoinhibition, 108–109
PHT. See Polyhydroxyalkanoates (PHT)
Phthalates, 249
PHV. See Polyhydroxy valerate (PHV)
Phycocyanin, 233–234
Phycoerythrobilin, 233–234
Physical activation, 365
  reaction conditions, 367t
Physical adsorption, 318
Physical pretreatment methods, 42–43
Physicochemical characteristics, 294, 297
Physicochemical conversion, 363
Physicochemical pretreatment, 43–47, 49
  acid pretreatment, 43–44
  alkali pretreatment, 44
  deep eutectic solvent pretreatment, 45
  ionic liquid pretreatment, 44–45
  process, 24, 47, 49
  steam explosion, 45–46
  supercritical CO_2, 46
Physisorption methods, 369
pI. See Isoelectric points (pI)
Pichia stipites, 142, 203
Pigments, 110, 233–234
  challenges and future aspects, 241
  commercially available pigments from agrowaste and drawbacks of using agrowastes, 240–241
  methods, 236–237
    biological methods, 237
    chemical methods, 237
    physical methods, 236–237
  process of pigment production from agrowaste, 240, 240f
  produced from agrowastes, 235–236, 236t
  production of pigments
    from agrowastes without using microorganisms, 237
    from microorganisms using agrowastes as substrate, 238–239

Pilot plant, 142
Pilot study, 396–399
PILs. See Protic ionic liquids (PILs)
Pinus pinaster, 347
Pistachio shells–based AC, 370
PLA. See Polylactic acid (PLA)
Planococcus sp. TRC1, 238
Plant cell wall, 24
Plastic waste, 245
  as feedstock for potential product formation, 257–258
  modification of plastics for waste management, 253–254
  pre-treatment by thermochemical depolymerization, 255
  renewal process for utilization of multiple use plastics waste, 254–255
  strategies for management, 250, 250f
  synergistic approach for plastic waste conversion, 258–259
Plasticizers, 249
Plastics, 245
  biodegradable plastics, 249–250
  and classification, 246–248
  conversion strategies using biotechnology, 254
  ecology and environment, 259
  enzymes in plastic reduction, 252
  hydrothermal treatment, 256–257
  incineration, 257
  mealworm in plastic reduction, 252
  micro-organisms in plastic reduction, 251
  polylactic acid-based polymer, 253
  polymers and properties, 247t
  problems associated with conventional plastic usage and disposals, 249
  pyrolysis, 255–256
  reduction strategies, 251
  replacement of conventional plastics using biotechnology, 252
  solvothermal process, 257
  starch-based polymer, 253
Platform molecules, 263
Pleurotus ostreatus, 138
PMMA. See Poly methyl methacrylate (PMMA)
Poly (butylene adipate-co-terephthalate) (PBAT), 245–246
Poly (butylene glycol adipate), 253
Poly (p-phenylene) (PPT), 328
Poly methyl methacrylate (PMMA), 221
Poly-ether ketone, 254
Poly-ethylene-co-tetrafluoroethylene (ETFE), 254
Polyaramid membranes, 195–196
Polyaromatic hydrocarbons, 315–316, 320–321
Polybrominated biphenyls (PBB), 257
Polybrominated diphenyl ethers (PDBE), 249
Polycyclic aromatic hydrocarbons (PAHs), 249
Polyelectrolytes, 318
Polyethylene (PE), 245–246, 392
Polyethylene terephthalate (PET), 245–246
Polyhydroxy butyrate (PHB), 251
Polyhydroxy hexanoate (PHH), 251

Polyhydroxy octanoate (PHC), 251
Polyhydroxy valerate (PHV), 251
Polyhydroxyalkanoate (PHA), 245–246, 328
Polylactic acid (PLA), 245–246, 328
  polylactic acid-based polymer, 253
Polymer-derived carbon, 364
Polymerase chain reaction (PCR), 205
Polymeric membranes, 195–196
Polymers, 364
Polypropylene (PP), 246
Polypyrrole (PPy), 370
Polysaccharide, 201–202
Polystyrene (PS), 246
Polystyrene sulfonic acid (PSSA), 254
Polytetrafluoroethylene (PTFE), 246
Polyunsaturated fatty acids, 108
Polyvinyl alcohol (PVA), 209
Polyvinyl chloride (PVC), 246
Porous graphite biochar (PGBC), 372
Posidonia oceanica, 350
Potassium (K), 276–278, 307–308, 320–321
Potassium sulfate (K_2SO_4), 269–273
Potassium-ion battery (KIB), 297–299
  comparative electrochemical performance of bio-based carbon material anodes for, 299t
Potato peel waste (PPW), 226
Powdered activated carbon (PAC), 318, 365
PP. See Polypropylene (PP)
PPT. See Poly (p-phenylene) (PPT)
PPW. See Potato peel waste (PPW)
Preaeration grit tanks, 313–314
Precipitation, 318
Prehydrolysis and Simultaneous Saccharification and Fermentation (PSSF), 142
Preliminary sewage treatment, 313–314
Pressure swing adsorption technology (PSA technology), 193–194
Pretreatment
  cons, 48–50
    biological pretreatment, 49–50
    mechanical pretreatment, 48–49
    physicochemical pretreatment, 49
  enzymes for pretreatment of lignocellulosic biomass, 140
  in generating bioenergy from waste biomass, 41–42
  methods of biomass for 2G bioethanol, 137
  by organic acids, 34
  of organic wastes, 13
  pros, 47–48
    biological pretreatment, 47–48
    mechanical pretreatment, 47
    physicochemical pretreatment, 47
  with sodium hydroxide, calcium hydroxides, and sodium carbonate, 31
  using alcohol as organic solvent, 33–34
  of waste biomass, 42–47
    alkaline pretreatment process, 30–32
    biological pretreatment, 23, 46–47
    biomass torrefaction, 25–26
    cellulose, 24
    chemical pretreatment process, 23
    different pretreatment processes, 23–24

  goals of pretreatment process, 24
  hemicelluloses, 24
  ionic liquid pretreatment process, 29–30, 29t
  lignin, 24
  mechanical pretreatment, 42–43
  organic solvent pretreatment process, 32–35
  ozonolysis pretreatment process, 26–28
  physical pretreatment process, 23
  physicochemical pretreatment, 24, 43–46
  steam explosion pretreatment process, 35–37
  types of biomasses, 21–23
  typical biomass component, 24
Prices paid for individual units of electricity, 401
Primary clarifiers, 314
Primary effluent filtration (PEF), 314–315
Primary sewage treatment, 314–315
Process feasibility analysis, 281
Process to further reduce pathogens (PFRP), 308
Process to significantly reduce pathogens (PSRP), 308
Prodigiosin, 233–234, 238
Production capacity (PC), 159
Proteases, 140, 202–203, 210
Proteins, 263–264
Proteobacteria, 315
Protic ionic liquids (PILs), 44
Protozoa, 308
PS. See Polystyrene (PS)
PSA technology. See Pressure swing adsorption technology (PSA technology)
Pseudocapacitance (PC), 290
Pseudocapacitive processes, 290–291
Pseudomonas, 315
  P. stutzeri, 77
Pseudomonas aeruginosa. See Rhamnolipid (*Pseudomonas aeruginosa*)
PSRP. See Process to significantly reduce pathogens (PSRP)
PSSA. See Polystyrene sulfonic acid (PSSA)
PSSF. See Prehydrolysis and Simultaneous Saccharification and Fermentation (PSSF)
Psychrophilic enzymes, 205–206
Psychrophilic microorganisms, 205–206
PTFE. See Polytetrafluoroethylene (PTFE)
Pulse electric field (PEF), 13
Puolanka Municipality, 409
PVC. See Polyvinyl chloride (PVC)
Pyrochars, 373
Pyrolysis, 4, 13, 63, 89, 120, 125–126, 157, 188, 250, 255–256, 264–265, 278, 285, 307, 321, 363–364
Pyrolytic carbon, 364

Q

Quercus ilex, 347

R

Radar diagram, 414, 414f
Radioactive wastes, 74

Radiophiles, 206
Radiophilic enzymes, 206–207
Ralstonia eutropha H16, 258–259
Rapid treatment, 275
Rate of organic loading (OLR), 168
RBC. *See* Rotating biological contractor (RBC)
Reactive oxygen species (ROS), 370
RECENT project. *See* Renewable Energy Empowerment in Northern territories project (RECENT project)
Recombinant DNA technology, 17, 201
Recovery rate (RR), 276–278
Recycle, redesign, remanufacture, recovered, reuse, and reduce principle (6R principle), 259
Recycling of plastic, 245–246, 250
Reduction strategies, 251
Renewable energy, 50, 57–58, 115, 121, 144, 185
  demand of, 385–386
Renewable Energy Empowerment in Northern territories project (RECENT project), 407
Renewable energy sources (RESs), 15, 165–166
Renewable material, 185–187
Renewable sources, 83, 190, 263
Renewal process for utilization of multiple use plastics waste, 254–255
Reovirus, 309
Research institutes, 385–386
Residual material, 167
Resistance temperature detector sensors (RTD sensors), 174–175
Resource recovery methods, 71–72
RESs. *See* Renewable energy sources (RESs)
Return on investment (ROI), 159
Rhamnolipid (*Pseudomonas aeruginosa*), 17
Rhizomucor miehei, 16
Rhodamine B, 370
Rhodococcus
  R. erythropolis, 17, 77
  R. opacus, 77
Rhodotorula, 203
Riboflavin, 233–234
Rice straw (RS), 138, 345–346
Rice waste (RW), 171–172, 390
Robur, 347
ROI. *See* Return on investment (ROI)
ROS. *See* Reactive oxygen species (ROS)
Rotating biological contractor (RBC), 305–306
Rotavirus, 309
RR. *See* Recovery rate (RR)
RS. *See* Rice straw (RS)
RTD sensors. *See* Resistance temperature detector sensors (RTD sensors)
RW. *See* Rice waste (RW)

S

SA. *See* Succinic acid (SA)
SAA. *See* Soaking Aqueous Ammonia (SAA)
Saccharification, 140–141
  biological hydrolysis, 141
  chemical hydrolysis, 140–141
Saccharomyces
  S. carlsbergensis, 203
  S. cerevisiae, 142, 146, 201–202, 258–259
  S. lypolytica, 216–217
Salmonella, 308–309
  S. typhi, 309
Sanitary landfilling, 64–65
Saprophytes, 139
Sargassum natans, 348
Saturated liquid, 46
SBPL. *See* Sugar beet pulp lixiviation (SBPL)
SC. *See* Seed of chia (SC)
Scale-up
  of biodiesel production and technoeconomic analysis, 158–159
    biodiesel and scaling-up, 158
    technoeconomic analysis, 158–159
  studies, 158
Scaling of impact categories, 409t
Scanning electron microscopy (SEM), 369
Schinopsis, 347
SCWG. *See* Supercritical water gasification (SCWG)
SDS. *See* Sustainable development scenario (SDS)
SEAI materials. *See* Sustainable Energy Authority of Ireland materials (SEAI materials)
Second-generation (2G)
  biodiesel, 152, 160
  bioethanol
    pretreatment methods of biomass for, 137
    production, 143, 145–146
  biofuels, 85, 120
  lignocellulosic technology, 51
Secondary sewage treatment, 315–317
  aerobic treatment, 315–317
Sedimentation tanks, 314
Seed of chia (SC), 373
SEM. *See* Scanning electron microscopy (SEM)
Sensitivity analysis, 181, 401–402
Separate hydrolysis and cofermentation (SHCF), 124
Separate Hydrolysis and Fermentation (SHF), 124, 142, 227
Septic tanks, 314
Serratia marcescens UCP 1549 and BWL 1001, 238
Sewage microalgae, 112
Sewage sludge (SS), 155–156, 166, 305–306, 390
  applications, 329–331
  biological properties, 309–310
  chemical properties, 309
  global production, 310
  laws and regulations associated with sewage sludge utilization, 310–313
    Australia and New Zealand, 312
    European Union, 312–313
    Japan, 311–312
    USA, 312
  physical properties, 308–309
  physicochemical properties, 308–310
  value-added products from sewage sludge, 319–329
Sewage systems, 305–306
Sewage treatment, strategies for, 313
Sewage waste, 215, 313–314
  composition of sewage sludge obtained from wastewater, 306f
  preliminary sewage treatment, 313–314
  primary sewage treatment, 314–315
  regulatory limits of heavy metals around different nations in world, 307t
  secondary sewage treatment, 315–317
  tertiary treatment, 317–318
SFE. *See* Supercritical fluid extraction (SFE)
SHCF. *See* Separate hydrolysis and cofermentation (SHCF)
Shewanella oneidensis, 76
SHF. *See* Separate Hydrolysis and Fermentation (SHF)
Shigella, 308–309
  S. boydii, 309
  S. dysenteriae, 309
  S. flexneri, 309
  S. sonnei, 309
Silicone oil, 122–123
Simultaneous saccharification and cofermentation (SSCF), 124, 142
Simultaneous saccharification and fermentation (SSF), 124, 142, 144, 226
Single atom catalyst, conversion process utilizing, 269
Single-use plastics, 246
Singlet oxygen ($^1O_2$), 370
Skimming, 313–314
Slow pyrolysis process, 321, 364
Slurry, 35–36, 175
Small-scale biogas facilities, 386–387
Small-scale digestion, 411
SNG. *See* Synthetic natural gas (SNG)
Soaking Aqueous Ammonia (SAA), 32
Social sustainability, 406
Sodium carbonate, 31
  pretreatment with, 31
Sodium chloride (NaCl), 269–273
Sodium hydroxide, pretreatment with, 31
Sodium-ion batteries (NIBs), 292–297, 296t
Softwood lignin, 137
Soil Contamination Countermeasures Law (1968), 311–312
Soil organic matter, 61–62
Solar energy, 115
Solgels, 209
Solid fuels, 120, 126–127
  biochar, 120, 126
  drop-in fuels, 120, 126–127
Solid garbage, 9
Solid organic waste (SOW), 55, 275
  classification of technologies for processing of, 58–65
    biological conversion techniques, 60–62
    sanitary landfilling, 64–65
    thermal conversion techniques, 62–64
  existing technologies on nutrient recovery from, 276–278, 278f
    AD, 278

composting, 276–278
　pyrolysis, 278
　and major constituents, 277t
　physicochemical characterization, 277t
　qualitative and quantitative analysis, 276
　sources of, 56–58
　　agricultural waste, 58
　　food waste, 56–57
　　garden/yard waste, 57
　　livestock waste, 57–58
Solid substrates, 44
Solid-state fermentation (SSF), 238, 266
Solids retention time (SRT), 168–169
"Solution-precipitation" mechanism, 289
Solvents, 221–222
　extraction method, 155–156, 237
　production from microbial cell factories using waste biomass, 222t
Solvothermal process, 257
Sonication, 43
Sophorolipid (*Candida bombicola*), 17
Sorghum straw, 342
Sorption enhanced steam reforming, 188
Sorption-enhanced chemical looping, 188
SOW. *See* Solid organic waste (SOW)
Soxhlet extraction, 237
Soy stalks, 345–346
Specific surface areas (SSAs), 285–286, 365
Sphingomonas, 315
Sporotrichum pulverulentum, 123
SRT. *See* Solids retention time (SRT)
SS. *See* Sewage sludge (SS)
SSAs. *See* Specific surface areas (SSAs)
SSCF. *See* Simultaneous saccharification and cofermentation (SSCF)
SSF. *See* Simultaneous saccharification and fermentation (SSF); Solid-state fermentation (SSF)
Starch, 202, 344
　starch-based polymer, 253
Stated Policies Scenario (STEPS), 165
Steam activation, 365
Steam explosion, 45–46
　autohydrolysis process, 36
　different approaches for steam explosion treatment, 36–37
　　acid-catalyzed steam explosion process, 37
　　ammonia fiber explosion process, 37
　　two-step pretreatment process, 36
　　wet explosion process, 37
　pretreatment process, 35–37, 35t
　variables effecting steam explosion pretreatment process, 36
　　effect of particle size and water content, 36
　　effect of temperature and retention period, 36
Steam reforming techniques, 188
STEPS. *See* Stated Policies Scenario (STEPS)
Straws, 342, 364
Strength, weakness, opportunity, and threat analysis (SWOT analysis), 55, 65, 66t
Struvite, 350
Suber, 347

Succinate, 218
Succinic acid (SA), 217–220
　biosynthetic production, 218
　global market with applications, 218–220
　metabolic pathway, 218, 219f
　physicochemical properties and brief history, 217
　production from microbial cell factories using renewable waste biomass, 220t
　structure, 218f
Sugar beet pulp lixiviation (SBPL), 390
Sugarcane, 85
Sulfadimidine, 370
Sulfation, 348
Sulfolobus
　S. shibate, 205
　S. solfataricus, 205
Sulfur, 83–84, 368–369
Sulfur oxide (SO_2), 63
Sulfuric acid (H_2SO_4), 140–141
Sunflower meal, 269–273
Supercapacitors, biomass-based materials for, 290–292
Supercritical CO_2 (scCO_2), 46
Supercritical fluid (SC fluid), 14, 46
Supercritical fluid extraction (SFE), 14, 236
Supercritical methanol
　process, 157
　transesterification, 156–157
Supercritical water gasification (SCWG), 63
Superoxide anion radicals ($O_2^{\bullet -}$), 370
Supply chain management, 144
　of lignocellulosic bioethanol, 144–145
Surface oxides, 368–369
Sustainability, 406–409
　assessment, 406
　　of case study, 414–415
　method, 405, 407–409
　results, 413
　scenarios, 410–411
　as equilibrium of human needs and ecosystem needs, 406f
Sustainable biomass processing, 1
Sustainable chemicals for advance treatment, 349–350
Sustainable development, 406
Sustainable development scenario (SDS), 386
Sustainable energy, 159–160
　sources, 30
Sustainable Energy Authority of Ireland materials (SEAI materials), 169
Sustainable hydrogen production, 185, 187
Sustainable value teaching, 408t
SWOT analysis. *See* Strength, weakness, opportunity, and threat analysis (SWOT analysis)
Synergy advantages, 407t
Syngas, 330–331
Synthetic biology approaches, 76
Synthetic dyes, 233–234
Synthetic graphite, 289
Synthetic natural gas (SNG), 187
Synthetic plastics, 251
Syntrophobacter wolinii, 167

T
Taenia saginata, 308
Tagetes erecta, 240
Talaromyces purpureogenus CFRM02, 238
Tandem mass spectrometry, 109–110
Tannins, 347
Tapeworms, 310
TBBPA. *See* Tetrabromobisphenol A (TBBPA)
TCA cycle. *See* Tricarboxylic acid cycle (TCA cycle)
TDSs. *See* Total dissolved solids (TDSs)
TEA. *See* Technoeconomic analysis (TEA)
Technoeconomic analysis (TEA), 120, 158–159
　scale-up of biodiesel production and, 158–159
Technoeconomic feasibility analysis of waste bioprocessing
　benefits of biogas plants, 387
　biogas plants for rural economy, 387
　challenges in development of biogas plants, 388
　costs involved in biogas generation, 388
　demand of renewable energy, 385–386
　　global primary energy supply scenario, 385f
　importance of biogas plants, 386–387
　materials and methods, 389–396
　　experimental setup and procedure, 392–394
　　mathematical modeling, 394
　　physical and chemical characteristics of organic wastes, 390–391
　　selection of organic wastes, 389–390
　　technoeconomic analysis of biogas generation, 394–396
　　types of biogas plants and selection for study, 391–392
　motivation and objectives of study, 388–389
　potential of biogas as future energy, 386
　results, 396–402
　　economic analysis of biogas production, 399
　　LCC, 400
　　net present value, 400
　　pilot study, 396–399
　　prices paid for individual units of electricity, 401
　　sensitivity analysis, 401–402
　　verification of theoretical predictions, 396
Technology readiness level, 350–351
Tectona grandis, 233–234
TEM. *See* Transmission electron microscopy (TEM)
Temperature and retention period, effect of, 36
Temperature-Phased Anaerobic Sequencing Batch Reactor (TPASBR), 392
Temperature-programmed desorption (TPD), 369
Tenebrio molitor, 252
Termobifida fusca, 252
Terra Preta soils, 321

430 Index

Tertiary treatment, 317–318
Tetrabromobisphenol A (TBBPA), 257
Tetracycline, 370
Thermal conversion techniques, 62–64
   gasification, 63
   hydrothermal treatment, 63
   incineration, 62–63
   pyrolysis, 63
Thermal decomposition, 157
Thermal degradation, 125
Thermal digester, 279–280, 279f
Thermal digestion process, 63–64, 275, 279–281
   comparative assessment with technologies, 281–282
   existing technologies on nutrient recovery from solid organic waste, 276–278, 278f
   optimization of process parameters, 279–280
   process description and design of digester, 279
   process feasibility analysis, 281
   qualitative and quantitative analysis of SOW, 276
   transformation of nutrients in thermal digestion, 280
Thermal graphitization, 289, 289f
Thermal methods, 1, 4
Thermal oxidation, 189
Thermal plasma technology, 74
Thermal pretreatment process, 25
Thermochemical biomass conversion, 363
Thermochemical conversion, 285
   process, 13, 264–265
   of waste biomass to bioenergy, 124–126
     combustion, 125
     hydrothermal liquefaction, 125–126
     pyrolysis, 125
Thermochemical depolymerization, 246
Thermochemical methods, 12–13, 89–90, 92–96, 98–99
   gasification, 13, 90
   hydrothermal liquefaction, 13, 90
   incineration for heat and electricity, 13
   microwave-assisted drying, 92–94
   microwave-assisted hydrothermal pretreatment, 94–96
   microwave-assisted torrefaction, 94
   pyrolysis, 89
Thermochemical transformation process, 26
Thermococcus litoralis, 205
Thermometer, 175–177
Thermophiles, 205
Thermophilic enzymes, 205
Thermoplastic starch (TPS), 245–246
Thermoplastics, 246
Thermosets, 246
Thermotoga maritima, 205
Thermus
   T. aquaticus, 205
   T. thermophilus, 205
3G biodiesel, 155
Titanium dioxide (TiO_2), 192
TOC. *See* Total Organic Carbon (TOC)
Torrefaction, 25–26, 92, 94, 322

Total dissolved solids (TDSs), 106
Total Organic Carbon (TOC), 309
Total solids (TS), 167
Total suspended solids (TSSs), 106, 314
Toulene, 257
Tower reactor, 317
Toxic organic chemicals, 307
TPASBR. *See* Temperature-Phased Anaerobic Sequencing Batch Reactor (TPASBR)
TPD. *See* Temperature-programmed desorption (TPD)
TPS. *See* Thermoplastic starch (TPS)
Traditional GW disposal methods, 57
Transesterification, 16, 156–157
Transition metals, 289
Transmission electron microscopy (TEM), 369
Transportable bioenergy, 126
   gaseous fuel, 128–129
   liquid fuel, 127–128
   production of, 126–129
   solid fuel, 126–127
Tree residues, 188
Triacylglycerols, 108, 269–273
Tricarboxylic acid cycle (TCA cycle), 215–218
Trichoderma
   T. koningii, 46, 123
   T. longibrachiatum, 46
   T. ressei, 46, 123
   T. viride, 46
Trickling filters, 315–316
TS. *See* Total solids (TS)
TSSs. *See* Total suspended solids (TSSs)
Two-step pretreatment process, 36

U

UAE. *See* Ultrasound-assisted extraction (UAE)
UASB reactors. *See* Up-low anaerobic sludge blanket reactors (UASB reactors)
UCO. *See* Used cooking oil (UCO)
UIG. *See* Upper-income group (UIG)
Ultrasonication, 237
Ultrasound irradiation. *See* Sonication
Ultrasound-assisted extraction (UAE), 14
Ultraviolet light (UV), 249–250
   UV-vis spectrophotometer, 118
United States (US), 310
Up-low anaerobic sludge blanket reactors (UASB reactors), 317, 388
Upcycling approaches, 259
Upper-income group (UIG), 276
US. *See* United States (US)
US Department of Energy (US DOE), 145–146, 217
US DOE. *See* US Department of Energy (US DOE)
Used cooking oil (UCO), 152–153
Ustilago
   U. maydis, 221
   U. vetiveriae, 221
UV. *See* Ultraviolet light (UV)

V

Vacuum pyrolysis, 322
Valorization, 58, 264
   methods, 1
   of waste biomass, 71, 342f
Value-added products
   from sewage sludge, 319–329
     biochar, 319–323
     bioplastics, 328–329
     biosurfactants, 327–328
     CDs, 327
     hydrochar, 323–327
   techniques for extraction of value-added products from biowaste, 11–15
Vegetable waste (VW), 56, 171–172, 235, 390
Vegetables, 223, 234
Vermicomposting, 60–61
VFAs. *See* Volatile fatty acids (VFAs)
Vibrio cholera, 309
Violacein, 233–234
Virgin oils, 116–117
Viruses, 308–309
Vitamins, 110
VOCs. *See* Volatile organic compounds (VOCs)
Volatile fatty acids (VFAs), 12, 168, 328, 392
Volatile organic compounds (VOCs), 188–189
Volatile solids (VSs), 167, 178–179
VSs. *See* Volatile solids (VSs)
VW. *See* Vegetable waste (VW)

W

Wallemia ichthyophaga, 204–205
Walnut shell, 372
WAS. *See* Waste Activated Sludge (WAS)
Waste, 85–86
Waste Activated Sludge (WAS), 388
Waste biomass, 1, 5, 41–42, 116–118
   advance pretreatment in biorefineries, 50
   biochemical conversion of waste biomass to bioenergy, 123–124
   bioeconomy and biorefinery, 129–130
   bioenergy in combating climate change and greenhouse effects, 115
   catalytic conversion of waste biomass to bioenergy, 122–123
   characterization of waste biomass, 116–119
     determination of cellulose, hemicellulose, and lignin content, 118–119
     determination of degree of polymerization, 119
     determination of moisture and ash content, 119
     XRD, BET-BJH, and particle size analysis, 119
   classification of waste biomass, 116–119, 117f
   combustion, 4
   conversion into biochemicals via valorization techniques, 343t
   gasification, 4
   as potential substrate for biorefinery, 41
   pretreatment in generating bioenergy from waste biomass, 41–42

pretreatment of waste biomass for bioenergy production, 42–47
pretreatment technologies for waste biomass
  cons, 48–50
  pros, 47–48
processing, 5
production of transportable bioenergy from waste biomass, 126–129
sources and classification of bioenergy obtained from waste biomass, 119–122
sources of waste biomass, 116–119
technological upgradation and scale-up, 51, 129–130
thermochemical conversion of waste biomass to bioenergy, 124–126
valorization, 1–4
  biochemical methods, 3
  challenges of, 5
  functional materials from biomass waste, 6
  improving treatment and processing of waste biomass, 5
  methods, 2t
  opportunities of, 5–6
  thermal methods, 4
waste biomass as potential source of bioenergy, 115–116
Waste cooking oil (WCO), 156
Waste disposal, 71–72
Waste generation and compositional variation in household organic waste, 56
  compositional variation in household organic waste, 57t
Waste lignocellulosic biomass, 42, 117–118
  composition of, 118f
Waste management, 71–72, 249
  system, 58
  techniques, 64–65
Waste Management and Public Cleaning Law (1970), 311–312
Waste materials, 83–84, 120
Waste oil, 156
Waste remedial approaches and associated challenges, 74
Waste remediation approach, 74–75
Waste treatment approaches, 74
Waste valorization, 266
  advances in molecular biotechnology tools for, 75–77, 75f
    microbe-based bionanomining, 77
    microbial fuel cell technology, 76
    omics-based and gene editing technologies, 76–77
    synthetic biology approaches, 76
  methods, 75
  sources and types of wastes, 72–74, 72f
    agricultural and livestock waste, 73
    industrial waste, 73
    municipal and electronic waste, 73–74
  status of waste generation in India and world, 71–72
  as valuable tool for resource recovery, 74–75
  waste bioprocessing as valuable tool for resource recovery, 74–75
  waste remedial approaches and associated challenges, 74
Waste-to-energy (W2E), 407
  technologies, 9, 158–159, 187
  transitions, 74–75
Wastes, 215
Wastewater, 105, 155–156
  production of biofuel from using algal strains, 203
  treatment, 105
  wastewater-grown microalgae, 203
Water content, effect of particle size and, 36
Water reclamation, 364
  application of biomass-derived catalysts for water treatment, 370–377
  biomass conversion methods, 363–364
  biomass-derived catalyst materials, 364–368
  carbon-based catalyst preparation and properties, 368–369
  challenges and future perspectives, 377–378
Water vapor, 35
Water-soluble anionic kraft lignin, 345
WCO. *See* Waste cooking oil (WCO)
Wet explosion process, 37
Wet torrefaction (WT), 25–26
Wheat straw, 345–346
Wheat straw pellets (WSP), 372
Wind energy, 115
WLP. *See* Wood-Ljungdhal pathway (WLP)
Wood, 364
  chips, 84–85
  wood-derived ashes, 350
Wood-Ljungdhal pathway (WLP), 258
Woody biomass, 84
WSP. *See* Wheat straw pellets (WSP)
WT. *See* Wet torrefaction (WT)

X

X-ray photoelectron spectroscopy (XPS), 369
Xenobiotics, 320–321
XPS. *See* X-ray photoelectron spectroscopy (XPS)
XRD, 119
Xylan, 24

Y

Yard waste, 57
Yarrowia lipolytica, 216–218
Yeast enzymes, 203

Z

ZCP composite, 370
Zetasizer nanoparticle analyzer, 119
Zinc oxide (ZnO), 370
Zymomonas mobilis, 146

Printed in the United States
by Baker & Taylor Publisher Services